W9-ARP-876

Geophysical Monograph Series

Including

IUGG Volumes
Maurice Ewing Volumes
Mineral Physics Volumes

Geophysical Monograph Series

77 **The Mesozoic Pacific: Geology, Tectonics, and Volcanism** *Malcolm S. Pringle, William W. Sager, William V. Sliter, and Seth Stein (Eds.)*

78 **Climate Change in Continental Isotopic Records** *P. K. Swart, K. C. Lohmann, J. McKenzie, and S. Savin (Eds.)*

79 **The Tornado: Its Structure, Dynamics, Prediction, and Hazards** *C. Church, D. Burgess, C. Doswell, R. Davies-Jones (Eds.)*

80 **Auroral Plasma Dynamics** *R. L. Lysak (Ed.)*

81 **Solar Wind Sources of Magnetospheric Ultra-Low Frequency Waves** *M. J. Engebretson, K. Takahashi, and M. Scholer (Eds.)*

82 **Gravimetry and Space Techniques Applied to Geodynamics and Ocean Dynamics (IUGG Volume 17)** *Bob E. Schutz, Allen Anderson, Claude Froidevaux, and Michael Parke (Eds.)*

83 **Nonlinear Dynamics and Predictability of Geophysical Phenomena (IUGG Volume 18)** *William I. Newman, Andrei Gabrielov, and Donald L. Turcotte (Eds.)*

84 **Solar System Plasmas in Space and Time** *J. Burch, J. H. Waite, Jr. (Eds.)*

85 **The Polar Oceans and Their Role in Shaping the Global Environment** *O. M. Johannessen, R. D. Muench, and J. E. Overland (Eds.)*

86 **Space Plasmas: Coupling Between Small and Medium Scale Processes** *Maha Ashour-Abdalla, Tom Chang, and Paul Dusenbery (Eds.)*

87 **The Upper Mesosphere and Lower Thermosphere: A Review of Experiment and Theory** *R. M. Johnson and T. L. Killeen (Eds.)*

88 **Active Margins and Marginal Basins of the Western Pacific** *Brian Taylor and James Natland (Eds.)*

89 **Natural and Anthropogenic Influences in Fluvial Geomorphology** *John E. Costa, Andrew J. Miller, Kenneth W. Potter, and Peter R. Wilcock (Eds.)*

90 **Physics of the Magnetopause** *Paul Song, B.U.Ö. Sonnerup, and M.F. Thomsen (Eds.)*

91 **Seafloor Hydrothermal Systems: Physical, Chemical, Biological, and Geological Interactions** *Susan E. Humphris, Robert A. Zierenberg, Lauren S. Mullineaux, and Richard E. Thomson (Eds.)*

92 **Mauna Loa Revealed: Structure, Composition, History, and Hazards** *J. M. Rhodes and John P. Lockwood (Eds.)*

93 **Cross-Scale Coupling in Space Plasmas** *James L. Horwitz, Nagendra Singh, and James L. Burch (Eds.)*

94 **Double-Diffusive Convection** *Alan Brandt and H. J. S. Fernando (Eds.)*

95 **Earth Processes: Reading the Isoto pic Code** *Asish Basu and Stan Hart (Eds.)*

96 **Subduction Top to Bottom** *Gray E. Bebout, David Scholl, Stephen Kirby, and John Platt (Eds.)*

97 **Radiation Belts: Models and Standards** *J. F. Lemaire, D. Heynderickx, and D. N. Baker (Eds.)*

98 **Magnetic Storms** *Bruce T. Tsurutani, Walter D. Gonzalez, Yohsuke Kamide, and John K. Arballo (Eds.)*

99 **Coronal Mass Ejections** *Nancy Crooker, Jo Ann Joselyn, and Joan Feynman (Eds.)*

100 **Large Igneous Provinces** *John J. Mahoney and Millard F. Coffin (Eds.)*

101 **Properties of Earth and Planetary Materials at High Pressure and Temperature** *Murli Manghnani and Takehiki Yagi (Eds.)*

102 **Measurement Techniques in Space Plasmas: Particles** *Robert F. Pfaff, Joseph E. Borovsky, and David T. Young (Eds.)*

103 **Measurement Techniques in Space Plasmas: Fields** *Robert F. Pfaff, Joseph E. Borovsky, and David T. Young (Eds.)*

104 **Geospace Mass and Energy Flow: Results From the International Solar-Terrestrial Physics Program** *James L. Horwitz, Dennis L. Gallagher, and William K. Peterson (Eds.)*

105 **New Perspectives on the Earth's Magnetotail** *A. Nishida, D. N. Baker, and S. W. H. Cowley (Eds.)*

106 **Faulting and Magmatism at Mid-Ocean** *Ridges W. Roger Buck, Paul T. Delaney, Jeffrey A. Karson, and Yves Lagabrielle (Eds.)*

107 **Rivers Over Rock: Fluvial Processes in Bedrock Channels** *Keith J. Tinkler and Ellen E. Wohl (Eds.)*

108 **Assessment of Non-Point Source Pollution in the Vadose Zone** *Dennis L. Corwin, Keith Loague, and Timothy R. Ellsworth (Eds.)*

109 **Sun-Earth Plasma Connections** *J. L. Burch, R. L. Carovillano, and S. K. Antiochos (Eds.)*

110 **The Controlled Flood in Grand Canyon** *Robert H. Webb, John C. Schmidt, G. Richard Marzolf, and Richard A. Valdez (Eds.)*

111 **Magnetic Helicity in Space and Laboratory Plasmas** *Michael R. Brown, Richard C. Canfield, and Alexei A. Pevtsov (Eds.)*

Geophysical Monograph 112

Mechanisms of Global Climate Change at Millennial Time Scales

Peter U. Clark
Robert S. Webb
Lloyd D. Keigwin
Editors

American Geophysical Union
Washington, DC

Published under the aegis of the AGU Books Board

Library of Congress Cataloging-in-Publication Data

Mechanisms of global climate change at millennial time scales / Peter U. Clark, Robert S. Webb, Lloyd D. Keigwin, editors.
p. cm. -- (Geophysical Monograph ; 112)
Includes bibliographical references.
ISBN 0-87590-095-X
1. Climatic changes. I. Clark, Peter U., 1956- . II. Webb, Robert S., 1959- . III. Keigwin, Lloyd D. IV. Series.
QC981.8.C5M427 1999
551.5' 253--dc21 99-40787
CIP

ISBN 0-87590-095-X
ISSN 0065-8448

Cover (images clockwise from upper right):
The French R/V *Marion Dufresne* routinely retrieves marine piston cores of 30 to 50 m length. (Photograph by M.Le Coz/IFRTP.) Coring the southern shore of Walker Lake, southwestern Nevada. (Photograph by Larry Benson, WRD, USGS.) Figure showing Greenland ice-core oxygen isotope record (GRIP) and relative abundance of a polar foraminifera (*N. pachyderma*) in the North Atlantic sediment core V23-81 for the period 10 to 50 kyr BP, ("HL" denotes Heinrich Layer). (Figure from Duplessy and Overpeck, 1994, after Bond et al., 1993.) The GISP2 Drill Dome, central Greenland. (Photograph by Mark Twickler, GISP2 SMO, University of New Hampshire.)

Printed in the United States of America.

CONTENTS

CONTENTS

PREFACE

This volume is intended to serve as a single, comprehensive resource for understanding the mechanisms of millennial-scale global climate change. Given the maturity of the science and understanding of the climate processes, we believe that the time is right for such a compilation of results. This monograph provides the paleoclimatology and climate dynamics communities with a comprehensive overview of current evidence and understanding of climate variability and abrupt climate change between orbital and interannual time scales.

One of the major uncertainties in global climate change research relates to the mechanisms and causes of abrupt climate change at millennial to submillennial time scales. Over the last 25 years, an increasing number of high quality geologic records of climate change with high-precision age control have confirmed the basic premise of Mitchell (1976) that there exists a range of climate variability across a continuum of time scales. Many climate variations at the high- and lower-frequency bands have been associated with distinct and interrelated driving processes. Within the Quaternary, orbital changes play a central role in driving the pace of climate changes in concert with associated feedbacks of global scale processes at periods of 10^4 to10^5 years. At the other end of the climate variability spectrum, there has been a growing understanding of the mechanisms of seasonal to interannual climate change and development of predictive models to simulate stable modes of ENSO variability in the late 1980s. Situated between the orbital forced climate change and the seasonal to interannual climate variability is the less well understood decade to millennial climate variability. Despite the growing abundance of evidence for decade to millennial-scale variability, there has remained a deficiency in the mechanistic understanding of paleoclimate change at these frequencies.

This monograph focuses on the current understanding of abrupt climate variations or events that have occurred at millennial to submillennial frequencies (10^3-10^4 years). These abrupt climate variations or events are superimposed on gradual glacial to interglacial climate changes in response to slowly changing orbital forcing. Unrecognized as significant reoccurring events 10 years ago, millennial to submillennial climate variability is now recognized as a characteristic feature of the global climate during the last glaciation. Records from the Greenland Ice Sheet and the North Atlantic clearly demonstrate significant climatic variability on these time scales. Evidence from both land and sea now indicates that the climate of regions outside the North Atlantic also varied at millennial time scales, suggesting the potential for a global response to conditions observed in the North Atlantic. The challenge now is to assess whether events well known in the North Atlantic record are related to events being identified elsewhere.

As editors, we have identified a number of important questions related to our current understanding of millennial to submillennial scale climate change:

- What are the forcings, linkages, and feedbacks which produce millennial-scale climate change?
- What is the sensitivity of various components of the Earth's climate system to millennial-scale climate change?
- Do the periodic abrupt shifts observed during glacial, deglacial, and interglacial periods all occur at the same frequency, in response to the same forcings, and involve the same climate processes?
- Are these abrupt shifts initiated at high northern latitudes, at low latitudes, or at high southern latitudes?
- Are these abrupt shifts a response to external or to internal forcings?
- Are the globally distributed records of these abrupt climate changes synchronous or time transgressive, in phase or out of phase relative to each other?
- What processes allow abrupt changes to be transmitted from the northern to the southern hemisphere or vice versa?

The significant breadth of global paleoclimate knowledge presented in this monograph provides the critical information required to answer many of these questions and provides a road map to address the remaining outstanding ones. A majority of the papers in the monograph are derived from presentations given at the Chapman Conference titled "Mechanisms of Millennial-Scale Global Climate Change," convened by Peter U. Clark and Robert S. Webb in Snowbird, Utah, June 14-18, 1998. However, the monograph includes a number of additional papers contributed by authors who were unable to attend the meeting, which ensures a comprehensive presentation of the current understanding of the mechanisms of millennial-scale global climate change. The range of subjects covered by the authors of individual chapters include analysis of modern climate and ocean dynamics; paleoclimate reconstructions derived from the marine, terrestrial, and ice core records; and paleoclimate modeling studies.

Robert S. Webb
NOAA, National Geophysical Data Center

Peter U. Clark
Oregon State University

Lloyd D. Keigwin
Woods Hole Oceanographic Institution

We'd like to thank the National Science Foundatuon for support of the Chapman Conferemce, the contributing authors to this volume, and the following reviewers for their efforts in reviewing the papers in this volume:

REVIEWERS

Richard Alley
John Andrews
Edouard Bard
Thomas Blunier
Anthony J. Broccoli
Mark A. Cane
Mark Chandler
Chris Charles
Julia Cole
Thomas Crowley
William Curry
Henry Diaz
Peter DeMenocal
Art Dyke
Peter Fawcett
Eric C. Grimm
Sveinung Hagen
Gerald Haug
Chris Hewitt
Konrad Hughen
Scott Lehman
Doug MacAyeal
Vera Markgraf
Jerry McManus
James Miller
Alan Mix
Robert Oglesby
Charles G. Oviatt
Tad W. Pfeffer
Stefan Rahmstorf
David H. Rind
Dan Shrag
Eric J. Steig
Jean Lynch-Steiglitz
Jozef Syktus
Robert S. Thompson
Robert Thunell
J. R. Toggweiler
James W. C. White
James D. Wright

Some aspects of ocean heat transport by the shallow, intermediate and deep overturning circulations

Lynne D. Talley

Scripps Institution of Oceanography, UCSD, La Jolla, CA

The ocean's overturning circulation can be divided into contributions from: (1) shallow overturning in the subtropical gyres to the base of thermocline, (2) overturning into the intermediate depth layer (500 to 2000 meters) in the North Atlantic, North Pacific and area around Drake Passage, and (3) overturning into the deep layer in the North Atlantic (Nordic Seas overflows) and around Antarctica. The associated water mass structures are briefly reviewed including presentation of a global map of proxy mixed layer depth. Based on the estimated temperature difference between the warm source and colder newly-formed intermediate waters, and the formation rate for each water mass, the net heat transport associated with all intermediate water formation is estimated at 1.0-1.2 PetaWatts (1 PW = 10^{15} W), which is equivalent in size to that for deep water formation, 0.6-0.8 PW. The heat transport due to shallow overturn, calculated as the residual between published direct estimates of heat transport across subtropical latitudes and these heuristic estimates of the intermediate and deep overturning components, is about 0.5 PW northward for the North Pacific and North Atlantic subtropical gyres and 0.0 to 0.2 PW southward for each of the three southern hemisphere subtropical gyres, exclusive of the shallow overturn in the southern hemisphere gyres which is associated with Antarctic Intermediate Water and Southeast Indian Subantarctic Mode Water formation.

Direct estimates of meridional heat transport of 1.18 PW (North Atlantic) and 0.63 PW (North Pacific) at 24°N are calculated from *Reid's* [1994, 1997] geostrophic velocity analyses and are similar to previously published estimates using other methods. The new direct estimates are decomposed into portions associated with shallow, intermediate and deep overturn, confirming the heuristic estimate for the North Pacific, where the shallow gyre overturning heat transport accounts for about 75% of the total and intermediate water formation for the remainder. The direct estimate for the North Atlantic indicates the opposite - about 75% of the total heat transport is associated with intermediate and deep water formation, split approximately

Mechanisms of Global Climate Change at Millennial Time Scales
Geophysical Monograph 112

equally, with the remainder associated with the shallow gyre overturn. the difference from the heuristic estimate for the North Atlantic suggests that the source waters for the intermediate and deep water overturn originate within the Gulf Stream at an average temperature warmer than 14°C.

1. INTRODUCTION

Meridional heat transport in the ocean is associated with heat gain and loss at the sea surface and hence transformation of surface water properties. The ocean and atmosphere together transport approximately 5-6 PW (1 PW = 1 PetaWatt = $10^{15}W$) of heat poleward in each hemisphere on an annual average; of this, approximately one-third to one-half is carried by the oceans [*Oort and Von der Haar,* 1976; *Talley,* 1984; *Hsiung,* 1985; *Keith,* 1995; *Josey et al.,* 1996]. The distribution of meridional ocean heat transport depends on ocean basin and the distribution of water mass transformation. The major elements of the net transfer of heat from the ocean to the atmosphere include heating throughout the tropics and large heat loss in the Gulf Stream and Kuroshio where warm northward boundary currents meet cold, dry continental air [Figure 1 after *Hsiung,* 1985]. [For recent maps, with similar patterns, see *da Silva et al.,* 1994; *Barnier et al.,* 1995; *Josey et al.,* 1997.] An asymmetry between the North Atlantic and North Pacific is evident in the large areas of heat loss extending up into the Nordic Seas in the North Atlantic. In the southern hemisphere, the more recent maps which extend to Antarctica show regions of heat loss in the poleward western boundary currents, with lower amplitudes than in the Kuroshio and Gulf Stream. In all versions of air-sea heat flux, large heat loss is not observed in the primary formation regions of most intermediate and deep water masses (described below). The absence of high surface flux may reflect a fair weather bias in data coverage but more likely is an indication that buoyancy loss associated with these water mass transformations is a gradual, broad-scale, cumulative process, culminating in a geographically-limited formation of the denser water.

Superimposed on Figure 1 are various published direct estimates of ocean heat transport at subtropical latitudes based on *in situ* temperature measurements and velocity estimates. All basins, except the South Atlantic, transport heat poleward across their subtropical gyres. The South Atlantic transports heat equatorward to the North Atlantic, feeding the northern hemisphere deep water formation.

Because ocean heat transport is linked to water mass transformation, it is useful to estimate the relative contribution of each of the types of water mass transformation to the overall heat transport. *Speer and Tziperman* [1992] and *Speer* [1992] related surface heat fluxes to water mass formation in isopycnal outcrop regions and were able to identify some especially marked transformations, particularly those associated with Subtropical and Subpolar Mode Waters at midlatitudes. Deep and intermediate water formation are not easily captured using Speer and Tziperman's method due to the localized and in some cases, subsurface, processes involved.

In situ temperature measurements and direct velocity estimates, where available for the full water column, can be used to estimate the relative contribution of different water mass formations to the heat transport. For the North Atlantic and North Pacific respectively, *Hall and Bryden* [1982] and *Bryden et al.* [1991] decomposed the direct heat transport into a portion carried by the Ekman transport, the total baroclinic contribution (overall overturn), and a horizontal circulation (due to the temperature difference in northward and southward flows at the same depth). They also considered the contribution to the net heat transport of different parts of the gyre and water column. Hall and Bryden concluded that in the North Atlantic at 24°N almost all of the heat transport is carried by the conversion of warm northward-flowing waters into deep/intermediate waters. *Roemmich and Wunsch* [1985] came to the same conclusion using a 1980 reoccupation of the 1957 section used by Hall and Bryden. In contrast, for the North Pacific at 24°N, *Bryden et al.* [1991] showed that about half of the heat transport is due to shallow overturn in the subtropical gyre and the other half is due to intermediate water formation.

Roemmich and Wunsch [1985] (North Atlantic), *Rintoul* [1991] (South Atlantic), *Roemmich and McCallister* [1989] (North Pacific), *Wunsch et al.* [1983] (South Pacific) and *Toole and Warren* [1993]/*Robbins and Toole* [1997] (Indian) presented total mass transport in isopycnal, isothermal or constant pressure layers along zonal sections which completely cross the ocean basins at 24°N and 30°S. These presentations clearly show the distribution of net northward and southward flow as a function of depth and hence depict overturn (upwelling and downwelling).

In section 2, the principal water masses involved in the overturning are described, including the apparent

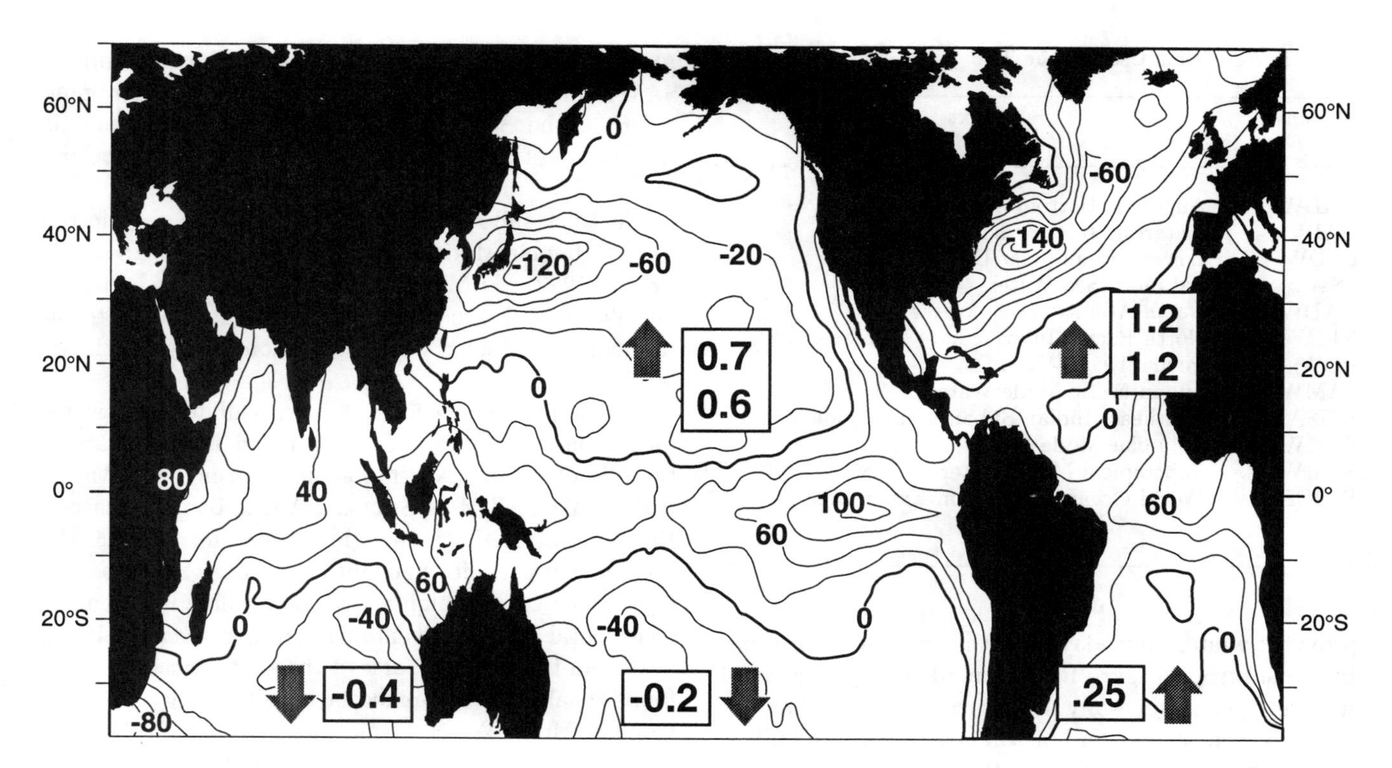

Figure 1. Ocean net heat gain (Wm^{-2}) after *Hsiung* [1985]. Positive numbers indicate heat gained by the ocean from the atmosphere. Direct estimates and directions of meridional heat transport are indicated for each subtropical gyre. The upper number in each box is from: North Pacific - *Bryden et al.* [1991]; North Atlantic - *Hall and Bryden* [1983]; Indian - *Robbins and Toole* [1997], including Indonesian throughflow; South Pacific - *Wunsch et al.* [1983]; South Atlantic - *Rintoul* [1991]. The lower number in the northern hemisphere boxes is calculated herein (section 3.2) using *Reid's* [1994, 1998] velocities and standard calculations for the Ekman layer transport.

precursors to the new water masses. Table 1 lists the acronyms used for water mass names. In section 3.1, heat transports associated with formation of each intermediate and deep water mass are estimated based on their formation rates and temperature change from their source waters. In section 3.2, *Reid's* [1994, 1997] velocity analyses at 24°N in the Pacific and Atlantic are used to calculate the heat transport due to shallow, intermediate and deep overturn. The net heat transports using Reid's velocities are comparable to those shown in Figure 1. The heuristic estimates of section 3.1 are then compared with the direct estimates of section 3.2, for the northern hemisphere only.

2. MAJOR OVERTURNING COMPONENTS: REVIEW OF WATER MASS DISTRIBUTIONS

Water mass formation can be thought of primarily as the densifying branch (downwelling) part of the large-scale overturn. Two processes dominate: open ocean buoyancy loss leading to mixed layer deepening, densification and convection (subtropical and subpolar mode water formation, Labrador and Greenland Sea intermediate water formation, Antarctic Intermediate Water formation), and buoyancy loss due to brine rejection under ice formation (North Pacific Intermediate Water formation, Antarctic Bottom Water formation, and possibly some aspects of Greenland Sea intermediate water formation). Upwelling and diffusion alter water mass properties and can be volumetrically significant even for narrowly-defined water masses such as the North Atlantic Deep Water where Antarctic Bottom Water upwells into it.

2.1. Winter Mixed Layer Depth

Water mass formation due to open ocean surface buoyancy loss is not a local process since mixed layer properties are cumulative along a flow path. Winter mixed layer depth is a useful indicator of vigorous surface layer processes and preconditioning for overturn. For the North Pacific, *Reid* [1982] showed that a useful

Table 1. Acronyms Used in the Text

Acronym	Text
AABW	Antarctic Bottom Water
AAIW	Antarctic Intermediate Water
LCDW	Lower Circumpolar Deep Water
LSW	Labrador Sea Water
NADW	North Atlantic Deep Water
NPIW	North Pacific Intermediate Water
PDW	Pacific Deep Water
SAMW	Subantarctic Mode Water
SEISAMW	Southeast Indian Subantarctic Mode Water
SPMW	Subpolar Mode Water
STMW	Subtropical Mode Water
WOCE	World Ocean Circulation Experiment

proxy for winter mixed layer depth is oxygen saturation. In the subtropical gyre Reid showed that the seasonal surface layer is generally supersaturated and that the 100% oxygen saturation horizon is a reasonable indicator of winter mixed layer depth. Because the winter surface layer in the subpolar region is undersaturated due to vigorous overturn, Reid found that the 94% saturation depth was a more useful proxy in the North Pacific's subpolar gyre.

In lieu of a complete analysis of the correspondence in each basin between oxygen saturation and winter mixed layer depth, the 95% oxygen saturation depth is used here (Plate 1). A global hydrographic data set comprised of all discrete bottle stations available from the one-time survey of the World Ocean Circulation Experiment (WOCE) hydrographic program and high quality hydrographic data from the National Oceanographic Data Center selected by J. Reid and A. Mantyla (personal communication) were used. Oxygen profiles were interpolated to 10 meter depths using an Akima cubic spline; the 95% saturation depth was then found through linear interpolation.

The deepest winter mixed layers, as indicated by depth of the 95% oxygen saturation, are close to and north of the Antarctic Circumpolar Current and are in the northern North Atlantic. The deep Southern hemisphere mixed layers are referred to as Subantarctic Mode Water [*McCartney,* 1977] and are the major precursor to Antarctic Intermediate Water formation. The Subantarctic Mode Waters (SAMW) are thickest in the southeast Indian Ocean and across the South Pacific. The easternmost SAMW in the Indian Ocean is the densest outcropping water in the combined South Atlantic and Indian subtropical gyre. The easternmost SAMW in the South Pacific is the densest outcropping water in the South Pacific's subtropical gyre and is the source of Antarctic Intermediate Water [*McCartney,* 1977].

The thick subpolar North Atlantic layers are referred to as Subpolar Mode Water (SPMW) [*McCartney and Talley,* 1982]. SPMW is a primary input to Labrador Sea Water and Greenland Sea Water and hence to intermediate and deep water formation in the North Atlantic.

Two important regions of intermediate/deep water formation do not have a deep mixed layer signature: the Okhotsk Sea (North Pacific Intermediate Water) and the Weddell/Ross Seas and Adelie Land (Antarctic Bottom Water). In these regions sea ice formation is the dominant process for buoyancy loss and is accompanied by production of a highly stratified, saline layer on the continental shelf which enters the deeper sea as a plume.

Within the western parts of the subtropical gyres are found locally thicker mixed layers. These are the Subtropical Mode Water formation sites and are an important component of the shallow overturn [*Speer and Tziperman,* 1992], although not the densest part of it. In the next set of subsections, the upper ocean, intermediate depth and abyssal overturning water masses are described briefly.

2.2. Upper Ocean - Subtropical Gyres

Shallow subtropical overturning has several components: poleward flow of warm water in the western boundary current, major buoyancy loss associated with formation and spreading of Subtropical Mode Water, and continued buoyancy loss leading to denser surface waters in the poleward, eastern gyre region, followed by a split into equatorward subduction beneath less dense waters and poleward flow into the subpolar regions. The subtropical circulations with their poleward western boundary currents and equatorward interior flows are indicated very schematically in Figure 2.

The thick layers of relatively homogenized water in the poleward-western corners of the subtropical gyres, adjacent to the poleward western boundary currents and their eastward, separated extensions (medium shaded regions in Figure 2) are the Subtropical Mode Waters (STMW). In the North Atlantic and North Pacific, STMW is associated with the largest surface buoyancy loss (Figure 1). Because the vertical stratification in the upper ocean in the North Atlantic is considerably weaker than in the North Pacific, the North Atlantic STMW is thicker and deeper. The southern hemi-

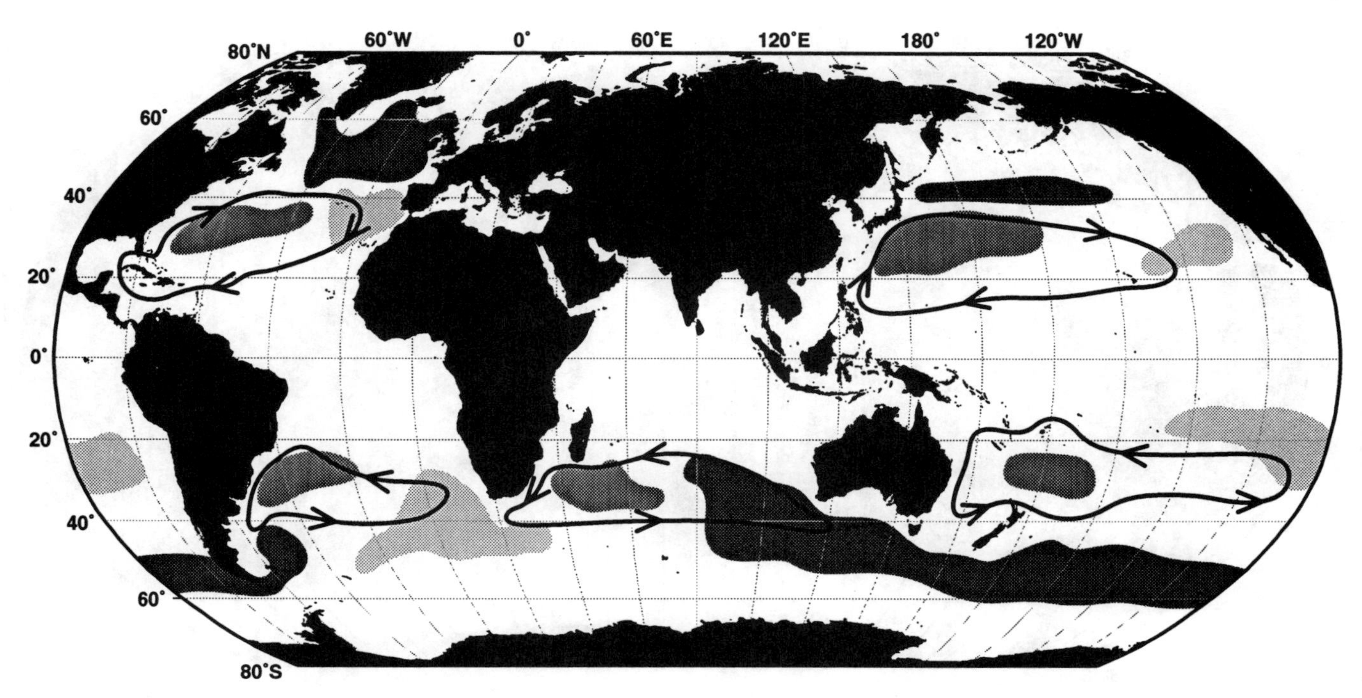

Figure 2. Subtropical mode water locations (medium shading). Low density mode waters of the eastern subtropical gyres (light shading). Highest density mode waters which subduct into the subtropical gyres (dark shading): Subpolar Mode Water in the North Atlantic, North Pacific Central Mode Water, and Subantarctic Mode Water in the southern hemisphere. Cartoons of subtropical gyre circulations are superimposed.

sphere STMW's, which are the group of thicker layers near 30°S, east of Brazil [*Tsuchiya et al.,* 1994], east of Madagascar [*Gordon et al.,* 1987; *Toole and Warren,* 1993] and north of New Zealand [*Roemmich and Cornuelle,* 1992], are somewhat less remarkable in thickness.

Warm, thick surface layers (light shading in Figure 2 and Plate 1) are also found on the eastern sides of each of the subtropical gyres. These layers are associated with low density mode waters which have not been as completely described as other mode waters. These low density mode waters are at 20 - 30° latitude in the North Pacific [*Hautala et al.,* 1998], the Madeira Mode Water in the North Atlantic [*Käse et al.,* 1985], and mode waters in the South Atlantic, Indian and South Pacific at 20 - 40°S. These thick mixed layers are not associated with particularly large local buoyancy loss (Figure 1). All are subducted equatorward in their respective gyres.

The densest surface waters in each subtropical gyre are found at the poleward, eastern boundary. These also are relatively thick "mode waters" (dark shading in Figure 2). When subducted equatorward, they form the bottom of the directly ventilated subtropical thermocline. There is nothing remarkable about the surface buoyancy loss (Figure 1) in the area where these poleward, eastern mode waters are found. Rather they appear to be a dynamical feature of the gyre with perhaps a slight enhancement due to surface buoyancy loss. These poleward, eastern mode waters are: the southern portion of the Subpolar Mode Water in the North Atlantic [*McCartney and Talley,* 1982], the North Pacific Central Mode Water [*Suga et al.,* 1997], the southeast Indian Subantarctic Mode Water [*McCartney,* 1977, 1982], and the southeast Pacific Subantarctic Mode Water or Antarctic Intermediate Water [*McCartney,* 1977, 1982].

In section 3.2 below, the maximum vertical extent of the shallow overturning is taken to be the maximum subtropical gyre surface density in winter and includes all of the subtropical mode waters. The poleward edge of the subtropical gyre is given by the zero of the annual average Sverdrup transport streamfunction. The nominal maximum potential densities relative to the sea surface in the North Pacific and North Atlantic are $\sigma_\theta = 26.2$ and 27.3, respectively [see *Yuan and Talley,* 1992 and *McCartney,* 1982, respectively]. In the southern hemisphere, the subtropical circulations are divided here at New Zealand where the eastward flow of the Antarctic Circumpolar Current is especially constricted. The maximum potential density of

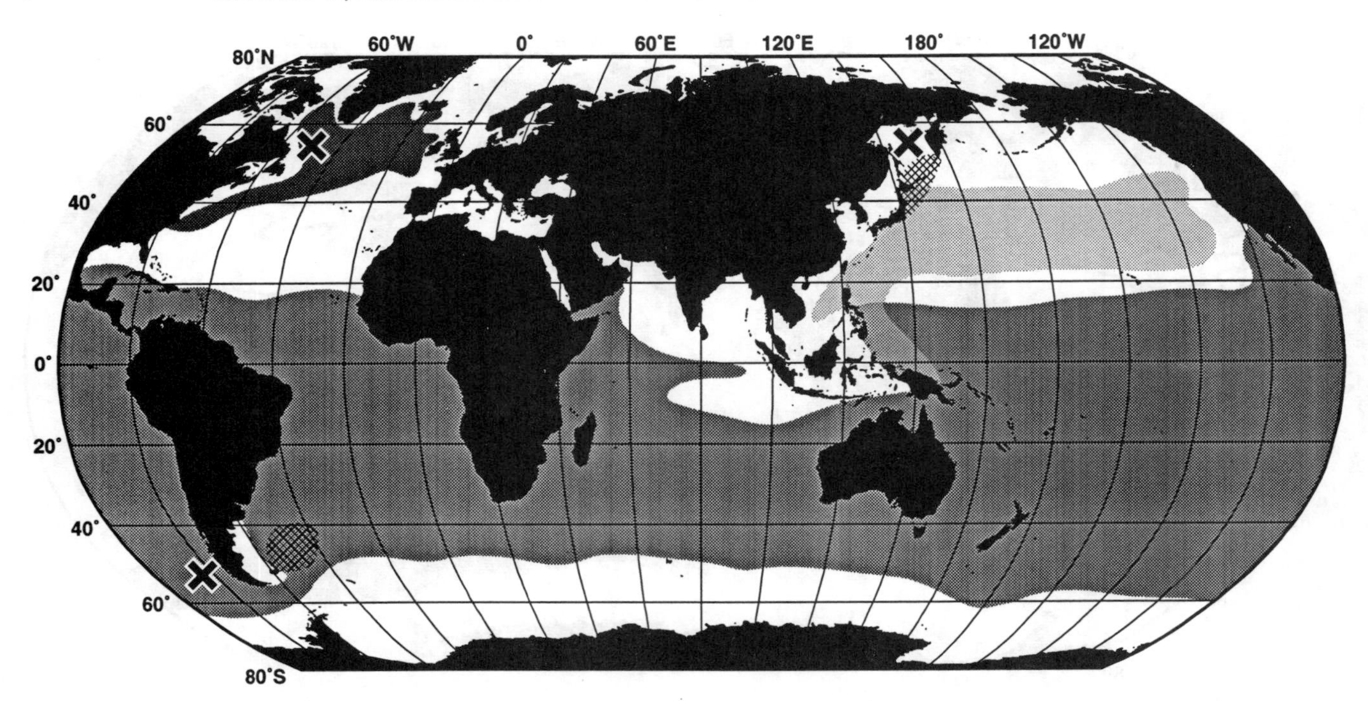

Figure 3. Locations of the intermediate water salinity minima Labrador Sea Water (LSW) (dark) [*Talley and McCartney,* 1982], North Pacific Intermediate Water (NPIW) (light) [*Talley,* 1993], and Antarctic Intermediate Water (AAIW) (medium) [*Talley,* 1998]. The total geographic influence of the newly-renewed intermediate waters extends well beyond these salinity minimum regions. (For instance, renewed water NPIW fills the North Pacific's subpolar gyre, but is not a salinity minimum due to the lower salinity surface waters [*Talley,* 1993]; the high oxygen signature of LSW extends into the South Atlantic as Middle North Atlantic Deep Water [*Wüst,* 1935]; high silica associated with AAIW extends to the subpolar North Atlantic [*Tsuchiya,* 1989].) Primary formation areas for the intermediate waters are indicated with X's. Hatching indicates regions where mixing is essential for setting properties of the intermediate waters near their sources. (Clearly mixing occurs globally throughout the water mass extent, causing gradual changes in properties and erosion of vertical property extrema.) The NPIW and AAIW salinity minima overlap in the western tropical Pacific.

the combined South Atlantic/Indian subtropical gyre is $\sigma_\theta = 26.8$ to 26.9, corresponding to the southeast Indian SAMW (SEISAMW) found south of Australia (see below). The maximum potential density of the South Pacific subtropical gyre is about $\sigma_\theta = 27.1$, corresponding to SAMW of the southeast Pacific, which is identical with Antarctic Intermediate Water (AAIW) there. Thus formally speaking AAIW in the South Pacific is the densest shallow overturning water mass.

2.3. Intermediate Water Formation

The intermediate layer of a given subtropical and tropical ocean can be defined to lie between the maximum subducted density in the subtropical gyre and the deep water described in the next subsection, that is, from about 500 to 2000 meters depth, depending on basin and location. The sources of intermediate waters result in salinity signatures. The fresh intermediate waters (Figure 3), originating from freshened surface subpolar waters, are the Labrador Sea Water (LSW), North Pacific Intermediate Water (NPIW), and Antarctic Intermediate Water (AAIW). Saline intermediate waters are formed in the Mediterranean and Red Seas but are not considered in section 3 since their total formation rate, temperature change and overturning heat transport are low.

The fresh intermediate waters are injected to intermediate depth at specific locations indicated in Figure 3. The water masses are: the Labrador Sea for the LSW [*Lazier,* 1980; *Clarke and Gascard,* 1983; *Talley and McCartney,* 1982], the Okhotsk Sea for the NPIW [*Talley,* 1991, 1993], and the region around southern South America for AAIW [*McCartney,* 1977; *Talley,* 1996a].

Each of the fresh intermediate waters is formed in a different manner. LSW is formed by intermediate depth

convection, to about 1500 m, acting on the thick surface layer of Subpolar Mode Water (Plate 1 and dark shading in Figure 2) whose density and thickness increase as it circulates cyclonically around the subpolar gyre [*McCartney and Talley,* 1982]. NPIW forms through brine rejection during sea ice formation in the northwestern Okhotsk Sea [*Kitani,* 1973; *Talley,* 1991], followed by strong tidal mixing [*Talley,* 1991], and possibly including deep convection in the southern Okhotsk Sea [*Wakatsuchi and Martin,* 1990; *Freeland et al.,* 1998]. NPIW enters the subtropical gyre in the North Pacific through strong interaction between the Oyashio and Kuroshio [*Talley,* 1993; *Talley et al.,* 1995]. AAIW in the South Pacific is the densest Subantarctic Mode Water (thick layers off Chile in Plate 1 and dark shading in Figure 2) and is subducted northward into the subtropical gyre [*McCartney,* 1977; *Talley,* 1996a]. Pacific AAIW is thus actually part of the shallow overturn. AAIW in the Atlantic and Indian Oceans derives from advection of the southeast Pacific AAIW through Drake Passage with additional modification in that region. AAIW enters the Atlantic's subtropical gyre through strong interactions between the Falkland and Brazil Currents [*Talley,* 1996a].

The total extent of the subpolar-origin intermediate water is of course much greater than depicted in Figure 3, which shows only the locations of the actual salinity minima. Although the salinity extrema erode away beyond these regions, a significant amount of the water at the intermediate densities nevertheless has originated in the intermediate water formation sites. For instance the influence of AAIW in the North Atlantic has been tracked through a high silica signal [*Tsuchiya,* 1989], and the influence of LSW in the South Atlantic is evident as an oxygen maximum [Middle North Atlantic Deep Water in *Wüst,* 1935].

The primary source waters of the fresh intermediate waters within their gyre regions are the surface waters in the southwestern regions of the northern hemisphere subpolar gyres (for NPIW and LSW) and the surface waters east of the Falkland/Brazil confluence for the AAIW [*McCartney,* 1977]. The source water temperatures are taken to be 12°to 15°C (Table 2, used in section 3.1).

2.4. Deep Water Formation

Deep water lies below about 2000 meters. There are two surface sources for the deep waters of the main ocean basins: (1) the intermediate waters of the Nordic Seas [*Swift et al.,* 1980] which overflow into the North Atlantic producing the densest part of the North Atlantic Deep Water (NADW), and (2) dense water formed around Antarctica, primarily in the Weddell and Ross Seas and along Adelie Land [*Rintoul,* 1998], producing bottom/deep waters known variously as Antarctic Bottom Water (AABW) or Lower Circumpolar Deep Water (LCDW).

In the North Atlantic, Nordic Sea overflow water is joined by the intermediate-depth, outflowing Labrador and Mediterranean Sea Waters, amalgamated into a single low nutrient, high oxygen water mass by the time of passage into the subtropical South Atlantic [*Wüst,* 1935]. The global influence of the high salinity NADW was demonstrated by *Reid and Lynn* [1971].

Various authors have shown that the temperature of the northward-flowing water in the South Atlantic that feeds NADW formation must be at least 12°C [*Gordon,* 1986; review in *Talley,* 1996b]. A temperature of 14°C is used herein, which is the winter surface water temperature in the North Atlantic Current region, assuming that this water is the principal warm water source of NADW [*McCartney and Talley,* 1984].

The southern hemisphere deep and bottom waters are primarily formed through ice processes on the shelves. The densest of these waters do not stray far from the Antarctic basins, confined by topography. The "Antarctic Bottom Water" observed at mid-latitudes originates at shallow to intermediate levels south of the Antarctic Circumpolar Current [e.g., *Reid,* 1994]. A rough depiction of the influence of AABW is shown in Figure 4 based on the location of waters at $\sigma_4 = 45.92$ kg/m^3. At this potential density relative to 4000 dbar, water of Nordic Sea origin is confined north of Newfoundland (light shading in Figure 4), separated from water of Antarctic origin (medium shading). At slightly lower potential densities (e.g. $\sigma_4 = 45.91$ kg/m^3), water from the two sources merge, and so although there is Antarctic influence, it is confounded with Nordic Sea influence.

The warm water source of Antarctic Bottom Water is most likely NADW that passes southward through the fronts of the Antarctic Circumpolar Current in the eastern South Atlantic and Indian Oceans. NADW upwells south of the Antarctic Circumpolar Current to supply most of the upper layer that is modified to produce AABW. Therefore the NADW temperature is chosen in section 3.1 for the source of AABW (Table 2).

3. HEAT TRANSPORTS AND WATER MASSES

Net heat transport is a useful concept when the mass balance is closed, hence reflecting the temper-

Table 2. Heuristic Heat Transports for Deep and Intermediate Waters

	Temperature change	Volume Transport	Heat Transport
		Deep Waters	
NADW	14°to 2°C	8 Sv	0.4 PW
AABW	2°to -1°C	10 Sv (30 Sv[a])	0.12 PW (0.36 PW[a])
Total	...	...	0.52 PW (0.76 PW)
		Intermediate Waters	
LSW	14°to 3°C	8 Sv	0.3 PW
NPIW	12°to 5°C	5 Sv	0.16 PW
AAIW (Atlantic-Indian)	14°to 4°C	10 Sv	0.4 PW
AAIW (Pacific)	14°to 4°C	4 Sv	0.2 PW
SEISAMW	14°to 9°C	5 Sv (10 Sv[b])	0.1 PW (0.2 PW[b])
Total	...	...	1.2 PW (1.3 PW)

[a] *Schmitz* [1995]
[b] Calculation here.

ature difference between incoming and outgoing flow. For instance, the commonly-calculated meridional heat transports across subtropical latitudes use mass balance through the entire vertical section. Meridional heat transports in the South Pacific and South Indian Oceans must contend with the net flow around Australia and therefore "complete" sections include the Indonesian throughflow region.

Heat transport can also be computed meaningfully for portions of the circulation which conserve mass. In the present treatment, the overturning circulation is separated into shallow, intermediate and deep components, each of which conserves mass with equal inflow (at some high temperature) and outflow (at a lower temperature). When, on the other hand, a quantity like heat transport is calculated for a portion of the circulation which on its own does not conserve mass, for instance the transport of the Gulf Stream or the Ekman layer transport, it is common to use degrees Celsius and refer to the quantity as "temperature transport" which is a misnomer since the units are Watts. However, the phrase does at least highlight the meaninglessness of the transport until it is taken together with another portion of the circulation which allows mass to be balanced.

Two approaches are taken here to computing heat transport associated with different parts of the overturn. First (section 3.1), heuristic estimates of the contributions of all intermediate and deep waters to overturn are calculated, based on formation rates and source and final temperatures for each water mass. Published direct heat transport estimates for sections at 30-32°S and 24°N are then used to calculate the overturning heat transport due to the shallow subductive subtropical gyre overturn. This calculation is done for the whole globe. Secondly (section 3.2) meridional heat transport is calculated directly for each of the overturning elements across 24°N using *Reid's* [1994, 1997] total velocity fields. This second approach is being extended now to the southern hemisphere where interpretation is more difficult because of ambiguities in defining shallow and intermediate water overturn and because of the connections between the three subtropical gyres through the Indonesian passages and south of Africa.

Neither of these two methods is concerned with the details of the transformation of waters from source to final properties: the heuristic method is much too simple, assuming a single temperature for the source and final properties, while the direct method looks only at northward and southward flow across a single latitude. The details do matter, however, when one considers the response of heat transport to changes in surface forcing.

3.1. Heuristic Structure of the Overturning Heat Transport

The relative sizes of heat transport associated with deep and intermediate water formation can be estimated from their source and final temperatures and formation rates. Each deep and fresh intermediate water mass is examined in the next few paragraphs.

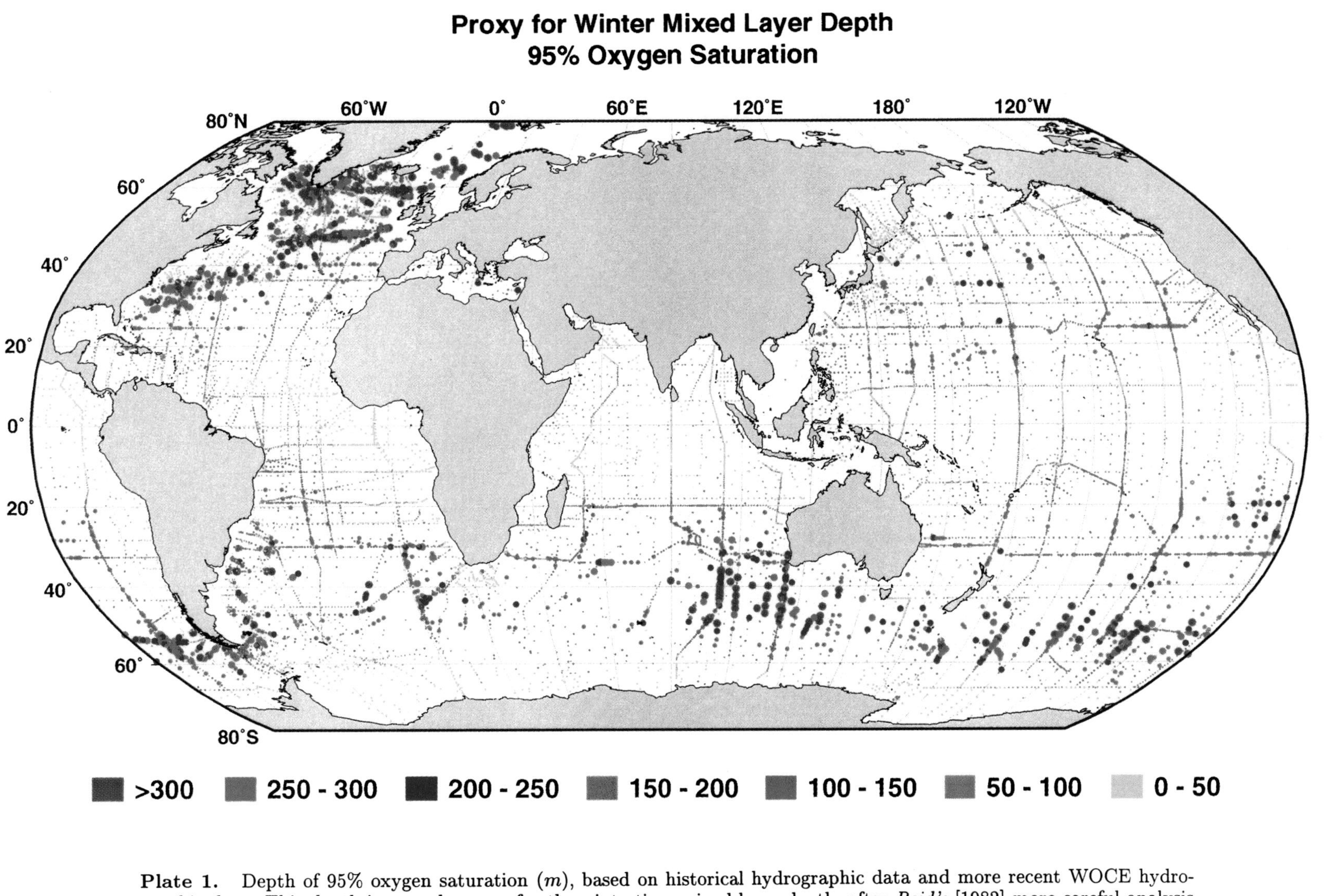

Plate 1. Depth of 95% oxygen saturation (m), based on historical hydrographic data and more recent WOCE hydrographic data. This depth is a crude proxy for the wintertime mixed layer depth, after *Reid's* [1982] more careful analysis for the North Pacific.

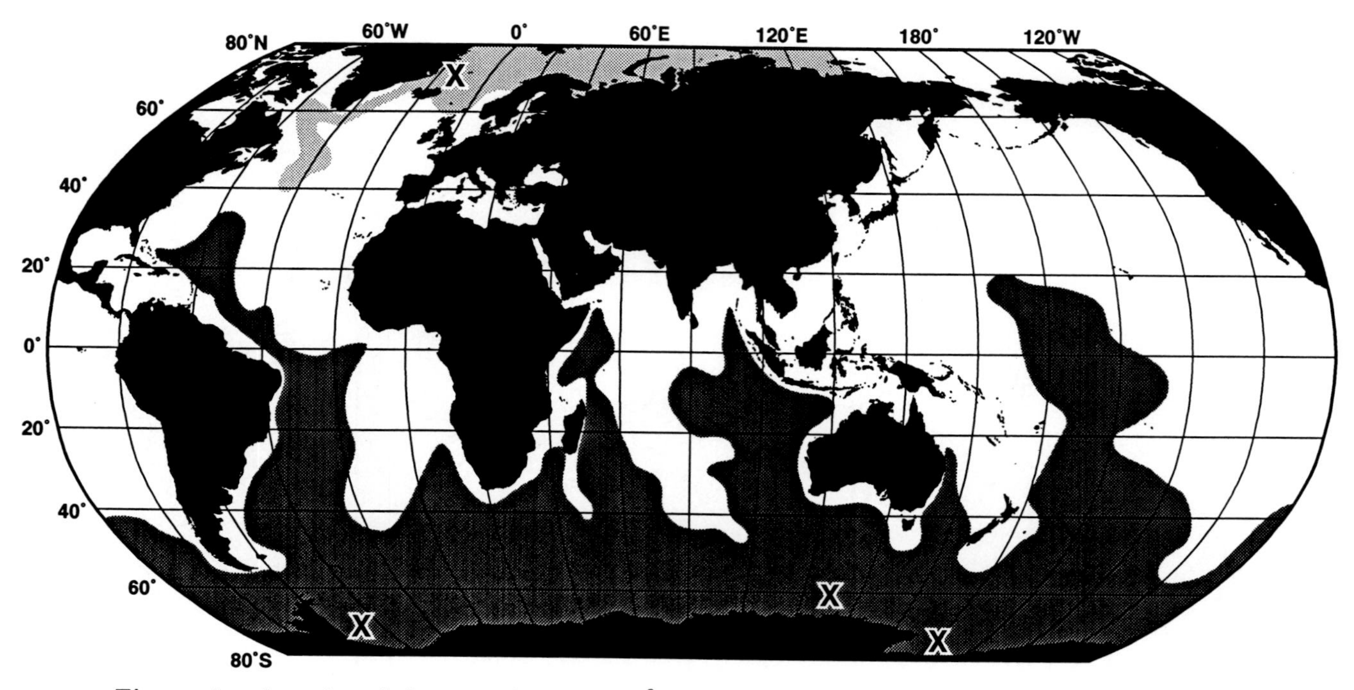

Figure 4. Location of the $\sigma_4 = 45.92$ kg m^{-3} isopycnal, representing the northward extent of the shallowest deep waters of Antarctic origin (dark shading) which can be unambiguously separated from those of North Atlantic origin (light shading). X's are formation regions for dense Antarctic waters. At lower densities, Antarctic water may be present but is not separated laterally from NADW in the North Atlantic.

As described in section 2, in the northern North Atlantic a thick layer of water of approximately 14°C is found east of the North Atlantic Current. *McCartney and Talley* [1982] define this as the warmest type of Subpolar Mode Water (SPMW), which is at the beginning of the process of cyclonic flow and gentle cooling around the subpolar gyre that eventually results in significant input into the Nordic Seas component of North Atlantic Deep Water (NADW) and production of Labrador Sea Water (LSW). The 14°C water cools as it is advected eastward; a portion splits northward into the subpolar gyre with the remainder staying in the subtropical gyre. For the heat budget, the details of this transformation are not important. A rough heat budget associated with NADW (Table 2) assumes an input temperature of 14°C, an outflow temperature of 2°C and a production rate of 8 Sv (1 Sv = $10^6 m^3/sec$) [*Worthington,* 1976; *McCartney and Talley,* 1984; *Dickson and Brown,* 1994; *Schmitz,* 1995]. Hence the estimated heat transport for this closed NADW mass balance is 0.4 PW. A rough heat budget for LSW assumes an input temperature of 14°C, an output temperature of 3°C, and a production rate of 8 Sv [*McCartney and Talley,* 1984; *Schmitz,* 1995], with an estimated heat transport of 0.3 PW. (Heat transports are quoted to tenths in the text although for calculations and in the tables they are carried in hundredths. An error estimate is difficult for such a rough calculation, but is expected to be on the order of 0.1 to 0.2 PW.)

If the 14°C SPMW, assumed as the main source of NADW and LSW, arises directly from water of about this temperature in the mid-subtropical gyre (hence in the Gulf Stream), with no input from warmer Gulf Stream waters, then the total heat transport of 1.2 PW which has been estimated directly across 24°N [*Hall and Bryden,* 1981; *Roemmich and Wunsch,* 1985; *Parilla et al.,* 1994] would be composed of about 0.7 PW due to NADW and LSW formation with the remaining 0.5 PW due to shallow overturn within the subtropical gyre. In section 3.2 below, the maximum heat transport associated with the shallow gyre overturn at 24°N is shown to be less than 0.3 PW, and thus the LSW/NADW heat transport is greater than 0.9 PW. Much of the water which feeds the NADW originates from Gulf Stream waters warmer than 14°C, with transformation to a lower temperature within the North Atlantic's subtropical (Gulf Stream and North Atlantic Current) circulation.

Table 3. Residual Shallow Gyre Overturn Heat Transports

Ocean	Direct Heat Transport Reference	Total Direct Heat Transport Estimate	Deep and Intermediate Water Heat Transports from Table 2	Residual Shallow Heat Transport
N. Atlantic	*Hall and Bryden* [1985]	1.2 PW	0.73 PW	0.47 PW
N. Atlantic	*Reid* [1994] and §3.2	1.18 PW		0.45 PW
N. Pacific	*Bryden et al.* [1991]	0.75 PW	0.16 PW	0.6 PW
N. Pacific	*Reid* [1998] and §3.2	0.63 PW		0.5 PW
S. Atlantic	*Rintoul* [1991]	0.25 PW	0.73-0.24 = 0.5 PW	-0.25 PW
S. Atlantic	*Rintoul* [1991]	0.25 PW	0.73-0.33 = 0.4 PW	-0.15 PW
S. Pacific	*Wunsch et al.* [1983]	-0.2 PW	0.2-0.24 = -0.04 PW	-0.16 PW
S. Pacific	*Wunsch et al.* [1983]	-0.2 PW	0.2-0.33 = -0.13 PW	-0.07 PW
Indian	*Robbins and Toole* [1995]	-0.4 PW	-0.35 PW	-0.05 PW
Indian	*Robbins and Toole* [1995]	-0.4 PW	-0.53 PW	0.13 PW

These heat transports (negative southward) are based on Table 2 and published direct heat transport estimates for 24°N and 30°S. South Atlantic and South Pacific deep/intermediate transports are the sum of their northern hemisphere deep/intermediate waters with an apportionment for southern hemisphere AABW and AAIW; upper and lower lines use minimum and maximum southern hemisphere transports from Table 2. The various use of one and two-place precision reflects the various estimates that go into the calculation, and should not be taken too literally.

The North Pacific forms intermediate water but not deep water. The warm water source for North Pacific Intermediate Water (NPIW) could be considered to be the thick layer of 10 - 12°C water found north of the Kuroshio. As in the North Atlantic, this thick layer advects eastward and a portion turns northward into the subpolar gyre, with maximum density occurring in the Okhotsk Sea due to brine rejection under sea ice formation and hence near the freezing point. The input to NPIW, however, is a mixture of the new dense water due to sea ice production and North Pacific waters produced through very strong tidal mixing at the Kuril Straits [*Talley,* 1991]. Thus 5°C, rather than a much colder temperature, is assigned here to the NPIW. Using an NPIW production rate of 5 Sv [*Talley,* 1997], the associated overturning heat transport is 0.2 PW (Table 2).

A heat transport of 0.75 PW across 24°N in the Pacific has been directly estimated by *Bryden et al.* [1991] and *Roemmich and McCallister* [1989] using a 1985 CTD section. (In section 3.2, the same data set with *Reid's* [1998] velocity analysis yields a direct heat transport estimate of 0.63 PW.) Thus the heat transport across this latitude due to shallow overturning might be 0.5-0.6 PW (Table 3). The complete calculation in section 3.2 below substantiates this heuristic argument, as does the analysis of *Bryden et al.* [1991].

In the southern hemisphere, Antarctic Intermediate Water is considered here to begin as a thick layer of Subantarctic Mode Water with an approximate temperature of 14°C east of the Brazil/Falkland Current confluence [*McCartney,* 1977]. Its final temperature just west of Chile is 4°C. Its formation rate is taken at 14 Sv [*Schmitz,* 1995], of which 4 Sv remains in the Pacific and 10 Sv passes through Drake Passage to enter the Atlantic/Indian gyre. The heat transport associated with AAIW formation is thus 0.6 PW (Table 2).

Southeast Indian Subantarctic Mode Water (SEISAMW) is similar dynamically to Pacific Antarctic Intermediate Water, and so a separate accounting for its heat transport is included here (see section 2.2 above). Its temperature is about 9°C and its formation rate is about 5 Sv according to *Schmitz* [1995]. The new WOCE sections which enclose the southeastern Indian Ocean can be used to compute SEISAMW export to the west and north in the Indian Ocean, with both a box at 32°S and 95°E and a closed section at 110°E. Both of these closed sections yield about 12 Sv of flow in the potential density range σ_θ = 26.7-26.9 into the Indian Ocean, with about half in the σ_θ = 26.7-26.8 and the other half in the σ_θ = 26.8-26.9 range. Therefore a larger transport of 10 Sv is used. If it is assumed that SEISAMW, like AAIW, originates at about 14°C, then the associated overturning heat transport is 0.1 PW or 0.2 PW for 5 Sv or 10 Sv, respectively (Table 2).

Antarctic Bottom Water (AABW) likely originates from NADW/Circumpolar Deep Water that upwells south of the Antarctic Circumpolar Current. The input temperature is about 2°C. The outflow temperature is about -1°C. With a production rate of 10 Sv, the associated overturning heat transport is about 0.1 PW (Table 2). However, *Schmitz* [1995] shows a net production rate of about 30 Sv, when both the Weddell and Ross Sea sources are considered; a source along Adelie Land has also been substantiated [*Rintoul,* 1998]. With this large net export rate of 30 Sv of AABW, the net overturning heat transport associated with its production could be as much as 0.4 PW (Table 2).

Partitioning the overturning heat transports across the southern hemisphere basins at 30°S where direct estimates of heat transport have been made is difficult because the basins are connected. The first assumption is that the overturning heat transports calculated across the northern hemisphere lines for NPIW, NADW and LSW are assigned to the southern hemisphere sections, and hence that the 12°- 14°C water that is the northern hemisphere source originates at the same temperature in the southern hemisphere. Secondly, it is assumed here that the overturning heat transport associated with AAIW and AABW formation is divided equally between the three southern hemisphere oceans.

The direct heat transport estimate of 0.25 PW northward across 32°S in the South Atlantic [*Rintoul,* 1991] is thus divided here into 0.73 PW northward due to LSW and NADW formation, -0.2 PW (southward) due to AAIW and -0.1 PW due AABW formation. The remaining -0.2 PW of southward transport is then ascribed to shallow overturn in the subtropical gyre (Table 3).

The direct heat transport of -0.2 PW southward across 28°S in the South Pacific [*Wunsch et al.,* 1983] was based on a circulation with only a small Indonesian throughflow. The South Pacific heat transport is partitioned here into 0.2 PW northward due to NPIW formation, -0.2 PW (southward) due to AAIW formation in the Pacific and -0.1 PW due to AABW formation. These estimates leave -0.1 PW southward for the remaining shallow gyre overturn in the South Pacific, exclusive of the AAIW formation there. Note that *Wunsch et al.* [1983] indicated that their heat transport is indistinguishable from 0 PW, and so the non-AAIW shallow overturn could also be indistinguishable from 0.

In the Indian Ocean, the heat gain of -0.4 PW north of 32°S [*Robbins and Toole,* 1997] is balanced by export of heat across the two-part section composed of the 32°S section and one across the Indonesian throughflow. This net export of 0.4 PW is divided here into -0.2 PW southward due to AAIW formation, -0.1 PW due to AABW formation, and -0.1 to -0.2 PW southward due to SEISAMW formation (Table 3). This budget leaves 0.03 to -0.07 PW, indistinguishable from 0, for additional shallow gyre overturn, suggesting that SEISAMW formation dominates the shallow overturn.

Despite the limitations of these calculations, which are only partially supported by the direct heat transport estimates of section 3.2 below, one robust conclusion can be drawn: the global meridional heat transport associated with the formation of deep and bottom waters (NADW and AABW) is not significantly larger than the heat transport associated with formation of intermediate waters (LSW, NPIW, AAIW and SEISAMW). Even if only LSW and NPIW are considered as intermediate waters, the net heat transport associated with them is comparable to that for the deep waters. The input waters for each of these water masses, with the exception of AABW, are warm, on the order of 12 to 14°C, and the outflowing waters are on the order of 2 to 5°C. The temperature differences between inflow and outflow are of the same order. The total transformation rates are of the same order. Therefore the overturning heat transports associated with the deep and intermediate waters are about the same size.

The meridional heat transport associated with the shallow overturning gyres appears to be larger in the northern hemisphere oceans than in the southern hemisphere oceans, although as seen below, actually only the North Pacific's shallow overturn is significant. Weaker gyre strength and weaker surface buoyancy loss in the southern hemisphere subtropical gyres might be the cause of the difference. Or it could be that AAIW and SEISAMW dominate shallow overturn in southern hemisphere, although they were counted here as intermediate water formation.

3.2. Direct Estimates of Heat Transport in the Northern Hemisphere

Ocean heat transport has been estimated directly from ocean velocity (geostrophic and Ekman) and temperature information along zonal sections crossing each of the subtropical gyres. Absolute geostrophic velocities for the sections represented by arrows in Figure 1 have been estimated using inverse models with varying constraints [*MacDonald and Wunsch,* 1996 for all oceans; *Roemmich and McCallister,* 1989 for the North Pacific; *Roemmich and Wunsch,* 1985 for the North Atlantic; *Wunsch et al.,* 1983 for the South Pacific; *Rintoul,* 1991 for the South Atlantic; *Robbins and Toole,* 1997 for the Indian Ocean]. Geostrophic reference velocities were selected in a more traditional way based on water mass analysis and relative flow patterns for several of the same sections [*Bryden et al.,* 1991 for the North Pacific; *Toole and Warren,* 1993 for the Indian; *Reid,* 1994 for the Atlantic sections; *Reid,* 1997 for the Pacific sections]. The method employed by *Hall and Bryden* [1982] for the North Atlantic zonal section took advantage of the zonal uniformity of vertically-averaged potential temperature along 24°N, allowing them to calculate the heat transport using only the baroclinic velocities rather than the absolute velocities.

The direct estimates presented here include only the two northern hemisphere subtropical gyres and are based on *Reid's* [1994, 1997] adjusted geostrophic velocities. CTD stations were occupied in 1981 along 24°N in the North Atlantic [*Roemmich and Wunsch,* 1985] and in 1985 along 24°N in the North Pacific [*Roemmich et al.,* 1991]. The North Atlantic section ends at the Bahamas in the west and is completed across the Gulf Stream at 26°N. The data (pressure, temperature, salinity) resolution is 2 dbar in the vertical. *Reid* [1994, 1997] included these two sections in his basin-wide flow analyses. He graciously provided files with his velocity solution at the deepest common level for each station pair. Reid assumed geostrophic mass balance across each section. For this study, Reid's velocities were adjusted to allow total mass balance to include Ekman transport, as described below.

The goal here is to estimate the contribution to the overturning heat transport due to shallow, intermediate and deep components. The shallow overturn is defined to extend down to the maximum density which is found at the sea surface in winter at the subtropical/subpolar gyre boundary, as defined by the Sverdrup transport. It is assumed that this is the maximum density of subducted overturn in the subtropical gyre (*Stommel's* [1979] Ekman demon concept). This assumption seems reasonable since even if there is baroclinic exchange across the gyre boundary, it most likely takes the form of southward Ekman transport of surface water (hence the densest surface water at the gyre boundary) and northward return flow of underlying water. Thus the shallow overturn is confined to the subducted portion of the subtropical gyre. This approach was taken by *Roemmich and Wunsch* [1985] for the North Atlantic 24°N section.

Once the maximum potential density for subduction is selected ($\sigma_\theta = 26.2$ for the North Pacific and $\sigma_\theta = 27.3$ for the North Atlantic), the mass transport across the subtropical zonal section is computed for this uppermost layer, including the geostrophic and Ekman components. The transport is divided between the northward western boundary current (Gulf Stream/Kuroshio) and the southward interior flow. Since the zero of northward Ekman transport lies at about 30°N and since all (northward) Ekman transport at 24°N is at much lower density than the maximum subducted density, it is assumed that all Ekman flow at 24°N loses buoyancy and joins the southward geostrophic flow of the subtropical gyre, with none continuing into the subpolar gyre. The mass balance required for a net heat transport for the shallow overturn is taken to be: (1) all northward Ekman transport at 24°N, (2) all southward geostrophic flow in the interior, and (3) the portion of the northward geostrophic flow in the western boundary current in this shallow layer that is required to balance mass, that is,

$$V_{wbcshall} = V_{ek} + V_{geointshall} \tag{1}$$

The remaining (geostrophic) transport in the warm water layer of the western boundary current enters the subpolar gyre and returns as intermediate or deep water. This result is similar to *Roemmich and Wunsch's* [1985] approach to what they labelled as "shallow wind-driven circulation" (which is described here as the thermohaline overturning associated with the subducting subtropical part of the wind-driven circulation).

For heat transport in the shallow overturning associated with (1), the annually-averaged temperature at 30 m from *Levitus et al.* [1994] is used with the Ekman transport, and the measured temperature averaged between adjacent CTD stations is used with the adjusted geostrophic velocities. The western boundary current portion of the heat transport is assumed to be of the least dense (warmest) water down to the depth required for mass balance. This choice maximizes the meridional heat transport assigned to the shallow overturn. A reasonable minimum value is obtained by using the average temperature of the whole shallow layer of the western boundary current, which reduces the northward temperature transport assumed for the western boundary current component of the shallow overturn, as was done by *Roemmich and Wunsch* [1985].

The intermediate and deep water overturning contributions to the heat transport across 24°N are described in the individual calculations below, where it is assumed that up and downwelling occur to the closest possible density for mass balance.

3.2.1. North Pacific. Hydrographic station data at 24°N in the Pacific were collected in 1985 [*Roemmich et al.,* 1991]. Meridional heat transport was calculated using an inverse method by *Roemmich and McCallister* [1989], who obtained 0.75 PW northward. The same heat transport was also calculated using adjusted reference velocities by *Bryden et al.* [1991]. (In comparison, *MacDonald and Wunsch* [1996] included the section in a global inverse model, and obtained 0.45 PW. Because regional attention was paid to initial reference velocity choices in the earlier papers which treated only the Pacific 24°N section, the earlier estimates were used in the heuristic treatment of section 3.1 above.)

Reid [1997] used the 24°N section in his Pacific circulation analysis and provided his set of bottom velocities (adjustments) for the present calculation. He assumed a net geostrophic mass balance across the section. Ekman transport across 24°N from *Hellerman and Rosenstein's* [1983] annually-averaged winds is 10.3 Sv northward. To adjust Reid's geostrophic transport to 10.3 Sv southward, a southward velocity correction of -0.0135 cm/sec was applied at all points of the section so that the geostrophic and Ekman transports sum to 0. (A similarly ad hoc adjustment was made by *Bryden et al.* [1991] for their estimate of heat transport at 24°N in the North Pacific.) The total, top-to-bottom, meridional heat transport across 24°N, using Reid's velocities adjusted as described and the Ekman transport, is (Table 4 and Figure 5)

$$\begin{aligned} H_{tot} = 0.63PW &= H_{ek} + H_{geostrophic} \\ &= 0.94PW - 0.31PW \end{aligned} \tag{2}$$

where the "heat" transports on the right hand side are relative to 0°C, and are referred to henceforth as "temperature transports", following *Hall and Bryden* [1982]. The heat transport on the left-hand side is associated with a complete mass balance and hence is a true net heat transport.

The maximum potential density of shallow overturn is $\sigma_\theta = 26.2$, based on the density at the zero of Sverdrup transport in the North Pacific [e.g., *Yuan and Talley,* 1992]. The shallow overturning heat transport, depicted in Figure 5b, is associated with the mass (volume) balance (Table 4):

$$\begin{aligned} V_{shallow} = 0Sv &= V_{wbcshall} + V_{ek} + V_{geointshall} \\ &= 21.5Sv + 10.3Sv - 31.8Sv \end{aligned} \tag{3}$$

$$\begin{aligned} H_{shallow} = 0.57PW &= H_{wbcshall} + H_{ek} + H_{geointshall} \\ &= 1.87PW + 0.94PW - 2.24PW. \end{aligned} \tag{4}$$

Again the heat transport on the left side is a true, net heat transport involving mass balance, while terms on the right side are temperature transports relative to 0°C. The Kuroshio (western boundary current) transport of 21.5 Sv required to balance the interior and Ekman transports is assumed to be of the least dense portion of the current, which is calculated from the data set to be all water of potential density less than $\sigma_\theta = 26.0$. This shallow heat balance leaves a small residual (lower bound) heat transport of 0.09 PW associated with intermediate water formation and deep overturn, which is now examined in detail.

The vertical mass overturn of the North Pacific was depicted by *Roemmich and McCallister* [1989] and *Bryden et al.* [1991] using zonally-integrated mass transports in pressure and potential temperature layers, respectively. The zonally integrated mass transport in potential density layers using *Reid's* [1997] velocities plus the correction made here for Ekman transport is shown in Figure 5c. The layers here were chosen to contain commonly-defined water masses. The layer boundaries are: the sea surface, $\sigma_\theta = 26.2$ for the maximum ventilated subtropical potential density as described above, $\sigma_\theta = 27.0$ lying between North Pacific Intermediate Water (NPIW) and Antarctic Intermediate Water (AAIW), $\sigma_\theta = 27.6$ between AAIW and Pacific Deep Water (PDW), $\sigma_2 = 36.96$ and $\sigma_4 = 45.84$ within the PDW, and $\sigma_4 = 45.88$ between the PDW and Lower Circumpolar Deep Water (LCDW, also known as Antarctic Bottom Water).

The main features of the mass transports here are similar to the previous depictions: northward Ekman transport, southward transport in the upper ocean, southward transport in the NPIW layer, northward transport in the AAIW layer, southward transport in the PDW, which is highest in the layer just above the LCDW, and northward transport of LCDW. The upper two PDW layers have very little transport, a feature also seen in the previous analyses.

The 3.1 Sv of northward-flowing bottom water must upwell and return southward across the section. This is assumed to be into the next layer up, of Pacific Deep Water, where the transport is -5.9 Sv. Assume that the temperature transport in this second layer is $\frac{3.1}{5.9}$ of the total temperature transport, given no information on exactly where the return of the deepest water occurs. Then the southward heat transport associated with this deep upwelling is

$$\begin{aligned} H_{bottom} = -0.002PW &= H_{LCDW} + \tfrac{3.1}{5.9}H_{PDW3} \\ &= 0.012PW - 0.014PW \end{aligned} \tag{5}$$

where LCDW refers to the bottom layer and PDW3 refers to the deepest of the three Pacific Deep Water layers.

Table 4a. Overall Volume, "Temperature" (T) and Heat Transports for 24°N

	North Pacific		North Atlantic	
	Volume (Sv)	Heat (PW)	Volume (Sv)	Heat (PW)
Ekman	10.3 Sv	0.94 PW (T)	5.5 Sv	0.43 PW (T)
Total Geostrophic	-10.3 Sv	-0.31 PW (T)	-5.5 Sv	0.75 PW (T)
Total	0.0 Sv	0.63 PW	0.0 Sv	1.18 PW

These estimates use *Reid's* [1994, 1998] velocity analyses, adjusted here to balance Ekman transport. (T) indicates temperature transport, as defined in the text.

Table 4b. Shallow and Intermediate/Deep Volume and Heat Transports for 24N

	North Pacific		North Atlantic	
	Volume (Sv)	Heat (PW)	Volume (Sv)	Heat (PW)
Ekman	10.3 Sv	0.94 PW (T)	5.5 Sv	0.43 PW (T)
Shallow Interior Geostrophic	-31.8 Sv	-2.24 PW (T)	-18.7 Sv	-1.54 PW (T)
Shallow Western Boundary Current Compensating Ekman and Shallow Interior	21.5 Sv	1.87 PW (T)	13.2 Sv	1.40 PW (T)
Shallow Layer Overturn Total	0.0 Sv	0.57 PW	0.0 Sv	0.29 PW
Intermediate/Deep Geostrophic	-1.6 Sv	-0.03 PW (T)	-15.8 Sv	-0.12 PW (T)
Shallow Western Boundary Current Compensating Intermediate and Deep Interior	1.6 Sv	0.09 PW (T)	15.8 Sv	1.01 PW (T)
Intermediate/Deep Overturn Total	0.0 Sv	0.06 PW	0.0 Sv	0.89 PW

These estimates use *Reid's* [1994, 1998] velocity analyses, adjusted here to balance Ekman transport. (T) indicates temperature transport, as defined in the text.

A downwelling of 2.8 Sv from the AAIW layer ($\sigma_\theta =$ 27.0-27.6) to the near-bottom layer is required to balance deep mass in this scheme. In *Roemmich and McCallister* [1989] there was almost complete mass balance in the bottommost two layers (northward at the bottom and compensating southward just above). Using Reid's analysis with the adjustment for Ekman transport, there is more southward flow just above the bottom layer than needed to compensate for the northward bottom layer. This excess flow requires downwelling from intermediate depths to the near-bottom layer to balance deep mass. A similar problem, requiring downwelling, was noted by *Bryden et al.* [1991] in their net transports for this same section for which they suggested mixing as a solution, but which they also noted was somewhat inconsistent with the mass balance. Without an obvious physical mechanism for so much downwelling, there is likely an error in the

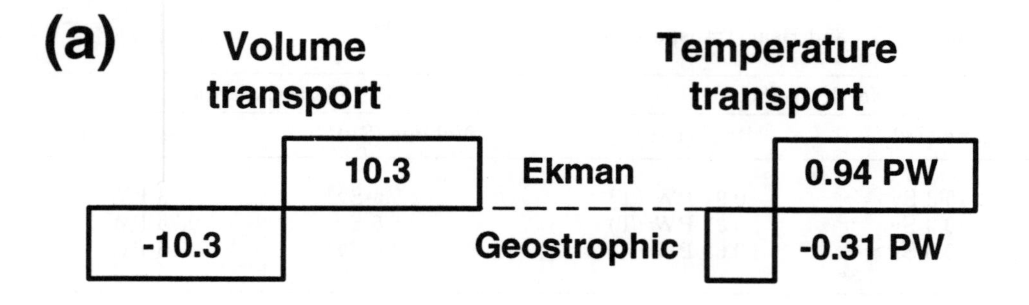

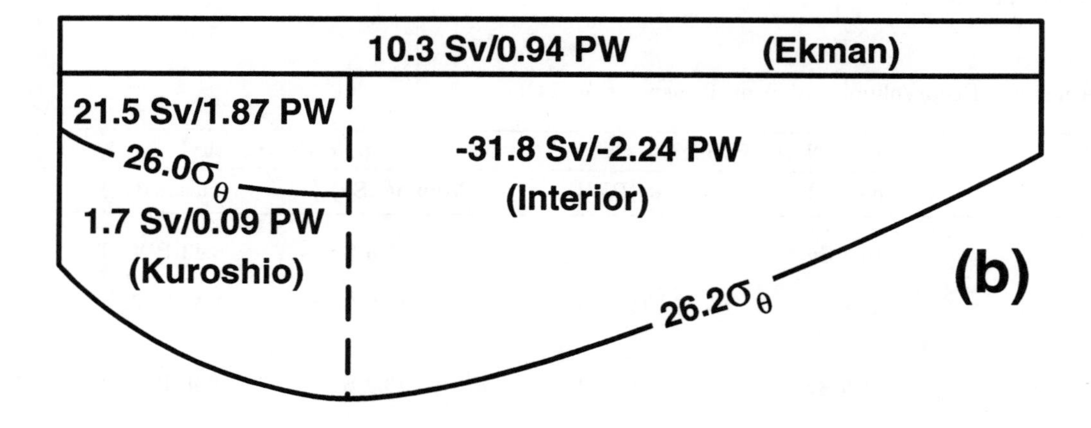

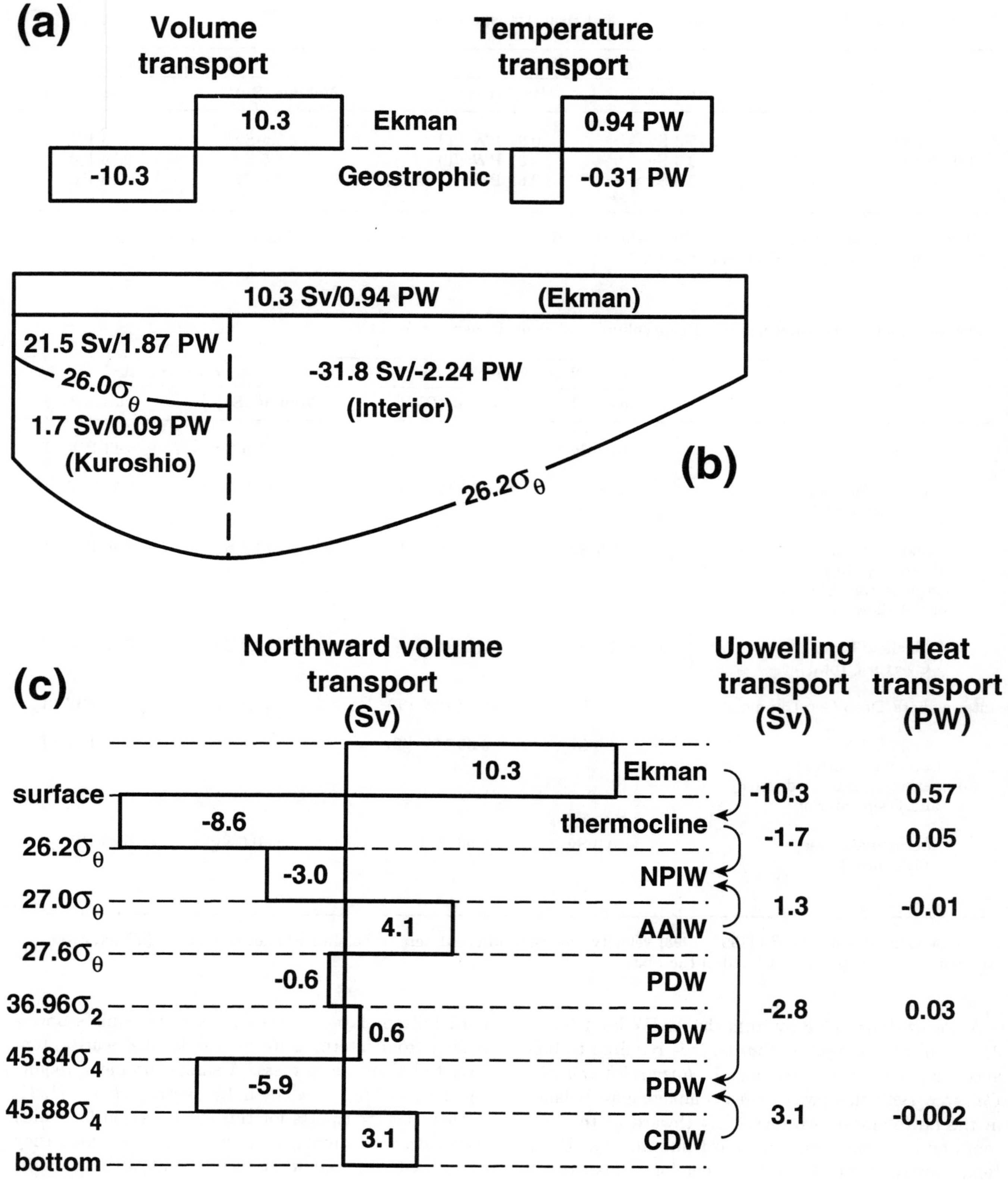

(c)

Northward volume transport (Sv)

Upwelling transport (Sv)

Heat transport (PW)

Layer (upper boundary–lower boundary)	Water mass	Northward volume transport (Sv)	Upwelling transport (Sv)	Heat transport (PW)
	Ekman	10.3	-10.3	0.57
surface – $26.2\sigma_\theta$	thermocline	-8.6	-1.7	0.05
$26.2\sigma_\theta$ – $27.0\sigma_\theta$	NPIW	-3.0	1.3	-0.01
$27.0\sigma_\theta$ – $27.6\sigma_\theta$	AAIW	4.1		
$27.6\sigma_\theta$ – $36.96\sigma_2$	PDW	-0.6	-2.8	0.03
$36.96\sigma_2$ – $45.84\sigma_4$	PDW	0.6		
$45.84\sigma_4$ – $45.88\sigma_4$	PDW	-5.9	3.1	-0.002
$45.88\sigma_4$ – bottom	CDW	3.1		

Figure 5. Direct heat transport estimate at 24°N in the Pacific, using *Reid's* [1997] absolute velocities, adjusted to zero mass balance using Ekman transport based on *Hellerman and Rosenstein's* [1983] annually-averaged winds. (a) Volume and temperature transports in the Ekman layer and total geostrophic flow. (b) Shallow overturning schematic, divided into the western boundary current, Ekman layer and southward interior geostrophic flow. The base of the shallow layer is at $\sigma_\theta = 26.2$ kg m^{-3}. The Kuroshio transport is divided in two: 21.5 Sv required to balance the Ekman and interior return flow in this layer, and a 1.7 Sv residual which flows on into the subpolar gyre. "Temperature" transports (heat transports without mass balance) are relative to 0°C. The net heat transport for this mass-balanced shallow overturn is 0.47 PW. (c) Zonally-integrated transports in isopycnal layers (bar graph). Vertical arrows and transports indicate assumed transfer of mass between layers, assuming that mass moves preferentially to the most adjacent layer. The heat transports at the right are based on the mass balances implied by the upwelling transports, and so are the net heat transports associated with each part of the overturn.

mass transports. (In Reid's original velocities, unadjusted for Ekman mass balance, there is near balance between these two deepest layers, suggesting that the excess flow and hence downwelling is an artifact of the *ad hoc* correction applied here.) Since deep temperature differences are small, such an error in deep transport has little effect on heat transport, as noted by *Bryden et al.* [1991]. For the downwelling transport, assume again that the temperature transport contribution from the AAIW layer is simply proportional to its total temperature transport. This downwelling carries

$$\tfrac{2.8}{4.1}H_{AAIW} + \tfrac{3.1}{5.9}H_{PDW3} = 0.04PW - 0.01PW = 0.03PW. \quad (6)$$

Because the temperature difference between these two layers is small despite their large vertical separation, this heat transport, even if incorrect, is small compared with the total northward heat transport of 0.63 PW for the section.

The remaining 1.3 Sv of the northward AAIW transport must upwell into the NPIW layer, partially compensating the 3.0 Sv of southward NPIW transport. This upwelling has associated southward heat transport of

$$\tfrac{1.3}{3.0}H_{NPIW} + \tfrac{1.3}{4.1}H_{AAIW} = -0.03PW + 0.02PW = -0.01PW. \quad (7)$$

The remaining 1.7 Sv of southward flow in the NPIW layer is assumed to come from the surface layer of the western boundary current, so the associated heat transport is the difference between the total heat transport of 0.63 PW and all of these layer transports, or 0.05 PW.

Thus the partition of northward heat transport across 24°N in the Pacific into the shallow overturn, NPIW overturn, and deep water upwelling is 0.57 PW, 0.09 PW and -0.002 PW, respectively, plus the putative 0.03 PW of downwelling heat transport from AAIW to PDW.

This partitioning between shallow subtropical gyre overturn and intermediate water overturn resembles the heuristic calculation of section 3.1 above (Table 2), which was based on rough assumptions about the intermediate water formation rate and associated temperature change from inflow to outflow.

3.2.2. North Atlantic. Hydrographic data have been collected along 24°N in the Atlantic four times - 1957, 1981, 1992 and 1998. *Hall and Bryden* [1981] calculated a net heat transport of 1.22 PW across the 1957 section based on temperature transports of 0.42 PW due to northward Ekman transport, 2.38 PW due to the northward current in Florida Straits, and -1.58 PW due to the southward interior flow.

Roemmich and Wunsch [1985] used an inverse method to calculate a meridional heat transport of 1.2 PW across the 1957 and 1981 sections. They divided this into 0.1 PW for the shallow "wind-driven" circulation using 27.4 σ_θ as the bottom of the subducted flow, and 1.1 PW for the "thermohaline" circulation. In their global inverse, *Macdonald and Wunsch* [1996] calculated a northward heat transport of 1.1 PW for the 1981 24°N section. *Parrilla et al.* [1994] found a net heat transport of 1.2 PW for the 1992 section, very similar in size to the calculations for the 1981 section, despite a temperature increase of 0.1 to 0.2°C from 1981 to 1992 at Labrador Sea Water depths. Heat transports are sensitive only to much grosser temperature changes, as well as changes in the mass transport distribution, which apparently differed little between 1981 and 1992.

Reid [1994] used the 1981 24°N section in his circulation analysis of the Atlantic, and supplied the geostrophic velocities for this calculation. The section is composed of two portions: a Gulf Stream crossing at 26°N and the main section at 24°N between the Bahamas and the eastern boundary. As done for the North Pacific in section 3.1 above, an Ekman transport of 5.5 Sv was calculated from the annually-averaged winds of *Hellerman and Rosenstein* [1983]. Since Reid's geostrophic mass was balanced, velocities here were adjusted at all points by -0.0118 cm/sec to balance the Ekman and geostrophic mass transports.

The total heat transport from Reid's geostrophic velocities with this correction applied, again using the 30 m annually-averaged temperature from *Levitus et al.* [1994] for the Ekman heat transport, is (Table 4a and Figure 6)

$$H_{tot} = 1.18PW = H_{ek} + H_{geostrophic} \\ = 0.43PW + 0.75PW \quad (8)$$

which is the same total as from *Hall and Bryden* [1981], *Roemmich and Wunsch* [1985] and *Parilla et al.* [1994]. Here the net geostrophic mass transport is southward (-5.5 Sv), but the associated temperature transport (relative to a temperature of 0°C) is northward, unlike for the North Pacific.

Shallow overturn in the North Atlantic is defined to be confined to densities lower than $\sigma_\theta = 27.3$, based on the maximum surface density in the subtropical gyre [e.g. *McCartney,* 1982]. In this upper ocean layer, the southward geostrophic transport east of the Bahamas is -18.7 Sv. With a northward Ekman transport across 24°N of 5.5 Sv, 13.2 Sv of the total of 29.0 Sv of the Florida Strait Gulf Stream's upper layer transport is required to balance mass. 15.8 Sv of the Gulf Stream's upper layer continue northward to become part of the intermediate/deep overturn. It is next assumed that the Gulf Stream transport of 13.2 Sv that remains in the subtropical gyre is the warmest, least dense part of the Gulf Stream or all water less dense than $\sigma_\theta =$ 25.8, with an associated temperature transport of 1.41 PW northward. The volume and heat balances for the subtropical subductive overturn, assuming that the Ekman transport and warmest Gulf Stream waters feed the southward interior flow, is then (Table 4b)

$$V_{shallow} = 0Sv = V_{wbcshall} + V_{ek} + V_{geointshall} \\ = 13.2Sv + 5.5Sv - 18.7Sv \quad (9)$$

$$H_{shallow} = 0.30PW = H_{wbc} + H_{ek} + H_{interior} \\ = 1.41PW + 0.43PW - 1.54PW. \quad (10)$$

as depicted in Figure 6b. This heat transport is somewhat less than the heuristic estimate (section 3.1) of 0.5 PW for the North Atlantic, and is likely an upper bound on the shallow heat transport, since it is assumed that the warmest Gulf Stream water stays in the subtropical gyre while the colder water in this shallow layer passes to the subpolar gyre. It is equally possible that some of the colder shallow water remains in the gyre; if it is assumed that the temperature transport for this surface to the $\sigma_\theta = 27.3$ layer is just a simple proportion of the total temperature transport in the Gulf Stream [as in *Roemmich and Wunsch,* 1983], the total heat transport for the shallow gyre overturn is reduced to about 0.0 PW. Similarly low heat transport for the shallow gyre was calculated by *Hall and Bryden* [1981] and *Roemmich and Wunsch* [1983].

The water column is next divided according to water masses to estimate the relative contribution of the remaining parts of the overturn to the total heat transport. The vertical distribution of zonally-integrated transport (Figure 6c) based on *Reid's* [1994] velocities is in keeping with that of *Roemmich and Wunsch* [1983], with a northward flux of 6.9 Sv of Antarctic Bottom Water (AABW), a southward flux of about 19 Sv of North Atlantic Deep Water (NADW), a southward flux of about 7 Sv of Labrador Sea Water (yielding a total southward flux of NADW plus LSW of about 26 Sv), and a net northward transport at the Antarctic Intermediate Water level and in the upper ocean.

Starting from the deepest layer, all 6.9 Sv of AABW must upwell, assumed to be into the southward-flowing NADW lying just above. The (southward) heat transport associated with this abyssal upwelling is -0.02 PW, which is small because of the small temperature difference between AABW and NADW.

The remaining 11.9 Sv of southward-flowing NADW and 6.5 Sv of LSW must be fed by downwelling from the surface and AAIW layers (water mass formation). The overturning heat transport associated with LSW is 0.43 PW and with NADW is 0.48 PW.

Thus the North Atlantic's meridional heat transport of 1.18 PW at 24°N is dominated by conversion of upper layer water into LSW and NADW, which contribute together 0.91 PW. In contrast, the North Pacific is dominated by the shallow overturning, as already concluded by *Hall and Bryden* [1981], *Roemmich and Wunsch* [1985] and *Bryden et al.* [1991]. The LSW overturning heat transport is a substantial fraction of the whole. However, the upper bound on the strength of the subtropical shallow overturn as calculated here is greater than in Roemmich and Wunsch and in Hall and Bryden, due to the assumption here that the Gulf Stream transport which feeds the shallow overturn is of the warmest waters. The upper bound on shallow overturning heat transport of 0.30 PW is lower than the

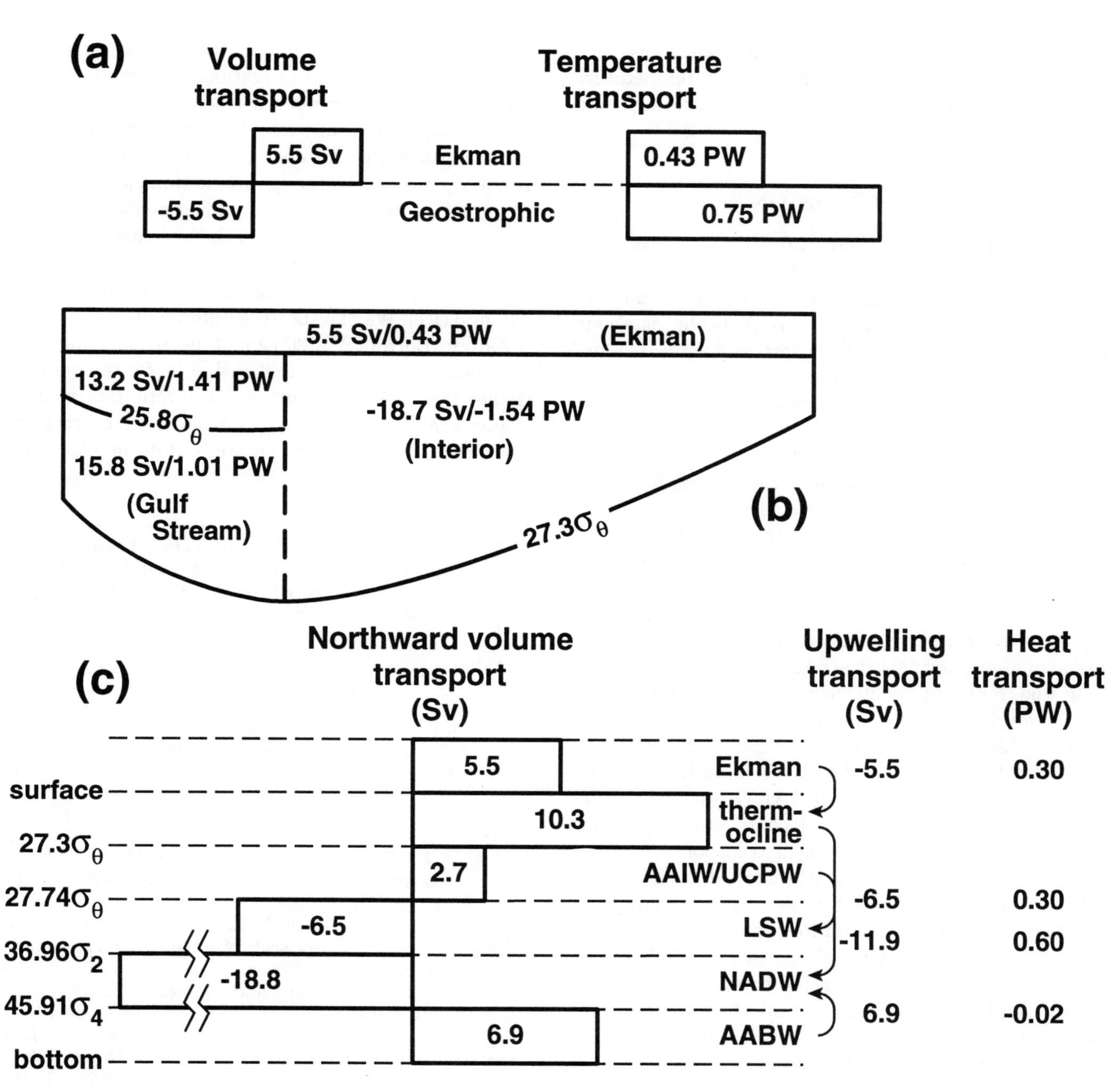

Figure 6. Direct heat transport estimate at 24°N in the Atlantic, using *Reid's* [1997] absolute velocities, adjusted to zero mass balance using Ekman transport based on *Hellerman and Rosenstein's* [1983] annually-averaged winds. (a) Volume and temperature transports in the Ekman layer and total geostrophic flow. (b) Shallow overturning schematic, divided into the western boundary current, Ekman layer and southward interior geostrophic flow. The base of the shallow layer is at $\sigma_\theta = 27.3$ kg m^{-3}. The Gulf Stream upper layer transport is divided in two, as in *Roemmich and Wunsch* [1985]: 22.1 Sv required to balance the Ekman and interior return flow in this layer, and 15.9 Sv residual which flows on into the subpolar gyre. "Temperature" transports (heat transports without mass balance) are relative to 0°C. The net heat transport for this mass-balanced upper layer overturn is 0.03 PW. (c) Zonally-integrated transports in isopycnal layers (bar graph). Vertical arrows and transports indicate assumed transfer of mass between layers, assuming that mass moves preferentially to the most adjacent layer. The heat transports at the right are based on the mass balances implied by the upwelling transports, and so are the net heat transports associated with each part of the overturn.

heuristic estimate of 0.47 PW for the shallow overturn (section 3.1), implying that the average temperature of surface/thermocline water that is transformed into LSW and NADW is greater than 14°C at 24°N.

4. DISCUSSION

Both the heuristic and more precise computations of the relative contribution of intermediate and deep waters to the global overturning heat transport reveal that intermediate water formation is responsible for about the same amount of heat transport as deep water formation. In the North Atlantic, intermediate and deep water formation are associated with at least 75% of the meridional heat transport across the subtropical gyre. The large heat loss near the Gulf Stream that is associated with Subtropical Mode Water formation functions as a step towards the complete and much denser total water mass formation of the northern North Atlantic and Nordic Seas. The North Pacific situation is exactly the opposite, with only weak intermediate water formation and no deep water formation. North Pacific shallow gyre overturn is responsible for about 75% of the meridional heat transport across the subtropical gyre.

In the southern hemisphere, Antarctic Intermediate Water (AAIW) formation involves a temperature change which is about three times larger than that for Antarctic Bottom Water (AABW) formation. If the formation rates of AAIW and AABW are roughly the same, AAIW formation would contribute more than AABW to global overturning heat transport.

AAIW formation in the southeastern Pacific is actually more akin to shallow subtropical gyre overturn than it is to the intermediate water formation of the northern hemisphere - it lies in the poleward-eastern side of the gyre and much of it is subducted equatorward around the subtropical gyre. Therefore, the shallow gyre overturn of the South Pacific is likely dominated by this water mass formation, with formation of less dense waters in the subtropical gyre mainly as precursors. The Subantarctic Mode Water of the southeastern Indian Ocean (SEISAMW) is dynamically and volumetrically similar to the southeast Pacific AAIW, and carries most of the shallow overturning heat transport of the Indian's subtropical gyre. The total contribution of SEISAMW to the global overturning transport is rather small since its temperature change from input to output water is about half that of the AAIW.

The good agreement between the heuristic and the direct estimates of shallow, intermediate and deep overturn in the northern hemisphere suggests that the crude assumptions made about the principal water masses, their temperatures and formation rates are reasonable and that these water masses actually do dominate the overturning heat transport. It is concluded that shallow, intermediate and deep overturning all contribute relatively equally to the global meridional heat transports. At a minimum, any simplified model or concept of the present or past states of the overturning circulation must include intermediate water formation. Even if Labrador Sea Water is considered to be part of North Atlantic Deep Water, Antarctic Intermediate Water and Antarctic Bottom Water cannot be so combined. In fact, because it is associated with a larger temperature change from source to outflow, Antarctic Intermediate Water formation likely contributes more to overturning heat transport than does Antarctic Bottom Water and cannot be ignored in a global analysis.

Acknowledgments
This study was supported by the National Science Foundation through grants OCE-9712209 and OCE-9529584. The absolute velocities used for the direct heat transport calculations were graciously supplied by J.L. Reid.

REFERENCES

Barnier, B., L. Siefridt and P. Marchesiello, Thermal forcing for a global ocean circulation model using a three-year climatology of ECMWF analyses, *J. Mar. Systems, 6,* 363-380, 1995.

Broecker W., The great ocean conveyor, *Oceanography, 4,* 79-89, 1991.

Bryden, H. L., D. H. Roemmich and J. A. Church, Ocean heat transport across 24-degrees-N in the Pacific, *Deep-Sea Res., 38,* 297-324, 1991.

Clarke, A., and J.-C. Gascard, The formation of Labrador Sea Water, Part I - Large scale processes, *J. Phys. Oceanogr., 13,* 1764-1778, 1983.

da Silva, A. M., C. C. Young and S. Levitus, Atlas of Surface Marine Data 1994. Vol. 3: Anomalies of Heat and Momentum Fluxes, NOAA Atlas NESDIS 8, U. S. Department of Commerce, NOAA, NESDIS, 413 pp., 1994.

Dickson, R. R. and J. Brown, The production of North Atlantic Deep Water: sources, rates and pathways, *J. Geophys. Res., 99,* 12319-12341, 1994.

Freeland, H. J., A. S. Bychkov, F. Whitney, C. Taylor, C. S. Wong and G. I. Yurasov, WOCE section P1W in the Sea of Okhotsk - 1. Oceanographic data description, *J. Geophys. Res., 103,* 15613-15623, 1998.

Gordon, A. L., Interocean exchange of thermocline water, *J. Geophys. Res., 91,* 5037-5046, 1986.

Gordon, A. L., J. R. E. Lutjeharms and M. J. Grundlingh, Stratification and circulation at the Agulhas Retroflection, *Deep-Sea Res., 34,* 565-600, 1987.

Hall, M. M. and H. L. Bryden, Direct estimates and mechanisms of ocean heat transport, *Deep-Sea Res., 29,* 339-359, 1982.

Hautala, S. L. and D. H. Roemmich, Subtropical mode water in the northeast Pacific basin, *J. Geophys. Res., 103,* 13055-13067, 1998.

Hsiung, J., Estimates of global oceanic meridional heat transport, *J. Phys. Oceanogr., 15,* 1405-1413, 1985.

Josey, S. A., E. C. Kent, D. Oakley and P. K. Taylor, A new global air-sea and momentum flux climatology, *Int'l. WOCE Newsletter, 24,* 3-5, 1996.

Käse, R., W. Zenk, T. B. Sanford, and W. Hiller, Currents, fronts and eddy fluxes in the Canary Basin, *Prog. Oceanogr., 14,* 231-257, 1985.

Keith, D. W., Meridional energy transport: uncertainty in zonal means, *Tellus, 47A,* 30-44, 1995.

Kitani, K., An oceanographic study of the Okhotsk Sea - particularly in regard to cold waters, *Bull. Far Seas Fish. Res. Lab., 9,* 45-76, 1973.

Lazier, J., Oceanographic conditions at Ocean Weather Ship Bravo, 1964-1974, *Atmos. Ocean., 18,* 227-238, 1980.

Levitus, S., and T. Boyer, World Ocean Atlas 1994, Vol. 4: Temperature, *NOAA Atlas NESDIS 4,* U.S. Department of Commerce, NOAA, NESDIS, Washington, D.C., 1994.

Luyten, J. R., J. Pedlosky, and H. Stommel, The ventilated thermocline, *J. Phys. Oceanogr., 13,* 292-309, 1983.

MacDonald, A. M., Property fluxes at 30°S and their implications for the Pacific-Indian throughflow and the global heat budget, *J. Geophys. Res., 98,* 6851-6868, 1993.

MacDonald, A. M. and C. Wunsch, An estimate of global ocean circulation and heat fluxes, *Nature, 382,* 436-439, 1996.

Mantyla, A. W., On the potential temperature in the abyssal Pacific Ocean, *J. Mar. Res., 33,* 341-354, 1975.

Mantyla, A. W., and J. L. Reid, Abyssal characteristics of the world ocean waters, *Deep-Sea Res., Part A, 30,* 805-833, 1983.

McCartney, M. S., Subantarctic Mode Water, in *A Voyage of Discovery, George Deacon 70th Anniversary Volume,* edited by M. Angel, pp. 103-119, Pergamon, Elmsford, N.Y., 1977.

McCartney, M. S., The subtropical recirculation of Mode Waters, *J. Mar. Res., 40,* suppl., 427-464, 1982.

McCartney, M. S. and L. D. Talley, The Subpolar Mode Water of the North Atlantic Ocean, *J. Phys. Oceanogr., 12,* 1169-1188, 1982.

McCartney, M. S. and L. D. Talley, Warm water to cold water conversion in the northern North Atlantic Ocean, *J. Phys. Oceanogr., 14,* 922-935, 1984.

Oort, A. H. and T. H. Vonder Haar, On the observed annual cycle in the ocean-atmosphere heat balance over the Northern Hemisphere, *J. Phys. Oceanogr., 6,* 781-800, 1976.

Parrilla, G. A., A. Lavín, H. Bryden, M. García and R. Millard, Rising temperatures in the subtropical North Atlantic Ocean over the past 35 years, *Nature, 369,* 48-51, 1994.

Reid, J. L., On the use of dissolved oxygen concentration as an indicator of winter convection, *Naval Research Reviews, 34,* 28-39, 1982.

Reid, J. L., On the total geostrophic circulation of the South Pacific Ocean: Flow patterns, tracers and transports, *Prog. Oceanogr., 16,* 1-61, 1986.

Reid, J. L., On the total geostrophic circulation of the North Atlantic Ocean: flow patterns, tracers and transports, *Progr. Oceanogr., 33,* 1-92, 1994.

Reid, J. L., On the total geostrophic circulation of the Pacific Ocean: Flow patterns, tracers and transports, *Prog. Oceanogr., 39* 263-352, 1997.

Reid, J.L. and R. J. Lynn, On the influence of the Norwegian-Greenland and Weddell seas upon the bottom waters of the Indian and Pacific oceans, *Deep-Sea Res., 18,* 1063-1088, 1971.

Rintoul, S. R., South Atlantic interbasin exchange, *J. Geophys. Res., 96,* 2675-2692, 1991.

Rintoul, S. R., On the origin and influence of Adelie Land bottom water, *Ocean, Ice and Atm. Interactions at the Antarctic Continental Margin, Ant. Res. Ser., 75,* 151-171, 1998.

Robbins, P. E. and J. M. Toole, The dissolved silica budget as a constraint on the meridional overturning circulation of the Indian Ocean, *Deep-Sea Res., 44,* 879-906, 1997.

Roemmich, D., Estimation of meridional heat flux in the North Atlantic by inverse methods, *J. Phys. Oceanogr., 10,* 1972-1983, 1980.

Roemmich, D., T. McCallister and J. Swift, A transpacific hydrographic section along latitude 24°N; the distribution of properties in the subtropical gyre, *Deep-Sea Res., 38 (Suppl.),* S1-S20, 1991.

Roemmich, D., and B. Cornuelle, The subtropical mode waters of the South Pacific Ocean, *J. Phys. Oceanogr., 22,* 1178-1187, 1992.

Roemmich, D. and T. McCallister, Large scale circulation of the North Pacific Ocean, *Prog. Oceanogr., 22,* 171-204, 1989.

Roemmich, D. and C. Wunsch, Two transatlantic sections: meridional circulation and heat flux in the subtropical North Atlantic Ocean, *Deep-Sea Res., 32,* 619-664, 1985.

Schmitz, W. J., On the interbasin-scale thermohaline circulation, *Rev. Geophys., 33,* 151-173, 1995.

Speer, K., Water mass formation from revised COADS data, *J. Phys. Oceanogr., 25,* 2445-2457, 1992.

Speer, K., and E. Tziperman, Rates of water mass formation in the North Atlantic Ocean, *J. Phys. Oceanogr., 22,* 93-104, 1992.

Stommel, H., Determination of water mass properties of water pumped down from the Ekman layer to the geostrophic flow below, *Proc. Nat. Acad. Sci., 76,* 3051-3055, 1979.

Suga, T., Y. Takei, and K. Hanawa, Thermostad distribution in the North Pacific subtropical gyre: The central mode water and the subtropical mode water, *J. Phys. Oceanogr., 27,* 140-152, 1997.

Swift, J. H., K. Aagaard, and S.-A. Malmberg, The contribution of the Denmark Strait overflow to the deep North Atlantic, *Deep-Sea Res., 27,* 29-42, 1980.

Talley, L. D., Meridional heat transport in the Pacific Ocean, *J. Phys. Oceanogr., 14,* 231-241, 1984.

Talley, L. D., Ventilation of the subtropical North Pacific: the shallow salinity minimum, *J. Phys. Oceanogr., 15,* 633-649, 1985.

Talley, L. D., An Okhotsk Sea water anomaly: implications for sub-thermocline ventilation in the North Pacific, *Deep-Sea Res., 38,* S171-190, 1991.

Talley, L. D., Distribution and formation of North Pacific Intermediate Water, *J. Phys. Oceanogr., 23,* 517-537, 1993.

Talley, L. D., Antarctic Intermediate Water in the South Atlantic, *The South Atlantic: Present and Past Circulation,* edited by G. Wefer, W.H. Berger, G. Siedler and D. Webb, Springer-Verlag, New York, 219-238, 1996a.

Talley, L. D., North Atlantic circulation, reviewed for the CNLS conference, *Physica D, 98,* 625-646, 1996b.

Talley, L. D., North Pacific intermediate water transports in the mixed water region, *J. Phys. Oceanogr., 27,* 1795-1803, 1997.

Talley, L. D., and M. S. McCartney, Distribution and circulation of Labrador Sea Water, *J. Phys. Oceanogr., 12,* 1189-1205, 1982.

Talley, L. D., Y. Nagata, M. Fujimura, T. Iwao, K. Tokihiro, D. Inagake, M. Hirai and K. Okuda, North Pacific Intermediate Water in the Kuroshio/Oyashio mixed water region in Spring, 1989, *J. Phys. Oceanogr., 25,* 475-501, 1995.

Toole, J. M. and B. A. Warren, A hydrographic section across the subtropical South Indian Ocean, *Deep-Sea Res., 40,* 1973-2019, 1993.

Tsuchiya, M., Circulation of Antarctic Intermediate Water in the North Atlantic Ocean, *J. Mar. Res., 47,* 747-755, 1989.

Tsuchiya, M., L. D. Talley and M. S. McCartney, Water mass distributions in the western Atlantic: a section from South Georgia Island (54S) northward across the equator, *J. Mar. Res., 52,* 55-81, 1994.

Wacongne, S., and R. Pacanowski, Seasonal heat transport in a primitive equation model of the tropical Indian Ocean, *J. Phys. Oceanogr., 26,* 2666-2699, 1996.

Wakatsuchi, M., and S. Martin, Satellite observations of the ice cover of the Kuril basin region of the Okhotsk Sea and its relation to the regional oceanography, *J. Geophys. Res., 95,* 13393-13410, 1990.

Worthington, L. V., On the North Atlantic circulation, *Johns Hopkins Oceanogr. Studies, 6,* 110 pp., 1976.

Wunsch, C., D. Hu and B. Grant, Mass, heat and nutrient fluxes in the South Pacific Ocean, *J. Phys. Oceanogr., 13,* 725-753, 1983.

Wüst, G., The stratosphere of the Atlantic Ocean, *Wiss. Erg. der Deutsch. Atl. Exp. auf dem Vermessungs- und Forschungsschiff Meteor 1925-1927, 6,* 109-288, 1935.

Yuan, X., and L. D. Talley, Shallow salinity minima in the North Pacific, *J. Phys. Oceanogr., 22,* 1302-1316, 1992.

L. D. Talley, Scripps Institution of Oceanography, 9500 Gilman Dr., 0230, La Jolla, CA 92093. (e-mail: ltalley@ucsd.edu)

Errors in Generating Time-Series and in Dating Events at Late Quaternary Millenial (Radiocarbon) Time-Scales: Examples From Baffin Bay, NW Labrador Sea, and East Greenland

John T. Andrews, Donald C. Barber, and Anne E. Jennings

INSTAAR and Department of Geological Sciences, University of Colorado, Boulder, CO

A crucial question to understand the climate system at millennial time-scales is whether we can detect leads and lags. We examine errors on downcore age data sets resulting from the application of two depth/radiocarbon age models: 1) interpolation between dates, versus 2) ordinary least squares (OLS) regression. In areas affected by changes in sediment accumulation rates, interpolation between dates on facies boundaries would seem most appropriate, whereas in areas of constant sedimentation OLS regression would seem appropriate. We estimated the ages of 50 and 21 data points respectively, from cores HU87033-009 and HU93030-007 (both with rates of sediment accumulation ca. 30-40 cm/ky) using the two age models. Both cores show intervals of increased sediment accumulation associated with iceberg rafting events possibly coeval with H-1, -2 and -3(?). The two age models produced average age differences of 0.88 ky in HU87033-009 and 0.48 ky in HU93030-007. Monte Carlo simulation experiments indicated that the interpolation method consistently resulted in the larger errors. We then examine the age distribution for the basal ages of two detrital carbonate (DC) "events" in Baffin Bay and the NW Labrador Sea, including H-1, and show that errors on dating the onset of these events are considerable (~± 300 yrs). We conclude that, when dealing with the generation of millennial time-series and correlation of abrupt events, more attention needs to be given to appropriate age/depth models and attendant errors. This analysis does not take into account the calibration of the radiocarbon time-scale nor bioturbation, both of which can introduce additional errors.

1. INTRODUCTION

The present interest in resolving millennial-scale changes in the global climate system [e.g. *Bond et al.*, 1992, 1993] requires attention to two issues, namely: what are the errors involved in developing a time-series of events, and how well can we date and hence correlate discrete abrupt changes in our proxy records? This paper presents an illustration of the problems involved with these two questions, using data from marine cores with radiocarbon dates that span the latter part of Marine Isotope Stage (MIS) 3 and extending to the latter part of MIS2. Errors in estimating the ages of individual data points must also have an impact on the resolution of events in the frequency domain, but this problem is not investigated in this note.

2. BACKGROUND

In three recent papers we have described the stratigraphy and chronology of late Quaternary marine sediments from the areas of 1) Baffin Bay, 2) the SE Baffin Island slope/NW Labrador Sea, and 3) the East Greenland slope immediately south of Denmark Strait [*Andrews et al.*, 1998a, 1998b, 1998c]. These three

Mechanisms of Global Climate Change at Millennial Time Scales
Geophysical Monograph 112

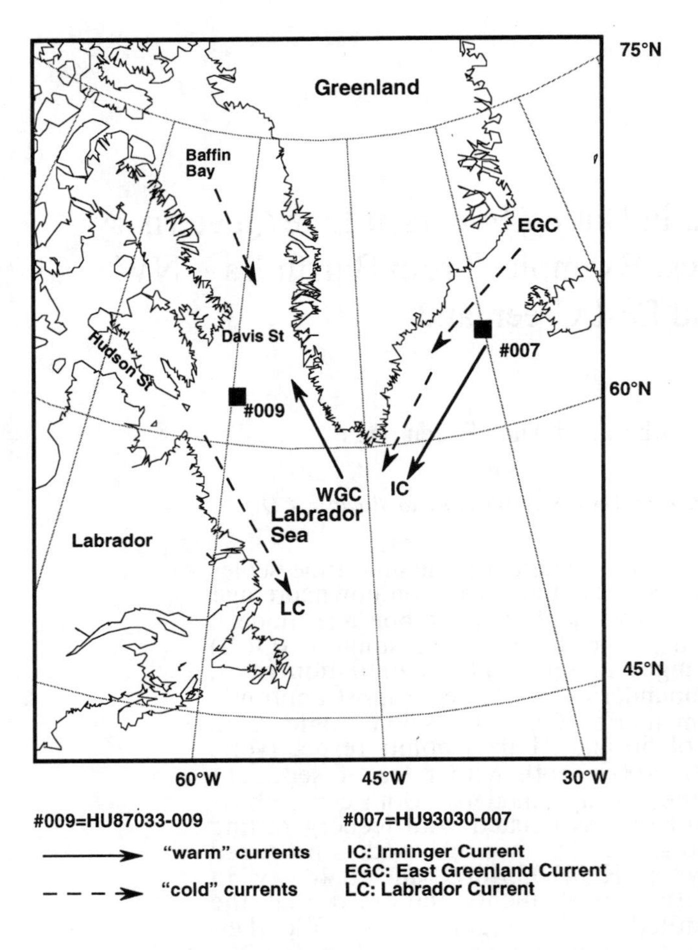

Figure 1. Location Map of the area between Iceland and the Eastern canadian Arctic showing the major present-day surface currents and the location of cores and sites noted in the text.

areas received sediment from different sectors of late Quaternary Northern Hemisphere ice sheets (Figure 1). Baffin Bay received some sediment from Greenland and by erosion of the channels and islands of the Canadian High Arctic [*Aksu,* 1981, 1985; *Aksu and Piper,* 1987; *Hiscott et al.,* 1989]. The SE Baffin slope was positioned to receive sediments from large ice streams flowing from the NE sector of the Laurentide Ice Sheet (LIS) toward the Labrador Sea, principally (N-->S) Cumberland Sound, Frobisher Bay, and Hudson Strait [*Andrews et al.*, 1993; *Jennings,* 1993; *Jennings et al.,* 1996]. Finally, the western slope of Denmark Strait could receive sediment from an East Greenland ice sheet margin advanced onto the shelf and/or sediments derived from sites farther north and brought across the Denmark Strait sill by icebergs as there is evidence for ice-scouring on the Denmark Strait slopes [*Andrews et al.,* 1996, 1997; *Stein,* 1996; *Stein et al.,* 1996; *Syvitski et al.,* in press].

A vital scientific question is: how confidently can we correlate events and time-series between these areas (Figure 1)? This dictates our ability to assess whether abrupt changes in the Northern Hemisphere ice sheet system during the late Quaternary was driven by, or drove climate, and whether we can confidently speak of leads and lags in ice sheet behavior at millennial time-scales. Although this paper is focused on data from our own research, we nevertheless feel that the issues addressed should be of considerable general interest and concern.

In this short paper we discuss important issues about the strategies and errors involved in a) developing a depth/age model (Figure 2), b) proceeding from that to the generation of a time-series for measurements on all the intervening data points (e.g. Figure 3), and finally considering the question of how well we can presently constrain the timing of an abrupt event, such as a Heinrich event [*Bond et al.,* 1993, 1992; *Broecker,* 1994; *Broecker et al.,* 1992; *Heinrich,* 1988].

3. DEPTH / AGE MODELS

During the late Quaternary in the areas where we work, the processes of sediment delivery varied greatly. Cores from these high latitude areas (Figure 1) have intervals of hemipelagic sediments interrupted by intervals of marked change in such properties as sediment color, composition, and amount of iceberg rafted detritus (IRD). The two cores we consider explicitly, HU87033-009 and HU93030-007 (Figure 1), are from the slopes off SE Baffin Island and East Greenland respectively. Changes in sediment properties and in the rates of sediment accumulation are principally driven by changes in the location of ice sheet margins on the adjacent shelf [e.g. *Jennings et al.,* 1996; *Andrews et al.,* 1998a]. Sediment accumulation rates (SAR, cm/ky), as determined by a series of AMS ^{14}C dates on foraminifera, are such that 5 to 10 cm sampling intervals result in century-scale temporal resolutions---thus we are potentially able to resolve millennial changes in our various proxies of ice sheet/ocean interaction. However, in order to generate a time-series from a depth/age (radiocarbon-dated) relation (e.g. Figure 2) we have to implicitly adopt a model of how SAR changes through time. Table 1, for example, shows the results of OLS regression for some of the cores we have studied (Figure 1).

Two approaches are frequently used to proceed from a depth/age relationship (Figure 2) to the presentation of a

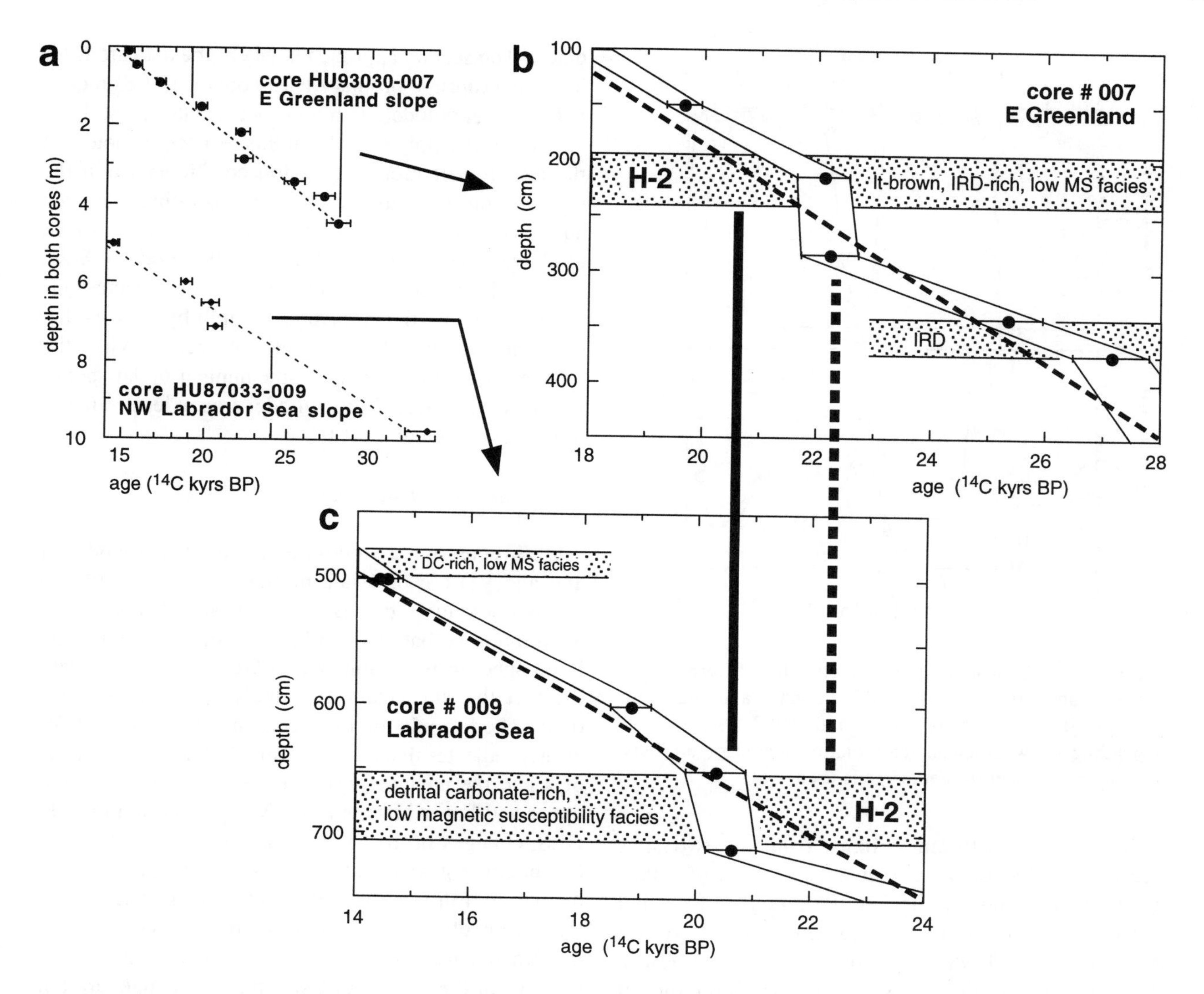

Figure 2. (A) Depth/age scatterplots for HU87033-009 and HU93030-007 (Figure 1). Dashed line shows OLS-fit regression. (B) Illustration of the differences between an ordinary least squares regression (OLS) fit and an interpolated fit for HU93030-007. (C) As in B, for HU87033-009. Note that 2B and C are arranged so that we can directly compare the onset and duration of H-2 at the two sites based on the OLS fit and on the interpolated fit. Thick solid vertical line shows onset of H-2 in HU87033-009 from interpolated fit; dashed vertical line shows H-2 onset from the OLS fit. Note also that the OLS fit results in an increased estimate of the duration of H-2, and that a significant off-set for the start of the event results from comparing the interpolated fit models.

downcore time-series (Figure 3), where the age at any depth in the core is obtained by some algorithm. In the first approach, a series of dates are obtained at approximately equal intervals downcore and some regression method is used to fit this relationship and to predict the ages of any depths. An example would be the data from HU93030-007 (#007 henceforth) from the East Greenland slope (Figures 1 and 2) [*Andrews et al.*, 1998a]. A contrasting method is to concentrate on dating the base and tops of major lithofacies changes and to develop time-series by interpolating between these dates. An example would be HU87033-009 (#009 henceforth) from the NW Labrador Sea [*Jennings et al.*, 1996] (Figures 1 and 2). We use as measures of ice sheet/ocean interaction (Figure 3) the following proxies: the amount of detrital carbonate as measured by the Chittick apparatus [*Dreimanis*, 1962]; and rock magnetic susceptibility [*Thompson and Oldfield*, 1986]. Both these cores have marked changes in lithofacies which mark major changes in regional ice sheet/ocean interactions [*Jennings et al.*, 1996; *Andrews et al.*, 1998a].

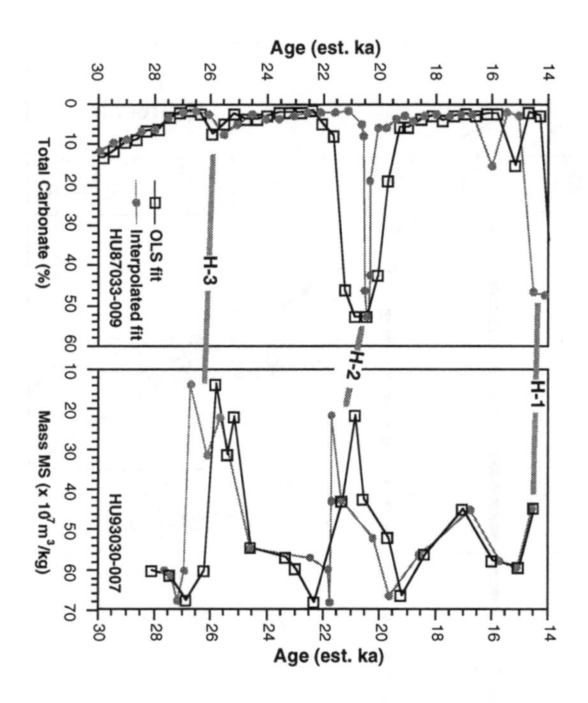

Figure 3. Downcore plot showing the differences in location and structure of total carbonate and magnetic susceptibility events in #009 and #007 respectively depending on whether an interpolation model or ordinary least squares model for depth/age is used.

We consider different proxies from the two sites because the characteristic sediment signatures of abrupt events in both cores, such as the amount of detrital carbonate, differ profoundly [*Jennings et al.,* 1996; *Andrews et al.,* 1998a]. Furthermore, our concern in this paper is with the errors involved in chronological correlations and not with developing appropriate models of ice sheet/ocean interactions for the two areas.

4. ERRORS IN LATE QUATERNARY RADIOCARBON TIME-SERIES

The purpose of this section is to examine the precision of late Quaternary chronostratigraphies derived from ^{14}C dates. In the marine environment an ideal number of dates for millennial-scale investigations would be ≥ 1 date per ky. Usually the dating density is less than that with 1 date per 2 ky or fewer (e.g. Figure 2).

Time-series of data values for a particular proxy are generated by interpolation and extrapolation using two common approaches as noted above. The first is simple linear interpolation between dated levels where the implicit assumption is that SAR is constant between dates. The second approach is where the data are fitted by an appropriate ordinary least-squares (OLS) model, either a linear model, with two coefficients (Table 1), or a higher order polynomial. In this respect we note that the addition of more regression coefficients into the equation must be justified by demonstrating that each added coefficient makes a genuine contribution to the total explained variance offered by the model [*Davis,* 1986]. This requires using analysis of variance and testing each coefficient against the null hypothesis that it is not significantly different from zero. When the number of coefficients equals the number of data points we have an interpolation model not a least squares model, even though r^2=1.0.

4.1. Some Statistical Issues

OLS regression assumes the distributions (depth {x} and age{y}) are bivariate normal, yet in virtually all depth/age studies this is not the case. A reasonable assumption is that age is a function of depth, but there is no specific provision within OLS to accommodate the fact that the errors on sample age increase with time. The formulation of forecasting errors [*Till,* 1974] clearly indicates that errors are smallest around the mean of x and the mean of y, and increase toward both the upper and lower bounds of the y (age) distribution. In OLS, changes in the extreme data points, i.e. those at the beginning and end, have most "weight" on the determination of the intercept and slope. This weighting effect is particularly true in virtually all depth/age regressions because there usually is no cluster of dates in the middle of the core, i.e. the data are not normally distributed but rather have a rectangular distribution.

The alternative approach of interpolation between successive pairs of dates assumes that the slope between each pair of dates is fixed, and thus this model can be considered as a series of linear regressions with r=1.0 in all cases.

Results of OLS regression for four cores from the northern North Atlantic (Figure 1) are given in Table 1. Notice that for these cores, with the number of data points varying between 4 and 9, the regression lines are clearly highly significant in terms of rejection of the null hypothesis, H_o (r=0.0, p=0.05), but the ± 1 standard error about each of the coefficients is large, and further, that the RMSE (obtained from observed date minus predicted date) is in the range of 1 - 2 ky. Data like these should be reported routinely with all age models if a regression approach is used.

Table 1. Data on ordinary least squares fits between depth and radiocarbon age of some of the cores located on Figure 1. Description of dates are included in *Kirby* [1996], *Jennings et al.* [1996], *Andrews et al.* [1998a, b, c].

Core Id: Number of dates	Intercept (a ±)[a]	slope (b±)	r^2	significance[b]	RMS Error
HU93030-007 9	14.29 (0.125)	0.0307 (0.0006)	0.97	0.000	0.7
HU87033-009 5	-5.69 (2.45)	0.0396 (.003)	0.97	0.0015	1.23
HU75009-IV-056 8	5.93 (0.65)	0.08 (0.003)	0.98	0.0000	1.04
HU75009-IV-054 4	14.55 (1.8)	0.084 (0.008)	0.98	0.008	2.12

a errors are ± one standard error

b usual level for rejection of null hypothesis H_0 is $p = 0.05$, reject H_0 in all cases.

5. COMPARISON OF METHODS

We use data from HU93030-007 (#007), and HU87033-009 (#009) (Figures 2 and 3) [*Andrews et al.*, 1998a; *Jennings et al.*, 1996] as examples of the differences in estimated ages that arise by application of one or other of the two approaches noted above to the development of time-series. We briefly discuss issues associated with other errors (such as calibrated radiocarbon dates, [*Bard*, 1998]) later in the paper. The density of radiocarbon dates for these two cores is considered "average" for late Quaternary environments [*van Weering et al.*, 1996]. The first core, #009, has 6 dates between 14 and 33 ka, and the second one has 9 dates between ~14 and 28 ka (Figure 2). In both instances, an ordinary least squares (OLS) model explains the data well, with r^2 values > 0.9 and rejection of the null hypothesis that r is not significantly different from 0 (Table 1).

Our data sets for #007 consist of 21 measurements on mass magnetic susceptibility, whereas in #009 there are 50 measurements on total carbonate weight percent. As noted earlier, the issue of interest here is not the proxy records *per se* but the conversion of this depth/proxy data into an age/proxy time series. On Figure 3 we plot the two derived time-series data sets for #009 and #007 against the two proxies of millennial-scale changes in North Atlantic paleoceanography, i.e. total carbonate % [*Andrews and Tedesco,* 1992], and mass magnetic susceptibility [*Andrews and Tedesco,* 1992; *Robinson et al.*, 1995; *Stoner et al.*, 1995] (Figure 3) and then derive the estimated ages of the 21 and 50 data points. Notice in Figure 4 that the distribution of sampling intervals (years per sample) varies depending on which model is used, and also of course on whether or not the core was sampled predominantly at equally spaced depth intervals. For example, note the difference in years/sample due to the different sampling strategies in #009 (nearly all equally spaced samples) and #007 (variable sampling interval).

In Figure 5 we plot the distribution of root mean square error (RMSE) for ages obtained by interpolation and those estimated from the linear regression model [*Stata,* 1996] (see Figure 3). For #009, with 5 dated levels, the RMSE for the estimated age for the 50 data levels is 0.88 ± 0.4 k; for #007, with 9 dates and 21 measurements, the RMSE is 0.48 ± 0.32 ky (Figure 5). In both cores these errors are of the same magnitude as the average sampling resolution (Figures 4 and 5). These RMS errors (Figure 5), involving the differences in derived ages between two common approaches, are substantial but they are not usually considered when multiple downcore time-series are compared. Further, these errors are not equivalent to drawing an envelope around the radiocarbon dates (Figure 2). One assumes that the differences noted in Figures 3 and 5 will decrease in proportion to the number of radiocarbon dates per unit time, and the number of rapid changes in SAR.

Figure 3 indicates that different conclusions can be drawn depending upon which method has been used to generate the time-series. In #009 dates were obtained as close to lithofacies boundaries (Heinrich layers) as possible, thus interpolated ages are able to accommodate changes in SAR. This is evident in the comparison of the duration of H-2 in #009 (Figure 3) where ages derived from the well-fitted linear model (i.e. significant

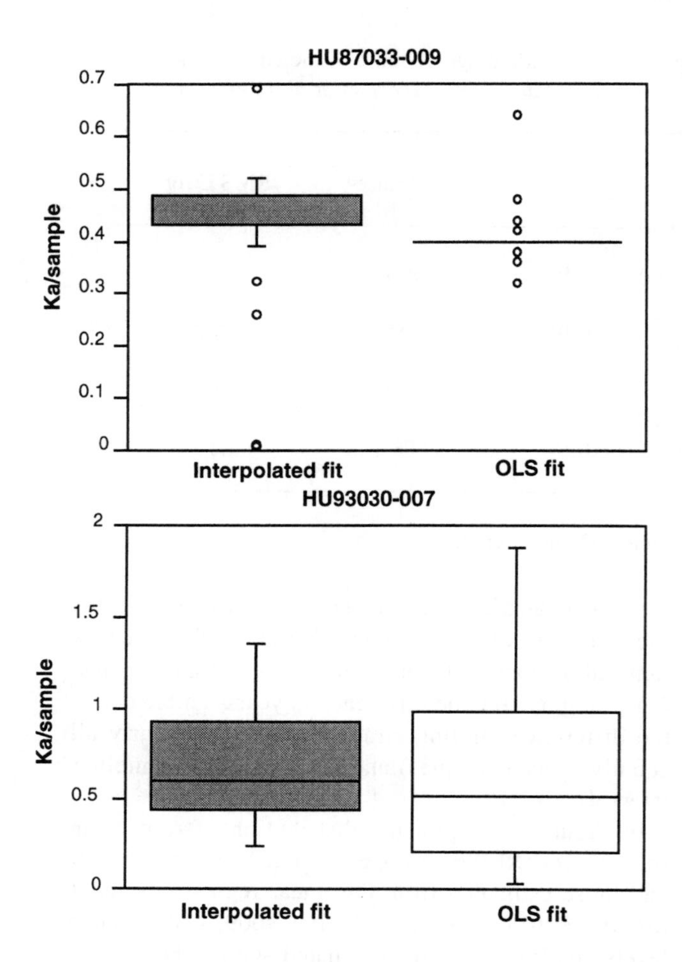

Figure 4. Box plots [*Velleman and Hoaglin,* 1981] of the distribution of sampling intervals (ka/sample) between HU87033-009 and HU93030-007 (Fig. 2). Variability is due to whether an OLS or interpolation depth/age model is used and whether the sampling is undertaken at equal intervals.

r^2, Table 1) give a duration for H-2 of ~ 3 ky compared to the more likely figure of ≤ 0.5 ky from the interpolated model (Figure 3). The data from #007 include dates from within IRD events (Figure 2) and the OLS approach would appear more suited to this core. If we compare the timing of two possible coeval events in #009 and #007 (H-2?) the events appear coeval with the application of the OLS linear models but offset by ~ 1.3 ky if interpolation is used (Figures 2 and 3).

5.1. Empirical Experiments

The results discussed above (e.g. Figure 3) assume that the reported dates are accurate. However, we know that there are errors associated with radiocarbon dates that are not always explicitly considered in the generation of a radiocarbon-based time-series. There are at least three sources of error in estimating the "true" radiocarbon ages of marine samples. These errors in dates are: 1) the quoted error in the laboratory analysis, usually of the order of ± 1% of the age at the 1 sigma level; 2) the error (mainly unknown) from variations in the ocean reservoir correction (400 ± yr) in time and space [*Bard et al.,* 1994; *Stocker and Wright,* 1998]; and 3) the "geological" error associated with dating the same level in the core but at another site 1 to 1000's m distant (also largely unknown). It is commonly assumed that errors are random variables and can cancel. The maximum assumed error is the sum of the error components, whereas the probable error is often derived from:

$$\mathrm{sqrt}\{e_1{}^2 + e_2{}^2 + e_3{}^2\} \qquad [\textit{Till}, 1974]$$

Commonly we wish to be at least 95% certain that an age estimate lies within certain confidence limits. As we are dealing with events in the range of 10 to 30 ka (Figure 2) the probable errors on the dates (± 3%?) may range from ± 300 to ± 1500 yrs.

Other sources of error which we do not explicitly evaluate include the calibration of radiocarbon dates, and the influence of bioturbation [e.g. *Manighetti and McCave,* 1995]. The question of calibrating radiocarbon years to sidereal years [*Stuiver and Reimer,* 1993; *Bard,* 1998] is a separate problem and involves additional uncertainties. Generally, the sense of the calibration is to increase the age of the data points,

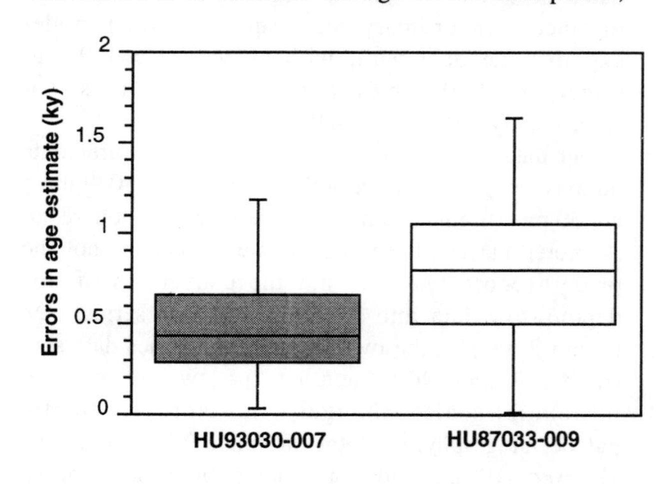

Figure 5. Box plot [*Velleman and Hoaglin,* 1981] of absolute error between linear and interpolated age models for #009 and #007. The area within the box represents 50% of the sample distributions.

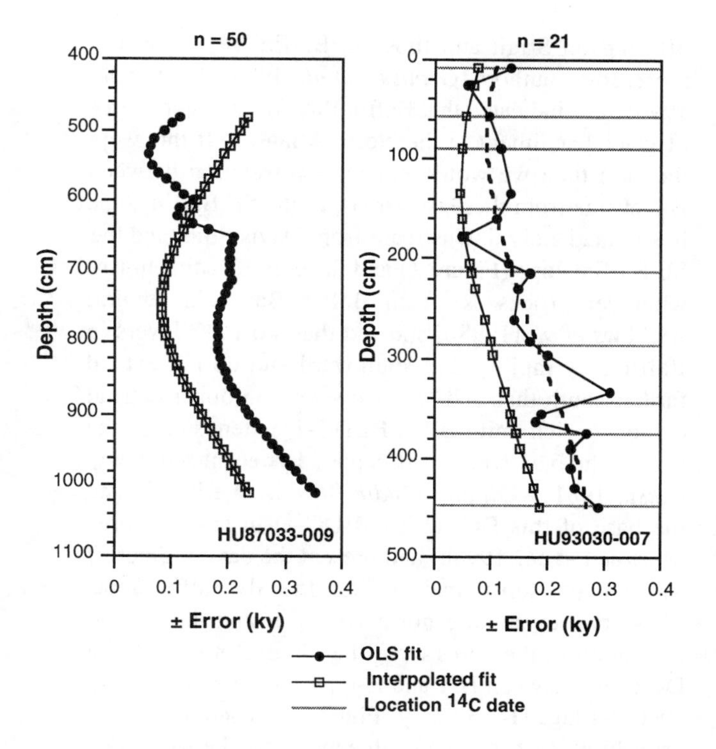

Figure 6. Plot of the downcore variations in the standard deviation of age estimates for the two cores (#009 and #007, Fig. 1) based on OLS and interpolation models of depth versus age. For #007 there are nine dates and 20 levels of data, whereas in #009 there are five dates and 50 levels of data. The dashed line for #007 is a smooth fit to the interpolation errors. Notice that is parallels the OLS curve.

which leads to a reduction in the OLS slope and in the individual slopes in interpolation. One obvious result of this will be to increase uncertainties in estimating the true duration of abrupt events (Figure 2). Because the SAR averaged over 30 cm/ky in our cores (Table 1, Figure 2) we feel justified in ignoring bioturbation as a major problem in the context of our paper [*Manighetti and McCave,* 1995].

We use only the laboratory counting errors on the radiocarbon dates from #007 and #009 (Figure 2) in a Monte Carlo experiment to investigate the variations in time-series derived from "typical" radiocarbon-dated marine cores (and without additional information derived from correlation with say the $\delta^{18}O$ time-series of SPECMAP). We generated a series of dates which varied randomly about the reported age for each of the dated levels in #007 (9 dates) and #009 (5 dates), using the routines in the statistical software package "Stata™"[*Stata,* 1996]. A new "seed" was set at each running of the quasi-random number generator. Our results are based on ten randomly generated depth/age sets per core from which we next generated ten interpolation models, and ten OLS linear models. We then calculated the ages of the proxy data points in both cores (n=21 and 50 respectively) for each of the 10 depth/age models and then derived the range of dates and the standard deviations for each level.

Figure 6 shows the downcore variations in the standard deviation of the age estimates for each of the 21 and 50 data points. There are differences in detail, as might be expected, but some generalizations can be drawn, namely: 1) the errors are larger for the interpolation method (by about a factor of two); 2) as expected, the errors generally decrease with age, because of the general decrease in reported error with age; 3) OLS errors follow a smooth curve whereas those from interpolation are more varied but have an underlying structure. Errors from OLS will theoretically be least in the region of the "average" age and will increase toward both limits, producing a U-shaped distribution. However, as the errors on the radiocarbon dates increase with age, the shape of the error distribution is not symmetrical (see Figure 6). In the interpolation method each pair of dates defines a regression line, thus the oscillations in the standard deviation downcore define a series of small-scale U-shaped changes.

One simple but relatively costly approach to the issue of dating variability for any event/depth, would be to obtain several (n) AMS ^{14}C dates at the same level in a core; this would reduce the ± standard error on the average age by sqrt(n). This has rarely been done. We did such an experiment on acid-insoluble organics from the Ross Sea Antarctica (*Andrews et al.,* in press) and found that samples had reported mean ages differing between 300 and 600 yrs (reported errors on the dates of ca. ± 50 yrs). We have replicated dates on planktonic foraminfera (450 yr ocean reservoir correction applied) from the base of H-1 in #009 and obtained results of 14,400 ± 205 and 14,530 ± 90. In a core from Baffin Bay we also replicated a date from another core [*Andrews et al.,* 1998b] with ages of 17,540 ± 110 and 17,490 ± 210. These two examples, although limited, are at least encouraging.

We illustrate on Figure 2B and C how the use of an appropriate age model can have a dramatic influence on the duration of discrete events, especially when modelled by an OLS procedure as compared to the interpolation method (Figure 2B and C). Thus the various discussions about the duration of H-events [cf. *Andrews,* 1998] should take into account how the estimates of duration where derived. Thus we would suggest that the rather long durations of H-events estimated by *Elliot et al.* [1998] might be partly explained by this comparison (Figure 2B and C), especially since those estimates are significantly longer than those derived from our "upstream" study on #007 with a significantly faster SAR (Figure 2A).

6. REGIONAL VARIATIONS IN AGES OF EVENTS

A different question is what is the regional variability in radiocarbon dates for a single, well-defined event. We take the base of H-1 in the NW Labrador Sea as one example, and the base of a possibly correlative DC-event in Baffin Bay, BBDC-1 as another.

We already noted that two dates from the base of H-1 in #009 are close to 14,500 BP. Other dates on what has been inferred to be the base of H-1 from cores in this region of the NW North Atlantic [*Andrews et al.,* 1994, 1998c; *Kirby,* 1996] include 14,225 ± 155 (I77-6-2), 13,730 ± 205 (I77-1-2), and 14,560 ± 105 (HU75-55). The uncalibrated, normalized distribution of radiocarbon dates [*Stolk et al.,* 1994] for this sample set (n = 5) indicates that there is a 75% probability that the base of H-1 in the NW Labrador Sea dates between 14.1 and 14.9 ka (Figure 6) . However, the date of 13,730 ± was not obtained at the base of a DC-layer but its position was judged on the basis of the $\delta^{18}O$ planktic curve [*Andrews et al.,* 1994]. If this date is omitted, the probability that the "true" age of the base of H-1 lies somewhere between 14.3 and 14.7 is 0.75 (Figure 7).

Our next question concerns the chronology of events in Baffin Bay versus those around Hudson Strait and the SE Baffin slope (Figure 1). *Aksu* [1981] described a series of DC-facies from a transect of cores along the axis of Baffin Bay. At the ODP site 645 in Baffin Bay [in *Littleton,* 1987] a series of sedimentary cycles were described and later associated with the glacial and deglacial processes in Baffin Bay [*Andrews et al.,* 1998b; *Hiscott et al.,* 1989]. Observations and modelling suggest that the channels (floored with Paleozoic limestones and dolostones) leading into northern Baffin Bay were the sites for ice streams [*Marshall and Clarke,* 1996]. These ice streams probably originated from the Innuitian Ice Sheet located over the islands of the Canadian Archipelago [*Dyke,* in press]. Although the Innuitian Ice Sheet was a separate ice sheet from the Laurentide Ice Sheet to the south [*Dyke and Prest,* 1987], it might be reasonable to expect some degree of correlation between DC events off Hudson Strait and those in Baffin Bay. Furthermore, some authors [*Grousset et al.,* 1993; *Revel et al.,* 1996] have indicated that Baffin Bay was a major source of glacial sediment to the North Atlantic. If this were the case then we would expect a correlation between DC-events from these two areas, as the SE Baffin slope lies immediately downstream from Davis Strait and the SE Baffin slope (Figure 1) and the exit of sediment (by whatever processes) from Baffin Bay. In general *Andrews et al.* [1998b] showed that most DC layers in Baffin Bay rapidly thin southward and do not extend farther south than ~70°N. However, an upper detrital carbonate-rich unit (called BBDC-1) extends along the axis of the bay, where it is capped by recent sediments [*Aksu,* 1981; *Aksu and Mudie,* 1985]. We have dated the base of this DC- unit (BBDC-1) in several cores [*Andrews et al.,* 1998b] and present the data in Figure 7 as a comparison with H-1 from the SE Baffin slope. These two events are not correlative, and there is no indication in the cores from the SE Baffin slope for a DC event between H-1 and H-0 [*Andrews et al.,* 1995]; BBDC-1 lags H-1 by 2 ky (Figure 7). There is an 85% probability that the age of this event lies between 12.0 and 12.8 ka. There are no other major DC-events in the Baffin Bay cores which might be coeval with H-1, and indeed *Andrews* [1998] argued [after Hiscott et al., 1989] that other DC-events in Baffin Bay lagged H- events in the North Atlantic, and that the Greenland ice core interstadial events were coeval with the Baffin Bay DC events [*Andrews et al.,* 1998b]. This observation indicates that the deglaciation of the northern regions of Baffin Bay was associated with increased advection of modified Atlantic Water into Baffin Bay via the West Greenland Current (which in turn is associated with the Irminger Current) [*Hiscott et al.,* 1989].

7. COMMENTS AND CONCLUSIONS

This paper has examined two issues associated with millennial scale variations in ice sheet/ocean interactions at a series of sites from the northern North Atlantic (Figure 1). The emphasis has not been so much on the correlation of events but rather on some of the difficulties of correlating and dating these records during the late Quaternary, but still within the interval when radiocarbon dating is a viable chronological tool (< 30 ka BP). The results of our simple analyses indicate that we should exercise caution in the development of time-series and answering questions of correlation, leads, or lags. In areas close to glaciated continental margins, where rapid changes in SAR are

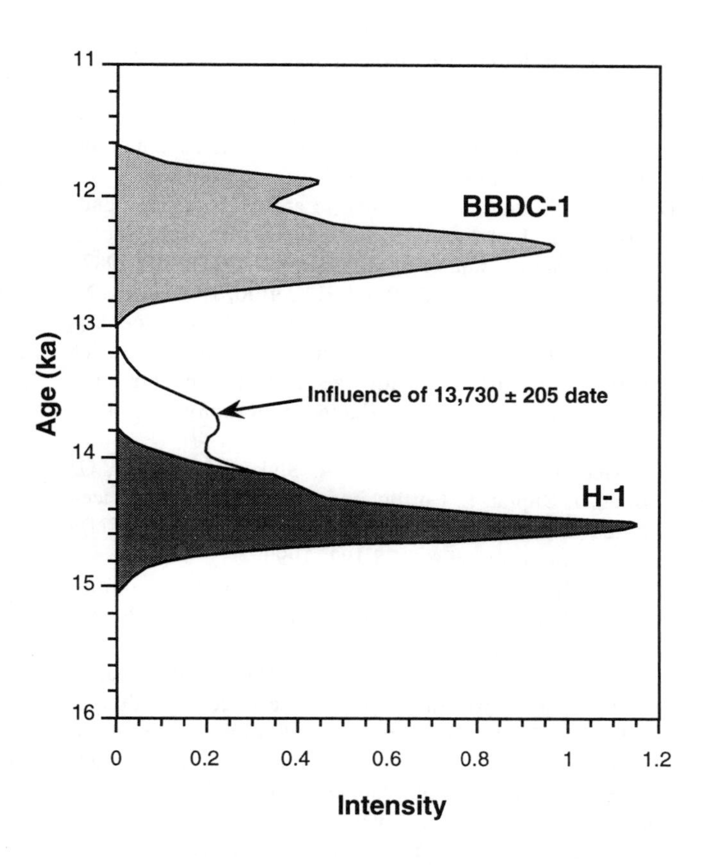

Figure 7. Probability plot of basal ages for H-1 based on radiocarbon AMS dates from the NW Labrador Sea [*Stolk et al.,* 1994], and for the base of the uppermost Baffin Bay DC-even). X-axis is referred to as intensity and the age scale is in radiocarbon time.

expected (Figure 2), it would seem that a depth/age model based on interpolation between dates on lithofacies boundaries would be most appropriate, whereas in areas where the supply of sediment has been more constant a linear or polynomial model should be considered. Our empirical study indicates that the errors generated by interpolation between dated levels generally have larger standard deviations than estimates based on OLS (Figure 5). This conclusion is based on simulated time-series from two cores where SAR varied significantly between intervals of hemipelagic sedimentation versus intervals associated with IRD and Heinrich events. The errors noted on Figure 6 are minimum errors because no allowance has been made for errors associated with lateral variability of dates on an event (e.g. Figure 7), nor of spatial variations in the ocean reservoir correction [*Bard et al.,* 1994; *Hughen et al.,* 1998]. Consideration of these issues indicates that problems of precise correlation (say ± 100 yrs at the 95% confidence level) during MIS 2 and 3 will be difficult to achieve.

Acknowledgements. Our research on these cores has been supported since 1990 by various grants from the National Science Foundation, Office of Polar Programs and most recently by OPP-9321135 and OPP-9707161, as well as NOAA Award NA47GP0188. We appreciate both the critical and helpful comments of Dr. E. Bard and another reviewer, and we thank them for their assistance.

REFERENCES

Aksu, A. E., Late Quaternary stratigraphy, paleo-environmentology, and sedimentation history of Baffin Bay and Davis Strait, PhD, 771 pp thesis, Dalhousie University, Halifax, N.S., 1981.

Aksu, A. E., Climatic and oceanographic changes over the past 400,000 years: Evidence from deep-sea cores on Baffin Bay and David Strait, in *Quaternary Environments: Eastern Canadian Arctic, Baffin Bay and Western Greenland*, edited by J. T. Andrews, pp. 181-209, Allen and Unwin, Boston, 1985.

Aksu, A. E., and P. J. Mudie, Late Quaternary stratigraphy and paleoecology of Northwest Labrador Sea, *Mar. Micropaleontol., 9*, 537-557, 1985.

Aksu, A. E., and D. J. W. Piper, Late Quaternary sedimentation in Baffin Bay, *Canadian Journal Earth Sciences, 24*, 1833-1846, 1987.

Andrews, J. T., Abrupt changes (Heinrich events) in late Quaternary North Atlantic marine environments: a history and review of data and concepts, *Journal Quaternary Science, 13*, 3-16, 1998.

Andrews, J. T., and K. Tedesco, Detrital carbonate-rich sediments, northwestern Labrador Sea: Implications for ice-sheet dynamics and iceberg rafting (Heinrich) events in the North Atlantic, *Geology, 20*, 1087-1090, 1992.

Andrews, J. T., K. Tedesco, and A. E. Jennings, Heinrich events: Chronology and processes, east-central Laurentide Ice Sheet and NW Labrador Sea, in *Ice in the Climate System*, edited by W. R. Peltier, pp. 167-186, Springer-Verlag, Berlin Heidelberg, 1993.

Andrews, J. T., H. Erlenkeuser, K. Tedesco, A. Aksu, and A. J. T. Jull, Late Quaternary (Stage 2 and 3) Meltwater and Heinrich events, NW Labrador Sea, *Quaternary Research, 41*, 26-34, 1994.

Andrews, J. T., G. Bond, A. E. Jennings, M. Kerwin, M. Kirby, B. MacLean, W. Manley, and G. H. Miller, A Heinrich like-event, H-0 (DC-0): Source(s) for Detrital Carbonate in the North Atlantic During the Younger Dryas chronozone, *Paleoceanography, 10*, 943-952, 1995.

Andrews, J. T., A. E. Jennings, T. Cooper, K. M. Williams, and J. Mienert, Late Quaternary sedimentation along a fjord to shelf (trough) transect, East Greenland (ca. 68°N), in *Late Quaternary Paleoceanography of North Atlantic Margins*, edited by J. T. Andrews, W. Austin, H. Bergsten and A. E. Jennings, pp. 153-166, Geological Society of London, London, 1996.

Andrews, J. T., L. M. Smith, R. Preston, T. Cooper, and A. E. Jennings, Spatial and temporal patterns of iceberg rafting (IRD) along the East Greenland margin, ca. 68 N, over the last 14 cal.ka, *Journal of Quaternary Science, 12*, 1-13, 1997.

Andrews, J. T., T. A. Cooper, A. E. Jennings, A. B. Stein, and H. Erlenkeuser, Late Quaternary IRD events on the Denmark Strait/Southeast Greenland continental slope (~65° N): Related to North Atlantic Heinrich Events?, *Marine Geology*, *149,* 211-228, 1998a.

Andrews, J. T., M. Kirby, A. Aksu, D. C. Barber, and D. Meese, Late Quaternary Detrital Carbonate (DC-) events in Baffin Bay (67° - 74° N): Do they correlate with and contribute to Heinrich Events in the North Atlantic?, *Quaternary Science Review*, 1998b.

Andrews, J. T., M. Kirby, A. E. Jennings, and D. C. Barber, Late Quaternary stratigraphy, chronology, and depositional processes on the Slope of SE Baffin Island, N.W.T.: DC- and H- events and implications for onshore glacial history, *Geophysique physique et Quaternaire, 52,* 91-105, 1998c.

Andrews, J. T., A. J. T. Jull, and A. Leventer, Replication of AMS ^{14}C dates on the acid-insoluble fraction of Ross Sea surface sediments. *Antarctic Journal of the US.*, in press.

Bard, E., Geochemical and geophysical implications of the radiocarbon calibration, *Geochimica Commochimica Acta , 62,* 2025-2038, 1998.

Bard, E., M. Arnold, J. Mangerud, M. Paterne, L. Labeyrie, J. Duprat, M.-A. Melieres, E. Sonstegaard, and J.-C. Duplessy, The North Atlantic atmosphere-sea surface ^{14}C gradient during the Younger Dryas climatic event, *Earth Planet Sci. Lett., 126*, 275-287, 1994.

Bond, G., H. Heinrich, W. S. Broecker, L. Labeyrie, J. McManus, J. T. Andrews, S. Huon, R. Jantschik, S. Clasen, C. Simet, K. Tedesco, M. Klas, G. Bonani, and S. Ivy, Evidence for massive discharges of icebergs into the glacial Northern Atlantic, *Nature, 360*, 245-249, 1992.

Bond, G., W. S. Broecker, S. Johnsen, J. McManus, L. Labeyrie, J. Jouzel, and G. Bonani, Correlations between climate records from North Atlantic sediments and Greenland ice, *Nature, 365*, 143-147, 1993.

Broecker, W. S., Massive iceberg discharges as triggers for global climate change, *Nature, 372*, 421-424, 1994.

Broecker, W. S., G. Bond, J. McManus, M. Klas, and E. Clark, Origin of the Northern Atlantic's Heinrich events, *Climatic Dynamics, 6*, 265-273, 1992.

Davis, J. C., *Statistics and data analysis in Geology*, 646 pp., John Wiley & Sons, New York, 1986.

Dreimanis, A., Quantitative gasometric determination of calcite and dolomite by using Chittick apparatus, *J. Sed. Petrol., 32*, 520-529, 1962.

Dyke, A. S. Last Glacial Maximum and deglaciation of Devon Island, Arctic Canada: Support for an Innuitian Ice Sheet, *Quaternary Science Reviews,* in press.

Dyke, A. S., and V. K. Prest, Late Wisconsinan and Holocene history of the Laurentide Ice Sheet, *Geographie physique et Quaternaire, 41*, 237-263, 1987.

Elliot, M., L. Labeyrie, G. Bond, E. Cortijo, J-L. Turon, N. Tisnerat, J-C. Duplessy, Millennial-scale iceberg discharges in the Irminger Basin during the last glacial period: relationship with the Heinrich events and environmental settings, *Paleoceanography, 13,* 433-447, 1998.

Grousset, F. E., L. Labeyrie, J. A. Sinko, M. Cremer, G. Bond, J. Duprat, E. Cortijo, and S. Huon, Patterns of ice-rafted detritus in the glacial North Atlantic (40-55°N), *Paleoceanography, 8*, 175-192, 1993.

Heinrich, H., Origin and consequences of cyclic ice rafting in the Northeast Atlantic Ocean during the past 130,000 years, *Quat. Res., 29*, 143-152, 1988.

Hiscott, R. N., A. E. Aksu, and O. B. Nielsen, Provenance and dispersal patterns, Pliocene-Pleistocene section at Site 645, Baffin Bay, in *Proceedings ODP, Scientific Results, 105*, edited by S. K. Stewart, pp. 31-52, Ocean Drilling Program, College Station, TX, 1989.

Hughen, K. A., J. T. Overpeck, S. J. Lehman, M. Kashgarian, J. Southon, L. C. Peterson, R. Alley, and D. M. Sigman, Deglacial changes in ocean circulation from an extended radiocarbon calibration, *Nature, 391*, 65-68, 1998.

Jennings, A. E., Late Quaternary chronology of deglaciation, Cumberland Sound, N.W.T., *Geographie physique et Quaternaire, 47*, 21-42, 1993.

Jennings, A. E., K. A. Tedesco, J. T. Andrews, and M. E. Kirby, Shelf erosion and glacial ice proximity in the Labrador Sea during and after Heinrich events (H-3 or 4 to H-0) as shown by foraminifera, in *Late Quaternary Palaeoceanography of the North Atlantic Margins*, edited by J. T. Andrews, W. E. N. Austin, H. Bergsten and A. E. Jennings, pp. 29-49, Geological Society Special Publications, 1996.

Kirby, M., Mid to Late Wisconsin Ice Sheet/Ocean Interactions in the Northwest Labrador Sea: Sedimentology, Provenance, Process, and Chronology, MSc thesis, University of Colorado, Boulder, 194 pp., 1996.

Littleton, M. R. (editor), *Part A-Initial Reports Sites 645-647 Baffin Bay and Labrador Sea*, 917 pp., Ocean Drilling Program Texas A & M University,Washington, D.C., 1987.

Manighetti, B., and I. N. McCave, Chronology for climate change: Developing age models for biogeochemical ocean flux. *Paleooceanography,* 10, 513-525, 1995.

Marshall, S. J., and G. K. C. Clarke, Geologic and

topographic controls on fast flow in the Laurentide and Cordilleran Ice Sheets, *J. Geophys. Res., 101* (B8), 17,827-17,839, 1996.

Revel, M., J. A. Sinko, F. E. Grousset, and P. E. Biscaye, Sr and Nd isotopes as tracers of North Atlantic lithic particles: Paleoclimatic implications, *Paleoceanography, 11*, 95-113, 1996.

Robinson, S. G., M. A. Maslin, and N. McCave, Magnetic Susceptibility Variations in Upper Pleistocene Deep-Sea Sediments of the N.E. Atlantic: Implications for Ice Rafting and Palaeocirculation at the Last Glacial Maximum, *Paleoceanography, 10*, 221-250, 1995.

Stata, User's Guide. Release 5. Four volumes. Stata Press, College Station, TX, 1996.

Stein, A. B., Seismic stratigraphy and seafloor morphology of the Kangerlugssuaq Region, East Greenland: Evidence for Glaciations to the Continental shelf break during the Late Weischelian and earlier, MSc, 293 pp thesis, University of Colorado, Boulder, 1996.

Stein, R., S.-I. Nam, H. Grobe, and H. Hubberten, Late Quaternary glacial history and short-term ice-rafted debris fluctuations along the East Greenland continental margin, in *Late Quaternary paleoceanography of North Atlantic margins*, edited by J. T. Andrews, W. A. Austen, H. Bergsetn and A. E. Jennings, pp. Geological Society, London, 1996.

Stocker, T. F. and D. G. Wright, The effect of a succession of ocean ventilation changes on ^{14}C, *Radiocarbon, 40*, 359-366, 1998.

Stolk, A. D., T. E. Tornqvist, K. P. V. Hekhuis, H. J. A. Berendsen, and J. Van Der Plicht, Calibration of ^{14}C Histograms: A Comparison of Methods, *Radiocarbon, 36*, 1-10, 1994.

Stoner, J. S., J. E. T. Channell, and C. Hillaire-Marcel, Magnetic properties of deep-sea sediments off southwest Greenland: Evidence for major differences between the last two deglaciations, *Geology, 23*, 241-244, 1995.

Stuiver, M., and P. J. Reimer, Extended ^{14}C data base and revised CALIB 3.0 ^{14}C age calibration program, *Radiocarbon, 35*, 215-230, 1993.

Syvistki, J. P. M., Jennings, A. E., and Andrews, J. T., High-Resolution Seismic Evidence for Multiple Glaciation across the SW Iceland Shelf, *Arctic and Alpine Research,* in press.

Thompson, R., and F. Oldfield, *Environmental Magnetism*, 227 pp., Allen & Unwin, Winchester, Mass., 1986.

Till, R. Statistical methods for Earth Scientist. John Wiley & Sons, New York, 159 pp.

van Weering, T. C. E., G. Ganssen, and L. Labeyrie (eds), Paleoceanography of the North Atlantic Region, *Mar. Geol., 131* (April), 1-3, 1996.

Velleman, P. F., and D. C. Hoaglin, *Applications, Basics, and Computing of Exploratory Data Analysis*, 354 pp., Duxbury, Boston, 1981.

J. T. Andrews (corresponding author), D. C. Barber, and A. E. Jennings, Institute of Arctic and Alpine Research and Department of Geological Sciences, Campus Box 450, University of Colorado, Boulder, CO 80309. (andrewsj@spot.colorado.edu, barberdc@ucsu.colorado.edu, jenninga@spot.colorado.edu)

The North Atlantic's 1-2 kyr Climate Rhythm: Relation to Heinrich Events, Dansgaard/Oeschger Cycles and the Little Ice Age

Gerard C. Bond[1], William Showers[2], Mary Elliot[3], Michael Evans[1], Rusty Lotti[1], Irka Hajdas[4], Georges Bonani[4], and Sigfus Johnson[5,6]

New evidence from deep-sea sediment cores in the subpolar North Atlantic demonstrates that a significant component of sub-Milankovitch climate variability occurs in distinct 1-2 kyr cycles. We have traced that cyclicity from the present to within marine isotope stage 5, an interval spanning more than 80 kyrs. The most robust indicators of the cycle are repeated increases in the percentages of two petrologic tracers, Icelandic glass and hematite-stained grains. Both are sensitive measures of ice rafting episodes associated with ocean surface coolings. The petrologic tracers exhibit a consistent relation to Heinrich events, implying that mechanisms forcing Heinrich events were closely linked to those forcing the cyclicity. Our records further suggest that Dansgaard/Oeschger events may be amplifications of the cycle brought about by the impact of iceberg (fresh water) discharges on North Atlantic thermohaline circulation. The tendency of thermohaline circulation to undergo threshold behavior only when fresh water input is relatively large may explain the absence of Dansgaard/Oeschger events in the Holocene and their long pacings (thousands of years) in the early part of the glaciation. Finally, evidence from cores near Newfoundland confirms previous suggestions that the Little Ice Age was the most recent cold phase of the 1-2 kyr cycle and that the North Atlantic tended to oscillate in a muted Dansgaard/Oeschger-like mode during the Holocene.

1. INTRODUCTION

The dominant modes of abrupt climate change in the North Atlantic are Dansgaard/Oeschger (D/O) cycles and Heinrich events. Both were recognized more than a decade ago, yet their underlying causes remain a matter of debate. D/O cycles are a series of large, abrupt temperature variations confined to the glacial portion of the Greenland ice core record (e.g., Dansgaard et al., 1993). They are regarded as a consequence of switches in the North Atlantic's thermohaline circulation that, as revealed by modeling efforts, could have been triggered by inputs of fresh water (Broecker et al., 1994). Heinrich events are massive ice rafting episodes thought to have originated through internally forced collapses of Laurentide ice in eastern Canada (MacAyeal, 1993).

Bond et al. (1993) demonstrated that Heinrich events and D/O cycles were closely coupled. That finding led to the idea that a potential source of the fresh water needed to trigger D/O cycles was the melting of icebergs. It was argued by analogy with Heinrich events that the iceberg production could have been forced by the internal oscillatory dynamics of ice sheets.

Bond et al. (1992), however, pointed out that each Heinrich event appeared to be preceded by cooling of the ocean surface, and hence, Heinrich events might have been

[1] Lamont-Doherty Earth Observatory of Columbia University, Palisades, NY
[2] Department of Marine Earth and Atmospheric Sciences, North Carolina State University, Raleigh, NC
[3] Laboratoire des Sciences du Climat et de l'Environment, Laboratoire mixte CNRS-CEA, Gif sur Yvette, France
[4] AMS 14 C Lab, ITP Eidgenössische Technische Hochschule (ETH) Honeggerberg, CH-8093 Zurich, Switzerland
[5] The Niehls Bohr Institute, Department of Geophysics, University of Copenhagen, Denmark
[6] Also at Science Institute, Department of Geophysics, University of Iceland, Reykjavik, Iceland

Mechanisms of Global Climate Change at Millennial Time Scales
Geophysical Monograph 112

a response to, rather than a cause of, climate change. That argument was strengthened by evidence that icebergs were discharged from multiple sources during all six Heinrich events (Revel et al., 1996).

Subsequently, evidence from North Atlantic sediments of climate shifts within the Holocene led Bond et al. (1997) to argue that internal ice sheet dynamics may not have been the underlying cause of the D/O cycles either. They found that the pacings of the Holocene variations and those of D/O cycles in marine isotope stage 3, where D/O cycles are best developed, are statistically the same with a mean of 1,470± 532 (1σ) years. Those findings led them to suggest that both glacial and interglacial climates were punctuated by a persistent climate cycle with a pacing of between 1 and 2 kyrs. If that is true they argued, D/O variations might have been an amplification of the 1-2 kyr cycle.

One objective of this paper is to further test the influence of climate on Heinrich events and D/O cycles by tracing the 1-2 kyr cyclicity through the last glaciation to about 80 ka (all ages are in calendar kyrs BP unless indicated otherwise). We will focus on three questions: Did the 1-2 kyr cycle persist through the early part of the glaciation where the D/O variations have much longer pacings? Were Heinrich events triggered by cold phases of the cycle as suggested by Bond and Lotti (1995)? If D/O variations were an amplification of the climate cycle, what caused the amplification, and why is the D/O pacing different from that of the cycle in parts of the glacial record?

We also address the question of whether the Little Ice Age (LIA), a climate deterioration that peaked about 300 to 400 years ago (Grove, 1987), was the most recent cold phase of the 1-2 kyr cycle. As Denton and Karlén (1973) originally pointed out, certain features of the LIA suggest it may have been a scaled-down version of a D/O event. If so, linking the LIA to the cycle will strengthen the argument by Bond et al. (1997) that the North Atlantic tended to oscillate in a muted D/O-like mode during the Holocene. Bond et al. (1997) could not establish that link because core top ages in their study predated the LIA. We now have high-resolution records through the LIA from new cores taken using multicoring techniques.

2. SOURCES OF DATA AND METHODS

Our glacial records come from three cores in the eastern North Atlantic (Fig. 1), VM23-81 (54°15'N, 16°50'W; 2393 m depth), DSDP site 609 (49°53'N, 24°14'W; 3884 m depth) and SU90-24 (62°40'N, 37°22'W; 2100 m depth). Previous work demonstrated that the cores are of high quality and have sedimentation rates of 6 to 17 cm/kyrs (Bond et al., 1992; Elliot et al., 1998).

In earlier work on late glacial North Atlantic sediment, Bond and Lotti (1995) and Bond et al. (1997) demonstrated that among the diverse constituents of IRD (ice rafted debris), the percentages of three grain types, which we refer to as petrologic tracers, were particularly useful indicators of millennial-scale climate variations. Two of the tracers, the percentages of fresh volcanic glass derived from Iceland and the percentages of hematite-stained grains derived from redbed-rich sedimentary rocks, were robust and consistent indicators of the 1-2 kyr cycle. A third, detrital carbonate derived mostly from eastern Canada, marked the North Atlantic's Heinrich events.

Following their findings and using the same methods as in Bond et al. (1997), we extended the record of all three tracers in VM23-81 and DSDP site 609 to the early part of the glacial cycle at a resolution of 100 to 200 years. We also measured Icelandic glass and detrital carbonate within the middle to late glacial interval of SU90-24 at a resolution of 100 to 300 years. Combined with our previous petrologic data, our findings reveal a series of statistically significant peaks, defined as consisting of more than two points and rising above the 2σ error of ±2%, punctuating nearly all of the last glacial cycle (Figs. 2a, b).

Our new Holocene records are from isotopic and petrologic measurements in two cores east of Newfoundland (Fig. 1; EW93-GGC36 at 44°58'N, 46°25'W; 3982 m depth and KN98-MC21 at 44°18'N, 46°16'W; 3958 m depth). KN98-MC21, a multicore, has a modern top (that is, a corrected core top radiocarbon age of -255 years) and a sedimentation rate of about 15 cm/ka. In that core we have radiocarbon-dated sediments corresponding to the Little Ice Age.

To standardize the petrologic measurements, we did all counting in the 63 to 150 micron grain size fraction. We adopted that convention because even in Holocene sediment that size fraction contains enough grains for counting (>200), and more importantly, limiting counts to that narrow size range reduces variability caused by the relation between grain size and grain composition. The percentages of both non-Icelandic tracers (detrital carbonate and hematite-stained grains) are reported on an ash-free basis because in many cores, especially those near Iceland, the abundance of ash tends to mask all other IRD components.

We always express the tracers in percentages rather than concentrations. Tracer concentrations may record only the influence of ocean surface temperature on the survival or melting rate of drifting icebergs (which is true for total IRD concentrations as well). The tracer percentages, though, potentially are evidence of additional mechanisms such as changes in circulation and/or changes in rates of iceberg discharge from specific sources. Moreover, the percentage make up of the tracers is not likely to be influenced by changes in velocities of bottom currents, and as elaborated in a following section, although the concentration signal is weak during warm climates the percentage signal remains robust throughout the entire record.

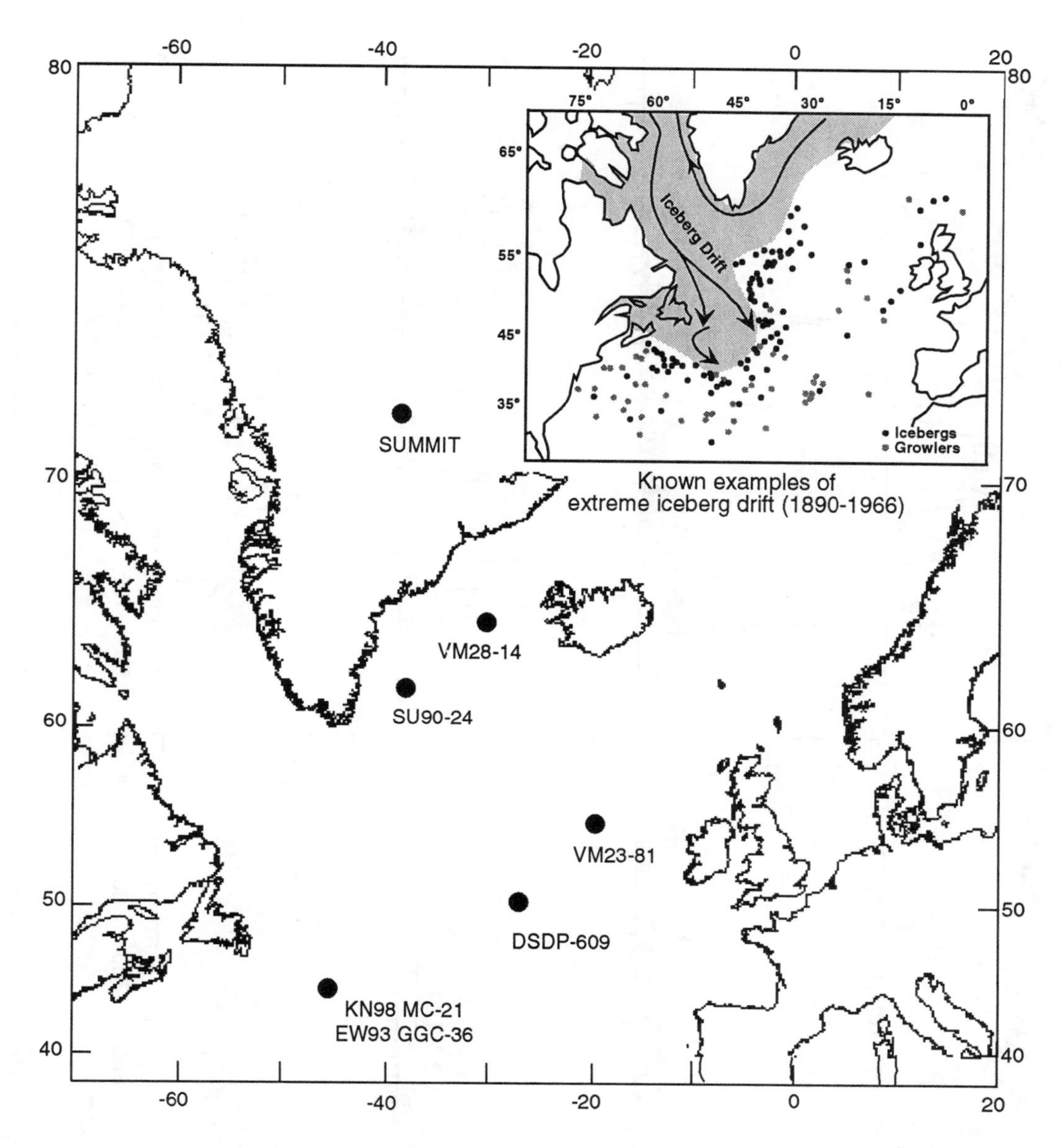

Figure 1. Location of cores cited in this study. Inset depicts modern iceberg discharges from the Labrador Sea. Shaded area is locus of most drifting icebergs. Dots indicate extreme examples of iceberg drift since the late 1800's based on ship captain's reports compiled in Oceanographic Atlas of the North Atlantic (1968). Growlers are less than 5 meters in height and less than 15 meters in length.

3. CHRONOLOGIES

AMS (accelerator mass spectrometer) radiocarbon ages for the late glacial are from Bond et al. (1993), Bond and Lotti (1995), Bond et al. (1997), and Elliot et al. (1998). Calibrations to 26 ka (radiocarbon) are from Bard (1993). Holocene chronologies are from AMS radiocarbon ages in Bond et al. (1997) and new measurements from Bond et al. (submitted), calibrated using CALIB 4.1 (Stuiver and Reimer, 1993; Stuiver et al., 1998).

To construct calendar age models for records older than 26 ka, we transferred the Meese/Sowers GISP2 time scale (Meese et al., 1994; Bender et al., 1994) into the *N. pachyderma* (s.) records from VM23-81 and DSDP site 609. For tie points we used prominent stadial-interstadial shifts that the ice core and marine records have in common (Fig. 3; Bond et al., 1993; see also Voelker et al., 1998). Ages between the tie points were obtained by interpolation. Using the calibrated radiocarbon ages for V23-81 thus obtained, we produced a calendar age model for SU90-24.

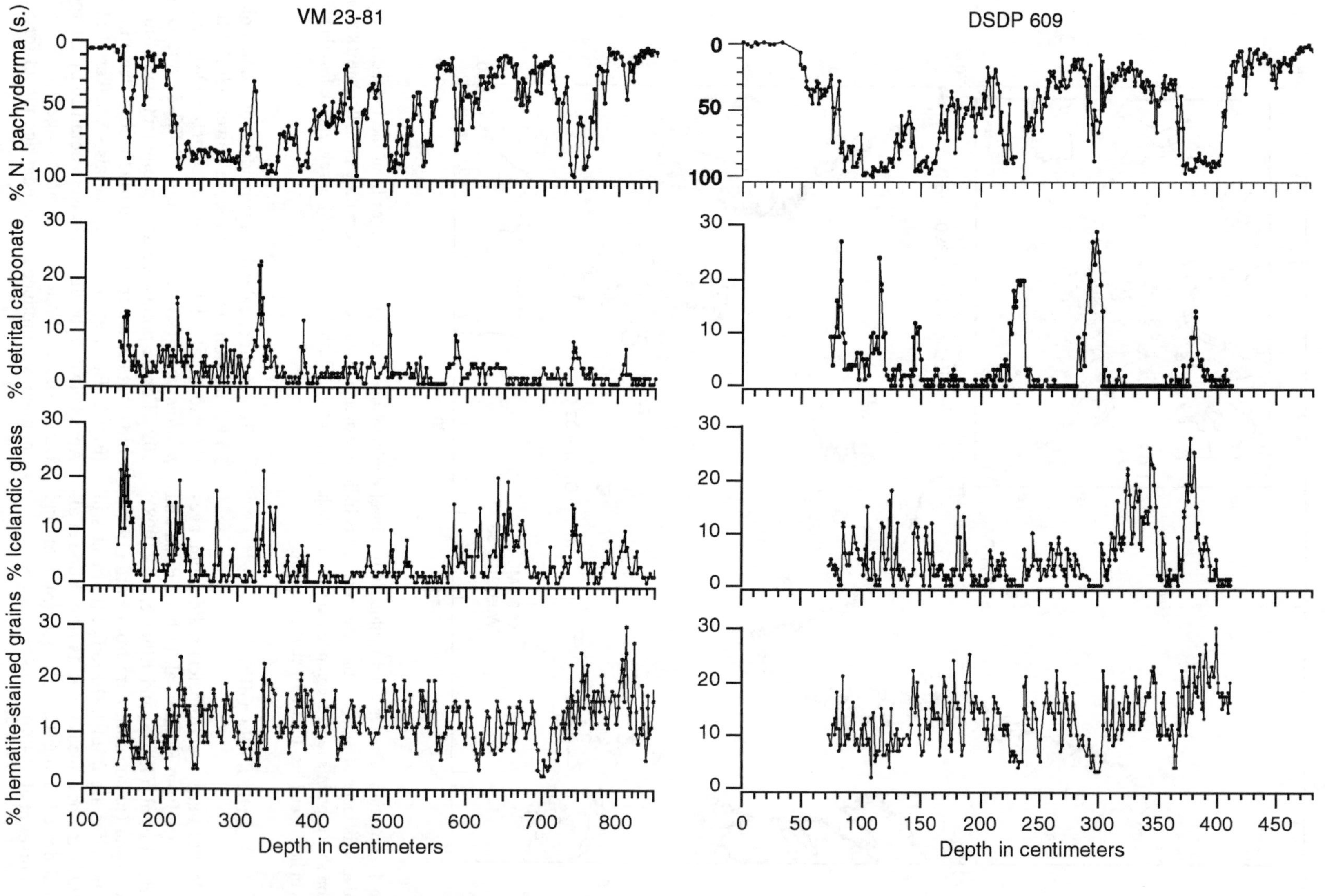

Figure 2. (a) Our petrologic and faunal measurements in VM23-81 and DSDP site 609. The measurements were made following the procedures in Bond et al. (1997).

In spite of the fewer number of tie points in DSDP site 609, the correlations among the ice core and marine records are quite good (r = 0.76 to 0.84; Fig. 3).

Errors in our calendar age models are difficult to estimate. Alley et al. (1997) concluded that the reproducibility of the GISP2 chronology is about 5% to 55 ka and decreases with increasing age and depth. The error is compounded in our age models by the likelihood that the ice core-marine tie points are not exactly the same age.

Even so, at least to about 55 ka our age models are almost certainly better than the only other alternative, SPECMAP (Martinson, et al., 1987), which has an estimated average error of ± 5 kyrs. The radiocarbon-calendar age offset between 12 ka and 38 ka in our marine records (Fig. 4) is similar to that reported elsewhere (e.g., Kitagawa et al., 1998). In addition, Bond et al. (1997) found that both the GISP2 age model and a calibrated radiocarbon chronology for the benthic $\delta^{18}O$ record in VM19-30 gave the same age for the marine isotope stage 2-3 boundary. That age is about 30 ka and is 6 kyrs older than estimated by SPECMAP.

Beyond 55 ka, the Meese/Sowers age model is based on comparing ice core atmospheric O_2 with the marine record. For that interval, our age model errors must be on the order of those estimated by SPECMAP.

Placed on our calendar age models, the records of the petrologic events from the 3 cores correlate reasonably well (Fig. 5). Peak to peak age differences are mostly less than 500 years, and with only two exceptions the Heinrich events are always separated by the same number of petrologic cycles.

4. RESULTS AND DISCUSSION

4.1. Tracing The 1-2 Kyr Climate Cycle To 80 Ka

Our records of Icelandic glass and especially hematite-stained grains reveal no statistical difference between the narrow range of event pacings that Bond et al. (1997) identified between 1 and 30 ka and those measured here in the earlier part of the glaciation (Figs. 5, 6; Table 1). For the Holocene and glacial intervals combined, the average pacing is 1469±514 years (1σ) in VM23-81 and 1476±585 years (1σ) in DSDP site 609. We also find that the shorter mean pacing of 1.3 to 1.4 kyrs is not unique to the Holocene; within the glaciation, intervals with similar durations to that of the Holocene (i.e., 12 kyrs) have similar mean values (Table 1).

Our results also confirm earlier arguments that the percentage shifts in Icelandic glass and hematite-stained grains must have been climatic in origin (e.g. Bond and Lotti, 1995). As the iceberg armadas carrying the two tracers come from different sources, the covariance of the two tracers through the entire record, particularly in DSDP site 609 (Figs. 2, 5), is evidence of the influence of climate

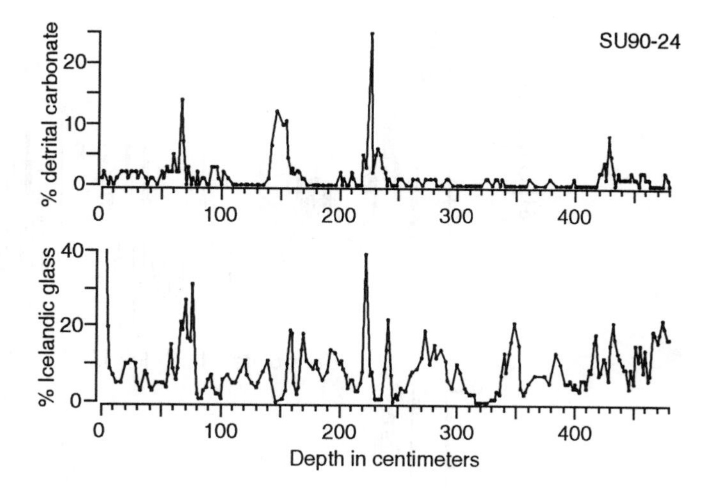

Figure 2. (b) Our petrologic measurements in SU90-24. The measurements were made following the procedures in Bond et al. (1997). See Elliot et al. (1998) for isotopic, lithic and faunal measurements in the core.

on unstable ice rather than of forcing by internal ice sheet instabilities. The evidence is robust as the tracer covariance is documented within each core and is independent of age models.

In addition, Bond et al. (1997) found that from the present to about 30 ka, tracer peaks coincided with ocean surface coolings. Our new records reveal the same relation for the early glaciation, at least where ocean surface coolings are documented by abundance peaks in *N. pachyderma* (s.) (Fig. 2).

The discontinuous record of Icelandic glass below 400 cm (about 32 ka) in VM23-81 (Fig. 2) probably reflects the location of the core at the downstream edge of the Ruddiman IRD belt (Ruddiman, 1977). Prior to the main phase of the glaciation, that site at times may have been out of reach of ash-bearing icebergs.

Thus, our petrologic records demonstrate that the 1-2 kyr climate cycle persisted through the shifting climate modes of the last 80 kyrs. The cycle punctuated the Holocene, a number of prolonged interstadials such as 1, 12 and 14, and several stadials, such as those between interstadials 2 and 3, 4 and 5, and 17 and 18 (Figs. 6, 7). Our results clearly reveal the sharp contrast between the irregular cyclicity of the D/O events and the much more regular pattern of cycles in our petrologic records (Fig. 6c).

Others have suggested a similar cyclicity in the North Atlantic's climate, although from less consistent records than we describe here. Keigwin and Jones (1989) concluded that North Atlantic Holocene and deglacial climates underwent abrupt shifts about every 2 kyrs, but found no evidence of that variability in the first 6 kyrs of the Holocene. Similarly, Pisias et al. (1973) inferred

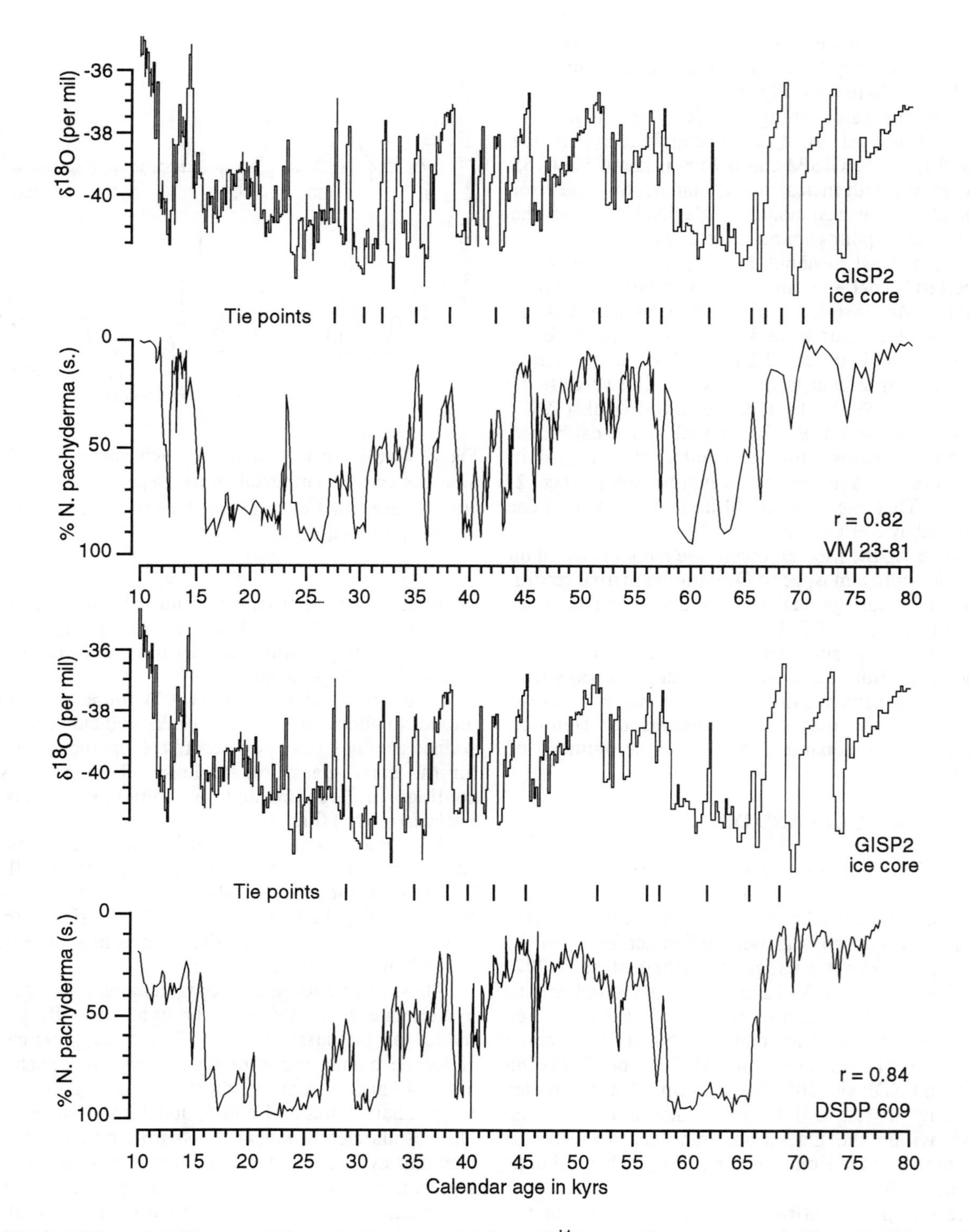

Figure 3. Procedure for constructing calendar age models beyond ^{14}C age calibrations. Correlations of ice core and marine records from Bond et al. (1993); GISP2 time scale was transferred into the *N. pachyderma* (s.) records using tie points shown as short bars. The correlation coefficient of the two *N. pachyderma* (s.) records is 0.76. Correlation coefficients of marine records and ice core shown on the Figure. GISP2 time scale is from (Meese et al. 1994; Bender et al. 1994, available from National Snow and Ice Data Center, University of Colorado at Boulder, and the World Data Center-A for Paleoclimatology, National Geophysical Data Center, Boulder, Colorado).

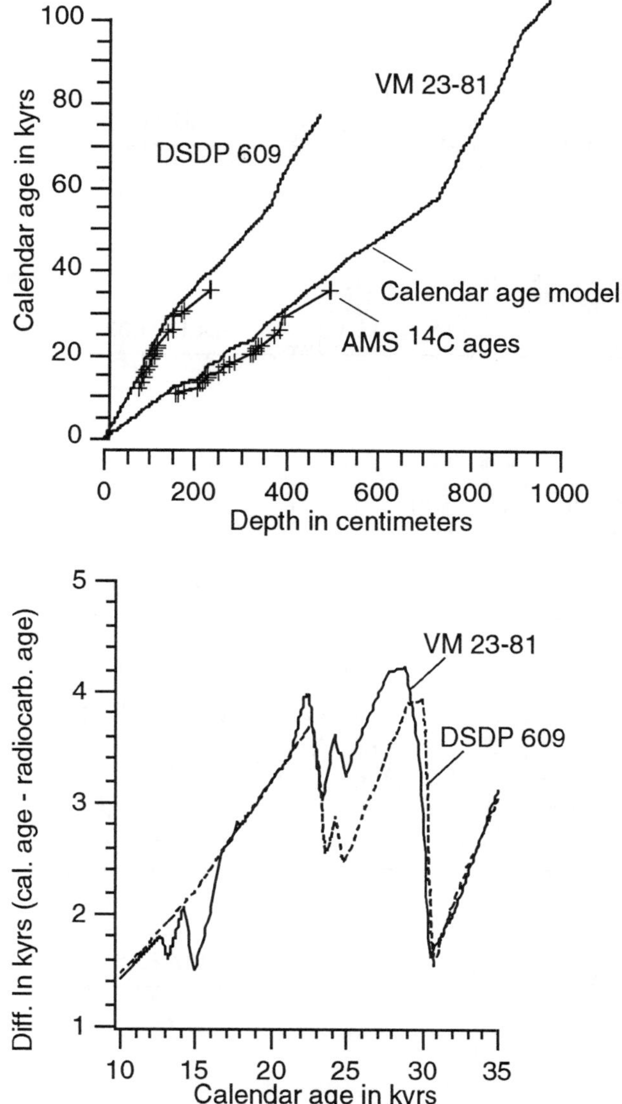

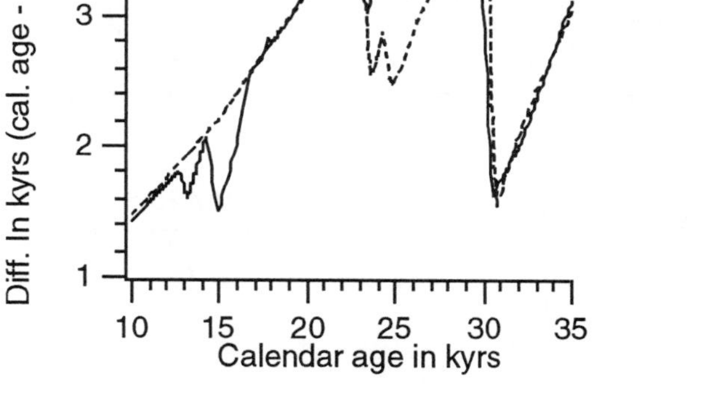

Figure 4. Above is the age-depth relation from the calendar age models in Fig. 3. Below is the difference between calendar and radiocarbon ages for the two cores, based on the age models in Fig. 3.

periodicities of 1.3 kyrs and 2.6 kyrs from the frequency makeup of North Atlantic climate records spanning the last 130 kyrs. The time series of the proxies they measured, however, revealed little evidence of such rapid variations, especially during the Holocene.

4.2. Strength And Consistency Of The Petrologic Signal

Why do robust petrologic variations in North Atlantic IRD punctuate both warm and cold climate states so consistently? Perhaps part of the answer is that the petrologic signal is not subject to biological overprints that tend to complicate other proxies. A more important reason though, lies in a rather unique combination of the sources of IRD, the sources of drifting ice, and the North Atlantic's surface circulation.

The circulation mechanism Bond et al. (1997) proposed for the petrologic variations during the warm climates of the Holocene, for example, arguably applies as well to the long, Holocene-like interstadials of the early glaciation. The Holocene signal was produced by recurring southward advections of cooler, ice bearing surface waters from north of Iceland. Petrologic analyses of core top sediments demonstrated that drift ice in those waters carried relatively large amounts of Icelandic glass that was produced mainly by eruptions onto ice drifting near the island. The drift ice also carried hematite-stained grains that were most likely derived from redbeds in East Greenland and Svalbard and/or from redbeds around the Arctic Ocean. Thus, during relatively warm climates with few calving glaciers, circulation changes alone were capable of producing robust peaks in percentages of the two tracers at least as far south as VM23-81.

That drift ice reached so far south during interstadials and interglacials is not as extraordinary as it might seem. Iceberg sightings reported by ship captains over the last century document extreme iceberg drift as far south as Bermuda and eastward over the VM23-81 and DSDP 609 coring sites (Fig. 1; Oceanographic Atlas of the North Atlantic, 1968). Those examples are minimum estimates as they reflect only what was observed in shipping lanes.

For the colder phases of the glaciation, however, the circulation mechanism does not explain all of our petrologic data. Compositional measurements of IRD in the likely paths of iceberg drift suggested that during marine isotope stages 2 and 3, D/O coolings forced iceberg discharges from redbed sources in the Gulf of St. Lawrence and perhaps from Iceland as well (Bond and Lotti, 1995).

Subsequent studies have confirmed that at least during the LGM and deglaciation icebergs were discharged on millennial time scales from the Gulf of St. Lawrence and elsewhere along the southern edge of the Laurentide ice sheet. Piper and Skene (1998) correlated a reddish, hematite-rich layer in deep-sea sediment near Newfoundland with the precursor to H1 in DSDP 609 and VM23-81 (Fig. 5). They concluded that both the reddish layer and the precursor were produced by rapid iceberg discharge from the Gulf of St. Lawrence. Stea et al. (1998) documented advances of ice in Nova Scotia (also a source of hematite-stained grains) during H2 and during the Younger Dryas.

Thus, we have identified two petrologic tracers whose percentages increased by similar amounts during both warm and cold climates. During the Holocene and the long interstadials of the early glaciation, changes in surface

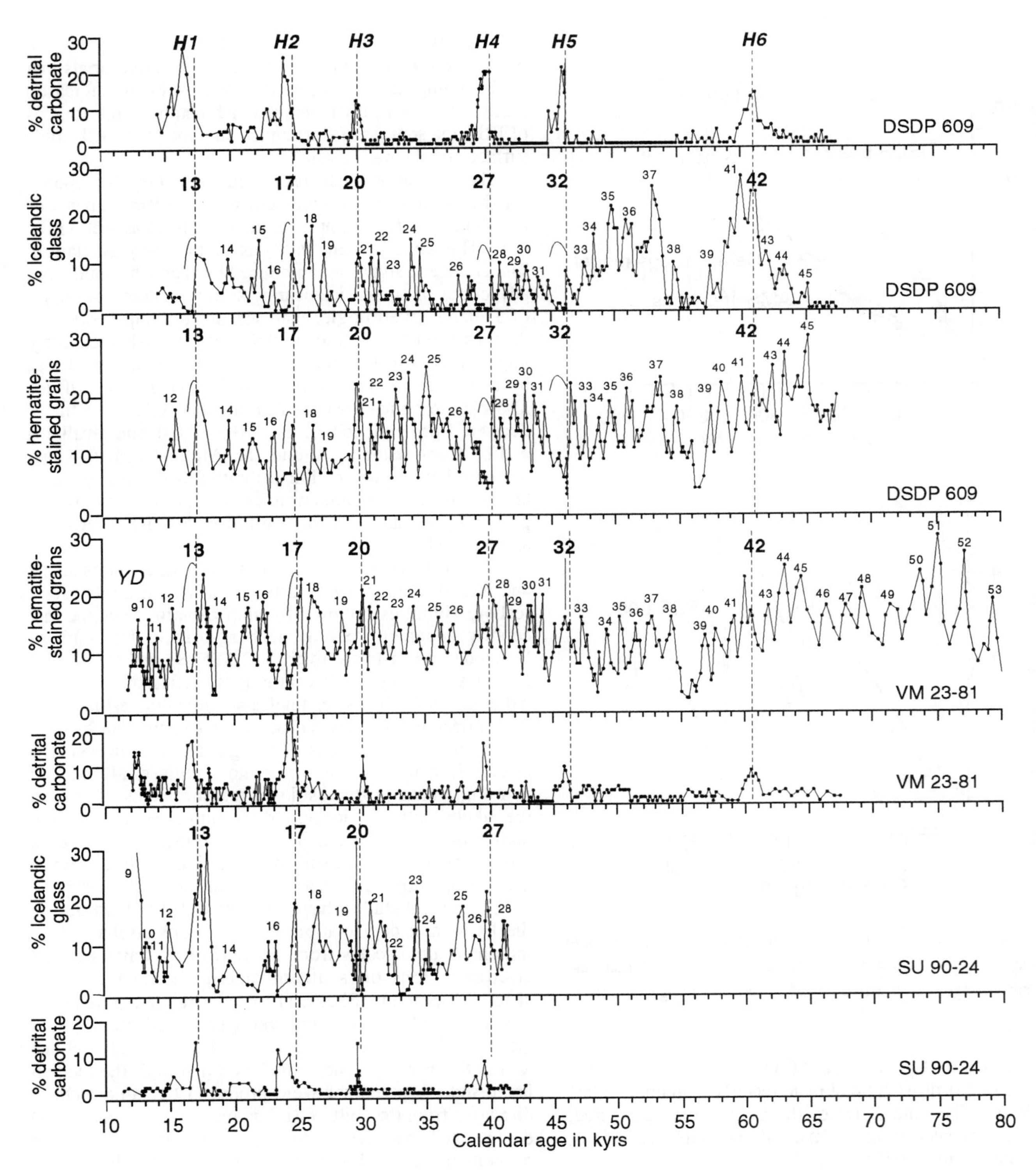

Figure 5. Our petrologic records on calendar time scales from Fig. 3. Short curved lines mark dilution of tracers by IRD in H events. Small numbers identify statistically significant peaks as defined in text; numbering begins with Little Ice Age = 0. Boldface numbers are H events. Dashed lines for H1, H2, H4 in all three cores and H5 in DSDP 609 mark boundaries between increases in detrital carbonate (H events) and petrologic precursors (unnumbered). Dashed lines for H3 and H6 mark synchronous increases in all petrologic tracers. YD = Younger Dryas.

Table 1. Cycle Pacing

	Time Interval in cal. kyrs.	Mean Pacing ±1 sigma in yrs.
	VM 23-81	
Holocene	1.5 - 12ka	1374±502
Late Glacial	12 - 32ka	1537±558
Early Glacial	32 - 75ka	1478±458
~12kyr step	13 - 24ka	1494±624
~12kyr step	22 - 34ka	1631±511
~12kyr step	31 - 43ka	1328±539
~12kyr step	43 - 55ka	1350±302
~12kyr step	53 - 64ka	1443±470
~12kyr step	64 - 79ka	1795±425
Holo. + Glacial	0 - 79ka	1469±514
	DSDP 609	
Glacial	15 - 65ka	1476±585

circulation controlled the tracer percentages, while during stadials a combination of changes in circulation and increases in iceberg discharges probably predominated.

In marked contrast to the strong attenuation of IRD concentrations during warmer climates, therefore, shifts in percentages of both tracers, and particularly those of the hematite-stained grains, are robust and continuous through the entire glacial-interglacial record. The amplitude of the tracer signal though is not related in any simple way to the amplitude of the climatic forcing.

Our inference that coolings forced iceberg discharges from certain glaciers does not conflict with the well-documented climate insensitivity of many glaciers and especially large ice sheets (e.g. Alley, 1991; Sturm et al., 1991; Mann, 1986; Wiles et al., 1995). In maritime settings where large amounts of moisture are available, such as the Gulf of St. Lawrence region and Iceland, glaciers and small ice caps undergo advances directly in response to rapid climate coolings (e.g., Oerlemans, 1992; Sigurdsson and Jonsson, 1995; Fleming et al., 1997).

Moreover, as Porter (1989) and Motyka and Begét (1996) concluded from glacial histories in southeastern Alaska, it is entirely reasonable to expect, on millennial time scales at least, that falling temperatures forced increases in calving rates in some (but certainly not all) tidewater glaciers.

We note, however, that the two petrologic tracers together constitute less than half of the total IRD in any sample. Stochastic, non-climatic processes may well have forced iceberg discharges from other sources we have not yet identified.

4.3. Spectral Analysis Of The Petrologic Signal

Spectral analysis of the record of percent hematite-stained grains in VM23-81 reveals power within a broad peak centered at ~1.8 kyr (0.55±0.15 cycles/kyr). The peak is significant relative to spectra from a red noise time series with similar memory to that of the proxy data (Fig. 8a). The frequency window explains 25% of variance in the raw time series and is used to filter that time series (Fig. 8b). Hence, even though our age models must contain a certain amount of error, our spectral data together with the persistent and narrow range of event pacings (Fig. 6c) suggest that the cycle may be periodic.

Bond et al. (1997) found evidence of spectral power at ~1.8 kyr in a shorter record of percent hematite-stained grains, constrained in that case entirely by calibrated radiocarbon ages (Bond et al., 1997). In addition, Mayewski et al. (1997) documented a weak 1.4 to 1.5 kyr spectral peak in major ion time series from the GISP2 ice core, and Bianchi and McCave found spectral power at 1.5 kyr in time series of sediment grain sizes from south of Iceland. Those spectra are statistically indistinguishable from the broad spectral peak in our records.

4.4. A Caveat On Evidence From Other Marine Proxies

We found no compelling evidence of a 1-2 kyr cycle in high resolution North Atlantic records of gray scale, of the percentages of *N. pachyderma* (s.) and of the $\delta^{18}O$ in that foraminifera (Fig. 9). We can suggest a number of reasons for those results. Gray scale (or color) mainly reflects changes in the composition and concentration of fine sediment (<62 microns), which typically constitutes more than 90% of the core sediment weight. Those changes are influenced not only by depositional rates of ice rafted sediments, but also by sediment discharges from rivers and by speeds of surface and bottom currents.

The abundances of *N. pachyderma* (s.) do not reflect summer temperature changes above about 15°C, which is its maximum growth temperature, and below about 5°C at which the species saturates the planktic population (Pflaumann et al., 1996). Moreover, by shifting growth seasons or depth habitats the species can maintain constant, optimum growth temperatures (Sarnthein et al., 1995; Wu and Hillaire-Marcel, 1994). It is also possible that during each cooling episode, freshening of surface waters by melting icebergs and/or advection of fresher surface waters from more northerly latitudes offset temperature-driven

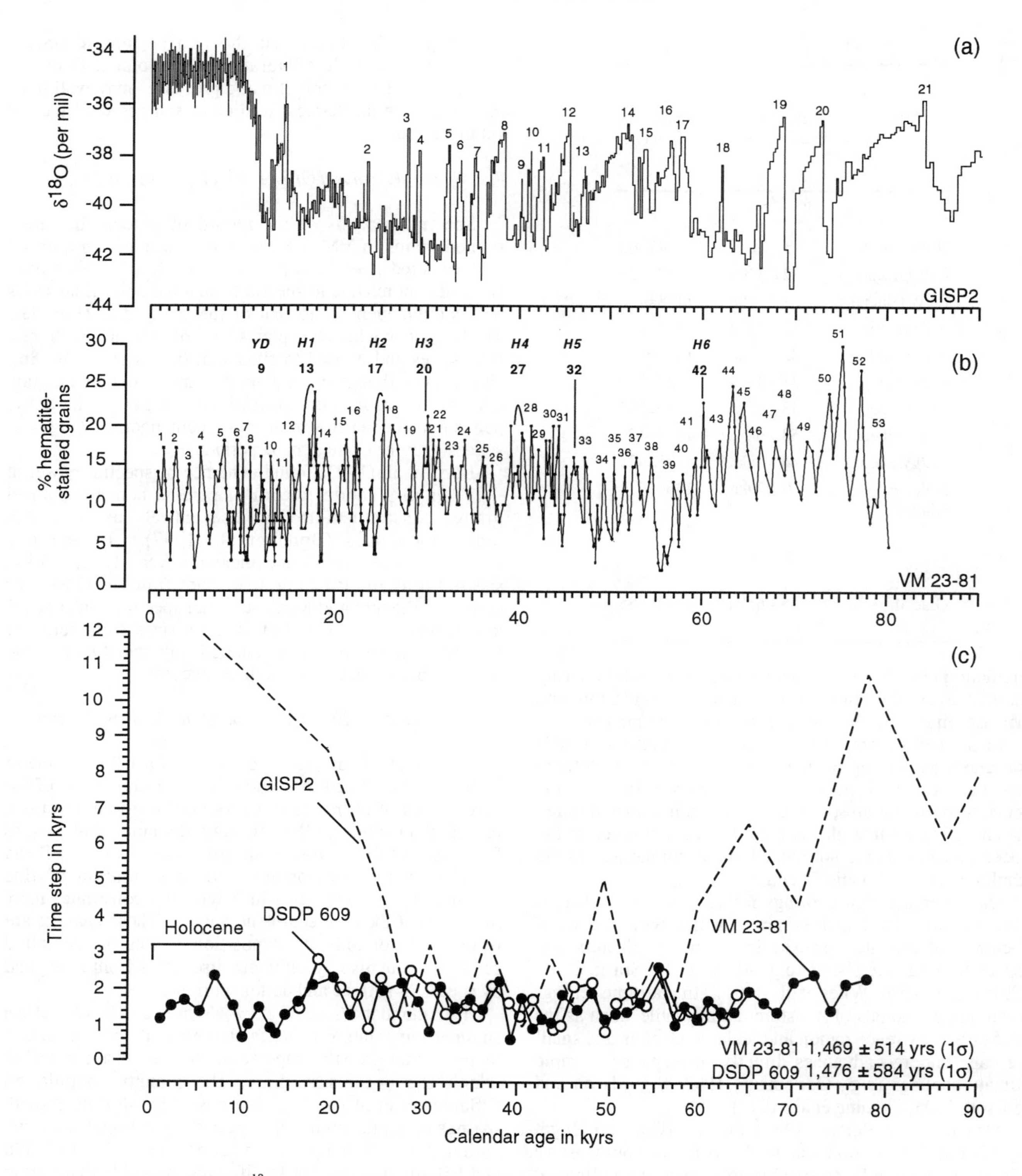

Figure 6. (a) GISP2 $\delta^{18}O$ record; numbered interstadials from Dansgaard et al. (1993). (b) VM23-81 record of hematite-stained grains; numbering as in Fig. 5. (c) Dots are pacings of petrologic events, defined as time separations of statistically significant peaks placed at midpoints between the peaks. Dashed line is same calculation for numbered GISP2 interstadials. In (c), regular pacing of petrologic events contrasts sharply with irregular pacings of D/O events. Holocene record of hematite-stained grains from Bond et al. (1997) was added for comparison with glacial records.

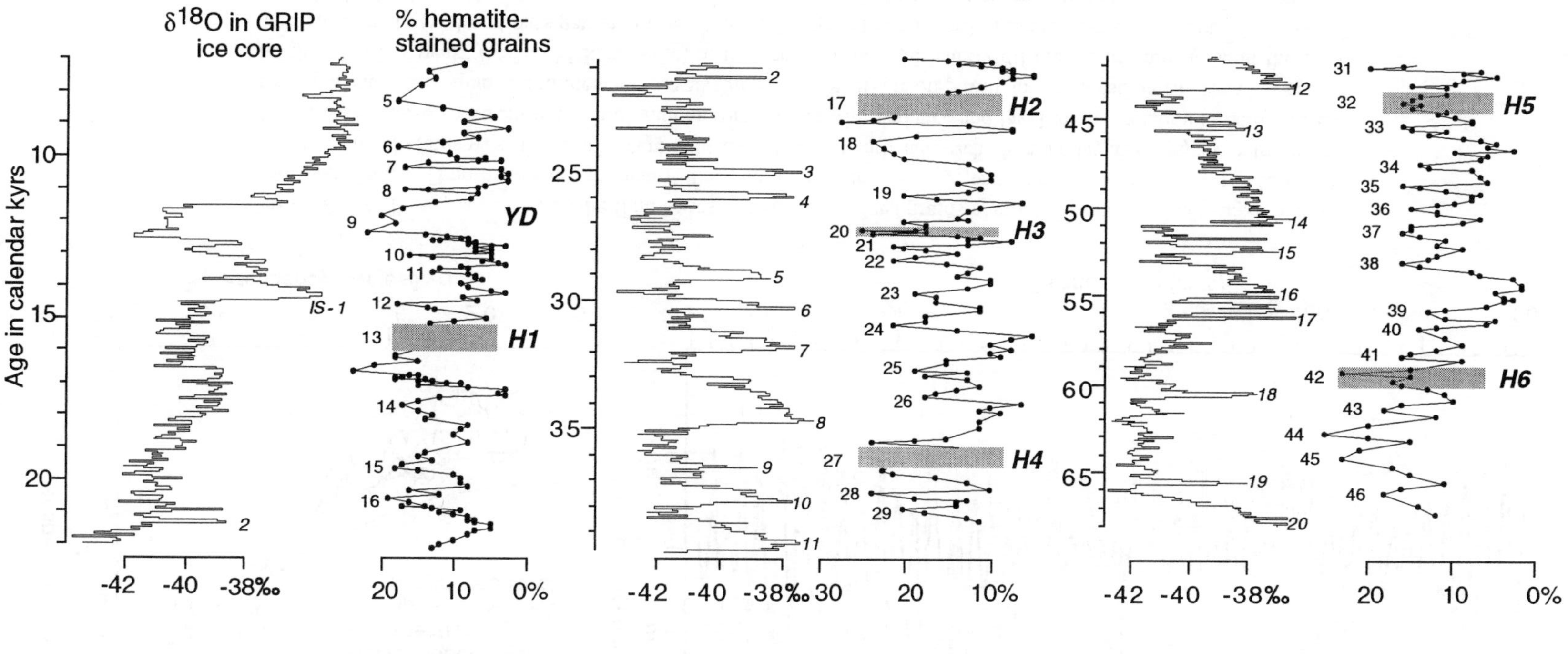

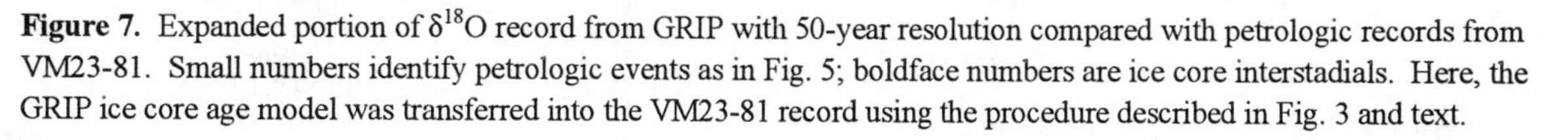
Figure 7. Expanded portion of $\delta^{18}O$ record from GRIP with 50-year resolution compared with petrologic records from VM23-81. Small numbers identify petrologic events as in Fig. 5; boldface numbers are ice core interstadials. Here, the GRIP ice core age model was transferred into the VM23-81 record using the procedure described in Fig. 3 and text.

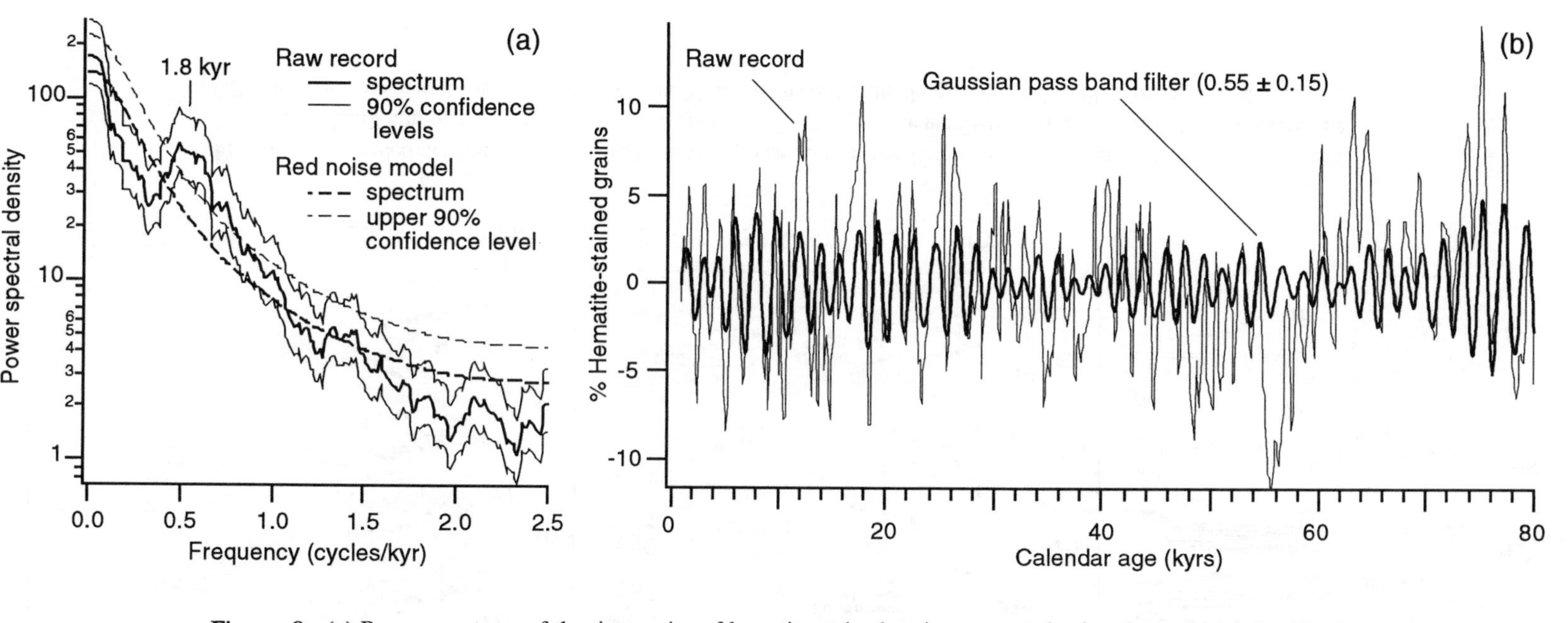

Figure. 8. (a) Power spectrum of the time series of hematite-stained grains computed using the multi-taper method (Thomson, 1982). Time bandwidth product = 4.5; number of windows = 9; bandwidth = 0.5 cycles/kyr. Record of hematite-stained grains was "filled in" where diluted by debris in carbonate-rich Heinrich layers (compare with Fig. 5). The mean red noise spectrum is for 1000 random time series with variance and lag-1 autocorrelation structure of the petrologic data. Chi-squared confidence intervals (90 percent level) are on power spectrum estimates. Results suggest that the ~1.8 kyr peak (0.55±0.15 kyrs/cycle) in the spectrum is significantly different from red noise at the 90% confidence level. (b) Band pass gaussian filter (0.55±0.15 kyr) of raw record of hematite-stained grains. The record was "filled in" where diluted by debris in carbonate-rich Heinrich layers (compare with Fig. 5). The filter explains about 25% of the variance in the raw record.

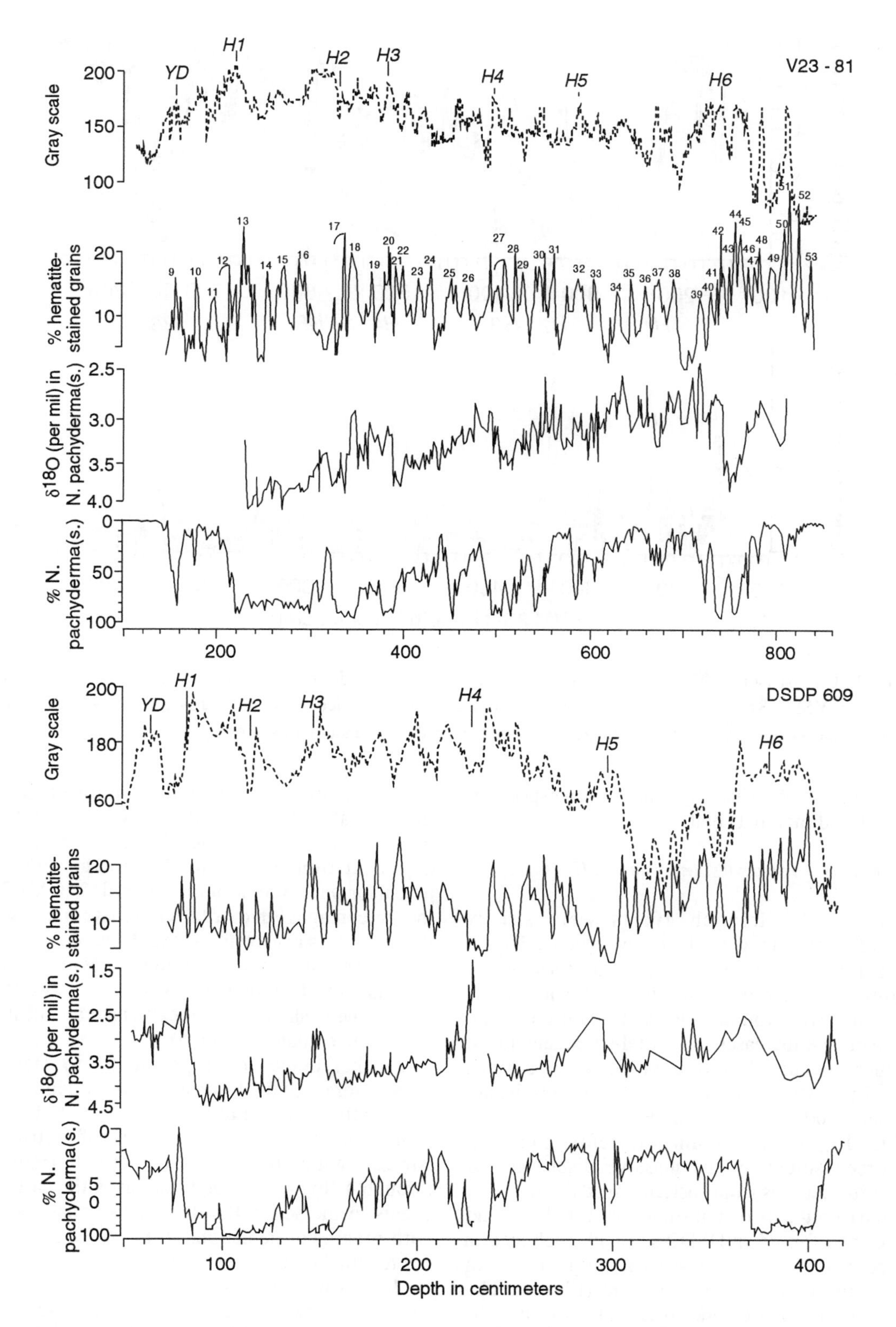

Figure 9. Comparison of records of hematite-stained grains with gray scale, *N. pachyderma* $\delta^{18}O$, and *N. pachyderma* (s.) abundances in VM23-81 and in DSDP site 609. Isotopic measurements in DSDP site 609 from Bond et al. (1992) and in VM23-81 from Labeyrie et al. (1995). Peaks in hematite-stained grains numbered as in Fig. 5.

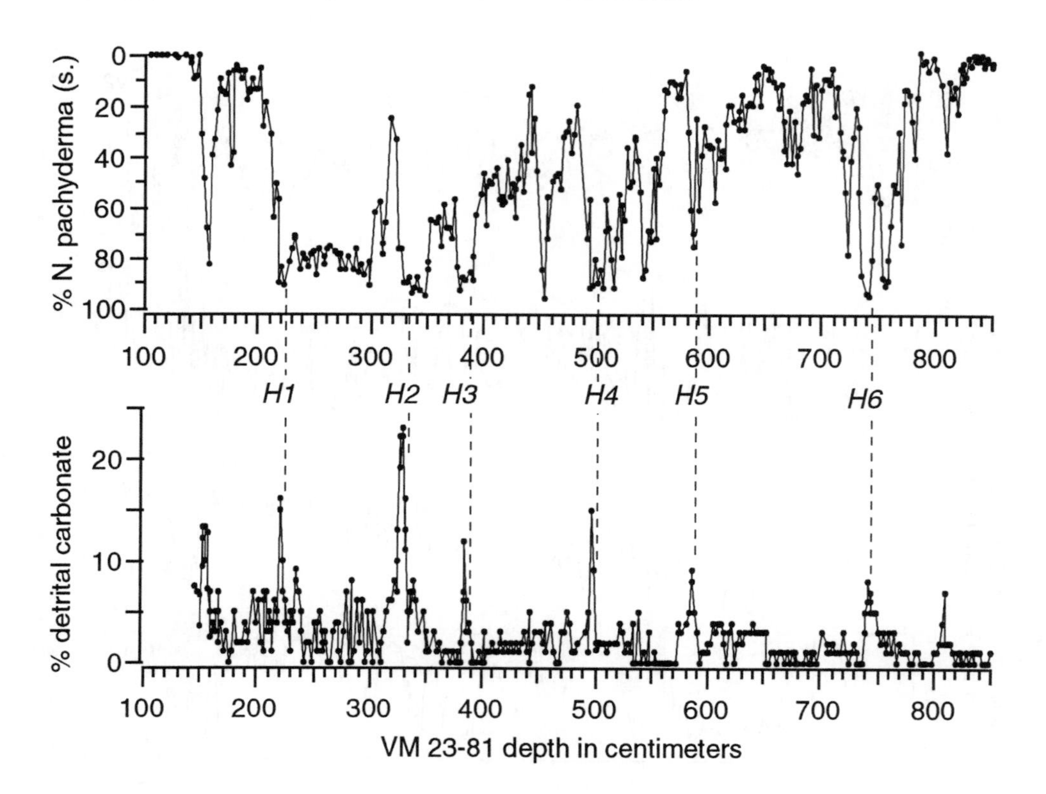

Figure 10. Comparison of *N. pachyderma* (s.) and Heinrich events as defined by increases in percentages of detrital carbonate in VM23-81. Dashed lines demonstrate that cooling, as indicated by an increase in *N. pachyderma* (s.) percentage, leads each increase in percentage of detrital carbonate above ambient values.

planktic $\delta^{18}O$ enrichment. That might have been especially common in the Ruddiman IRD belt.

4.5. Heinrich Events, Precursors And The 1-2 Kyr Cycle

Precursors. If Heinrich events were strictly consequences of stochastic, glaciologically driven collapses of ice in Hudson Strait (e.g. MacAyeal, 1993), they should exhibit no consistent relation to changes in climate. Bond et al. (1992), however, found that each Heinrich event, defined as percentage increases in detrital carbonate above ambient values, occurred during cold ocean surface temperatures (Fig. 10). In each case, the cooling began between 500 and 1,000 years before the increase in detrital carbonate (Fig. 10), and the coolings therefore could not have been forced solely by the massive discharges of carbonate-bearing icebergs from eastern Canada.

Our new measurements demonstrate that for H1, 2, and 4 in both cores and for H5 in DSDP site 609, each lead in ocean surface cooling coincides with an increase in percentages of the two petrologic tracers (Figs. 5, 11). Those results confirm and extend to Heinrich event 5 the evidence of "precursors" to Heinrich events and the link of those precursors to the climate cycle first noted by Bond and Lotti (1995).

The abrupt decreases in the two tracers at the levels of the detrital carbonate peaks in H1, 2, and 4 (Fig. 5) are most likely due to dilution by massive amounts of IRD deposited from the Heinrich icebergs. For example, a precursor but no dilution occurs at H5 in VM23-81 where the percentages of detrital carbonate are low relative to those for the same event in DSDP site 609.

On the other hand, we found no evidence of petrologic precursors to Heinrich events 3 and 6. For both Heinrich events, the peaks in detrital carbonate and peaks in the two tracers are coincident (Fig. 5).

Further evidence of precursors to Heinrich events comes from SU90-24, located several degrees north of the main path of the Heinrich icebergs (Fig. 1). In that core prominent peaks in percentages of detrital carbonate correlate within the error of dating with Heinrich events 1 through 4 (Fig. 5). In addition, as is the case for Heinrich events to the south, the detrital carbonate peaks coincide with prominent depletions in the $\delta^{18}O$ of *N. pachyderma* (s.) reported by Elliot et al. (1998).

Just as we found for the precursors in the cores to the south, increases in Icelandic glass that are part of the 1-2 kyr cycle in SU90-24 precede detrital carbonate peaks at H1, H2 and H4 and are coincident with a detrital carbonate peak at H3 (Fig. 5). The relation of the Heinrich events to

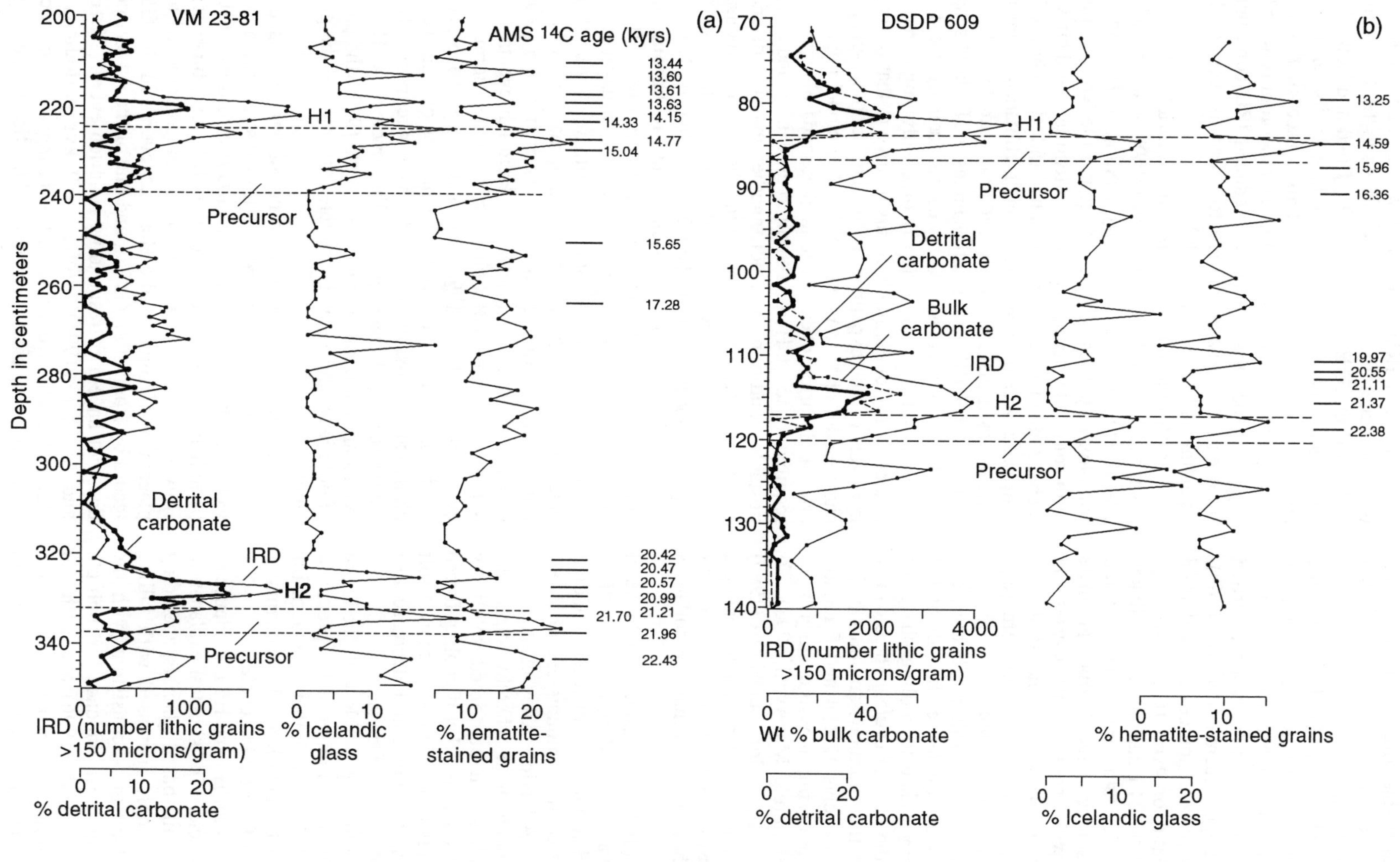

Figure 11. (a) Evidence of petrologic precursors to H1 and H2 in VM23-81. Precursors are defined as increases in percentages of Icelandic glass and hematite-stained grains during initial increases of IRD and initial shifts to cooler ocean surface temperatures (see also Fig. 9). AMS radiocarbon ages from Bond and Lotti (1995). (b) Evidence of petrologic precursors to H1 and H2 in DSDP site 609 defined as in (a). Also shown is the percentage of fine fraction carbonate, which, based on microscopic analysis is entirely non-biogenic. Sharp bases of this material coincide precisely with sharp bases of the two Heinrich layers as revealed by x-radiographs. AMS radiocarbon ages are from Bond et al. (1992).

precursors is especially clear in this core because the small amounts of IRD from Heinrich icebergs did not significantly dilute the other tracers.

The detrital carbonate peaks in SU90-24 probably are evidence of extreme northward drift of Heinrich icebergs in the western North Atlantic, although even today small numbers of icebergs exiting the Labrador Sea are carried toward the site in the Irminger Current (Murry, 1969). Although we cannot firmly rule out sources of the detrital carbonate in East Greenland or even the Arctic (e.g. Bischof et al., 1996), the SU90-24 peaks are clearly linked by age, composition and $\delta^{18}O$ signature to the Heinrich events farther south. The source of the detrital carbonate perhaps can be confirmed with geochemical analyses and dating of ice-rafted grains, as has been demonstrated by provenance studies of Heinrich layers elsewhere (e.g. Gwiazda, et al., 1996; Hemming et al., 1998).

Relation to the 1-2 kyr cycle. Taken together, our findings in the three cores are evidence that all of the Heinrich events were closely linked to cooling phases of certain oscillations within the 1-2 kyr cycle. There can be no question, however, that the enormous amounts of ice discharged during Heinrich events 1, 2, 4 and 5 require glaciological processes involving massive surging or collapse of ice in Hudson Strait (e.g. MacAyeal, 1993). To reconcile both pieces of evidence, Bond and Lotti (1995) suggested that if ice in Hudson Strait were unstable enough, coolings and/or sea level rises accompanying the precursor iceberg discharges might have induced massive collapses of ice.

More recently, Hulbe (1997) suggested that lowered temperatures alone might have induced rapid growth of an ice shelf in the Labrador Sea which, after a few hundred years, would have discharged large amounts of carbonate-bearing ice into the North Atlantic. Alley et al. (1996) proposed an intriguing mechanism whereby surge-like behavior of ice sheet lobes can be forced solely by atmospheric cooling. They suggested that if cooling forced ice with basal layers frozen in bedrock to advance over soft deforming sediment, basal friction could decrease fast enough to induce a surge-like loss of ice mass.

All of those alternatives are compatible with the lag of the Heinrich events relative to the onset of the precursor coolings. Heinrich events 1, 2, 4 and 5, therefore, arguably were the consequences of glaciological mechanisms triggered by cooling phases of the 1-2 kyr cycle.

The absence of precursors for H3 and H6 is evidence of a different relation to the cycle. The simplest explanation is that calving rates of marine-based glaciers, including ones in eastern Canada, simply increased with falling temperatures. That mechanism is consistent with the increasing amount of geochemical evidence for multiple IRD sources in H3 (Revel et al., 1996) and with our evidence of nearly synchronous increases in all three petrologic tracers during both H3 and H6 (Fig. 5). It is also consistent with the relatively small contribution of carbonate-rich IRD from eastern Canadian sources and with the generally low concentrations of IRD in both H3 and H6 (Fig. 5; Bond et al., 1992; Bond and Lotti, 1995; Stoner and Channell, 1998). The lack of distinct isotopic evidence of an eastern Canadian source in those two Heinrich events, so prominent in the other four (Gwiazda et al., 1996), perhaps reflects the reduced IRD contribution from eastern Canada relative to other sources, such as Scandinavia (Revel et al., 1996).

A different interpretation of Heinrich events was advanced by McCabe and Clark (1998) who argued that Laurentide ice sheet instability leading to H1 triggered complex responses of glaciers around the North Atlantic, with some advancing and some retreating. In their view, iceberg discharges from Iceland and the Gulf of St. Lawrence were the response of small, climatically sensitive glaciers to a glaciologically induced climate shift. Our evidence that IRD from both sources leads the discharge from Laurentide ice, however, weakens their argument.

As Bond and Lotti (1995) originally suggested, the longer pacing of Heinrich events relative to that of the 1-2 kyr cycle can be explained in terms of the binge-purge ice sheet model of MacAyeal (1993). The model implies that after a Heinrich event, several thousands of years were required for ice in Hudson Strait to grow back and reach an unstable condition in which another massive discharge could have been triggered. During those several thousands of years, ice in Hudson Strait would be insensitive to triggering mechanisms of the 1-2 kyr cycle (Fig. 12).

The tendency of Heinrich events to become more closely spaced with decreasing time (e.g. Fig. 5; Bond et al., 1992) may have been a direct consequence of intensification of the glacial climate. As climate deteriorated from stage 3 to stage 2, it may have taken less time for ice in Hudson Strait to recover from the previous collapse (Fig. 12). In addition, Heinrich events always appear within the coldest phases of the longer-term cooling cycles (Bond et al., 1993) when the largest ice shelves might have grown or when the highest calving rates in climatically sensitive ice might have occurred. To the extent that the longer-term cooling cycles and Heinrich events are interrelated (e.g. MacAyeal, 1993), the pacings of both may have been tied to intensification of the glacial climate.

The timing of Heinrich events in our records is not consistent with previous conclusions that Heinrich events were periodic. Mayewski et al. (1997) estimated a 6.1 kyr periodicity for Heinrich events in GISP2 ionic times series. However, the 6.1 kyr cycle does not coincide with the placement of some of the Heinrich events in the GISP2 record by Bond et al. (1993), and one of the 6.1 kyr peaks does not have a corresponding Heinrich event. McIntyre and Molfino (1996) inferred a 9 kyr periodicity for Heinrich events from a record of changes in thermocline depths from the eastern equatorial Atlantic. However, they assigned a calendar age of 34 ka to H3 which is dated at

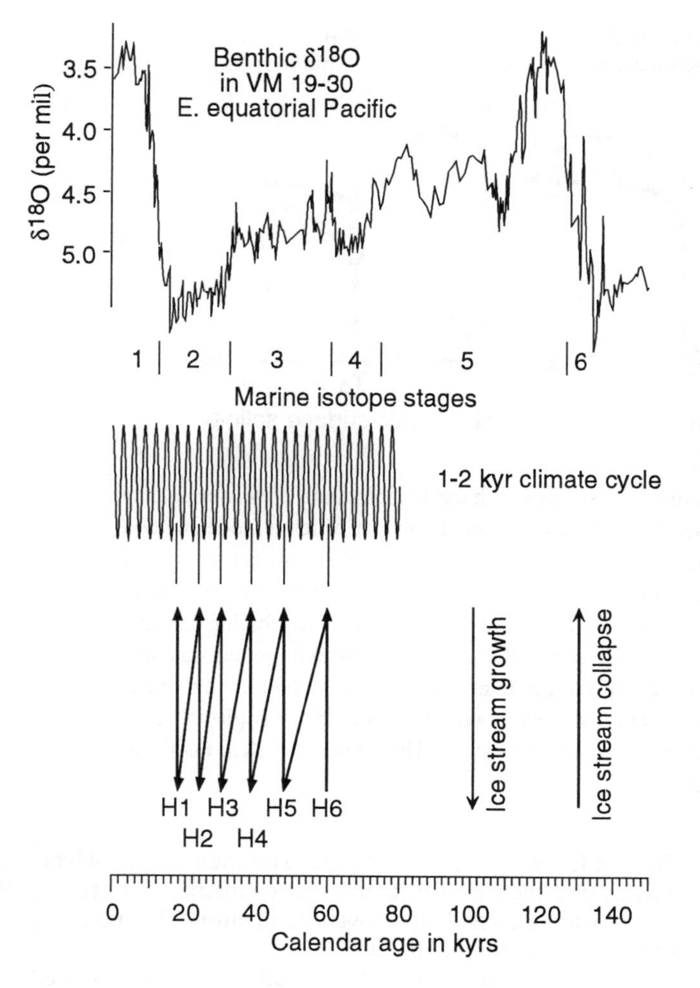

Figure 12. Hypothetical example of how the change in pacing of Heinrich events with time might have been related to intensification of the glacial cycle (see text for explanation). VM19-30 record is from Shackleton (1983).

about 30 ka in our records (Fig. 5; see also Bond and Lotti, 1995 and Bond et al., 1997).

4.6. Dansgaard/Oeschger Events And The 1-2 Kyr Cycle

Coupled ocean-atmosphere models have demonstrated that D/O coolings could have been initiated by the injection of a sufficient amount of fresh water into convecting regions of the Labrador or GIN Seas to weaken or shut-down North Atlantic thermohaline circulation (e.g. Rahmstorf, 1994, 1995; Stocker, 1996; Schiller et al., 1997). The models further imply that if the fresh water perturbation was a transient phenomena, evaporation and convection would quickly remove the fresh water anomaly and the circulation would return to its initial, switched-on state (Schiller et al., 1997).

The repetition of D/O events over time has been more difficult to model. Broecker et al. (1990) and Birchfield and Broecker (1990) suggested that the recurrence of D/O cycles might reflect the impact of a salt oscillator on North Atlantic thermohaline circulation. Sakai and Peltier (1995) suggested from modeling experiments that a large overall increase in surface fresh water flux to the glacial North Atlantic would force thermohaline circulation to oscillate on millennial time scales. The models have been questioned (Rahmstorf, 1996), and neither in their present formulation simulates the finding of a D/O-like cyclicity in the Holocene.

We suggest that an alternative mechanism for the recurrence of D/O events is the 1-2 kyr cycle itself. That is plausible if, as our records imply, cooling phases of the cycle freshened surface waters of the glacial North Atlantic through increased rates of iceberg discharges and perhaps by southward advections of fresher surface waters from north of Iceland as well. Indeed, during most of stage 3 when D/O cycles were best developed, a D/O event coincides with nearly every IRD event in our records (e.g. Figs. 6, 7; Bond and Lotti, 1995).

One way that mechanism might have operated is suggested by the tendency of non-linear systems to exhibit threshold behavior. In that context, a D/O event would occur each time the climate cycle forced enough iceberg (fresh water) discharge to exceed a threshold value (e.g. Y_o in Fig. 13a). The increase in fresh water flux need not have been large if the glacial North Atlantic was already in an initial state close to the threshold value. That could well have been the case. $\delta^{18}O$ anomalies of 0.6‰ to > -1‰ estimated by Duplessy et al. (1991) for surface water in the glacial subpolar North Atlantic suggest that those waters overall were fresher than mean modern surface water. Provided that the system returned to its initial state during the subsequent warm phase of the cycle as the fresh water input decreased, a D/O shift could be induced during the next cold phase of the cycle, and so on.

The much-diminished amplitudes of the Holocene cycles, therefore, could reflect a different response to perturbations induced by the cycle such that the threshold value was never exceeded (Fig. 13b). The initial ocean state may have been farther removed from the threshold value (i.e., more stable) than during the glacial owing to a higher salt content (Duplessey et al., 1991; 1992). With few glaciers calving into the North Atlantic, climate-driven changes in iceberg discharges would have been small, as the records of Holocene ice rafting imply (Bond et al., 1997). Thus, even though increases in drifting ice in the North Atlantic produced robust changes in IRD petrology during the Holocene, fresh water fluxes associated with those changes may have been insufficient to trigger large shifts in thermohaline circulation.

The long, Holocene-like interstadials of the early glaciation could reflect a similar response to the climate cycle. IRD concentrations at the start of the interstadials

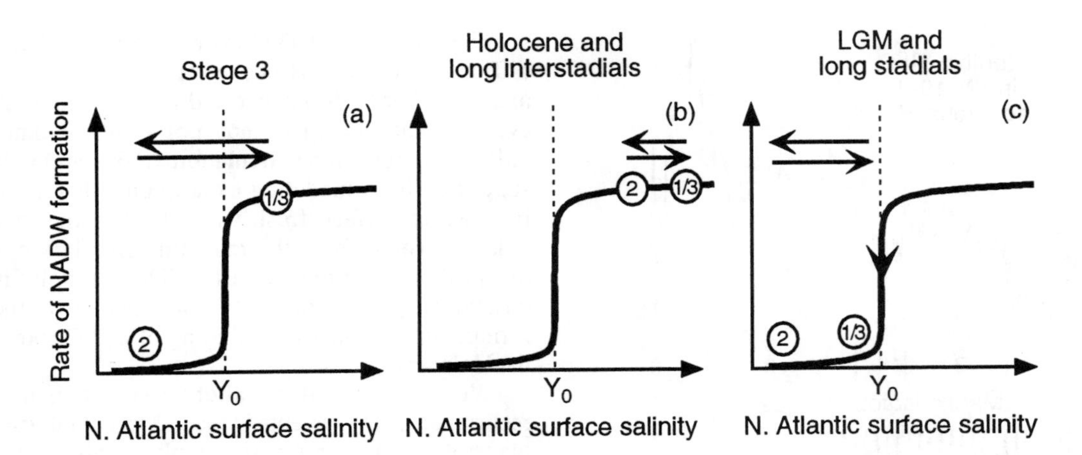

Figure 13. Threshold diagrams (after Stocker, 1996) depicting one way in which the 1-2 kyr climate cycle might have been amplified into Dansgaard/Oeschger cycles of the glaciation. Y_o is a threshold value at which a mode switch in circulation occurs. Parallel arrows give a hypothetical range of changes in surface salinities as might result from increases and decreases in fresh water input regulated by the climate cycle. 1 is an initial state, 3 is a final state and 2 is a new equilibrium state of the system. (a) Increases in fresh water fluxes exceed the threshold value, forcing recurring non-linear responses. Here, fresh water injection is a transient process and the system returns to its initial state after each mode switch. (b) Recurring shifts in fresh water fluxes during warmer intervals are small and do not exceed the threshold value. That mode might have characterized the Holocene and the long interstadials of the early glaciation. (c) A low overall surface salinity prevents the system from returning to the initial state. That mode may have dominated the long stadials of the glaciation. See text for further explanation.

are nearly as low as during the Holocene (Rasmussen et al., 1996; Bond et al., 1992). It may have taken several thousands of years of ice growth, which would have occurred during the transitional coolings (Fig. 7), before the climate cycle could have forced enough fresh water discharge to trigger a full D/O event. Rasmussen et al. (1996) proposed a similar explanation for a series of interstadial events within stage 3 that they identified in high sedimentation rate cores from near the Færoe Islands. It is also possible that during the long transitional coolings, increasing iceberg discharges and fresh water fluxes from the growing ice margins shifted the system progressively closer to a threshold value.

A mechanism for reinitiating the long interstadials of the early glaciation was suggested by Bond et al. (1993) and Paillard and Labeyrie (1991). They argued that during Heinrich events enough glacial ice was discharged that ice fronts shifted rapidly landward, thereby reducing fresh water input to the ocean and initiating vigorous resumption of North Atlantic Deep Water (NADW) production. Similarly, during each of the early stage 3 D/O events, relatively large amounts of IRD were deposited, and each IRD peak was followed immediately by prominent ocean surface warmings (e.g. Bond and Lotti, 1995). We suggest that large iceberg discharges accompanying early stage 3 D/O shifts were in a sense "mini" Heinrich events, which, just as was the case for their larger counterparts, triggered abrupt warmings and rapid switches to strong thermohaline overturn in the North Atlantic.

The threshold mechanism plausibly can be extended one step further to explain the presence of more than one petrologic shift within long stadials, such as between interstadials 1 and 2, 4 and 5, 17 and 18, and 18 and 19 (Fig. 7). At those times the overall surface salinity may have remained at values low enough to prevent the system from returning to the initial state during warmer (and presumably somewhat saltier) phases of the cycle (Fig. 13c). That is consistent with the near saturation of *N. pachyderma* (s.) during those stadials (> 90%; Fig. 2), indicating stabilization of the polar front south of the coring sites.

Our threshold mechanism is not contradicted by the lack of evidence for a 1-2 kyr cycle in parts of the ice core $\delta^{18}O$ record. The cycle would be muted during the Holocene and long interstadials if the precipitation site and the oceanic water vapor source underwent similar changes in temperature (e.g., Stuiver et al. 1997). In fact, ice core interstadials 8 and 14 are punctuated by subtle shifts in $\delta^{18}O$ that may have correlatives in the petrologic signal (Fig. 7). In addition, as Bond et al. (1997) suggested, during the long stadials air masses carrying the climate signal may have been blocked from the interior of Greenland by expansion of polar atmospheric circulation.

4.7. The Little Ice Age And The 1-2 Kyr Cycle

The LIA was a relatively severe climate episode characterized by a discontinuous series of unusually cold years between early 1600 and late 1800 A.D. (e.g. Grove, 1987; Lamb, 1995). It is best documented in Europe but known to have occurred elsewhere. Borehole temperatures from Summit Greenland (Dahl-Jensen et al., 1998) and $\delta^{18}O$ in ice cores from Northern Greenland (Fischer et al., 1998) indicate an ~1°C drop in Greenland air temperatures during the LIA. At the same time in the North Atlantic the polar front shifted southeastward, sea ice nearly surrounded Iceland, and the ocean surface may have cooled by 5°C near the Færoes and by 1° to 2°C east of Bermuda (Lamb, 1995; Olgivie, 1991, 1992, Keigwin, 1996).

Our new records from KN98-MC21 provide additional evidence of the impact of the LIA in the North Atlantic. In that core we have found robust increases in percentages of two petrologic tracers within the time interval corresponding to the LIA (Fig. 14). One of the tracers, detrital carbonate, probably was transported by ice from carbonate sources in the Arctic Islands, West Greenland, or if by sea ice alone, from the Labrador coast. The other, Icelandic glass, must have been carried to the site by icebergs and/or sea ice originating from near Iceland. With one possible exception, there are no documented examples of tidewater glaciers in Iceland during the LIA (Grove, 1987); hence, the peaks in rafted glass most likely reflect enhanced survival of ash-bearing icebergs as the ocean surface cooled. Associated with the petrologic peaks is an ~10% increase in concentration of IRD (Bond et al., submitted).

It is interesting to note that the increases in the petrologic tracers may have begun as early as 700 years ago (Fig. 14). That age is consistent with historical accounts that as early as the mid-1300s ice had become a problem for navigation around Iceland and between Iceland and Greenland (e.g. Jones, 1968; Barlow et al., 1997; Olgivie, 1991).

We also found a significant deep-water signal associated with the LIA in KN98-MC21. Measurements of $\delta^{18}O$ in the benthic foraminifera *C. wuellerstorfi* reveal a 0.2 to 0.3 per mil enrichment accompanying the peaks in the two petrologic tracers (Fig. 14). The $\delta^{18}O$ enrichment is too large to be explained by glacier expansion. As the coring site lies within the modern southward flowing NADW or Deep Western Boundary Current of McCartney (1992), the $\delta^{18}O$ enrichment likely reflects a drop in the temperature of that water mass.

If there was no ice volume change during the LIA and the enrichment was due entirely to temperature, the cooling would be on the order of 1° to 2°C (Epstein et al., 1953). The lowered deep-water temperature may reflect cooling of the surface waters north of Iceland where NADW forms, consistent with the evidence of substantially colder temperatures in those waters during the LIA.

Two results of our analyses confirm that the LIA is the most recent cold phase of the 1-2 kyr cycle. First, the time step between the coldest years of the LIA and the preceding cold event, dated at about 1.4 kyrs B.P. (Fig. 14), is between 1 and 1.3 kyrs. It clearly falls within the range of the pacing of the cycle. Second, deeper in KN98-MC21 and in a nearby core, EW93-GGC36, we have identified a series of prominent peaks in the same two petrologic tracers that characterize the LIA, Icelandic glass and detrital carbonate (Fig. 14). Within the limits of error in the calibrated radiocarbon ages, those events are correlative with the Holocene to Younger Dryas events in VM28-14 and VM23-81 (Fig. 14). In those two cores, however, the petrologic peaks within the Holocene have a different petrology, conspicuously lacking detrital carbonate. Thus, at the sites east of Newfoundland, a distinctive petrologic signature ties the LIA to the earlier cold phases in the Holocene part of the 1-2 kyr cycle.

5. SUMMARY AND CONCLUSIONS

The percentages of petrologic tracers we have identified within IRD are highly sensitive indicators of climate change, at least in the subpolar North Atlantic. They have produced robust evidence of a 1 to 2 kyr cycle punctuating virtually the full range of glacial to interglacial climates since 80 ka.

We have also demonstrated that all of the North Atlantic's Heinrich events are closely linked to cooling phases of the 1-2 kyr cycle. We regard those results as evidence that climate forcing played a fundamental role in the origin of Heinrich events, most likely operating principally as a trigger of massive ice sheet collapses.

The most compelling evidence that D/O variations are an amplification of the 1-2 kyr cycle is the striking convergence of the D/O and petrologic cycle pacings through nearly all of stage 3, a nearly 30-kyr interval during which D/O cyclicity is best developed (e.g. Fig. 6). The threshold mechanism we propose here provides an explanation for both the amplification of the 1-2 kyr cycle into full D/O events and the irregular pacing of those events through the glaciation. That mechanism also is consistent with the relation of D/O events to the intensity of the glacial cycle. D/O cycles with long interstadials occurred in late stage 5 and early stage 3 when ice volumes were relatively low. The well-developed 1-2 kyr D/O cycles

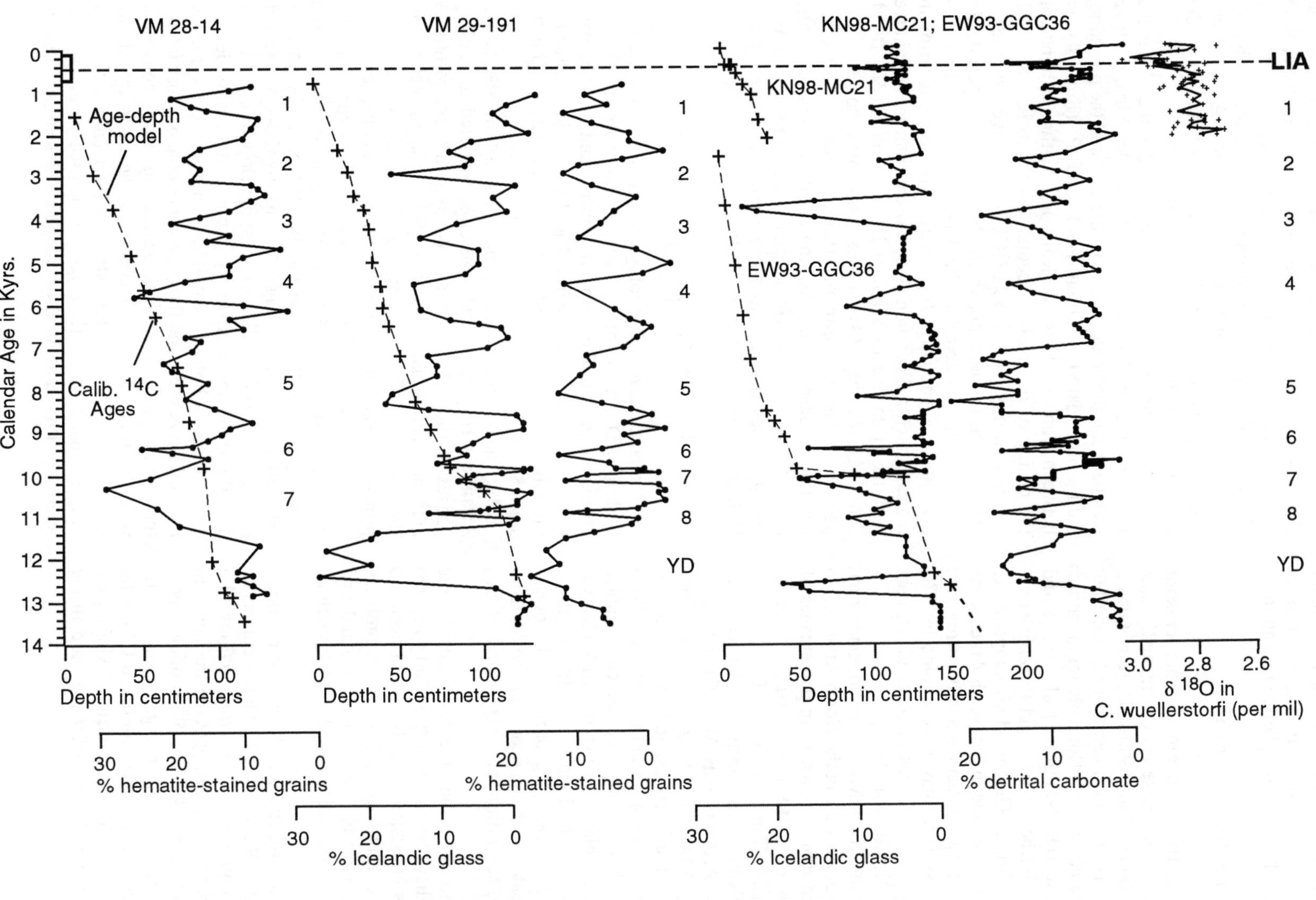

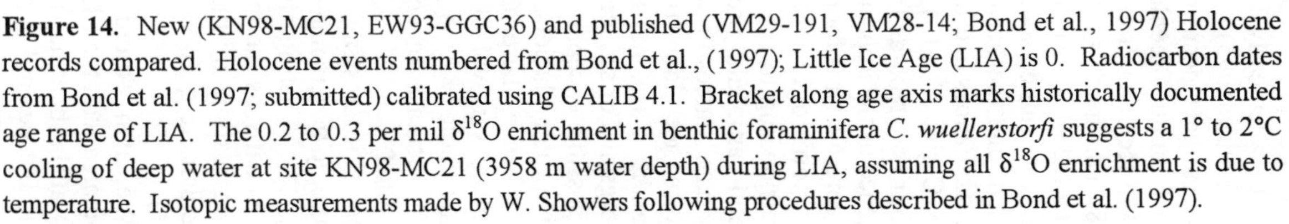

Figure 14. New (KN98-MC21, EW93-GGC36) and published (VM29-191, VM28-14; Bond et al., 1997) Holocene records compared. Holocene events numbered from Bond et al., (1997); Little Ice Age (LIA) is 0. Radiocarbon dates from Bond et al. (1997; submitted) calibrated using CALIB 4.1. Bracket along age axis marks historically documented age range of LIA. The 0.2 to 0.3 per mil $\delta^{18}O$ enrichment in benthic foraminifera *C. wuellerstorfi* suggests a 1° to 2°C cooling of deep water at site KN98-MC21 (3958 m water depth) during LIA, assuming all $\delta^{18}O$ enrichment is due to temperature. Isotopic measurements made by W. Showers following procedures described in Bond et al. (1997).

occurred in mid to late stage 3 at times of intermediate ice volumes. D/O events with prolonged stadials dominated stages 2 and 4 when ice volumes were large and the North Atlantic's polar front was stabilized at its full glacial position.

Our findings, therefore, are evidence that millennial-scale climate variability documented in Greenland ice cores and in North Atlantic sediments through the last glaciation was not forced by ice sheet instabilities, but instead arose through modulation of a pervasive 1-2 kyr cycle. The persistence of the cycle through virtually the entire 80 kyr record points to a single forcing mechanism that operated independently of the glacial-interglacial climate states.

We have no new insight here, though, on the nature of the forcing mechanism beyond what has been suggested already (Mayewski et al., 1997; Bond et al., 1997; Bianchi and McCave, 1999). Those suggestions include climatic responses at harmonics of the orbital frequencies, internally-forced oscillatory behavior of the ocean-atmospheric system, and solar forcing, which has been linked to the LIA (Stuiver et al., 1993). None of those alternatives is supported thus far by particularly compelling evidence.

Our coupling of the LIA to the series of earlier Holocene cycles corroborates previous arguments that the North Atlantic's Holocene climate tended to oscillate in a D/O-like mode (e.g. Bond et al., 1997). Cold phases of D/O cycles were marked by deep southward penetrations of the polar front, increased amounts of drifting ice and severe coolings in western Europe and in the North Atlantic. The same features, although with smaller amplitudes, are particularly well documented for the LIA. Moreover, both the LIA and D/O coolings lasted for a few hundred years, and both had bi-hemispheric imprints. If Bianchi and McCave (1999) are correct, the LIA also was accompanied by a reduction of NADW formation, just as appears to have been the case during D/O coolings.

If the LIA is indeed the most recent cold phase of the climate cycle, then the origin of the LIA must be linked fundamentally to the origin of the cycle. That may rule out mechanisms such as changes in global volcanic activity (e.g. Jones and Bradley, 1992) which are not likely to vary in a continuous and regular manner.

Finally, if we are correct that the 1-2 kyr cycle is a persistent feature of climate, at least in the North Atlantic, then one conclusion seems inescapable. Independent of any anthropogenic forcing, the North Atlantic's climate eventually will shift (or in fact may be shifting now) toward the warm phase of the cycle. That shift will be superimposed upon and, therefore, may modulate higher frequency climate variability in the North Atlantic such as the North Atlantic Oscillation. If that is true, efforts to predict future climate trends in the North Atlantic must take into account the nature and origin of the 1-2 kyr climate cycle.

Acknowledgments. We thank Christina Bruckner, Mazie Cheseby, and Nichole Anest for sample preparation, census counts of foraminiferal populations, and hand picking foraminifera for radiocarbon and isotopic analyses. Comments by Peter Clark and Robin Webb helped improve the manuscript. New radiocarbon dating in core EW98-MC21 was done at ETH, Zurich. This work was supported by NSF grants OCE 95-09997, OCE 96-18641 and OCE 97-12456 and NOAA grant UCSIO P.O. 10156283, and grants to LDEO Deep-Sea Sample Repository by NSF OCE97-11316 and ONR N00014-96-1-0186.

REFERENCES

Alley, R.B., Sedimentary processes may cause fluctuations of tidewater glaciers, *Annals of Glaciol.* 15, 119-124, 1991.

Alley, R.B., C.A. Shuman, D.A. Meese, A.J. Gow, K.C. Taylor, K.M. Cuffey, J.J. Fitzpatrick, P.M. Grootes, G.A. Zielinski, M. Ram, G. Spinelli, and B. Elder, Visual-stratigraphic dating of the GISP2 ice core: Basis, reproducibility, and application, *Jour. Geophys. Res.* 102, C12, 26,367-26,381, 1997.

Alley, R.B., S. Anadakrishnan, K.M. Cuffy, Subglacial sediment transport and ice-stream behavior, *Ant. J. U.S. 31*, 81-81, 1996.

Alley, R.B., and P.U. Clark, The deglaciation of the northern hemisphere: a global perspective, *Ann. Rev. Earth Planet. Sci.*, 27, 149-182, 1999.

Bard, E., M. Arnold, R.G. Fairbanks, B. Hamelin, ^{230}Th-^{234}U and ^{14}C ages obtained by mass spectrometry on corals, *Radiocarbon* 35, 191-199, 1993.

Barlow, L.K., J.P. Sadler, A.E.J. Ogilvie, P.C. Buckland, T. Amorosi, J.H. Ingimundarson, P. Skidmore, A.J. Dugmore, and T.H. McGovern, Interdisciplinary investigations of the end of the Norse Western Settlement in Greenland, *The Holocene* 7, 4, 489-499, 1997.

Bender, M.J., T. Sowers, M-L. Dickson, J. Orchardo, P. Grootes, P. Mayewski, and D. Meese, Climate correlations between Greenland and Antarctica during the past 100,000 years, *Nature*, 372,663-666, 1994.

Bianchi, G.G. and I.N. McCave, Holocene periodicity in North Atlantic climate and deep-ocean flow south of Iceland, *Nature* 397, 515-517, 1999.

Birchfield, G.E., and W.S. Broecker, A salt oscillator in the glacial northern Atlantic? Part II: A 'Scale Analysis' Model, *Paleoceanography* 5, 835-843, 1990.

Bischof, J., D.L. Clark, and J.-S. Vincent, Origin of ice-rafted debris: Pleistocene paleoceanography in the western Arctic Ocean, *Paleoceanography* 11, 6, 743-756, 1996.

Bond, G.C. and R. Lotti, Iceberg discharges into the North Atlantic on millennial time scales during the last glaciation, *Science*, v. 267, p. 1005-1010, 1995.

Bond, G.C., W. Broecker, S. Johnsen, J. McManus, L. Labeyrie, J. Jouzel, and G. Bonani, Correlations between climate records from North Atlantic sediments and Greenland ice, *Nature*, v. 365, 143-147, 1993.

Bond, G.C., Heinrich, H., Broecker, W., Labeyrie, L., McManus, J., Andrews, J., Huon, S., Jantschik, R., Clasen, S., Simet, C.,

Tedesco, K., Klas, M., Bonani, G., and Ivy., S., Evidence for massive discharges of icebergs into the North Atlantic ocean during the last glacial, *Nature*, v. 360, 245-249, 1992.

Bond, G.C., W. Showers, M. Cheseby, R. Lotti, P. Almasi, P. deMenocal, P. Priore, H. Cullen, I. Hajdas, and G. Bonani, A pervasive millennial-scale cycle in North Atlantic Holocene and Glacial climates, *Science* 278, 1257-1266, 1997.

Broecker, W.S., G. Bond, M. Klas, G. Bonani, and W. Wolfli, A salt oscillator in the glacial North Atlantic?: 1. The Concept, *Paleoceanography* 5, 469-477, 1990.

Broecker, W.S., Massive iceberg discharges as triggers for global climate change, *Nature* 372, 421-424, 1994.

Broecker, W.S., Paleocean circulation during the last deglaciation: A bipolar seesaw?, *Paleoceanography* 13, 119-121, 1998.

Crowley, T.J., Northern Atlantic deep water cools the Southern Hemisphere, *Paleoceanography* 7, 489-497, 1992.

Dahl-Jensen, D., K. Mosegaard, N. Gundestrup, G.D. Clow, S.J. Johnsen, A.W. Hansen, and N. Balling, Past temperatures directly from the Greenland ice sheet, *Science* 282, 268-271, 1998.

Dansgaard, W., S. Johnsen, H.B. Clausen, D. Dahl-Jensen, N.S. Gundestrup, C.U. Hammer, C.S. Hvidberg, J. Steffensen, A.E. Sveinbjörnsdottir, J. Jouzel, G. Bond, Evidence for general instability of past climate from a 250-kyr ice core record, *Nature*, v. 364, 218-220, 1993.

Denton, G.H. and W. Karlén, Holocene climatic variations — Their pattern and possible cause, *Quat. Res.* 3, 153-205, 1973.

Duplessy, J.C., L. Labeyrie, M. Arnold, M. Paterne, J. Duprat, and T.C.E. van Weering, Changes in surface salinity of the North Atlantic Ocean during the last deglaciation, *Nature* 358, 485-488, 1992.

Duplessy, J.-C., L. Labeyrie, A. Juillet-Leclerc, F. Maitre, J. Duprat, and M. Sarnthein, Surface salinity reconstruction of the North Atlantic Ocean during the last glacial maximum, *Oceanologica Acta* 14, 4, 311-324, 2991.

Elliot, M., L. Labeyrie, G. Bond, E. Cortijo, J.-L. Turon, N. Tisnerat, and J-C. Duplessy, Millennial-scale iceberg discharges in the Irminger Basin during the last glacial period: Relationship with the Heinrich events and environmental settings, *Paleoceanography* 13, 433-446, 1998.

Epstein, S.R., R. Buchsbaum, H.A. Lowenstam, and H.C. Urey, Revised carbonate-water isotopic temperature scale, *Geol. Soc. Am. Bull.*, *64*, 1315-1325, 1953.

Fleming, K.M., J.A. Dowdeswell, J. Oerlemans, Modelling the mass balance of northwest Spitsbergen glaciers and responses to climate change, *Annals of Glaciol.* 24, 203-210, 1997.

Fischer, H., M. Werner, D. Wagenbach, M. Schwager, T. Thorsteinnson, F. Wilhelms, J. Kipfstuhl, S. Sommer, Little ice age clearly recorded in northern Greenland ice cores, *Geophys. Res. Letts.* 25, 10, 1749-1752, 1998.

Grootes, P.M., and M. Stuiver, Oxygen 18/16 variability in Greenland snow and ice with 10^{-3} to 10^{-5}-year time resolution, *Jour. Geophys. Res.* 102, C12, 26,455-26,470, 1997.

Grove, J.M., *The Little Ice Age*, Methuen, 494 pp., London and New York, 1987.

Gwiazda, R.H., S. Hemming and W. Broecker, Tracking icebergs sources with lead isotopes: The provenance of ice-rafted debris in Heinrich layer 2, *Paleoceanography* 11, 1, 77-94, 1996.

Hemming, S.R. Broecker, W.S., Sharp, W.D., Bond, G.C., Gwiazda, R.H., McManus, J.F., Klas, M., Hajdas, I., Provenance of Heinrich layers in core V28-82, northeastern Atlantic: 40Ar/39Ar ages of ice-rafted hornblende, Pb isotopes in feldspar grains, and Nd-Sr-Pb isotopes in the fine sediment fraction, *Earth and Planetary Science Letters*, 164, 317-333, 1998.

Hulbe, C.L., An ice shelf mechanism for Heinrich layer production, *Paleoceanography* 12, 711-717, 1997.

Jones, P.D. and R.S. Bradley, Climatic variations over the last 500 years, in *Climate Since A.D. 1500*, edited by R.S. Bradley and P.D. Jones, pp. 649-679, Routledge, London and New York, 1995.

Jones, G., *A History of the Vikings*, 504 pp., Oxford University Press, Ely House, London, 1968.

Kitagawa, H., and J. van der Plicht, Atmospheric radiocarbon calibration to 45,000 yr B.P.: Late Glacial fluctuations and cosmogenic isotope production, *Science* 279, 1187-1190.

Keigwin, L.D., The Little Ice Age and Medieval Warm Period in the Sargasso Sea, *Science*, 274, 1504-1508, 1996.

Keigwin, L.D., and G.A. Jones, Glacial-Holocene stratigraphy, chronology, and paleoceanographic observations on some North Atlantic sediment drifts, *Deep-Sea Res.* 36, 6, 845-867, 1989.

Lamb, H.H., Climate, *History and the Modern World*, 431 pp., Routledge, London and New York, 1995.

MacAyeal, D.R., Binge/purge oscillations of the Laurentide Ice Sheet as a cause of the North Atlantics Heinrich events, *Paleoeanography* 8, 775-784, 1993.

Mann, D.H., Reliability of a Fjord glacier's fluctuations for paleoclimatic reconstructions, *Quat. Res.* 25, 10-24, 1986.

Martinson, D.G., Pisias, N.G., Hays, J.D., Imbrie, J., Moore, T.C Jr., Shackleton, N.J., Age dating and the orbital theory of the Ice Ages: Development of a high-resolution 0 to 300,000 year chronostratigraphy, *Quat. Res.* 27, 1-29, 1987.

Mayewski, P.A., L.D. Meeker, M.S. Twickler, S. Whitlow, Q. Yang, W.B. Lyons, and M. Prentice, Major features and forcing of high-latitude northern hemisphere atmospheric circulation using a 110,000 year-long glaciochemical series, *J. Geophys. Res.* 102, 26,345-26,366, 1997.

McCabe, A.M., and P.U. Clark, Ice-sheet variability around the North Atlantic Ocean during the last deglaciation, *Nature* 392, 373-377, 1998.

McCartney, M.S., Recirculating components to the deep boundary current of the northern North Atlantic, *Prog. Oceanog.* 29, 283-383, 1992.

McIntyre, A. and B. Molfino, Forcing of Atlantic equatorial and subpolar millennial cycles by precession, *Science* 274, 1867-1870, 1996.

Meese, D.A., R.B. Alley, R.J. Fiacco, M.S. Germani, A.J. Gow, P.M. Grootes, M. Illing, P.A. Mayewski, M.C. Morrison, M. Ram, K.C. Taylor, Q. Yang, and G.A. Zielinski, Preliminary depth-age scale of the GISP2 ice core, *Special CRREL Report 94-1*, US, 1994.

Motyka, R.J., and J.W. Begér, Taku Glacier, Southeast Alaska, U.S.A.: Late Holocene history of a tidewater glacier, *Arctic and Alpine Res.* 28, 1, 42-51, 1996.

Murry, J.E., The drift, deterioration and distribution of icebergs in the North Atlantic Ocean, in *Ice Seminar, Special Volume 10*, edited by V.E. Bohme, Canadian Institute of Mining and Metallurgy, Washington, D.C., 3-18, 1969.

Oceanographic Atlas of the North Atlantic: U.S. Naval Oceanographic Office, Washington, D.C, 1968.

Oerlemans, J., Climate sensitivity of glaciers in southern Norway; application of an energy-balance model to Nigardsbreen, Hellstugubreen and Alfotbreen, *Jour. Glaciol.* 38, 129, 223-232, 1992.

Ogilvie, A.E.J., Climatic Changaes in Iceland A.D. c 865 to 1598, in G.F. Bigelow, presenter, *The Norse of the North Atlantic, Acta Archaeologica 61*, pp. 233-251, Munskgaarad, Copenhaagen, 1991.

Ogilvie, A.E.J., Documentary evidence for changes in the climate of Iceland, A.D. 1500 to 1800, in *Climate Since A.D. 1500*, edited by R.S. Bradley and P.D. Jones, pp. 92-117, Routledge, London and New York, 1995.

Paillard, D. and L. Labeyrie, Role of the thermohaline circulation in the abrupt warming after Heinrich events, *Nature* 372, 162-164, 1994.

Pflaumann, U., J. Duprat, C. Pujol, and L. Labeyrie, SIMMAX, a modern analog technique to deduce Atlantic sea surface temperatures from Planktonic foraminifera in deep sea sediments, *Paleoceanography* 11, 15-35, 1996.

Piper, D.J.W. and K.I. Skene, Latest Pleistocene ice-rafting events on the Scotian Margin (eastern Canada) and their relationship to Heinrich events, *Paleoceanograhy* 13, 205-214, 1998.

Porter, S.C., Late Holocene fluctuations of the fiord glacier system in Icy Bay, Alaska, U.S.A., *Arctic and Alpine Res.* 21, 4, 364-379, 1989.

Porter, S.C., Pattern and forcing of Northern Hemisphere glacier variations during the last Millennium, *Quat. Res.* 26, 27-48, 1986.

Pisias, N.G., J.P. Dauphin, and C. Sancetta, Spectral analysis of Late Pleistocene-Holocene sediments, *Quat. Res.* 3, 3-9, 2973.

Rahmstorf, S., Bifurcations of the Atlantic thermohaline circulation in response to changes in the hydrological cycle, *Nature* 378, 145-149, 1995.

Rahmstorf, S., Rapid climate transitions in a coupled ocean-atmosphere model, *Nature* 372, 3, 82-85, 1994.

Rahmstorf, S., On the freshwater forcing and transport of the Atlantic thermohaline circulation, *Climate Dynamics* 12, 799-811, 1996.

Rasmussen, T.L., E. Thomsen, T.C.E. van Weering, and L. Labeyrie, Rapid changes in surface and deep water conditions at the Færoe margin during the last 58,000 years, *Paleoceanography* 11, 6, 757-771, 1996.

Revel, M., J.A. Sinko and F.E. Grousset, Sr and Nd isotopes as tracers of North Atlantic lithic particles: Paleoclimatic implications: *Paleoceanography* 11, 1, 95-113, 1996.

Ruddiman, W.F., Late Quaternary deposition of ice-rafted sand in the subpolar North Atlantic (lat 40° to 65°N), *Geol. Soc. America Bull. 88*, 1813-1827, 1977.

Sakai, K. and W.R. Peltier, A simple model of the Atlantic thermohaline circulation: Internal and forced variability with paleoclimatological implications, *J. Geophys. Res.* 100, 13455-13479, 1995.

Sarnthein, M., E. Jansen, M. Weinelt, M. Arnold, J.C. Duplessy, H. Erlenkeuser, A. Flatoy, G. Johannessen, T. Johannessen, S. Jung, N. Koc, L. Labeyrie, M. Maslin, U. Pflaumann, and H. Schulz, Variations in Atlantic surface ocean paleoceanography, 50°-80°N: a time-slice record of the last 30,000 years, *Paleoceanography* 10, 1063-1094, 1995.

Schiller, A., U. Mikolajewicz, and R. Voss, The stability of the North Atlantic thermohaline circulation in a coupled ocean-atmosphere general circulation model, *Climate Dynamics* 13, 325-347, 1997.

Shackleton, N.J., J. Imbrie, and M.A. Hall, Oxygen and carbon isotope record of East Pacific core V19-30: implications for the formation of deep water in the late Pleistocene North Atlantic, *Earth and Planetary Science Letters* 65, 233-244, 1983.

Sigurdsson, O., T. Jonsson, Relation of glacier variations to climate changes in Iceland, *Annals of Glaciol.* 21, 263-270, 1995.

Stea, R.R., Piper, D.J.W., Fader, G.B.J., Boyd, R., Wisconsinan glacial and sea-level history of Maritime Canada and the adjacent continental shelf: A correlation of land and sea events, *Geol. Soc. Am. Bull.*, 110, 821-845, 1998.

1953.Steig, E.J., E.J. Brook, J.W.C. White, C.M. Sucher, M.L. Bender, S.J. Lehman, D.L. Morse, E.D. Waddington, G.D. Clow, Synchronous climate changes in Antarctica and the North Atlantic, *Science* 282, 92-96.

Stocker, T.F., The ocean in the climate system: Observing and modeling its variability, in *Topics in Atmospheric and Interstellar Physics and Chemistry, European Research Course on Atmospheres, Volume 2*, edited by C.F. Boutron, pp. 39-90, Les Editions de Physique, Les Ulis, France, 1996.

Stoner, J.S, J.E.T. Channell, and C. Hillaire-Marcel, A 200 kyr geomagnetic stratigraphy for the Labrador Sea: Indirect correlation of the sediment record to SPECMAP, *Earth Planet. Sci. Lett* 159, 165-181, 1998.

Stuiver, M. and P.J. Reimer, Extended ^{14}C data base and revised CALIB 3.0 ^{14}C age calibration program, in *Calibration 1993*, edited by M. Stuiver and R.S. Kra, pp. 215-230, Radiocarbon 35, 1, 1993.

Stuiver, M., Braziunas, T.F., Sun ocean, climate, and atmospheric $^{14}CO_2$: an evaluation of causal and spectral relationships, *The Holocene*, 3, 289-305, 1993.

Stuiver, M., Braziunas, T.F., Grootes, P.M., Is there evidence for solar forcing of climate in the GISP2 oxygen isotope record?, *Quat. Res.* 48, 259-266, 1997.

Stuiver, M., P.J. Reimer, E. Bard, J.W. Beck, G.S. Burr, K.A. Hughen, B. Kromer, G. McCormac, J. van der Plicht, and M. Spurk, INTCAL98 Radiocarbon age calibration, 24,000-0 cal BP, *Radiocarbon* 40, 3, 1041-1083, 1998.

Sturm, M., D.K. Hall, C.S. Benson, and W.O. Field, Non-climatic control of glacier-terminus fluctuations in the Wrangell and Chugach mountains, Alaska, U.S.A., *Jour. Glaciol.* 37, 348-356, 1991.

Voelker, A.H.L., M. Sarnthein, P.M. Grootes, H. Erlenkeuser, C. Laj, A. Mazaud, M.-J. Nadeau, and M. Schleicher, Correlation of marine 14C ages from the Nordic Seas with the GISP2 isotope record: Implications for ^{14}C calibration beyond 25 ka BP, *Radiocarbon* 40, 517-534, 1998.

Wiles, G.C., P.E. Calkin, A. Post, Glacier fluctuations in the Kenai Fjords, Alaska, U.S.A.: An evaluation of controls on iceberg-calving glaciers, *Arctic and Alpine Res.* 27, 3, 234-245, 1995.

Wu, G. and C. Hillaire-Marcel, Oxygen isotope compositions of *sinistral Neogloboquadrina pachyderma* tests in surface sediments, North Atlantic Ocean: *Geochimica et Cosmochimica Acta* 58, 1303-1312, 1994.

Gerard C. Bond, Michael Evans, and Rusty Lotti, Lamont-Doherty Earth Observatory of Columbia University, Rt. 9W, Palisades, NY 10964

Mary Elliot, Laboratiore des Sciences du Climat et de l'Environment, Laboratoire mixte CNRS-CEA, Gif sur Yvette, France

Sigfus Johnson, The Niehls Bohr Institute, Department of Geophysics, University of Copenhagen, Haraldsgade 6, DK-2200, Copenhagen N, Denmark; also at Science Institute, Department of Geophysics, University of Iceland, Dunghaga 3, IS-107 Reykjavik, Iceland. Also at Science Institute, Department of Geophysics, University of Iceland, Reykjavik, Iceland.

William Showers, Department of Marine, Earth and Atmospheric Sciences, North Carolina State University, 1125 Jordan Hall, Raleigh, NC 27695

Millennial-scale Changes in Ventilation of the Thermocline, Intermediate, and Deep Waters of the Glacial North Atlantic

W. B. Curry, T. M. Marchitto, J. F. McManus, D. W. Oppo and K. L. Laarkamp

Woods Hole Oceanographic Institution, Woods Hole, MA

Changes in North Atlantic thermohaline circulation on millennial time scales were analogous to those observed on Milankovitch time scales: production of glacial intermediate water increased at the expense of the deep water. During the first half of isotope stage 3 (55 to 30 ka), NADW production was operating at nearly full strength when its production was interrupted by the iceberg discharges associated with Heinrich events. Reduced production of NADW began before each Heinrich event and continued well after the event, suggesting that the iceberg discharges were not the immediate cause of the change in thermohaline overturning. After 30 ka, when Milankovitch-forced reductions in NADW raised the water mass boundary to above 3200 m, reduced benthic foraminiferal $\delta^{13}C$ associated with Heinrich events was more pronounced in shallower and more northerly cores. In the western subtropical gyre, high values of benthic foraminiferal $\delta^{13}C$ and low values of Cd/Ca imply that the upper water column was well ventilated throughout marine isotope stages 3 and 2 with no evidence for decreased ventilation during Heinrich events. Intermediate waters (1849 m) in the subpolar North Atlantic were well ventilated during most of the Heinrich events, although mixing with southern ocean waters is evident for Heinrich events 1 and 2.

Enhanced intermediate water production during the last glacial period was associated with oscillating temperatures within the subtropical gyre. Heinrich events H2, H3 and H4 were local warmings of the thermocline, while H1 and H5 were cold events. Two factors compete to cause these oscillations: 1) a redistribution of heat caused by variations in the intensity of thermohaline circulation in the North Atlantic [*Manabe and Stouffer*, 1997] and 2) variations in the temperature of water advected along isopycnals in the ventilated portion of the thermocline [*Gu and Philander*, 1997; *Bush and Philander*, 1998].

1. INTRODUCTION

It is becoming increasingly clear that large changes in the ocean's thermohaline overturning played a role in the rapid climate changes that occurred during the last ice age. There exists a growing body of geochem-

Mechanisms of Global Climate Change at Millennial Time Scales
Geophysical Monograph 112

ical evidence for decreases in the production of North Atlantic Deep Water (NADW) during short-term cold periods like the Younger Dryas [*Boyle and Keigwin,* 1987], the earlier Heinrich events during marine isotope stages 2 and 3 [*Keigwin and Lehman,* 1994; *Oppo and Lehman,* 1995; *Vidal et al.,* 1997; *Zahn,* 1997; *Curry and Oppo,* 1997] and, in the highest sedimentation rate cores, during the cold phases of Dansgaard-Oeschger cycles that occur interspersed among the Heinrich events [*Oppo and Lehman,* 1995; *Keigwin and Boyle,* 1998, in press]. Much less is known about the response of intermediate and thermocline waters to these rapid climate events. Using cores from the Bahamas, *Marchitto et al.* [1998] showed that decreased nutrient concentration and increased ventilation of the lower thermocline accompanied a decrease in production of NADW during the Younger Dryas, in a manner analogous to the glacial-interglacial differences that have been observed for the Atlantic Ocean. *Zahn et al.* [1997], using a core from 1099 m water depth, have attributed large decreases in benthic foraminiferal δ^{13}C during Heinrich events to reduced thermohaline overturning at all depths in the eastern North Atlantic.

Because of the important role of thermohaline overturning in the North Atlantic to the global heat budget, it is important to determine the extent to which reduced production of deep water in the North Atlantic may have been compensated for by enhanced production of intermediate water, maintaining a strong net northward oceanic heat transport into the North Atlantic [*Webb et al.,* 1997]. The goal of this paper is to evaluate the changes in thermocline, intermediate and deep water circulation during the short-term climate oscillations that occurred during the last glacial interval.

2. STRATEGY

Our strategy derives from studies of circulation on glacial-interglacial and Milankovitch time scales, where there is evidence for enhanced production of intermediate water compensating for reduced production of deep water [*Boyle and Keigwin,* 1987; *Oppo and Fairbanks,* 1987; *Curry et al.,* 1988; *Duplessy et al.,* 1988; *Oppo and Lehman,* 1993; *Sarnthein et al.,* 1994]. The observation of nutrient-depleted waters above ~2000 m in the North Atlantic reflects the presence of a shallow water mass (named Glacial North Atlantic Intermediate Water (GNAIW) by *Boyle and Keigwin* [1987]), produced in the North Atlantic south of Iceland, and with a density less than that of present-day NADW. Northward penetration of higher-nutrient deep waters from the southern ocean compensated for the decreased volume of NADW; penetration as far as 60°N was observed in the eastern Atlantic Ocean [*Oppo and Lehman,* 1993; *Sarnthein et al.,* 1994].

This understanding of the glacial circulation required a comprehensive geographic and bathymetric reconstruction of the distribution of benthic foraminiferal δ^{13}C and trace metal compositions - many time series of benthic foraminiferal chemistry from a variety of locations and water depths in the Atlantic. A similar type of reconstruction will be needed to identify the changes that occurred on millennial and shorter time scales.

With that need in mind our strategy is to document the synoptic, millennial-scale variations in water mass indices in several key locations and water masses in the North Atlantic region, including deep water, intermediate water, and the thermocline of the North Atlantic subtropical gyre. New and published cores from the Ceara Rise and the Bahamas will be compared to records from the North Atlantic [*Lehman and Oppo,* in manuscript] near the isopycnal outcrop regions for the deeper parts of the modern thermocline (Figure 1, Table 1). In this way we will evaluate millennial-scale changes in deep and intermediate water production and their effects on the heat distribution in the North Atlantic.

We will evaluate two competing processes which can affect this heat distribution. First, when deep water production is reduced, there is a warming of the upper water column, most prominent at thermocline depths, throughout the tropical regions of the Atlantic basin [*Manabe and Stouffer,* 1997]. The warming, which reaches more than 3°C, results in part from the reduced advection of heat into the North Atlantic where deep water convection normally occurs. Thus heat which is normally lost to the atmosphere during periods of active thermohaline overturning in the North Atlantic, is redistributed within the upper water column. Competing with this process is the effect of cooling on the northern regions of the North Atlantic where the thermocline is ventilated [*Bush and Philander,* 1998]. Because a large-scale cooling of North Atlantic surface waters occurred during the glacial period, overall the thermocline must have been colder than today. Superimposed on this were millennial-scale, warm-cold oscillations of North Atlantic surface waters, which must have affected thermocline temperatures as well. Since reduced deep water production (which would have warmed the tropical thermocline) occurred during short-term periods of cold North Atlantic sea surface temperatures (which would tend to cool thermocline temperatures), the two effects are in opposition.

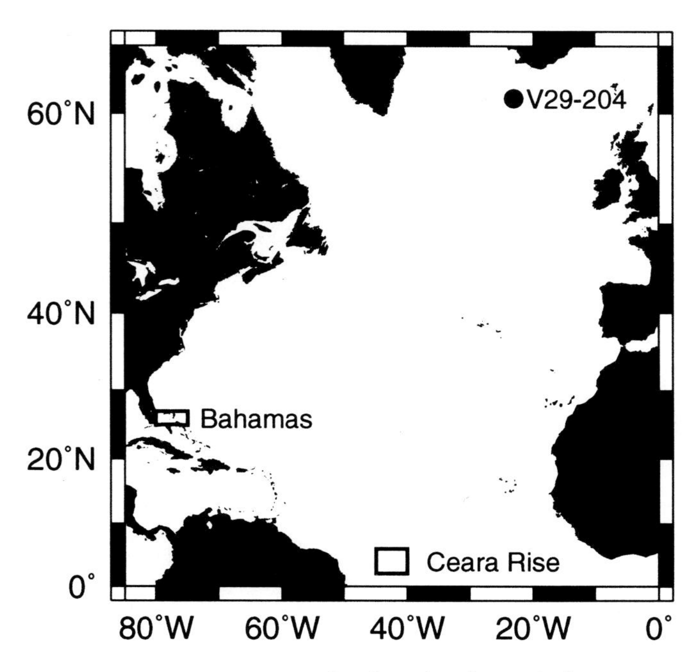

Figure 1. Location map showing the three study areas: Ceara Rise, Bahamas and V29-204 in the North Atlantic.

The core locations are well suited for this study. The Ceara Rise is an aseismic feature in the western equatorial Atlantic that intersects the present transition from NADW above 4000 m to Antarctic Bottom Water (AABW) below. A series of deep sea piston cores collected there in 1992 will be used to monitor changes in the depth of the NADW-AABW water mass boundary during the last glacial period. The eastern side of the Bahamas intersects the water column from the surface down to about 1500 m, the full depth range of the permanent thermocline of the subtropical gyre. A core collected in 1988 from these well-sedimented slopes will provide a record from the lower thermocline/intermediate water. V29-204 is located near the region where deeper portions of the thermocline are ventilated today. Its water depth (1849 m) places it within GNAIW during marine isotope stage (MIS) 2. Using these cores, we will apply indicators of nutrient chemistry (δ^{13}C,Cd/Ca) and physical properties (δ^{18}O) to describe the geometry and properties of deep and intermediate water masses and thermocline waters across several events of rapid climate change during the last glacial period.

3. METHODS

The isotopic data were acquired on a Finnigan MAT 252 mass spectrometer equipped with the "Kiel" automated carbonate device. Tests of *Cibicidoides wuellerstorfi*, *Cibicidoides kullenbergi*, *Planulina ariminensis* or other *Cibicidoides* or *Planulina* species were selected for isotopic analysis. Individual specimens of each taxon were measured. For the Bahamas core, additional specimens of *Hoeglundina elegans* were analyzed. The procedures for the isotopic analysis of carbonate are detailed in *Curry* [1996] and *Ostermann and Curry* [in manuscript]. Conversion to PDB is through the intermediate standard NBS-19; replicate analyses (n >2000) of this standard yield external precisions of ±0.03 ‰ for δ^{13}C and ±0.07 ‰ for δ^{18}O. The data from one core, EW9209-1JPC, was published in *Curry and Oppo* [1997]. Some of the data from OC205-103GGC were previously published in *Slowey and Curry* [1995] and *Marchitto et al.* [1998]. Revisions to our interlaboratory calibration to PDB have been applied to the data in this paper; the effect is to increase the δ^{18}O values by about 0.1 ‰ from the original *Slowey and Curry* [1995] and *Curry and Oppo* [1997] publications. The isotopic data from V29-204 were produced following the same procedures and using the same mass spectrometer system [*Lehman and Oppo,* in manuscript].

Cd/Ca measurements for core OC205-103GGC were produced using the procedures outlined in *Boyle and Keigwin* [1985/86] as modified by *Boyle et al.* [1995]. The benthic species *H. elegans* was selected for analysis. The Cd/Ca values were converted to estimates of

Table 1. Core Locations, Water Depths and Glacial Sedimentation Rates

Core	Latitude	Longitude	Depth (m)	Rate (cm kyr^{-1})
EW9209-1JPC	5.907	44.195	4056	5.3
EW9209-2JPC	5.635	44.470	3528	6.1
EW9209-3JPC	5.313	44.260	3288	6.2
OC205-103GGC	26.070	78.056	965	4.5
V29-204	61.180	23.020	1849	16.5

Table 2. Radiocarbon Ages for Samples of *G. sacculifer* Picked From EW9209 Cores and OC205-103GGC.

Depth in Core (m)	Radiocarbon Age and Error	Calendar Age, years	Accession Number
EW9209-1JPC			
0.03	3,880 ± 40	3,830	OS-4543 (WHOI)
0.13	6,540 ± 35	7,010	OS-4860 (WHOI)
0.20	9,600 ± 45	10,290	OS-4542 (WHOI)
0.33	13,350 ± 50	15,360	OS-4858 (WHOI)
0.43	16,600 ± 310	19,100	OS-4857 (WHOI)
0.60	18,750 ± 70	21,900	OS-4859 (WHOI)
0.83	23,700 ± 120	27,500	OS-4541 (WHOI)
0.97	26,500 ± 240	30,500	OS-4540 (WHOI)
1.17	29,900 ± 210	34,100	OS-4856 (WHOI)
1.33	32,400 ± 180	36,700	OS-4545 (WHOI)
1.53	38,800 ± 340	42,900	OS-4544 (WHOI)
EW9209-2JPC			
0.03	3,890 ± 35	3,837	OS-8347 (WHOI)
0.10	4,070 ± 80	4,084	OS-8348 (WHOI)
0.20	8,080 ± 100	8,482	OS-8349 (WHOI)
0.37	13,050 ± 100	14,874	OS-8350 (WHOI)
0.67	16,900 ± 140	19,423	OS-8351 (WHOI)
0.93	19,950 ± 100	23,100	OS-8352 (WHOI)
1.20	24,500 ± 120	28,300	OS-8353 (WHOI)
1.43	28,100 ± 190	32,200	OS-8354 (WHOI)
1.60	30,000 ± 170	34,200	OS-8355 (WHOI)
1.77	32,800 ± 200	37,100	OS-8356 (WHOI)
1.90	34,100 ± 270	38,400	OS-8357 (WHOI)
EW9209-3JPC			
0.03	6,210 ± 35	6,653	OS-8361 (WHOI)
0.17	8,090 ± 55	8,490	OS-8362 (WHOI)
0.37	5,230 ± 40	5,585	OS-8363 (WHOI)
0.57	13,150 ± 50	15,033	OS-8364 (WHOI)
0.80	15,950 ± 70	18,455	OS-8365 (WHOI)
0.93	16,500 ± 70	18,975	OS-8366 (WHOI)
1.20	20,400 ± 80	23,700	OS-8367 (WHOI)
1.40	22,400 ± 80	25,900	OS-8656 (WHOI)
1.58	26,700 ± 170	30,700	OS-8658 (WHOI)
1.82	29,300 ± 170	33,500	OS-8657 (WHOI)

sea water Cd using a distribution coefficient of 1; this estimate is referred to by the term CdWH in the figures and text. A subset of this data was published in *Marchitto et al.* [1998]. Here we extend the record through most of MIS 3.

AMS radiocarbon measurements were performed on *Globigerinoides sacculifer* for EW9209-3JPC, EW9209-2JPC, EW9209-1JPC and OC205-103GGC and are presented in Table 2. The AMS radiocarbon measurements for V29-204 were performed on *N. pachyderma* (l) and *G. bulloides* [*Lehman and Oppo,* in manuscript]. The radiocarbon ages were converted to calendar ages (Table 2) and then the geochemical records for each core were placed on the time scale by linear interpolation. None of the cores in this study come from the main region of ice rafting that formed the Heinrich layers, so comparing events in our records to the Heinrich events is based solely on quality of the chronologies we have developed and the accuracy of the ages of the Heinrich events in other well dated cores (Table 3).

The isotopic and trace metal data are available at the National Geophysical Data Center, World Data Center

Table 2. (continued)

Depth in Core (m)	Radiocarbon Age and Error	Calendar Age, years	Accession Number
OC205-103GGC			
0.10	920 ± 35	510	OS-10523 (WHOI)
0.62	5,290 ± 45	5,640	OS-10524 (WHOI)
0.88	7,630 ± 45	7,930	OS-15376 (WHOI)
1.13	11,000 ± 50	12,500	OS-10526 (WHOI)
1.21	12,200 ± 55	13,800	OS-10525 (WHOI)
1.34	17,100 ± 100	19,700	OS-10527 (WHOI)
1.51	20,200 ± 85	23,400	OS-10528 (WHOI)
1.71	25,900 ± 120	29,800	OS-10529 (WHOI)
2.00	31,500 ± 170	35,800	OS-10530 (WHOI)
2.20	37,400 ± 360	41,600	OS-10645 (WHOI)
2.40	39,600 ± 390	43,300	OS-15377 (WHOI)
2.67	45,700 ± 500	49,300	OS-15378 (WHOI)
2.70	39,500 ± 480	N/A	OS-10531 (WHOI)

The $^{14}C/C$ ages are reported based on a half-life of 5568 years and without reservoir corrections or conversion to calendar years. Calendar ages have been calculated using a 400 year reservoir correction and applying the *Stuiver and Braziunas* [1993] calibration curve for the younger ages and a U/Th calibration curve for the older ages [*Bard et al.*, 1992]. The age reversal in the upper part of EW9209-3JPC was not used to develop the chronology for this core. The reversal does not affect our discussion because it occurs in the Holocene section of the core, which is not discussed in this paper. The upper section of this EW9209-3JPC suffered from coring disturbance because of a collapsed core liner, which is the likely cause for the age reversal. The 2.70 m sample from OC205-103GGC was not used the chronology.

A for Paleoclimatology (www.ngdc.noaa.gov) or by ftp from mahi.whoi.edu (128.128.16.21).

4. RESULTS

4.1. Deep Water Variability - Ceara Rise

The $\delta^{13}C$ variability from three cores at the Ceara Rise (3288, 3528, and 4056 m), shows evidence for both Milankovitch and shorter-term variations in the depth of the NADW:AABW transition (Figure 2). Today the transition at 4000 m is marked by colder, fresher AABW overlain by warmer, saltier NADW. NADW has significantly higher $\delta^{13}C$ of ΣCO_2 because it is formed from surface waters in the North Atlantic that have been stripped of nutrients and enriched in ^{13}C. Short-term lows in the benthic foraminiferal $\delta^{13}C$ records indicate the volumetric expansion of AABW at the expense of NADW since the mean ocean $\delta^{13}C$ does not exhibit large, correlative changes on millennial time scales. The records for the early part of MIS 3 show several reductions in $\delta^{13}C$ that are nearly synchronous with the large input of ice rafted debris associated with Heinrich events 3, 4 and 5 (Table 3). These reductions in $\delta^{13}C$ imply that the water mass transition shoaled by at least 800 m. Thus, the thickness of AABW, which is about 500 m today at Ceara Rise, more than doubled during the Heinrich events. Each minimum in $\delta^{13}C$ was much longer than a Heinrich event, beginning several thousand years before and ending several thousand years after the event. The reduced production of NADW was therefore not directly caused by the ice rafting; rather the ice rafting event occurred within an interval of already reduced NADW production. Following the reduced production associated with Heinrich events 4 and 5, $\delta^{13}C$ values in each core rebounded, suggesting that NADW production resumed at a Holocene-like level.

Following Heinrich event 3, however, $\delta^{13}C$ values in all three cores remained very low for about 15 kyr until the deglaciation. Furthermore there was far less coherent short-term variability. The very low $\delta^{13}C$ values approach the values observed in the Pacific Ocean [*Curry and Oppo*, 1997], suggesting that from the time of Heinrich 3 until deglaciation (approximately 30 to 15 ka) NADW was not present at or below about 3200 m in the equatorial Atlantic. Shallower cores which are closer to NADW source regions exhibit millennial

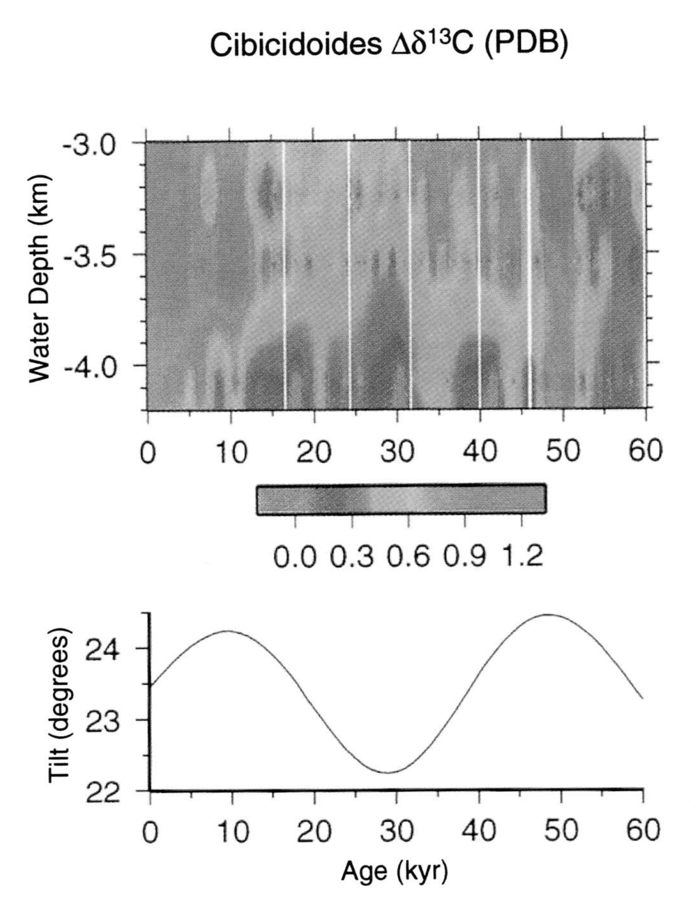

Plate 1. Time-depth variations in *Cibicidoides* $\Delta\delta^{13}$C document the strong penetration of southern ocean water into the Atlantic Ocean during the last glaciation. $\Delta\delta^{13}$C is calculated as the carbon isotopic difference between the measured δ^{13}C value in the core and the synchronous δ^{13}C value in Pacific Ocean core V19-30 [*Shackleton et al.* , 1983] (corrected to an equivalent *Cibicidoides* δ^{13}C value). This parameter normalizes for whole ocean changes in δ^{13}C caused by variations in the ocean's carbon budget. Maxima in $\Delta\delta^{13}$C (in orange and red) imply the presence of NADW; minima imply the presence of southern-ocean water with δ^{13}C values very much like those in the Pacific Ocean. The white lines show the timing of Heinrich events in the North Atlantic from Table 3. During the second half of stage 3, low values of $\Delta\delta^{13}$C shoal to above 3200 m, implying a volumetric reduction of NADW and at least a doubling of the volume of southern ocean water. Between the Heinrich events during early stage 3, NADW production appears to reach Holocene-like levels. Variations in deep water production that occur on millennial time scales are superimposed on a large, orbitally-forced record of variations in NADW production which appears to parallel the earth's tilt cycle.

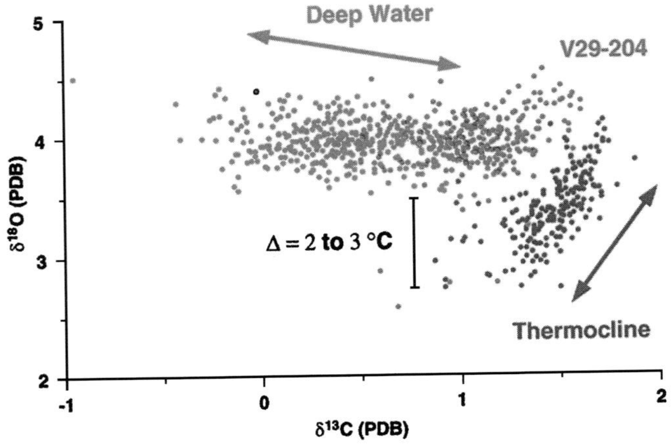

Plate 2. Covariation of δ^{18}O and δ^{13}C for the three locations used in this study. The data plotted are those which fall between 24 and 55 ka, the MIS 3 interval. The deep water (red symbols) records display negative covariation, implying mixing between colder, lower δ^{13}C water mass from the south with a warmer, higher δ^{13}C water mass from the north. The thermocline record (blue symbols) displays positive covariation; when δ^{18}O is lower (warmer or less salty), ventilation is poorer. The V29-204 record (green symbols) is a hybrid of the two records. At times this core is overlain by southern ocean water; at other times it is overlain by GNAIW. The range of δ^{18}O values in the thermocline is much larger than in the deep and intermediate waters, implying 2 to 3°C greater variability in temperature during the glacial period.

Table 3. Ages of the Heinrich Layers

HL	Vidal *et al.*		GISP2 Age	This paper
	$^{14}C/C$	Calendar		
1	13.8	16.5	15.8-16.5	16.5
2	21.0	24.3	24.0	24.3
3	27.15	31.6	30.2	31.6
4	35.2	39.9	38.5-39.6	39.9
5	43.7	47.8	45.5-46.2	46.0

The ages of the Heinrich layers used in this paper come from *Vidal et al.* [1997] radiocarbon ages and correlation to the ice core records. First, we converted the $^{14}C/C$ ages in *Vidal et al.* [1997] to calendar ages. Then we compared these ages to the correlative stadial events in the GISP2 ice core $\delta^{18}O$ record. We used the correlations obtained by *Bond et al.* [1993] and the new layer-counted age model for GISP2 [*Grootes and Stuiver*, 1997]. The age estimates sometimes disagree by up to 2 kyr. For Heinrich layers 1 to 4, we chose to use the radiocarbon ages from the deep sea record since they directly date the layers. For Heinrich layer 5, the $^{14}C/C$ ages for this event are at the limits of the technique, so we chose to use the estimated age for H5 based on its correlation to the ice core record.

scale variations implying that production of this water mass continued to fluctuate on short time scales and synchronous with the Heinrich events [*Oppo and Lehman*, 1995; *Vidal et al.*, 1997]. However, between ~15 and 30 ka these fluctuations occurred in a water mass that was already reduced by about one third of its equivalent Holocene volume. Plate 1 presents the time-depth variability of benthic foraminiferal $\delta^{13}C$ for the last 60,000 years, plotted as the isotopic enrichment with respect to the Pacific Ocean ($\Delta\delta^{13}C$), illustrating both the millennial and orbital-scale variations in deep water circulation. Millennial-scale minima in $\Delta\delta^{13}C$ are evident for Heinrich events 3, 4 and 5 during MIS 3, but are obscured by the very large shoaling of the water mass boundary that follows Heinrich event 3. This shoaling appears to be driven to first order by the Milankovitch tilt cycle and the shoaling precludes observing millennial-scale decreases in $\Delta\delta^{13}C$ for Heinrich events 1 and 2 because the location is already covered by southern ocean water.

The changes in deep water production are correlated with variations in the $\delta^{18}O$ of *G. sacculifer* (Figure 3), a mixed layer-dwelling planktonic foraminifer; highest $\delta^{18}O$ values (colder surface waters) occurred during periods of reduced NADW volume. *Curry and Oppo* [1997] attributed these $\delta^{18}O$ increases to local coolings of sur-

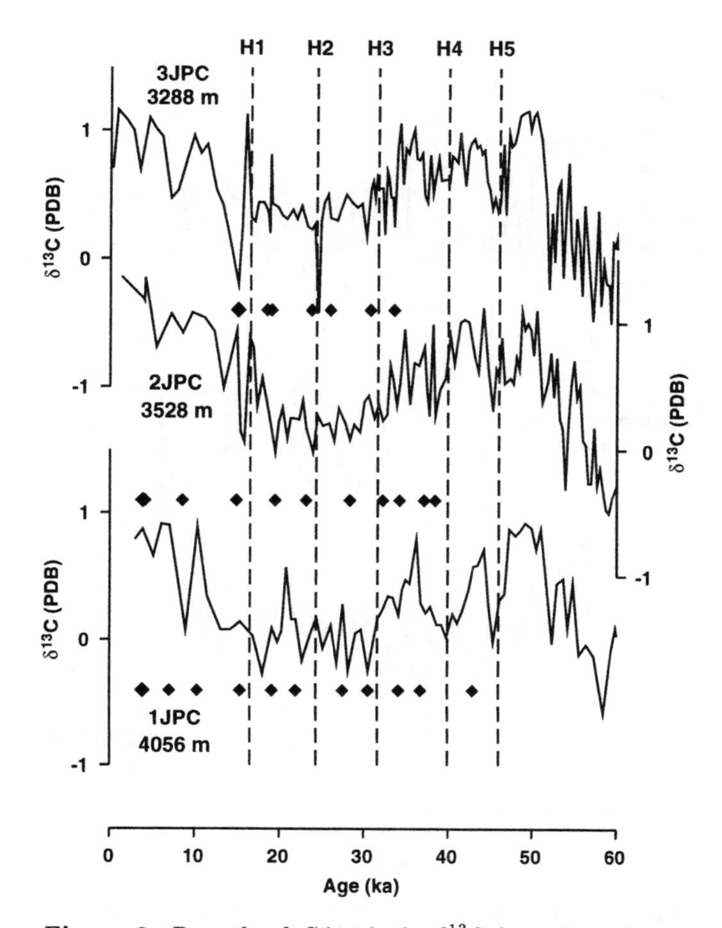

Figure 2. Records of *Cibicidoides* $\delta^{13}C$ from three Ceara Rise cores which span the modern transition from NADW to AABW. The data for EW9209-1JPC are from *Curry and Oppo* [1997]. The lines represent the average isotopic composition at each stratigraphic level. Each average value is based on several measurements of individual tests. The filled diamonds are the levels of the AMS radiocarbon ages (converted to calendar age) from Table 2. Ages between these dates were interpolated. Short-term decreases in $\delta^{13}C$, implying reduced volume of NADW, are synchronous with the Heinrich ice rafting events in the North Atlantic. These short-term decreases in deep water production are superimposed on a large, Milankovitch-forced variation in deep water production.

face waters of 2 to 3°C. Thus there appears to be a link between convection in the North Atlantic and surface water temperatures in the western equatorial Atlantic. The cause of this relationship is unclear, however, since variations in heat transport associated with NADW formation would be unlikely to cause large changes in temperature near the equator [*Manabe and Stouffer*, 1988, 1997; *Crowley*, 1992; *Rahmstorf*, 1994]. Later in this paper we will explore one possible explanation for this variability in SST: advection of cold waters from high

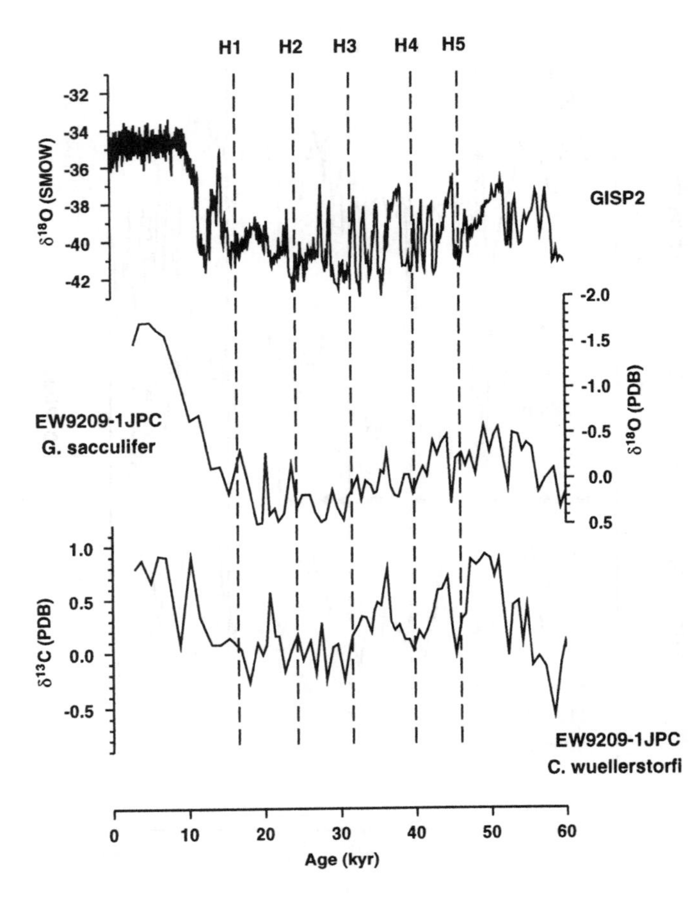

Figure 3. Oscillations in deep water production appear to be synchronous with the longer D-O oscillations in air temperature over Greenland and with surface water temperature in the western equatorial Atlantic. a) the GISP2 δ^{18}O record using the layer counted chronology; b) δ^{18}O record for *G. sacculifer* from EW9209-1JPC [*Curry and Oppo,* 1997]; and c) δ^{13}C record for *Cibicidoides* from EW9209-1JPC [*Curry and Oppo,* 1997]. Warmings in air temperature over Greenland correlate with warming at the equator and maxima in NADW production.

to low latitudes along isopycnals within the thermocline and enhanced upwelling at low latitudes.

4.2. Thermocline and Intermediate Water Variability - Bahamas

On glacial-interglacial time scales, significant changes occurred in the physical and chemical properties of the thermocline [*Slowey and Curry,* 1992, 1995; *Marchitto et al.,* 1998]. During glacial maxima, large δ^{18}O increases in benthic foraminifera imply that the thermocline cooled by more than 4°C. At the same time, δ^{13}C values of the benthic foraminifera increased by 0.3 to 0.6 ‰; the largest increases occurred at the present depth of the oxygen minimum zone. *Marchitto et al.* [1998] documented that glacial sea water Cd values (and by implication [PO_4]) were about one half of Holocene values, confirming that the increased δ^{13}C values during the glacial maximum were due to reduced nutrient concentrations. Taken together these results suggest that ventilation of the lower thermocline occurred in a region of Ekman pumping where surface water temperatures were 4°C colder. Thus a colder thermocline and enhanced ventilation of the upper water column occurred during the glacial maximum, a period when NADW production was significantly reduced.

Cooling of the thermocline, though, is not the expected result of reduced deep water production in the North Atlantic. If variations in thermohaline circulation were the only forcing, the thermocline should warm during periods of reduced overturning. *Manabe and Stouffer* [1997] show that short-term reductions in thermohaline circulation in the North Atlantic cause the tropical thermocline (the northward-flowing, warm water portion of the meridional overturning cell) to warm by 3°C when thermohaline circulation decreases from 16 Sv to 4 Sv. The warming results from heat redistribution; heat which is normally advected to the north and lost to the atmosphere during deep water production, remains in the upper water column during periods of reduced thermohaline circulation. *Rahmstorf* [1994] produced a much smaller warming of the tropical gyre when reduced deep water production was replaced by enhanced glacial intermediate water production. Interestingly, neither study produces large surface water temperatures changes outside of the subpolar North Atlantic; the large changes in temperature are restricted to the thermocline. That thermocline temperatures were 4°C colder during the glacial maximum implies that cold waters were advected into the thermocline along isopycnals from their outcrop locations. This process dominated the heat redistribution caused by a weakening in thermohaline circulation.

The variations in glacial surface properties (temperature and wind stress, for instance) were responding to the change in climate state caused by orbital forcing and the result was a significantly colder surface ocean in the regions where the thermocline is ventilated. However, superimposed on these longer-term (10,000yr) climate changes may be shorter-term (millennial-scale) oscillations in thermocline temperatures resulting from variations in thermohaline circulation. Short-term warmings (which would be δ^{18}O lows at the Bahamas) may be coincident with the evidence for reduced deep water production (δ^{13}C lows in the deep water records). Using this conceptual model, we will try to evaluate changes in upper water column properties on millennial time scales.

Previous work by *Marchitto et al.* [1998] showed that during the initial stages of the deglaciation, Cd concentrations in the aragonitic benthic foraminifer *H. elegans* from the base of the North Atlantic subtropical gyre thermocline began to increase at the same time as Cd concentrations in the deep North Atlantic began to decrease [*Boyle and Keigwin,* 1987]. During the Younger Dryas, both shallow and deep water records exhibit reversals in their deglacial trends, the shallower site exhibiting a shift toward lower Cd (better ventilation) and the deeper site a shift toward higher Cd (reduced NADW). After the Younger Dryas, both records continued on their deglacial trends toward Holocene values, with active production of NADW leading the reduction in ventilation of the lower thermocline.

To examine events prior to the Younger Dryas, we have extended the records of benthic foraminiferal $\delta^{13}C$ and Cd concentration to span the upper part of marine isotope stage 3 (Figure 4). *H. elegans* Cd concentration and *Cibicidoides* $\delta^{13}C$ each exhibit little variability through the record prior to the deglaciation. Low Cd and high $\delta^{13}C$ values together imply that nutrient concentrations were reduced throughout most of isotope stages 3 and 2. Sea water Cd values remained well below Holocene levels for the entire interval from 50 to 10 ka. *Cibicidoides* $\delta^{13}C$ values, on the other hand, increased from Holocene-like values at about 50 ka to maximum values at about 20 ka, which may reflect changes in non-nutrient factors which affect sea water $\delta^{13}C$ of ΣCO_2 [*Charles et al.,* 1993]. Although small variations in these nutrient proxies may have occurred on millennial time scales, the variations are incoherent and do not appear to be synchronous with the known ages of the Heinrich events. The persistence of very low Cd values throughout stages 3 and 2 implies that ventilation was strong at the times of most Heinrich events. Unfortunately the sedimentation rate in this interval (4.5 cm kyr^{-1}) compromises our ability to observe truly short-term events. The median sample spacing for the Cd measurements is 800 years, but it is longer during the glacial period. The median sample spacing for $\delta^{13}C$ measurements is 400 years, but increases to more than 600 years during stage 3. Resolution of events shorter than 1000 years would be difficult because of bioturbation and sample spacing.

In contrast to the small variations seen in the benthic foraminiferal nutrient proxies, much larger changes occur in benthic foraminiferal $\delta^{18}O$, implying that there were variations in water column temperature, $\delta^{18}O$ of

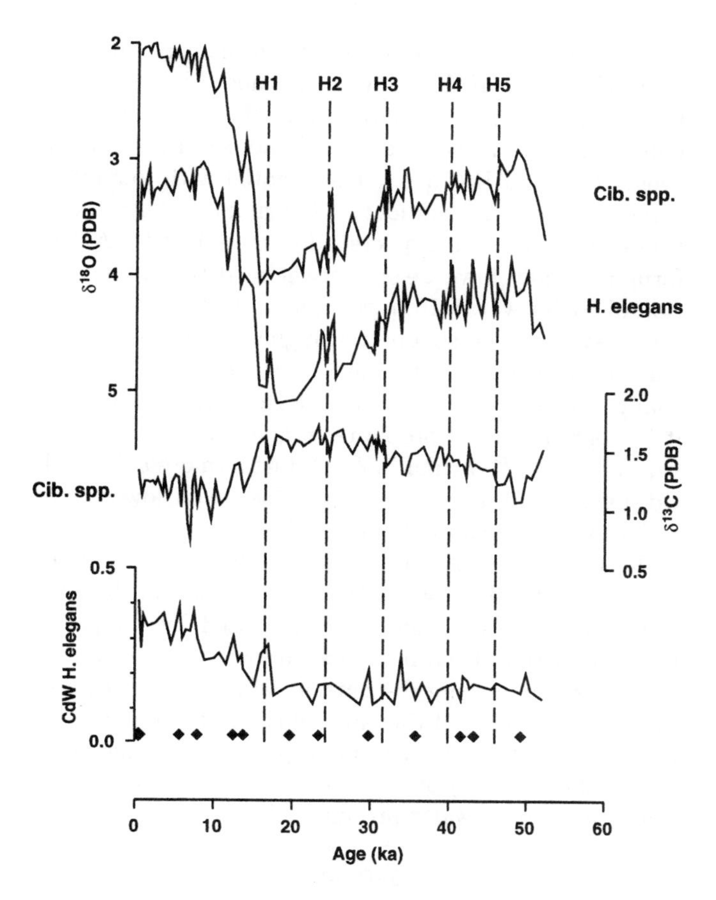

Figure 4. Variations in benthic foraminiferal chemistry for OC205-103GGC. The upper panels are $\delta^{18}O$ records for *Cibicidoides* spp. and *H. elegans.* The next record is the $\delta^{13}C$ record for *Cibicidoides.* The lines represent the average isotopic composition at each stratigraphic level. Each average value is based on several measurements of individual tests. The lower panel is the CdWH, the estimated Cd concentration of sea water based on Cd/Ca measurements of *H. elegans* using a distribution coefficient of 1 [*Boyle et al.,* 1995]. The filled diamonds are the levels of the AMS radiocarbon ages (converted to calendar age) from Table 2. Ages between these dates were interpolated. The nutrient proxies (CdWH and $\delta^{13}C$) exhibit little variability throughout MIS 2 and 3; both imply that nutrient concentrations at the Bahamas were reduced. $\delta^{18}O$ variability is larger than for $\delta^{13}C$, suggesting that there were short-term changes in water column temperature during stage 3.

sea water or both. For both the *Cibicidoides* and *H. elegans* records, the glacial-interglacial difference in $\delta^{18}O$ exceeds 2.0 ‰. On shorter time scales there are numerous oscillations in $\delta^{18}O$ of the benthic foraminifera that exceed 0.5 ‰ and are common to both species. Many lows in $\delta^{18}O$, both short-term spikes and broad, multi-millennial peaks, occur between the Heinrich events. For several events (H2 and perhaps H3 and H4) low $\delta^{18}O$ values occur synchronously with the Heinrich events. If these lows were caused by increased temperature, they may be the result of reduced thermohaline circulation. On the other hand, several Heinrich events (H1 and H5) are not associated with low $\delta^{18}O$ values, but are rather local $\delta^{18}O$ maxima. At these times, any warming expected due to reduced NADW formation may have been overwhelmed by the advection of colder waters from the north. The $\delta^{18}O$ records are noisy, however, so a better, higher-resolution record will be needed to determine the exact relationship of changes in the thermocline to the Heinrich events. But it is clear that the isotopic records exhibit much more variability in $\delta^{18}O$ than in $\delta^{13}C$; furthermore $\delta^{13}C$ and Cd suggest that nutrient concentrations were lower than Holocene values throughout all of MIS 2 and 3.

Because many of the minima in benthic foraminiferal $\delta^{18}O$ occur between Heinrich events, we believe that variations were more likely caused by temperature rather than from mixing with low $\delta^{18}O$ meltwater caused by the iceberg discharges. Whatever the cause, though, the source of the variability most likely was to the north in the isopycnal outcrop region (for the ventilated thermocline) or where convection was occurring (for intermediate water production).

4.3. North Atlantic Surface and Intermediate Waters

Records of surface water properties and intermediate water chemistry are available from *Lehman and Oppo* [in manuscript] for a high sedimentation rate core (V29-204, 1849 m, 16 cm kyr^{-1}) from the subpolar North Atlantic, in the region where the deeper parts of the subtropical gyre were probably ventilated during much of the last glacial period. Large, frequent variations in surface water temperatures at this location are synchronous with air temperature records from Greenland ice cores and there are indications of decreased nutrient concentration of the intermediate water during several periods of reduced deep water production [*Lehman and Oppo,* 1995].

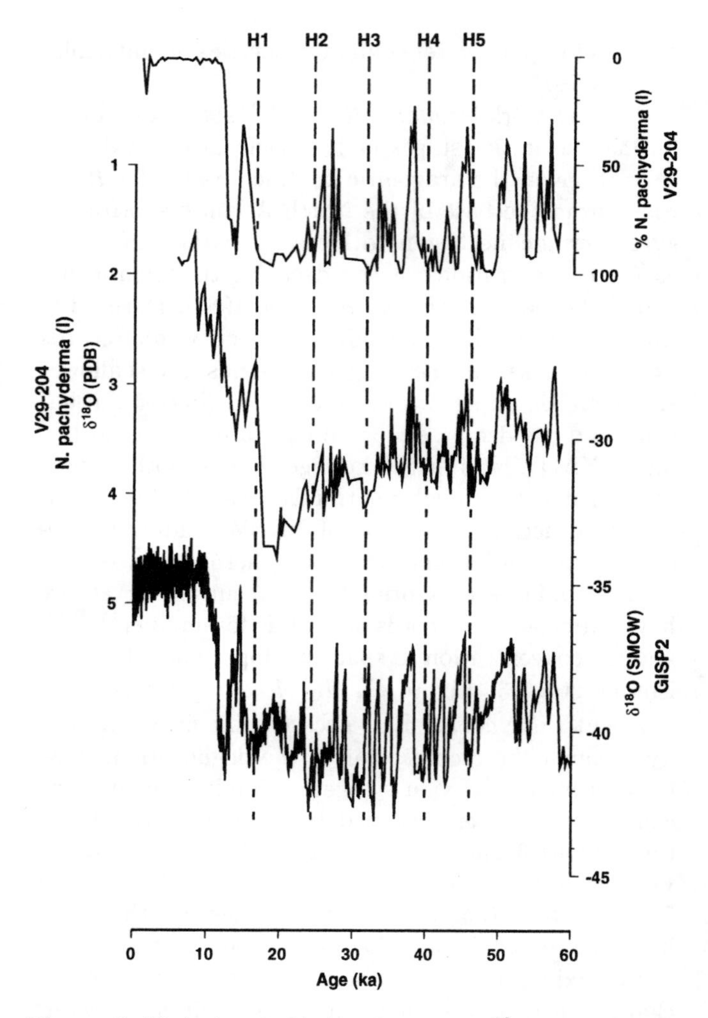

Figure 5. Variations in % abundance and $\delta^{18}O$ of *N. pachyderma* (l) in V29-204 [*Lehman and Oppo,* in manuscript] compared to the record of $\delta^{18}O$ at GISP2. The covariation of faunal abundances and $\delta^{18}O$ indicate large changes in surface water temperatures at this North Atlantic location. The variations in surface conditions are synchronous with changes in air temperature over Greenland. The absence of ice rafted debris during minima in $\delta^{18}O$ in this record indicate that melt water did not contribute to the $\delta^{18}O$ lows in V29-204.

The surface water proxies for V29-204 are presented in Figure 5 and include $\delta^{18}O$ and the abundance variations of *N. pachyderma* (l). These records exhibit, large, synchronous oscillations; when $\delta^{18}O$ values of *N. pachyderma* (l) are high, the relative abundance of the left coiling (polar) variety also increases. The records suggest that that millennial-scale oscillations of sea surface temperature in this region were large: 4°C on the basis of the *N. pachyderma* (l) $\delta^{18}O$ values and 4 to 6°C on the basis of the *N. pachyderma* (l) abundance. This well dated record also suggests that the periods associated with Heinrich events were cold at this location and that any melt water effects on the *N. pachyderma* (l) $\delta^{18}O$ record were dominated by larger changes in temperature. This location was minimally affected by meltwater because V29-204 is north of the main salinity anomaly associated with the Heinrich layers (40 to 55°N). Within the errors of the chronology, the records are also synchronous with variations in the air temperature over Greenland [*Lehman and Oppo,* 1995, in manuscript], as might be expected at this location which is near to and down wind of southern Greenland (see Figure 1).

The benthic foraminiferal $\delta^{13}C$ record also exhibits variations on millennial time scales, but with a pattern very different from those seen in the deeper waters (Figure 6). At this shallower depth, the $\delta^{13}C$ variations are smaller than in the deep records and the oscillations are sometimes of opposite sign. Several of the Heinrich events during isotope stage 3 are local maxima in $\delta^{13}C$ (H3, H4, H5), while others appear to be local minima (H1, H2). Unlike most records of $\delta^{13}C$ from deeper water, there is little difference in mean carbon isotopic composition between the Holocene and the glacial period. This highly resolved record (median sample spacing of 250 years) probably captured events of reduced thermohaline circulation which were longer than about 500 years.

For several of the Heinrich events, benthic foraminifera in V29-204 have lower $\delta^{13}C$ values than coeval benthic foraminifera at the Bahamas (Figure 6). For instance, $\delta^{13}C$ values during H1 are 0.6 ‰ lower than at the Bahamas, within well dated and high sedimentation rate intervals of both cores. $\delta^{13}C$ values during H2 may be as much as 0.4 ‰ lower at V29-204. In contrast, $\delta^{13}C$ values for H3 and H5 are the same at both locations. For H4, $\delta^{13}C$ values are lower at V29-204 by about 0.2-0.3‰. These gradients in $\delta^{13}C$ between sites may be caused by different aging of the water masses, different surface equilibration with atmospheric CO_2 or, as we believe is most likely, different fractional mixing with southern ocean waters.

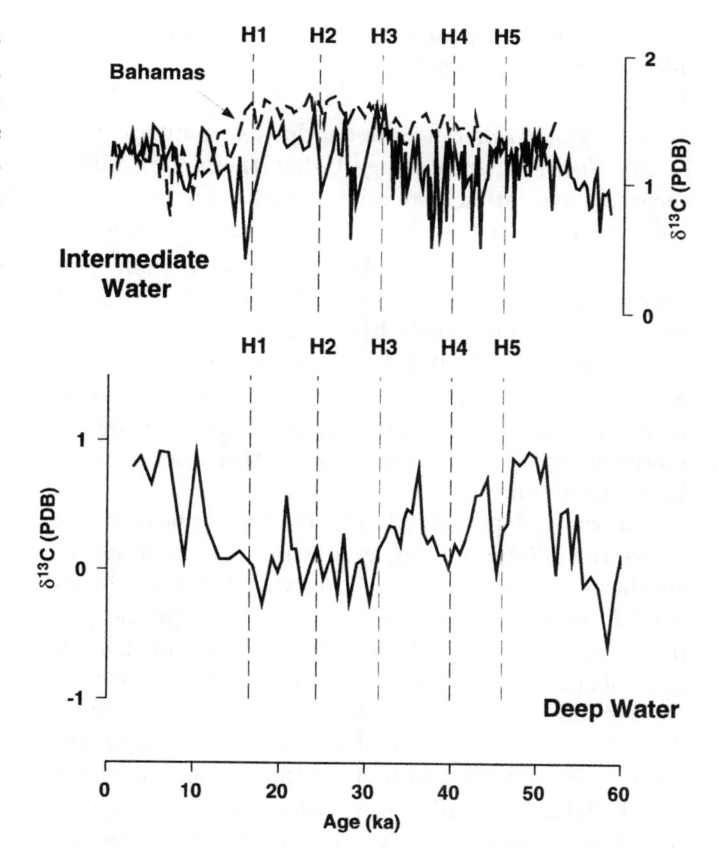

Figure 6. $\delta^{13}C$ variations within thermocline (dashed line), intermediate and deep waters during the last glacial period. While the deep water site exhibits a reduction in $\delta^{13}C$ during the Heinrich events and throughout the glacial maximum, the shallower sites generally do not. The Bahamas record is always the highest in $\delta^{13}C$ during MIS 2 and 3. For several short intervals, V29-204 has the same values (Heinrich events 3 and 5) as the Bahamas; for other events it has much lower $\delta^{13}C$ values. We suggest that the very low $\delta^{13}C$ values in V29-204 coincident with H1 and H2 were caused by mixing with southern ocean water.

5. DISCUSSION AND IMPLICATIONS

There are several patterns in these data which result from variations in the intensity of circulation of the subtropical gyre, and intermediate and deep waters:

- the deep water records display evidence for nutrient enrichment during Heinrich events;
- at intermediate depths (1849 m) there is a mixed signal for short-term oscillations in $\delta^{13}C$: for Heinrich events 3, 4 and 5 there are increases in $\delta^{13}C$ while for Heinrich events 1 and 2 there are decreases;
- $\delta^{13}C$ and sea water Cd estimates in the thermocline imply low nutrient concentrations and enhanced ventilation throughout the glacial period;
- glacial benthic foraminiferal $\delta^{18}O$ variability in the thermocline is large, $\delta^{13}C$ variability is small;
- at low latitudes, surface waters exhibit short-term coolings that are synchronous with Heinrich events and decreased deep water production.

Below we attempt to explain these disparate observations with a model for glacial circulation that contains a vigorous gyre circulation throughout MIS 3 and 2 interacting with a North Atlantic thermohaline circulation that is varying in intensity of deep and intermediate water production. When the intermediate water mass is active, it is cold, well ventilated, and low in nutrients, affecting the chemical and thermal structure of the upper water column.

5.1. Ventilation of the Upper Water Column and Mixing with Southern Ocean Water

The regional differences in benthic foraminiferal $\delta^{13}C$ in the North Atlantic suggest that there were differences in ventilation and mixing with southern ocean water that affected the lateral and vertical distribution of $\delta^{13}C$ of ΣCO_2 during the last glacial period. The lower thermocline waters at the Bahamas during much of the glacial most likely had a source region near the surface waters at V29-204. In contrast, the intermediate water at V29-204 had a source region farther north; its $\delta^{13}C$ record reflects variations in the preformed $\delta^{13}C$ modified by varying amounts of southern ocean water in the local mixture.

The water depth of V29-204 (1849 m) places it just above the 2000 m mixing zone which separated glacial southern ocean water from overlying GNAIW during MIS 2, based on the water column $\delta^{13}C$ profile produced by *Oppo and Lehman* [1993]. They documented that benthic foraminiferal $\delta^{13}C$ values decreased from ~1.2 to 0.6‰ over the depth interval 2063 m to 2209 m. The low $\delta^{13}C$ values were of southern ocean origin and even lower values were observed throughout the glacial North Atlantic at all depths below this sharp boundary. The higher-resolution record of V29-204 shows that minimum $\delta^{13}C$ values for *Cibicidoides* during Heinrich event 1 were 0.4‰ even as shallow as 1849 m. The very low values are most easily obtained by mixing with the low $\delta^{13}C$, southern ocean water mass. A short-term shoaling of this glacial water mass boundary would account for the low in $\delta^{13}C$ during Heinrich event 1. Decreased production of GNAIW would be implied by this process.

The differences in $\delta^{13}C$ observed between V29-204 and the Bahamas reflect their relative positions in the water column. OC205-103GGC is located today at about 965 m in water with a density (σ_θ) of 27.5, while the water mass bathing V29-204 has a density of 27.9. This difference is reflected in their different Holocene benthic foraminiferal $\delta^{18}O$ values: *Cibicidoides* $\delta^{18}O$ values at the Bahamas are about 0.5‰ lower than at V29-204. During the glacial maximum, this $\delta^{18}O$ difference decreased to about 0.3‰, a smaller difference but one large enough to preclude mixing across isopycnals between these sites.

Consequently southern ocean water would influence the $\delta^{13}C$ record of V29-204 before it would affect cores as shallow as OC205-103GGC; this is the likely explanation for the large $\delta^{13}C$ difference observed between the two cores during Heinrich event 1 and at other times during the glacial period. We suggest that variations in mixing of southern ocean water caused the the other low $\delta^{13}C$ values in V29-204. At the water depth of the Bahamas core, however, there are no short term decreases in $\delta^{13}C$ that would imply decreased ventilation or increased mixing with southern ocean water. The difference between the $\delta^{13}C$ records at the Bahamas and V29-204 is a measure of the degree to which thermohaline circulation in the North Atlantic decreased: the greatest reduction in thermohaline circulation occurred during Heinrich events 1 and 2, where GNAIW was maintained only at very shallow water depths; Heinrich events 3 and 5 show no difference in $\delta^{13}C$ between the Bahamas and V29-204, implying a larger volume of GNAIW.

Thus at times V29-204 was bathed by GNAIW with low nutrients and high $\delta^{13}C$ of ΣCO_2, while at other times by southern ocean deep water with higher nutrients and lower $\delta^{13}C$ of ΣCO_2. The case for this becomes stronger by comparing the $\delta^{13}C$ and $\delta^{18}O$ isotopic data from stage 3 for deep, intermediate and thermocline cores in the North Atlantic (Plate 2). The $\delta^{18}O$ and $\delta^{13}C$ values within the deep water records from the Ceara Rise exhibit a negative covariation. This pattern reflects the replacement of NADW (with higher temperature and higher $\delta^{13}C$ of ΣCO_2) with southern ocean water (with lower temperature and lower $\delta^{13}C$ of ΣCO_2). For the thermocline record at the Bahamas there is a strong positive covariation of the isotopic data which implies a relationship much like that seen in the Holocene-glacial maximum comparison of *Slowey and Curry* [1995]: higher $\delta^{18}O$ values implying a colder thermocline co-occurring with higher $\delta^{13}C$ values implying nutrient reduction, air-sea exchange at colder temperatures and better ventilation. The V29-204 data are a hybrid of the thermocline and deep water trends. For parts of the record, there is negative covariation of $\delta^{13}C$ and $\delta^{18}O$ values which parallels the deep water trend; for other parts of the record there is positive covariation, which parallels the thermocline trend. Thus V29-204, located at a water depth near to the boundary between GNAIW and southern ocean water during much of the glaciation, was affected by both water masses at various times.

A potential weakness in our conclusion comes from the relatively low sedimentation rate during isotope stage 3 (4.5 cm kyr^{-1}) in the Bahamas $\delta^{13}C$ record, which may mean that we have missed low $\delta^{13}C$ events of very short duration. This may explain why there are no low $\delta^{13}C$ events during some Heinrich events at the Bahamas, unlike the V29-204 record or the eastern Atlantic record of *Zahn et al.* [1997]. However, we do not believe that sedimentation rate alone can explain the differences we observe.

The isotopic record of *Zahn et al.* [1997], from 1099 m water depth on the Portugal continental margin, exhibits decreases in δ^{13}C beginning before the ice rafting events and a gradual increase following, much like the changes in δ^{13}C values seen in our deep water record (1JPC). They documented that decreases in benthic foraminiferal δ^{13}C began 1.5 to 2.5 kyr prior to the ice rafting event and that the rebound in δ^{13}C postdated the ice rafting by 1 to 3 kyr. With the Heinrich events lasting 300 to 400 years, the inferred decrease in thermohaline overturning must have lasted 3000 years or more [*Zahn et al.,* 1997]. Events of this duration should have been detected in V29-204, and also in the Bahamas records. The lack of any significant minima in benthic foraminiferal δ^{13}C or maxima in CdWH at the Bahamas imply that events of reduced thermohaline circulation at this time must have been short and occurring within periods of generally enhanced ventilation of the upper water column. On the other hand, δ^{18}O values at the Bahamas exhibit variations that are consistent with reduced thermohaline overturning during Heinrich events 1 and 2, two very low δ^{13}C events in the *Zahn et al.* [1997] record. For each of these events, lower δ^{18}O values imply a short-term warming of the upper water column because of reduced northward heat transport. This warming should have been associated with a small decrease in benthic foraminiferal δ^{13}C values, based on covariation observed in Plate 2, but the magnitude of the δ^{13}C anomaly would have been about 0.1 ‰, too small to be observed with confidence.

Our observations do not preclude regional differences in ventilation, particularly differences between a vigorous main flow in western basins compared to sluggish recirculated flow in the eastern basins. Indeed we believe, but cannot yet prove, that the differences in δ^{13}C between the Bahamas and the *Zahn et al.* [1997] record may have resulted from an east-west hydrographic contrast within the North Atlantic for long intervals encompassing the Heinrich events. Only a very high sedimentation rate core from the Bahamas or other western basin location will show unequivocally if thermohaline overturning shutdown completely during Heinrich events. It is unlikely, though, that a 3,000 year shutdown in thermohaline circulation was missed in the Bahamas records.

5.2. Oscillations in Temperature or Sea water δ^{18}O in the Thermocline

Oxygen isotopic variations in foraminifera are affected by both changing temperature and sea water oxygen isotopic composition; without some other indicator of temperature or δ^{18}O, it is difficult to uniquely differentiate among the two possibilities. For the planktonic foraminiferal δ^{18}O in V29-204 (Figure 5), the case is strong that the variations are driven by temperature for two reasons. First, the δ^{18}O variations covary with the abundance of left coiling *N. pachyderma.* Second, minima in δ^{18}O occur within intervals of diminished iceberg rafting, implying that meltwater input was not causing the δ^{18}O changes [*Lehman and Oppo,* in manuscript]. The benthic foraminiferal δ^{18}O variations in V29-204 also exhibit lower values between the intervals of high ice rafting (not shown in Figure 5), although the amplitude of the δ^{18}O variations (0.5 ‰ or lower) is smaller than for *N. pachyderma* (l) [*Lehman and Oppo,* in manuscript]. If these variations were solely due to temperature, then the water mass bathing V29-204 fluctuated by about 2°C. The association of low benthic and planktonic foraminiferal δ^{18}O with minima in ice rafting and minima in abundance of *N. pachyderma* (l) is strong indication that they were not caused by variations in fresh water input.

There are several reasons to believe that the δ^{18}O variations at the Bahamas were also driven by temperature and not variations in fresh water input. Many low values in δ^{18}O for *Cibicidoides* and *H. elegans* occur within the warmer intervals between the Heinrich events, periods for which there is little evidence of significant meltwater injection into the North Atlantic. There are some similarities of the variations in δ^{18}O at the Bahamas with the planktonic foraminiferal δ^{18}O records for higher and lower latitudes (Figure 7), which makes us lean toward an interpretation of synchronous temperature variability of the gyre as a whole. At the Bahamas, however, we cannot determine the timing of events with great certainty, which limits our interpretation. For instance, if the low δ^{18}O values falling between Heinrich events were synchronous with intervening cold phases of the Dansgaard-Oeschger cycles rather than the warm, our interpretation of the record would be different. Unfortunately the poorer chronology of the Bahamas record and the low sedimentation rates preclude coming to a definite conclusion based on the timing of the changes in δ^{18}O. (Our search for a higher sedimentation rate core from this area continues.)

The δ^{18}O variations in the Bahamas benthic foraminiferal records were too large to have been caused by meltwater injection and be maintained, after mixing and recirculation, some 3000 km away from the region of subduction. Depending on the δ^{18}O value for meltwater (-20 to -40 ‰), the thermocline at the Bahamas would be required to contain from 1.5 to 3% meltwater to cause a 0.6 ‰ decrease in δ^{18}O of the

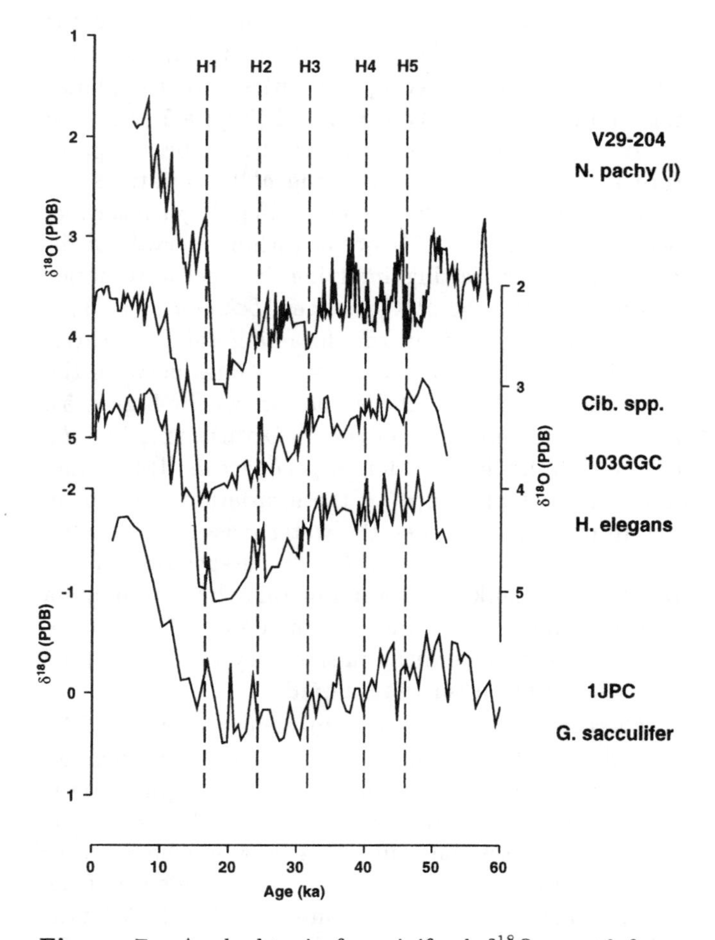

Figure 7. A planktonic foraminiferal $\delta^{18}O$ record from the high latitude North Atlantic (V29-204), a benthic foraminiferal $\delta^{18}O$ record from within the glacial thermocline (OC205-103GGC), and a planktonic foraminiferal $\delta^{18}O$ record from the western equatorial Atlantic (EW9209-1JPC). The pattern of low $\delta^{18}O$ values between the Heinrich events is clear in V29-204 and 1JPC. At the Bahamas, the pattern is less clear. For several events (H2, H3, H4) $\delta^{18}O$ values appear to decrease, while $\delta^{18}O$ values increase during other events (H1 and H5). These different responses may reflect the competing influences of a rapidly circulating cold gyre and the heat redistribution associated with reduced thermohaline circulation in the North Atlantic. Unfortunately the sedimentation rate in the Bahamas core may be too low to resolve with certainty the chronology of the events.

benthic foraminifera. Local salinity would be reduced by more than 0.5 ‰, which would have had a large effect on the density structure and stability of the water column. Mixing with meltwater affected locations farther to the north [*Rasmussen et al.*, 1996; *Vidal et al.*, 1998; *Dokken et al.*, 1998], where overflows from the Norwegian-Greenland Sea influenced the records and the minima in benthic foraminiferal $\delta^{18}O$ are synchronous with the ice rafting. However, these records are from depths and densities that would be unlikely to mix with waters that eventually end up in the subtropical gyre thermocline: their benthic foraminiferal $\delta^{18}O$ values are about 1 ‰ greater than the Bahamas benthic foraminifera, implying significantly colder and much denser water than can be found at the Bahamas. Indeed the low benthic foraminiferal $\delta^{18}O$ values in these cores are thought to be derived from entrainment of meltwater into deep water through brine rejection [*Vidal et al.,* 1998; *Dokken et al.,* 1998]. As such the water temperature must be close to freezing. It is unlikely that any process like this is affecting the source of the subtropical gyre waters.

On the other hand, the $\delta^{18}O$ values of benthic foraminifera at V29-204 are similar to the heaviest values we observe at the Bahamas. Thus it is quite possible that during periods when their $\delta^{13}C$ and $\delta^{18}O$ values were the same, V29-204 and the base of the thermocline at the Bahamas were bathed by the same, cold water mass.

Consequently we believe that the primary cause of the variations in benthic foraminiferal $\delta^{18}O$ at the Bahamas was temperature variability. The fluctuations were about 2 to 3°C, but the timing of the fluctuations is not consistently related to the temperature oscillations in the North Atlantic. Several Heinrich events exhibit $\delta^{18}O$ minima in one or both of the $\delta^{18}O$ records at the Bahamas, with the most prominent being $\delta^{18}O$ minima during H2 (in both benthic species), H3 (in *Cibicidoides* only) and H4 (in *H. elegans* only). $\delta^{18}O$ values appear to be maxima for H5 and for H1. (The small $\delta^{18}O$ low near H1 is caused by a single measurement of an *H. elegans* that we believe was mixed downward by bioturbation.)

What was the rest of the North Atlantic gyre doing at this time? Planktonic foraminiferal $\delta^{18}O$ records from the Bahamas imply that surface waters were at 2 to 4°C colder throughout the glacial period [*Curry,* unpublished data]. There were oscillations in the temperature of surface waters on the northern margin of the gyre all during isotope stage 3 with colder surface water temperatures synchronous with cold phases of the Dansgaard-Oeschger cycles and Heinrich events [*Keigwin and Boyle,* 1998, in press; *Sachs and Lehman,* 1998]. Furthermore, the colder surface waters were synchronous with reduced deep water production on millennial time scales [*Keigwin and Boyle,* 1998, in press]. At Ceara Rise, on the southern margin of the gyre, short-term increases in planktonic foraminiferal oxygen isotopic composition were synchronous with decreased production of NADW. Thus, on average, the gyre was colder than today (by ~2-4°C) and the millennial-scale oscillations were large on its northern (4°C) and southern (2 to 3°C) margins. Only within the thermocline is there evidence for warming during periods of reduced thermohaline overturning, during the short-term δ^{18}O minima associated with Heinrich events 2, 3 and 4. These events appear to be superimposed on a gyre that is otherwise very cold. A cause other than shutting down thermohaline circulation is needed to explain this observed widespread cooling of the North Atlantic gyre.

5.3. Isopycnal Mixing and Cooling of the Tropics

If we are correct that the δ^{18}O variations at the Bahamas are primarily caused by temperature changes, then perhaps we are observing in our δ^{18}O records the interaction of two competing processes: mixing of cold water along isopycnals in a cold, vigorously circulating gyre and the heat redistribution that accompanies a change in North Atlantic thermohaline circulation. The isopycnal mixing of cold water may provide a mechanism for cooling tropical thermocline and surface waters and for forcing changes in the tropics from an extra-tropical location (Figure 8). Superimposed on this overall cooling may be short-term warmings which result from variations in the production of NADW. The premise for our argument is based on the modeling studies of *Gu and Philander* [1997] and *Bush and Philander* [1998], which focus on the effects of wind-driven processes in the gyre. Enhanced trade wind activity, in response to a change in pole-to-equator thermal gradient advects surface water to the north. Equatorward flow of colder water along isopycnals balances poleward flow of surface water, causing a reduction in temperature of the upwelled water near the equator and further cooling of the tropical surface water. The two modeling studies focussed on the shallow, wind driven region of the thermocline; here we extend the idea by analogy to the thermocline as a whole.

In Figure 8 we present the distribution of isopycnals in the western North Atlantic for the GEOSECS data set and we summarize the location and magnitude of temperature oscillations within the gyre system. At the region of isopycnal outcrop (V29-204) δ^{18}O oscillations reach 1 ‰, indicating that sea surface temperature is oscillating by more than 4°C. Within the intermediate waters in the North Atlantic (V29-204 benthics), temperatures appear to be oscillating by about 2°C. In the Bermuda Rise region, *Keigwin and Boyle* [1998, in press] (using planktonic foraminiferal δ^{18}O) and *Sachs and Lehman* [1998] (using alkenone SST

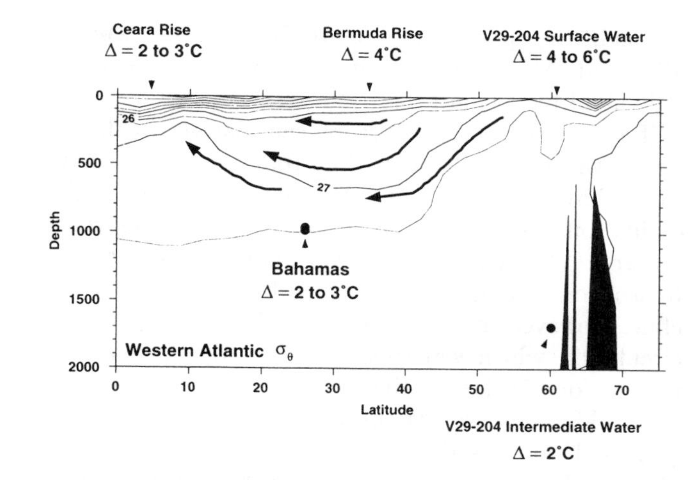

Figure 8. A schematic representation of the warm-cold oscillations we believe are affecting the gyre during the last glacial period. There is evidence for surface water temperature oscillations at V29-204 and near the equator. *Keigwin and Boyle* [1998, in press] and *Sachs and Lehman* [1998] present evidence for surface temperature oscillations northern margin of the subtropical gyre. Average temperatures within the thermocline were colder than today and superimposed were oscillations of about 2 to 3°C. Taken together, these data suggest that the gyre was significantly colder during most of MIS 3 and 2 and that propagation of cold water along isopycnals was responsible. If this is true, then cold water would have been advected to the equator to within a few hundred meters of the surface, providing a source of cold water to an upwelling system that was probably more active at that time. Competing with this process, however, were short term warmings associated with reduced thermohaline circulation. For several of the Heinrich events (H2, H3 and H4), the effects of reduced thermohaline circulating may have dominated and the thermocline warmed while the surface waters of the gyre remained cold.

proxies) record oscillations in sea surface temperature of about 4°C. Near the equator, *Curry and Oppo* [1997] calculated that the millennial-scale oscillations in δ^{18}O were caused by 2 to 3°C oscillations in sea surface temperature. Finally, at about 1000 m water depth in the Bahamas we observe δ^{18}O variability that we believe is most easily explained by 2 to 3°C oscillations in thermocline temperature. Thus at several locations within the gyre, the outcrop region in the north, the surface waters of the gyre, and at the surface near the equator, short-term, coolings were synchronous (or nearly so) with cold phases in the North Atlantic. Within the thermocline, however, the timing of the oscillations is less consistent: while there are several intervals for which the δ^{18}O maxima line up, there are as many examples of out-of-phase behavior as in-phase behavior. The magnitude of the variability is clear, showing up as "noisy" 0.5 ‰ variations in the time series (Figure 7), but as systematic $\delta^{18}O$ and $\delta^{13}C$ covariation in the measurements of the individual tests (Plate 2).

If the mean temperature of the glacial North Atlantic was 2 to 4°C colder than today, the entire gyre would have been starting from this colder state. The oscillations in temperature in the region of isopycnal outcrop (V29-204) would be transmitted into the gyre by isopycnal mixing, which cooled the tropical thermocline. However, during periods of reduced thermohaline circulation, which were times of very cold surface waters in the North Atlantic, the gyre thermocline may have warmed in response. At times this warming dominated (H2, and perhaps H3 and H4), but at other times it did not (H1 and H5). In addition, the large coolings of higher latitude surface waters increased the equator-to-pole temperature gradient, enhancing the low latitude winds and increasing upwelling near the equator. Isopycnal mixing of colder gyre waters toward the south makes the upwelled water colder on average, enhancing the cooling along the equator. Isopycnal mixing then provides a mechanism to enhance cooling at the equator during the glacial maximum and throughout much of isotope stage 3 when gyre temperatures were on average colder than today. The fluctuations in temperature would be enhanced by this process when the increased upwelling co-occurs with colder waters advected to the south in a vigorous gyre circulation. Thus the isopycnal mixing is an extra-tropical feedback which may help to explain the unexpectedly large variations observed by *Curry and Oppo* [1997] in planktonic foraminiferal $\delta^{18}O$ near the equator.

Unfortunately, there is contradictory microfossil evidence for enhanced equatorial upwelling occurring during Heinrich events. *McIntyre and Molfino* [1996] used the abundance of the coccolith *Florisphaera profunda* to show that reduced upwelling occurred during Heinrich events in the eastern equatorial Atlantic. In contrast, *Curry and Oppo* [1997] demonstrated that maxima in the upwelling sensitive planktonic foraminifer *Neogloboquadrina dutertrei* occurred synchronously with maxima in *G. sacculifer* and *Globigerinoides ruber* $\delta^{18}O$ during Heinrich event 4 in the western equatorial Atlantic. From the existing data, it is not possible to provide a general description of enhanced or decreased equatorial upwelling during Heinrich events.

6. SUMMARY

1. Synchronous with the Heinrich event pulses of ice rafting were short-term decreases in the volume of NADW suggesting that deep thermohaline overturning in the North Atlantic decreased during each event. The millennial-scale decreases in NADW production observed during Heinrich events 1, 2 and 3 occurred to a water mass that was already operating at only about 2/3 of its Holocene efficiency because of Milankovitch-forced reductions in deep water production.

2. The oscillations in deep water production were synchronous with changes in surface water temperature in the western equatorial Atlantic. Surface waters cooled by about 2 to 3°C at times when deep water production in the North Atlantic decreased.

3. Throughout most of isotope stage 3, benthic foraminiferal nutrient proxies imply that the upper water column (965 m) remained low in nutrient concentration and that there was little variability on millennial time scales. Benthic foraminiferal $\delta^{18}O$, on the other hand, exhibited significant variability (>0.5 ‰). Low $\delta^{18}O$ values appear to be associated with Heinrich events 2, 3 and 4. Additional short-term lows occurs between the Heinrich events. In contrast, Heinrich events 1 and 5 were associated with high $\delta^{18}O$ values.

4. Surface water conditions in the subpolar region exhibited large changes in temperature. These oscillations appear to be synchronous with changes in air temperature over Greenland. The region is north of the main area of iceberg discharge associated with the Heinrich events and is thus not in a region with a melt water signal in planktonic foraminiferal $\delta^{18}O$.

5. The oscillations in $\delta^{18}O$ observed within the thermocline appear to be caused by a combination of changes in temperature at the source region which may be superimposed with short-term fluctuations in temperature caused by variations in North Atlantic thermohaline circulation. When the thermocline is colder, $\delta^{13}C$ values are higher, implying enhanced ventilation. During Heinrich events, the surface waters of the northern gyre and the western equatorial region are each several degrees colder, which may be caused by enhanced isopycnal mixing of cold water in the gyre coupled with enhanced upwelling in the western equatorial Atlantic. However, a coincident reduction in thermohaline circulation opposed this cooling and at times (H2, H3 and H4) this process may have dominated and caused a short-term warming in the thermocline at the Bahamas. Our speculation remains to be tested in detail both with better cores from the subtropical gyre thermocline and with a more thorough analysis of the history of equatorial upwelling during MIS 2 and 3.

6. The variations in benthic foraminiferal $\delta^{13}C$ in mid-depth core V29-204 (1849 m) appear to be a hy-

brid of the variations seen in the deep water and thermocline records. Negative covariation of benthic foraminiferal $\delta^{18}O$ and $\delta^{13}C$ (the deep water pattern) imply a stronger input of southern ocean water during cold events; positive covariations (the upper water pattern) imply a stronger influence of GNAIW during cold events. Strong variability related to water depth should thus be expected in evaluating the relationship of benthic foraminiferal $\delta^{13}C$ to Heinrich events because the lower boundary between GNAIW and southern ocean water is not the same for each event. Based on the differences in $\delta^{13}C$ between V29-204 and the Bahamas, the volume of GNAIW was largest during Heinrich events 5 and 3, smaller during event 4 and smallest during events 1 and 2.

Acknowledgments. Jean Lynch-Stieglitz and Jim Wright reviewed this manuscript and their comments stimulated us to present our results in a more balanced manner. We thank Peter Clark and Robert Webb for inviting us to participate in the Chapman Conference. We thank Lloyd Keigwin for providing his unpublished data. D. R. Ostermann, M. Jeglinski, S. Prew, L. Zou and E. Joynt provided technical support. Funding was provide by the NSF and NOAA. The views expressed herein are those of the authors and do not necessarily reflect those of NOAA or its subagencies. Graduate fellowship support was provided to TMM by the MIT/WHOI Joint Program in Oceanography and the Joint Oceanographic Institutions, and to KLL by the Office of Naval Research. A Postdoctoral Scholarship to JFM was provided by WHOI. The WHOI and LDEO core repositories are supported by grants from NSF and ONR. This is WHOI Contribution Number 9898.

REFERENCES

Bard, E., M. Arnold, and B. Hamlin, Present status of the radiocarbon calibration for the late Pleistocene (abstract), Fourth International Conference on Paleoceanography, *GEOMAR Report*, *15*, 52-53, 1992.

Bond, G., H. Heinrich, W. Broecker, L. Labeyrie, J. McManus, J. Andrews, S. Huon, R. Jantschik, S. Clasen, C. Simet, K. Tedesco, M. Klas, G. Bonani, and S. Ivy, Evidence for massive discharges of icebergs into the North Atlantic during the last glacial period, *Nature*, *360*, 245-249, 1992.

Bond, G., W. Broecker, S. Johnsen, J. McManus, L. Labeyrie, J. Jouzel, and G. Bonani, Correlations between climate records from North Atlantic sediments and Greenland ice, *Nature*, *365*, 143-147, 1993.

Boyle, E. A., and L. D. Keigwin, Comparison of Atlantic and Pacific paleochemical records for the last 15,000 years: changes in deep ocean circulation and chemical inventories, *Earth and Planetary Science Letters*, *76*, 135-150, 1985/86.

Boyle, E. A. and L. D. Keigwin, North Atlantic thermohaline circulation during the past 20,000 years linked to high-latitude surface temperature, *Nature*, *330*, 35-40, 1987.

Boyle, E. A., L. Labeyrie, and J.-C. Duplessy, Calcitic foraminiferal data confirmed by cadmium in aragonitic *Hoeglundina*: Application to the last glacial maximum in the northern Indian Ocean, *Paleoceanography*, *10*, 881-900, 1995.

Bush, A. B. G. and S. G. H. Philander, The role of ocean-atmosphere interactions in tropical cooling during the last glacial maximum, *Science*, *279*, 1341-1344, 1998.

Charles, C. D, J. D. Wright, and R. G. Fairbanks, Thermodynamic influences on the marine carbon isotope record, *Paleoceanography*, *8*, 691-697, 1993.

Crowley, T. J., North Atlantic Deep Water cools the southern hemisphere, *Paleoceanography*, *7*, 489-497, 1992.

Curry, W. B., Late Quaternary deep circulation in the western equatorial Atlantic, in *The South Atlantic: Present and Past Circulation*, edited by G. Wefer, W. H. Berger, G. Siedler, and D. Webb, Springer-Verlag, New York, 1996.

Curry, W. B., J.-C. Duplessy, L. Labeyrie and N. J. Shackleton, Quaternary deep-water circulation changes in the distribution of $\delta^{13}C$ of deep water ΣCO_2 between the last glaciation and the Holocene, *Paleoceanography*, 3, 317-341, 1988.

Curry, W. B. and D. W. Oppo, Synchronous, high-frequency oscillations in tropical sea surface temperature and North Atlantic Deep Water production during the last glacial cycle, *Paleoceanography*, *12*, 1-14, 1997.

Dansgaard, W., et al., Evidence for general instability of past climate from a 250-kyr ice-core record, *Nature*, *364*, 218-220, 1993.

Dokken, T., E. Jansen, and J. Adkins, Rapid climatic changes documented from the Nordic Seas during the last glacial cycle: timing and possible mechanisms, Abstracts of the Chapman Conference "Mechanisms of Millennial-scale Climate Change", Snowbird, Utah, 17, 1998.

Duplessy, J.-C., N. J. Shackleton, R. G. Fairbanks, L. Labeyrie, D. Oppo, and N. Kallel, Deepwater source variations during the last climatic cycle and their impact on the global deepwater circulation, *Paleoceanography*, *3*, 343-360, 1988., 1988.

Grootes, P.M., and M. Stuiver, Oxygen 18/16 variability in Greenland snow and ice with 10^3 to 10^5-year time resolution. *Journal of Geophysical Research*, *102*, 26455-26470, 1997.

Gu, D. and S. G. H. Philander, Interdecadal climate fluctuations that depend on exchanges between the tropics and the extratropics, *Science*, *275*, 805-807, 1997.

Keigwin, L. D., and E. A. Boyle, Rapid climate oscillations of the last glacial cycle in the western North Atlantic, Abstracts of the Chapman Conference "Mechanisms of Millennial-scale Climate Change", Snowbird, Utah, 11, 1998.

Keigwin, L. D., and E. A. Boyle, Surface and deep ocean variability in the northern Sargasso Sea during marine isotope stage 3, *Paleoceanography*, in press .

Keigwin, L. D., and S. J. Lehman, Deep circulation change linked to Heinrich event 1 and Younger Dryas in a mid-depth North Atlantic core, *Paleoceanography*, *9*, 185-194, 1994.

Lehmann, S. J. and D. W. Oppo, Mode switches in abyssal circulation linked to millennial-scale climate change in Greenland and Antarctica, *EOS*, *76*, F309, 1995.

Manabe, S., and R. J. Stouffer, Two stable equilibria of a coupled ocean-atmosphere model, *J. Clim.*, *1*, 841-866, 1988.

Manabe, S. and R. J. Stouffer, Coupled ocean-atmosphere model response to freshwater input: Comparison to the Younger Dryas event, *Paleoceanography*, *12*, 321-336, 1997.

Marchitto, T. M., W. B. Curry, and D. W. Oppo, Millennial-scale changes in North Atlantic circulation since the last glaciation, Nature, *393*, 557-561, 1998.

McIntyre, A. and B. Molfino, Forcing of Atlantic equatorial and subpolar millennial cycles by precession, *Science*, *274*, 1867-1870, 1996.

Oppo, D. W. and R. G. Fairbanks, Variability in the deep and intermediate water circulation of the Atlantic Ocean during the past 25,000 years: Northern hemisphere modulation of the Southern Ocean, *Earth Planet. Sci. Lett.*, *86*, 1-15, 1987.

Oppo, D. W. and S. J. Lehman, Mid-depth circulation of the subpolar North Atlantic during the last glacial maximum, *Science*, *259*, 1148-1152, 1993.

Oppo, D.W., and S. J. Lehman, Suborbital timescale variability of North Atlantic Deep Water during the past 135,000 years, *Paleoceanography*, *10*, 901-910, 1995.

Ostermann, D. R. and W. B. Curry, A heavy δ^{18}O standard for source-mixing detection, linearity and correction in a stable isotope mass spectrometer, *Paleoceanography*, in manuscript.

Rahmstorf, S., Rapid climate transitions in a coupled ocean-atmosphere model, *Nature*, *372*, 82-85, 1994.

Rasmussen, T. L., E. Thomsen, T. C. E. van Weering, and L. Labeyrie, Rapid changes in surface and deep water conditions at the Faeroe Margin during the last 58,000 years, *Paleoceanography*, *11*, 757-771, 1996.

Sachs, J. P. and S. J. Lehmann, A new late Pleistocene SST record with 50-100 year resolution for the subtropical North Atlantic, Abstracts fro the Sixth International Conference on Paleoceanography, Lisbon, 197, 1998.

Sarnthein, M. A., K. Winn, S. J. A. Jung, J.-C. Duplessy, L. Labeyrie, H. Erlenkeuser, G. Ganssen, Changes in east Atlantic deepwater circulation over the last 30,000 years: Eight time slice reconstructions, *Paleoceanography*, *9*, 209-267, 1994.

Shackleton, N. J., J. Imbrie, and M. A. Hall, Oxygen and carbon isotope record of East Pacific core V19-30: Implications for the formation of deep water in the late Pleistocene North Atlantic, *Earth Planet. Sci. Lett.*, *65*, 233-244, 1983.

Slowey, N. C. and W. B. Curry, Enhanced ventilation of the North Atlantic subtropical gyre during the last glaciation, *Nature*, *358*, 665-668, 1992.

Slowey, N. C. and W. B. Curry, Glacial-interglacial differences in circulation and carbon cycling within the upper western North Atlantic, *Paleoceanography*, *10*, 715-732, 1995.

Stuiver, M., and T. F. Braziunas, Modeling atmospheric ^{14}C influences and ^{14}C ages of marine samples to 10,000 B.C., *Radiocarbon*, *35*, 137-189, 1993.

Vidal, L., L. Labeyrie, E. Cortijo, M. Arnold, J. C. Duplessy, E. Michel, S. Becque, and T. C. E. van Weering, Evidence for changes in the North Atlantic Deep Water linked to meltwater surges during Heinrich events, *Earth and Planetary Science Letters*, *146*, 13-27, 1997.

Vidal, L., L. Labeyrie, and T. C. E. van Weering, Benthic δ^{18}O records in the North Atlantic over the las glacial period (60-10kyr): Evidence for brine formation, *Paleoceanography*, *13*, 245-251, 1998.

Webb, R. S., S. J. Lehman, D. H. Rind, and R. J. Healy, Ocean heat transports and LGM cooling (abstract), *EOS Trans. AGU*, *75*(44), Fall Meet. Suppl., 381, 1994.

Webb, R. S., D. H. Rind, S. J. Lehmann, R. J. Healy, and D. Sigman, Influence of ocean heat transport on the climate of the last glacial maximum, *Nature*, *385*, 695-699, 1997.

Zahn, R., J. Schonfield, H.-R. Kudrass, M.-H. Park, H. Erlenkeuser, and P. Grootes, Thermohaline instability in the North Atlantic during meltwater events: Stable isotope and ice-rafted detritus records from core SO75-26KL, Portuguese margin, *Paleoceanography*, *12*, 696-710, 1997.

W. B. Curry,K. L. Laarkamp, T. M. Marchitto, J. F. McManus, D. W. Oppo, Department of Geology and Geophysics, Woods Hole Oceanographic Institution, 368 Woods Hole Rd., Woods Hole, MA 02543. (e-mail: wcurry@whoi.edu; klaarkamp@whoi.edu; tmarchitto@whoi.edu; jmcmanus@whoi.edu; doppo@whoi.edu)

Temporal Variability of the Surface and Deep Waters of the North West Atlantic Ocean at Orbital and Millenial Scales

Laurent Labeyrie[1], Heloïse Leclaire, Claire Waelbroeck, Elsa Cortijo, Jean-Claude Duplessy, Laurence Vidal[2], Mary Elliot, and Brigitte Le Coat

Laboratoire des Sciences du Climat et de l'Environnement, Unité mixte CEA-CNRS Domaine du CNRS 91198 Gif/Yvette cedex, France

Gérard Auffret

IFREMER, Technopole Brest-Iroise, 29280 Plouzané cedex, France

The millennial variability of the North Atlantic Current (NAC) and the deep NADW has been studied in Core CH 69-K09 in relation with the evolution of the Laurentide ice sheet, the Heinrich events (HE) and the orbitally driven changes of insolation. This core is located at the base of the Newfoundland margin, near the boundary between NAC and the Labrador current. Changes in sea surface temperature, *G. bulloides* and *N. pachyderma s.* $\delta^{18}O$ indicate that both the northward oceanic heat transport by NAC and the Polar Front oscillated at millennial frequency in parallel with the oscillations of the Northern ice sheets. The benthic foraminifera $\delta^{13}C$ record, used as proxy for deep North Atlantic water ventilation (4100m water depth), and thus the thermohaline conveyor belt, varied in phase with NAC. The millennial evolution exhibits analogies with the chaining of events occurring under the influence of insolation changes associated with the Earth precession variations. NAC transport of warm water to the Northern Atlantic Ocean goes on for several kyr, as high latitude polar water cool when summer solstice drift out of perihelie. Thermohaline circulation follows NAC activity as an efficient positive feedback. The large temperature gradients favored snow accumulation and fast growth of ice sheets. Cold phases initiate as ice sheets expend to the ocean and collapse pseudo-periodically, thus reducing

(1) Also at Département de Géologie, Université d'Orsay, 91405 Orsay cedex, France
(2) Now at Geowissenschaften, Bremen University, 28334 Bremen, Germany

Mechanisms of Global Climate Change at Millennial Time Scales
Geophysical Monograph 112

high latitude surface salinity and stopping deep water convection. Besides HE, icebergs invaded frequently the northern seas, with associated melting and salinity drop. These invasions occurred also during warm periods of ice sheets retreat and within the Holocene. Simultaneously the Polar Front shifted southward and deep Atlantic ventilation decreased.

1. INTRODUCTION

The millennial oscillations of climate recorded in the Greenland ice cores during the last glacial period [Dansgaard et al., 1984; Taylor et al., 1993; Grootes et al., 1993] have stirred a considerable amount of research in the paleo-ceanographic community. North Atlantic high latitude records of ice rafted detritus and the associated oscillations of the polar front have been correlated in details with the Greenland air temperature proxies [Bond et al., 1993; Elliot et al., 1998; Fronval et al., 1995]. The North Atlantic thermohaline circulation was affected by a similar variability during that period [Keigwin and Lehman, 1994; Labeyrie et al., 1995; Rasmussen et al., 1996; Vidal et al., 1997; Zahn et al., 1997]. Other records indicate nearly synchronous responses outside of the North Atlantic, such as the oxygen minimum intermediate waters of the eastern tropical Pacific [Behl and Kennett, 1996] and the northern Indian Ocean [Schulz et al., 1998]. But the causes and consequences of these climatic changes, the relationship with the evolution of the northern ice sheets, the dynamics of the atmosphere and the surface and deep oceans are still not clearly identified.

These interactive processes are better understood for the climate variability occurring at the Milankovitch time scale, when external forcings are known (the changes in temporal and mean zonal distribution of insolation). Ruddiman and Mc Intyre [1979, 1981] have discussed, in particular, the strong positive feedback of the delayed response of the North Atlantic thermohaline circulation during periods of Northern summer insolation decay, which allows rapid accumulation of snow over the growing ice sheets. Imbrie et al. [1992,1993], have precised in a conceptual model the links between the ice sheets evolution and the different modes of thermohaline circulation along the glacial-interglacial cycles. For the coupled ocean-atmosphere system, which operates on time scales of days to centuries, the background forcing of the ice sheets is a slow process. We feel therefore justify to compare the dynamic of the north Atlantic ocean and climate over the Milankovitch and millenial time scales, in the same core and with the same proxies.

We had the chance for that study to be able to sample a high sedimentation rate core from a key location of the NW Atlantic (CH 69-K09, 41°45'4 N, 47°21' W, 4100 m water depth), at the boundary between the warm North Atlantic current (NAC) and the polar Labrador current (Figure 1.1). Changes in planktic foraminifera associations and oxygen isotopic ratios monitor the North-South oscillations of tropical waters during warm periods and subpolar waters during cold periods. Simultaneously, the benthic foraminifera isotopic ratios (oxygen and carbon) monitor the hydrological changes in North Atlantic deep waters. The variability of the iceberg input is recorded by the amount of ice rafted detritus. Among those, detrital carbonate may be used to record ice shelves outputs from the Laurentide ice sheet [Bond et al., 1992], and volcanic detritus to track the output from the Icelandic ice [Bond and Lotti, 1995; Elliot et al., 1998]. Results suggest that the processes which drive the variability of the Northern Hemisphere climate and North Atlantic hydrology at Milankovitch time scales and those which drive it at millennial time scale, although different, operate through a similar system of retroactions.

2. MATERIAL, METHODS AND DATA

The core CH 69-K09 was recovered from an area of smooth topography, at the foot of the Newfoundland margin. The 3.5 khz profiles in the area show a very regular succession of alternating transparent and soft reflectors, with no evidence of slumping or turbidites [Labeyrie et al., 1996]. This site is swept by the lower North Atlantic Deep Water (NADW) contour current. The Laurentide ice sheet was not far to the north-west during the last glacial period, with icebergs flowing from the North transported by the Labrador current (Figure 1.2). This place is thus appropriate to study the temporal relationship between Laurentide ice sheet, northward surface heat and salt transfers, and evolution of the thermohaline activity. Pastouret et al. [1975] initially studied core CH 69-K09 and demonstrated, by a careful sedimentological and micro-paleontological study, that this 14.95 m core covers at least the last glacial-interglacial cycle. The record contains a succession of oscillations of the "Gulf Stream" (in fact the NAC [Rossby, 1996]), alternating with periods of increased input of "ice-rafted debris and relatively coarse turbidite-type beds", which correspond to the now recognized "Heinrich events" [Heinrich, 1988]. Dissolution is significant at some levels within the core (as indicated by large abundance of broken foraminifera shells), especially where diatom frustula are abundant. However, the core has remained permanently above the carbonate compensation depth [Labeyrie et al., 1996] and both benthic and planktic foraminifera are present throughout the entire core.

Initially, the working half of the core was divided within oriented 10 cm long sections and about 2 cm has been sampled at the top of each sections by Pastouret et al.

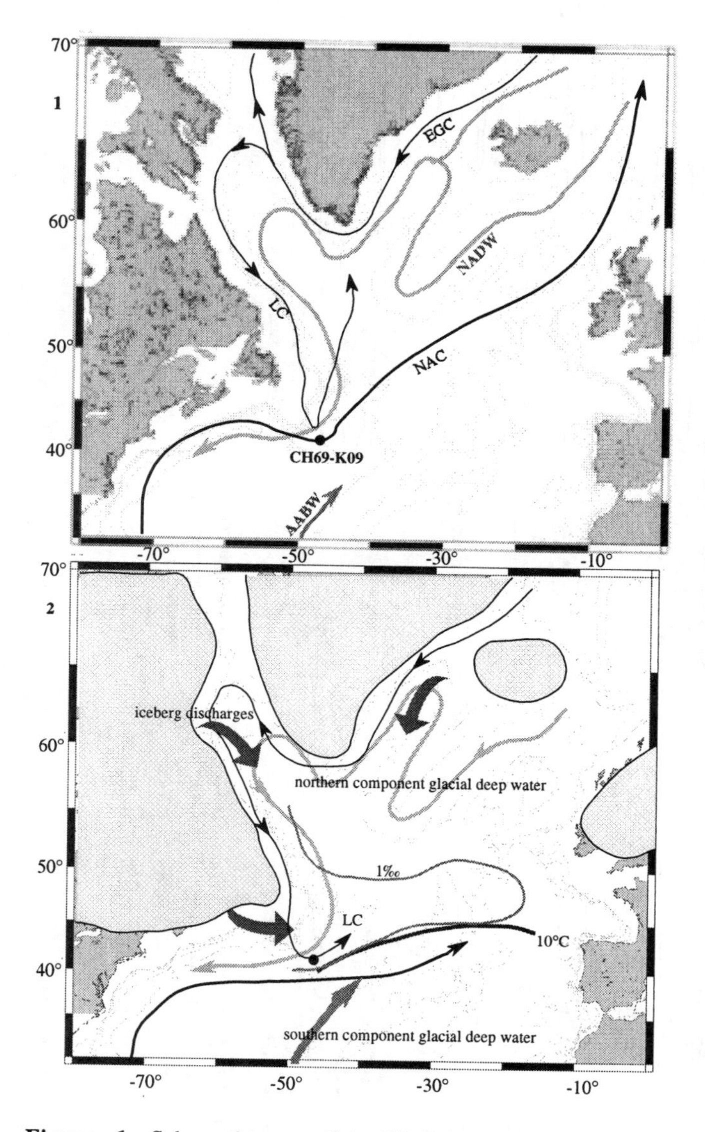

Figure 1. Schematic map of the North Atlantic ocean, with the location of sediment core CH 69-K09.
1.1 : modern setting, with indications of the main surface water currents (NAC: North Atlantic Current ; EGC East Greenland Current; LC: Labrador Current), North Atlantic deep water flow (NADW), and Antarctic Bottom Water flow (AABW). 1.2: Last glacial period, with approximate location of ice sheets (from CLIMAP [1981]), zones of iceberg discharges, and currents. The location of the 10°C summer surface sea water isotherm is extracted from the reconstruction of the Northern Atlantic hydrology during Heinrich event 4 (35 kyr BP) [Cortijo et al., 1997]. The 1‰ $\delta^{18}O$ isoline shows the zone of excess iceberg melting, for the same period) [Cortijo et al., 1997].

[1975]. Our study was done on the other part of each section and completed for few levels by the samples sieved by Pastouret. The present study is based on 346 different sediment levels sampled for micropaleontology and isotopic analysis.

2.1 Planktic foraminifera species distribution

Foraminifera were counted following the methodology and species classification of CLIMAP (except for the class p.d. intergrade, which was lumped with *Neogloboquadrina pachyderma* right coiling), with a minimum of 350 individual shells counted at each level.

Because core CH 69-K09 is located along the NAC path, its modern foraminiferal assemblage has the distinct feature of comprising species living in the warm waters of the NAC together with species living in the underlying colder waters. The core top contains about 10% of tropical and subtropical species (*Orbulina universa*, *Globigerinoides ruber* pink and white, *Globigerinoides sacculifer*), 60% of transitional / subarctic species (*Globigerina bulloides* and *N. pachyderma* right coiling) and 30% of arctic species (*Globigerina quinqueloba* and *N. pachyderma* left coiling). A very sharp hydrographic gradient separates the NAC from the cold Labrador current water which flows to the south along the Grand Banks, few hundred miles to the west, with numerous eddies transporting surface waters from one water mass to the other. As productivity is higher in the colder waters, the modern fauna is shifted towards transitional / subarctic associations. The Imbrie and Kipp method of SST reconstruction by statistical transfer functions [CLIMAP, 1981] is not adapted to such mixing of populations in places of large hydrographic gradients. Communality between the observed mixed populations and the factors defined from the core tops data base are low, especially during the Holocene. The Modern Analog Technique, or MAT [Prell, 1985] is not adapted either. The core-top SST estimated by MAT is too cold (by about 2°C in winter and 2.5°C in summer). As shown by Waelbroeck et al. [1998], this bias can be attributed to the fixed number of best analogs retained and lack of reference core tops located in the area of core CH 69-K09. We therefore used the revised analog method (RAM) to reconstruct SSTs in the present study. This method is a modified version of MAT in which an objective selection criterion of the best analogs, and the indirect approach are implemented [Waelbroeck et al., 1998].

The indirect approach consists in a remapping of the data in the environmental space before searching for analogs. The reference database used in the computations is a subset from the Atlantic database [Pflaumann et al., 1996] consisting of 615 core tops located between 5° and 80°N, and the corresponding winter Tc (cold) and summer Tw (warm) SSTs interpolated from the 1° x 1° WOA94 data set [Levitus, 1994]. The SSTs reconstructed by RAM for core CH 69-K09 surface samples (12.83 ± 0.04 °C in winter; 20.94 ± 0.03°C in summer) are equal to the atlas values. RAM's good reconstruction results from the use of the larger and more homogenous database obtained by interpolation and from the optimized selection of modern analogs.

Results are reported Figure 2-1 (Tw) and 2-2 (Tc). Dissimilarity coefficients are always lower than 0.13. For

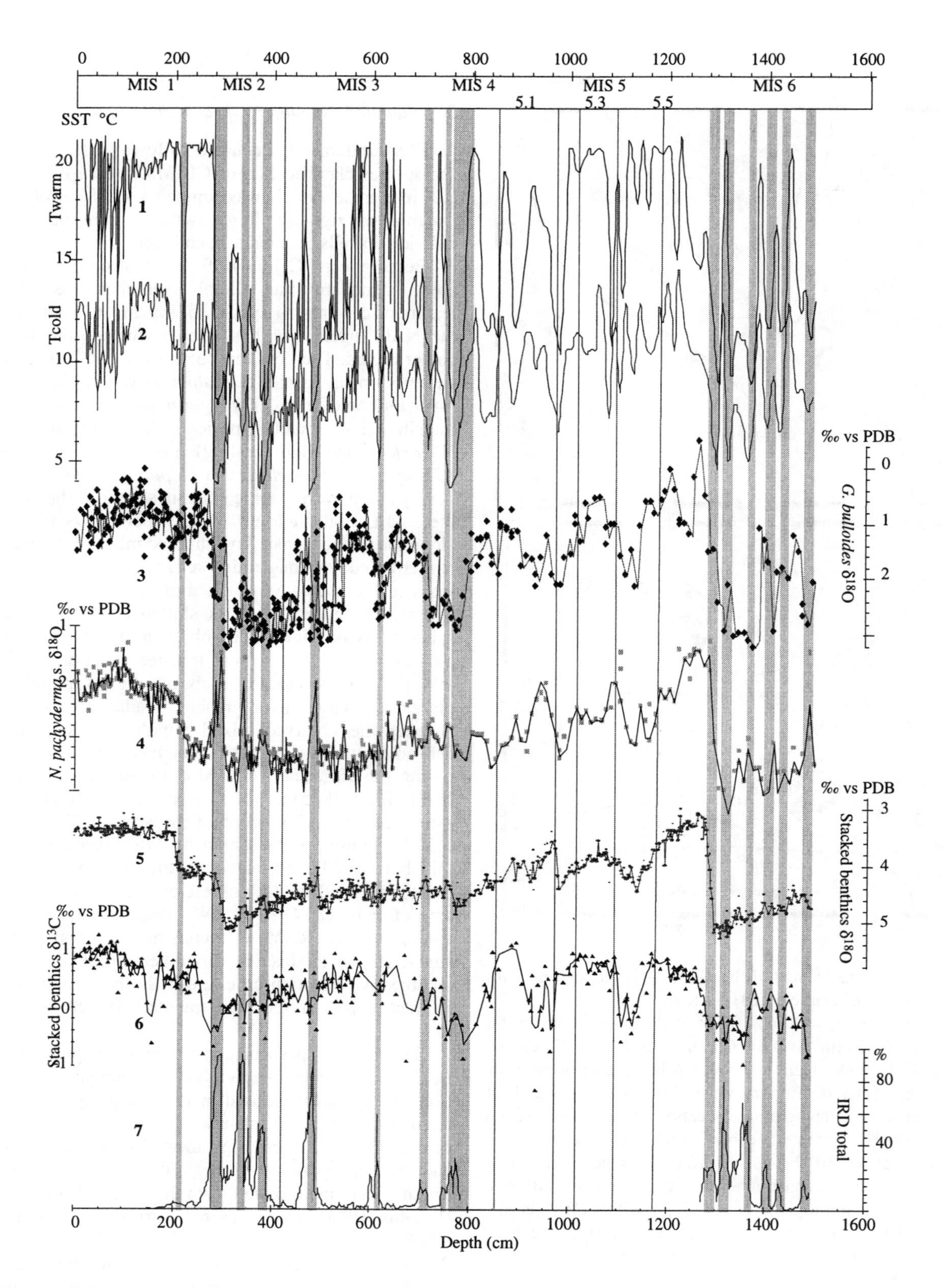

Figure 2. The main parameters analyzed in core CH 69-K09, expressed vs. depth in the core : 1- Sea Surface Temperature (SST) for summer (Twarm) ; 2- idem for winter (Tcold); 3- *G. bulloides* $\delta^{18}O$; 4- Neogloboquadrina pachyderma s. (N. pac. S.) $\delta^{18}O$; 5- Stacked benthic foraminifera $\delta^{18}O$ (with individual values and the 5 points running mean line); 6- Stacked benthic foraminifera $\delta^{13}C$ (with individual values and the 3 points running mean line); 7- proportion of ice rafted detritus (IRD) in the >150μm size fraction.

most of the warm periods (the 189 levels for which Tw values range between 12 and 20.2°C), standard deviations of Tc and Tw estimates are very large : up to 3.9 °C with a mean of 1.9°C for Tc and up to 5.7°C with a mean of 2.9°C for Tw. Standard deviations decrease to a mean value of 0.7°C for both Tw and Tc at lower temperatures. The periods of high standard deviations in the estimated SSTs are likely linked to the presence of the frontal zone in the area of the core. Conversely, the stability of the warmest SST around the value found at the core top (Tw about 21°C) would indicate that the present hydrological structure corresponds to the maximum northward penetration of the warm NAC/Gulf Stream system, but also that this Northern position was relatively stable in the past, as predicted by Rossby [1996].

The down-core variability is much higher for Tw than Tc. Tw total amplitude covers about 15°C, but the signal oscillates mainly (as an on/off system) between about 20°C and about 11°C, probably as the subtropical front moved north or south of the core location. Tc changes much more regularly, as the subtropical front was south of the core location during winter.

2.2 Isotopic studies

The oxygen isotopic ratio of foraminifera is a function of the growth temperature and the water isotopic ratio [Emiliani, 1966]. Because of their limited ecological range and the hydrological variability at the core location, we had to analyze different species of planktonic foraminifera: *N. pachyderma* left coiling was picked to monitor cold (< 7°C water) surface or subsurface waters (about 7 shells per sample), and *G. bulloides* (4 to 6 shells per sample) to monitor transitional and subtropical waters (about 7 to 18°C water) [Bé, 1971; Duplessy et al., 1991]. To limit ontogenic effects, the size range was limited to the fraction 250 to 315 μm wherever possible. Results are reported as $\delta^{18}O$ vs. PDB, in Figure 2-3 and 2-4. Values vs PDB are calculated in reference to the standard calibration of NBS 19 [Coplen, 1988], with additional corrections for source non-linearity of the MAT 251 mass spectrometer.

The isotopic records are typical of the North Atlantic, with well defined glacial/interglacial changes. However, several systematic differences appear between the records. They are attributed to the different ecological ranges of these species. The *G. bulloides* $\delta^{18}O$ and Tc signals exhibit strong analogies. Within the temperature range of transitional /subtropical waters (between about 7°C and 18°C), *G. bulloides* lives in its optimal ecological environment and its $\delta^{18}O$ is well adapted to record surface water temperature variations [Duplessy et al., 1991]. By contrast, the tropical summer SSTs reconstructed in many portions of the core, including the core-top, are not imprinted in the *G. bulloides* $\delta^{18}O$ record. In fact, the temperature of deposition of calcite derived from the core-top *G. bulloides* $\delta^{18}O$ value is 12.6°C, whereas the measured modern summer SST is 20.9°C [Levitus, 1994]. This species tends thus to live in cooler deeper waters or to develop earlier in the year. Temperature and salinity conditions, consistent with the *G. bulloides* $\delta^{18}O$ measured in the core-top sample are encountered in summer between 50 and 100 m water depth (temperature ≈ 13°-14°C and salinity ≈ 35.5), or at the surface during winter (SST around 12.8°C, [Levitus, 1994]).

The *N. pachyderma* s. $\delta^{18}O$ record follows less regularly the SST signal. At the location of core CH 69-K09, *N. pachyderma* s. is an exotic cold species, linked to winter (or thermocline) injections of Labrador polar waters. Its growth temperature (calculated from core top shell $\delta^{18}O$) correspond to about 6-8°C, about 4-6°C colder that Tc. This species, which is limited to polar waters does not even reflect mean winter conditions at the core site, but the advection of colder waters brought by the Labrador Current.

Four different species of benthic foraminifera were sampled, as none of the species could be sampled all along. *Cibicides wuellerstorfi* (321 levels) and *Nuttalides umbonifera* (129 levels) are common during the Interglacials (Marine Isotopic Stages MIS 1 and 5) and during the warm parts of MIS 3. *Uvigerina peregrina* (284 levels) are more frequent during the glacial periods. This species is generally considered in the deep Atlantic ocean as better adapted to an environment corresponding to lower NADW activity [Streeter and Shackleton, 1979]. *Melonis barleanum* (analyzed at 249 levels) is relatively abundant during transition periods. Most of the sedimentary levels were analyzed in duplicates, or more, with one or several independently picked species and 2 to 5 shells per sample.

A stacked benthic $\delta^{18}O$ record was obtained by merging all benthic foraminifera isotopic values after correction for specific isotope fractionation ($\delta^{18}O$ +0.64‰ for *C. wuellerstorfi,* 0 for *U. peregrina,* +0.40 for *M. barleanum* and +0.50 ‰ for *N. umbonifera*) (Figure 2-5). MIS 1 to 6 have been defined based on this record, which is very similar to that of the reference deep Pacific ocean signal from core V19-30 [Shackleton et al., 1983]. The mean $\delta^{18}O$ difference between cores CH 69-K09 and V19-30 is -0.1‰ (±0.18‰ at one sigma). This small difference implies that the deep Atlantic (below 4000 m) was fed during most of the last 180 kyr by water with similar physical properties that the deep Pacific Ocean.

The *C. wuellerstorfi* $\delta^{13}C$ is classically considered a good proxy for active North Atlantic deep water ventilation [Boyle, 1986; Boyle and Keigwin, 1982; Duplessy et al., 1988; Duplessy et al., 1984]. Progressive addition of CO_2 derived from oxidized organic matter ($\delta^{13}C$ around -25‰) decreases the $\delta^{13}C$ of the total dissolved CO_2 in proportion to the consumption of dissolved oxygen [Kroopnick, 1984]. However, *C. wuellerstorfi* disappears at levels which are important for monitoring glacial deep water variability. We have tested the other benthic species, analyzed in the course of our study, to fill at least in part these gaps. *U. peregrina*

$\delta^{13}C$ has been taken in the Pacific Ocean as proxy for deep water ventilation [Shackleton et al., 1983], but not in the Atlantic ocean. Zahn et al. [1986] have demonstrated that, because of its infaunal habitat, the $\delta^{13}C$ of this species is affected by the low $\delta^{13}C$ of the intersticial water. The mean difference $\delta^{13}C$ *C. wuellerstorfi-U. peregrina,* for the 216 analyses done at the same levels, is 0.99‰ ± 0.33. The same measurements done for 170 $\delta^{13}C$ *C. wuellerstorfi-M. barleanum* pairs gives 1.09 ± 0.60, confirming that *M. barleanum,* because of its infaunal habitat and strong dependence on high fluxes of organic matter to the sediment [Thomas et al., 1995], cannot be used for dissolved CO_2 $\delta^{13}C$ paleoreconstruction. *N. umbonifera – C. wuellerstorfi* pairs from the same levels differ only by 0.44‰, with a mean standard deviation of ±0.16‰, the same standard deviation that for paired *C. wuellerstorfi* analyses. We consider therefore that *N. umbonifera* may also be used as proxy for deep water ventilation.

The record with stacked $\delta^{13}C$ analyses of *C. wuellerstorfi* and *N. umbonifera* (corrected by +0.44‰) is reported in Figure 2-6. The signal is relatively noisy, with a mean absolute difference of 0.28‰ for two successive samples, to be compared to 0.14‰ for the equivalent $\delta^{18}O$ pairs. However, the record shows the typical trends observed in other deep Atlantic cores [Labeyrie and Duplessy, 1985; Mix and Fairbanks, 1985], with glacial values (MIS 2, 4, 5.2, 5.4 and 6) around -0.5 to 0‰, and interglacial values around +1‰. The large amplitude changes are explained by the replacement of well-ventilated NADW by less ventilated, nutrient-rich deep water of southern origin during each cold period [Boyle and Keigwin, 1985; Duplessy et al., 1988]. As the core is located only few hundred meters above the modern transition zone between NADW and AABW [Levitus, 1994], the high frequency signal may also record short term variability in North Atlantic deep water activity.

2.3 Ice Rafted Detritus and Heinrich events

We follow Heinrich [1988] and subsequent studies in using the amount of ice-rafted coarse grain detritus (IRD) to identify the periods of major ice sheet collapse. Pastouret et al. [1975] already observed a succession of levels with large amount of coarse grain material in core CH 69-K09, but attributed them to ice-triggered "turbidite deposits". To separate IRD from wind blown material, we counted IRD only as grains over 150µm size. We therefore do not consider the larger amount of fine silt and clay which is also transported by icebergs. In the present study, IRD were quantitatively estimated only for the glacial periods, between 150 and 780 cm, and below 1270 cm down-core. Counts are reported as grains/g dry sediment and as proportion of IRD to the total number of elements larger than 15µm (Figure 2-f).

Heinrich events (HE) 1 to 6 are easily identified for the last glacial period, as large IRD peaks rich in detrital carbonate [Bond et al. 1992]. At least 3 similar events are also present during MIS 6. Smaller IRD increases (mostly characterized by volcanic residues) are apparent between HE's, and during the cold event of the Younger Dryas, at 210 cm. Gray bands identify in Figure 2 the corresponding depths across the other records. They all correspond to large drops in SST and *N. pachyderma* s. $\delta^{18}O$, as described in numerous high resolution northern Atlantic records [Bond et al., 1993; Bond et al., 1992; Cortijo et al., 1997; Labeyrie et al., 1995]. Small decreases in the benthic foraminifera $\delta^{18}O$ signal (by about 0.2 ‰) may also be evidenced during most of the HE. This result confirms the observations of Rasmussen et al. [1996] and Vidal et al. [1998], done at shallower water depths, that the freshwater with low $\delta^{18}O$ produced by the melting icebergs during the HE was mixed rapidly in the whole ocean. Low benthic foraminifera $\delta^{13}C$ values are also observed during most of the HE and smaller events, as observed by Sarnthein et al. [1994] and Vidal et al.[1997] between 2000 and 3500 m water depth. This implies that HE resulted in the invasion to shallower depths of the poorly ventilated deep water of southern origin which was usually present at the core location during the glacial period, in agreement with model simulations [Fichefet et al., 1994].

2.4 AMS ^{14}C dating and the time scale of core CH 69-K09

23 different planktic foraminifera samples were dated on the AMS ^{14}C Tandetron of Gif-sur-Yvette [Duplessy et al., 1986]. Following the strategy proposed by Bard [1987] to limit the effects of bioturbation, dating was done on monospecific samples at the levels of maximum relative abundance. Four different foraminifera species were analyzed : *N. pachyderma* s., *N. pachyderma* d., *G. bulloides*, and *G. inflata*. Three different samples from the same sediment depth (454 cm) were analyzed to test external reproducibility at ages around 31 kyrBP. The two *G. bulloides* samples have dates 0.5 kyr apart, which is acceptable, taking into account the mean error at one sigma for the ages obtained in that range, about 0.3 to 0.4 kyr Their ages are however significantly younger (by about 1.5 kyr) than the age of *N. pachyderma* d, sampled at the same depth or 0.6 kyr younger that *N. pachyderma* s sampled 24 cm above (see Table 1). Such difference may not be attributed to bioturbation mixing. Both species probably develop, during that period, within waters with very different ^{14}C ventilation exchange with the atmosphere. The more polar species *N. pachyderma* would correspond to the older ventilation age. The modern range of ventilation ages for surface waters exceeds 600 yr [Bard, 1988]. Our result would indicate that during periods of excess input of fresh water at the surface and increased seasonal sea-ice, stratification and ventilation age of the surface and subsurface high latitude waters could be significantly increased. Bard [1994] evidenced a similar trend during the Younger Dryas, with a larger that 1 kyr difference between the subpolar and polar surface waters. However, We have not enough data to correct the ^{14}C ages

Table 1. Construction of the age scale: Results of the AMS ^{14}C dating , with the depth of sampling (cm), the species analysed (*Neogloboquadrina pachyderma d., N. pachyderma s., Globigerina inflata, Globigerina bulloides*), the standard measured age (kyr), the corrected age for the "reservoir effect", the error at 1 sigma, the calendar age, using the Stuiver et al. [1995] correction and the Laj et al. [1996] correction.

	AMS ^{14}C datings				Calendar ages	
Depth	species	Age	Age corr.	Error (1σ)	Correction	
cm		ka	Res. -0.4ka	ka	Stuiver	Laj
0	*N. pachyderma* d.	1.15	0.75	0.06	0.67	-
20	*N. pachyderma* d.	2.24	1.84	0.04	1.74	-
33	*G.inflata*	3.13	2.73	0.045	2.79	-
52	*G.bulloides*	4.77	4.37	0.07	4.87	-
85	*G.bulloides*	6.66	6.26	0.05	7.18	-
130	*G.inflata*	8.84	8.44	0.06	9.44	-
175	*G.inflata*	9.69	9.29	0.07	10.25	-
215	*G.inflata*	11.14	10.74	0.07	12.67	-
230	*G.inflata*	12.33	11.93	0.08	13.91	-
280	*N. pachyderma* s.	13.63	13.23	0.11	15.79	-
300	*N. pachyderma* s.	14.97	14.57	0.12	17.45	-
320	*G.bulloides*	18.02	17.62	0.14	20.99	-
	interpolated		18		21.48	21.48
365	*N. pachyderma* s.	23.01	22.61	0.21	-	26.48
370	*N. pachyderma* s.	24.81	24.41	0.23	-	28.44
390	*N. pachyderma* s.	28.01	27.61	0.32	-	31.71
430	*N. pachyderma* s.	31.56	31.16	0.41	-	34.85
454	*N. pachyderma* d.	32.13	31.73	0.38	-	35.39
454	*G.bulloides*	30.65	30.25	0.29	excluded	-
454	*G.bulloides*	31.14	30.74	0.37	excluded	-
475	*N. pachyderma* s.	33.58	33.18	0.46	-	36.84
490	*N. pachyderma* s.	34.81	34.41	0.56	excluded	-
490	*G.bulloides*	34.94	34.54	0.55	-	38.20
510	*N. pachyderma* s.	39.09	38.69	0.81	-	42.11
535	*N. pachyderma* s.	42.4	42	1.1	-	44.46
618	*N. pachyderma* s.	46.9	46.5	1.9	-	47.96
1215	*G.bulloides*	53.6	53.2		background	-
1270	*N. pachyderma* d.	49.6	49.2	2.5	background	-

for such changes in surface water ventilation age. Thus, we used a constant ventilation age model to construct the age scale : all the measurements were corrected by 0.4 kyr, knowing that the correction may be underestimated during cold periods by as much as 1 kyr [Bard, 1988].

To facilitate comparison with the Greenland isotopic records, the ages have been converted to the calendar scale using Stuiver and Reimer [1993] and Bard [1993] calibration until 18 kyr BP ^{14}C years. Beyond, we assumed that the first order correction to the ^{14}C dates was linked to the changes in ^{14}C production which derived from the past evolution of the Earth magnetic field. We applied the Laj et al. [1996] model to estimate that correction. For ages older that 40 kyr BP, we used the SPECMAP age model for the $\delta^{18}O$ record [Martinson et al., 1987]. The resulting age model is given in Table 2 and Figure 3. All correlations and age calculations were done using the Analyseries software of Paillard et al. [1996]. The bottom of the core (older that 150 kyr BP) is poorly constrained in ages, and the corresponding results will not be discussed.

The ^{14}C stratigraphy and resulting dated proxy records compare well with other published records from the North Atlantic for the last 25 kyr BP [Bard and Broecker, 1992; Bond et al., 1993; Sarnthein et al., 1994], especially if we take into consideration the changes in surface ventilation age [Bard et al., 1994]. Comparison is more difficult for older periods. While H4 is dated between 32 and 35 kyr BP (^{14}C age, i.e. 36-39 kyr in the calendar scale) [Cortijo et al., 1997; Elliot et al., 1998; Zahn et al., 1997], we obtained a mean ^{14}C age of 33.2 kyr BP for that event (36.8 kyr calendar). A direct correlation has been done between the Tc record of core CH 69-K09 and the GISP 2 $\delta^{18}O$ record

[Grootes and Stuiver, 1997; Grootes et al., 1993; Stuiver et al., 1995]. The cold event which corresponds to H4 in the marine records, (between interstadials 8 and 9) is dated around 39 kyr BP in GISP2, thus about 2 kyr older (Figure 3). Both age models are similar (within the uncertainties of correlation) for ages younger than 30 kyr BP. By contrast, the difference increases between 45 and 70 kyr BP, a period with no independent age control for the marine record.

3. DISCUSSION

The aim of our paper is to study the variability of the North West Atlantic ocean and its connections with global climate, ice sheets dynamic and ocean circulation, at millennial time scale, in comparison with its reactions to the changes in insolation in the precessionnal band (20 kyr periodicity). The discussion is based on the analogies between the evolution of the different forcing functions (as insolation) and responses of part of the climatic system and rests on the usually untold assumptions that : (1) the climatic system is so complex that the degree of similarity between different signals should directly correspond to their degree of linkage; (2) the different components of the climatic system evolve at their own internal pace driven by the distribution of energy and mass (hours to days for the atmosphere, months to hundreds of years for the ocean, millennia to 10^5 years for ice sheets, atmospheric pCO_2 and insolation changes). The slower reacting components act as "external forcing" for the faster components. The only real external forcing for the climate system derives from the orbital changes of insolation distribution. We neglect, for lack of information, all other factors which modulate the solar flux to the Earth surface.

During the last several hundred thousand years, period of oscillation of the large ice sheets, most of the paleoclimatic information in ice and sediment records (foraminiferal $\delta^{18}O$, SST, atmospheric pCO_2, high latitude air temperature, sea level) have a typical common first order signature which is modulated by the 100 kyr periodicity. Correlation coefficients between these signals are high whatever the interrelationships, because these parameters respond first to the evolution of the large continental ice sheets [Imbrie et al., 1993]. This is one of the reasons why we did not consider here the changes in atmospheric pCO_2. They have strong analogies with the changes in ice volume, and their respective signatures in the proxy records cannot be easily distinguished within the time scale of our study.

The 20 kyr^{-1} frequency band of climate variability is one of the easiest to study, as changes in insolation distribution are the overwhelming forcing factor, and most paleo-proxies present a near linear response in that band [Imbrie et al., 1993; Imbrie et al., 1992].

Proxy records of core CH 69-K09 present both high and low frequency variability. For simplicity, we shall discuss independently the variability which results directly from the astronomical insolation changes in the precessionnal band and the variability corresponding to the frequency band 1 kyr^{-1} to 1/5 kyr^{-1}, generally associated with oscillations within the interactive ocean-ice-atmosphere system. Sampling resolution is not sufficient to explore higher frequency bands (centennial variability). We will not discuss the 1/100 kyr^{-1} "excentricity" band either, as the records are too short (180 kyr coverage, thus less than 2 periods). We used both the temporal records and the power distribution of the variability over the frequency domain (calculated using the Blackman-Tuckey, cross Blackman-Tuckey and Multitaper Method (MTM) spectral methods as included in the Analyseries software [Paillard et al., 1996].

Table 2. Final list of constraints between depth (cm) and calendar age (kyr).

Age Scale	
Depth cm	Calendar ky.
0	0.67
20	1.74
33	2.79
52	4.87
85	7.18
130	9.44
175	10.25
215	12.67
230	13.91
280	15.79
300	17.45
320	20.99
365	26.48
370	28.44
390	31.70
400	32.50
430	34.84
454	35.39
475	36.84
490	38.21
510	42.10
530	43.99
620	48.11
700	54.67
750	57.60
970	84.13
1035	96.38
1115	107.55
1200	118.69
1280	126.58
1295	135.10
1338	141.33
1410	162.79

3.1. The long term variability and the precession band

The precession band is the main insolation forcing band on long term changes of climate. The characteristic variability around 1/21 kyr^{-1} is recognized in most paleo-climatic records, and is generally linked to low latitude processes. At high latitudes, changes in solar obliquity play a large role in the modulation of the incoming solar energy [Imbrie et al., 1993; Imbrie et al., 1992]. The different proxy records are reported versus age in Figure 4, after least square smoothing (5 points running) and sampling each kyr, to filter out the high frequencies which will be studied separately. The inter-relations between the insolation and each temporal set has been studied by cross Blackman-Tuckey analysis [Jenkins and Watts, 1968]. Results are reported Table 3. The spectral responses of the characteristic proxies are reported in Figure 5 and the filtered responses in the precession band are reported in Figure 6.

3.1.1. Surface water proxies

Sub-tropical/transitional waters and the NAC. The variability of the summer and winter SST signals is relatively small within the precession band (10% of the integrated variance for Tc, 8% for Tw), compared to the 1/100 and 1/41 ky^{-1} , which indicates a large influence of the changes in ice volume at that location [Imbrie et al., 1992]. The warmest conditions (for Tc and Tw) lag the minimum ice volume (benthic foraminifera $\delta^{18}O$) by 6 kyr and the June perihelion by 10-12.7 kyr (Table 2).Thus, warmest conditions appear in phase with minimum northern hemisphere summer insolation, or maximum winter insolation. Our results follow Ruddiman and McIntyre [1981] observations from a set of cores from the central North Atlantic. They attributed this apparent inverse relationship between summer insolation and sea surface temperature to the low latitude insolation forcing of the heat accumulation in the intertropical eastern Atlantic and the associated wind driven transport and forcing on Gulf Stream. Villanueva et al. [1998] have observed a similar phase for a precessionnally driven change in open ocean productivity recorded in core SU 90-08, eastwards of core CH 69-K09. They attribute the observed productivity changes to an increased wind driven mixing of the upper thermocline. These observations are in agreement with the hypothesis that winter storms play a major role for transferring the latent heat accumulated in the western equatorial Atlantic to the high latitude ice sheets [Ruddiman and McIntyre, 1981; McIntyre and Molfino, 1996]. The phase relationship would imply that a large part of the solar heat transfer to the oceans which occurs in the Southern Hemisphere in summer is transferred to the Northern Hemisphere through the Equatorial trade wind driven circulation.

The planktic foraminifera $\delta^{18}O$ records (Figures 4.4 and 6.1) integrate the evolution of the global sea water $\delta^{18}O$, growth temperature (an SST increase by 4°C corresponds to a shell $\delta^{18}O$ decrease by 1‰ [Emiliani, 1966; Duplessy et al., 1991]) and spatial distribution of salinity. In the modern North Atlantic Ocean, 1‰ increase in sea surface $\delta^{18}O$ corresponds approximately to a 2‰ increase in salinity [Craig and Gordon, 1965]. Warmer waters are also saltier (the 12°C SST gradient between tropical and transitional surface waters corresponds approximately to a 1‰ salinity gradient). We did not try to separate SST and salinity in the $\delta^{18}O$ record [Duplessy et al. 1991], because likely the growth season would have extended toward summer during glacial periods and we have little constraint on growth temperature to estimate salinity.

The *N. pachyderma* s. $\delta^{18}O$ signal leads by more that 5 kyr the *G. bulloides* signal, during the large oscillations of MIS 5 and 1 (Figure 4.4). Differences in the $\delta^{18}O$ records of both species have already been described and linked to their respective ecological limitations [Duplessy et al., 1992]. For instance, the higher sensitivity of the *N. pachyderma* s. $\delta^{18}O$ to the Heinrich meltwater events is attributed to its polar surface water habitat [Labeyrie et al., 1995]. In fact, the *N. pachyderma* s. $\delta^{18}O$ record exhibits strong analogies with the benthic $\delta^{18}O$ signal, but with a different dynamic, being lighter during interglacials and the warm oscillations of stage 5.1 and stage 5.3. *N. pachyderma* s. $\delta^{18}O$ decreased faster and increased earlier than the benthic $\delta^{18}O$ at most oscillations.

Results of the spectral analysis of the $\delta^{18}O$ data series, using the cross Blackman-Tuckey method [Jenkins and Watts, 1968] against insolation, are reported Table 3. In these analyses, following Imbrie et al. [1992], the $\delta^{18}O$ are multiplied by (-1) as proxy for ice volume and ocean temperature. Both *N. pachyderma* s. $\delta^{18}O$ and benthic $\delta^{18}O$ lag June perihelion by about 4 to 4.5 kyr. This lag, similar to the one calculated by Imbrie and co-authors [1992], has been attributed by them to the delay of continental ice sheet waxing and waning in response to insolation changes. However, the *N. pachyderma s.* $\delta^{18}O$ record does not show the early surface water temperature response predicted by Imbrie et al. [1992] for high latitude polar waters and observed by Cortijo et al. [1994, 1999] in Norwegian Sea at 75°N. The *G. bulloides* $\delta^{18}O$ signal, which should represent better the evolution of the sub-tropical waters does not present in the precessional band a variability with significant coherence with insolation, but the corresponding filtered signal is lagging the *N. pachyderma s.* and benthic $\delta^{18}O$ signals by about 5 kyrs (Figure 6) in agreement with the evolution observed during MIS 5. It presents thus a lag similar to SST when compared to insolation.

3.1.2. Deep water proxy : the benthic foraminifera $\delta^{13}C$ record

The water depth of core CH 69K-09 (4100 m) corresponds at present to the deep NADW. During glacial periods, NADW is replaced by poorly ventilated water

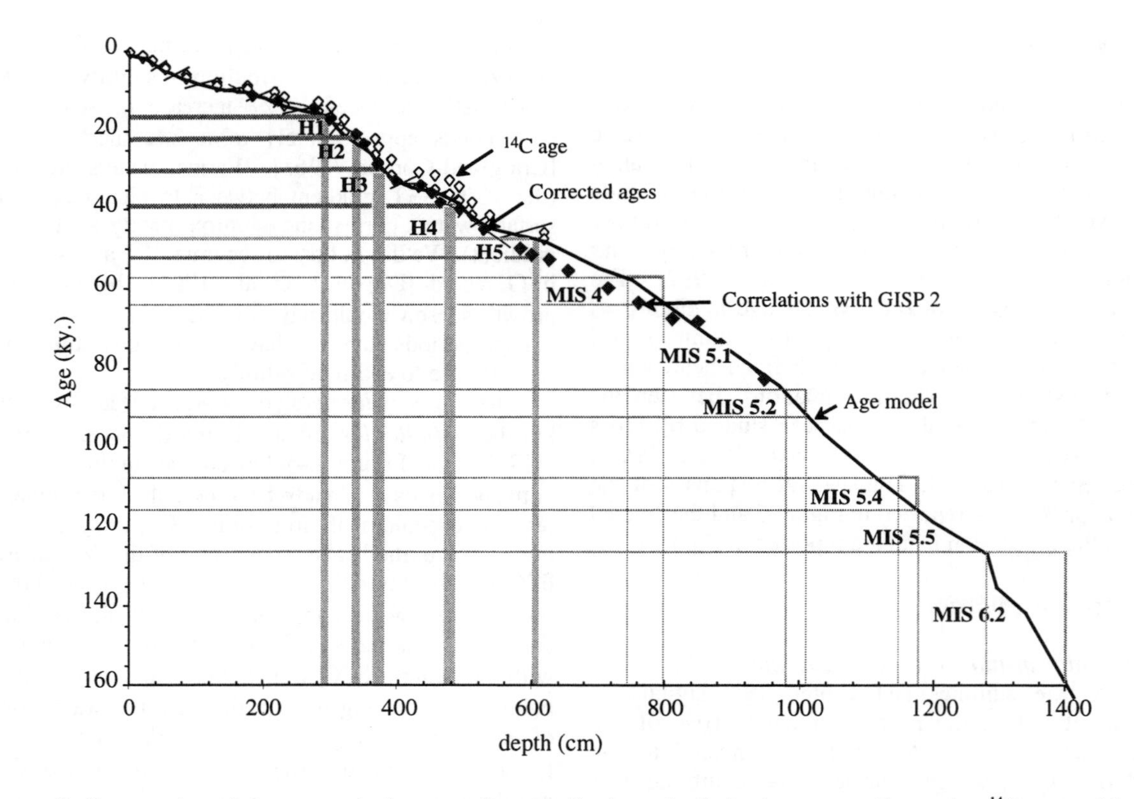

Figure 3. Construction of the age scale (expressed as age (kyr) vs. depth (cm). - empty diamonds : ^{14}C ages; -black diamonds ^{14}C ages corrected as calendar ages; grey diamonds : points of correlation with GISP2; black line : age model.

originating from the Southern Hemisphere [Duplessy et al., 1988]. The benthic foraminifera $\delta^{13}C$ record shows interesting similarities with Tw (Figure 4). Cross spectral analysis (Blackman-Tuckey) shows high coherence with insignificant phasing for the insolation forcing bands. We obtain the same results using $\delta^{13}C$, obtained after subtraction from the CH 69-K09 benthic $\delta^{13}C$ signal the mean oceanic $\delta^{13}C$ changes [Raymo et al. 1990]. For that calculation we follow Imbrie et al. [1992], using the benthic $\delta^{13}C$ record of Pacific core V 19-30 [Shackleton and Pisias 1985] as reference for global ocean $\delta^{13}C$. Our results differ from those of Imbrie et al. [1992]. Their $\delta^{13}C$ signal follows ice volume with a lag of about 2.5 kyr However, they used as proxy for NADW ventilation the benthic record from ODP 607, at 41°N and 3427 m depth [Ruddiman et al., 1989]. This record presents a rather low resolution, and does not sample adequately the $\delta^{13}C$ variability of the deep North Atlantic waters. For core CH 69-K09, a well define zero phase (±1 kyr) and high coherence (above 0.7) are apparent for both the obliquity and precession bands in the cross Blakman-Tuckey analyses between benthic $\delta^{13}C$ or $\delta^{13}C$ and Tc or Tw. Thus December perihelion (heat accumulation in the Gulf of Mexico?), North West Atlantic SST (ocean heat transport through the Gulf Stream and NAC?) and deep water ventilation at 4100 m (activity of NADW?) covary in phase. We may also consider that the 5 kyr lag of these parameters with ice volume (represented by the benthic $\delta^{18}O$) corresponds in fact to an evolution of the intensity of the thermohaline conveyor belt approximately in phase with the derivative of the benthic $\delta^{18}O$, ie. the growth and decay rate of the ice sheet) at least in the precession band.

We are inclined to see a cause and effect relationship in this narrow correlation, with maximum accumulation of heat and evaporation at low latitudes in phase with maximum transport of salt by surface circulation and vapor by the atmosphere during the periods of minimum summer insolation, thus the most favorable for maximum ice sheet growth. These observations are in agreement with the hypothesis of Ruddiman and McIntyre [1979] that surface waters maintained high temperatures during ice-sheet growth. We propose that it is the whole thermohaline conveyor belt circulation which acts as a strong positive feedback for ice sheet growth.

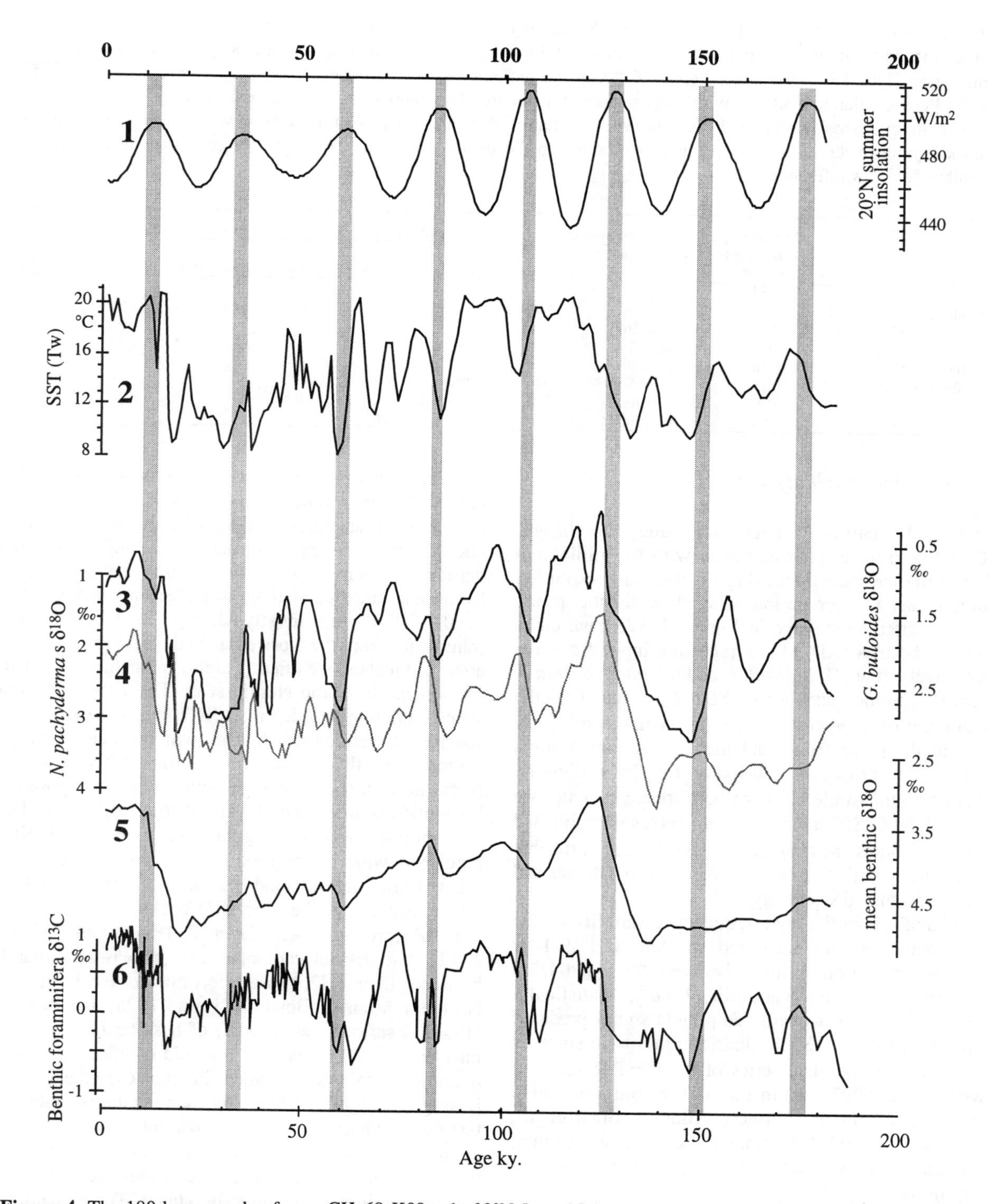

Figure 4. The 180 kyr records of core CH 69-K09 : 1- 20°N June 15 insolation [Berger, 1978](W/m^2); 2-Summer sea surface temperature (SST Tw); 3- *G. bulloides* δ^{18}O; 4- *N. pachyderma s.* δ^{18}O; 5- stacked benthic foraminifera δ^{18}O ; 6-stacked benthic foraminifera δ^{13}C. Grey bands correspond to the periods of maximum summer insolation modulated by Earth precession.

Table 3. Results of the Spectral analysis (Blackman-Tuckey with band width 0.016, frequency domain 0 to 0.2 ky^{-1}). In the first column : proportion of variance within the precession band (0.048 ± 0.008 ky^{-1}) respectively for insolation (June 15, 65°N), stack benthic $\delta^{18}O$, for *G. bulloides* $\delta^{18}O$, *N. pachyderma s.* $\delta^{18}O$, stack benthic $\delta^{13}C$, $\delta^{13}C$ Pacific-Atlantic gradient, Winter sea surface temperature (Tc), Summer sea surface temperature (Tw). In the two next columns, cross Blackman-Tuckey between insolation and each variable; following two columns idem with the stack benthic $\delta^{18}O$ as reference (analysis limited to the benthic $\delta^{13}C$ and Tw); last two columns, idem with the benthic ?$\delta^{13}C$ as reference.

	%variance spectral band 0.048±0.008 $ky.^{-1}$	vs. Insolation (June 15, 65°N)		vs. benthic $\delta^{18}O$		vs. benthic $\delta^{13}C$		vs. Δbenthic $\delta^{13}C$	
		Coh.	phase (ky.)	Coh.	phase (ky.)	Coh.	phase (ky.)	Coh.	phase (ky.)
Insolation	0.57	-	-	-	-	-	-	-	-
Benthics $\delta^{18}O$(*-1)	0.1	0.74	4.2	-	-	-	-	-	-
G.Bull $\delta^{18}O$(*-1)	0.1	0.36	-	-	-	-	-	-	-
N.pach $\delta^{18}O$(*-1)	0.1	0.7	4.4	-	-	-	-	-	-
Benthics $\delta^{13}C$	0.18	0.6	11.3	-	-	-	-	-	-
Δbenthic $\delta^{13}C$	0.26	0.8	13.8	0.63	6.4	-	-	-	-
Tc	0.1	0.5 (>0.55)	11.4	-	-	0.82	0	0.77	0
Tw	0.08	0.5 (>0.55)	12.7	0.56	5.8	-	-	0.73	1

3.2. The millennial variability

Our aim in that part of the paper is to study the changes in NAC and deep water proxies linked with the millennial evolution of the ice sheets, focusing on the high resolution study of the last 60 kyr period. The characteristic proxy signals are reported versus age in Figure 7. We have taken into account, besides IRD which traces the input from the main ice sheets, the *G. bulloides* and *N. pachyderma s.* $\delta^{18}O$ records, summer and winter SST (Tw and Tc), the relative amount of *N. pachyderma* s., as an indicator of the proximity of the polar front, and the benthic foraminifera $\delta^{18}O$ and $\delta^{13}C$ records. As a reference for the millennial variability of high latitude northern hemisphere climate, we have reported the GISP2 ice $\delta^{18}O$ record versus its published time scale [Grootes and Stuiver, 1997], and probable correlative events with the ocean records. The Glacial and Holocene are discussed separately.

Spectral analysis of the high frequency variability of the different climatic proxies measured in core CH 69-K09 exhibits an accumulation of power between 1/1.1 and 1/2.5 ky^{-1} (Figure 5). This high variability band is statistically significant with a F-Test on the amplitude signal produced by the MTM analyses. Such distribution of variance is observed in the geochemical series of the GISP 2 ice core [Mayewski et al., 1997] and in the high resolution marine sediment proxies from the northern Atlantic [Bond et al., 1997; Cortijo et al., 1995]. We may therefore consider that, at least statistically, we adequately sampled the millennial variability in the core.

3.2.1. The glacial world

The IRD record. All the Heinrich events are clearly defined, with a large increase in IRD. Peaks of detrital carbonate are observed for each HE, including H3, but also, as a more or less continuous background, during the Last Glacial Maximum, between 30 and 15 kyr BP. Other events (Younger Dryas, before and after H2, between H5 and H6 at about 55 kyr) are apparent, but with a small contribution of detrital carbonates. This indicates that icebergs originating from Laurentide ice shelves were melting in the area [Bond et al., 1992], but mostly during the HE. Basaltic IRD transported from the Norwegian and Greenland seas volcanic areas and Iceland are often present, indicating a contribution of icebergs from the Nordic seas. The Fenno-Scandian ice sheet probably provided also IRD, but we did not analyse specific mineralogical tracers which could identify that source. We will consider the basaltic IRD as a proxy for a North-Eastern (NE) icebergs source, to be opposed to the Laurentide source traced by the detrital carbonate. That NE contribution does not appear, in core CH 69-K09, as regular as what is observed in the Northern Atlantic nearer the volcanic areas [Bond and Lotti, 1995; Elliot et al., 1998]. Significant basaltic IRD peaks are found around 55 kyr BP (probably Ash layer 2 [Ruddiman and Glover, 1972], but detrital carbonate also increases at that level), between H3 and H2, and at the time of Ash layer 1. In the Northern Atlantic, Bond and Lotti [1995] and Elliot et al. [1998], described a succession of basaltic IRD peaks which may correspond to each of the cold oscillations (Dansgaard /Oeschger stadials) recorded in the Greenland ice sheet [Dansgaard et al., 1984]. The sedimentation rate in core CH 69-K09 and the analytical resolution may be not large enough to individualize each peak, but the record does show that the general evolution of the Laurentide and of the NE ice sheets might not be directly related. There is no support in core CH 69-K09 for the idea that each event was marked first by an input of basaltic material, and then by the carbonate IRD [Bond and Lotti, 1995]. Basaltic grains decrease in proportion during most of the HE, and for some of them, before the detrital carbonate (H2, in particular) but the lower basaltic grain proportion within the HE itself may derive from an increased dilution during periods of increased flux of IRD from the Laurentide ice shelves.

The surface water proxies. A large part of the interstadial/stadial oscillations may be recognized in the records of the "warm" surface water proxies, as lower *G. bulloides* $\delta^{18}O$ and/or increased SST during the interstadials. This is specially characteristic for the warmest interstadials, immediately after the HE. During the HE's, SST decreases to the coldest recorded temperatures (5°C for Tc and 8°C for Tw). *G. bulloides* $\delta^{18}O$ also decreases during H4, H2 and H1 , which proves that iceberg melted in the area during these events. The record is rather different for the polar *N. pachyderma* s. $\delta^{18}O$, which does not show the light isotopic values associated with the interstadials, even for the warmest post HE interstadials. The proportion of *N. pachyderma* s. in the fauna, however, decreases significantly. Thus, polar waters retreated north during the interstadials, but their hydrology did not change significantly. At the opposite, large peaks of low *N. pachyderma* s. $\delta^{18}O$ are observed during all the HE (including H3), in phase with low SST, high proportion of *N. pachyderma* s., and high IRD. HE correspond therefore to a sudden invasion of polar waters just north of the area (a *N. pachyderma* s. proportion of 70% is typical of the polar front waters [Bé and Tolderlund, 1971; Pflaumann et al., 1996]), with lower salinity and iceberg meltwater. A characteristic isotopic signature of the events is the convergence of $\delta^{18}O$ values for *G. bulloides* and *N. pachyderma* s. from the same age.

A similar, but smaller amplitude convergence of isotopic values between both species may be seen during the large stadials (by example just prior to H5 (49 kyr BP) and at 45, 43, 39, 35 and 25.5 kyr BP), in phase with small IRD events. Each are marked by decreases in the *N. pachyderma* s. $\delta^{18}O$ and not in the *G. bulloides.* Such trend would be explained by the opposite isotopic effect of SST and salinity. Cooling of NAC waters (where *G. bulloides* develops) was probably much larger during the stadials and HE that the cooling of polar waters, already very cold. Neglecting temperature changes for polar waters, we may estimate the order of magnitude of the changes in polar water salinity corresponding to these events: a 1±0.5‰? $\delta^{18}O$ change for *N. pachyderma* s. would correspond to an equivalent decrease in salinity (about 1 ‰), if the source of freshwater is melting icebergs with a $\delta^{18}O$ of -35‰ vs. SMOW.

The benthic records. The stacked benthic $\delta^{18}O$ record shows small decreases (~0.1 to 0.3‰) synchronous with the *N. pachyderma* s. signal for most of the HE. We probably did not sample the true variability of the signal, as amounts of benthic foraminifera decrease drastically during the periods of high IRD flux. The $\delta^{18}O$ decreases correspond to the penetration at depth of the low $\delta^{18}O$ water which derived from the iceberg melting [Vidal et al., 1998]. What happens during the smaller D/O cycles is less clear, for lack of resolution and a low signal to noise ratio. The benthic foraminifera $\delta^{13}C$ is also noisy. However, as observed in shallower benthic $\delta^{13}C$ records [Sarnthein et al., 1994; Vidal et al., 1997], $\delta^{13}C$ drops within most of the HE, and at least for part of the stadials as by example around 50, 44.5, 43 kyr BP, indicating a general decrease of NADW formation for each of these events. Although in limit of resolution, $\delta^{13}C$ drops (0.1 to 0.3 ‰) are approximately in phase with the stadial SST drops or southward shift of the NAC indicated by increasing *N. pachyderma* s. %. Other $\delta^{13}C$ drops correspond to the warm interstadials post HE. Indications of high latitude ice melting, associated with these warm events have been provided by Elliot et al. [1998]. Lowering of the deeper water ventilation would therefore be also associated with the earlier phase of the interstadials. Vidal et al. [1997] provided however evidences that at shallower depth, ventilation resumed immediately after the HE. These events would thus not affect significantly the thermohaline conveyor belt, contrarily to what happens during HE and major stadials. Observations by Keigwin and Jones [1994] on the variability of the input of carbonate in high sedimentation rate cores from the Bermuda rise support the idea that deep water dynamics was significantly affected by the D/O oscillations. These results confirm the study done in a higher sedimentation rate core from the Southern Norwegian sea [Rasmussen et al., 1996], both the Heinrich events and the smaller stadials are associated there with cooling and lowered surface salinity.

Spectral analysis of the benthic $\delta^{18}O$ and $\delta^{13}C$, and of the Tc signals for the period 10 to 60 kyr PB in the frequency domain 0.1 to 1 ky^{-1} confirms the interrelationship observed in the temporal domain. The peaks of maximum power observed for SST are also observed for the benthic $\delta^{18}O$ and $\delta^{13}C$, and cross-Blackman-Tuckey analysis show high coherence with negligible phase. Thus, our proxies for surface and deep water circulation oscillate in a coherent way within the frequency ranges of the Heinrich and D/O events. This is a further indication that the evolution of the thermohaline conveyor belt is linked to the short term evolutions of high latitude surface water hydrology and ice sheets.

3.2.2. The Holocene

Core CH 69-K09 presents a significant variability of its surface water proxies during the Holocene, with also a peak in variance around 1/1.4 kyr^{-1}. We observe trace input of ice rafted detritus (few grains/g) around 0.7, 1.5, 2.7, 4.8 kyr BP, and larger IRD peaks (more than 10 grains/g) at 7.5-8.5 kyr BP and around 12 kyr BP (YD). Tc becomes cooler by 2°C or more and Tw up to 5°C at about the same periods. The number of IRD grains is however very small for the last 5 kyr period, and cooling is marked by only one to two consecutive points. Some of these dates are similar to those described by Bond et al. [1997] in the North East Atlantic cores, and by O'Brien et al. [1995] in the GISP2 geochemical signal.

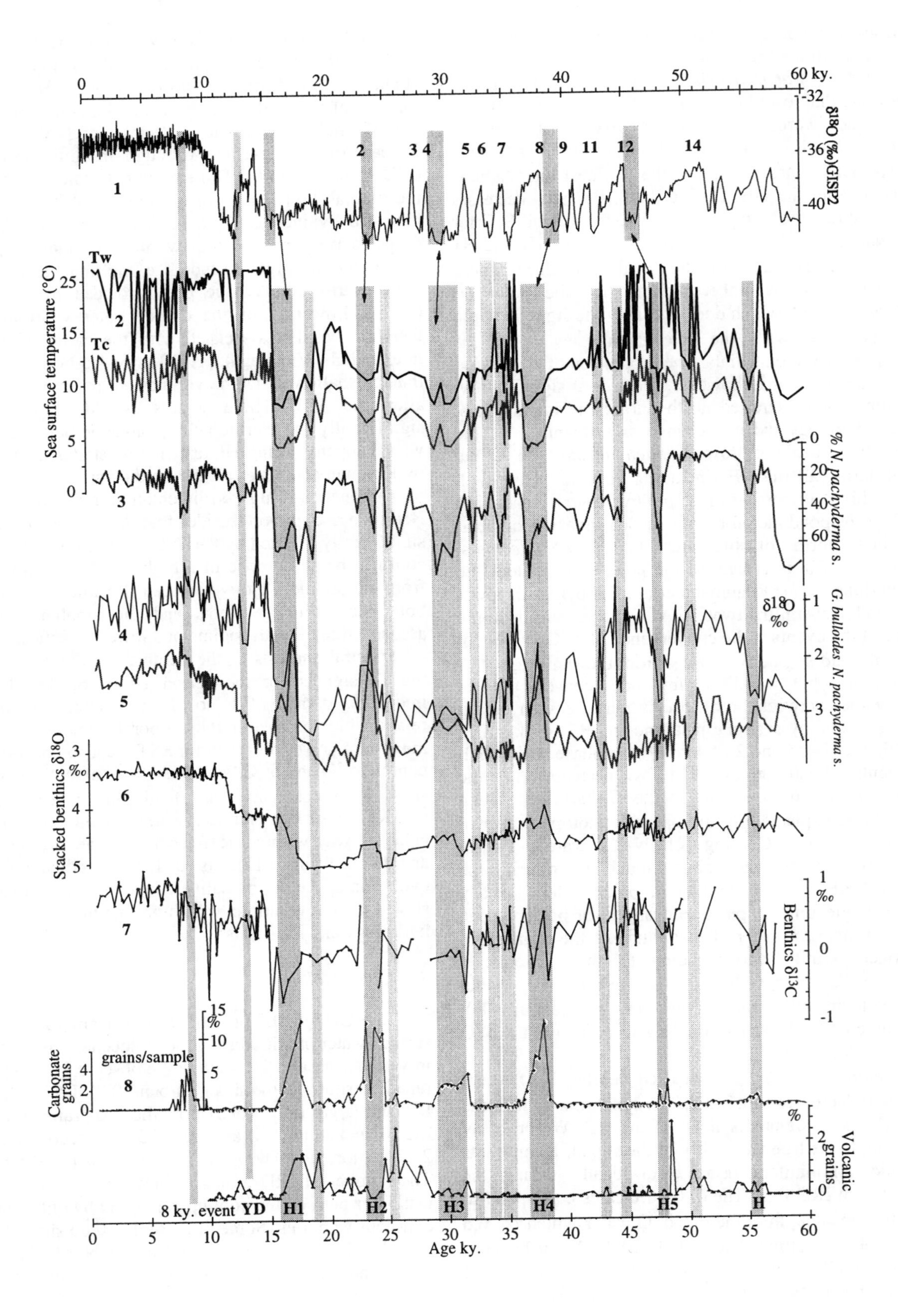

60 ky.
δ18O (‰) GISP2
Tw
Tc
Sea surface temperature (°C)
% N. pachyderma s.
δ18O ‰
G. bulloides N. pachyderma s.
Stacked benthics δ18O ‰
Benthics δ13C ‰
Carbonate grains
grains/sample
%
Volcanic grains
8 ky. event
YD
H1
H2
H3
H4
H5
H
Age ky.

The 8 kyr event (8.2 kyr in the GISP2 chronology) has generated quite a lot of excitation among the paleo-climatology community in recent years [Alley et al., 1997; Bond et al., 1997]. This event corresponds in core CH 69-K09 to an abrupt 2.5°C winter temperature cooling, an invasion of polar *N. pachyderma* s. foraminifera to the site of the core, and an associated low $\delta^{18}O$ fresh water event : sea surface salinity decrease, calculated following Duplessy et al. [1991] from the shift in *G. bulloides* $\delta^{18}O$ (but with a growth temperature taken as T cold), is larger that 1 ‰. The ice rafted detritus include detrital carbonates, which implies that the residual northern Canadian Laurentide was a source of icebergs. This cold phase has therefore all the characteristics of the abrupt cold and low salinity events observed for each Heinrich event during the glacial period [Labeyrie et al., 1995]. However there was only residual ice over the Northern continents during that period, and thus the 8 kyr event could not be the result of the major ice sheet collapse mechanism proposed for the Last Glacial events [MacAyeal, 1993].

We observed a similar evolution with a larger amplitude, during the Younger Dryas, 12 to 13 kyr BP: a cooling larger than 3°C for Tc and 8°C for Tw, a 3 ‰ drop in sea surface salinity and a significant increase of ice rafted detritus (including detrital carbonates from the Laurentide source of icebergs). Seen at the location of core CH 69-K09, there are therefore strong reasons to think that both the YD and the 8 kyr cold events are similar to Heinrich events, which resulted in surface water cooling and decrease in the NAC northward surface water transport.

If we focus on the long term trends throughout Holocene, we find, at a smaller scale, a similar evolution to what is observed for the other warm periods. *N. pachyderma* s. $\delta^{18}O$ decreases regularly until the 8 kyr BP event, and then progressively increases to below the top of the core (by about 0.5 ‰), while *G. bulloides* $\delta^{18}O$ increases during the cold spells between 4 and 8 kyr BP, and decreases again during the following warm period. The benthic foraminifera $\delta^{18}O$ is stable after the big jump of the second melt water pulse (11 kyr BP.), with a small tendency for increasing values since 8 kyr BP, but the range is small, less than 0.1 ‰. The benthic $\delta^{13}C$ signal presents a large variability, but not specifically linked to the surface water events. A first maximum is observed just after the warm Bölling-Allerod period. YD is marked by a very short low $\delta^{13}C$ peak at its beginning. Other low $\delta^{13}C$ peaks punctuate the progressive $\delta^{13}C$ increase apparent until about 6 kyr BP. The 8 kyr BP event does not appear in the benthic record.

The only major event within upper Holocene is around 4 kyr BP and is marked in the *G. bulloides* $\delta^{18}O$ and SST records as a time of large variability. The deep water ventilation was also quite variable during that period, but further work is needed to quantify more precisely its links with changes in surface and deep water hydrology.

4. CONCLUDING REMARKS

Core CH 69-K09 presents exceptional characteristics for the study of climate's millennial-scale variability, with its regular sedimentation at a rate larger that 8 cm/kyr, its coverage of the last 180 kyr BP, its location, at the North-West boundary of the NAC, at the point where this warm water current leaves the American coast towards the North East Atlantic Ocean and Norwegian Sea. The simultaneous records, at around 400 years resolution, of the Heinrich and other ice rafted events, changes in sea surface temperature and salinity and deep water properties offer a sound basis for the study of the causes and consequences of abrupt climatic changes.

On Glacial /Interglacial time scales, the ice sheets, the North Atlantic climatology and the thermohaline activity are driven for a significant part by insolation changes modulated by the precession of the equinoxes. Low latitude winter northern hemisphere and summer southern hemisphere insolation seem to exert a direct influence on warm water transport to the North, and probably on associated winter snow storminess, which plays a major role in ice sheet growth. The thermohaline conveyor belt works at maximum efficiency during these periods. With the development of ice sheets, southern expansion of low salinity arctic waters and iceberg production contribute to a change in the mode of thermohaline circulation. NAC,

Figure 5. Spectral response of some of the parameters studied in core CH 69-K09 :
1-*G. bulloides* $\delta^{18}O$, power distribution vs. frequency (0 to 0.1 ky^{-1}) using Blakman-Tuckey (black line, band width 0.02) and Multi Tapper Method (MTM) (grey line, band width 0.06), with at the bottom an F test on the amplitude spectrum. Peaks are considered significant if F> 0.8. The precession band corresponds to the grey box. 2- idem for summer Sea Surface Temperature (SST Tw). 3- same analysis, for winter sea surface temperature (Tc) but on the frequency range 0 to 1.5 kyr^{-1}. Are reported on the same diagram the results from 3 different spectral analyses methods available in the Paillard et al. [1996] package : Blackman-Tuckey, Maximum Entropy and MTM. The upper diagram gives the power distribution of the variance vs. frequency. The lower diagram gives the amplitude distribution resulting from the MTM analysis, with the F-Test. Significant periods are individualized (kyr) .

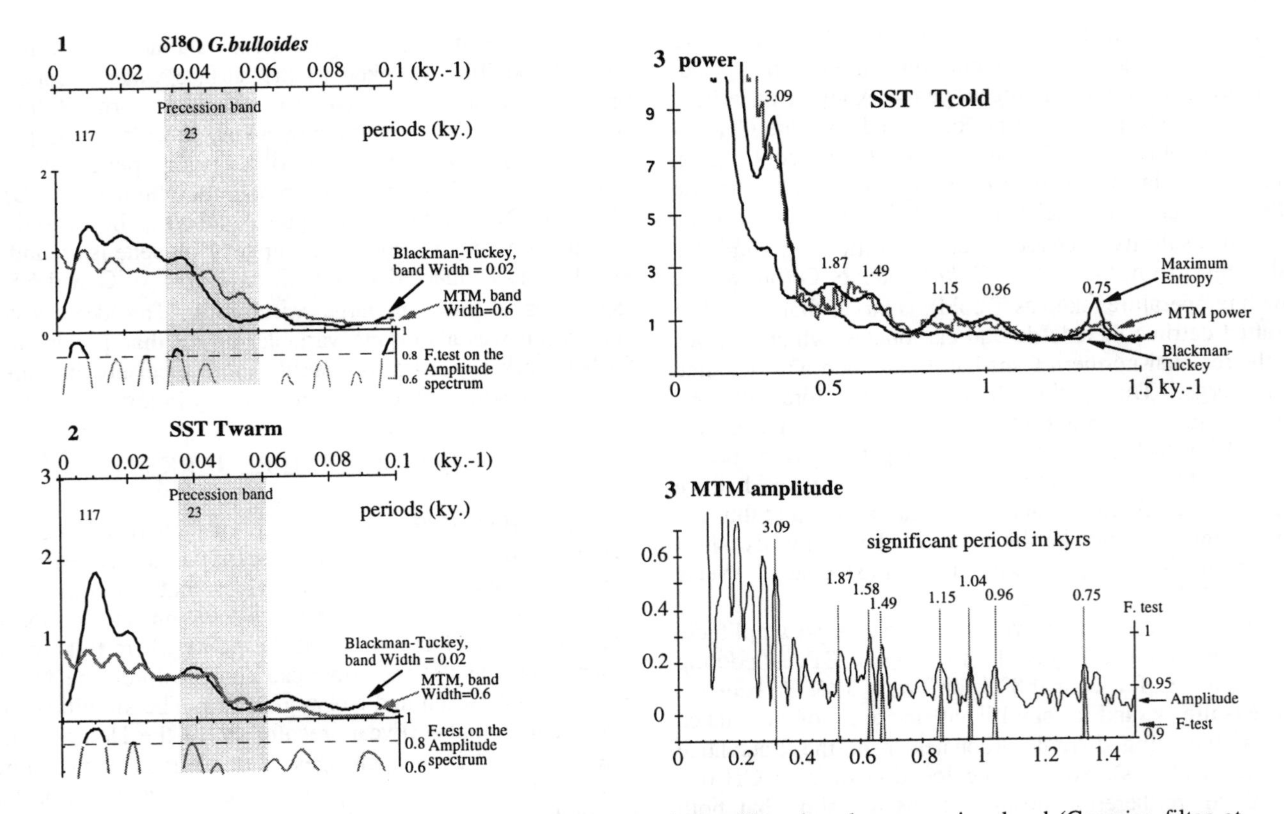

Figure 6. Characteristics signals for core CH 69-K09, band-pass filtered at the precession band (Gaussian filter at 0.05± 0.02), over the last 180 kyr.
1- summer insolation June 15, 65°N [Berger 1978], not filtered, in W/m^2; *2- N. pachyderma s.* δ^{18}O; 3- Mean benthic δ^{18}O; *4- G. bulloides* δ^{18}O; 5- Summer sea surface temperature (Tw); 6- Benthic foraminifera δ^{13}C; 7- Winter insolation December 15 [Berger 1978], not filtered.

although cooler that during interglacials, still transport to the North-East a significant amount of warm waters, flowing out of Newfoundland not far from the Southern boundary of the Laurentide ice sheet. Vidal et al. [1997] hypothesized that intermediate or deep water was probably actively forming during that period in southern Labrador Sea, not far north of core CH 69-K09 location.

For the millennial variability, our results confirm that both HE and stadial cooling are associated with southern invasion of polar waters, disruption of the NAC circulation, and decrease in deep water ventilation. The major ice surges (HE) occurred with a spacing which decreased progressively during the long term development of the glacial cycle, from about 15 kyrs between H6 and H5 to 6 kyr between H2 and H1 (Figure 7). Each HE dramatically increased the area covered by the arctic low salinity cold water lid (to about 40°N [Cortijo et al., 1997, 1999; Labeyrie et al., 1995], and more or less stopped deep and intermediate water formation in the high latitude North Atlantic. Warm and saline surface water re-invaded the North Atlantic, and some thermohaline circulation resumed instantaneously after the end of the HE [Paillard and Labeyrie, 1994]. However, at the depth of core CH 69K09, ventilation resumed only several hundred years after the HE's. This is attributed to the extension of a low salinity meltwater lead at high northern latitudes associated with the warm interstadial peaks which followed each HE [Elliot et al., 1998]. The evolution of the surface and deep water characteristics during HE and D/O cycles is thus rather complex, with probably shifts between several modes and zones of deep water formation in few hundred to thousand years, depending on when and where was expending the polar low salinity surface water.

The smaller amplitude millennial climatic variability is pervasive through glacial and interglacial periods, with similar periodicity (around 1.5 kyr). Stadials correspond to surface water cooling and southward extension of the cold and low salinity arctic waters. Surface and deep water thermohaline transport is decreased. Significant input of ice rafted debris are recorded in different locations of the Northern Atlantic, including at the location of core CH 69-K09. The corresponding signature of continental ice low δ^{18}O meltwater is observed for both surface and deep

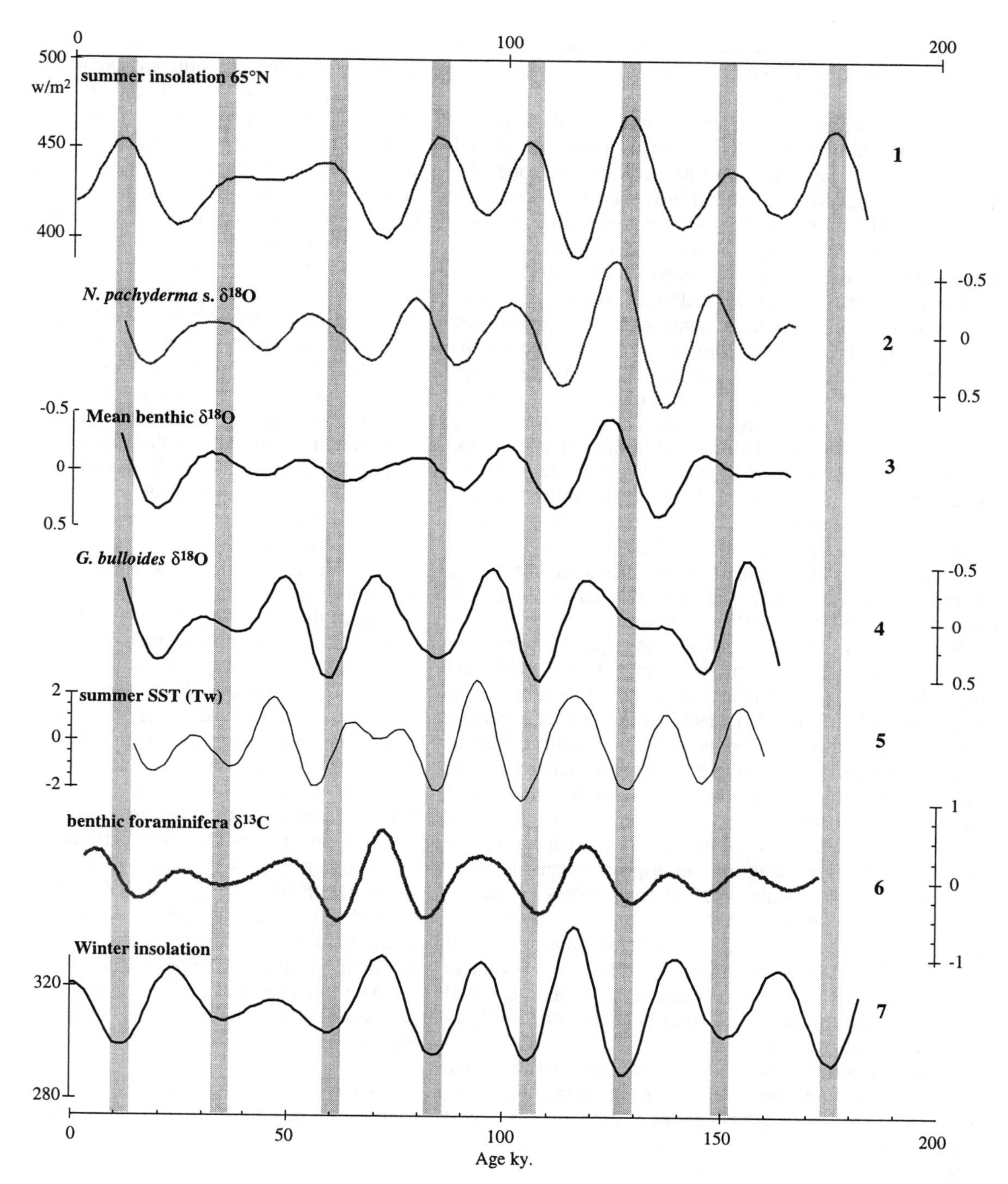

Figure 7. Detailed records for the last 60 kyr : 1- the ice $\delta^{18}O$ in the GISP2 Greenland ice core, a proxy for air temperature; 2- Summer (Tw) and Winter (Tc) sea surface temperature; 3- relative proportion of polar planktic foraminifera *N. pachyderma* s. ; 4- *G. bulloides* $\delta^{18}O$; 5- *N. pachyderma s.* $\delta^{18}O$; 6- stacked benthic foraminifera $\delta^{18}O$; 7- stacked benthic foraminifera $\delta^{13}C$; 8- proportion of carbonate grains (IRD) within the >150μm size range. For the upper part (Holocene), the total number of grains per sample is reported; 9- proportion of basaltic grains (IRD) within the >150μm size range. Age scale for GISP2 from Grootes and Stuiver [1997]; age scale for core CH 69-K09 in calendar years (see text).

Norwegian Sea waters [Rasmussen et al., 1996]. A higher analytical resolution would be necessary to confirm that result in core CH 69-K09.

Sakai and Peltier [1997] have shown that a global thermohaline circulation model, placed in a mode of glacial circulation (with low thermohaline activity, controlled by the flux of surface fresh water and low temperatures at high latitudes) exhibits free oscillations at millennial frequencies. The same result is obtained if the model is coupled to a simple energy balance model. Bond et al. [1997], who demonstrated that the North Atlantic ocean hydrology oscillates also during inter-glacials at those frequencies, favor the hypothesis of a "climatic oscillator" within the coupled ocean-atmosphere system to explain the millennium oscillations. The Dansgaard/Oeschger cycles in the Greenland ice records would correspond to a reinforcement of the oscillations during the glacial periods, perhaps through a "thermohaline oscillator" as described by Sakai and Peltier [1997], but without direct forcing from the ice sheets. The peaks in IRD observed south of Iceland would be a response of the surface circulation, with southern invasion of the polar waters (increased strength of the East Greenland current) and an associated larger iceberg transport to the Northern Atlantic Ocean.

We find however peculiar the quasi permanent association, even at the latitude of core CH 69-K09 and during the Holocene, of an input of ice rafted detritus during the cold episodes. Bond et al. [1997] consider it as a residual semi permanent flux of icebergs, with their transport along the East Greenland and Labrador currents modulated by the internal oscillations of the ocean circulation. We suggest that coastal or marine based part of ice sheets, even if they cover a small area, could contribute as a feedback to the oscillations of the coupled atmosphere-ocean system, bringing excess meltwater both during surges or during warming periods. Both would affect areas of deep water convection and decrease the thermohaline conveyor belt activity and associated northward heat transfer. As these reactions would lag by tens or hundreds of years the initial forcing mechanism, they could contribute very efficiently to a sustained oscillatory behavior.

If we consider now the connections between the evolution in the Earth precession frequency band and in the millennial band, our results do not support any direct forcing of the Heinrich events by low latitude insolation, as proposed by McIntyre and Molfino [1996]. The lag time between each Heinrich events decreases as the volume of the large ice sheets increase along the 60 kyr period of the last Glacial, independently of the insolation forcing. We prefer an indirect connection between HE and insolation: higher insolation at low latitude in winter (December perihelion) increases atmospheric water transport to high latitudes through winter storms. This accelerates ice sheet development. The larger the ice sheet, the more sensitive it becomes to that excess accumulation, with increases in both the frequency of surges and southward penetration of the polar surface water low salinity lid. Cortijo et al. [1994] have shown that this southward expansion lags by only 1 to 2 kyr the decrease in high latitude summer insolation. However, the North Atlantic Current (the surface water component of the thermohaline conveyor belt) is not affected before 5 to 10 kyr, which leave enough time to pursue ice sheets growth and development of ice surges. Conversely, during periods of increased northern summer insolation, meltwater events, or surges from the northern parts of the ice sheets affected NAC and decrease the conveyor belt activity for a limited duration. The thermohaline conveyor belt was rapidly reactivated with its associated atmospheric transport of water vapor to high latitudes [Cortijo et al. 1997; Paillard and Labeyrie, 1994]. As long as the rate of ice sheet destruction (through semi-periodic HE) stays below the rate of ice sheet growth, the system will go on with the development of the ice age. The Heinrich events would thus be the product of several effects, among which rapid ice sheet growth and overflow to the ocean are of major importance. We have seen that the main ice sheet concerned by these events is the Laurentide.

A similar evolution is observed for the millennial variability during periods of large ice sheets. There, the source of the events appears mostly limited to the ice sheets around Norwegian and Greenland seas (this work and Elliot et al. [1998]). Initial high latitude warmth during active conveyor belt periods favors ice accumulation, through increased high latitude water vapor transport and winter snow accumulation. After some lag, coastal ice sheets would become unstable, and increase the fresh water load over the northern Atlantic and Norwegian/Greenland seas, thus decreasing convection and heat transport. But these small events at about 1.5 kyr periodicity would affect only on a minor way NAC activity, which would pursue its contribution to the growth of the ice sheets. Thus, immediately after dissipation of the excess low salinity surface water load, thermohaline convection would invade again the Northern Atlantic, and a new D/O cycle would be on the way. The inertia of the largely continental Laurentide ice sheet would make it less sensitive to these rapid oscillations.

Acknowledgment. This study has been initiated in the course of a collaboration with the D.G.O. in Bordeaux, we acknowledge useful discussions with M. Labracherie, J.L. Turon and F. Grousset. Sampling of the core CH 69-K09 has been possible thanks to the quality of preservation by R. Kerbras (IFREMER). V. Bout-Roumazeille, A. Boelaert, D. Paillard, C. Kissel, D. Blamart, E. Michel, C. Lalou and F. Lemoine have all significantly contributed to our research on the causes and consequences of the millennial climate variability. J. Tessier's efficiency with the mass spectrometers is also thanked. Financial support from the French CEA, CNRS, and INSU (PNEDC) and EU ENV4-CT95-0131 and EV 95-117 have permitted the thousands of costly isotopic analyses needed for that study. P. Clark gave the impetus for writing the paper. R.

Webb, A. Mix and an anonymous reviewer helped to improve considerably the original text. This is an IMAGES contribution and LSCE contribution number 264.

REFERENCES

Alley, R.B., P.A. Mayewski, T. Sowers, M. Stuiver, K.C. Taylor, and P.U. Clark, Holocene climatic instability: a prominent, widespread event 8200 yr ago, Geology, 25, 483-486, 1997.

Bard, E., M. Arnold, J. Moyes, and J.-C. Duplessy, Reconstruction of the last deglaciation: deconvolved records of $\delta^{18}O$ profiles, micropaleontological variations, and accelerator ^{14}C dating, Clim. Dynam., 1, 101-112, 1987.

Bard, E., Correction of accelerator mass spectrometry ^{14}C ages measured in planktonic foraminifera : paleoceanographic implications, Paleoceanography, 3, 635-645, 1988.

Bard, E., and W.S. Broecker (Ed.), The Last deglaciation : Absolute and Radiocarbon Chronologies., NATO ASI series I , vol. 2, 344 pp., Springer-Verlag, 1992.

Bard, E., M. Arnold, and R.G. Fairbanks, ^{230}Th-^{234}U and ^{14}C ages obtained by mass spectrometry on corals, Radiocarbon, 35, 191-199, 1993.

Bard, E., M. Arnold, J. Mangerud, M. Paterne, L. Labeyrie, J. Duprat, M.A. Mélières, E. Sonstegaard, and J.C. Duplessy, The North Atlantic atmosphere-sea surface ^{14}C gradient during the Younger Dryas climatic event, EPSL, 126, 275-287, 1994.

Bé, A.W.H., Plankton abundance in the North Atlantic Ocean, in Fertility of the Sea, edited by J.D. Costlow, pp. 17-50, Gordon & Breach Science Publishers, New York, 1971.

Bé, A.W.H., and D.S. Tolderlund, Distribution and ecology of living planktonic foraminifera in surface waters of the Atlantic and Indian Oceans, in Micropaleontology of Oceans, edited by B.M. Funnell and W.R. Riedel, pp. 105-149, Cambridge Univ. Press, London, UK, 1971.

Behl, R.J., and J.P. Kennett, Brief intrestadial events in the Santa Barbara basin, NE Pacific during the past 60 kyr, Nature, 379, 243-246, 1996.

Berger, A.L., Long-term variations of daily insolation and Quaternary climatic change, J. Atmos. Sci., 35, 2362-2367, 1978.

Bond, G., H. Heinrich, W. Broecker, L. Labeyrie, J. McManus, J. Andrews, S. Huon, R. Jantschik, C. Clasen, C. Simet, K. Tedesco, M. Klas, and G. Bonani, Evidence for massive discharges of icebergs into the glacial north Atlantic, Nature, 360, 245-249, 1992.

Bond, G., W. Broecker, S. Johnsen, J. McManus, L. Labeyrie, J. Jouzel, and G. Bonani, Correlations between climate records from north Atlantic sediments and Greenland ice, Nature, 365, 143-147, 1993.

Bond, G., and R. Lotti, Iceberg discharges into the North Atlantic on Millenial time scales during the last glaciation, Science, 267, 1005-1010, 1995.

Bond, G., W. Showers, M. Cheseby, R. Lotti, P. Almasi, P. deMenocal, P. Priore, H. Cullen, I. Hajdas, and G. Bonani, A pervasive millenial-scale cycle in North Atlantic Holocene and glacial climates, Science, 278, 1257-1266, 1997.

Boyle, E.A., Deep ocean circulation, preformed nutrients, and atmospheric carbon dioxide: theories and evidence from oceanic sediments, in Mesozoic and Cenozoic Oceans, Geodynamics edited by K.J. Hsü, pp. 49-59, AGU, Washington, DC, 1986.

Boyle, E.A., and L.D. Keigwin, Deep circulation of the North Atlantic over the last 200,000 years: Geochemical evidence, Science, 218, 784-787, 1982.

Boyle, E.A., and L.D. Keigwin, Comparison of Atlantic and Pacific paleochemical records for the last 250,000 years: changes in deep ocean circulation and chemical inventories, Earth Planet. Sci. Lett., 76, 135-150, 1985.

CLIMAP, Seasonal reconstructions of the Earth's surface at the last glacial maximum, GSA Map and Chart Ser., MC-36, Geol. Soc. Am., Boulder, CO, 1981.

Coplen, T.B., Normalization of oxygen and hydrogen isotope data., Chem. Geol. (isotope Geosci. Sect.), 72, 293-297, 1988.

Cortijo, E., J.C. Duplessy, L. Labeyrie, H. Leclaire, J. Duprat, and T.C.E. van Weering, Eeemian cooling in the Norwegian Sea and North Atlantic preceding continental ice sheet growth, Nature, 372, 446-449, 1994.

Cortijo, E., P. Yiou, L. Labeyrie, and M. Cremer, Sedimentary record of rapid climatic variability in the North Atlantic ocean during the last glacial cycle, Paleoceanography, 10, 921-926, 1995.

Cortijo, E., L. Labeyrie, L. Vidal, M. Vautravers, M. Chapman, J.C. Duplessy, M. Elliot, M. Arnold, J.L. Turon, and G. Auffret, Changes in sea surface hydrology associated with Heinrich event 4 in the North Atlantic Ocean between 40° and 60°N, E.P.S.L., 146, 29-45, 1997.

Cortijo, E., S. Lehman, L. Keigwin, M. Chapman, D. Paillard, and L. Labeyrie, Changes in meridional temperature and salinity gradients in the North Atlantic Ocean (30° to 72°N) during the Last Interglacial Period, Paleoceanography, 14, 23-33, 1999.

Craig, H., and A. Gordon, Deuterium and Oxygen 18 variations in the ocean and the marine atmosphere, in Stable isotopes in Oceanic Studies and Paleotemperatures, edited by E. Tongiorgi, pp. 9-130, C.N.R. Pisa, Spoleto, 1965.

Dansgaard, W., S.J. Johnsen, H.B. Clausen, D. Dahl-Jensen, N. Gundestrup, and C.U. Hammer, North Atlantic climatic oscillations revealed by deep Greenland ice cores, in Climate Processes and Climate Sensitivity, Geophysical Monograph Series edited by J.E. Hansen and T. Takahashi, pp. 288-298, American Geophysical Union, Washington, D.C., 1984.

Duplessy, J.C., N.J. Shackleton, R.K. Matthews, W.L. Prell, W.F. Ruddiman, M. Caralp, and C. Hendy, ^{13}C record of benthic foraminifera in the last interglacial ocean: implications for the carbon cycle and the global deep water circulation, Quat. Res., 21, 225-243, 1984.

Duplessy, J.C., M. Arnold, P. Maurice, E. Bard, J. Duprat, and J. Moyes, Direct dating of the oxygen-isotope record of the last deglaciation by ^{14}C accelerator mass spectrometry, Nature, 320, 350-352, 1986.

Duplessy, J.C., N.J. Shackleton, R.G. Fairbanks, L.D. Labeyrie, D. Oppo, and N. Kallel, Deepwater source variations during the last climatic cycle and their impact on the global deepwater circulation, Paleoceanography, 3, 343-360, 1988.

Duplessy, J.C., L.D. Labeyrie, A. Juillet-Leclerc, F. Maitre, J. Duprat, and M. Sarnthein, Surface salinity reconstruction of the north Atlantic ocean during the last glacial maximum, Oceanologica Acta, 14, 311-324, 1991.

Duplessy, J.C., L.D. Labeyrie, M. Arnold, M. Paterne, J. Duprat, and T.C.E. van Weering, Changes in surface salinity of the North Atlantic Ocean during the last deglaciation, Nature, 358, 485-488, 1992.

Elliot, M., L. Labeyrie, G. Bond, E. Cortijo, J.L. Turon, N. Tisnerat, and J.C. Duplessy, Millenial scale iceberg discharges in the Irminger Basin during the last glacial period: relationship with the Heinrich events and environmental settings, Paleoceanography, 13, 443-446, 1998.

Emiliani, C., Paleotemperature analysis of caribbean cores P6304-8 and P6304-9 and a generalized temperature curve for the past 425,000 years, Jour. Geology, 74, 109-126, 1966.

Fichefet, T., S. Hovine, and J.C. Duplessy, A model study of the Atlantic thermohaline circulation during the last glacial maximum, Nature, 372, 252-255, 1994.

Fronval, T., E. Jansen, J. Bloemendal, and S. Johnsen, Oceanic evidence for coherent fluctuations in Fennoscandian and Laurentide ice sheets on millenium time scales, Nature, 374, 443-446, 1995.

Grootes, P.M., M. Stuiver, J.W.C. White, S. Johnsen, and J. Jouzel, Comparison of oxygen isotopes records from the GISP2 and GRIP Greenland ice cores, Nature, 466, 552-554, 1993.

Grootes, P.M., and M. Stuiver, Oxygen 18/16 variability in Greenland snow and ice with 13-3 to 105-year resolution, Journal of Geophysical Research, 102, 26455-26470, 1997.

Heinrich, H., Origin and consequences of cyclic ice rafting in the Northeast Atlantic Ocean during the past 130,000 years, Quat. Res., 29, 142-152, 1988.

Imbrie, J., E. Boyle, S. Clemens, A. Duffy, W. Howard, G. Kukla, J. Kutzbach, D. Martinson, A. McIntyre, A. Mix, B. Molfino, J. Morley, L. Peterson, N. Pisias, W. Prell, M. Raymo, N. Shackleton, and J. Toggweiler, On the structure and origin of major glaciation cycles, 1 linear responses to Milankovitch forcing, Paleoceanography, 7, 701-738, 1992.

Imbrie, J., A. Berger, E. Boyle, S. Clemens, A. Duffy, W. Howard, G. Kukla, J. Kutzbach, D. Martinson, A. McIntyre, A. Mix, B. Molfino, J. Morley, L. Peterson, N. Pisias, W. Prell, M. Raymo, N. Shackleton, and J. Toggweiler, On the structure and origin of major glaciation cycles, 2 The 100,00 years cycle, Paleoceanography, 8, 699-735, 1993.

Jenkins, G.M., and D.G. Watts, Spectral analysis and its applications, 525 pp., Holden Day, San Francisco, CA, 1968.

Keigwin, L.D., and G.A. Jones, Western North Atlantic evidence for millenial-scale changes in ocean circulation and climate, Journal of Geophysiscal Research, 99, 12397-12410, 1994.

Keigwin, L.D., and S.J. Lehman, Deep circulation change linked th Heinrich event 1 and Younger Dryas in a middepth North Atlantic core, Paleoceanography, 9, 185-194, 1994.

Kroopnick, P., Distribution of ^{13}C and ΣCO_2 in the world oceans, Deep-Sea Res., 32, 57-77, 1984.

Labeyrie, L.D., and J.C. Duplessy, Changes in the oceanic $^{13}C/^{12}C$ ratio during the last 140,000 years: high-latitude surface records, Paleogeogr., Paleoclimatol., Paleoecol., 50, 217-240, 1985.

Labeyrie, L.D., J.C. Duplessy, and P.L. Blanc, Variations in mode of formation and temperature of oceanic deep waters over the past 125,000 years, Nature, 327, 477-482, 1987.

Labeyrie, L.D., J.C. Duplessy, J. Duprat, A.J. Juillet-Leclerc, J. Moyes, E. Michel, N. Kallel, and N.J. Shackleton, Changes in the vertical structure of the north Atlantic ocean between glacial and modern times, Quaternary Science Reviews, 11, 401-413, 1992.

Labeyrie, L., L. Vidal, E. Cortijo, M. Paterne, M. Arnold, J.C. Duplessy, M. Vautravers, M. Labracherie, J. Duprat, J.L. Turon, F. Grousset, and T. van Weering, Surface and deep hydrography of the Northern Atlantic Ocean during the last 150 kyr, Phil. Trans. R. Soc. Lond. B, 348, 255-264, 1995.

Labeyrie, L., Y. Lancelot, J.L. Turon, F. Bassinot, and Y. Balut, Le programme IMAGES et la campagne de carottages géants du Marion-Dufresne en Atlantique nord, Rep.Rapport annuel pp 35-39, IFRTP, 1996.

Laj, C., A. Mazaud, and J.C. Duplessy, Geomagnetic intensity and ^{14}C abundance in the atmosphere and ocean during the past 50 kyr, Geophysical Research Letters, 23, 2045-2048, 1996.

Levitus, S., World Ocean Atlas, 81, NOAA, 1994.

MacAyeal, D.R., Binge/purge oscillations of the Laurentide Ice sheet as a cause of the North Atlantic's Heinrich events, Paleoceanography, 8, 775-784, 1993.

Martinson, D.G., N.G. Pisias, J.D. Hays, J. Imbrie, T.C. Moore, and N.J. Shackleton, Age dating and the orbital theory of the ice ages: development of a high-resolution 0-300,000 year chronostratigraphy, Quat. Res., 27, 1-30, 1987.

Mayewski, P.A., L.D. Meeker, M.S. Twickler, S. Whitlow, Q. Yang, W.B. Lyons, and M. Prentice, Major features and forcing of high-latitude northern hemiphere atmospheric circulation using a 110 000-year-long glaciochemical series, Journal of Geophysical Research, 102, 26345-26366, 1997.

McIntyre, A., and B. Molfino, Forcing of Atlantic Equatorial and Subpolar Millenial Cycles by Precession, Science, 274, 1867-1870, 1996.

Mix, A.C., and R.G. Fairbanks, North Atlantic surface control of Pleistocene deep-ocean circulation, Earth and Planet. Sci. Lett., 73, 231-243, 1985.

O'Brien, S.R., P.A. Mayewski, L.D. Meeker, D.A. Meese, M.S. Twickler, and S.I. Whitlow, Complexity of Holocene climate as reconstructed from a Greenland ice core, Science, 270, 1962-1964, 1995.

Paillard, D., and L.D. Labeyrie, Role of the thermohaline circulation in the abrupt warming after Heinrich events, Nature, 372, 162-164, 1994.

Paillard, D., L.D. Labeyrie, and P. Yiou, AnalySeries 1.0: a Macintosh software for the analysis of geophysical time-series, E.O.S., 77, 379, 1996.

Pastouret, L., G. Auffret, M. Hoffert, M. Melguen, H.D. Needham, and C. Latouche, Sédimentation sur la ride de Terre-Neuve, Canadian Journal Earth Science, 12, 1019-1035, 1975.

Pflaumann, U., J. Duprat, C. Pujol, and L. Labeyrie, SIMMAX, a modern analog technique to deduce Atlantic sea surface temperatures from Planktonic foraminifera in deep sea sediments, Paleoceanography, 11, 15-35, 1996.

Prell, W.L., The stability of low-latitude sea-surface temperatures: an evaluation of the CLIMAP reconstruction with emphasis on the positive SST anomalies, Rep.TR 025, U.S. Department of Energy, Washington, DC, 1985.

Rasmussen, T.L., T.C.E. Van Weering, and L. Labeyrie, Climatic instability, ice sheets and ocean dynamics at high northern latitudes during the last glacial period (58-10 ka BP), Quaternary Science Reviews, 16, 71-80, 1996.

Raymo, M.E., W.F. Ruddiman, N.J. Shackleton, and D.W. Oppo, Evolution of Atlantic-Pacific ^{13}C gradients over the last 2.5 m.y., Earth Planet. Sci. Lett., 97, 353-368, 1990.

Rossby, T., The North Atlantic current and surrounding waters : At the crossroads, Reviews of Geophysics, 34, 463-481, 1996.

Ruddiman, W.F., and L.K. Glover, Vertical mixing of ice-rafted volcanic ash in North Atlantic sediments, Geological Society of America Bulletin, 83, 2817-2836, 1972.

Ruddiman, W.F., and A. McIntyre, Warmth of the subpolar north Atlantic ocean during northern hemisphere ice-sheet growth, Science, 204, 173-175, 1979.

Ruddiman, W.F., and A. McIntyre, Oceanic mechanism for amplification of the 23,000-year ice-volume cycle, Science, 212, 617-627, 1981.

Ruddiman, W.F., M.E. Raymo, D.G. Martinson, B.M. Clement, and J. Backman, Pleistocene evolution of Northern Hemisphere ice sheets and North Atlantic ocean, Paleoceanography, 4, 353-412, 1989.

Sakai, K., and W.R. Peltier, Dansgaard-Oeschger oscillations in a coupled atmosphere-ocean climate model, American Meteorological Society, 10, 949-970, 1997.

Sarnthein, M., K. Winn, S.J.A. Jung, J.C. Duplessy, L.D. Labeyrie, H. Erlenkeuser, and G. Ganssen, Changes in East Atlantic deep water circulation over the last 30,000 years : Eight time slice reconstructions, Paleoceanography, 9, 209-267, 1994.

Schulz, H., U. von Rad, and H. Erlenkeuser, Correlation between Arabian Sea and Greenland climate oscillations of the past 110 000 years, Nature, 393, 54-57, 1998.

Shackleton, N.J., J. Imbrie, and M.A. Hall, Oxygen and carbon isotope record of East Pacific core V19-30: implications for the formation of deep water in the late Pleistocene North Atlantic, Earth Planet. Sci. Lett., 65, 233-244, 1983.

Shackleton, N.J., and N.G. Pisias, Atmospheric carbon dioxide, orbital forcing, and climate, in The Carbon Cycle and Atmospheric CO_2: Natural Variations Archean to Present, Geophys. Monogr. Ser. 32, edited by E. Sundquist and W.S. Broecker, pp. 303-317, AGU, Washington, D.C., 1985.

Streeter, S.S., and N.J. Shackleton, Paleocirculation of the deep north Atlantic: 150,000-year record of benthic foraminifera and oxygen-18, Science, 203, 168-171, 1979.

Stuiver, M., and P.J. Reimer, Extended ^{14}C data base and revised CALIB 3.0 ^{14}C age calibration program, Radiocarbon, 35, 215-230, 1993.

Stuiver, M., P.M. Grootes, and T.F. Braziunas, The GISP2 $\delta^{18}O$ climate record of the past 16 500 years and the role of sun, ocean, and volcanoes, Qusternary Research, 44, 341-354, 1995.

Taylor K. et al., The "flicking switch" of late Pleistocene climate change, Nature, 361, 432-436, 1993.

Thomas, E., L. Booth, M. Maslin, and N.J. Shackleton, Northeastern Atlantic benthic foraminifera during the last 45000 years: changes in productivity seen from the bottom up, Paleoceanography, 10, 545-562, 1995.

Vidal, L., L. Labeyrie, E. Cortijo, M. Arnold, J.C. Duplessy, E. Michel, S. Becqué, and T.C.E. van Weering, Evidence for changes in the North Atlantic Deep water linked to meltwater surges during the Heinrich events, E.P.S.L., 146, 13-27, 1997.

Vidal, L., L. Labeyrie, and T.C.E. van Weering, benthic $\delta^{18}O$ records in the North Atlantic over the last glacial period (60-10 ka) : evidence for brine formation, Paleoceanography, 13, 1998.

Villanueva, J., J. Grimalt, L. Labeyrie, E. Cortijo, L. Vidal, and J.L. Turon, Precessional forcing of productivity in the North Atlantic Ocean, Paleoceanography, 13, 561-571, 1998.

Waelbroeck, C., L. Labeyrie, J.C. Duplessy, J. Guiot, M. Labracherie, H. Leclaire, and J. Duprat, Improving paleoSST estimates based on planktonic fossil faunas, Paleoceanography, 13, 272-283, 1998.

Zahn, R., K. Winn, and M. Sarthein, Benthic foraminiferal $\delta^{13}C$ and accumulation rates of organic carbon: Uvigerina peregrina group and Cibicidoides wuellerstorfi, Paleoceanography, 1, 27-42, 1986.

Zahn, R., J. Schonfeld, H.R. Kudrass, M.H. Park, H. Erlenkeuser, and P. Grootes, Thermohaline instability in the North Atlantic during meltwater events: Stable isotope and ice-rafted detritus records from core SO75-26KL, Portugese margin, Paleoceanography, 12, 696-710, 1997.

Elsa Cortijo Laboratoire des Sciences du Climat et de l'Environnement (LSCE), Unité mixte CEA-CNRS Domaine du CNRS 91198 Gif/Yvette cedex, France

Jean Claude Duplessy, LSCE, Unité mixte CEA-CNRS Domaine du CNRS 91198 Gif/Yvette cedex, France

Mary Elliot, LSCE, Unité mixte CEA-CNRS Domaine du CNRS 91198 Gif/Yvette cedex, France

Laurent Labeyrie, LSCE, Unité mixte CEA-CNRS Domaine du CNRS 91198 Gif/Yvette cedex, France; Also at Département de Géologie, Université d'Orsay, 91405 Orsay cedex, France

Heloise Leclaire, LSCE, Unité mixte CEA-CNRS Domaine du CNRS 91198 Gif/Yvette cedex, France

Brigitte Lecoat, LSCE, Unité mixte CEA-CNRS Domaine du CNRS 91198 Gif/Yvette cedex, France

Claire Waelbroeck, LSCE, Unité mixte CEA-CNRS Domaine du CNRS 91198 Gif/Yvette cedex, France

Laurence Vidal now at Geowissenschaften, Bremen University, 28334 Bremen, Germany

Gerard Auffret; IFREMER, Technopole Brest-Iroise, 29280 Plouzané cedex, France

Origin of Global Millennial Scale Climate Events: Constraints from the Southern Ocean Deep Sea Sedimentary Record

Ulysses S. Ninnemann

Graduate Department, Scripps Institution of Oceanography, La Jolla, CA

Christopher D. Charles

Geosciences Research Division, Scripps Institution of Oceanography, La Jolla, CA

David A. Hodell

Department of Geology, University of Florida, Gainesville, FL

A select set of high deposition rate deep sea sedimentary sequences from the Subantarctic Atlantic and Pacific allows both a detailed description of the global thermohaline circulation changes over the entire last glacial cycle and a simultaneous evaluation of the possible climatic effects of this deep ocean variability. From the benthic foraminiferal carbon isotope records we infer that significant and abrupt fluctuations in the strength of the North Atlantic thermohaline circulation occurred throughout the course of the last glacial cycle--including during the peak of the last interglacial period (120 ka). The new observations from interglacial Stage 5 suggest that the mechanisms of thermohaline instability need not involve glacial meltwater into the North Atlantic as a necessary feedback. From the planktonic foraminiferal isotopic records, we outline the specific upper ocean changes that accompanied the deep ocean changes. The sense of the deep/surface comparison is that indications of greater poleward extent of warmer surface waters usually (but not always) occur in conjunction with indications of reduced North Atlantic Deep Water production. Carbon isotopes of planktonic foraminifera suggest that this relationship cannot simply be a measure of changes in surface heat transport, but rather must involve changes in the actual southward mixing of subtropical waters, ultimately a product of variable wind stress. The phase relationship between thermohaline circulation changes and Southern Hemisphere climate varies over different regions of the climate spectrum, and,

Mechanisms of Global Climate Change at Millennial Time Scales
Geophysical Monograph 112

as a result, it is not straightforward to assess whether the millennial scale climate transitions in the high latitudes of both hemispheres were strictly in antiphase. In all, the specific physical effects inferred from these deep sea sediment records may be quite distinct from those formulated in climate model experiments featuring thermohaline instability.

INTRODUCTION

The parallel behavior of ice core records [e.g. *Dansgaard et al.* 1993] and marine core records throughout the circum-Atlantic region [e.g., *Keigwin and Jones*, 1994; *Marchitto et al*, 1997] demonstrates that variability in the North Atlantic thermohaline circulation must have been an important factor governing the characteristics of the late Pleistocene millennial scale climate oscillations. These results are satisfying, because North Atlantic Deep Water (NADW) formation obviously influences the regional heat budget of the high latitude Northern Hemisphere in the current climate, and models of thermohaline instability clearly suggest the capacity for abrupt changes over millennial timescales [*Broecker et al*, 1990; *Rahmstorf*, 1995]. Yet, despite this consistency of theory and observations, two more general (and more complex) issues remain unresolved: the origin of the thermohaline instability itself has not been established, and the ultimate global climate effects of North Atlantic Deep Water variability are not at all obvious. These questions--which are at the heart of understanding abrupt climate variability and are therefore among the most important issues for future climate concerns--have not yet been answered, in part because they require observations of the relative timing and magnitude of millennial scale variability from the global oceans.

Here we attack these two issues by examining the evidence from several high-deposition rate Southern Ocean deep sea sediment records. The first objective is to construct a detailed record of the changes in North Atlantic Deep Water flux to the Southern Hemisphere over the entire last glacial cycle (140,000 kyrs). Previous work on piston cores suggested that Circumpolar Deep Water (CDW) chemistry is sensitive to the changes in NADW input to the Southern Ocean, on both orbital timescales (the inference from low sedimentation rate cores; e.g. *Oppo and Fairbanks*, [1987]) and millennial timescales (as evidenced from expanded sedimentary sequences covering the last 80 kyr; e.g. *Charles et al.* [1996]). However, using recently acquired ocean drilled cores, we can now extend the analysis of abrupt thermohaline variability into the previous interglacial periods, beyond the range of piston cores [*Gersonde et al.*, in press]. This extension therefore allows the comparison of millennial scale variability between glacial and interglacial states. In the absence of broad geographic coverage, this evolutive approach is one of the best available means for testing possible mechanisms for thermohaline circulation changes such as the "salt oscillator" model [*Broecker et al.*, 1990] or meltwater-induced collapses in thermohaline overturning [*Manabe and Stouffer*, 1997]. For example, if the strength of the thermohaline circulation fluctuated significantly in previous interglacial periods, then this observation would practically eliminate meltwater from Northern Hemisphere ice sheets as a necessary precondition for North Atlantic convective instability.

In the same deep sea sediments that monitor deep water fluctuations, we can also describe the climatic variability occurring on millennial timescales in the surface Southern Ocean. This surface-to-deep comparison offers a unique view of the relationship between deep water production in one hemisphere and climate change in the opposite hemisphere. Our strategy here will be to apply this comparison to model predictions for the global climate effects of NADW variability. For example, recent coupled ocean/atmosphere modeling experiments with variable thermohaline circulation suggest that, rather than warming the Southern Ocean (as originally proposed for glacial-interglacial timescales [*Weyl*, 1968]), vigorous NADW production actually has a net cooling effect on the Southern Hemisphere--presumably a result of the reduction in northward oceanic heat transport in the Atlantic [*Crowley, 1992*; *Schiller et.al.* 1997] (Figure 1). If these models are correct, then upon the onset of thermohaline circulation changes, one might predict a weak sea surface temperature (SST) response in the high latitude South Atlantic that is in anti-phase with the stronger temperature response in the North Atlantic .

Such predictions for the Southern Hemisphere should be readily testable, and therefore, the description of the amplitude, geographic distribution, and phasing of past Southern Ocean surface climate oscillations imposes significant constraints on the mechanisms for interhemispheric climate change.

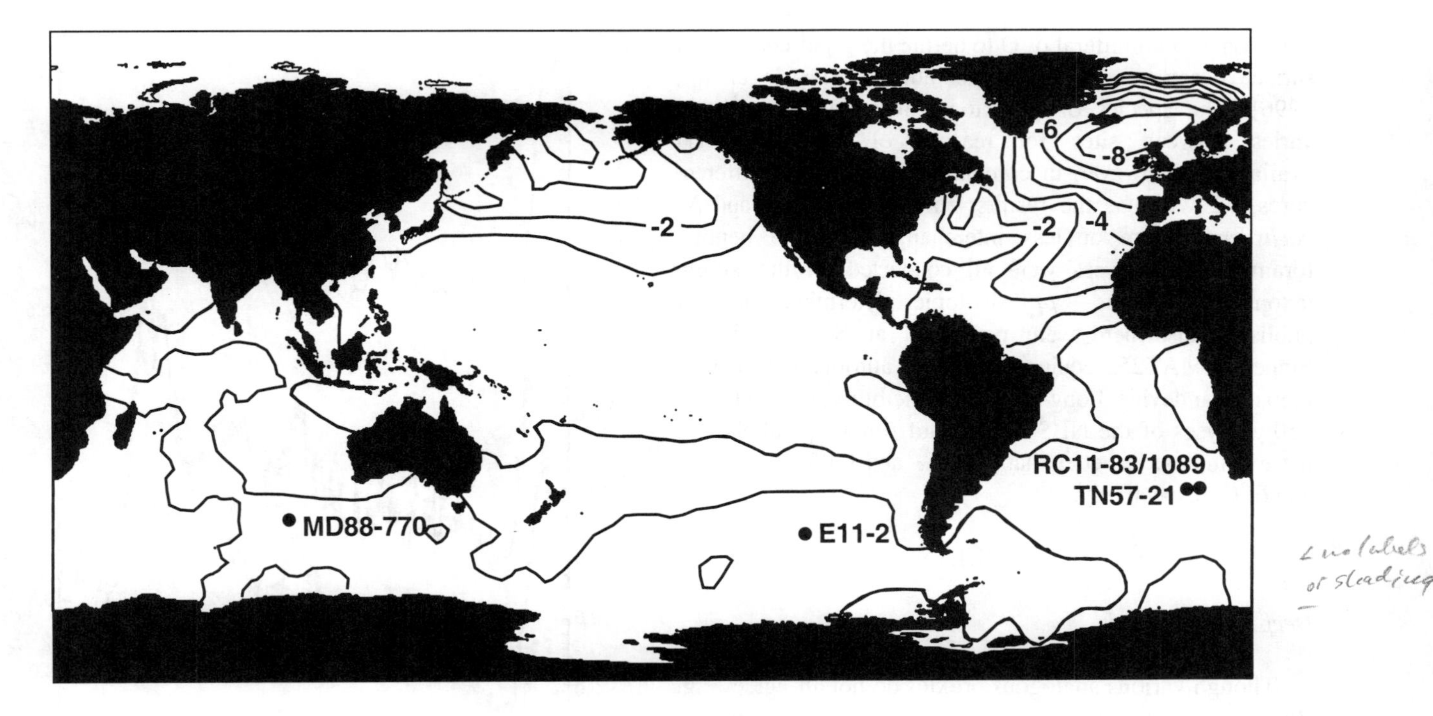

Figure 1. Model results showing the surface ocean temperature response to a meltwater-induced shutdown of NADW [Schiller et al., 1997]. Although the high latitude Northern Hemisphere cools, most of the Southern Hemisphere experiences a slight (0-1°C) warming. The contour interval is 1°C and regions with a positive temperature change are shaded. Also plotted are the locations of cores referenced in the text.

METHODS AND MATERIALS

To approach the issues of millennial scale variability in the Southern Ocean as we have defined them requires deep sea sediment sequences that are (i) characterized by sedimentation rates exceeding 10 cm/kyr; (ii) demonstrably continuous through at least the last 100 kyr, and; (iii) capable of yielding complete isotopic records as stratigraphic tools and as proxies of climate. These requirements currently limit the number of records available for consideration to a select few. From the Indian and Pacific sectors, respectively, we use the published isotopic records from core MD88-770 [*Labeyrie et al*., 1996] and E11-2 [*Ninnemann and Charles*, 1997]. From the Atlantic sector, we present new data from ODP Site 1089, taken from the location of piston core RC11-83 [*Charles et al.*, 1996]. To form a complete Atlantic section through 140 ka, we have spliced the data from RC11-83 and Site 1089. The details of this process will be presented elsewhere, but for our purposes here, we have constructed the splice at 69 kyr, essentially the base of core RC11-83. We also demonstrate the degree of reproducibility of the RC11-83/1089 isotopic time series using nearby core TN057-21.

The age scales for those core sequences not published previously (i.e. Site 1089 and TN057-21) derive from visual matching of the benthic $\delta^{18}O$ records to the SPECMAP chronology [*Martinson et al.* 1987]. In the case of TN057-21, more detailed age assignments were made by aligning the fine scale features of the isotopic records with those of RC11-83. Figure 2 shows the resulting benthic $\delta^{18}O$ time series generated using the various chronologies, along with the *Martinson et al.* [1987] stacked record. We estimate that the age assignments for any given point in the time series are accurate to within several percent of the total age. This level of accuracy is certainly not adequate to demonstrate the relative phasing of the millennial scale variability among the different cores. Therefore, any arguments concerning the phase of abrupt changes must rely on comparison between different variables within a single core.

We limit our consideration of proxy variables in these cores to benthic and planktonic foraminiferal isotopic records. Following established principles, we take the benthic foraminiferal $\delta^{13}C$ as an index of deep ocean circulation changes [*Curry et. al*, 1988], and we use planktonic foraminiferal $\delta^{18}O$ to define the rapid changes in surface ocean sea surface temperature [*Labeyrie et al.*, 1996]. The species of planktonic foraminifera analyzed varies between sites for reasons of abundance and availability. However, in comparing results from different cores, we will use the results from *G. bulloides* and *N. pachyderma* more or less interchangeably. The benthic foraminiferal analyses were all conducted on the single taxon *Cibicidoides, spp.* Isotopic determinations not published elsewhere were performed at S.I.O. using a Finnegan MAT252 equipped with an automatic carbonate preparation device. Long-term reproducibility, deduced from 150 analyses of the NBS-19 standard run in parallel with these samples, is better than 0.08‰ and 0.06‰ for $\delta^{18}O$ and $\delta^{13}C$, respectively.

RESULTS

Deep Water Records

Though various analogous proxies do not all agree [e.g. *Boyle*, 1992; *Yu et al*, 1997], we believe that Southern Ocean benthic foraminiferal $\delta^{13}C$ provides a relatively faithful reflection of global scale changes in thermohaline circulation. Accordingly, we expand here on the empirical evidence supporting a direct tie between conditions in the North Atlantic and the deep Southern Ocean, especially on millennial timescales.

Figure 3 shows the comparison between the RC11-83/1089 record of Southern Ocean deep water chemistry and Greenland ice core record of climate over the last 140 kyr. Warm periods in the North Atlantic region (evidenced by increased $\delta^{18}O$ of Greenland ice) coincide generally with well ventilated Lower Circumpolar Deep Water (LCDW) in the Southern Ocean (evidenced by high benthic foraminiferal $\delta^{13}C$). New results from isotope Stage 5 in Site 1089 show that this correlation between the Greenland ice core record and the Southern Ocean benthic foraminiferal $\delta^{13}C$ persists throughout much of the entire last climate cycle. This agreement even includes some of the controversial Eemian interval (120-125 kyr B.P.) [*Chappellaz et al.*, 1997], where the benthic $\delta^{13}C$ drops significantly (by 0.6‰) in the middle of isotopic Stage 5e. This strong $\delta^{13}C$ minimum occurs firmly in the interglacial, because it appears several thousand years before the first increase in benthic $\delta^{18}O$ that signifies the abrupt end to the last interglacial period in the North Atlantic sediment cores [*Adkins et al.*, 1997]. The GRIP ice core is obviously disturbed in sections of the last interglacial period, and therefore, we do not attach any particular significance to the correlation in the Eemian. But

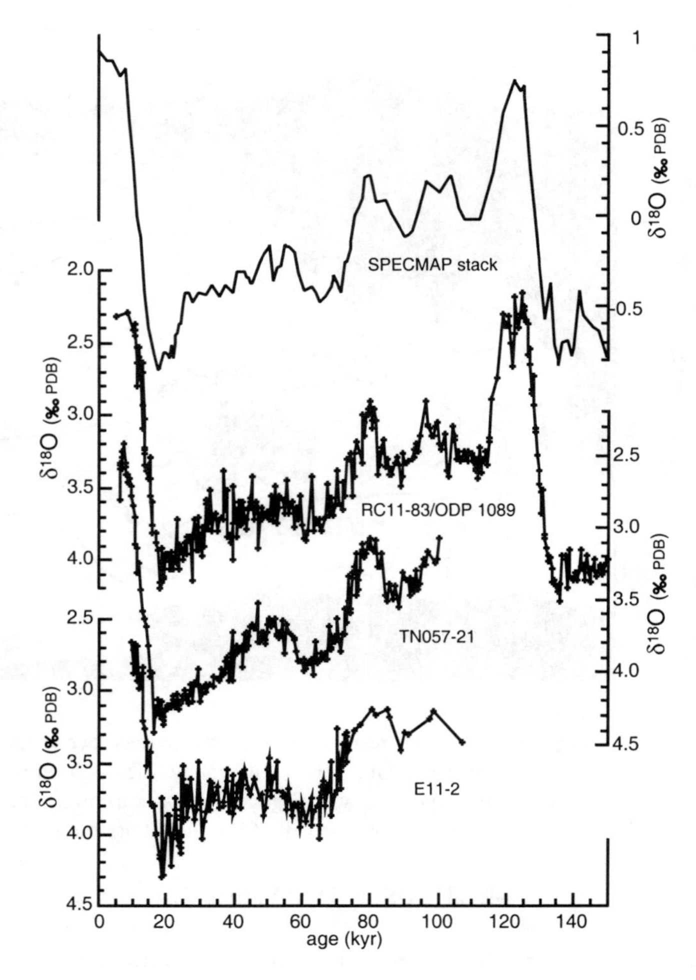

Figure 2. The time series of benthic $\delta^{18}O$ derived from splicing piston core RC11-83 and ODP Site 1089 (40°56'S, 9°54'E, 4624m) is plotted with the benthic $\delta^{18}O$ in nearby core TN057-21 (41°08'S, 7°49E, 4981m) and the Pacific core E11-2 (56°S, 115°W, 3109m). The SPECMAP benthic $\delta^{18}O$ stack [Martinson et al., 1987] used as a reference curve to generate the chronologies is also plotted. For an indication of sedimentation rates, the depth of the last interglacial in the spliced RC11-83/Site 1089 record is 21 meters and the base of TN057-21 is 13.50 meters.

regardless of whether the ice core record through Stage 5 is valid, we have no reason to suspect that our data is the product of local artifacts. At face value, it suggests that abrupt fluctuations in deep water chemistry occurred during both fully glacial and fully interglacial climate states.

Thermohaline circulation is the only possible common influence between South Atlantic deep water nutrient proxy records and the Northern Hemisphere climate recorded in Greenland $\delta^{18}O$ ice; therefore, the similarity between the records over millennial timescales is most plausibly the result of changes in the production (and/or) southward flux

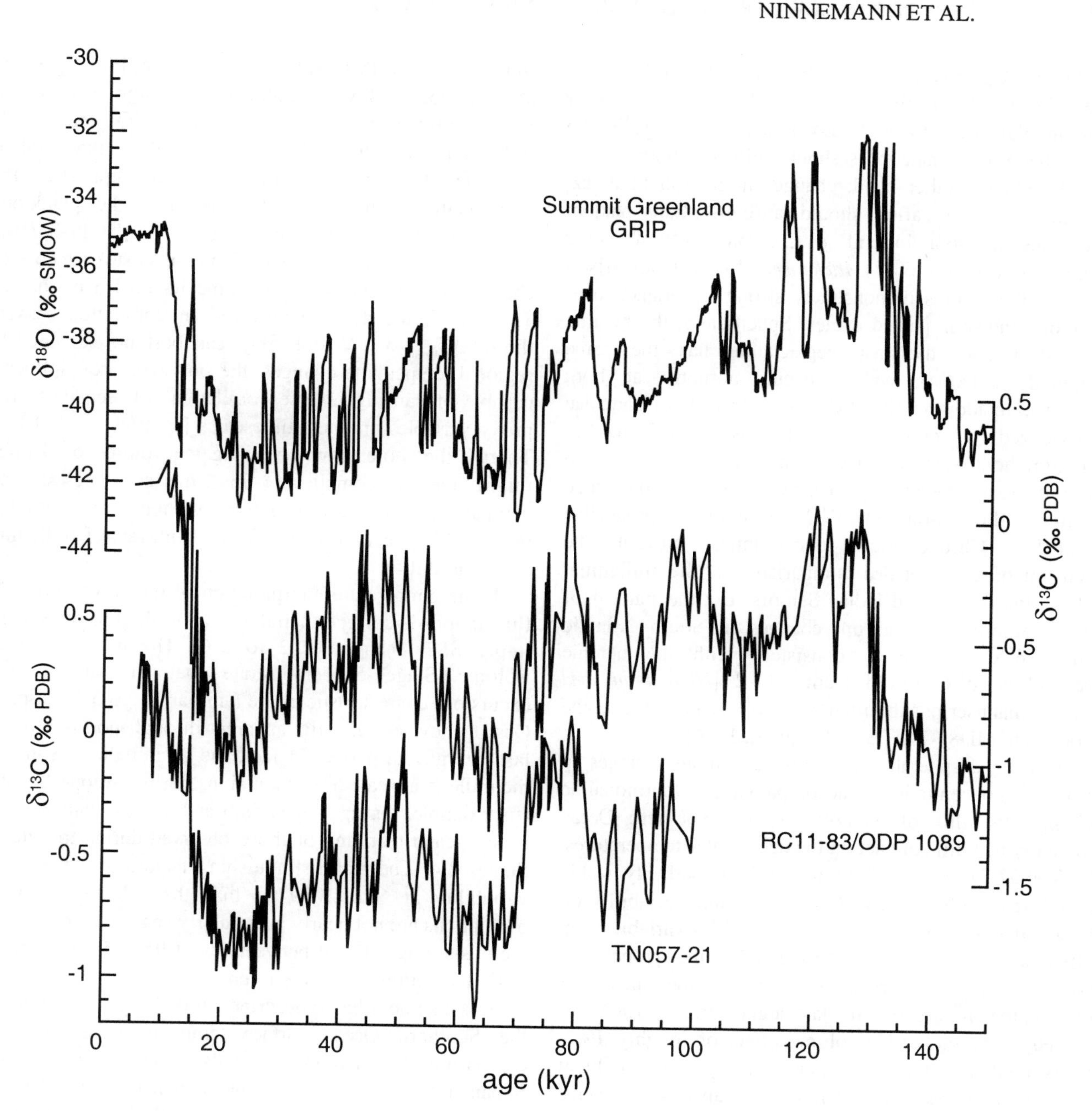

Figure 3. The GRIP ice core δ^{18}Oice record, along with two time series of benthic foraminiferal δ^{13}C: one derived from the same splice (RC11-83/ODP 1089) shown in Figure 2, and the other from core TN057-21. (For illustration purposes only, we created the 140 kyr Greenland record by using the Bender et al. [1994] age model in the section after 71 kyr and the Dansgaard et al., [1993] age model prior to 71 kyr) Both benthic δ^{13}C series show strikingly similarities to the δ^{18}O of Greenland ice, arguing for significant changes in NADW flux throughout much of the last glacial cycle.

of NADW. Still, there are legitimate questions regarding the thermohaline circulation interpretation of Southern Ocean benthic foraminiferal δ^{13}C, because the amplitude of the fluctuations is larger than is expected from water mass mixing alone. In particular, the changing intensity of surface productivity [*Mackensen et al.*, 1993] or changes in alkalinity [*Spero et al.* 1997] have been suggested as possible confounding influences for benthic foraminiferal δ^{13}C. Thus, to use benthic δ^{13}C as a circulation proxy, it is necessary, at the very least, to demonstrate that millennial scale δ^{13}C oscillations are expressed consistently in different sediment cores.

Figure 3 also compares the benthic $\delta^{13}C$ in RC11-83 and TN057-21. Both sites are bathed by Lower Circumpolar Deep Water (LCDW), and any change in the chemistry of this water mass should influence both records equally. On the other hand, changes in microhabitat and organic flux to the seafloor should cause essentially random deviations in foraminiferal $\delta^{13}C$ from bottom water $\delta^{13}C_{\Sigma CO2}$ of up to 0.6‰ [*Mackensen*, 1993]. Comparison between the records demonstrates strong covariance over both long and short period cycles. Specifically, the benthic $\delta^{13}C$ shift over the last deglaciation has the same magnitude (>1.0‰), and it occurs abruptly at both locations. In addition, the $\delta^{13}C$ records from both cores are characterized by millennial oscillations (of 0.3 to 0.7 ‰).throughout isotopic Stage 3. Thus, we see no evidence for a random productivity overprint, because the large amplitude fluctuations in $\delta^{13}C$ are highly reproducible regionally. Chronological uncertainties prevent the extension of such detailed comparison on the millennial scale to the Pacific and Indian Sectors. But the pattern, if not the amplitude, of abrupt changes in South Atlantic benthic $\delta^{13}C$ is entirely consistent with the benthic foraminiferal $\delta^{13}C$ record from E11-2 [*Ninnemann and Charles*, manuscript submitted] and, to a certain extent, the record from MD88-770 [*Labeyrie et al.*, 1996].

However, despite the seemingly enormous changes in nutrient concentration accompanying thermohaline switches, the flux of NADW into the Southern Ocean apparently had little impact on the deep water temperatures. The TN057-21 benthic $\delta^{18}O$ record is remarkably stable through isotope Stage 3, with no discernable variability on millennial scales. This lack of benthic $\delta^{18}O$ variability on millennial timescales is underscored by the pronounced "extra" amplitude observed over orbital timescales. For example, the amplitude of the last deglaciation is 1.8‰ (as opposed to the global ice volume effect of roughly 1‰), suggesting that the temperature and salinity of the ambient deep water at these sites did have the capacity to change substantially over glacial cycles. Thus, the thermohaline circulation effects on Southern Ocean nutrient chemistry and temperature were decoupled, either because the direct temperature effect of NADW is too small in the Southern Ocean [*Manabe and Stouffer*,1997], or because the NADW-related temperature changes were somehow compensated by Antarctic deep water formation processes.

Surface Water Records

Unlike their benthic counterparts in the same sediment cores, planktonic foraminifera $\delta^{18}O$ time series from the Subantarctic South Atlantic are characterized by significant short period (1-4kyr) variability. Although a variety of factors influence the foraminiferal $\delta^{18}O$ record, the millennial scale fluctuations are most likely the result of sea surface temperature changes. Empirical support for this interpretation comes from the fact that the planktonic foraminiferal $\delta^{18}O$ time series from the RC11-83/1089 splice and the deuterium record from the Vostok ice core are characterized by many of the same features over the last 150 kyr (Figure 4). The observed anticorrelation between the $\delta^{18}O_{pl.foram}$ and Vostok δD_{ice} can best be explained by regional temperature changes, the only common influence capable of causing inverse correlation between these two proxy variables on short timescales [*Charles et al*, 1996]. Though the more slowly varying ice volume $\delta^{18}O$ effect might clearly complicate this marine core-ice core comparison in any given interval, we make no attempt to remove this effect artificially for our analysis of millennial scale variability.

If the temperature interpretation of the short term $\delta^{18}O$ fluctuations is correct, then the $\delta^{18}O$ oscillations in Stage 3 imply SST changes of 2 to 3°C. The new data from isotopic Stage 5 demonstrate that the abrupt $\delta^{18}O$ excursions characteristic of the familiar Stage 3 "Dansgaard Oeschger cycles" are still apparent through portions of the last interglacial period. In fact, the amplitude of some of the millennial scale $\delta^{18}O$ excursions during isotope Stage 5 (for example, during Stages 5a,b and e) was equivalent to, if not greater than any of those observed during the glacial Stages 2-4. The real distinguishing feature of the Stage 5 variability at Site 1089 is that these large amplitude oscillations are not nearly as regularly spaced as in Stage 3; there is a long interval between about 95-115 ka when the $\delta^{18}O$ fluctuations were less intense.

It is also possible to describe a broad regional pattern of the Southern Ocean surface temperature response by comparing the isotopic signals from cores in the Subantarctic Pacific (E11-2) and Indian (MD88-770). For this purpose, it is important that all these cores have comparable resolution. The planktonic $\delta^{18}O$ time series in both the Subantarctic Pacific and Indian generally show smaller amplitude variability through isotope Stage 3 than is observed in RC11-83/1089 (Figure 5), with one notable exception: the $\delta^{18}O$ anomaly centered at 25 kyr, which is probably the result of glacial meltwater [*Labeyrie et al.*, 1986; *Shemesh et al.*, 1995]. In fact, there is little room for significant SST variability in the Pacific $\delta^{18}O$ signal, because the amplitude of the fluctuations is no greater than those observed in benthic $\delta^{18}O$ records [e.g. *Shackleton and Pisias*, 1985]. At most, if the Stage 3 planktonic $\delta^{18}O$ oscillations in the Subantarctic Pacific are entirely

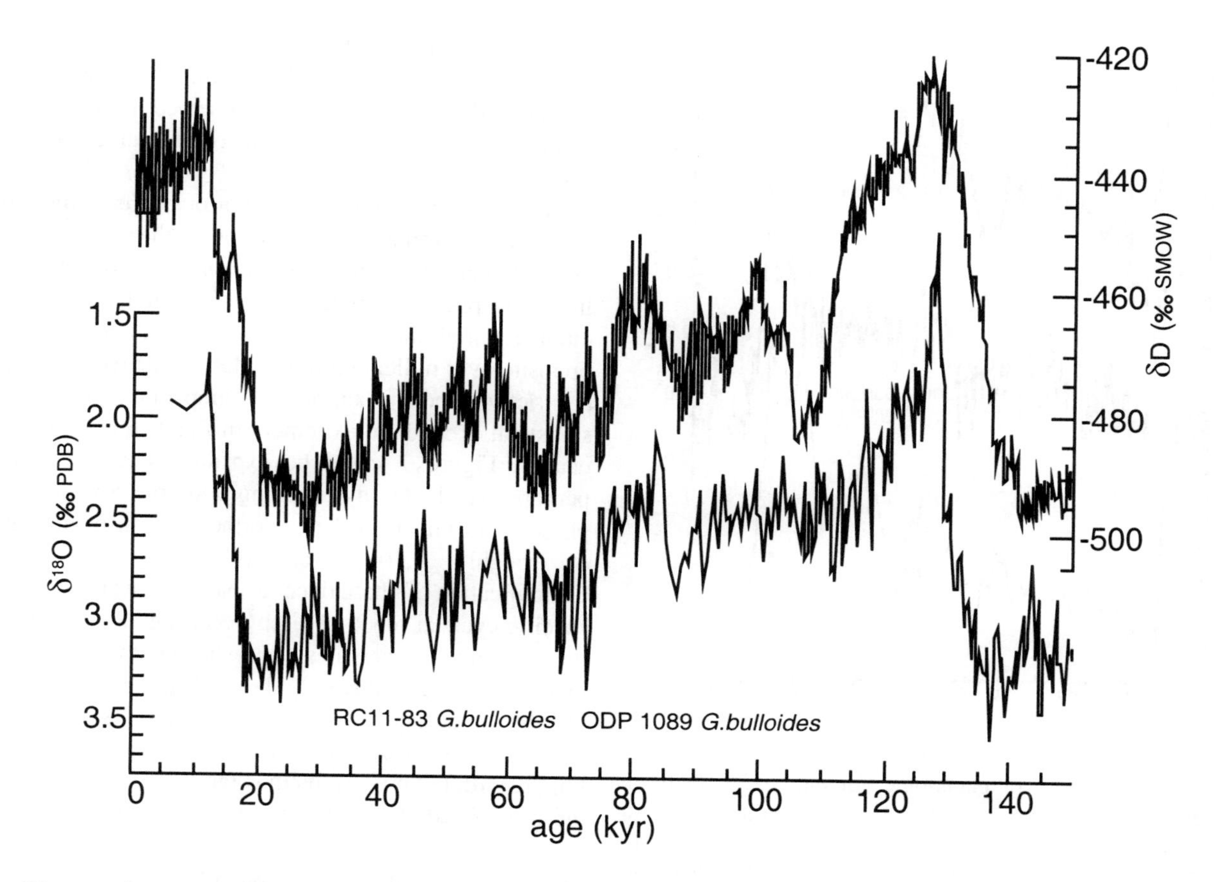

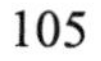

Figure 4. Time series of planktonic foraminifera $\delta^{18}O$ (*G.bulloides*) derived from the same spliced sequence RC11-83/1089, with the δDice time series from Vostok [Jouzel et al., 1987], using the Bender et al. [1994] age model.

temperature-related, then they are still only one third as large as their Subantarctic Atlantic counterparts. These various comparisons emphasize the fact that the more rapid climate oscillations are not manifested uniformly in the Southern Hemisphere SST's: the South Atlantic sector was apparently particularly sensitive to the mechanisms forcing millennial scale variability.

While the geographic pattern of change provides some general constraints on the physical processes that produced the millennial-scale planktonic $\delta^{18}O$ oscillations, the planktonic foraminiferal carbon isotope records may yield more direct clues. Figure 6 shows the $\delta^{13}C$ and $\delta^{18}O$ time series of planktonic foraminifera (*G.bulloides*) in the RC11-83/1089 sequence. In contrast to the lower frequency (orbital) variability, the higher frequency (millennial) isotopic variability appears to be positively correlated: $\delta^{18}O$ minima (warm intervals) correspond to $\delta^{13}C$ minima over the past 150 kyr. The clearest examples of this relationship occur during termination I and termination II. Significantly, this relationship extends through most of the last interglacial period, and therefore, it is a general feature of the millennial scale oscillations in the Subantarctic. In fact, it is perhaps the most obvious pattern of change in planktonic foraminiferal records throughout the South Atlantic [*Charles and Fairbanks*, 1990; *Hodell et al*., in press].

This behavior of the carbon isotopic composition of planktonic foraminifera is important, because modern foraminiferal $\delta^{13}C$ and $\delta^{18}O$ values in surface waters of the Southern Ocean are positively correlated over the Subantarctic region. The inset of Figure 6b shows this positive correlation between planktonic $\delta^{18}O$ and $\delta^{13}C$ in Southern Ocean core-tops spanning 40° to 50° S in the Atlantic. The surface water $\delta^{13}C$ (and therefore planktonic foraminiferal $\delta^{13}C$) in this region is controlled by a complex set of processes [*Gruber et al*., in press], but the mixing line established in the core top transect suggests a very simple interpretation of the millennial scale fluctuations in both $\delta^{18}O$ and $\delta^{13}C$: the tandem variability most likely results from a change in the proportion of polar (vs. subtropical) water in the surface layer overlying northern Subantarctic core sites.

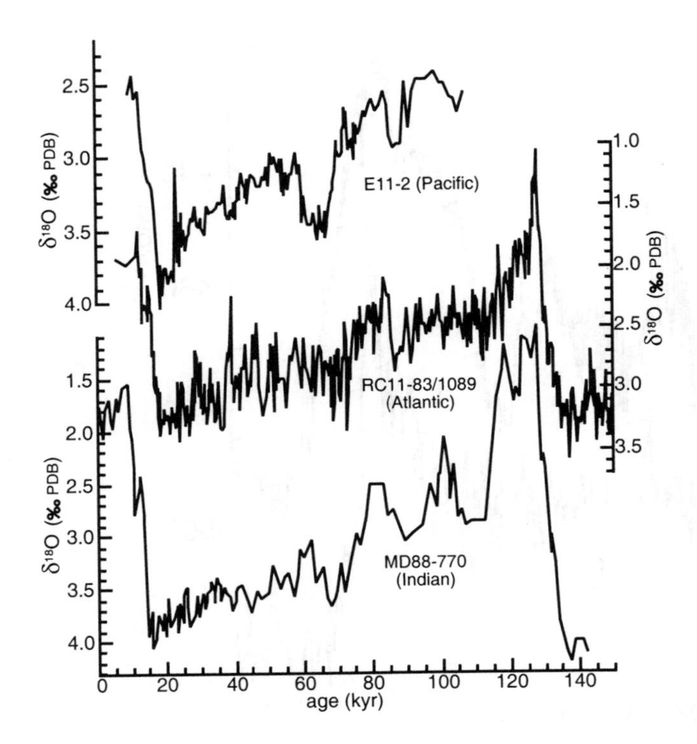

Figure 5. Time series of planktonic foraminifera $\delta^{18}O$ from the Subantarctic Atlantic core RC11-83, the Pacific core E11-2, and the Indian Ocean core MD88-770 [Labeyrie et al., 1996]. The Subantarctic Atlantic core RC11-83 shows the highest amplitude variability on millennial timescales.

Surface/Deep Comparison

The geographical pattern of planktonic isotopes and the interpretation of surface water mixing establishes a more specific context for the comparison to the record of deep ocean circulation (benthic $\delta^{13}C$). Figure 7 compares the benthic $\delta^{13}C$ and planktonic $\delta^{18}O$ time series from the RC11-83/Site 1089. This comparison is independent of chronological uncertainties, because the two records are derived from the same high-sedimentation-rate sequence. Over millennial timescales, most of the rapid benthic $\delta^{13}C$ increases (increased NADW flux) correspond with increases in planktonic $\delta^{18}O$ (cooler Subantarctic Atlantic).

However, this pattern is opposite to the long-period variability, where weak thermohaline circulation (inferred from benthic $\delta^{13}C$) apparently accompanies colder Subantarctic temperatures (inferred from planktonic $\delta^{18}O$) over glacial cycles. Also, there is at least one very prominent exception to the general millennial scale relationship between surface and deep water records. During the last interglacial period (130-125 kyr B.P.), increased benthic $\delta^{13}C$ corresponded with decreased planktonic $\delta^{18}O$ – an "in phase" relationship between the hemispheres, as we have defined the proxies. This example demonstrates that not all millennial oscillations observed in the Southern Ocean record were created as part of the same sequence of events.

The different responses over various parts of the climate spectrum (as well as the exceptions to the rule such as the last interglacial) complicate the analysis of phasing of interhemispheric climate change. Though the sense of climate variability between the hemispheres is usually opposite over millennial timescales in our records, proving a strictly bipolar "antiphase" temperature response on statistical grounds is extremely difficult. For example, the data in Figure 7 could be explained equally well, and perhaps even better, by invoking a simple Southern Ocean phase lead (with respect to Northern Hemisphere climate) of roughly 1,500 years [*Charles et al.*, 1996]. Quantification of phase depends partly on whether there is a specific cyclical mode of deep ocean circulation operating on millennial timescales, or whether the thermohaline instability is an inherently broad band process featuring variable frequencies over different intervals of time. To answer this kind of question conclusively requires a very long record, but our results thus far (Figure 3) suggest that the broad band characterization of the rapid oscillations in thermohaline circulation is more appropriate. Consequently, the usefulness of "frequency domain" analyses may be limited until longer records are available, and we prefer to emphasize the mechanisms operating over some of the most distinctive intervals. For example, it is interesting to note that the positive correlation between planktonic $\delta^{13}C$ and $\delta^{18}O$ (our measure of surface water mass mixing) remains intact over isotope Stage 5e, whereas the correlation between benthic $\delta^{13}C$ and planktonic $\delta^{18}O$ (our measure of the influence of NADW on Southern Ocean climate) does not.

DISCUSSION

While it may seem counterintuitive, there are several advantages to exploring the Southern Ocean record of North Atlantic thermohaline instability. The main benefit of this "downstream" approach is that it yields perhaps the clearest indication of global scale thermohaline effects. As a result, this approach also provides an avenue for linking ice core records from different hemispheres more explicitly. The extension of observations through the entire glacial cycle highlights both of these aspects of the Southern Ocean record (cf. Figure 3 and 4), while adding new dimensions to the debate over the triggers for thermohaline instability and the ultimate effects on global climate.

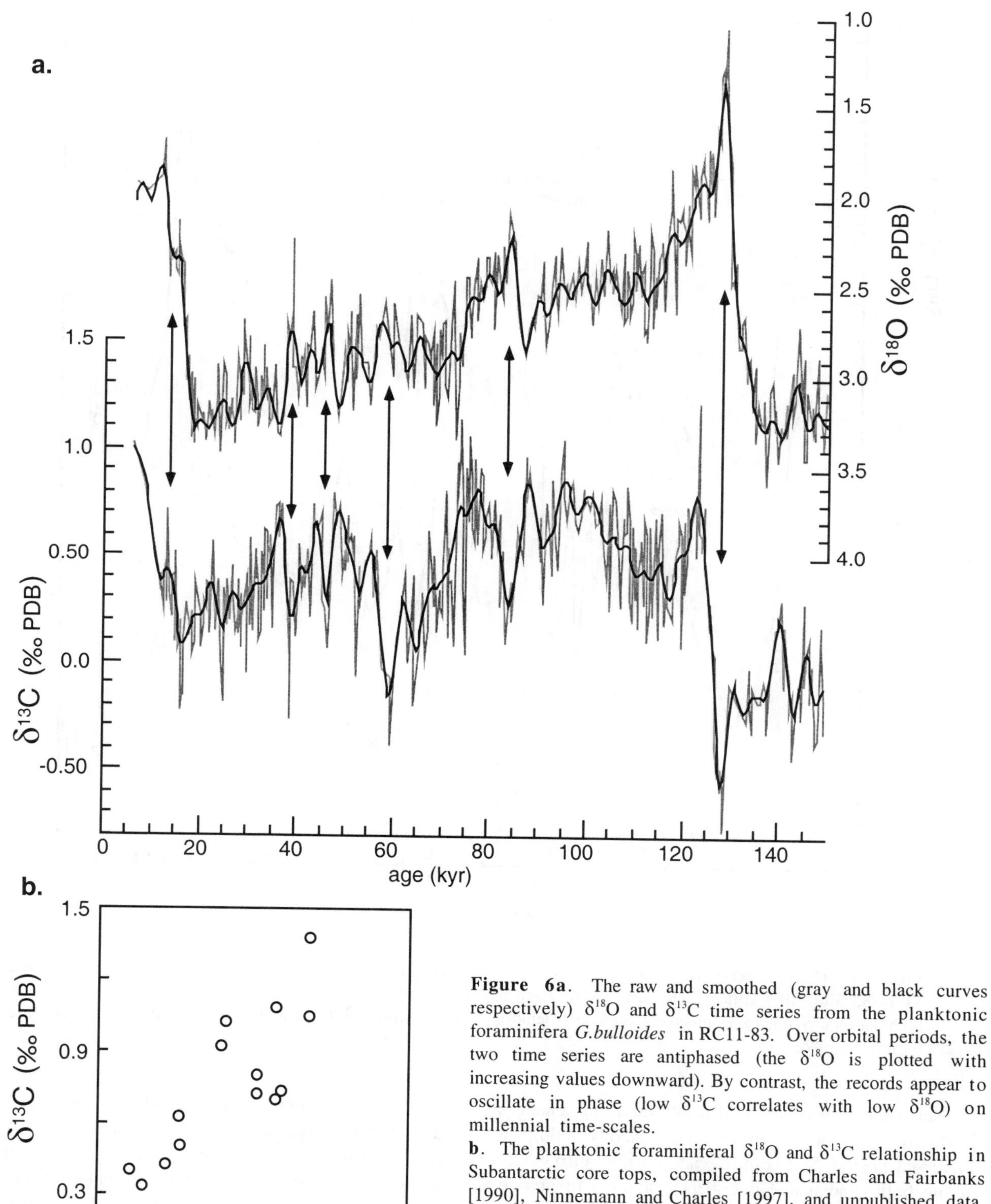

Figure 6a. The raw and smoothed (gray and black curves respectively) $\delta^{18}O$ and $\delta^{13}C$ time series from the planktonic foraminifera *G.bulloides* in RC11-83. Over orbital periods, the two time series are antiphased (the $\delta^{18}O$ is plotted with increasing values downward). By contrast, the records appear to oscillate in phase (low $\delta^{13}C$ correlates with low $\delta^{18}O$) on millennial time-scales.

b. The planktonic foraminiferal $\delta^{18}O$ and $\delta^{13}C$ relationship in Subantarctic core tops, compiled from Charles and Fairbanks [1990], Ninnemann and Charles [1997], and unpublished data. The positive correlation is opposite to that expected from nutrient-related $\delta^{13}C$ variability.

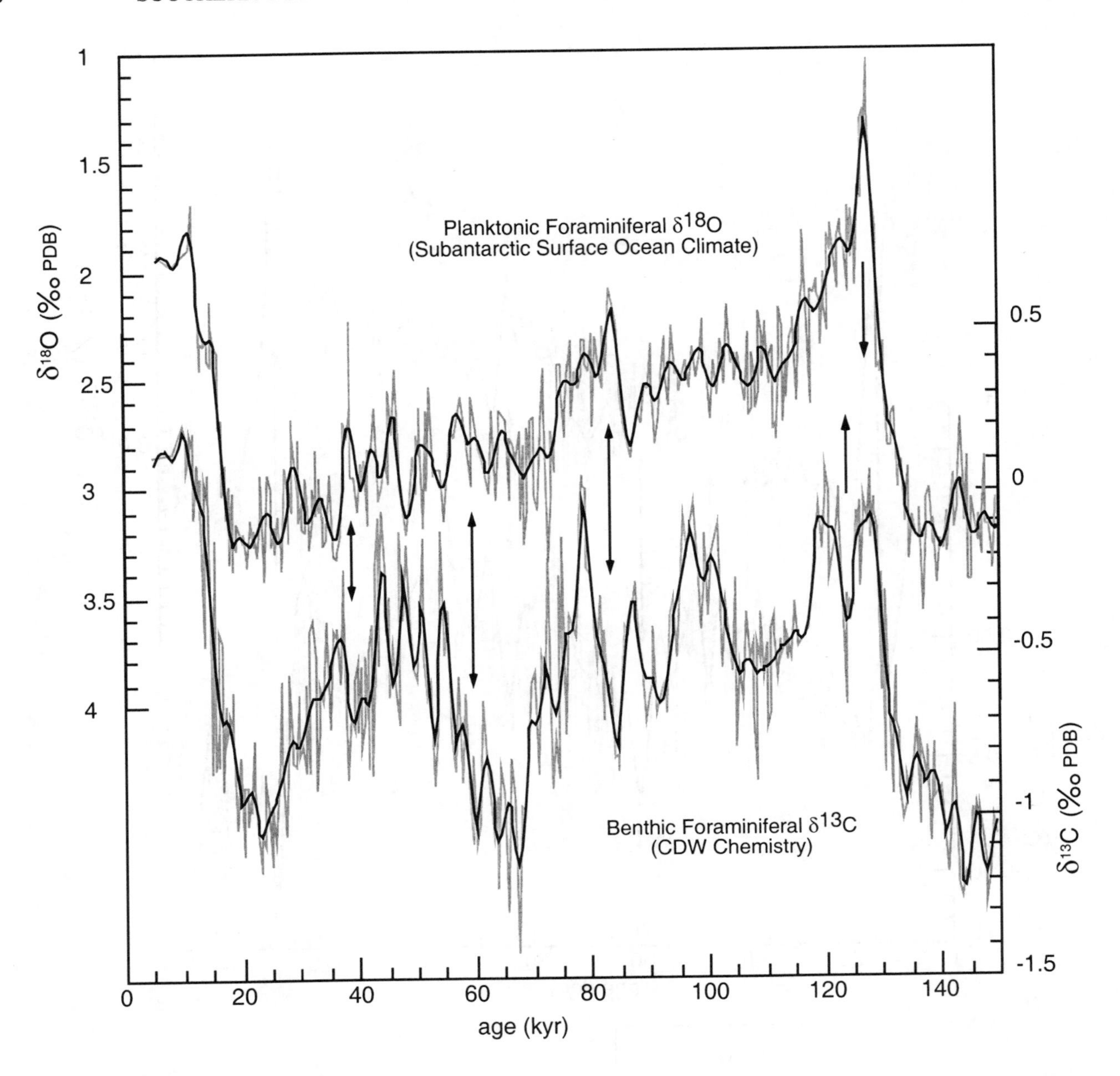

Figure 7. The time series of planktonic $\delta^{18}O$ (*G. bulloides*) plotted with the benthic $\delta^{13}C$ time series from the RC11-83/Site 1089 splice. The phasing between the records provides a direct comparison between changes in high latitude Southern Hemisphere climate (planktonic foraminiferal $\delta^{18}O$) and deep ocean circulation (benthic foraminiferal $\delta^{13}C$).

New data shows that significant and abrupt changes in Northern Hemisphere climate and NADW flux were expressed in the Southern Ocean nutrient proxy records during both glacial and interglacial states. In particular, our observations suggest that there were several prominent reductions in thermohaline circulation within isotope Stage 5. Though the estimates of global ice volume throughout Stage 5 are highly uncertain, one of the most prominent benthic $\delta^{13}C$ oscillations (circulation switches) occurs during the middle of the Eemian, presumably when ice sheet influence reached an absolute minimum. The inference that abrupt deep circulation changes were at times independent of ice volume implies that the original North Atlantic "salt oscillator" concept [*Broecker et al*, 1990] cannot apply to all cases of thermohaline instability, because this model requires significant glacial meltwater input (on the order of several tenths of a Sverdrup) to render deep water production unstable. Furthermore, the presence of a large amplitude benthic $\delta^{13}C$ excursion during the last interglacial suggests that even relatively warm climates, similar to our current state, are not immune to rapid thermohaline circulation changes. Our results therefore offer a means of reconciling the apparent paradox that, while some circulation model results call for a weakening of the thermohaline circulation in a future greenhouse world [*Manabe and Stouffer*, 1993; *Stocker and Schmittner*, 1997], the geological record has consistently suggested that thermohaline instability was a product mainly of glacial climates [*Raymo et al*, 1990]. It may be the case that, once appropriately resolved records of previous interglacial periods are examined, there will be more evidence of different kinds of instability--for example, both thermohaline "catastrophes" caused by rapid influx of glacial meltwater and conveyor shutdowns resulting from more gradual perturbations to the density of the upper layer of the North Atlantic.

However, before taking the interpretation of the last interglacial $\delta^{13}C$ observations too far, it is appropriate to consider their context. For example, it is probably significant that the most prominent benthic $\delta^{13}C$ excursions in RC11-83/1089 during isotope Stage 5 occur precisely in the middle of two large 41,000 cycles of this proxy variable (at 125 ka and 84 ka). Thus, unlike the "Dansgaard-Oeschger" events during Stage 3, the $\delta^{13}C$ changes observed in the last interglacial period could simply represent a harmonic response to the processes that drove the large amplitude 41 kyr cycle in this record. Since many records of North Atlantic conditions are remarkably stable throughout the last interglacial period [e.g. *Oppo and Lehman*, 1995] one might also conclude that these interglacial $\delta^{13}C$ excursions represent times when Southern Ocean deep water chemistry became uncoupled from North Atlantic climate. Recent work in the North Atlantic suggests that this deep water/surface climate decoupling may have been basin wide [*McManus et al.*, 1999]. In any case, regardless of whether the interglacial deep ocean excursions were forced or unforced, or whether they were of strictly North Atlantic origin, we can conclude that ice sheets are not required to drive abrupt oscillations in the thermohaline circulation.

Assuming that Southern Ocean benthic $\delta^{13}C$ does in fact monitor NADW variability faithfully (albeit in an amplified form), then we can also test for the direct climatic effect of North Atlantic mode switches in the Southern Hemisphere. The lack of a significant CDW temperature (benthic foraminiferal $\delta^{18}O$) response during these short period oscillations suggests that the deep ocean is not a direct conduit connecting high latitude temperature variability in both hemispheres. These observations support coupled ocean-atmosphere results that show little (0.5°C or less) temperature change in CDW in response to a shutdown in NADW production [*Manabe and Stouffer*, 1997]. Thus, if thermohaline reorganizations did drive Southern Ocean millennial scale climate events, the effect was not through the direct propagation of temperature anomalies in deep water masses and their influence on Southern Ocean sea ice distribution.

On the other hand, a variety of studies have predicted that NADW affects Southern Hemisphere SST's because of the cross-equatorial heat transport associated with thermohaline overturning [*Manabe and Stouffer*, 1988; *Crowley*, 1992; and *Schiller et al.*, 1997 among others], and we can now test these model predictions more completely. The most obvious test of the models is the phasing of climate between the hemispheres, because the models predict an anti-phase response: the Northern Hemisphere should warm in conjunction with a stronger conveyor, while the Southern Hemisphere should cool. In this regard, our data are generally in agreement with at least the sense of the model results. Over much of the last glacial cycle, one could infer cooler Subantarctic temperatures accompanying vigorous thermohaline circulation.

However, a more specific observation is that the last interglacial period shows a completely different pattern of behavior than the rest of the following 130,000 years. This observation is important, because if the relationship between deep ocean circulation and surface ocean climate evolves over glacial cycles, then one could conclude that North Atlantic Deep Water is not the primary mechanism driving rapid temperature changes in the Subantarctic.

A second test of the coupled ocean/atmosphere predictions for a weakened conveyor involves the amplitude and geographic pattern of millennial scale fluctuations. The models' results show a very subtle and fairly uniform zonal and meridional change in Southern Ocean SST, with only slightly higher sensitivity in the Atlantic sector. Although our core coverage is hardly ideal, the available records show that the Subantarctic Atlantic responds to millennial oscillations much more strongly than do the other sectors of the Subantarctic. Thus, simple changes in northward heat flux associated with NADW formation, as might be predicted in a simple energy balance models [e.g. *Crowley*, 1992] cannot be the only influence on Subantarctic millennial scale variability: some other process must amplify the response, at least in the Atlantic sector. This apparent discrepancy between the couple ocean atmosphere models and observations may be stretching the comparison past the point of legitimacy, since Southern Ocean deep water formation processes (and their controls on sea ice distribution) are notoriously poorly represented in current models. But, with these limitations in mind, it is reasonable to consider the exact physical processes that could drive Subantarctic temperature variability.

In this respect, the deep sea sedimentary evidence imposes a more explicit constraint: whatever mechanism was responsible for the observed planktonic $\delta^{18}O$ variability must also have influenced surface water $\delta^{13}C$ simultaneously. The simplest explanation for a correlative change in planktonic $\delta^{13}C$ and $\delta^{18}O$ observed over millennial cycles is that wind driven mixing processes altered the relative amounts of subtropical (vs. polar) waters at a given Subantarctic site. Of course, any number of specific mechanisms--from shifts in the fronts, to changes in the amount of meridional eddy mixing--could achieve this effect. But all possible mechanisms necessarily involve a shift in the position or strength of the westerlies. Westerly wind variability may also help explain why the surface temperature response is different throughout the various sectors of the Southern Ocean, because the effect of winds on ocean temperature and circulation is strongly dependent on basinal and continental geometry, and is perhaps sensitive to changes in vertical density gradients [*Klinck and Smith*, 1992]. In fact, modeling studies show that the SST response to changes in wind stress is quite heterogeneous among different sectors of the Southern Ocean [*Rahmstorf and England*, 1997].

Requiring changes in the wind field to drive millennial scale temperature variability in the Southern Ocean does not necessarily eliminate the thermohaline circulation as an important global climatic influence--obviously, any process that alters SST gradients would also likely affect the Southern Hemisphere westerlies. But this demand does bring other climatic variables into consideration, because thermohaline circulation is only one of many different processes that help establish surface temperature gradients. For example, observations of modern interannual climate variability demonstrate that tropical Pacific temperature anomalies are linked to downstream anomalies in the Southern Ocean surface wind, temperature, sea ice extent and surface pressure [*White and Peterson*, 1996], and it is likely that past changes in the mean tropical temperatures would have similar effects. Consistent with this mechanism, a Barbados coral record [*Guilderson et al.*, 1994] and rapidly accumulating sediment records [*Arz et al.*, 1999] suggest that oscillations in tropical Atlantic SST's were nearly synchronous with variations in RC11-83 $\delta^{18}O$ and Vostok δD_{ice} over the last deglaciation. In addition, *Curry and Oppo* [1997] showed evidence that tropical Atlantic SST's may have varied by 2-3°C, on timescales similar to those inferred from the RC11-83/1089 planktonic $\delta^{18}O$ record.

In summary, there is general agreement between observations of Southern Ocean millennial scale variability and model results for a bipolar antiphase climate response. This agreement offers some support for the idea that thermohaline circulation changes produce significant global climate variability. But we also emphasize that those model effects resembling simple changes in diffusive heat transport (accompanying thermohaline instability) may not adequately represent the actual changes observed in Southern Ocean deep sea sediments. In fact, at this point, it is equally likely that tropical climate instability [*Cane and Clement*, this volume] could have induced the patterns of variability in the planktonic foraminiferal record, which apparently were driven locally by changes in westerly winds.

Yet, these issues are more than just locally important, because the Vostok ice core deuterium record is so well correlated with the Subantarctic Atlantic oxygen isotope record. Thus, the mechanisms responsible for the variability in our records undoubtedly contributed to the Antarctic ice core records as well. Given this broad regional significance of Southern Ocean variability, and given the equivocal evidence for a North Atlantic driven thermohaline influence on this variability, an explicit consideration of the tie to the tropical oceans (through both modelling and observational studies) represents one logical aim for future exploration.

Acknowledgements. We are grateful to K. Ludwig for her collaboration on and development of the Stage 5 interval of ODP Site 1089. We also thank the scientific party and crew of Cruise TTN-057 and ODP Leg 177. Samples from Site 1089 were provided by the Ocean Drilling Program with sponsorship by NSF. Comments by L. Keigwin, J. McManus and an anonymous reviewer significantly improved the paper. This work was funded by NSF 0CE-9503817 and JOI-USSSP awards.

REFERENCES

Adkins, J.F., E.A. Boyle, L. Keigwin, and E. Cortijo, Variability of the North Atlantic thermohaline circulation during the last interglacial period., *Nature*, *390*, 154-156, 1997.

Arz, H.W., J. Pätzold, and G. Wefer, The deglacial history of the western tropical Atlantic as inferred from high resolution stable isotope records off northeastern Brazil, *Earth and Plant. Sci. Letters*, *167*, 105-117, 1999.

Bender, M., T. Sowers, M. Dickson, J. Orchardo, P. Grootes,

P. Mayewski, and D. Meese, Climate teleconnections between Greenland and Antarctica throughout the last 100,000 years., *Nature*, *372*, 663-666, 1994.

Bond, G., W. Broecker, S. Johnsen, J. McManus, L. Labeyrie, J. Jouzel, and G. Bonani, Correlations between climate records from North Atlantic Sediments and Greenland ice., *Nature*, *365*, 143-147, 1993.

Boyle, E.A., Cadmium and the $\delta^{13}C$ paleochemical ocean distributions during the stage 2 glacial maximum, *Annu. Rev. Earth Planet Sci*, *20*, 245-287, 1992.

Broecker, W.S. and G.G. Denton, The role of the ocean-atmosphere reorganizations in glacial cycles. *Geochimica et Cosmochimica Acta 53*, 2465-2501, 1989.

Broecker, W.S., G. Bond, M. Klas, G. Bonani and W. Wolfli, A salt oscillator in the glacial North Atlantic I, The concept. *Paleoceanography* 5, 469-477, 1990.

Cane, M., and A. Clement, A role for the Tropical Pacific Coupled Ocean-Atmosphere System on Milankovitch and Millennial Timescales, *this volume*, Part II.

Chappellaz, J., E. Brook, T., Blunier, and B. Malaize, CH4 and d18O of O2 records from Antarctic and Greenland ice: a clue for stratigraphic disturbance in the bottom part of the Greenland Ice Core Project and the Greenland Ice Sheep Project 2 ice cores., *J. Geophys. Res.*, *10*, 26,547-26,538, 1997.

Charles, C.D., J. Lynch-Stieglitz, U.S. Ninnemann, and R.G. Fairbanks, Climate connections between the hemispheres revealed by deep sea sediment core/ice core correlations., *Earth and Plantetary Science Letters*, *142*, 19-27, 1996.

Charles, C.D., and R.G. Fairbanks, Glacial to Interglacial Changes in the Isotopic Gradients of Southern Ocean Surface Water, in *geological History of the Polar Oceans: Arctic Versus Antarctic*, edited by U. Bleil, and J. Thiede, pp. 519-538, Kluwer Academic Publishers, Netherlands, 1990.

Crowley, T.J., North Atlantic Deep Water cools the southern hemisphere., *Paleoceanography*, *7*, 489-498, 1992.

Curry, W.B., J.C. Duplessy, L.D. Layberie, and N.J. Shackleton, Changes in the distribution of $\delta^{13}C$ of deep water ΣCO_2 between the last glaciation and the Holocene., *Paleoceanography*, *3*, no.3, 317-341, 1988.

Curry, W.B., and D.W. Oppo, Synchronous, high-frequency oscillations in tropical sea surface temperatures and North Atlantic Deep Water production during the last glacial cycle., *Paleoceanography*, *12*, 1-14, 1997.

Dansgaard, W., S.J. Johnson, H.B. Clausen, D. Dahl-Jensen, N. Gundestrup, C.U. Hammer, C.S. Hvidberg, J.P. Steffensen, A.E. Sveinbjornsdottir, J. Jouzel, and G. Bond, Evidence for general instability of past climate from a 250 kyr ice core record., *Nature*, *364*, 218-220, 1993.

Gersonde, R., D. Hodell, P. Blum et al., Proceedings of the Ocean Drilling Program, Initial Reports, 177: College Station, TX (Ocean Drilling Program), in press.

Guilderson, T.P., R.G. Fairbanks, and J.L. Rubenstone, Tropical temperature variations since 20,000 year ago: modulating interhemispheric climate change., *Science*, *263*, 663-665, 1994.

Gruber, N., C.D. Keeling, R.B. Bacastow, P.R. Guenther, T.J. Lueker, M.Wahlen, W.G. Mook, and T.F. Stocker Spatiotemporal patterns of carbon-13 in the global surface oceans and the oceanic Suess effect. *Global Biogeochemical Cycles,* in press

Hodell, D.A., C.D. Charles, and U.S. Ninnemann, Comparison of Interglacial Stages in the South Atlantic Sector fo the Southern Ocean for the past 450 kyrs: Implications for Marine Isotope Stage (MIS11), Global Planet. Change, in press.

Jouzel, J., C. Lorius, J.R. Petit, C. Genthon, N.I. Barkov, V.M. Kotlyakov, and V.M. Petrov, Vostok ice core: a continuous isotope temperature record over the last glacial climate cycle, *Nature*, *329*, 403-408, 1987.

Keigwin, L.D., and G.A. Jones, Western North Atlantic evidence for millennial-scale changes in ocean circulation and climate., *J. of Geophys. Res.*, *99*, 12397-12410, 1994.

Klinck, J.M., and D.A. Smith, Effect of wind changes during the last glacial maximum on the circulation in the Southern Ocean, *Paleoceanography*, *8*, no.4, 427-433, 1993.

Labeyrie, L.D., J.J. Pichon, M. Labracherie, P. Ippolito, J. Duprat, and J.C. Duplessy, Melting history of the Antarctica during the past 60,000 years., *Nature*, *322*, 701-706, 1986.

Labeyrie, L.D., M. Labracherie, N. Gorfti, J.J. Pichon, M. Vautravers, M. Arnold, J.C. Duplessy, M. Paterne, E. Michel, J. Duprat, M. Caralp, and J.L. Turon, Hydrographic changes of the Southern Ocean (southeast Indian sector) over the last 230 kyr., *Paleoceanography*, *11*, no. 1, 57-76, 1996.

Laskar, J., The chaotic motion of the solar system: A numerical estimate of the chaotic zones, Icarus, 88:266-291, 1990.

Mackensen, A., H.W. Hubberten, T. Bickert, G. Fischer, and D.K. Fütterer, The d13C in benthic foraminiferal tests on fontbotia wuellerstorfi (schwager) relative to the d13C of dissolved inorganic carbon in Southern Ocean deep water: Implications for glacial ocean circulation models., *Paleoceanography*, *8*, 586-610, 1993.

Marchitto, T.M. W.B Curry and D.W. Oppo,. Millennial-scale changes in North Atlantic circulation since the last glaciation. *Nature 393*, 557-561, 1998.

Martinson, D.G., N.G. Pisias, J.D. Hays, J. Imbrie, T.C. Moore, and N.J. Shackleton, Age Dating and the Orbital Theory of the Ice Ages: Development of a High-Resolution 0 to 300,000-Year Chronostratigraphy, *Quat.Res.*, *27*, 1-29, 1987.

Manabe, S., and R.J. Stouffer, Two stable equilibria of a coupled ocean-atmosphere model., *J. Clim.*, *1*, 841-866, 1988.

Manabe, S., and R.J. Stouffer, Century-scale effects of increased atmospheric CO2 on the ocean-atmosphere system, *Nature*, *364*, 215-218, 1993.

Manabe, S., and R.J. Stouffer, Coupled ocean-atmosphere model response to freshwter input: Comparison to Younger Dryas event., *Paleoceanography*, *12* (April), 321-336, 1997.

McManus, J.F., D.W. Oppo, and J.L. Cullen, A 0.5-Million-Year Record of Millennial-Scale Climate Variability in the North Atlantic, *Science*, *283*, 971-975, 1999.

Ninnemann, U.S., and C.D. Charles, Regional differences in Quaternary Subantarctic nutrient cycling: Link to intermediate and deep water ventilation., *Paleoceanography*, *12*, 560-567, 1997.

Oppo, D.W., and R.G. Fairbanks, Variability in the deep and intermediate water circulation of the Atlantic Ocean during the past 25,000 years: Northern Hemisphere modulation of the Southern Ocean, *Earth and Plantet. Sci. Lett.*, *86*, 1-15, 1987.

Oppo, D.W., and S.J. Lehman, Suborbital timescale variability of North Atlantic Deep Water during the past 200,000 years, *Paleoceanography*, *10*, 5, 901-910, 1995.

Rahmstorf, S., Bifurcations of the Atlantic thermohaline circulation in response to changes in the hydrological cycle., *Nature*, *378*, 145-149, 1995.

Rahmstorf, S., and M.H. England, Influence of Southern Hemisphere Winds on North Atlantic Deep Water Flow., *Journal of Physcial Oceanography*, *27*, 2040-2054, 1997.

Raymo, M.E., W.F. Ruddiman, N.J. Shackleton, and D.W. Oppo, Evolution of Atlantic-Pacific d13C gradients over the last 2.5 m.y., *Earth and Planetary Science Letters*, *97*, 353-368, 1990.

Schiller, A., U. Mikolajewicz, and R. Voss, The stability of the North Atlantic thermohaline circulation in a coupled ocean-atmosphere gerneral circulation model., *Climate Dynamics*, *13*, 325-347, 1997.

Shackleton, N.J., and N.G. Pisias, Atmospheric Carbon Dioxide, Orbital Forcing, and Climate, in *The Carbon Cycle and Atmospheric CO2: Natural Variations Archean to Present*, edited by E.T. Sundquist, and W.S. Broecker, pp. 303-317, 1985.

Shemesh, A., L.H. Burckle, and J.D. Hays, Late Pleistocene oxygen isotope records of biogenic silica from the Atlantic sector of the Southern Ocean., *Paleoceanography*, *10*, 179-196, 1995.

Spero, J.J., J. Bijma, D.W. Lea, and B. E Bemis, Effect of seawater carbonate concentration on foraminiferal carbon and oxygen istotopes, *Nature*, *390*, 497-500, 1997.

Stocker, T.F., and A. Schmittner, Influence of CO2 emission rates on the stability of the thermohaline circulation, *Nature*, *388*, 862-865, 1997.

Weyl, P.K., The role of the oceans in climate change: a theory of the ice ages., *Meteorolo. Monogr.*, *8*, 37-62, 1968.

White, W.B., and R.G. Peterson, An Antarctic circumpolar wave in surface pressure, wind, temperature and sea-ice extent., *Nature*, *380*, 699-702, 1996.

Yu, E.F., R. Francois, and M.P. Bacon, Similar rates of modern and last-glacial ocean thermohaline circulation inferred from radiochemical data, *Nature*, *379*, 689-694, 1996.

U.S. Ninnemann, Graduate Department, Scripps Institution of Oceanography, University of California San Diego, La Jolla, CA 92093-0208.

C. D. Charles, 3119 Sverdrup Hall, Scripps Institution of Oceanography, La Jolla, CA 92093-0220.

D. Hodell, Department of Geology, P.O. Box 117340, University of Florida, Gainesville, FL 32611.

High-frequency Oscillations of the Last 70,000 Years in the Tropical/Subtropical and Polar Climates

Frank Sirocko[1,2], Dirk Leuschner[1], Michael Staubwasser[1] & Jean Maley[3], Linda Heusser[4]

High-resolution pollen records from laminated sediments of the crater lake Barombi Mbo in the rainforest of tropical Africa, from laminated marine sediments of the Santa Barbara Basin off California, and eolian dust in deep-sea cores from the northern Indian Ocean are used to evaluate possible forcing mechanisms of abrupt climate change in the tropical/subtropical regions during the last 25,000 years. The three regions show a common series of century-scale abrupt climate extremes during the LGM and continuing during the deglacial into the Holocene. Further back in time during oxygen isotope stage 3 the Greenland ice cores reveal flickering oscillations, the Dansgaard-Oeschger cycles. These cycles also occur in the intermediate water ventilation of the equatorial Pacific. An equivalent for the Dansgaard-Oeschger cycles operates also in the low latitudes, reconstructed trom monsoon-related dust, upwelling and ventilation records in the Arabian Sea, northern Indian Ocean. Besides the teleconnection between the monsoonal and North Atlantic/Greenland climate we observe some resemblance between stages of stronger monsoon with intervals of warm air temperatures over the Antarctic. As the low latitudes, in particular the equatorial Pacific, receive the very most part of the incoming solar radiation, which is then exported to the high latitudes, we interpret the apparent array of teleconnections within the past global climate cycles to be probably driven by the El Niño/Southern Oscillation anomaly and associated changes in the position of the subtropical/subpolar jet stream.

INTRODUCTION

The geoid of the Earth (Fig. 1) has an area of $511*10^6 km^2$, 40 % ($204*10^6 km^2$) of which fall within the tropical latitudes (astronomically defined as the area where the angle of the sun reaches 90°), and 4.1 % ($21*10^6 km^2$) of which fall within each polar region (defined as the area between a pole and 66.5^o, i.e. where the sun does not eclipse during winter). The subtropics are not defined astronomically but by the poleward extent of the Hadley cell, which reach up to 40° during the seasonal climax, however with large regional variations. To calculate the surface area of tropics and subtropics together we used an annual average extent up to 30° N and S for the Hadley cell (Fig. 1); this brings the entire area of the low latitudes (used in this paper as tropics and subtropics together) to a value of $293*10^6 km^2$, or 57% of the Earth's surface.

This large area in the low latitudes receives the vast majority of solar insolation reaching the Earth; with an annual average of about 300 kcal/cm^2 in the low latitudes and about 130 kcal/cm^2 in the polar regions. In total, the annual heat gain of the tropics/subtropics is about ten times higher than in the polar regions. The mid latitudes are characterized by intermediate insolation values, but strong ocean surface- and atmospheric currents transporting heat from the low latitudes to the high latitudes.

The evolution of these heat-transporting currents during the Earth's history is forced by tectonic process on the

[1] GeoForschungsZentrum Potsdam, Germany
[2] Institut für Geowissenschaften, Johannes Gutenberg-Universität Mainz, Germany
[3] Paleoenvironnement & Palynologie (CNRS & ORSTOM) Univ. Montpellier-2, Montpellier, France
[4] Lamont Doherty Earth Observatory, Palisades, NY

Mechanisms of Global Climate Change at Millennial Time Scales
Geophysical Monograph 112

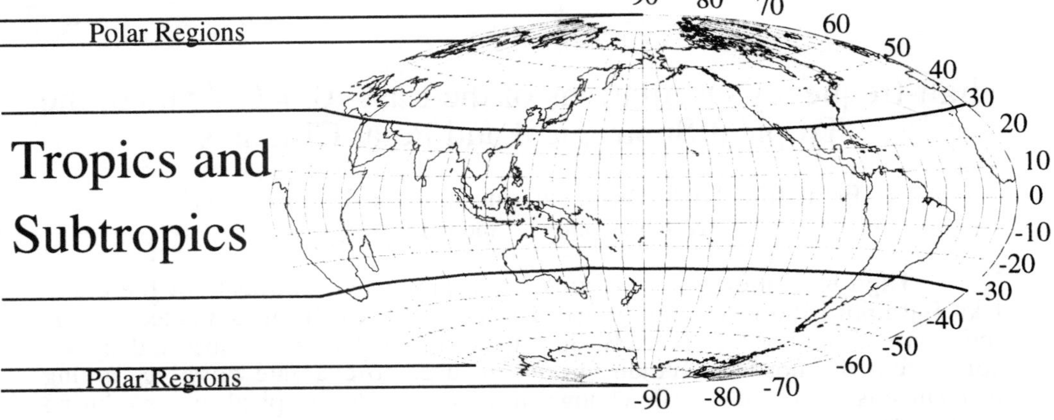

Figure 1. The climatic belts of the Earth in an equal area projection.

million-year time scale, e.g. *Prell and Kutzbach* [1992], by sinusoidal variations in the orbital constellation of the Earth and Sun on time scales of ten- to hundred thousand years [Milankovitch, 1939; Berger and Loutre, 1991; Imbrie et al., 1992]. Annual, seasonal and daily variations in these currents are a function of the local solar insolation on the ground, temperature of the Earth and surface ocean as well as associated feedback phenomenons like ocean current extent, atmospheric moisture transport, and vegetation.

Well dated climate records for the interval between the very long and very short cycles, i.e. the interval of decadal-, century-, and millennial-scale (10-1000 years) have become available only very recently. In 1993, the new Greenland ice cores GISP and GRIP corroborated the earlier results by *Dansgaard et al.* [1984] and stimulated a discussion concerning the cause of millennial-scale climate changes. Since then similar cycles have been reported from almost all over the globe, see examples for the Pacific [*Kotilainen and Shackleton*, 1995; *Behl and Kennett,* 1996], Atlantic [*McIntyre and Molfino*, 1996; *Bond et al.*, 1998], Indian Ocean [*Schulz et al.*, 1998; *Reichart et al.*, 1999; *Leuschner and Sirocko*, in press], Europe [*Thouveny et al.*, 1994], Asia [*Porter and Zisheng*, 1995], America [*Grimm et al.*, 1993; *Heusser*, 1998]. Unfortunately, there is no well dated record from Africa spanning the entire last 70,000 years.

A direct comparison of the Antarctic and Greenland climate has been done by synchronizing the Vostok and GISP2 ice core via trace gas content [*Bender et al.*, 1994]; however, this comparison is not finally concluded, as the age model for Vostok is still developing, see *Bender at al.*, this issue). In any case, similarities between the Dansgaard-Oeschger temperature cycles over Greenland and cycles in the Antarctic climate become apparent. The main question is why the abruptness of events in the Greenland climate is not resolved in the Vostok ice; is it a function of the low snow accumulation rate on the high Antarctic ice shield, or a genuine pattern of the Antarctic climate? A new, high-resolution ice core from the coastal Antarctic shows very similar fluctuations as over Greenland [*Steig et al.*, 1999]. In any case, the statement appears to be justified that millennial-scale climate variability is a dominating signal in well dated high-resolution paleoclimate records all over the world.

OBJECTIVE

If the Dansgaard-Oeschger Cycles (1450-year, 3000 year length) are a dominant global climate signal of the last glacial cycle, then it is necessary to look for globally working mechanisms linking/causing abrupt climate change. One of the most critical points to be considered is the phase relation between events that appear to be synchronous at first sight. Within the present time resolution of absolute dating, we have to accept an apparent error of centuries in the records of the last glacial, and even millennia in the records of marine oxygen isotope stage 3 and 4 (see for example a 3500 year mismatch in the chronologies of the nearby GRIP and GSIP cores for interstadial 8 [*Johnsen et al.*, 1992; *Grootes et al.*, 1993]). *Stocker et al.* [1992] used a coupled atmosphere/ocean model to show an out-of phase relation between the North Atlantic and the Antartic climate, and developed the concept of a hemispheric asymmetry of events, i.e. stadials on the northern hemisphere may represent interstadials on the southern hemisphere. Before the consistency of lead in the Antarctic is not finally resolved in long paleorecords (see work by *Blunier et al.* [1998]) any discussion on the forcing mechanism of climate oscillations must remain tentative.

It is the objective of this paper to add to this discussion by carefully comparing a few high-resolution records from the low latitudes (African tropical rain forest, Indian monsoon, Pacific ENSO (El Niño/Southern Oscillation), explore the possibility of teleconnections within the paleoclimate of the low latitudes, and put this into the perspective of the flickering climate oscillations as visible in the ice core records from both high latitudes.

1. THE LAST 25 Cal ka

1.1. The monsoonal climate, Arabian Sea (0-25 Cal ka)

Sirocko et al. [1993] published a series of several deglacial abrupt climate events in the monsoonal region of the northern Indian Ocean (Fig. 2), with the specific events at 11,6 Cal ka and 14, 5 Cal ka being teleconnected to the climate transitions of the high northern hemisphere, the earlier events being addressed to changes in the high southern latitudes [*Sirocko et al.*, 1996]. *Heusser & Sirocko* [1997] then realized that $\delta^{18}O$ spikes during the LGM, not accounted as significant events in the earlier work, were of the same nature as those during the termination and reveal a teleconnection between the Pacific ENSO and the Asian monsoon. These events near 17.0, 16.1, 14.3, 13.5, 13.1, 12.1, 9.9, 8.8, 7.3 ^{14}C ka in the Indian monsoon (Fig. 3 c), recently also reported from the South China Sea at 17.0, 15.9, 15.5, 14.7, 13.9, 13.5, 12.1, 11.5, 10.7 ^{14}C ka [*Wang et al.*, in press], are synchronous within the dating precision of ^{14}C. Unfortunately, these events could alternatively well be out of phase, given the large error bars of the ^{14}C technique during the termination. A synchroneity between these events over the Arabian Sea, South China Sea and east Pacific (see below) appears however rather likely, as the Asian monsoon and the Pacific El Niño systems are tightly coupled via land albedo, the atmosphere and ocean currents [*Shukla*, 1987; *Barnett et al.*, 1991; *Webster and Yang*, 1992; *Barnett et al.*, 1994; *Meehl*, 1994; *Tourre and White*, 1995; *Li and Yanai*, 1996; *Webster et al.*, 1998].

1.2 The low latitude east Pacific, Santa Barbara Basin (0-25 Cal ka)

The partly laminated sediments of ODP893 from the Santa Barbara Basin off southern California reveal an excellent record of the changes in hydrography of the low latitude Pacific [*Behl and Kennett*, 1996]. The pollen in these sediments are a direct function of vegetation (temperature and precipitation driven) in adjacent southern California. *Heusser* [1995], *Heusser and Sirocko* [1997], *Heusser* [1998] presented pollen curves, with a prominent feature being abrupt spikes in the abundance of *Pinus* (a tree growing in mountainous regions and well adapted to drought and bushfire), the series for 0-25 Cal ka being reproduced in Fig. 3 a. Each pine maximum (up to 60% of all tree pollens) is paralleled by a disappearance of the

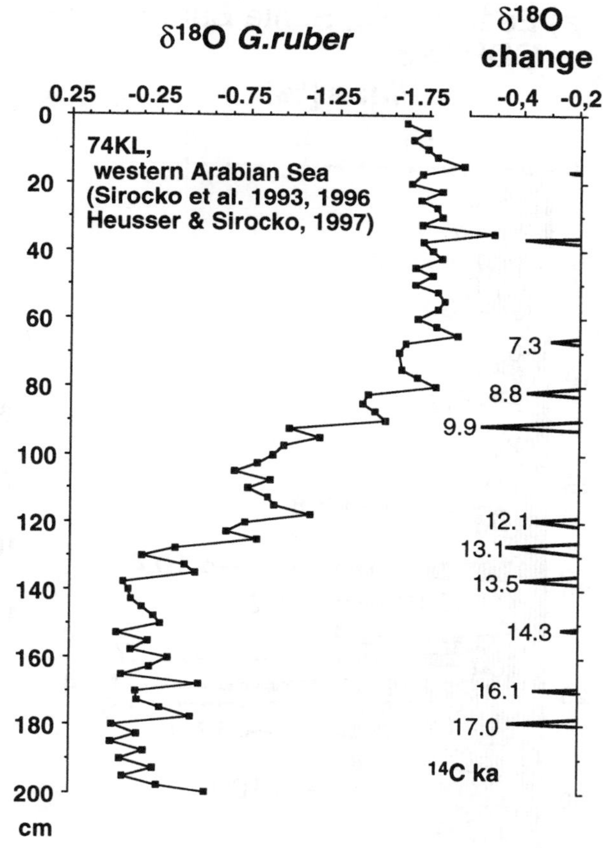

Figure 2. $\delta^{18}O$ of planktonic species *G.ruber* in core 74KL from the western Arabian Sea [*Sirocko et al.*, 1993, 1996], $\delta^{18}O$ change is the difference between a value and the value immediately below, highlighting the major changes in the curve (from *Heusser & Sirocko*, 1997). The event at 9.9 ^{14}C-kyrBP is equivalent to the prominent end of the Younger Dryas at 11.6 Cal-ka, the event at 12.1 ^{14}C-kyrBP being near to the Bolling transition at 14.5 Cal-ka.

juniper trees (decreasing from 60% to 10% during a pine maximum), a minimum in total pollen abundance, and maximum of charcoal fragments [*Heusser and Sirocko*, 1997]. The minimum of total pollen in combination with the charcoal maxima shows that the deglacial juniper forests were destroyed by fire, and pine survived or even spread at the expense of juniper. Accordingly, the *Pinus*-oscillations in Fig. 3 a are probably primarily a record of paleo-bushfires, i.e. drought and heat. Today, bushfires in southern California are a function of Pacific El Niño anomalies. The deglacial paleovegetation record from the laminated Santa Barbara sediments [*Heusser and Sirocko*, 1997], suggests that a similar mechanism as ENSO was operating at least periodically also during the Last Glacial Maximum.

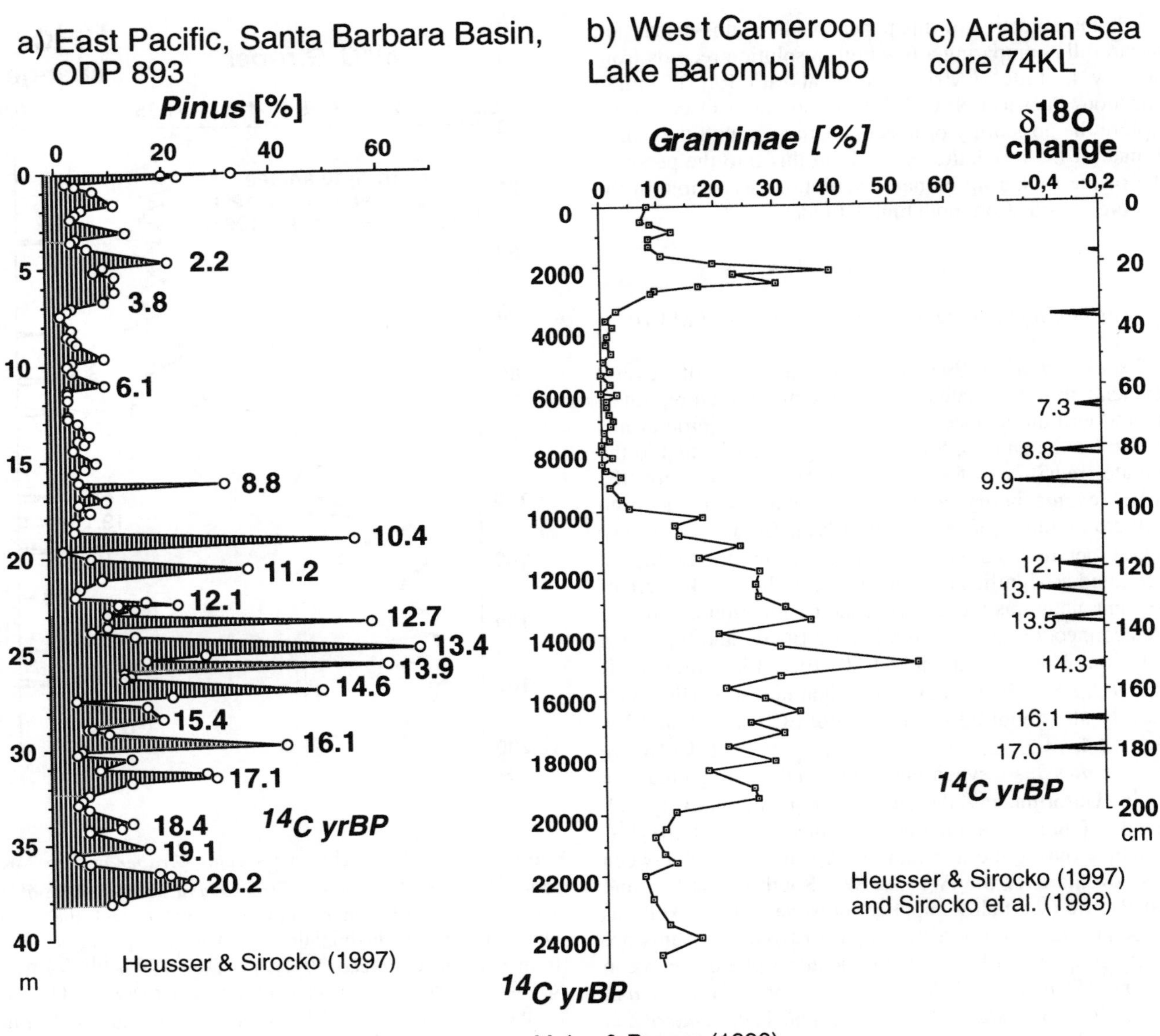

Figure 3. (a) Abundance (percent of total tree pollens) of pine pollens in ODP 893A Santa Barbara Basin off southern California, from *Heusser & Sirocko* [1997]. (b) Abundance (percent of total pollens) of Graminae (grasses) in lake Barombi Mbo, tropical rain forest of West Africa-Cameroon, from *Maley and Brenac* [1998]. (c) $\delta^{18}O$ change in core 74KL, western Arabian Sea, from Fig. 2.

1.3. Tropical central Africa, the equatorial rain forest Domain, Lake Barombi Mbo in SW Cameroon (0-25 Cal ka).

In tropical Africa few pollen records go back before 20 Cal ka BP, the LGM period. Several detailed studies were done in East Africa (see a recent synthesis in *Jolly et al.* [1997]) and one in the forest Domain of central Africa in south-west Cameroon, the lake Barombi Mbo [*Maley*, 1997]. Located 80 km from the sea (4°40'N-9°24'E) and at an altitude of ca. 300 m, the lake Barombi Mbo is a 1 Myr old crater with a diameter of ca. 2 km and a maximum depth of 110 m where a 23.5 m sediment core was obtained. The laminated sediments give a rather regularly dated sequence of pollen variability during the last 28,000 ^{14}C-yrBP [*Giresse et al.*, 1991; *Maley and Brenac*, 1998]. In Fig. 3 b we present the curve of Graminae (grasses), with high Graminae abundance representing arid phases and low Graminae abundances represent times of tropical rainforest domination. The sampling resolution for the pollen curve is of ca. 250 yr in the Holocene and ca. 400 yr during the LGM. Accordingly, spikes of 200 year duration as in Fig. 3 a cannot be resolved in Fig. 3 b. The early and middle Holocene is completely dominated by arboreal pollen indicating a maximum extension and density of the rain forest. After the LGM the transition period corresponding to the fast colonization phase of the forest intervened between ca. 10,000 and 9500 ^{14}C-yrBP, just after the end of the Younger Dryas time period. During the glacial and deglacial from 20,000 to 10,000 ^{14}C-yrBP, the savannas, mostly characterized by grasses, expanded greatly in this region. However the landscape was of a mosaic of forest and savanna in which the patches of forest were the largest, as was confirmed by isotopic studies [*Giresse et al.*, 1994]. A short maximum of aridification was dated for 15,000 - 14,500 ^{14}Cyr BP, almost the same time when icebergs from the initial decay of the Laurentian ice sheet formed "Heinrich Layers" in the North Atlantic [*Bond and Lotti,* 1995].

In tropical Africa the late Holocene period started at ca. 3800 ^{14}Cyr BP initiating major climatic changes [*Maley*, 1997]. The equatorial West (Bosumtwi) and East Africa (Taganyika and other nearby lakes) suffered a general and dramatic lacustrine regression between 3800 and 2800 ^{14}Cyr BP, but the evergreen rain forest Domain of West Cameroon (Barombi Mbo) was simultaneously favoured by a maximum of humidity [*Maley and Brenac*, 1998]. This patterned response with opposite trends was typical for the entire tropical belt during the period between 3800 and 2800 ^{14}C-yrBP, with strong drying in India and northern Australia, but with a large increase of rains in the Andes of Southern America [*Maley*, 1997], as being documented in a 20 m transgressive jump at 3800 ^{14}C-yrBP in the lake Titicaca [*Martin et al.*, 1993]. Such pattern characterized by alternating sectors with opposite changes corresponds typically to ENSO phenomenons related to Sea Surface Temperature (SST) variations [*Gunn*, 1991; *Rodbell et al.*, 1999]. Major changes were observed also in Antarctica at this date [*Mosley-Thompson et al.*, 1993], and also in the Santa Barbara Basin sediments with strong increase in pine-dominated vegetation at 3800 ^{14}C-yrBP (Fig. 3) or in the Euro-Asian continent [*Fontes et al.*, 1996] indicating a global phenomenon [*Kelts*, 1997].

1.4. Summary on tropical/subtropical climate variability (0-25 Cal ka)

The pollen curves from southern California and south-west Cameroon (on opposite sites of the globe, but in the same latitudinal band) look remarkably similar, and share their most extreme phases with similar structures in the evolution of the monsoon (Fig. 2, 3). All three regions are characterized by spikes (not all of them being fully developed in each record) of climate extremes lasting for not more than 200 years. Within the dating precision of ^{14}C these anomalies could all be exactly in phase, being synchronous, or they could be out of phase, but in either case, they show the same structure, a 1450- and 1100-year cyclicity [*Sirocko et al.*, 1996] (Fig. 6). These are exactly the same periodicities that *Grootes and Stuiver* [1998] observed in the Greenland ice core $\delta^{18}O$. Accordingly, low and high northern latitudes appear to follow the same principal forcing mechanism, with the pollen record from the Santa Barbara Basin revealing a distinct spike-like pattern, whereas other regions/records show a step-like evolution. Within the dating precision, these spikes and steps could be synchronous.

2. THE LAST 70,000 YEARS

2.1. The monsoonal climate, Arabian Sea (20-70 Cal ka)

Schulz et al. [1998] recently published a well dated, mostly laminated record from the oxygen minimum zone at the continental slope of Pakistan that reflect a tight correlation of the monsoonal interstadial/stadial sequence with the Greenland ice cores and sedimentation processes in the Arabian Sea, see also *Reichart et al.* [1999]. Corroborating these results, we show in Fig. 4 a carbonate curve for core 70KL from the central Arabian Sea, i.e. from a site on the Indus fan. Turbidites (today only frequent in the eastern sector of the Indus fan) were not detected in the entire section of core 70KL. Thus, despite its location on the fan, eolian dust appears to be the major contributor of clastic sediments. Palygorskite abundances of > 5% in the clay mineral fraction during the LGM section clearly reflect that the source of the 70KL sediments is from Arabia where palygorskite is abundant, whereas this mineral is absent on the Indian continent and in Indus fan turbidites [*Sirocko and Lange*, 1991]. Accordingly, the sediments are eolian, being transported by the northwesterly winds that dominate dust dispersal over the Arabian Sea [*Sirocko and Sarnthein,* 1989]. Indeed, eolian deposition rates from a global numerical modelling study reveal very similar values to the accumulation rates of clastic components in core 70KL [*Mahowald et al.*, subm.]. Carbonate content in the Arabian Sea is mainly a function of dilution with clastic eolian components, as revealed by the negative correlation of $CaCO_3$ percent and clastic accumulation rate in core tops [*Sirocko et al.*, in press]. At the site of 70KL we might expect secondary influences on the carbonate abundance from changing upwelling intensity off Arabia (however the core is outside the modern upwelling area) and carbonate dissolution (the core is from 3810 m waterdepth, i.e. at depth of the CCD [*Kolla et al.*, 1976]). The core has at the moment eleven AMS^{14}C dates on a mix of the planktonic species *G.ruber* and *G.sacculifer*.

Measurements of carbonate concentrations have a resolution of 1 cm between samples (Fig. 4). The difference between the long term average (30-point running mean, grey curve in Fig. 4 a) and a specific $CaCO_3$ value reaches up to 8 %. This difference is presented in Fig. 5

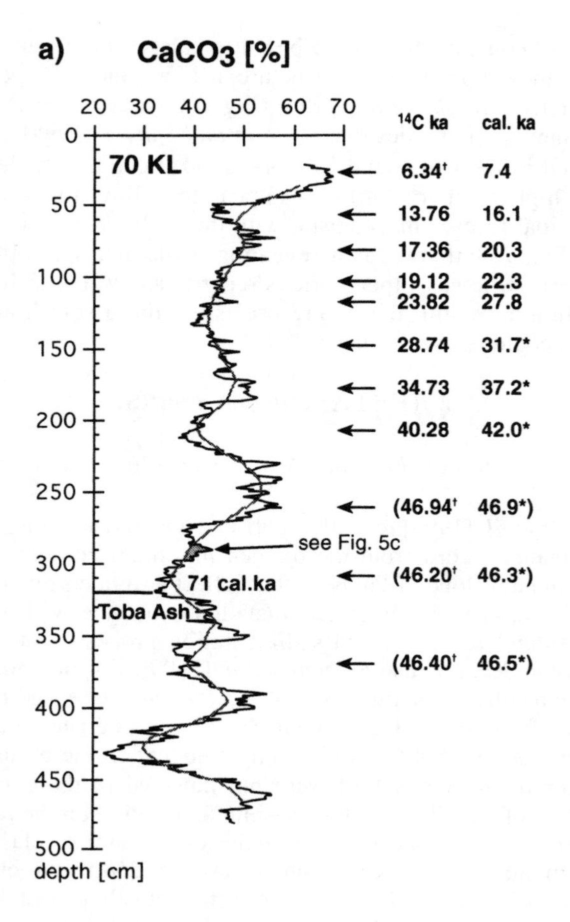

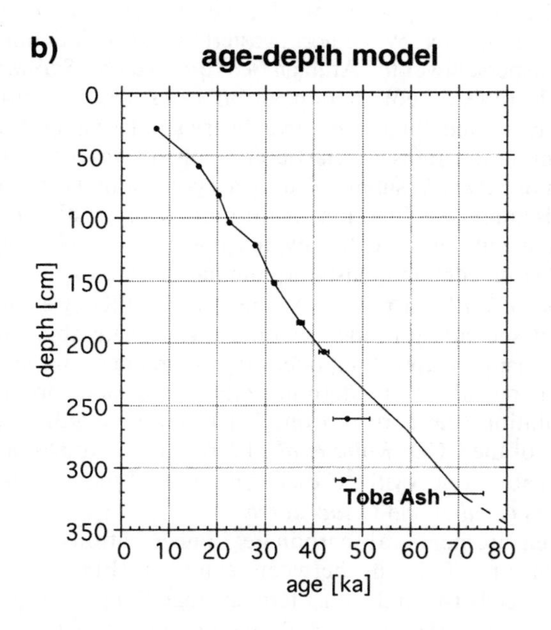

and represents a very simple high-pass filter. The variations of this filter output show a succession of humid/arid intervals very similar to the interstadial /stadial flickering of the $\delta^{18}O$ in the Greenland ice. Based on ^{14}C-AMS ages and the calibration routine of *Voelker et al.* [1998] we developed an age model for core 70KL (Fig. 4 b). Beyond ^{14}C control we used the Toba ash as a time marker. This ashlayer, resulting from a volcanic eruption on Sumatra at 71 Cal ka [*Zielinski et al.*, 1996; *Zielinski et al.*, 1997] was recently detected in the laminated records from Pakistan [*Schulz et al.*, 1998]. Between 40 ka and 70 ka we have no direct age control and had to interpolate ages assuming constant sedimentation rates (which is certainly not the case). ^{14}C-ages from this interval appear suspect (Fig. 4, constant ages near 46 ka between 220 and 370 cm coredepth), as we are too near to the natural limit of the technique and any minor contamination would result in a large effect on the calculated ^{14}C-age. Accordingly, because of these stratigraphic limitations we cannot compare the monsoon history directly to the climate evolution of the high latitudes.

Nevertheless, looking at the record between 20-40 ka (Fig. 5) and based on the findings of *Schulz et al.* [1998] it appears that the European/North Atlantic stadials (i.e. cold events) were associated with drought in the Arabian desert, resulting in stronger dust flux to the ocean via the northwestly winds that dominate eolian transport in this region.

Figure 4. Core 70KL was recovered during cruise "Sonne42". Located in the Arabian Sea (17 30,69°N; 61 41,82°E) on the Indus-fan at a waterdepth of 3810 m. A) The $CaCO_3$-content (black line) was calculated from the measurement of inorganic CO_3 with a Ströhlein Culomat. Samples were spaced in 1 cm intervals. The grey line represents the long term trend of the $CaCO_3$ record (30-point running mean). A simple high-pass filter was calculated by the difference of the trend and the original data. ^{14}C-dates on the planktonic foraminifera G.ruber (samples markd with a cross are a mix of *G.ruber* and *G.sacc.)* are calibrated after E. Bard (pers.comm. during the EGS conference at Aquafredda) and *Voelker et al.* [1998] (samples marked by asterix) including a 400 year seawater correction. The Toba ash layer (compare *Schulz et al.* [1998]) is represented by a layer of glass shards up to 100 μm diameter. B) Age depth model for core 70KL. Error Bars for the Toba Ash Layer represent the published dates between 68 ka and 75 ka BP by *Ninkovich et al.* [1978], *Chesner et al.* [1991] and *Westgate et al.* [1997] using K/Ar, Ar-Ar and fission track methods. The two deepest ^{14}C ages are not used for the age-model as they are very near to the limit of applicability of the ^{14}C-method and in conflict with the Toba Ash date. More information on the stratigraphy of core 70KL is given in *Leuschner and Sirocko* [in press].

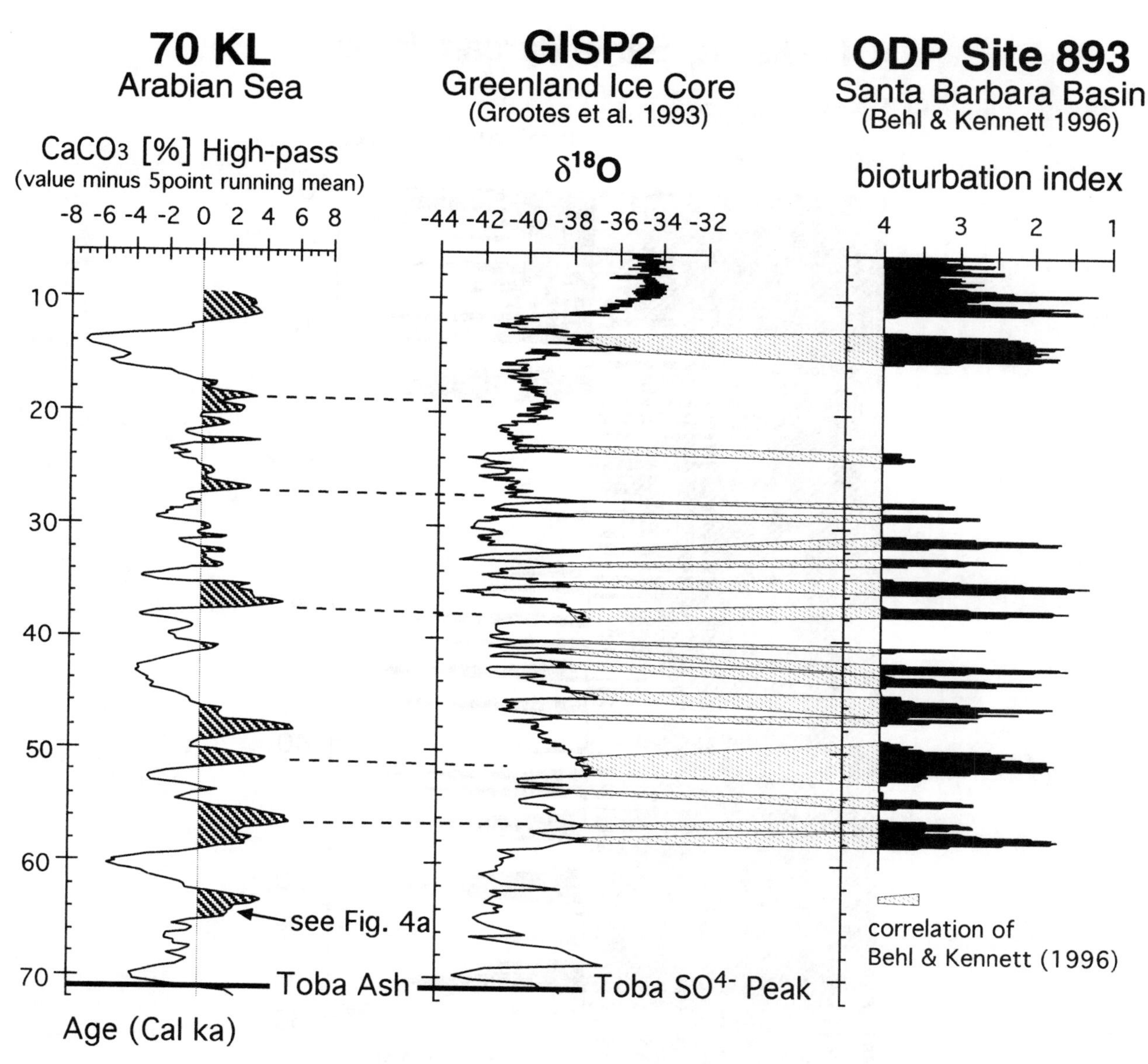

Figure 5. Dansgaard-Oeschger-scale variability in 70KL, GISP2 Greenland ice and low latitude Pacific. GISP2 oxygen isotopes [*Grootes et al.* 1993]. Santa Barbara Basin bioturbation index, showing non-bioturbated, laminated sediments during interstadials. The correlation to GISP2 interstadials was already made by *Behl & Kennett* [1996]. High-frequency variations in the $CaCO_3$ record of 70KL (see Fig. 4).

2.2 The low latitude east Pacific, Santa Barbara Basin (20-70 Cal ka)

In Fig. 5 we also show the relation between the high-frequency events in the Arabian Sea, the bioturbation index from ODP Hole 893A in the Santa Barbara Basin [*Behl and Kennett*, 1996] and the GISP2 $\delta^{18}O$ records (Fig. 5). *Heusser* [1998] note that the distribution, duration and amplitude of warm (anoxic, laminated) events identified in the Santa Barbara Basin corresponds with similar events in the relative abundance of oak, the diagnostic component of xeric southern ocean woodland (oak curve not reproduced in this paper). The high-frequency variability of the *Pinus* record during the last 70,000 years has in contrast a rather different relation to the Santa Barbara Basin lamination index. In Fig. 6 we compare a high-pass signal (value minus 5-point running mean) of the *pinus* curve from Heusser (1998) with the lamination index. The *pinus*

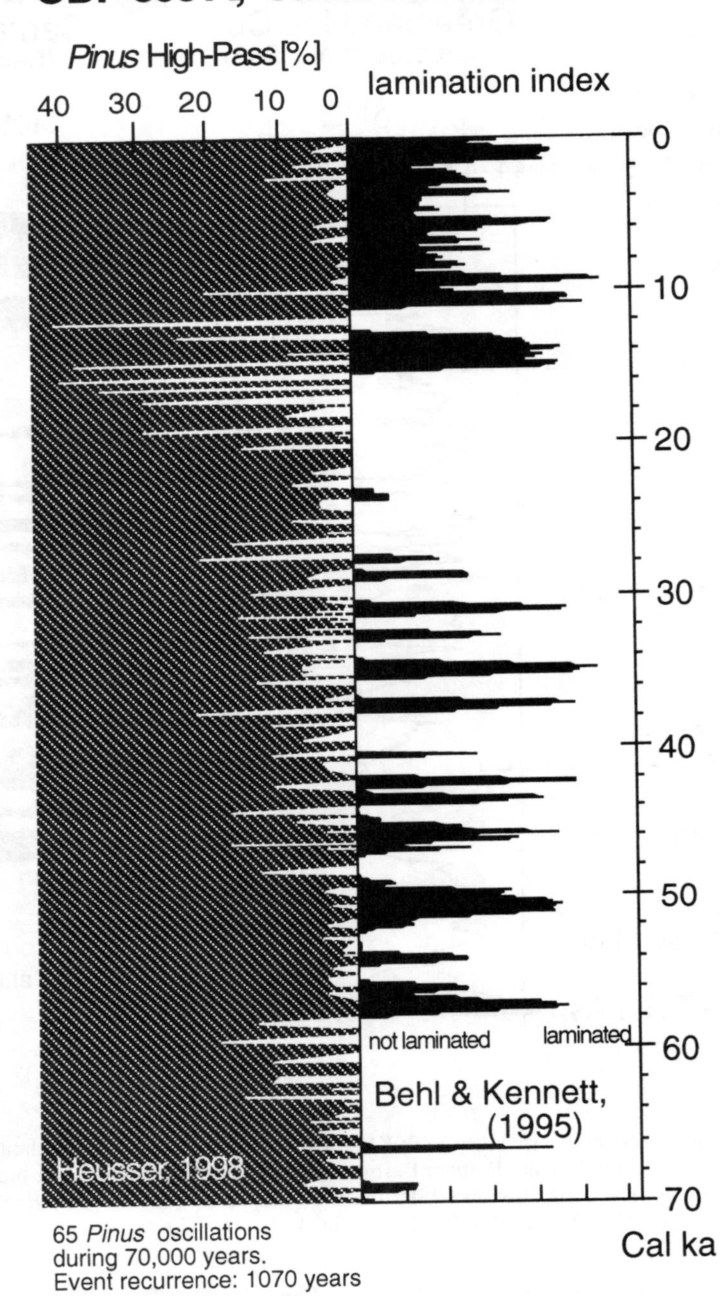

Figure 6. High pass filter (value minus 5-point running mean) for *pinus* abundance in the Santa Barbara Basin, ODP 893A (see complete *pinus* curve in *Heusser*, 1998) in comparison to the lamination index of *Behl & Kennett* [1995] for the same core.

maxima occur very regular with an average spacing of 1070 years. A periodicity of 1070 years is identical to the value reported by *Grootes and Stuiver* [1998] for the $\delta^{18}O$ of the GISP2 ice core during the Holocene. *Heusser & Sirocko* [1997] attributed the *pinus* variations to the Pacific ENSO, because the *pinus* maxima were associated with bushfire since 13.000 yrBP (see above), and bushfire in southern California is most frequent during strong El Niño anomalies. Unfortunately, charcoal abundance was not determined for the time 25-70 Cal ka in the Santa Barbara pollen record. The very regularity of the signal from stage 4, to stage 3, to stage 2, to stage 1 suggest that the same process operating during the Holocene and deglacial was operating continuously also during the past (Fig. 6).

The age model used for Fig. 6 is the same for both indices [*Ingram and Kennett*, 1995], the same as used by *Behl and Kennett* [1996] for the lamination index. Therefore, the *pinus* variations can be compared directly to the lamination index. All longer lamination phases in the Santa Barbara Basin sediments began with a pine maximum, reflecting a phase of active El Niño oscillations. Changes in the state of ventilation of the intermediate waters of the eastern Pacific apparently occurred during times of active El Niño. In like manner, the flickering oscillations in the Greenland ice could be seen in this context (see Behl & Kennetts original correlation between lamination intervals and Greenland interstadials, Fig. 5). Thus, pine maxima (interpreted as several decades/centuries with strong El Niño anomalies) appear to represent times of climate transitions, with effects on Pacific subsurface hydrology and global atmospheric circulation patterns.

2.3. The Antarctic ice cores (20-70 Cal ka)

J. Jouzel published the high-resolution δD record for the Vostok ice core in 1993 using the *Lorius et al.* [1985] time scale. This time scale was later modified by *Jouzel et al.* [1996], *Sowers et al.* [1993] and *Bender et al.* [1994] (last agemodel version see this issue). In Fig. 7 we plot the original Vostok δD on the time scale of *Sowers et al.* [1993] and compare it with the Greenland GISP2 and 70KL monsoon record. Keeping in mind that we had to interpolate the 70KL record between 40 and 70 ka, the carbonate maxima (stronger monsoons) could be matched with the five Antarctic warm phases during this time interval, but even this must remain a very tentative interpretation and we are far from looking at phase relations between the climate evolution of these widely separated areas.

An important observation in this context is the finding by *Blunier et al.* [1998] that the Antarctic warming events are leading the Greenland interstadials. The Toba ash could be an excellent time marker to further compare the phasing not only between high northern and southern latitudes, but between high and low latitudes; unfortunately, SO_4 peaks in the Antarctic ice are not published yet for this time interval. Any possible link between the high Antarctic (Vostok) and the monsoonal latitudes (either with or without a phase lag) terminate at least near 20 Cal ka when northern ice sheets reached their maximum extent and the Vostok temperature evolution already starts into the deglacial (Fig. 7), see however *Steig et al.* [1998] for a contrasting Antarctic record. Within the LGM the Asian monsoon appears to be more tightly connected with the slow northern hemisphere deglaciation. But still within the LGM the shortlasting warming pulses of several decades/centuries of the low latitude climate (Fig. 3) are largely independent from the high northern climate, which apparently is controlled by the massive, slowly reacting continental ice sheets.

3. DISCUSSION AND CONCLUSIONS

3.1. Forcing mechanism for globally teleconnected millennial-scale climate oscillations

In this paper we look mainly at the structure of climate oscillations. Absolute age to age comparisons are still somewhat arbitrary, because the dating of ice-cores, lake sediments and ocean sediments is not sufficient to do this in detail for times earlier than the Holocene. *Chappellaz et al.* [1993], *Bender et al.* [1994], *Blunier et al.* [1998] using the isotopes of atmospheric O_2, ^{10}Be and methane are about to succeed in comparing at least the records from the high latitudes, but as shown in the introduction, these cores represent only about 8.2 % of the area of the globe. 57 % of the Earth's area is in the subtropics and tropics; thus, the majority of Earth's space and its central heating system around the equator would be excluded if explanations on the forcing mechanisms of millennial climate variations would be developed only from polar ice cores.

Adding the low latitude evidence to the ice-core-only comparison results in three main observations:

i) The North Atlantic Heinrich anomaly 1 is associated with short intervals of extreme drought in the continental deserts (evidence from core 74KL [*Sirocko et al.*, 1996], Santa Barbara Basin [*Heusser & Sirocko,* 1997] and Barombi Mbo [*Maley et al.*, 1998] in the present day rainforest Domain.

ii) Stadials of the Dansgaard Oeschger Oscillations are associated with drought in the continental deserts (evidence from core 70KL [this study] and Santa Barbara Basin [*Heusser, 1998*]).

iii) Dansgaard Oeschger Oscillations are associated with changes in the oxygenation of the intermediate water in the Pacific (evidence from the Santa Barbara Basin [*Behl and Kennett*, 1996]) northern Indian Ocean (Fig. 4, 5) and continental margin of Pakistan [*Schulz et al*, 1998].

Interpretation for the flickering behaviour of the northern hemisphere high latitude climate have been usually done by invoking changes in the thermohaline circulation of the deep ocean, affecting the transport of warm subtropical waters into the North Atlantic [*Broecker*, 1997]. Meteorological observation and model results have supported this concept and highlighted the dominating role of the North Atlantic SST for the temperature of Europe and northern Asia, e.g. *deMenocal and Rind* [1993]. None of these experiments however, has shown an influence of SST changes in the North Atlantic to effect tropical latitudes, the monsoon system or even the Indo-Pacific ENSO system. In contrast, it has become apparent that the Pacific El Niño is very capable of effecting even the remotest parts of our globe, northern and southern hemisphere and that Asian monsoon and Pacific El Niño

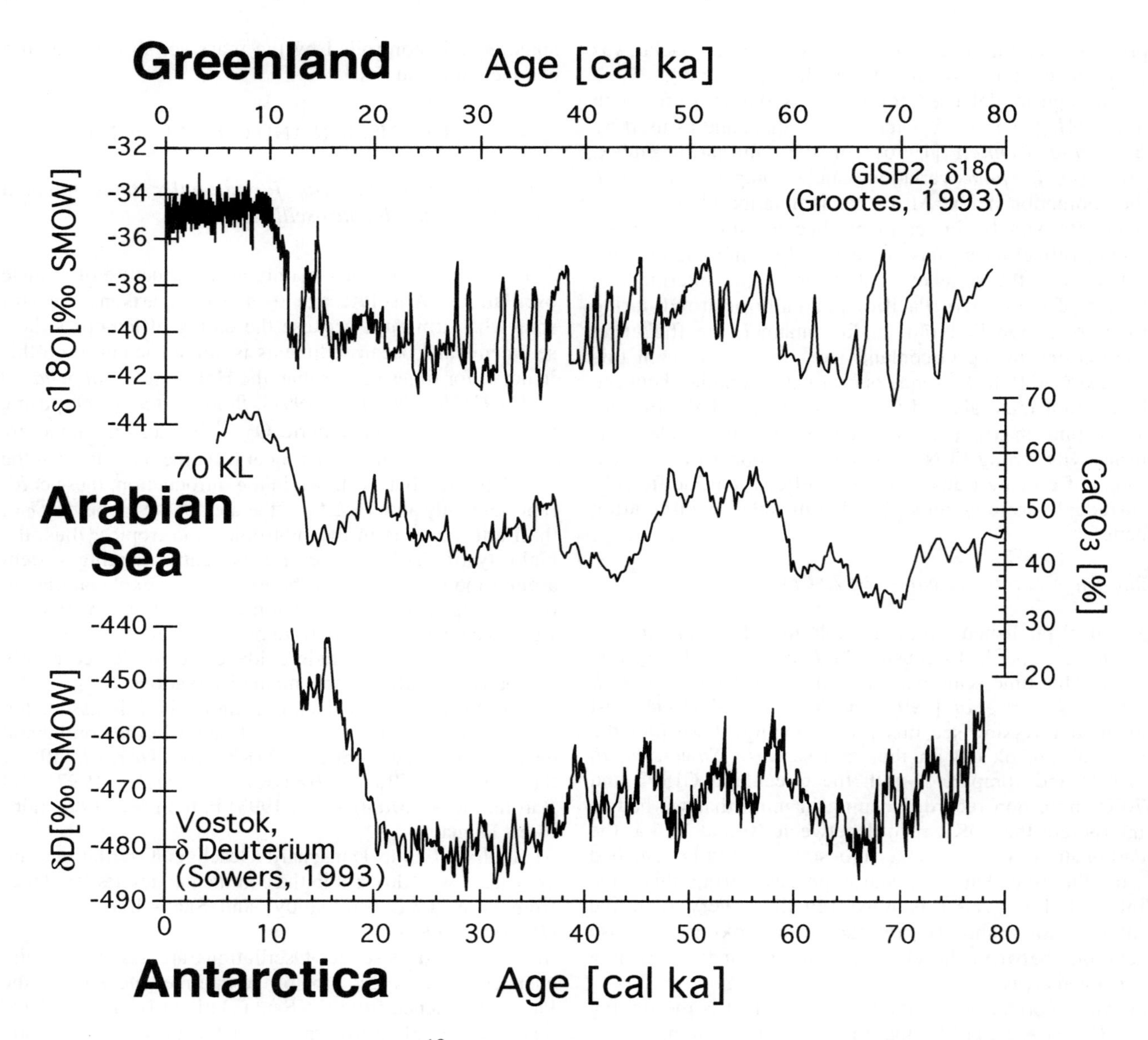

Figure 7. Comparison between GISP2 $\delta^{18}O$, 70KL $CaCO_3$-content and Vostok δDeuterium record [*Sowers et al.*, 1993].

are a coupled system [*Shukla*, 1987; *Barnett et al.*, 1991; *Webster and Yang*, 1992; *Barnett et al.*, 1994; *Meehl*, 1994; *Tourre and White*, 1995; *Li and Yanai*, 1996; *Webster et al.*, 1998].

Bjerknes [1966, 1969] already demonstrated the close connection between ENSO and the subtropical jetstream. This jetstream of the upper troposphere and lower stratosphere encompasses the entire globe (Fig. 8). It reaches deep into the tropics and subtropics during boreal winter, but stays in the north during boreal summer. Strong El Niño phases with warm tropical SST effect the strength/position/ and seasonal duration of this jet stream, which is thus the perfect link to teleconnect atmospheric signals all over the globe within weeks/months, as already shown by J. Bjerknes, who was the first to phrase global atmospheric "teleconnections".

Holocene/late glacial land records and deep-sea records (thousand year sampling resolution) from the equatorial Pacific have suggusted that El Niño oscillations were not active during glacial times [*Mc Glone et al.*, 1992]. However, short, century-scale El-Niño-like activity phases would not be visible in the existing records from sites of low sedimentation rates. The evidence from the pollen and charcoal records from the Santa Barbara Basin suggests that mechanisms like El Niño have been active during short time intervals of the last 70,000 years, a scenario that easily

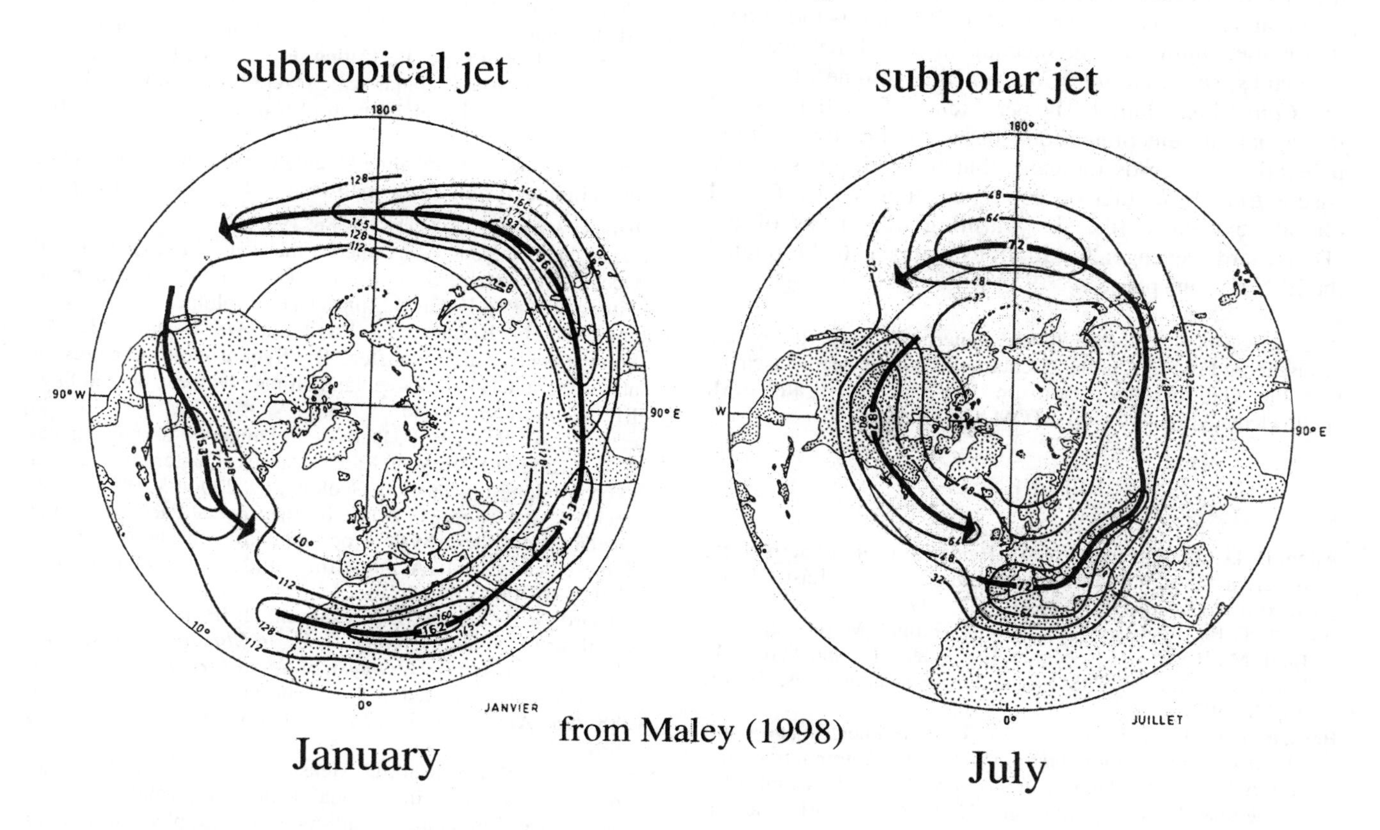

Figure 8. The Subtropical Jetstream: his present-day mean position and speed (km/hour) in the Northern Hemisphere in January and July, from *Maley* [in press] after *Estienne and Godard* [1979].

could explain the teleconnections between high and low latitudes, in particular if invoking changes in the strength and position of the subtropical jet.

The importance of the jetstream was already highlighted in the discussion on the origin of glacial precipitation anomalies in the US, as this jet, coming from the North Pacific must have circled southward around the Laurentian ice sheet, delivering precipitation to the inner continent, as during the "Trans US wet event" causing high lake levels in the glacial western US [*Allen and Anderson*, 1993; *Benson et al.*, 1997]. Two synchronous phases of high lake levels have been also reported (24-18.5 Cal ka, 28-14.5 Cal ka) for the high mountains of the Sahara (in particular the Tibesti, Hoggar, Air and Red Sea Hills). These high lake levels are restricted to mountainous regions above 2000 m height [*Maley*, in press]. Below this level, the Sahara was extremely dry during the LGM [*Sarnthein*, 1978; *Rognon and Coude-Gaussen*, 1996]. Apparently, the Subtropical Jet (Fig. 8) during the LGM and deglacial intensified precipitation not only in the US, but synchronously in the elevated regions of subtropical Africa. Accordingly, the low- and midtroposphere moisture budget show opposite behaviour in the low latitudes during the LGM. Most recent ice cores from elevated sites in Bolivia also indicate stronger snow accumulation during the LGM in the subtropics of the southern hemisphere. $\delta^{18}O$ of this snow reflects a succession of events, most similar to the succession of pinus maxima in the Santa Barbara Basin sediments (Fig. 3), indicating LGM teleconnections between the jetstreams of both hemispheres [*Thompson et al.*, 1998].

The other candidate to force high and low latitude climate to common variations would be primary variations in the intensity of the solar radiation or the amount of solar radiation reaching the Earth's surface, being controlled by greenhouse gases, clouds. In this case, warming/cooling signals should be in phase, and consistent globally. The example from the late Holocene (3800 ^{14}C-yrBP) revealed that this is not the observed pattern. ENSO related signals are however strongly patterned, with only the times of change being synchronous (slightly modified by the local feedback machanisms). The cause of strong El Niño events is, however, unknown, being discussed in the context of stochastic processes, internal changes in the heat-storage capacity of the Pacific ocean and global salt oscillator, or primary variations in the solar insolation strength. Despite the unknown cause of El Niño anomalies, this sea surface temperature anomaly of the equatorial Pacific is today one, if not the, main forcing mechanism of global weather anomalies, see *Cane and Clement* [this volume], *Clement and Cane* [this volume]. The persistence of oscillations and the apparent teleconnections between the low latitude paleoclimate records indicates that this system is in any case extremely regular on geological time scales. Global climate extremes, like the warming transitions of the "Dansgaard Oeschger Oscillations", appear to be related the El Niño "on" phases.

Acknowledgements: The study was generously supported by the "Heisenberg Program" of the "Deutsche Forschungsgemeinschaft" and the GeoForschungszentrum Potsdam. L.H. and J.M. acknowledge grants from ORSTOM, CNRS and NSF.

REFERENCES

Allen, B. D. and Anderson, R. Y., Evidence from western North America for rapid shifts in climate during the Last Glacial Maximum. *Science,* 260, 1920-1922, 1993.

Barnett, T. P., Bengtsson, L., Arpe, K., Flugel, M., Graham, N., Latif, M., Ritchie, J., Roeckner, E., Schlese, U., and Tyree, M., Forecasting global ENSO-related climate anomalies. *Tellus,* 46, 381-397, 1994.

Barnett, T. P., Dümenil, L., Schlese, U., Roeckner, E. and Latif, M., in *Teleconnections linking worldwide climate anomalies,* edited by M. H. Glantz et al., pp. 191-225, Cambridge University Press, Cambridge, New York, Port Chester, Melbourne, Sydney, 1991.

Behl, R. J. and Kennett, J. P., Brief interstadial events in the Santa Barbara Basin, NE Pacific, during the past 60 kyr. *Nature,* 379, 243-246, 1996.

Bender, M., Sowers, T., Dickson, M.-L., Orchardo, J., Grootes, P. M., Mayewski, P. A. and Meese, D. A., Climatic correlations between Greenland and Antarctica during the past 100,000 years. *Nature,* 372, 663-666, 1994.

Bender, M., Malaize, B., Orchardo, J., Sowers, T. and J. Jouzel, High precision correlations of Greenland and Antarctic ice core records over the last 100 kyr. this volume.

Benson, L., Burdett, J., Lund, S., Kashgarian, M. and Mensing, S., Nearly synchronous climate change in the Northern Hemisphere during the last termination. *Nature,* 388, 263-265, 1997.

Berger, A. and Loutre, M. F., Insolation values for the climate of the last 10 million years. *Quaternary Science Reviews,* 10, 297-317, 1991.

Bjerknes, J., A possible response of the atmospheric Hadley circulation to equatorial anomalies of ocean temperature. *Tellus,* 18, 820-829., 1966.

Bjerknes, J., Atmospheric teleconnections from the equatorial Pacific. *Monthly Weather Review,* 97, 163-172, 1969.

Blunier, T., Chappellaz, J., Schwander, J., Dällenbach, A., Stauffer, B., Stocker, T. F., Raynaud, D., Jouzel, J., Clausen, H. B., Hammer, C. U. and Johnsen, S. J., Asynchrony of Antarctic and Greenland climate change during the last glacial period. *Nature,* 394, 739-743, 1998.

Bond, G. and Lotti, R., Iceberg discharges into the North Atlantic on millennial time scales during the last glaciation. *Science,* 267, 1005-1010, 1995.

Bond, G., Showers, W., Cheseby, M., Lotti, R., Almasi, P., deMenocal, P., Priore, P., Cullen, H., Hajdas, I. and Bonani, G., A pervasive millennial-scale Cycle in the North Atlantic Holocene and Glacial Climates. *Nature,* 278, 1257-1266, 1998.

Broecker, W. S., Thermohaline Circulation, the Achilles Heel of our climate system: will man-made CO_2 upset the current balance? *Science,* 278, 1582-1588, 1997.

Cane, M. and Clement, A., A role for the tropical Pacific Coupled Ocean-Atmosphere System on Milankovitch and Millennial Timescales. Part II: Global Impacts. this volume.

Chappellaz, J., Blunier, T., Raynaud, D., Barnola, J. M., Schwander, J. and Stauffer, B., Synchronous changes in atmospheric CH4 and Greenland climate between 40 and 8kyr BP. *Nature,* 366, 443-445, 1993.

Chesner, C.A., Rose, W.I., Deino, A., Drake, R. and Westgate, J.A., Eruptive history of Earth's largest Quaternary caldera (Toba, Indonesia) clarified. Geology, 19, 200-203, 1991.

Clement, A. and Cane, M., A role for the tropical Pacific Coupled Ocean-Atmosphere System on Milankovitch and Millennial Timescales. Part I: A modelling study of tropical Pacific Variability. this volume.

Dansgaard, W., Johnsen, S. J., Clausen, H. B., Dahl-Jensen, D., Gundestrup, N. and Hammer, C. U., in *Climate processes and climate sensitivity. Geophysical Monograph 29, Maurice Ewing Volume 5*, edited by J. E. Hansen and T. Takahashi, pp. 288-298, American Geophysical Union, Washington D.C., 1984.

deMenocal, P. B. and Rind, D., Sensitivity of Asian and African climate to variations in seasonal insolation, glacial ice cover, sea surface temperature, and Asian Orography. *Journal of Geophysical Research,* 98, 7265-7287, 1993.

Estienne, P. and Godard, A., *Climatologie,* A. Colin, Paris, 1979.

Fontes, J. C., Gasse, F. and Gibert, E., Holocene environmental changes in Lake Bangong basin (Western Tibet). Part I: Chronology and stable isotopes of carbonates of a Holocene lacustrine core. *Palaeogeogr., Palaeoclimatol., Palaeoecol.,* 25-47., 1996.

Giresse, P., Maley, J. and Brenac, P., Late Quaternary palaeoenvironments in the Lake Barombi Mbo (West Cameroon) deduced from pollen and carbon isotopes of organic matter. *Palaeogeogr., Palaeoclimatol., Palaeoecol.,* 107, 65-78, 1994.

Giresse, P., Maley, J. and Kelts, K., Sedimentation and palaeoenvironment in crater lake Barombi Mbo, Cameroon, during the last 25,000 years. *Sedimentary Geology,* 71, 151-175, 1991.

Grimm, E. C., Jacobsen, G. L., Watts, W. A. and Hansen, B. C. S. M., K.A., A 50,000-year record of climate oscillations from Florida and its temporal correlation with the Heinrich events. *Science,* 261, 198-200, 1993.

Grootes, P. M. and Stuiver, M., Oxygen 18/16 variability of Greenland snow and ice with 10^{-3} to 10^5 year time resolution. *Journal of Geophysical Research,* 102, 26,455-26,470, 1998.

Grootes, P. M., Stuiver, M., White, J. W. C., Johnsen, S. and Jouzel, J., Comparison of oxygen isotope records from the GISP2 and GRIP Greenland ice cores. *Nature,* 366, 552-554, 1993.

Gunn, J., Influence of various forcing variables on global energy balance during the period of intensive instrumental observation (1958-1987) and their implications for paleoclimate. *Climatic Change,* 19, 393-420, 1991.

Heusser, L., Pollenstratigraphy and paleoecologic interpretation of the last 160 kyr from Santa Barbara Basin, ODP hole 893A. in *Proc. Ocean Drill. Program Sci. Results*, edited by J. Kennett et al., pp. 265-279, 1995.

Heusser, L., Direct correlation of millennial-scale changes in western North American vegetation and climate with changes in the California current system over the past 60,000 years. *Paleoceanography,* 13, 252-262, 1998.

Heusser, L. and Sirocko, F., Millennial Pulsing of vegetation change in southern California: A record of Indo-Pacific ENSO events from the past 24,000 years. *Geology,* 25, 243-246, 1997.

Imbrie, J., Boyle, E. A., Clemens, S. C., Duffy, D., Howard, W. R., Kukla, G., Kutzbach, J., Martinson, D. G., McIntyre, A., Mix, A. C., Molfino, B., Morley, J. J., Peterson, L. C., Pisias, N. G., Prell, W. L., Raymo, M. E., Shackleton, N. J. and Toggweiler, J. R., On the structure and origin of major glaciation cycles. 1. Linear responses to Milankovitch forcing. *Paleoceanography,* 7, 701-738, 1992.

Ingram, B. L. and Kennett, J. P., Radiocarbon chronology and planktonic-benthic foraminiferal age differences in Santa Barbara Basin sediments (ODP Hole893A), in *Proceedings of Ocean Drilling Program, Scientific results*, edited by J. P. Kennett et al., pp. 19-27, Ocean Drilling Program, College Station, 1995.

Johnsen, S. J., Clausen, H. B., Dansgaard, W., Fuhrer, K., Gundestrup, N., Hammer, C. U., Iversen, P., Jouzel, J., Staufer, B. and Steffensen, J. P., Irregular glacial interstadials recorded in a new Greenland ice core. *Nature,* 359, 311-313, 1992.

Jolly, D., Taylor, D., Marchant, R., Hamilton, A., Bonnefille, R., Buchet, G. and Riollet, G., Vegetation dynamics in central Africa since 18,000 yr BP : records from interlacustrine highlands of Burundi, Rwanda and western Uganda. *J.Biogeography,* 24, 495-512., 1997.

Jouzel, J., Waelbroeck, C., Malaize, B., Bender, M., Petit, J. R., Stievenard, M., Barkov, N. I., Barnola, J. M., King, T., Kotlyakov, V. M., Lipenkov, V., Lorius, C., Raynaud, D., Ritz, C. and Sowers, T., Climatic interpretation of the recently extended Vostok ice records. *Climate Dynamics,* 12, 513-521, 1996.

Kelts, K., Aquatic response signatures in lake core sequences as global evidence of rapid moisture balance shifts around 4000 years ago. *Terra Nova,* 9, Abstract 68/2A05., 1997.

Kolla, V., Bê, A. W. H. and Biscaye, P. E., Calcium carbonate distribution in the surface sediments of the Indian Ocean. *Journal of Geophysical Research,* 81, 2605-2616, 1976.

Kotilainen, A. T. and Shackleton, N. J., Rapid climate variability in the north Pacific Ocean during the past 95,000 years. *Nature,* 377, 323-326, 1995.

Leuschner, D. and Sirocko, F., The low latitude monsoon climate during Dansgaard-Oeschger cycles and Heinrich Events. *Quaternary Science Reviews*, in press.

Li, C. and Yanai, M., The onset and interannual variability of the Asian Summer Monsoon in relation to Land-Sea thermal contrast. *J. of Climate,* 9, 358-375, 1996.

Lorius, C., Jouzel, J., Ritz, C., Merlivat, L., Barkov, N. I., Korotkevich, Y. S. and Kotyakov, V. M., A 1550,000 -year climatic record from Antarctic ice. *Nature,* 316, 591-596, 1985.

Mahowald, N., Kohfeld, K. E., Hansson, M., Balkanski, Y., Harrison, S. P., Prentice, I. C., Schulz, M. and Rohde, H., Dust sources and deposition during the Last Glacial Maximum and current climate: A comparison of model results with paleodata from ice cores and marine sediments. *JGR,* subm.

Maley, J., in *Third millennium BC climate change and Old World collapse*, edited by H.N.Dalfes et al., pp. 611-640, Springer-Verlag, Berlin., 1997.

Maley, J., Last Glacial Maximum lacustrine and fluviatiles Formations in the Tibesti and other Saharan mountains, and large scale climatic teleconnections linked to the activity of the Subtropical Jetstream. *Global & Planetary Change,* in press.

Maley, J. and Brenac, P., Vegetation dynamics, palaeoenvironments and climatic changes in the forests of western Cameroon during the last 28,000 years. *Review of Palaeobotany and Palynology,* 99, 157-187, 1998.

Martin, L., Fournier, M., Mourguiart, P., Siffedine, A. and Turcq, B., Southern Oscillation signal in South American palaeoclimatic data of the last 7000 years. *Quat.Res.,* 39, 338-346., 1993.

Mc Glone, M. S., Kershaw, A. P. and Markgraf, V., in *El Niño, Historical and paleoclimatic aspects of the Southern Oscillation*, edited by H.F.Diaz and V.Markgraf, pp. 435-462, Cambridge Univ. Press, 1992.

McIntyre, A. and Molfino, B., Forcing of Atlantic Equatorial and Subpolar Millennial Cycles by precession. *Science,* 274, 1867-1870, 1996.

Meehl, G., Coupled land-ocean-atmosphere processes and South Asian monsoon variability. *Science,* 266, 263-267, 1994.

Milankovitch, M., in *Handbuch der Klimatologie*, edited by W. Köppen and R. Geiger, Gebrüder Bornträger, Berlin, 1939.

Mosley-Thompson, E., Thompson, L. G., Lin, P. N., Davis, M. E. and Dai, J., Rapid Holocene warming and cooling events in Central East Antarctica. *Am.Geoph.Union,* EOS Suppl., 1993.

Ninkovich, D., Shackleton, N.J., Abdel-Monem, A.A., Obradovich, J.D. and Izett, G., K-Ar age of the late Pleistocene eruption of Toba, north Sumatra *Nature,* 276, 574-577, 1978.

Porter, S. C. and Zisheng, A., Correlation between climate events in the North Atlantic and China during the last glaciation. *Nature,* 375, 305-307, 1995.

Prell, W. L. and Kutzbach, J. E., Sensitivity of the Indian monsoon to changes in orbital parameters, glacial and tectonic boundary conditions and atmospheric CO2 concentration. *Nature,* 360, 647, 1992.

Reichart, G. J., Lourens, L. J. and Zachariasse, W. J., Temporal variability in the northern Arabian Sea oxygen minimum zone (OMZ) during the last 225,000 years. *Paleoceanography,* 13, 607-621, 1999.

Rodbell, D. T., Seltzer, G. O., Anderson, D. M., Abbott, M. B., Enfield, D. B. and Newman, J. H., An 15,000-year record of El Nino-driven alluviation in Southwestern Ecuador. *Science,* 283, 516-520, 1999.

Rognon, P. and Coude-Gaussen, G., Paleoclimates off northwest Africa (28°- 35°N) about 18,000 C/14 yr BP based on continental eolian deposits. *Quaternary Research,* 46, 118-126., 1996.

Sarnthein, M., Sand deserts during glacial maximum and climatic optimum. *Nature,* 272, 43-46., 1978.

Schulz, H., von Rad, H. and Erlenkeuser, H., Correlation between

Arabian Sea and Greenland climate oscillations of the past 110,000 years. *Nature,* 393, 54-57, 1998.
Shukla, J., in *Monsoons*, edited by J. S. Fein and P. L. Stephens, pp. 399-464, John Wiley and Sons, 1987.
Sirocko, F., and Sarnthein, M., Wind-borne deposits in the Northwestern Indian Ocean: record of Holocene sediments versus modern satellite data, in: *Paleoclimatology and Paleometeorology: Modern and Past Patterns of Glacial Atmospheric Transport* edited by M. Leinen and M. Sarnthein, pp. 401-433. Kluwer Academic Publishers, Dordrecht, Boston, London, 1989.
Sirocko, F. and Lange, H., Clay mineral accumulation rates in the Arabian Sea during the Late Quaternary. *Marine Geology,* 97, 105-119, 1991.
Sirocko, F., Sarnthein, M., Erlenkeuser, H., Lange, H., Arnold, M. and Duplessy, J. C., Century-scale events in monsoonal climate over the past 24,000 years. *Nature,* 364, 322-324, 1993.
Sirocko, F., Garbe-Schönberg, D., McIntyre, A. and Molfino, B., Teleconnections between the subtropical monsoon and high latitude climates during the last deglaciation. *Science,* 272, 526-529, 1996.
Sirocko, F., Garbe-Schönberg, D. and Devey, C., Processes controlling trace element geochemistry of Arabian Sea sediments during the last 25,000 years. *Global and Planetary Change,* in press.
Sowers, T., Bender, M., Labeyrie, L., Martinson, D., Jouzel, J., Raynaud, D., Pichon, J. J. and Korotkevich, Y. S., A 135,000, year Vostok-SPECMAP common temporal framework. *Paleoceanography,* 8, 737-766, 1993.
Steig, E.J., Brook, E.J., White, J.W.C., Sucher, C.M., Bender, M.L., Lehman, S.J., Morse, D.L., Waddington, E.D. and Clow, G.D., Synchronous climate changes in Antarctica and the North Atlantic, *Science*, 282, 92- 95, 1998.
Stocker, T.F., Wright, D.G. and Mysak, L.A., A zonally avaraged, coupled ocean-atmosphere model for paleoclimate studies. *J.Clim.*, 5, 773-797, 1992.
Thompson, L. G., Davis, M. E., Mosley-Thompson, E., Sowers, T. A., Henderson, K. A., Zagorodnov, V. S., Lin, P.-N., Mikhalenko, V. N., Campen, R. K., Bolzan, J. F., Cole-Dai, J. and Francou, B., A 25,000-year tropical climate history from Bolivian Ice Cores. *Science,* 282, 1858-1864, 1998.
Thouveny, N., de Beaulieu, J.-L., Bonifay, E., Creer, K. M., Guiot, J., Icole, M., Johnsen, S., Jouzel, J., Reille, M., Williams, T. and Williamson, D., Climate variations in Europe over the past 140 kyr deduced from rock magnetism. *Nature,* 371, 503-506, 1994.
Tourre, Y. M. and White, W. B., ENSO signals in global upper-ocean temperature. *Journal of Physical Oceanography,* 25, 1317-1332, 1995.
Voelker, A. H. L., Sarnthein, M., Grootes, P. M., Erlenkeuser, H., Laj, C., Mazaud, A., Nadeau, M.-J. and Schleicher, M., Correlation of marine ^{14}C ages from the nordic seas with the GISP2 isotope record: implications for radiocarbon calibration beyond 25 ka BP. *Radiocarbon,* 1998 in press.
Wang, L., Sarnthein, M., Grootes, P. M. and Erlenkeuser, H., Millennial reoccurrence of century scale abrupt events of East Asian monsoon: a possible heat conveyor for the global deglaciation. *Paleoceanography,* in press.
Webster, P. J., Magana, V. O., Palmer, T. N., Shukla, J., Tomas, R. A., Yanai, M. and Yasunari, T., Monsoons: Processes, predictability, and the prospects for prediction. *J. Geoph. Res.,* 103, 14451-14510, 1998.
Webster, P. J. and Yang, S., Monsoon and ENSO: Selectively interactive systems. *Q.J.R.Meteorol.Soc.,* 118, 877-926, 1992.
Westgate, J., Sandhu, A. and Shane, P. Fission track dating. In: Taylor and Aitken, M. Chronometric dating in archeology. Plenum Press. New York, 127-158, 1997.
Zielinski, G. A., Mayewski, P. A., Meeker, L. D., Grönvold, K., Germani, M. S., Whitlow, S., Twickler, M. S. and Taylor, K., Volcanic aerosol records and tephrochronology of the Summit, Greenland, ice cores. *Journal of Geophysical research,* 102, 26625-26640, 1997.
Zielinski, G. A., Mayewski, P. A., Meeker, L. D., Whitlow, S., Twickler, M. S. and Taylor, K., Potential atmospheric impact of the Toba mega-eruption ≈71,000 years ago. *Geophysical Research Letters,* 23, 837-840, 1996.

Linda Heusser, 100 Clinton Road, Tuxedo, NY, USA
Dirk Leuschner, GeoForschungsZentrum, Telegraphenberg, 14473 Potsdam, Germany
Jean Maley, Paleoenvironnement & Palynologie (CNRS & ORSTOM) Univ. Montpellier-2, 34095 Montpellier, France
Frank Sirocko, Institut für Geowissenschaften, Johannes Gutenberg-Universität, 55099 Mainz, Germany. email: sirocko@mail.uni-mainz.de
Michael Staubwasser, GeoForschungsZentrum, Telegraphenberg, 14473 Potsdam, Germany

Rapid Climate Oscillations in the Northeast Pacific During the Last Deglaciation Reflect Northern and Southern Hemisphere Sources

Alan C. Mix[1], David C. Lund[2], Nicklas G. Pisias[1], Per Bodén[3], Lennart Bornmalm[4], Mitch Lyle[5], and Jennifer Pike[6]

Planktic foraminiferal species abundances, benthic and planktic foraminiferal stable isotopes, radiocarbon, and organic carbon contents of deep-sea cores off Oregon and Northern California reveal abrupt millennial-scale climate oscillations during the past 20,000 years. Changes in the near-surface ocean are essentially coincident with the Bølling-Allerød and Younger-Dryas climate oscillations observed in Greenland ice cores and North Atlantic sediments. This finding supports the concept of atmospheric transmission of climate signals between oceans within the Northern Hemisphere. Abrupt cooling of North Pacific surface waters occurred in mid-Holocene time, indicating that the warm events of the early Holocene and deglaciation are anomalous relative to modern climate. Higher export productivity is associated with warm events in the North Pacific. These biotic changes may have contributed to variations in the shallow (~400 m depth) oxygen minimum zone off California, and may in part explain the apparent coincidence of local anoxia with warming in Greenland. Benthic foraminiferal $\delta^{13}C$ and ^{14}C data from lower intermediate waters (980 m depth) suggests that higher ventilation (either faster formation or greater gas exchange) occurred during the Bølling-Allerød and early Holocene warm events. Synchronicity with surface ocean changes points to North Pacific source waters, and ventilation during warming leads to a hypothesis that salinity rather than temperature controls intermediate water formation at these times. In the deep North Pacific (2700 m depth) benthic foraminiferal $\delta^{18}O$ changes imply early warming roughly synchronous with warming of the Southern Ocean. Both $\delta^{13}C$

[1]College of Oceanic and Atmospheric Sciences, Oregon State University, Corvallis, Oregon
[2]Kennedy School of Government, Harvard University, Cambridge, Massachusetts
[3]Department of Geology and Geochemistry, Stockholm University, Stockholm, Sweden
[4]Department of Marine Geology, University of Goteborg, Goteborg, Sweden
[5]Center for Geophysical Investigation of the Shallow Subsurface, Boise State University, Boise, Idaho
[6]Department of Earth Sciences, University of Cardiff, Cardiff, Wales

Mechanisms of Global Climate Change at Millennial Time Scales
Geophysical Monograph 112

and ^{14}C suggest an abyssal ventilation event (either faster formation or greater gas exchange) during deglaciation at the same time as short-term cooling in Antarctica, pointing to a Southern Ocean source of variability in the deep Pacific. Thus, climate changes that characterize both northern and southern sources appear to propagate through the Pacific Ocean.

1. INTRODUCTION

Understanding climate variability on millennial time scales has become a major focus of the paleoclimate community. Starting in the 1970's, ice-core records clearly demonstrate variability on millennial time scales [*Dansgaard, et al.*, 1971]. Frequencies of variation similar to the ice core records were first documented in the North Atlantic by *Pisias et al.* [1973]. Major episodes of ice rafting, the so-called Heinrich events, have been documented in the North Atlantic [*Bond et al.*, 1993; *Bond and Lotti,* 1995] and may represent dramatic surges of former ice streams [*MacAyeal,* 1993]. These events are part of a package of high-frequency variations of climate in the North Atlantic and adjacent regions including deep-water circulation [*Oppo and Lehman,* 1995].

Recent evidence from both land and sea indicates that the climate of some regions outside of the North Atlantic also varied at millennial scales [*Porter and An,* 1995; *Thunell and Mortyn,* 1995; *Kotilainen and Shackleton,* 1995; *Kennett and Ingram,* 1995; *Lowell et al.,* 1995; *Charles et al.,* 1996; *Curry and Oppo,* 1997; *van Geen et al.,* 1997; *Lund and Mix,* 1998]. The potential exists for a widespread response to global forcing external to the climate system [*Broecker,* 1995], for a response arising within the climate system of the North Atlantic region and propagating elsewhere [*Clark and Bartlein,* 1995], or for an effect elsewhere within the climate system that is independent of North Atlantic variability.

The challenge now is to assess how millennial-scale events, reasonably well known in the North Atlantic record, are related to events identified globally, and more importantly to identify the processes that provide global linkage on these time scales. Are the linkages made through the atmosphere, the deep and surface ocean, or both? Does the rest of the world follow changes initiated in the North Atlantic, or does the chain of events start elsewhere?

Our strategy is to identify the timing of regional responses that yield insight into the dynamics of rapid climate changes. This approach follows that of establishing the phase sequence of regional paleoclimatic oscillations on longer (orbital) scales [*Imbrie et al.,* 1989]. Resolving sequences of events on the short timescales examined here presents special challenges, however, as it pushes the limits of chronologic uncertainty. We approach this uncertainty in two ways. First, we focus on an age range that can be dated precisely by radiocarbon, specifically on short-term climatic events of the last deglaciation as examples of millennial scale changes. Here we follow *Alley and Clark* [1999], who summarize current knowledge of the global distribution of such events and highlight the need for new data from poorly known areas such as the North Pacific. Second, we examine relationships between multiple proxies of different climatic systems within the same core samples. Leads or lags among such proxies within the same samples are reliable, even if the absolute chronology is uncertain.

2. WHY THE NORTHEAST PACIFIC?

Sediments from the Santa Barbara Basin off southern California and the Gulf of California already provide clear evidence of millennial-scale events in the shallow reaches of the Pacific Ocean. The patterns of change are remarkably similar to those of the North Atlantic and in Greenland ice cores [*Keigwin and Jones,* 1990; *Behl and Kennett,* 1996; *Cannariato et al.,* 1999; *Hendy and Kennett,* 1999], but the precise timing of events and processes driving such changes remain unclear.

Increases in the oxygen content of bottom waters of the Santa Barbara Basin appear to be associated with cool events in Greenland ice cores and North Atlantic sediments [*Behl and Kennett,* 1996; *Cannariato et al.,* 1999]. Plausible mechanisms to explain variations in oxygen here include changing ventilation of the North Pacific thermocline, shifts in intermediate water masses, and changes in large-scale biological productivity of the Pacific that modify the region's oxygen minimum zone [*Kennett and Ingram,* 1995].

Here, we examine variability of surface and deep waters and evidence for paleoproductivity in cores off Oregon and Northern California (Figure 1). This area is important for several reasons. Sediment cores here monitor: 1) the wind-driven position of the boundary between the subtropical and subpolar gyres of the North Pacific as reflected in the character of the California Current [*Lynn and Simpson,* 1987], 2) upwelling in a highly productive eastern boundary current in a region where much of the flux of organic matter to the deep North Pacific occurs [*Huyer,* 1983], and 3) lower intermediate and deep waters of the North Pacific, where world's "oldest", most nutrient-rich, and lowest $\delta^{13}C$ deep waters (i.e., those longest isolated from interaction with the sea surface) are found [*Reid,* 1965; *Mantyla and Reid,* 1983; *Broecker et al.,* 1985].

High hemipelagic sedimentation rates at two sites off Northern California and Southern Oregon, driven by rapid erosion of the continent and high biogenic productivity, facilitate high-resolution study of climate change. These two sites are particularly well placed for monitoring the upper and lower bounds of nutrient-rich North Pacific Deep Water (Figure 2). Core W8709A-13PC (42.117°N, 125.750°W, 2712 m water depth) lies within the gradient between low-$\delta^{13}C$ Pacific Deep Water and higher-$\delta^{13}C$ bottom waters that enter the Pacific basin from the south [*Kroopnick*, 1985]. Ocean Drilling Program (ODP) Site 1019 (41.682°N, 124.930°W, 980 m water depth) monitors the gradients above Pacific Deep Water where it mixes with overlying intermediate waters.

Stable isotope data from core W8709A-13PC were documented previously [*Lund and Mix*, 1998] but nine new radiocarbon dates reported here improve the chronology. This site, located 130 km from the Oregon coast, is under the advective core of the California Current outside of the coastal upwelling system. Planktonic foraminifera here record the strength and character of eastern boundary advection related to the position of the North Pacific Subpolar Front (Figure 2).

ODP Site 1019, 70 km offshore of Eureka, California, records upper-ocean variability and export productivity within the wind-driven coastal upwelling system. Because of its proximity to the coast, it has high sedimentation rates (averaging >50 cm/ka during the last deglaciation) and thus offers a highly detailed record of millennial-scale events during the last glacial cycle [*Lyle et al.*, 1997].

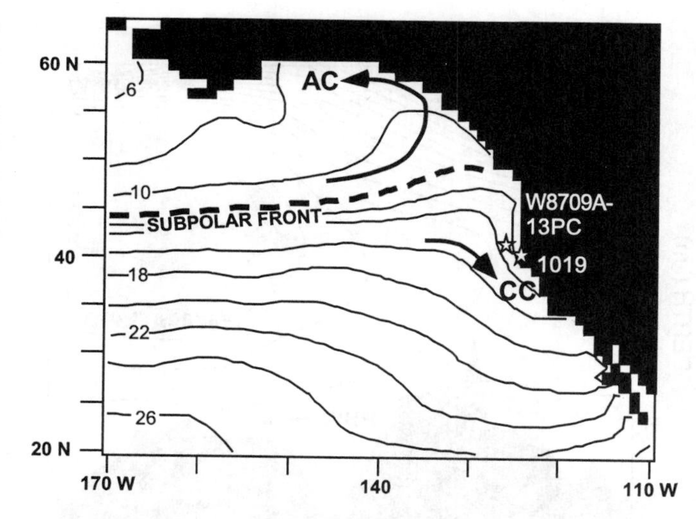

Figure 1. Core W8709A-13PC and ODP Site 1019 are under the northern California Current, south of the North Pacific Subpolar Front. Upper ocean proxies here are sensitive to both the position of the Subpolar Front and the heat content of the northern subtropical gyre. Site 1019 is under the coastal upwelling system, while W8709A-13PC is under the advective core of the California Current. Here, AC is the Alaska Current, and CC is the California Current.

3. HYPOTHESES

We explore several working hypotheses to explain millennial-scale climate oscillations observed in Pacific sediments, considering effects of upper ocean circulation and temperature, influences of biological productivity on the properties of thermocline and intermediate waters, and varying sources of mid-depth and deep water masses.

3.1. Upper Ocean Circulation and Temperature

Model studies suggest that atmospheric transmission of North Atlantic cooling to the North Pacific could be significant and rapid [*Manabe and Stouffer*, 1988; *Mikolajewicz et al.*, 1997]. A first working hypothesis based on such models is that changes in sea-surface temperatures of the North Pacific would be similar to those of the North Atlantic.

To test this hypothesis, we examine the timing of upper ocean circulation changes in the northeast Pacific. If changes in upper ocean temperatures or circulation in the North Pacific mimic those of the North Atlantic, perhaps with a small lag, then atmospheric transmission of a regional cooling is likely. *Hendy and Kennett* [1999] argue for such a connection based on variations in surface water conditions in Santa Barbara Basin.

If the North Pacific surface ocean record is substantially different from that of the North Atlantic, then other processes may be important. For example, a modeling study that includes the hypothesized effect of rapid reduction of Laurentide Ice Sheet elevation during Heinrich events predicts cooling in western North America at the same time as warming of the North Atlantic [*Hostetler et al.*, 1999].

3.2. North Pacific Intermediate Water

Cooling in the North Pacific has the potential to modify formation of North Pacific Intermediate Water (NPIW). At present, little or no deep water forms and intermediate water formation is limited in the North Pacific because the surface ocean here has relatively low salinity [*Warren*, 1983]. Salinity is low because evaporation rates are low. Evaporation rates are low because the sea surface is cool. To some extent, the sea surface is cool because there is little formation of intermediate or deep water, which would draw oceanic heat poleward from the subtropics. This loop of cause and effect suggests the possibility of dramatic change into a state with greater formation of intermediate or deep water which could be self sustaining, if first initiated by either extreme cooling or higher sea-surface salinity.

NPIW forms today in the Northwest Pacific and in the Sea of Okhotsk due to extreme cooling of the sea surface in winter [*Freeland et al.*, 1998]. Near its source, NPIW is detected in the upper 1000 m of the water column by its low salinities and high oxygen relative to adjacent water masses [*Yamanaka et al.*, 1998]. Interaction with the sea

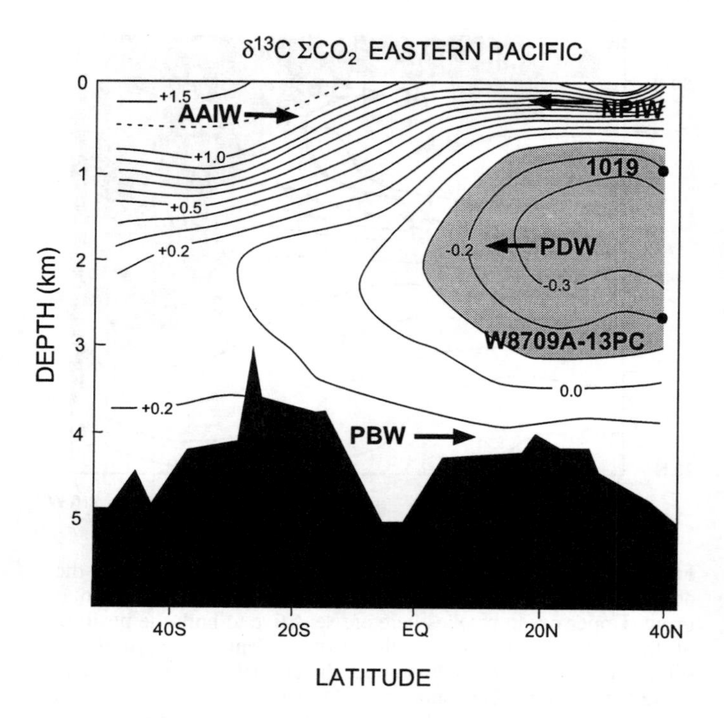

Figure 2. Benthic foraminifera in core W8709A-13PC, at 2700 m depth, monitor the mixing zone of low-$\delta^{13}C$, high-nutrient Pacific Deep Water (PDW, shaded), and higher-$\delta^{13}C$ Pacific Bottom Water (PBW). Benthic foraminifera in ODP Site 1019, at 980 m depth, monitor the mixing zone between PDW and relatively high-$\delta^{13}C$ North Pacific Intermediate Water (NPIW). AAIW is Antarctic Intermediate Water. Contours are $\delta^{13}C$ of ΣCO_2 in the water column, after Kroopnick [1985].

surface is limited, however, so oxygen levels are low in newly formed NPIW relative to other typical intermediate waters such as Antarctic Intermediate Water (AAIW). This property of relatively low oxygen in NPIW led to original (now discounted) hypotheses that NPIW formed by diapycnal mixing in the absence of substantial interaction with the atmosphere [*Reid,* 1965].

During the Last Glacial Maximum (LGM), increased formation of NPIW in the far northwest Pacific is suggested based on relatively high $\delta^{13}C$ of benthic foraminifera [*Keigwin,* 1998]. Its penetration appears limited to the upper 2 km of the water column within the Sea of Okhotsk. Enhanced formation of NPIW occurred in spite of apparently greater stratification and stability of the glacial water column in the subpolar North Pacific [*Zahn et al.,* 1991]. A likely mechanism is extreme winter cooling in the northwest Pacific driven by cold, dry winds coming off Asia.

More rapid formation of NPIW at the LGM suggests a hypothesis for the millennial scale, whereby North Pacific cooling, perhaps associated with atmospheric transmission of North Atlantic events such as the Younger Dryas cooling event, also enhanced NPIW formation. Although the timing of events is not firmly constrained, oxic conditions in the Santa Barbara Basin [*Behl and Kennett,* 1996; *Cannariato et al.,* 1999; *Hendy and Kennett,* 1999] are generally associated with cool events, as predicted by the atmospheric transmission hypothesis. A further test of this hypothesis requires establishing precise dates on changes in upper ocean properties of the North Pacific and the relationship of sea-surface cooling to oxygenation and apparent ventilation in intermediate waters.

3.3. Productivity and the Oxygen Minimum Zone

Another hypothesis to explain variations in anoxia along the California margin is that changes in the biological productivity of the North Pacific modify the intensity of the oxygen minimum zone (OMZ) and the properties of NPIW in the eastern Pacific. Such effects could operate independently of intermediate water formation.

At present, degradation of organic matter raining out of productive North Pacific waters depletes much of the original oxygen in NPIW in the eastern Pacific. Thus, NPIW in the northeast Pacific is detected presently by its density (at shallower depths than in the west) and to some extent by low salinity [*Talley,* 1993]. As a result of high North Pacific productivity and relatively weak ventilation of NPIW, the shallow OMZ of the eastern Pacific margin is strong, approaching anoxia in borderland basins off Southern California and in parts of the Gulf of California.

If millennial-scale intervals of anoxia on the California margin reflect high North Pacific productivity, we would predict seafloor evidence of higher organic rain associated with warmer intervals in the subpolar Pacific. We examine this hypothesis by comparing high-resolution records of organic carbon content in deep-sea sediments, which in the northeast Pacific appears to be driven mainly by changing export productivity [*Lyle et al.,* 1992], with isotopic and faunal estimates of sea surface properties.

3.4. Intermediate and deep waters

Preliminary data on $\delta^{13}C$ and trace metals in benthic foraminifera from deep northeast Pacific cores suggest millennial-scale variations in mid-depth and deep water properties [*van Geen et al.,* 1997; *Lund and Mix,* 1998]. However, given the limitations of published chronologies it remains uncertain whether the entire deep North Pacific changes in response to upper ocean processes, or whether different water masses record different sources of variability.

Changes in the northward flow rate of Pacific Bottom Water and return southward flow of Pacific Deep Water could modify properties of the deep Pacific from below [*Lund and Mix,* 1998]. Such an effect could be caused by: 1) buoyancy forcing in the Southern Ocean, with a prediction of more bottom water formation near Antarctica during cold intervals [*Rahmstorf,* 1995], 2) variations in westerly winds over the Southern Ocean, which help to maintain geostrophic ventilation of the ocean interior at mid depths [*Toggweiler and Samuels,* 1993], or 3) by Southern Ocean winds acting as a "flywheel". In this scenario, relatively constant wind-driven demand for total export of deep water to the Antarctic would favor turnover within the deep Pacific when North Atlantic Deep Water sources are reduced [*Lund and Mix,* 1998].

To complicate matters, intermediate water masses formed in either the North or South Pacific could penetrate more deeply into the ocean interior than they do at present. Available $\delta^{13}C$ evidence argues against formation of deep water in the North Pacific during glacial time [*Keigwin,* 1998], but the effects on millennial scales remain uncertain. A modeling study suggests the possibility that Antarctic Intermediate Water could be much stronger in the past, crossing the equator and ventilating the North Pacific at times [*Campin et al.,* 1999], and preliminary data from the Southern Ocean suggest potential changes in such sources [*Ninnemann et al.,* 1997].

Addressing these possibilities, we monitor the deep Pacific at a depth of 2700 m (core W8709A-13PC), in the mixing zone between modern North Pacific deep water and Pacific Bottom Water, and also at a depth of 980 m (Site 1019), in the mixing zone of North Pacific Deep Water and NPIW (Figure 2). If the entire North Pacific were flushed faster or slower, benthic foraminiferal $\delta^{13}C$ records from the two sites would covary. If only intermediate water formed in the North Pacific, then benthic $\delta^{13}C$ would increase at Site 1019 much more than at W8709A-13PC. If the flux rate of bottom waters from the Antarctic increased, the core W8709A-13PC would record the larger changes in deep water properties.

Atmospheric transmission of climatic effects from the North Atlantic to sources of deep water in the North Pacific would imply North Pacific ventilation at times of North Atlantic cooling, although with some lag in the deep Pacific [*Mikolajewicz et al.,* 1997].

Linkages of the deep North Pacific to Southern Ocean forcing would imply different timing of events. For example, buoyancy forcing in the Southern Ocean [*Rahmstorf,* 1995] would predict deep Pacific ventilation events associated with Antarctic cooling, which on millennial time scales has significantly different timing than in the North Atlantic [*Charles et al.,* 1996; *Blunier et al.,* 1998; *Broecker,* 1998].

A mechanism that drives ventilation of the deep interior of the Atlantic and Pacific by varying the strength of Southern Ocean winds [*Toggweiler and Samuels,* 1993] would imply broadly similar oscillations in deep water properties of both oceans. Finally, the "Antarctic Flywheel" effect, in which enhanced deep Pacific ventilation would compensate for reduced sources of North Atlantic Deep Water [*Lund and Mix,* 1998] would suggest opposite changes in the two oceans.

These different scenarios imply very different climate mechanisms for propagating rapid climate changes around the globe. They are not mutually exclusive -- all may operate in concert. At this early stage of research on Pacific climate variability we ask the question whether variations in the proxies for upper ocean circulation, export productivity, and intermediate and deep water properties all change together, or whether differences in signals reveal different linkages in the different systems.

4. METHODS

4.1. Stable Isotopes

All stable isotope data reported here were analyzed at the College of Oceanic and Atmospheric Sciences stable isotope laboratory at Oregon State University, using a Finnigan/MAT 251 mass spectrometer equipped with an Autoprep Systems single acid bath carbonate reaction device.

The benthic foraminifera *Uvigerina sp.* and *Cibicides wuellerstorfi* (3-10 specimens per sample) were hand picked from the >150 μm size fraction. Left-coiling specimens of the planktic foraminifera *Neogloboquadrina pachyderma* (40-50 specimens per sample) were selected from the 150-250 μm fraction (most were near 150 μm). Prior to stable isotopic analysis, foraminifera were cleaned ultrasonically in alcohol and roasted for 1 hr at 400°C under high vacuum to remove organic contaminants. Reactions of carbonates to produce CO_2 gas occurred in ~100% orthophosphoric acid at 90°C.

Precision of isotope analyses of the local OSU carbonate standard was ±0.06‰ for $\delta^{18}O$ and ±0.03‰ for $\delta^{13}C$, respectively. Calibration to the widely used Pee Dee Belemnite (PDB) scale was through the NBS-19 and NBS-20 standards provided by the U.S. National Institute of Standards and Technology.

4.2. Radiocarbon

All radiocarbon dates used here are based on accelerator mass spectrometry (AMS) of hand-picked foraminifera. Most of the dates of core W8709A-13PC were published previously [*Gardner et al.,* 1997; *Ortiz et al.,* 1997; *Lund and Mix,* 1998].

New dates presented here for core W8709A-13PC are based on 5-10 mg calcite of monospecific planktonic foraminifera or mixed-species of benthic foraminifera. The foraminifera were ultrasonically cleaned and etched in weak phosphoric acid solution, and then reacted in vaccuo in 100% phosphoric acid. Purified carbon dioxide gas was stored in acid cleaned, pre-roasted, glass tubes and transferred to Lawrence Livermore National Laboratory, where targets were made and analyzed for ^{14}C content (^{13}C corrected).

Radiocarbon dates for samples of ODP Site 1019 were analyzed at the Uppsala AMS facility at University of Uppsala, Sweden, using foraminifera isolated and cleaned at Stockholm University.

Radiocarbon corrections assume a modern reservoir age. At site 1019, inside the coastal upwelling zone, the reservoir age for planktic foraminifera is 800 years [*Robinson and Thompson* 1981; *Southon et al.*, 1990]. The modern reservoir age for planktic foraminifera in core W8709A-13PC, outside the coastal upwelling zone of the northeast Pacific, is 720 years [*Ortiz et al.*, 1997]. For benthic foraminifera, the modern reservoir age is estimated at 1750 years for site 1019, and 2310 years for core W8709A-13PC, based on measured water column radiocarbon contents near 980 m and 2700 m depths in the North Pacific, respectively [*Toggweiler et al.*, 1989].

Radiocarbon dates less than 20,000 reservoir-corrected ^{14}C years old were converted to calendar ages using version 4.1 of the CALIB radiocarbon software [*Stuiver and Reimer,* 1993], which has been updated with recent calibration data sets [*Stuiver et al.*, 1998]. Local reservoir corrections (less 400 years to account for nominal surface ocean radiocarbon ages assumed in CALIB) were entered as constant ΔR values in the CALIB program. An assumed error of ± 200 years was added to ΔR, to account for potential changes in regional reservoir ages. Based on this input, CALIB calculates a time-varying reservoir age and a model-generated history of upper-ocean radiocarbon content driven by changing radiocarbon production terms, as well as a corrected calendar age.

The CALIB software is more appropriately used for oceanic dates based on organisms living in near-surface waters that exchange carbon rapidly with the atmosphere (such as planktonic foraminifera) than it is for deep-sea organisms (such as benthic foraminifera) that may reflect large regional variations in local reservoir ages. Thus, although we used CALIB to calculate calendar ages based on benthic foraminifera, our inferred age models use the results from planktonic foraminifera where possible.

For ages older than 20,000 reservoir-corrected ^{14}C years, we corrected radiocarbon dates to calendar dates using the glacial polynomial algorithm of *Bard* [1998], which is calibrated based on dated corals between about 10,000 and 36,000 ^{14}C years before present.

Calculations of apparent ^{14}C ventilation ages of subsurface waters are made here by subtracting raw ^{14}C dates based on planktic foraminifera from those based on benthic foraminifera in the same samples. The rationale for such calculations is developed by *Broecker et al.* [1984].

4.3. Foraminiferal Species

Samples were hand picked for planktic foraminifera (>150 μm) using CLIMAP taxonomic categories [*Saito et al.*, 1981; *Parker,* 1962], with the exception that we do not recognize the "*pachyderma-dutertrei*" intergrade category of Kipp [1976] in this area. Qualitative estimates of sea-surface temperatures (SST's) are reported as percentages of left-coiling *N. pachyderma* relative to the total number of *N. pachyderma* specimens. An advantage of this method over a total faunal analysis in the northeast Pacific is that it is relatively insensitive to selective dissolution.

Where SST's are less than 5°C, the *N. pachyderma* core-top population is generally comprised of >95% left-coiling specimens. Where sea-surface temperatures are >15°C, the percentages of left-coiling specimens is generally less than 5%. This provides a sensitive faunal indicator of climate off Oregon, where mean annual temperatures range from <10°C (in the coastal upwelling systems) to >15°C (offshore in the California Current). Temperatures during the last glacial maximum are thought to have been 3-4°C cooler than at present in our study area [*Sabin and Pisias,* 1996; *Ortiz et al.*, 1997; *Doose et al.*, 1997].

4.4. Organic Carbon

Accurate measurements of organic carbon content are time consuming, and this limits the ability to generate high-resolution time series needed for assessing millennial-scale changes in export productivity. We get around this limitation by using an optical method, the SCAT system (Split Core Analysis Track), which provides information on sediment composition based on diffuse reflectance spectroscopy of sediment core surfaces [*Mix et al.*, 1992]. The principle is that different sedimentary assemblages have different spectral signatures

The SCAT analyzed 1024 wavelengths ranging from 250-950 nm (Ultraviolet, Visible, and near-Infrared wavelengths). This tool has proven useful in estimating calcium carbonate [*Mix et al.*, 1995; *Harris et al.*, 1997; *Ortiz et al.*, 1999], as well as goethite and hematite contents [*Harris and Mix,* 1999]. Here we use SCAT to estimate organic carbon content.

Calibration of regression equations with low-resolution chemical or mineralogical data yields high-resolution estimates of major changes in sedimentary content. The most reliable estimates off the Oregon and California margin are of organic carbon percentages [*A.C. Mix, unpublished data*, 1999], which can be estimated optically with a standard error of ±0.15%, approaching the analytical precision of some wet chemical methods [*Gardner et al.*, 1997], and sufficient for analysis of relatively large changes observed in Site 1019.

The optical estimates made here, calibrated with 303 measurements made by coulometry aboard D/V JOIDES Resolution in holes 1018A, 1018C, and 1019C [*Lyle et al.*, 1997], provide values of organic carbon content at Site 1019 with an average sample spacing of 150 years during the last deglaciation.

Organic carbon contents in core W8709A-13PC were analyzed by coulometry at Boise State University, with an estimated precision of ±0.05% by weight [*Lyle et al.*, 1999].

5. RESULTS

5.1 Radiocarbon Chronologies

5.1.1. Core W8709A-13PC. AMS radiocarbon dates from core W8709A-13PC are reported in Table 1, and illustrated as calendar corrected ages in Figure 3. Some of these dates were reported earlier [*Lyle et al.*, 1992; *Gardner et al.*, 1997; *Ortiz et al.*, 1997; *Lund and Mix*, 1998]. For clarity we have reproduced those dates in Table 1, correcting some small typographical errors in both age (referring where possible to the original laboratory report) and depth (by re-examining the original sediment core stored at the Oregon State University core laboratory) that have appeared in previous published versions.

There are a few age reversals in core W8709A-13PC. We have chosen to ignore the planktic foraminiferal date at 181 cm, which is younger than the date from 170.5 cm and inconsistent with the benthic date. The planktic date at 196.25 cm depth is older than adjacent dates, and it too is ignored in our inferred age model. Finally, the date on planktic foraminifera at 392.5 cm [*Lyle et al.*, 1992] is younger than adjacent dates and was culled. All three culled samples are included in Table 1, and shown in Figure 3.

Inferred ages in core W8709A-13PC pass through calendar corrected planktic foraminiferal dates in all cases except for the three samples noted above. In these cases, inferred dates pass through both calendar corrected benthic foraminiferal dates, and adjacent planktic foraminiferal dates. Rather than attempt a smooth polynomial fit through the array of dates, we have chosen an objective age model that interpolates linearly between our inferred dates (solid lines in Figure 3). We did not manipulate the age model to improve the correlation of other proxy data between cores or to events elsewhere. We note conservative minimum and maximum age ranges (dashed lines in Figure 3) which are based on the extremes of propagated errors of both benthic and planktic dates.

The new benthic and planktic dates at 170.5 cm and the new benthic date at 181 cm depth in core W8709A-13PC modify the inference of *Lund and Mix* [1998] that a deglacial radiocarbon "plateau" [*Edwards et al.*, 1993; *Hughen et al.*, 1998] was present in this core. Our modification of the age model changes the depth interval of the Younger Dryas interval in core W8709A-13PC (i.e., the depth interval equivalent to calendar ages 11.5-13.0 ka) [*Alley et al.*, 1993] from ~163-191 cm [*Lund and Mix*, 1998] to ~142-165 cm [*this paper*].

5.1.2. Site 1019. AMS radiocarbon data from ODP Site 1019 are reported in Table 2 and illustrated as calendar corrected dates in Figure 4. In most cases the calendar corrected radiocarbon dates from planktic and benthic foraminifera agree well. There are a few exceptions. The planktic foraminiferal date at 5.16 meters composite depth (mcd) in Hole 1019A is anomalously old relative to the benthic date. At 5.82 mcd, the benthic age is anomalously old (Figure 4b). These anomalies may in part reflect changes in reservoir ages through time, due either to changing upwelling intensity (which could change reservoir ages by a few hundred years at most), or to ventilation of thermocline waters with radiocarbon-rich surface waters (which might explain the data at 5.82 mcd), or to ventilation of local bottom waters (at 980 m water depth, this would require lower intermediate water to explain the young benthic date at 5.16 mcd).

Until more data confirm the benthic-planktic age difference patterns we make no attempt to adjust for varying reservoir ages through time (other than that included in the CALIB radiocarbon correction scheme). To avoid age reversals with depth in the core, our inferred age model (solid lines in Figure 4) passes through the calendar corrected benthic foraminiferal date at 5.16 mcd. In all other cases, our inferred age model passes through the calendar corrected planktic foraminiferal dates. As in core W8709A-13PC, we also define minimum and maximum ages (dashed lines in Figure 4) based on propagated errors in both benthic and planktic foraminiferal dates.

We retain all dates from Site 1019 for purposes of calculating benthic-planktic ^{14}C age differences.

5.2. Time Series of Changing Environments

5.2.1. Core W8709A-13PC. Lund and Mix [1998] presented benthic foraminiferal stable isotope data based mostly on the species *C. wuellerstorfi* from core W8709A-13PC. These data are reproduced in Figure 5 along with new data from the genus *Uvigerina* sp., and plotted using the new chronology developed here.

The variations in $\delta^{18}O$ in the two species are very similar but values from *C. wuellerstorfi* are lower by 0.66 ± 0.09 $^{o}/_{oo}$ (n=185) than those from *Uvigerina* sp. Consistent with previous workers, we added 0.64 to all $\delta^{18}O$ values from *C. wuellerstorfi*.

Changes in $\delta^{13}C$ based on *C. wuellerstorfi* (Figure 5b) suggest that an apparent event of stronger deep-sea ventilation, recorded by $\delta^{13}C$ about 0.2 to 0.3 $^{o}/_{oo}$ above background values (Figure 5b) occurs from 13.5-12.3 ka, near the beginning or slightly preceding the Younger Dryas interval (shaded following the Greenland Summit, GISP-2, Greenland Ice Sheet Program, ice core $\delta^{18}O$, Figure 5e) [*Grootes et al.*, 1993; *Alley et al.*, 1993].

Variations of $\delta^{13}C$ in *Uvigerina* sp. do not match those of *C. wuellerstorfi* (Figure 5b). *Uvigerina* $\delta^{13}C$ values are on average 0.70 $^{o}/_{oo}$ (±0.11, n=185) lower than those from *C. wuellerstorfi* in the same samples. In some cases the variations between species are nearly a mirror image. Anomalously low $\delta^{13}C$ values in *Uvigerina* sp. relative to other species are commonly attributed to high organic

Table 1. Radiocarbon data from core W8709A-13PC.

Sample Identifier	Depth (cm)	Sample Type	^{14}C Lab	Ref.	^{14}C age (ka ± ka)	Res.Cor. age (ka)	Cal. Age (ka ± ka)	Inferred Age (ka)
W8709A-13PC	27.50	Mixed Planktic	CAMS	1	7.00±0.23	6.28	7.22+0.32	7.2
W8709A-13PC	96.25	Mixed Benthic	CAMS	1	11.00±0.12	8.69	9.80±0.43	9.6
W8709A-13PC	98.75	Mixed Benthic	CAMS	1	11.02±0.24	8.71	9.80±0.57	9.7
W8709A-13PC	126.25	Mixed Planktic	CAMS	1	9.96±0.23	9.24	10.31±0.59	10.3
W8709A-13PC	126.25	Mixed Benthic	CAMS	1	11.58±0.17	9.27	10.33±0.56	10.3
W8709A-13PC	128.75	Mixed Benthic	CAMS	1	11.45±0.16	9.14	10.29±0.37	10.5
W8709A-13PC	139.00	Mixed Planktic	CAMS-39369	2	10.72±0.07	10.00	11.33±0.53	11.3
W8709A-13PC	139.00	Mixed Benthic	CAMS-45713	3	12.19±0.06	9.88	11.16±0.39	11.3
W8709A-13PC	154.00	*N.pachyderma*(L)	CAMS-39370	2	11.20±0.06	10.48	12.33±0.54	12.3
W8709A-13PC	154.00	Mixed Benthic	CAMS-45718	3	12.63±0.06	10.32	11.80±0.61	12.3
W8709A-13PC	170.50	Mixed Planktic	CAMS-45715	3	12.19±0.06	11.47	13.35±0.37	13.3
W8709A-13PC	170.50	Mixed Benthic	CAMS-45714	3	13.08±0.07	10.77	12.71±0.30	13.3
W8709A-13PC	181.00	*N.pachyderma*(L)	CAMS-39371	2	11.31±0.14	10.59	12.48±0.56	13.6
W8709A-13PC	181.00	Mixed Benthic	CAMS-45719	3	13.88±0.11	11.57	13.45±0.32	13.6
W8709A-13PC	191.50	*N.pachyderma*(L)	CAMS-44505	3	12.76±0.06	12.04	13.93±0.35	13.9
W8709A-13PC	191.50	Mixed Benthic	CAMS-44506	3	14.33±0.09	12.02	13.95±0.36	13.9
W8709A-13PC	196.25	Mixed Planktic	CAMS	1	13.86±0.10	13.14	15.68±0.32	14.1
W8709A-13PC	196.25	Mixed Benthic	CAMS	1	14.56±0.14	12.25	14.11±0.68	14.1
W8709A-13PC	198.75	Mixed Planktic	CAMS	1	13.00±0.09	12.28	14.13±0.62	14.2
W8709A-13PC	198.75	Mixed Benthic	CAMS	1	14.70±0.25	12.39	14.29±0.73	14.2
W8709A-13PC	212.50	*G. bulloides*	CAMS-44507	3	13.70±0.07	12.98	15.51±0.67	15.5
W8709A-13PC	212.50	Mixed Benthic	CAMS-44508	3	14.86±0.07	12.55	14.87±0.62	15.5
W8709A-13PC	221.25	Mixed Planktic	CAMS	1	14.05±0.14	13.33	15.90±0.36	15.9
W8709A-13PC	221.25	Mixed Benthic	CAMS	1	15.95±0.18	13.64	16.26±0.38	15.9
W8709A-13PC	223.75	Mixed Planktic	CAMS	1	14.02±0.14	13.30	15.87±0.35	16.0
W8709A-13PC	223.75	Mixed Benthic	CAMS	1	15.77±0.32	13.46	16.05±0.60	16.0
W8709A-13PC	227.5	Mixed Planktic	NZ	4	15.27±0.22	14.55	17.31±0.42	17.4
W8709A-13PC	301.25	Mixed Planktic	CAMS	1	16.71±0.12	15.99	18.96±0.40	19.0
W8709A-13PC	301.25	Mixed Benthic	CAMS	1	18.36±0.20	16.05	19.03±0.44	19.0
W8709A-13PC	303.75	Mixed Planktic	CAMS	1	17.03±0.15	16.31	19.33±0.42	20.0
W8709A-13PC	303.75	Mixed Benthic	CAMS	1	18.63±0.18	16.32	19.34±0.43	20.0
W8709A-13PC	332.5	Mixed Planktic	NZ	4	18.37+0.27	17.65	20.88+0.56	20.9
W8709A-13PC	382.00	*G. bulloides*	CAMS-39372	2	22.14±0.14	21.42	25.24±0.16	25.1
W8709A-13PC	392.5	Mixed Planktic	NZ	4	19.82+0.64	19.10	22.54+0.83	25.4
W8709A-13PC	401.25	Mixed Planktic	CAMS	1	22.00±0.25	21.28	25.07±0.29	25.7
W8709A-13PC	401.25	Mixed Benthic	CAMS	1	23.56±1.02	21.25	25.04±1.19	25.7
W8709A-13PC	403.75	Mixed Planktic	CAMS	1	21.91±0.24	21.19	24.97±0.28	25.8
W8709A-13PC	403.75	Mixed Benthic	CAMS	1	24.54±0.51	22.23	26.17±0.59	25.8
W8709A-13PC	443.00	*G. bulloides*	CAMS	2	24.78±0.14	24.06	28.28±0.16	28.3
W8709A-13PC	497.00	*G. bulloides*	CAMS	2	28.14±0.40	27.42	32.09±0.45	32.1
W8709A-13PC	557.00	*G. bulloides*	CAMS	2	32.00±0.32	31.28	36.38±0.35	36.4

1) Gardner et al. [1997], 2) Lund and Mix, [1998], 3) This study, 4) Lyle et al. [1992].

carbon rain and a preferred habitat within pore waters [*Zahn et al.*, 1986]. Such an explanation does not fit all the data here, as organic carbon content of the cores does not peak during times of maximum $\delta^{13}C$ offsets between species. If anything, the opposite relationship holds here. Large $\delta^{13}C$ differences between the taxa appear to be associated with lower organic carbon contents (Figure 5d).

The percentage of left-coiling *N. pachyderma* (Figure 5c) is high (essentially 100%, consistent with cold conditions) prior to 15 ka (calendar). Warmer conditions from 13 to 14 ka (50-60% left coiling *N. pachyderma*) are followed by a cold reversal (80-90% left coiling *N. pachyderma*) between 11.5 and 12.5 ka (calendar). A return to warmer conditions occurs at ages < 11.5 ka (calendar).

Given the precision of our age model (inferred ages and minimum-maximum ranges show as horizontal bars in Figure 5a), the oscillation in *N. pachyderma* coiling in core

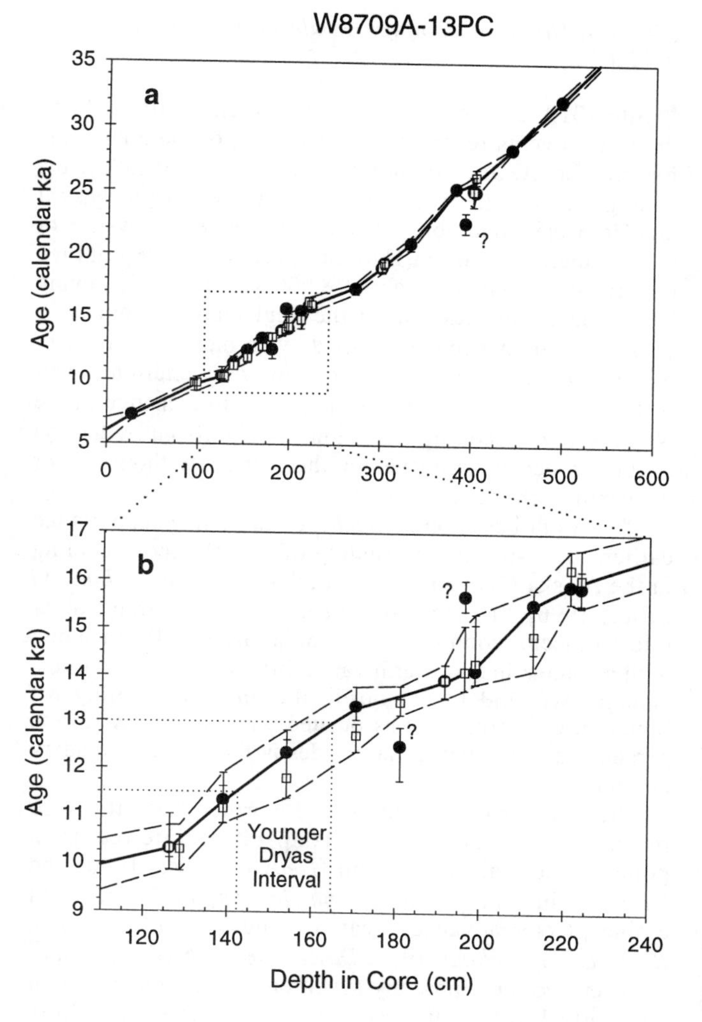

Figure 3. Age model constrained by AMS radiocarbon dates in core W8709A-13PC. All dates are shown as calendar corrected values with $\pm 1\sigma$ error bars (see text). Solid circles are dates based on planktonic foraminifera. Open squares are dates based on benthic foraminifera. Question marks indicate suspect dates. The solid line is the inferred age model, and dashed lines mark the upper and lower bounds of propagated errors. a) 5-35 ka (calendar), b) detail of 9-17 ka, illustrating the age control for the Younger-Dryas and Bølling-Allerød intervals.

W8709A-13PC is essentially in phase with the Younger Dryas temperature oscillations in Greenland as recorded by $\delta^{18}O$ in the Greenland Summit (GISP-2) ice core (Figure 5e). Planktic foraminifera are rare in core W8709A-13PC within the Holocene, and the top ~6000 years were not recovered in this core, so with this core we can not assess the extent of variability within the last 10 ka.

5.2.2. Site 1019. Within the coastal upwelling system at a shallower water depth and with high sedimentation rates, Site 1019 reveals additional details of the northeast Pacific variability during the last deglacial period.

The species *C. wuellerstorfi* is less common at Site 1019 than in core W8709A-13PC. The $\delta^{18}O$ of benthic foraminifera *C. wuellerstorfi* and *Uvigerina* sp. (Figure 6a) agree (with an offset between species of 0.68 ± 0.13‰, n=38, essentially identical to that in other cores) (Figure 6a).

The $\delta^{18}O$ of planktonic foraminifera *N. pachyderma* (left-coiling) display relatively large oscillations during the last deglaciation, with values about 0.5 to 0.8 ‰ greater (i.e., 2-3°C cooler if entirely attributed to temperature) during the Younger Dryas interval than before or after that event (Figure 6a). This pattern of change is essentially identical to the change in coiling direction of *N. pachyderma*, to more left-coiling specimens during the Younger Dryas event (Figure 6c). Within the limits of the chronology, deglacial warming and cooling at site 1019 are essentially synchronous with similar warming and cooling events recorded in the Greenland Summit (GISP-2) ice core (Figure 6e) [*Grootes et al.*, 1993; *Alley et al.*, 1993].

Site 1019 also records a return to cold conditions (more left-coiling *N. pachyderma* specimens) within the last 8.0 ka. The anomaly appears to be the relatively warm conditions (low percentages of left-coiling) that preceded and followed the Younger Dryas event. In terms of the *N. pachyderma* coiling, mid Holocene conditions appear roughly similar to those of the glacial maximum.

Benthic foraminiferal $\delta^{13}C$ values at Site 1019, based mostly on *Uvigerina sp.*, are highest relative to background values just prior to and again significantly after the Younger Dryas interval (Figure 6b). Values of $\delta^{13}C$ in *Uvigerina* sp. are on average 0.58 ± 0.27 ‰ (n=38) lower than those of *C. wuellerstorfi.* There is no apparent decrease in benthic $\delta^{13}C$ values associated with intervals of high organic carbon content (Figure 6d), so productivity influences on benthic $\delta^{13}C$ do not appear to be a major problem. To first approximation, $\delta^{13}C$ data from *Uvigerina* sp. and *C. wuellerstorfi* agree (with an offset), although *C. wuellerstorfi* is often absent, and always rare. Thus, for Site 1019 we use the $\delta^{13}C$ data from *Uvigerina* sp. as an indicator of bottom water properties. This is not ideal, but is supported by calculations of apparent ventilation ages based on benthic-planktonic differences in radiocarbon dates.

The total range of variation in benthic $\delta^{13}C$ at Site 1019 (~0.8 ‰) is large relative to global glacial-interglacial changes $\delta^{13}C$ of about 0.3 ‰ [*Curry et al.*, 1988]. Average $\delta^{13}C$ values of *Uvigerina* sp. near the glacial maximum (>18 ka calendar) approach those of the Holocene. Within the Holocene, a ~0.6 ‰ decrease in benthic $\delta^{13}C$ values occur near 8 ka (calendar), roughly coincident with apparent cooling documented by increases in the percentage of left coiling *N. pachyderma*.

Significant oscillations occur in the $\delta^{13}C$ of the planktic foraminifera *N. pachyderma* (Figure 6b), but these changes do not appear to be systematically related to warming or cool events, or organic carbon contents at Site 1019 (Figure 6c,d).

The range or organic carbon concentrations is quite high, from <1 to >4 % (Figure 6d). The organic carbon values are generally low during the cool events recorded by dominance of left coiling *N. pachyderma*. Large increases in organic carbon content occur during warm events recorded locally (Figure 6c) and in the Greenland ice core (Figure 6e). This pattern of higher inferred export production during warm events mimics the glacial-interglacial pattern of paleo-productivity in this region [*Lyle et al.*, 1992; *Ortiz et al.*, 1997].

The similarity of organic carbon percentages at Site 1019 and the temperature oscillations in the Greenland Summit (GISP-2) $\delta^{18}O$ record [*Grootes et al.*, 1993] is striking, especially prior to the Younger Dryas event during the Bølling-Allerød interstade. Although the apparent timing of the organic carbon peaks in Site 1019 appears to lead warm events in Greenland by a few hundred years, given the potential errors in the radiocarbon chronology illustrated by the horizontal bars in Figure 6a, major events in the two regions may be synchronous.

6. DISCUSSION

6.1. Upper ocean changes reflect atmospheric connections.

Both core W8709A-13PC under the offshore California Current and Site 1019 under the coastal upwelling system provide strong support for a cool event (dominance of left-coiling *N. pachyderma*) during deglaciation that is essentially synchronous with the Younger Dryas interval as documented in the Greenland Summit (GISP-2) ice core (Figure 7). This finding supports the concept of rapid atmospheric transmission of millennial-scale warming and cooling cycles between the Atlantic and the Pacific.

Based on the range of planktonic foraminiferal $\delta^{18}O$ values before, during, and after the Younger Dryas event (0.5-0.7 $^{o}/_{oo}$, Figure 6), we estimate the range of upper-ocean temperature changes at Site 1019 to be 2-3°C (assuming no other changes). This extent of change is smaller than surface water changes recorded in the Santa Barbara Basin, where $\delta^{18}O$ data from planktic foraminifera suggest rapid temperature oscillations of 4-8°C [*Hendy and Kennett,* 1999]. In both areas, observed changes in temperature are higher than those predicted by a coupled ocean-atmosphere model, ~1°C near the eastern boundary of the North Pacific [*Mikolajewicz et al.*, 1997].

Our finding of significant oscillations in near-surface conditions that are essentially synchronous presents an opportunity to use this oscillation as a stratigraphic marker. By comparing variations of other properties in the same samples in these northeast Pacific cores, we can place a range of processes, including variations in deep water properties, into a well constrained stratigraphy to examine dynamics of the North Pacific system during this interval.

6.2. Mid-Holocene cooling: An abrupt response to orbital forcing?

Site 1019 reveals that, for the northeast Pacific at least, the warm events recorded by the coiling of the planktonic foraminifera *N. pachyderma,* which precede and follow the Younger Dryas interval, are anomalous relative to typical late Holocene conditions. The coastal upwelling system of the Northern California Current appears to have returned toward its ice-age state about 8,000 years ago, following brief warm episodes during the deglaciation (Figure 7). This sense of change is supported by an increase in $\delta^{18}O$ of *N. pachyderma* at the same time; however a return to lower $\delta^{18}O$ values near 6,000 years ago, with little change in the *N. pachyderma* coiling ratio, implies significant reduction of upper ocean salinities in the region without major temperature change (Figure 6).

This result based on *N. pachyderma* coiling supports the earlier, lower-resolution, finding of mid-Holocene cooling in the Alaska Gyre based on radiolarian faunas [*Sabin and Pisias,* 1996]. To the south, *Pisias* [1979] inferred similar mid-Holocene cooling in the Santa Barbara Basin along with a change in short-term variability. Consistent with the changes we find to the north, the apparent cooling off Southern California was associated with an increase in the percentage of the fauna that resides in the eastern boundary current.

Mid-Holocene cooling of the northeast Pacific also matches reductions in land temperature inferred from pollen in coastal Washington [*Heusser et al.,* 1980] and British Columbia [*Mathewes and Heusser,* 1981], and from alpine glacier advances near 5.7 and 3.9 ka [^{14}C ages; *Burke and Birkeland,* 1983; *Davis,* 1988]. Analogous mid-Holocene cooling, possibly synchronous with the events of the North Pacific, also occurred in the Arctic and North Atlantic regions [*Kerwin et al.,* 1999]. Thus, the mid-Holocene cooling of the northeast Pacific documented here may be part of a widespread phenomenon in the high-latitude regions of the Northern Hemisphere.

The cause of such mid-Holocene cooling is unknown. One possibility is that gradual reduction in summer (and increase in winter) insolation through the Holocene, associated with well-known changes in Earth's orbit, may have resulted in expansion of polar waters in the Northern Hemisphere. Mechanisms related to orbital insolation change may include gradual reduction of direct summer heating of the subpolar North Pacific, which would cool the ocean directly, and weaker winter cooling, which could suppress formation of NPIW and the associated northward advection of upper ocean heat.

The relatively rapid change observed near 8000 years ago implies a threshold response to the gradual change in insolation. An abrupt shift might be expected from a mechanism that includes intermediate water formation during the early Holocene warm intervals. Enhanced intermediate water formation at these times would help to

Table 2. Radiocarbon data from ODP Site 1019 (all dates new to this paper).

Sample Identifier	DIC[1] (m)	MCD[2] (m)	Sample Type	^{14}C Lab Ref. No.	^{14}C age (ka ± ka)	Res Cor Age (ka)	Cal. Age (ka ± ka)	Inferred Age (ka)
1019A01H1 40-46 cm	0.43	2.84	bark	UA11171	6.03±0.15	6.03	6.82±0.24	6.8
1019A01H2 22-28 cm	1.75	4.17	Mixed Planktic	UA12954	9.95±0.11	9.15	10.29±0.37	10.1
1019A01H2 22-28 cm	1.75	4.17	Mixed Benthic	UA12955	10.81±0.12	9.06	10.27±0.16	10.1
1019A01H2 97-103 cm	2.51	4.92	Mixed Planktic	UA11246	10.21±0.12	9.41	10.45±0.41	10.6
1019A01H2 97-103 cm	2.51	4.92	Mixed Benthic	UA11662	11.13±0.08	9.38	10.48±0.41	10.6
1019A01H2 122-128 cm	2.75	5.16	Mixed Planktic	UA12956	11.41± 0.17	10.61	12.47±0.57	11.2
1019A01H2 122-128 cm	2.75	5.16	Mixed Benthic	UA12957	11.54±0.09	9.79	11.12±0.39	11.2
1019A01H3 38-44 cm	3.40	5.81	Mixed Planktic	UA12958	11.58±0.14	10.78	12.70±0.32	12.7
1019A01H3 38-44 cm	3.40	5.81	Mixed Benthic	UA12959	13.29±0.22	11.54	13.43±0.38	12.7
1019A01H3 78-84 cm	3.80	6.21	Mixed Planktic	UA12960	11.95±0.11	11.15	13.02±0.15	13.0
1019A01H3 78-84 cm	3.80	6.21	Mixed Benthic	UA12961	12.83±0.09	11.08	12.99±0.24	13.0
1019A01H4 24-30 cm	4.76	7.11	Mixed Planktic	UA11867	13.35±0.12	12.55	14.87±0.63	14.8
1019A01H4 24-30 cm	4.76	7.11	Mixed Benthic	UA11866	14.26±0.14	12.51	14.34±0.63	14.8
1019A01H4 128-134 cm	5.80	8.21	Mixed Planktic	UA12962	15.08±0.12	14.28	17.00±0.36	17.0
1019A01H4 128-134 cm	5.80	8.21	Mixed Benthic	UA12963	16.21±0.19	14.46	17.20±0.40	17.0
1019A01H5 110-116 cm	7.13	9.54	Mixed Planktic	UA11531	16.04±0.14	15.24	18.10±0.39	18.1
1019A01H5 110-116 cm	7.13	9.54	Mixed Benthic	UA11663	17.48±0.28	15.73	18.66±0.49	18.1
1019A01H6 60-66 cm	8.14	10.55	Mixed Planktic	UA12964	18.55±0.21	17.75	20.98±0.47	21.0
1019A01H6 60-66 cm	8.14	10.55	Mixed Benthic	UA12965	19.08±0.19	17.33	20.51±0.45	21.0
1019A01H7 71-77 cm	9.74	12.15	Mixed Planktic	UA11865	21.24±0.28	20.44	23.95±0.54	24.0
1019A01H7 71-77 cm	9.74	12.15	Mixed Benthic	UA11864	23.86±0.86	22.11	25.85±1.18	24.0

[1]DIC is depth in core. [2]MCD is meters composite depth [*Lyle et al.*, 1997].

amplify warming and sustain circulation by drawing warm and salty water northward from the subtropics. This concept is discussed in more detail in section 6.4 below.

Model experiments indicate that climates in North America, including extent of snow cover and potential for glaciation, are sensitive to sea surface temperatures in the North Pacific [*Peteet et al.*, 1997]. This is particularly interesting in light of inferences based on the phase of climate cycles in various regions, that the earliest cooling events in the global response to orbital forcing must be in the Northern Hemisphere [*Imbrie et al.*, 1989]. Whether the high latitude North Pacific plays a role in triggering these ice age cycles awaits the availability of longer time series of upper ocean temperatures in this region, and more detailed models of the downstream impacts of North Pacific cooling on climates in North America and elsewhere.

6.3. Variability in the North Pacific oxygen minimum zone: Productivity or intermediate water formation?

Two hypotheses are available to explain available data on the well known millennial-scale oscillations in oxygen minimum zone of the northeast Pacific [*Kennett and Ingram*, 1995; *Behl and Kennett*, 1996; *Cannariato et al.*, 1999]. First, enhanced formation of intermediate or thermocline waters in the North Pacific during cold events could change oxygenation in the Santa Barbara Basin and California Margin. Second, increased export productivity during warm intervals might have increased consumption of oxygen in the thermocline of the North Pacific, resulting in anoxia on the California margin.

This second hypothesis appears to be viable. Organic carbon burial at site 1019 is highest during warm events (essentially coincident with the Bølling-Allerød interstade and early Holocene warmth; Figure 7), which are also times of consistent anoxia in the OMZ [*Behl and Kennett*, 1996]. These events include the presence of diatom mats at Site 1019 [*Lyle et al.*, 1997], which are associated with high-export ecosystems, and also with increased abuncance of coastal upwelling diatom assemblages during early Holocene warmth [*Sancetta et al.*, 1992]

Of course, this inference begs a question of what drives the change in export productivity. At present, productivity of the North Pacific ecosystem is not limited by the major nutrients phosphate and nitrate. One possibility is iron limitation. A primary source of iron to the North Pacific is resuspension of sediments deposited on continental shelves and offshore advection in strong coastal upwelling systems [*Johnson et al.*, 1999]. In the northeast Pacific, such strong coastal upwelling occurs at present during summer [*Huyer*, 1983], and on longer time scales is associated with warm intervals. During cold climatic intervals, coastal upwelling was reduced due to suppression of northerly summer winds [*Lyle et al.*, 1992; *Ortiz et al.*, 1997]. A likely consequence of enhanced iron input associated with coastal upwelling is to favor productive ecosystems that include large diatoms which are more efficiently exported from the sea surface

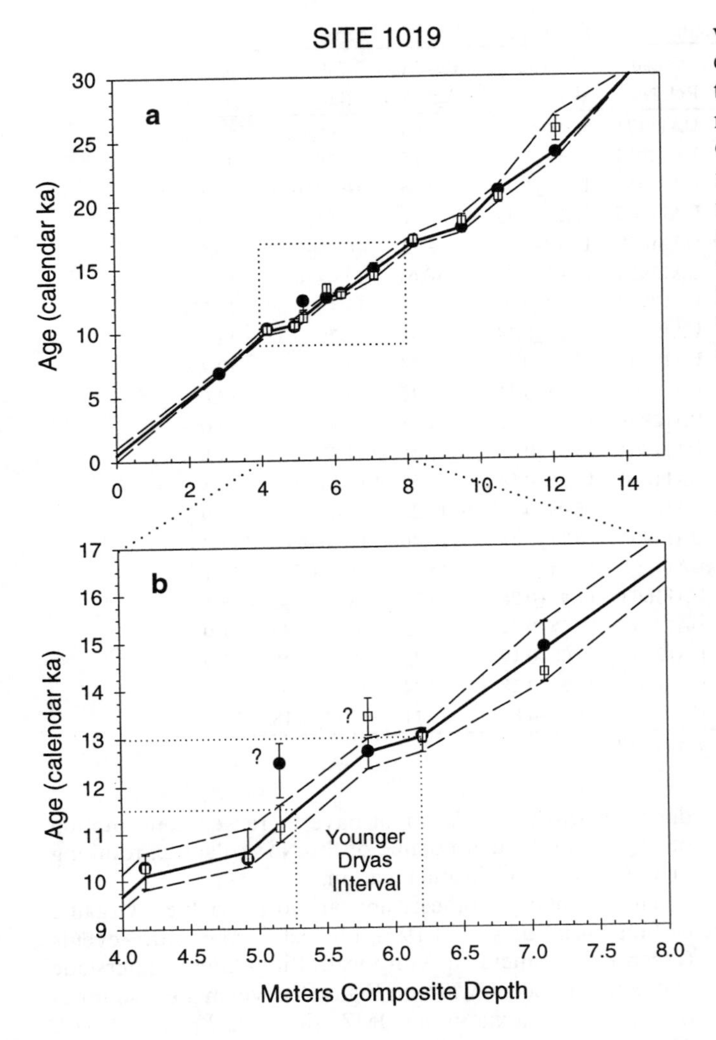

Figure 4. Age model constrained by AMS radiocarbon dates in ODP Site 1019. All dates are shown as calendar corrected values with $\pm 1\sigma$ error bars (see text). Solid circles are dates based on planktonic foraminifera. Open squares are dates based on benthic foraminifera. Question marks indicate suspect dates. The solid line is the inferred age model, and dashed lines mark the upper and lower bounds of propagated errors. a) 0-3 ka (calendar), b) detail of 9-17 ka, illustrating the age control for the Younger-Dryas and Bølling-Allerød intervals.

[*Hood et al.*, 1990; *Takeda*, 1998]. The net effect would be enhanced export of organic matter and depletion of oxygen in subsurface waters during intervals of warmth in the northeast Pacific. Such a linkage may help to explain the variations in anoxia in Santa Barbara Basin during the deglacial interval [*Behl and Kennett*, 1996; *Cannariato et al.*, 1999].

We can not exclude the possibility of greater ventilation of thermocline waters in the range of a few hundred meters water depth during cold intervals. Arguments in favor of such ventilation include an apparent reduction in the ^{14}C age difference between benthic and planktic foraminifera within Santa Barbara Basin during the Younger Dryas event [*Kennett and Ingram*, 1995]. During deglacial times, the sill depth of the Santa Barbara Basin was about 350-400 m. The modern density at these depths off southern California approaches that of the winter mixed layer in the far North Pacific [*Van Scoy et al.*, 1991]. Thus, it is likely that the Santa Barbara Basin and the oxygen minimum zone off California would have been more sensitive to deep winter mixing and formation of subpolar mode waters in the past. Below we suggest, however, that increases in water ventilation events during the Younger Dryas cooling did not necessarily include NPIW.

6.4. *Ventilating intermediate waters: A northern source?*

Increases in benthic foraminifera $\delta^{13}C$ and decreases in the radiocarbon age differences between benthic and planktic foraminifera are potential indicators of sub-surface ventilation in the North Pacific. Both tracers are compromised to some extent by gas exchange processes [*Broecker et al.*, 1984; *Broecker and Maier-Reimer,* 1992; *Campin et al.*, 1999]. Greater exchange with the atmosphere at high latitudes would increase the $\delta^{13}C$ value and decrease the preformed radiocarbon age of a cold, newly formed subsurface water mass, independent of changes in the rate of watermass formation. Thus, ventilation of the deep sea could be accomplished either by greater rates of watermass formation, or by increases in gas exchange.

At ~980m water depth, Site 1019 presently samples the mixing zone between NPIW and Pacific Deep Water (Figure 2). At this site, relatively high benthic foraminiferal $\delta^{13}C$ is closely associated with upper-ocean warmth in the Bølling-Allerød interstade, and in the early Holocene (Figure 8). Even though this record is based on the genus *Uvigerina* sp., which is sometimes compromised by productivity overprints on the $\delta^{13}C$ record, we tentatively accept it as an approximate record of watermass variability here. We have shown that the $\delta^{13}C$ record does not appear to be an artifact of productivity, because it changes in the opposite sense that would be expected based on organic carbon contents of the sediment.

An inference of greater ventilation of intermediate water during the early Holocene and deglacial warm intervals is supported by data on the radiocarbon age differences between benthic and planktic foraminifera. At site 1019, this difference has greater than modern values (implying lower ventilation) during the Younger Dryas cold interval, and less than modern values (implying more rapid ventilation) during the early Holocene warm interval (Figure 8a).

The relatively close linkage between benthic $\delta^{13}C$ at Site 1019 and the deglacial oscillations in the Greenland ice core (Figure 8) argues for atmospheric transmission of signals and a northern-hemisphere source of the watermass, but not in the sense first hypothesized. On the millennial

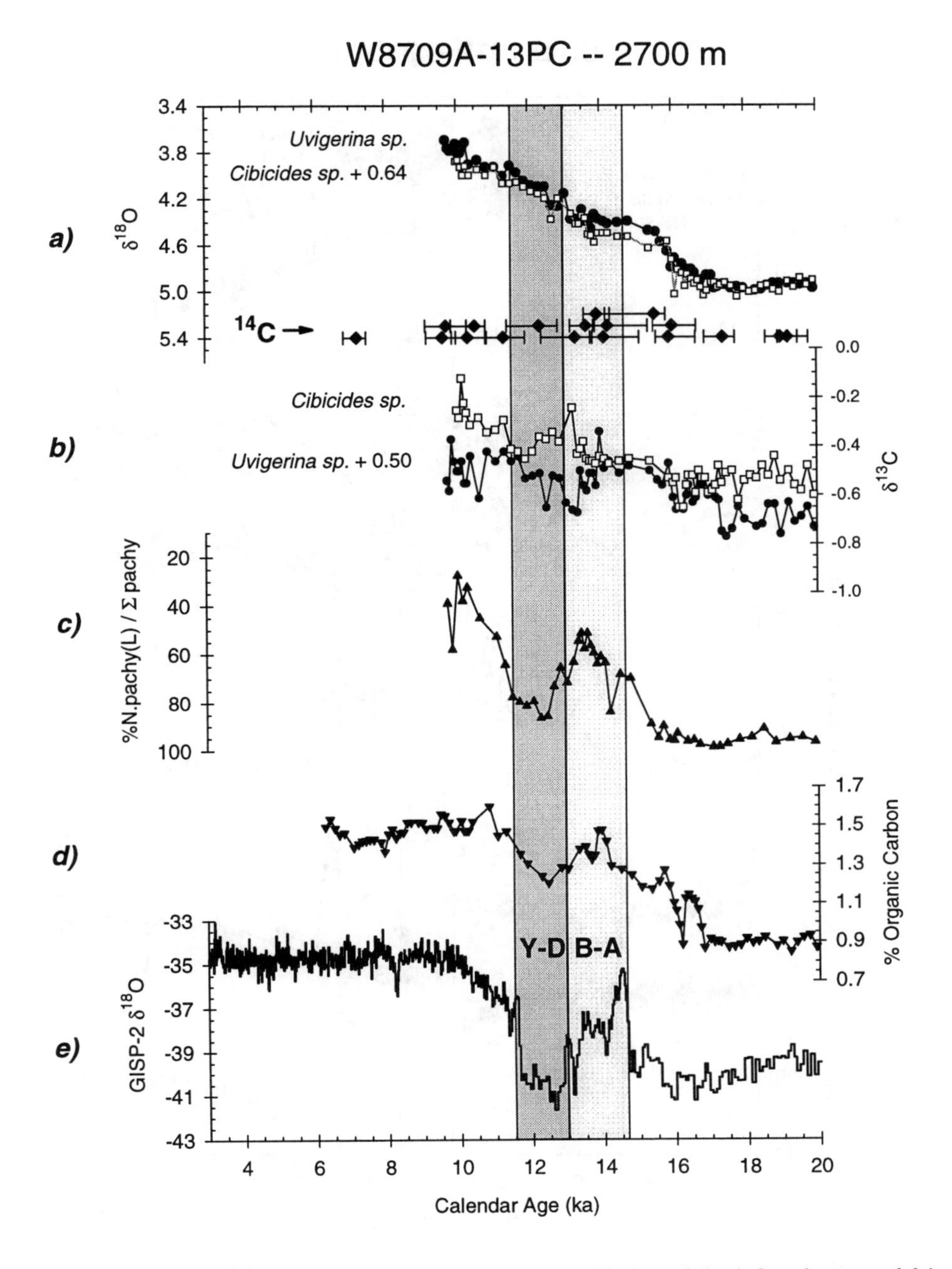

Figure 5. Time series in core W8709A-13PC, based on linear interpolation of the inferred age model in Figure 3. Shaded bars indicate the Younger Dryas (Y-D) and Bølling Allerød (B-A) intervals. a) Benthic foraminiferal $\delta^{18}O$. Open squares are *C. wuellerstorfi* + 0.64. Filled circles are *Uvigerina sp.* Filled diamonds are inferred ages. Horizontal bars indicate the range of the minimum and maximum age models at each inferred age datum. b) Benthic foraminiferal $\delta^{13}C$. Open squares are *C. wuellerstorfi*. Solid circles are *Uvigerina sp.* + 0.50. c) The percentage of left-coiling *N. pachyderma* relative to total *N. pachyderma*. d) Weight % organic carbon. e) Greenland Summit (GISP-2) ice core $\delta^{18}O$ ($^{o}/_{oo}$ relative to SMOW, data from Grootes et al., [1993], timescale from Alley et al., [1993]) for reference.

scale, our data suggest greater ventilation of mid-depth waters during warm, not cold events. To the extent that the $\delta^{13}C$ and ^{14}C data from Site 1019 record variations in NPIW formation rates, they would suggest that NPIW formation is not responsible for the apparent variations in the OMZ on the California margin.

The possibility of significant intermediate water coming from a North Pacific source during the early Holocene and deglacial warm intervals suggests that higher salinity, rather than cooling, may have provided the density required for formation. This is different from, but not necessarily in conflict with, the glacial maximum situation in which

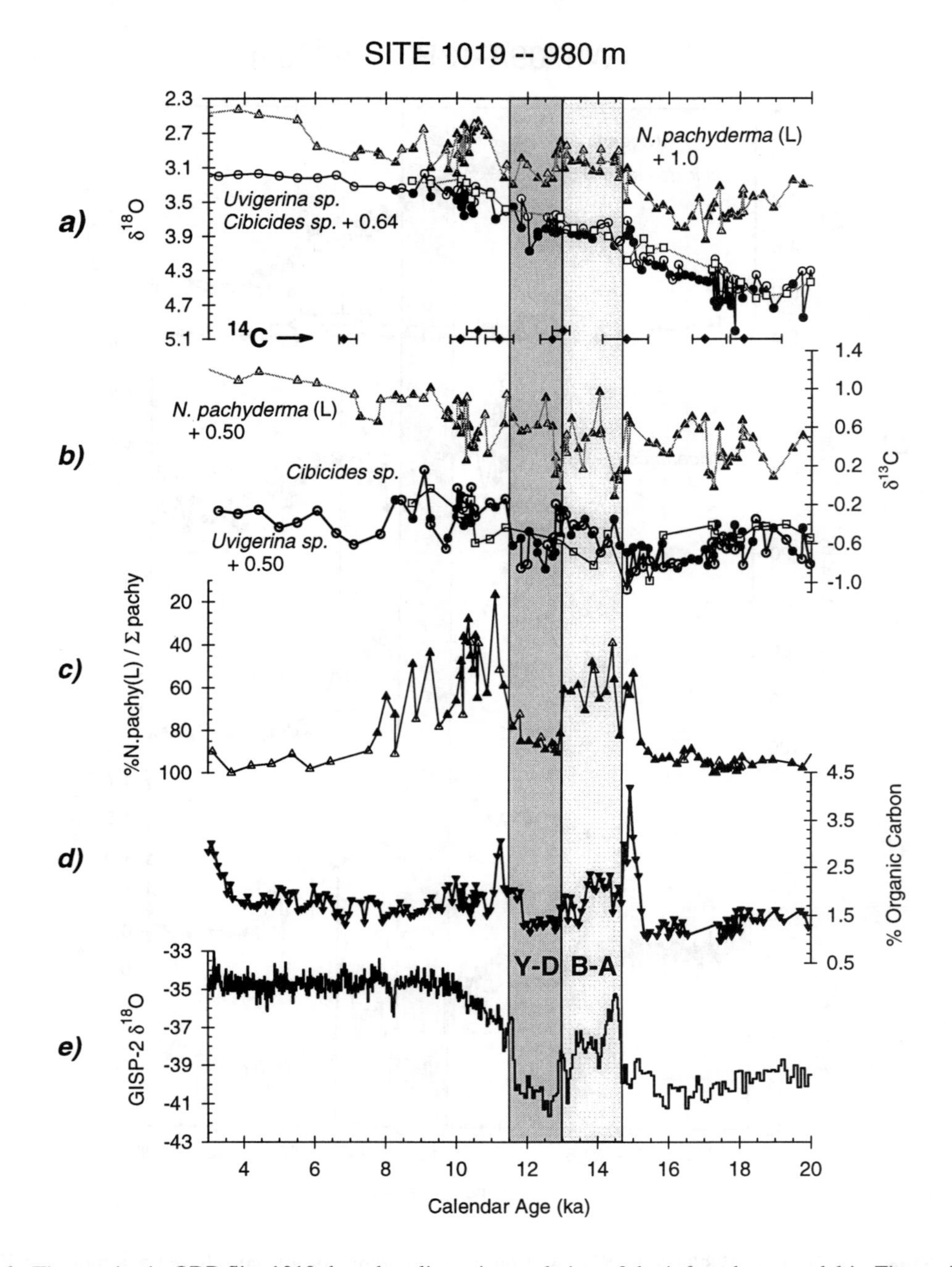

Figure 6. Time series in ODP Site 1019, based on linear interpolation of the inferred age model in Figure 4. Shaded bars indicate the Younger Dryas (Y-D) and Bølling Allerød (B-A) intervals. a) $\delta^{18}O$ ($^o/_{oo}$ relative to PDB). Open squares are *C. wuellerstorfi* + 0.64, from hole 1019c. Filled circles are *Uvigerina* sp. from hole 1019c. Open circles are *Uvigerina* sp. from hole 1019a. Filled triangles are left-coiling *N. pachyderma* + 1.0, from hole 1019a. Open triangles are left-coiling *N. pachyderma* + 1.0, from hole 1019a. Filled diamonds are inferred ages, and horizontal bars indicate the range of the minimum and maximum age models at each inferred age datum. b) $\delta^{13}C$. ($^o/_{oo}$ relative to PDB). Open squares are *C. wuellerstorfi*, from hole 1019c. Filled circles are *Uvigerina* sp. + 0.50 from hole 1019c. Open circles are *Uvigerina* sp. + 0.50 from hole 1019a. Filled triangles are left-coiling *N. pachyderma* + 0.50, from hole 1019a. Open triangles are left-coiling *N. pachyderma* + 0.50, from hole 1019a. c) The percentage of left-coiling *N. pachyderma* relative to total *N. pachyderma*. Solid triangles are from hole 1019a. Open triangles are from hole 1019c. d) Weight % organic carbon in hole 1019c. e) Greenland Summit (GISP-2) ice core $\delta^{18}O$ ($^o/_{oo}$ relative to SMOW, data from Grootes et al., [1993], timescale from Alley et al., [1993]).

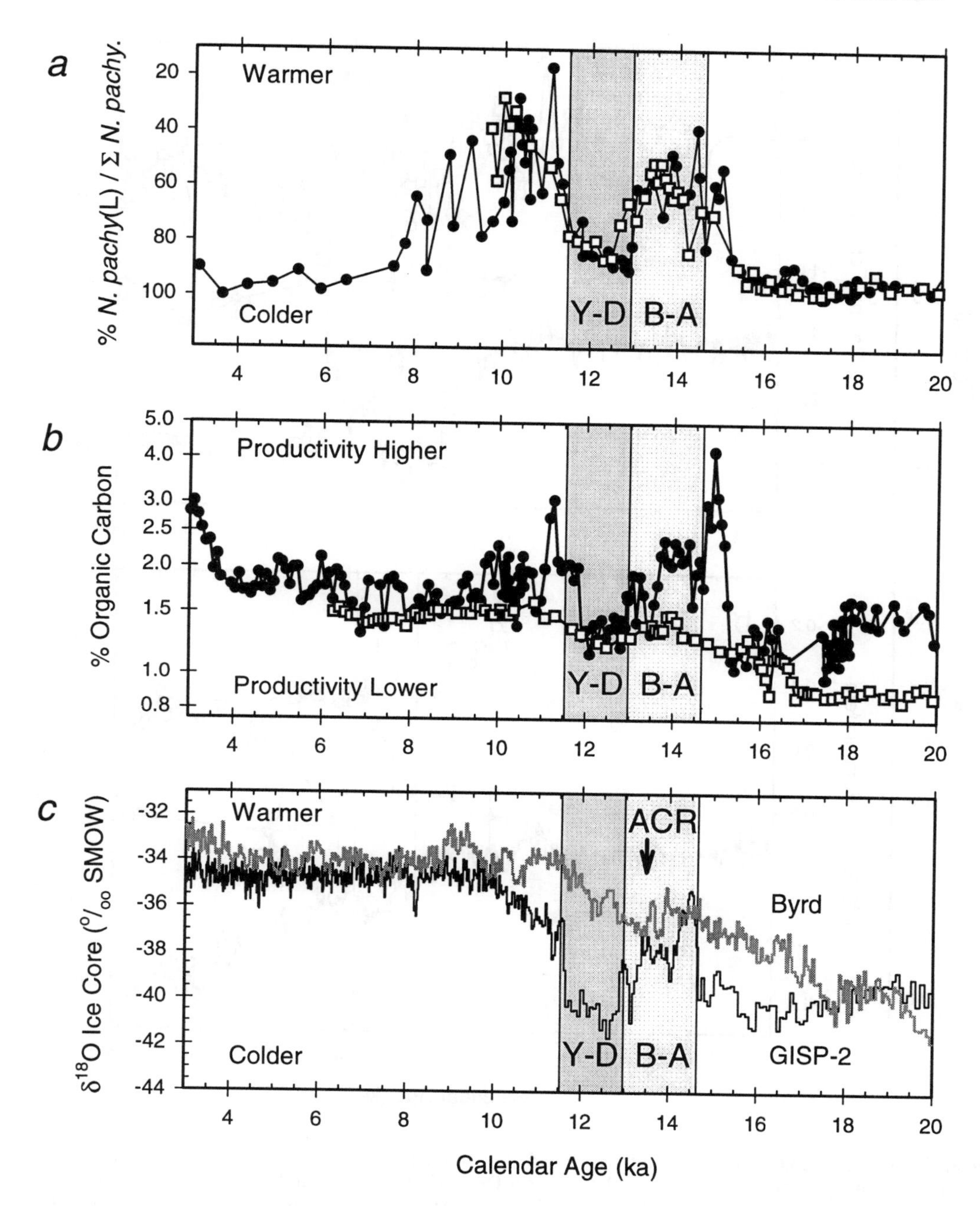

Figure 7. Upper ocean proxies in core W8709A-13PC and ODP Site 1019. a) The percentage of left-coiling *N. pachyderma* relative to total *N. pachyderma*. Filled circles are from Site 1019. Open circles are from core W8709A-13PC. Lower values (up on the graph) indicate warmer conditions. Note approximate synchrony between the two sites, the good match of cooling with the Younger Dryas interval, and apparent mid-Holocene cooling near 8 ka (calendar). b) Percentages of organic carbon in hole 1019C (filled circles) and core W8709A-13PC (open circles) plotted on a log scale. Higher values suggest higher export productivity. Note that high values correspond to warm intervals in (a). c) For comparison, Greenland Summit (GISP-2) ice core $\delta^{18}O$ ($^{o}/_{oo}$ relative to SMOW, black line, data from Grootes et al., [1993], timescale from Alley et al., [1993]) defines the ages of the Younger Dryas (Y-D) and Bølling-Allerød (B-A) intervals. Byrd Ice Core $\delta^{18}O$ ($^{o}/_{oo}$ relative to SMOW, gray line, data from Johnsen et al., [1972], timescale from Bender et al., [1999]) define the timing of the Antarctic Cold Reversal (ACR). The upper ocean proxies from the North Pacific are better correlated to $\delta^{18}O$ data from Greenland than to $\delta^{18}O$ data from Antarctica.

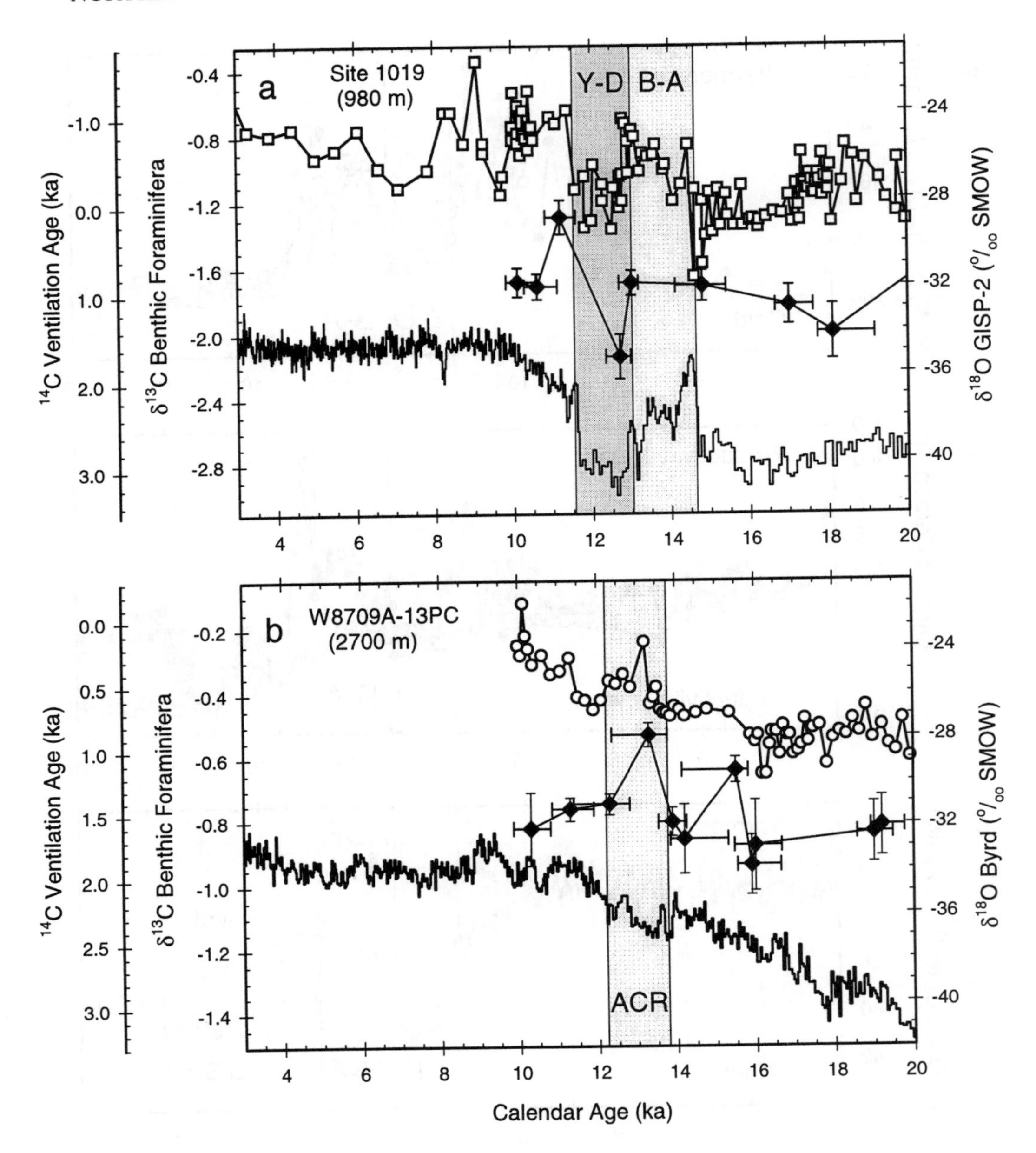

Figure 8. Deep ocean proxies from the northeast Pacific. a) In ODP Site 1019, benthic foraminiferal δ^{13}C (based on Uvigerina sp., open squares) are lower during the Younger Dryas (Y-D) interval and higher during the Bølling-Allerød and early Holocene intervals, suggesting greater ventilation at 980m depth during warm events in the North Pacific. This inferences is supported by estimates of ^{14}C ventilation age, the difference between benthic and planktonic ^{14}C dates (filled diamonds with error bars). Lower values (up on graph) indicate faster replacement of lower intermediate waters, or a greater $^{14}C/^{12}C$ of the water mass when it left the sea surface. Greenland Summit (GISP-2) ice core δ^{18}O ($^o/_{oo}$ relative to SMOW, black line, data from Grootes et al., [1993], timescale from Alley et al., [1993]) defines the ages of the Younger Dryas (Y-D) and Bølling-Allerød (B-A) intervals. b) In core W8709A-13PC benthic foraminiferal δ^{13}C (based on C. wuellerstorfi., open circles) are high relative to background values during the Antarctic Cold Reversal (ACR, 12.3-13.8 ka calendar) and near 10 ka (calendar), suggesting that greater ventilation at 2700 m depth is associated with cold events in the Southern Ocean. This inference based on δ^{13}C is supported by estimates of ^{14}C ventilation age, the difference between benthic and planktonic ^{14}C dates (filled diamonds with error bars). Lower values (up on graph) indicate faster replacement of deep or bottom waters, or a greater $^{14}C/^{12}C$ of the water mass when it left the sea surface. δ^{18}O in the Byrd Ice Core ($^o/_{oo}$ relative to SMOW, black line, data from Johnsen et al., [1972], timescale from Bender et al., [1999]) define the timing of the ACR.

stronger sources of intermediate water in the Sea of Okhotsk were associated with ice-age cooling [*Keigwin,* 1998].

Active formation of NPIW during millennial-scale warm events may provide a mechanism to explain the range of temperature changes observed in the California Current region [*Hendy and Kennett,* 1999; *this paper*], which is higher than that predicted by a coupled atmosphere-ocean model simulation [*Mikolajewicz et al.*, 1997]. Regional warmth in the North Pacific would increase evaporation rates and sea-surface salinities, possibly leading to intermediate water formation that would help to advect more oceanic heat (and salt) northward from the subtropics. This link essentially reverses the causal chain of *Warren* [1983] and provides a plausible feedback mechanism to amplify upper ocean temperature changes in the North Pacific. It also implies that the North Pacific may be susceptible to rapid shifts in its mode of circulation between two potentially stable states.

6.5. *Ventilating deep waters: A southern source?*

Deeper in the water column (~2700 m), benthic foraminiferal $\delta^{13}C$ from the species *C. wuellerstorfi* and benthic-planktic ^{14}C age differences suggest a somewhat different history of ventilation (Figure 8b). Here, in the mixing zone between northward flowing Pacific Bottom Water and southward flowing Pacific Deep Water, high apparent ventilation (inferred from high $\delta^{13}C$ and a low benthic-planktic ^{14}C age difference) occurs near 13 ka (calendar).

This ventilation event appears to be poorly correlated with the Younger Dryas oscillation of the Northern Hemisphere, but is essentially in phase with the Antarctic Cold Reversal (ACR) as documented by relatively low $\delta^{18}O$ in the Byrd Ice Core ($\delta^{18}O$ from *Johnsen et al.* [1972], timescale from *Bender et al.* [1999]). Additional high $\delta^{13}C$ values, suggesting greater ventilation of the deep Pacific, occur near 10 ka (calendar), coincident with another cooling event in Antarctica.

Thus, the pattern of isotopic variation in the deep North Pacific suggests the possibility of greater ventilation coming from the Southern Hemisphere, transmitted northward via bottom waters. This finding is based on limited data and remains somewhat uncertain based on the chronologic constraints. The inference is strengthened, however, by the stratigraphic relationship of benthic foraminiferal $\delta^{13}C$ and apparent Younger-Dryas cooling of surface waters recorded in the same core samples. If our inference is correct, it implies that the time signature of millennial-scale oscillations in the Northern and Southern Hemisphere, now referred to as a bi-polar seesaw [*Broecker,* 1998; *Charles et al.*, 1996], will be detected elsewhere at different depths within the Pacific.

6.6. *Deep-sea temperatures and Antarctic warming.*

Benthic foraminiferal $\delta^{18}O$ values in Site 1019 (all corrected to equivalent *Uvigerina* sp. values) vary from ~3.1 ‰ to ~4.8 ‰, a range of more than 1.7 ‰ (Figure 9). Although samples from W8709A-13PC do not reach the late Holocene, glacial values reach 5.0 ‰. Late Holocene values in nearby cores are about 3.3 ‰. This range of $\delta^{18}O$ of more than 1.6 ‰ is important, because if recent estimates are correct that the amplitude of the ice-volume driven change in mean oceanic $\delta^{18}O$ is about 1.0 ‰ [*Schrag et al.*, 1996] then the deep North Pacific (at 980m) may have cooled at the last glacial maximum by about 3°C. At present, the temperatures of water in contact with the sea floor at Site 1019 and at W8709A-13PC are 3.6°C and 1.6°C, respectively, so ice-age water temperatures here would have been <1°C at 980m depth and <-1°C at 2700 m depth.

The time series of benthic foraminiferal $\delta^{18}O$ are compared to sea level estimates from Barbados corals [*Fairbanks,* 1989], and to $\delta^{18}O$ data from the Byrd ice core in Antarctica [*Johnsen et al.,* 1972; age model of *Bender et al.,* 1999] in Figure 9. This plot overlays sea level on oceanic $\delta^{18}O$ with the scaling implied by Barbados coral data of 0.11‰ per 10 m [*Fairbanks and Matthews,* 1978]. If the *Schrag et al.* [1996] scaling of sea level relative to $\delta^{18}O$ was used, deviations between these records would be even larger.

Through much of the transition, benthic $\delta^{18}O$ and sea level are essentially coincident (within the limits of the chronology). A major deviation occurs between sea level and benthic $\delta^{18}O$ in both cores prior to ~15.5 ka (calendar). Although sea level data are sparse in this interval, it is unlikely that values are much lower than the three dated points between 18 and 19 ka (calendar) [*Fairbanks,* 1989]. Recent summaries indicate very little melting of Northern Hemisphere ice sheets in this interval, and suggest that much of the melting associated with the ~14 ka (calendar) sea level rise occurred in Antarctica [*Clark et al.*, 1996]. A likely conclusion based on the deviation between benthic $\delta^{18}O$ and sea level is that significant warming of the deep sea occurred prior to ~15.5 ka (calendar) without major change in sea level (and thus without major melting of ice sheets).

Comparison of the benthic foraminiferal $\delta^{18}O$ records to the Byrd ice core $\delta^{18}O$ suggests that early warming of the deep Pacific may be coincident with early warming in Antarctica (Figure 9). Similar early warming in surface waters have been observed in the subantarctic region [*Charles et al.*, 1996; *Broecker,* 1998]. If true, this implies that sea surface conditions near Antarctica are important in setting the heat budget of the deep sea, at least in regions where Antarctic Bottom Water is an important contributor to abyssal water masses.

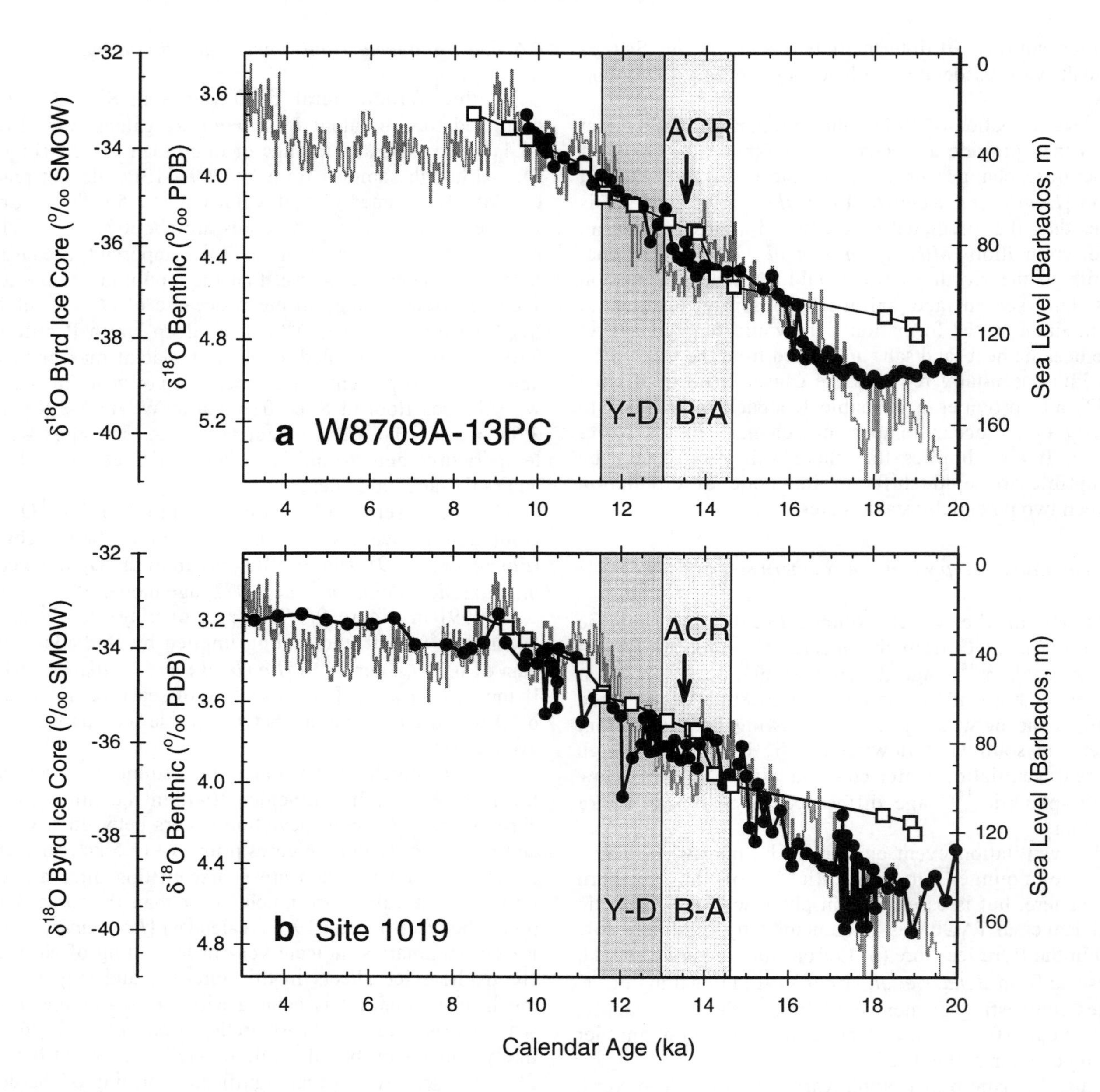

Figure 9. Benthic foraminifera $\delta^{18}O$ data (‰ PDB, filled circles), including data from *C. wuellerstorfi (+0.64)* and *Uvigerina* sp., are compared to sea level data (open squares) from Barbados [*Fairbanks,* 1989], and $\delta^{18}O$ in the Byrd Ice Core ($\delta^{18}O$ SMOW, black line, data from Johnsen et al., [1972], timescale from Bender et al., [1999]). Bars note the timing of the Younger Dryas (Y-D) and Bølling-Allerød (B-A) intervals [*Alley et al.,* 1993]. a) Core W8709A-13PC (2700 m water depth), b) ODP Site 1019 (980 m water depth).

7. CONCLUSIONS

Planktonic foraminiferal species abundances, benthic and planktonic foraminiferal stable isotopes, radiocarbon, and organic carbon content of deep-sea cores off Oregon and California reveal millennial-scale climate oscillations during the past 20,000 years. We conclude the following.

1) Surface ocean changes are essentially coincident with the Bølling-Allerød and Younger-Dryas climate oscillations observed in Greenland and the North Atlantic. This finding supports the concept of atmospheric transmission of climate signals between oceans in the Northern Hemisphere. Significant cooling of North Pacific surface waters also occurred in mid-Holocene time, and this event may be part of a widespread phenomenon in the high-latitude regions of the Northern Hemisphere.

2) Organic carbon contents of northeast Pacific sediments suggest that high export productivity was

associated with warm events. This enhanced organic flux probably contributed to depletion of oxygen along the margin of California, and could account for oxic/anoxic oscillations observed there near 400-m water depth, alligned with rapid climatic events observed in Greenland. Although we can not exclude the possibility of variations in thermocline ventilation, major changes in North Pacific Intermediate Water are not required to explain variations in the intensity of the oxygen minimum zone.

3) Benthic foraminiferal $\delta^{13}C$ and ^{14}C data from 980 m depth suggest that higher ventilation (either faster formation or greater gas exchange) of lower intermediate waters occurred during millennial-scale warm intervals, mirroring the Bølling-Allerød and Younger-Dryas climate oscillations observed in Greenland and the North Atlantic. As with the surface water tracers, atmospheric transmission of climate changes within the Northern Hemisphere is supported, but a tentative conclusion that higher intermediate water formation was associated with North Pacific warming suggests that increases in salinity would be required to compensate for effects of warming on water density. We speculate that a mode of circulation with enhanced formation of NPIW during warm events may be self-sustaining and may contribute to the observed range of temperature oscillations in the region.

4) In the deep North Pacific (2700 m), benthic foraminiferal $\delta^{13}C$ and ^{14}C suggest a ventilation event during deglaciation (either faster formation or greater gas exchange) that coincides with the Antarctic Cold Reversal in the Byrd Ice Core. Such a connection, if confirmed at additional sites with firm chronologic constraints, would support the concept of enhanced formation of Antarctic Bottom Water associated with cooling in the Southern Ocean.

5) Benthic foraminiferal $\delta^{18}O$ data from the deep Pacific, compared with available sea level data, suggest that warming of the deep ocean preceded major deglaciation and sea level rise. This warming may be linked to early warming in Antarctica.

6) Our findings of millennial-scale events in the Northeast Pacific, with a range of apparent timings in different proxies from the same sediment core samples, are important because they may reveal dynamic processes in a chain of events leading to natural climate oscillations at millennial and longer scales. The earliest changes appear to be in the deep sea, perhaps associated with events in Antarctica.

7) Research on millennial-scale climate changes in the Pacific region is in its infancy relative to the more heavily studied North Atlantic region. Limits to our findings come from possible chronologic errors and the difficulty in getting good high-resolution records from North Pacific sediments, in which carbonate fossils are relatively rare. To resolve some of these issues in the Pacific, arrays of high-sedimentation rate sites with well-constrained chronologies and a rich array of paleoclimatic proxies are needed.

Acknowledgments. We thank A. Morey, J. Wilson, and W. Rugh for technical assistance at OSU and M. Kashgarian at Lawrence Livermore National Laboratory for the radiocarbon dates in core W8709A-13PC. Reviews by L. Keigwin and G. Haug, and discussions with P. Clark and S. Hostetler improved the paper. This work was funded by NSF, USSAC, and the Swedish National Research Council. NSF funded sediment curation at Oregon State University.

REFERENCES

Alley, R.B., and P.U. Clark, The deglaciation of the Northern Hemisphere: A global perspective, *Ann. Rev. Earth Planet. Sci.*, 27, 149-182, 1999.

Alley, R.B., D.A. Meese, C.A. Shuman, A.J. Gow, K.C. Taylor, P.M. Grootes, J.W.C. White, M. Ram, E.D. Waddinton, P.A. Mayewski, and G.A. Zielinski, Abrupt increase in Greenland snow accumulation at the end of the Younger Dryas event, *Nature*, 362, 527-529, 1993.

Bard, E., Geochemical and geophysical implications of the radiocarbon calibration, *Geochim. Cosmochim. Acta*, 62, 2025-2038, 1998.

Behl, R.J. and J.P. Kennett, Brief interstadial events in the Santa Barbara Basin, NE Pacific, during the past 60 kyr, *Nature,* 379, 243-246, 1996.

Bender, M.L., B. Malize, J. Orchardo, T. Sowers and J. Jouzel, High precision correlations of Greenland and Antarctic ice core records over the last 100 kyr, in *Mechanisms of Millennial Scale Global Climate Change*, edited by P.U. Clark, R.S. Webb, and L.D. Keigwin, AGU, Washington DC, (in press 1999).

Blunier, T., J. Chappellaz, J. Schwander, A. Dällenbach, B. Stauffer, T.F. Stocker, D. Raynaud, J. Jouzel, H.B. Clausen, C.U. Hammer, and S.J. Johnsen, Asynchrony of Antarctic and Greenland climate change during the last glacial period, *Nature*, 394, 739-743, 1998.

Bond, G., W. Broecker, S. Johnsen, J. McManus, L. Labeyrie, J. Jouzel, and G. Bonani, Correlations between climate records from North Atlantic sediment and Greenland ice, *Nature*, 365, 143-147, 1993.

Bond, G. and R. Lotti, Iceberg discharges into the North Atlantic on millennial time scales during the last glaciation, *Science,* 267, 1005-1010, 1995.

Broecker, W.S., Chaotic climate, *Sci. Am.*, November, 62-68, 1995.

Broecker, W.S., Paleocean circulation during the last deglaciation: A bipolar seesaw? *Paleoceanography*, 13, 119-121, 1998.

Broecker, W.S. and E. Maier-Reimer, The influence of air and sea exchange on the carbon isotope distribution in the sea, *Global Biogeochem. Cycles*, 6, 315-320, 1992.

Broecker, W.S., A.C. Mix, M. Andree, and H. Oeschger, Radiocarbon measurements on coexisting benthic and planktic foraminifera shells: Potential for reconstructing ocean ventilation times over the past 20,000 years, *Nuclear Instruments and Methods in Physics Research*, B5, 331-339, 1984.

Broecker, W.S., T. Takahashi, and T. Takahashi, Sources and flow patterns of deep-ocean waters as deduced from potential temperature, salinity, and initial phosphate concentration, *J. Geophys. Res.* 90, 6925-6939, 1985.

Burke, R.M. and P.W. Birkeland, Holocene glaciation in the mountain ranges of the western United States, in *Late*

Quaternary Environments of the United States, edited by H.E. Wright, v.1, pp. 3-11, Univ. Minnesota Press, Minneapolis, MN, 1983.

Campin, J.-M., T. Fichefet, and J.-C. Duplessy, Problems with using radiocarbon to infer ocean ventilation rates for past and present climates, *Earth Planet. Sci. Lett.*, 165, 17-24, 1999.

Cannariato, K.G., J.P. Kennett, and R.J. Behl, Biotic response to late Quaternary rapid switches in Santa Barbara Basin: Ecological and evolutionary implications, *Geology*, 27, 63-66, 1999.

Charles, C.D., J. Lynch-Stieglitz, U.S. Ninnemann, and R.G. Fairbanks, Climate connections between the hemisphere revealed by deep sea sediment core / ice core correlations, *Earth Planet. Sci. Lett.*, 142, 19-27, 1996.

Clark, P.U., R.B. Alley, L.D. Keigwin, J.M. Licciardi, S.J. Johnsen, and H. Wang, Origin of the first global meltwater pulse following the last glacial maximum, *Paleoceanography,* 11, 563-577, 1996.

Clark, P.U. and P.J. Bartlein, Correlations of late Pleistocene glaciation in the western United States with North Atlantic Heinrich events, *Geology*, 23, 483-486, 1995.

Curry, W., J.-C. Duplessy, L.D. Labeyrie, and N.J. Shackleton, Changes in the distribution of $\delta^{13}C$ of deep water ΣCO_2 between the last glaciation and the Holocene. *Paleoceanography,* 3, 317-341, 1988.

Curry, W.B., and D.W. Oppo, Synchronous, high-frequency oscillations in tropical sea surface temperatures and North Atlantic Deep Water production during the last glacial cycle, *Paleoceanography,* 12, 1-14, 1997.

Dansgaard, W., S.J. Johnsen, H.G. Clausen, and C.C. Langway, Jr., Climatic record revealed by the Camp Century ice core, in *The Late Cenozoic Glacial Ages*, edited by K.K. Turekian, pp. 37-56, Yale Univ. Press, New Haven, CT, 1971.

Davis, P.T., Holocene glacier fluctuations in the American cordillera, *Quat. Sci. Rev.*, 7, 129-157, 1988.

Doose, H., F.G. Prahl, and M.W. Lyle, Biomarker temperature estimates for modern and last glacial surface waters of the California Current system between 33° and 42°N, *Paleoceanography*, 12, 615-622, 1997.

Edwards, R.L., J.W. Beck, G.S. Burr, D.J. Donahue, J.M.A. Chappell, A.L. Bloom, E.R.M. Druffel, and F.W. Taylor, A large drop in atmospheric $^{14}C/^{12}C$ and reduced melting in the Younger Dryas, documented with ^{230}Th ages of corals, *Science,* 260, 962-968, 1993.

Fairbanks, R.G., A 17,000-year glacio-eustatic sea level record: influence of glacial melting rates on the Younger Dryas event and deep-ocean circulation, *Nature*, 342, 637-642, 1989.

Fairbanks, R.G. and R.K. Matthews, The marine oxygen isotope record in Pleistocene coral, Barbados, West Indies, *Quat. Res.,* 10, 181-196, 1978.

Freeland, H.J., A.S. Bychkov, FF.Whitney, C. Taylor, C.S. Wong, and G.I Yurasov, WOCE section P1W in the Sea of Okhotsk 1: Oceanographic data description, *J. Geophys. Res.*, 103, 15,613-15,623, 1998.

Gardner, J.V., W.E. Dean, P. Dartnell, Biogenic sedimentation beneath the California Current system for the past 30 kyr and its paleoceanographic significance, *Paleoceanography*, 12, 207-225, 1997.

Grootes, P.M., M. Stuiver, J.W.C. White, S.J. Johnsen, and J. Jouzel, Comparison of oxygen isotope records from the GISP2 and GRIP Greenland ice cores, *Nature*, 366, 552-554, 1993.

Harris, S.E. and A.C. Mix, Pleistocene precipitation balance in the Amazon Basin recorded in deep-sea sediments, *Quat. Res.,* 51, 14-26, 1999.

Harris, S.E., A.C. Mix, and T. King, Biogenic and terrigenous sedimentation at Ceara Rise, western tropical Atlantic, supports Plio-Pleistocene deep-water linkage between hemispheres, in *Proceedings of the Ocean Drilling Program, Scientific Results*, v. 154, edited by N.J. Shackleton, W. Curry, and C. Richter, pp. 331-348, College Station, TX (ODP), 1997.

Hendy, I.L. and J. P. Kennett, Latest Quaternary North Pacific surface-water responses imply atmosphere-driven climate instability, *Geology*, 27, 291-294, 1999.

Heusser, C.J., L.E. Heusser, and S.S. Streeter, Quaternary temperatures and precipitation for the Northwest coast of North America, *Nature*, 286, 702-704, 1980.

Hood, R., M. Abbott, A. Huyer, and P.M. Kosro, Surface patterns in temperature, flow, phytoplankton biomass, and species composition in the coastal transition zone off Northern California, *J. Geophys. Res.*, 95, 18,081-18,094, 1990.

Hostetler, S.W., P.U. Clark, P.J. Bartlein, A.C. Mix, and N.J. Pisias, Atmospheric transmission of North Atlantic Heinrich events, *J. Geophys. Res.*, 104, 3947-3952, 1999.

Hughen, K.A., J.T. Overpeck, S.J. Lehman, M. Kashgarian, J. Southon, L.C. Peterson, R. Alley, and D.M. Sigman, Deglacial changes in ocean circulation from an extended radiocarbon calibration, *Nature*, 391, 65-68, 1998.

Huyer, A., Coastal upwelling in the California Current System. *Prog. Oceanography*, 12, 259-284, 1983.

Imbrie, J., A. McIntyre, and A.C. Mix, Oceanic response to orbital forcing in the late Quaternary: Observational and Experimental Strategies, in *Climate and Geosciences*, edited by A. Berger, S. Schneider, and J.-C. Duplessy, pp. 121-164, Kluwer Academic, Dordrecht, 1989.

Ingram, B.L. and J.P. Kennett, Radiocarbon chronology and planktonic-benthic foraminiferal ^{14}C age differences in Santa Barbara Basin sediments, Hole 893A, in *Proceedings of the Ocean Drilling Program, Scientific Results*, v. 146, edited by J.P. Kennett, J.G. Baldauf, and M. Lyle, pp. 19-27, College Station, TX (ODP), 1996.

Johnsen, S.J., W. Dansgaard, H.B. Claussen, and C.C. Langway, Oxygen isotope profiles through the Antarctic and Greenland ice sheets, *Nature,* 235, 429-434, 1972.

Johnson, K.S., F.P. Chavez, and G.E. Friederich, Continental-shelf sediment as a primary source of iron for coastal phytoplankton, *Nature,* 398, 697-700, 1999.

Keigwin, L.D., Glacial-age hydrography of the far northwest Pacific Ocean, *Paleoceanography*, 13, 323-339, 1998.

Keigwin, L.D. and G.A. Jones, Deglacial climatic oscillations in the Gulf of California, *Paleoceanography*, 5, 1009-1023, 1990.

Kennett, J.P. and B.L Ingram, A 20,000-year record of ocean circulation and climate change from the Santa Barbara Basin, *Nature*, 377, 510-514, 1995.

Kerwin, M.W., J.T. Overpeck, R.S. Webb, A. DeVernal, D.H. Rind, and R.J. Healy, The role of oceanic forcing in mid-Holocene Northern Hemisphere climate change, *Paleoceanography,* 14, 200-210, 1999.

Kipp, N.G., New transfer function for estimating past sea-surface conditions from sea-bed distribution of planktonic foraminiferal assemblages in the North Atlantic, in *Investigation of Late Quaternary Paleoceanography and Paleoclimatology*, edited by R.M. Cline, and J.D. Hays, Memoir 145, pp. 3-42, Geol. Soc. America, Boulder, CO, 1976.

Kotilainen, A.T., and N.J. Shackleton, Rapid climate variability in the North Pacific Ocean during the past 95,000 years, *Nature*, 377, 323-326, 1995.

Kroopnick, P.M., The distribution of ^{13}C of TCO_2 in the world oceans, *Deep-Sea Res.*, 32, 57-84, 1985.

Lowell, T.V., C.J. Heusser, B.G. Andersen, P.I. Moreno, A. Hauser, L.E. Heusser, C. Schluchter, D.R. Marchant, and G.H. Denton, Interhemispheric correlation of late Pleistocene glacial events, *Science*, 269, 1541-1549, 1995.

Lund, D.C. and A.C. Mix, Millennial-scale deep water oscillations: Reflections of the North Atlantic in the deep Pacific from 10 to 60 ka, *Paleoceanography*, 13, 10-19, 1998.

Lyle, M., I. Koizumi, C. Richter, and the shipboard scientific party, *Proceedings of the Ocean Drilling Program, Initial Reports*, (part 1), 167, 0-495pp, College Station, TX (ODP), 1997.

Lyle, M., A. Mix, C. Ravelo, D. Andreasen, L. Heusser, A. Olivarez, Kyr-scale $CaCO_3$ and C_{org} events along the northern and central California margin: Stratigraphy and origins, *Proceedings of the Ocean Drilling Program, Scientific Results*, 167, edited by M. Lyle, I. Koizumi, and C. Richter, College Station, TX (ODP), in press 1999.

Lyle, M., R. Zahn, F. Prahl, J. Dymond, R. Collier, N. Pisias, and E. Suess, Paleoproductivity and carbon burial across the California Current: The multitracers transect, 42°N, *Paleoceanography*, 7, 251-272, 1992.

Lynn, R.J. and J.J. Simpson, The California Current System: The Seasonal variability of its physical characteristics, *J. Geophys. Res.*, 92, 12947-12966, 1987

MacAyeal, D.R., Growth/Purge oscillations of the Laurentide ice sheet as a cause of the North Atlantic's Heinrich events. *Paleoceanography*, 8, 775-784, 1993.

Manabe, S., and R.J. Stouffer, Two stable equilibria of a coupled ocean-atmosphere model, *J. Climate*, 1, 841-866, 1988.

Mantyla, A.W., and J.L. Reid, Abyssal characteristics of the world ocean waters, *Deep-Sea Res.*, 30, 805-833, 1983.

Mathewes, R.W. and L.E. Heusser, A 10,000 year palynological record of temperature and precipitation trends in southwestern British Columbia, *Can. J. Bot.*, 59, 707-710, 1981.

Mikolajewicz, U., T.J. Crowley, A, Schiller, and R. Voss, Modeling teleconnections between the North Atlantic and North Pacific during the Younger Dryas, *Nature*, 387, 384-387, 1997.

Mix, A.C., S.E. Harris and T. Janecek,. Estimating lithology from nonintrusive reflectance spectra: ODP Leg 138, in *Proceedings of the Ocean Drilling Program, Scientific Results,* 138, edited by N.G. Pisias, L. Mayer, T. Janecek, A. Palmer-Julson, and T.H. van Andel, pp. 413-428, College Station, TX (ODP), 1995.

Mix, A.C., N.G. Pisias, W. Rugh, S. Veirs, Leg 138 Shipboard Sedimentologists, and the Leg 138 Scientific Party, Color reflectance spectroscopy: A tool for rapid characterization of deep-sea sediments. In: *Proceedings of the Ocean Drilling Program, Initial Repts*, 138 (volume 1), edited by L. Mayer, N. Pisias, and T. Janecek, pp. 67-77, College Station, TX, (ODP), 1992.

Ninnemann, U., and C.D. Charles, Regional differences in Quaternary Subantarctic nutrient cycling: Link to intermediate and deep water ventilation, *Paleoceanography* 12, 560-567, 1997.

Oppo, D.W. and S.J. Lehman, Suborbital timescale variability of North Atlantic Deep Water during the past 200,000 years, *Paleoceanography*, 10, 901-910, 1995.

Ortiz, J., A. Mix, S. Harris, and S. O'Connell, Diffuse spectral reflectance as a proxy for percent carbonate content in North Atlantic sediments. *Paleoceanography*. 14, 171-186, 1999.

Ortiz, J.D., A.C. Mix, S. Hostetler, and M. Kashgarian, The California Current of the last glacial maximum: Reconstruction at 42°N based on multiple proxies, *Paleoceanography*, 12, 191-205, 1997.

Parker, F., Planktonic foraminiferal species in Pacific sediments, *Micropaleontology*, 8, 219-254, 1962.

Peteet, D., A. Del Genio, and K.-W. Lo, Sensitivity of Northern Hemisphere air temperatures and snow expansion to North Pacific sea surface temperatures in the Goddard Institute for Space Studies general circulation model. *J. Geophys. Res.* 102, 23,781-23,791, 1997.

Pisias, N.G., Model for paleoceanographic reconstructions of the California Current during the last 8000 years. *Quat. Res.*, 11, 373-386, 1979.

Pisias, N., Sancetta, C., and Dauphin, P., Spectral analysis of late Pleistocene-Holocene sediments, *Quat. Res.*, 3, 3-9, 1973.

Porter, S.C. and Z. An, Correlation between climate events in the North Atlantic and China during the last glaciation, *Nature* 375, 305-308, 1995.

Rahmstorf, S., Bifurcations of the Atlantic thermohaline circulation in response to changes in the hydrological cycle, *Nature*, 378, 145-149, 1995.

Reid, J.L., Jr., *Intermediate waters of the Pacific Ocean*, 85 pp., Johns Hopkins Press, Baltimore, MD, 1965.

Robinson, S.W.and G. Thompson, Radiocarbon corrections for marine shell dates with application to southern Pacific Northwest Coast prehistory, *Syesis,* 14, 45-57, 1981.

Sabin, A.L. and N.G. Pisias, Sea surface temperature changes in the Northeast Pacific Ocean during the past 20,000 years and their relationship to climatic change in Northwestern North America, *Quat. Res.*, 46, 48-61, 1996.

Saito, T., P.R. Thompson and D. Breger, *Systematic Index of Recent and Pleistocene Planktonic Foraminifera*, Univ. Tokyo Press, Tokyo, Japan, 1981.

Sancetta, C., M. Lyle, L. Heusser, R. Zahn, and J.P. Bradbury, Late-glacial to Holocene changes in winds, upwelling, and seasonal production of the northern California Current system, *Quat. Res.*, 38, 359-370, 1992.

Schrag, D.P., G. Hampt, and D.W. Murray, The temperature and oxygen isotopic composition of the glacial ocean, *Science*, 272, 1930-1932, 1996.

Smith, A.B., Living planktonic foraminifera collected along an east-west traverse in the North Pacific, *Contrib. Cushman Foundation for. Foraminiferal Research*, 15, 131-134,1964.

Southon, J.R., D.E. Nelson, and J.S. Vogel, A record of past ocean-atmosphere radiocarbon differences from the northeast Pacific, *Paleoceanography*, 5, 197-206, 1990.

Stuiver, M. and P.J. Reimer, Extended ^{14}C database and revised CALIB radiocarbon calibration program, *Radiocarbon*, 35, 215-230, 1993.

Stuiver, M., P.J. Reimer, E. Bard, J.W. Beck, G.S. Burr, K.A. Hughen, B. Kromer, F.G. McCormac, J. vanderPlicht, and M. Spurk, INTCAL98 radiocarbon age calibration, 24,000-0 cal BP, *Radiocarbon* , 40, 1041-1083, 1998.

Takeda, S., Influence of iron availability on nutrient consumption ratio of diatoms in oceanic waters, *Nature* 393, 774-777, 1998.

Talley, L., Distribution and formation of the North Pacific Intermediate Water, *J. Phys. Ocean.*, 23, 517-537, 1993.

Thunell, R. and P.G. Mortyn, Glacial climate instability in the northeast Pacific Ocean, *Nature*, 376, 504-506, 1995.

Toggweiler, J.R., K. Dixon, and K. Bryan, Simulations of radiocarbon in a coarse-resolution world ocean model, 1, steady state prebomb distributions, *J. Geophys. Res.*, 94, 8217-8242, 1989.

Toggweiler, J.R., and B. Samuels, Effect of Drake Passage on the global thermohaline circulation, *Deep-Sea Res.* 42, 477-500, 1993.

van Geen, A., R.G. Fairbanks, P. Dartnell, M. McGann, J.V. Gardner, and M. Kashgarian, Ventilation changes in the northeast Pacific during the last deglaciation, *Paleoceanography*, 11, 519-528, 1997.

van Scoy, K.A., D.B. Olson, and R.A. Fine, Ventilation of the North Pacific Intermediate waters: The role of the Alaskan Gyre, *J. Geophys. Res.*, 96, 16,801-16,810, 1991.

Warren, B.A., Why is no deep water formed in the North Pacific? *J. Mar. Res.*, 41, 327-347, 1983.

Yamanaka, G., Y. Kitamura, and M. Endoh, Formation of North Pacific Intermediate Water in Meteorological Research Institute ocean general circulation model 1: Subgrid-scale mixing and marginal sea fresh water, *J. Geophys. Res.*, 103, 30,885-30,903, 1998.

Zahn, R., T.F. Pedersen, B.D. Bornhold, and A.C. Mix, Water mass conversion in the glacial subarctic Pacific (54°N, 148°W): Physical constraints and the benthic-planktonic stable isotope record, *Paleoceanography,* 6, 543-560, 1991.

Zahn, R., K. Winn, and M. Sarnthein, Benthic foraminiferal $\delta^{13}C$ and accumulation rates of organic carbon, Uvigerina peregrina group and Cibicidoides wuellerstorfi, *Paleoceanography*, 6, 543-560, 1986.

P. Bodén, Dept. of Geology and Geochemistry, Stockholm University, S-106 91, Stockholm, Sweden (per.boden@geo.su.se)

L. Bornmalm, Dept. of Marine Geology, Göteborg University, Box 460, S-403 Göteborg, Sweden (lennart@gvc.gu.se)

M. Lyle, CGISS, Boise State University, Boise, ID 83725 (mlyle@kihei.boisestate.edu)

D.C. Lund, Kennedy School of Government, 79 JFK Street, Harvard University, Cambridge, MA 02138 (dave_lund@harvard.edu)

A.C. Mix, College of Oceanic and Atmospheric Sciences, Oregon State University, Corvallis, OR 97331 (mix@oce.orst.edu)

J. Pike, Department of Earth Sciences, Cardiff University, PO Box 914, Cardiff, CF1 3YE, UK (pikej@cardiff.ac.uk)

N.G. Pisias, College of Oceanic and Atmospheric Sciences, Oregon State University, Corvallis, OR 97331 (pisias@oce.orst.edu)

High Precision Correlations of Greenland and Antarctic Ice Core Records Over the Last 100 kyr

Michael L. Bender[1,2], Bruno Malaize[1,3], Joseph Orchardo[1], Todd Sowers[4], and Jean Jouzel[3]

We have remeasured the $\delta^{18}O$ of O_2 and $\delta^{15}N$ of N_2 in trapped gases in the GISP2 and Vostok ice cores, and recalculated variations with depth in the $\delta^{18}O$ of paleoatmospheric O_2. The new records have improved precision and resolution, allowing us to correlate the two ice cores using $\delta^{18}O$ of O_2 as a time-stratigraphic marker with about half the previous uncertainty in relative ages. We transfer the relative timescales from the gas records to the ice by correcting for differences in the age of the ice and the trapped gas. We then compare records of isotopic temperature vs. time for the GISP2 and Vostok cores. Our comparison confirms that all 6 long (> 2 kyr) interstadial events found in GISP2 between 35-75 ka also occur at Vostok, and strongly suggests that the 9 short GISP2 interstadials occurring within this time period have counterparts at Vostok as well. On average, long and short interstadial events at Vostok and GISP2 are in phase. Various uncertainties, however, allow for corresponding events to be out of phase by as much as 1.4 kyr on average. The near synchroneity in climate change between GISP and Vostok during interstadial events contrasts with the climate phasing during Termination 1, when long, slow warmings as well as millennial events were largely out of phase.

INTRODUCTION

Climates have varied dramatically during the Pleistocene epoch because of the growth and decay of the great continental ice sheets and because of associated changes in albedo, ocean circulation, atmospheric circulation, and atmospheric chemistry. Pleistocene climate change is expressed most simply in records of the $\delta^{18}O$ of planktonic and benthic foraminifera in deep sea sediments. These records reflect some combination of global ice volume and local surface temperature or regional deep water temperature. Much of the variance in these $\delta^{18}O$ records is cyclic, with periods of 100 kyr, 41 kyr, and 23-19 kyr. These of course correspond to periods over which variations in Earth's orbit around the sun change the seasonal distribution of insolation at any point on the Earth's surface. Many aspects of Earth's resulting climate variations have much of their variance concentrated in orbital periods [*Imbrie et al.*, 1992], which are therefore believed to "pace" glacial - interglacial climate change. The amplitude of the climate changes requires that effects of insolation be amplified by variations in albedo,

1 Graduate School of Oceanography, University of Rhode Island, Kingston, RI

2 Department of Geosciences, Princeton University, Princeton, NJ

3 Laboratoire pour Modelisation de Climat et l'Environnement, CEA, Saclay, France

4 Department of Geosciences, Pennsylvania State University, State College, PA

Mechanisms of Global Climate Change at Millennial Time Scales
Geophysical Monograph 112

greenhouse gas forcing, ocean circulation and heat fluxes, and other effects [*Imbrie et al.*, 1992].

Data from Greenland ice cores demonstrate another mode of climate variability. In Fig. 1, we plot $\delta^{18}O$ of ice vs. age for the GISP2 ice core [*Grootes et al.*, 1993]. Heavier (less negative) values of $\delta^{18}O$ correspond to warmer temperatures. This figure shows the well-known sequence of 23 interstadial events during the last 100 ka [*GRIP*, 1993; *Grootes et al.*, 1993]. Going forward in time, individual events begin with a very rapid warming. A slow cooling follows, and a rapid cooling to baseline temperatures terminates each event. Interstadial events have a duration of several hundred to several thousand years, baseline-to-baseline.

While the interstadial events are most readily observed in the Greenland $\delta^{18}O_{ice}$ record, there is compelling evidence that these events were global in extent. *Bender et al.* [1994b] showed that interstadial events in Antarctica were associated with the longer (>2 kyr) events in Greenland. Uncertainties of about ±3 kyr in their relative dating made it impossible for them to determine the phasing of these events between Greenland and Antarctica. Recently *Blunier et al.* [1998] have presented CH_4 records confirming the association of long interstadial events in east and west Antarctica (Vostok and Byrd) with long events in Greenland. Their results, which correlate GRIP, Vostok, and Byrd from ~ 10-47 ka, suggest that interstadial climate change is out of phase between West Antarctica and Greenland. They also show that long warmings occur first in Antarctica.

Atmospheric CH_4 concentrations rose during interstadial times [*Chappellaz et al.*, 1993a; *Brook et al.*, 1996], suggesting that precipitation increased during these events, especially in the tropics [*Chappellaz et al.*, 1993b]. Equatorial Atlantic sea surface temperature rose at times of long interstadial events in Greenland [*Curry and Oppo*, 1997]. North Atlantic surface water was warmer during interstadial events [*Bond et al.*, 1993], ice rafting was restricted [*Bond and Lotti*, 1995], North Atlantic deep water formation was apparently enhanced [e.g., *Keigwin and Jones*, 1994; *Lehman and Oppo*, 1997], the Arctic atmospheric circulation cell extended further to the south [*Mayewski et al.*, 1994], and the North Pacific thermocline was less well ventilated [*Behl and Kennett*, 1996; *Kennett and Ingram*, 1995; *Adkins and Boyle*, 1997]. Interstadial events were thus of global scale, involving extensive changes in climate.

We are interested in improving our knowledge of the nature of interstadial events, and comparing climate changes during these events and the last termination. Here we present new $\delta^{18}O_{atm}$ data ($\delta^{18}O$ of atmospheric O_2) from the GISP2 and Vostok ice cores. These are more detailed and precise than those of *Bender et al.* [1994], allowing a more accurate correlation of Vostok and GISP2 climate records. We briefly discuss some attributes of interstadial climate change and its causes.

Finally, we consider the phasing of climate change during Termination 1, and we discuss how the interhemispheric phasing of climate change differed during interstadial events and the last termination.

EXPERIMENTAL PROCEDURES AND RESULTS

We measured $\delta^{18}O$ of O_2 and $\delta^{15}N$ of N_2 in over 300 samples from 189 depths of the Vostok and GISP2 ice cores. Measurements were made using the basic procedures outlined by *Sowers et al.* [1989]. We melt the ice in vacuum, refreeze the meltwater, and condense the released air into a stainless steel tube immersed in liquid helium. We then admit the sample into a Finnigan MAT 252 isotope ratio mass spectrometer and measure the $\delta^{18}O$ of O_2, $\delta^{15}N$ of N_2, and $\delta O_2/N_2$ (as 32/29) by double collection against a dry air standard. We measure peak heights of O_2, N_2, Ar, and CO_2 against the standard by peak jumping. To improve productivity, we modified our original procedure by omitting the second melting and refreezing step in the gas extraction procedure. This change may cause incomplete degassing and a small bias in O_2/N_2 and Ar/N_2 ratios. Nonetheless, we can say that O_2/N_2 and Ar/N_2 ratios of samples analyzed in this study were in the range previously measured for the Vostok and GISP2 cores [*Bender et al.*, 1995]. $\delta^{15}N$ and $\delta^{18}O$ in air samples extracted with a single melt/refreeze cycle are indistinguishable from values in air extracted using two melt/refreeze cycles. We correct the measured values of the $\delta^{18}O$ of O_2 for zero enrichment, fractionation in the ion source, and isobaric interferences as described previously [*Sowers et al.*, 1989]. We measured zero enrichments and the variation of $\delta^{18}O$ with $\delta O_2/N_2$ every day we analyzed samples, and corrected sample data according to average values of these properties during periods when values appeared to be stable. We calibrated the working standard against air approximately once every 6 months, and averaged the results over the duty cycle.

We compute the $\delta^{18}O$ of paleoatmospheric O_2, $\delta^{18}O_{atm}$, from the equation $\delta^{18}O_{atm} = \delta^{18}O - 2x\ \delta^{15}N$. The latter term corrects for gravitational fractionation, which enriches heavy isotopes with depth in the firn in an amount proportional to their mass difference with respect to light isotopes [*Craig et al.*, 1988; *Schwander*, 1989; *Sowers et al.*, 1989]. During periods of rapid climate change, temperature gradients between the atmosphere and the base

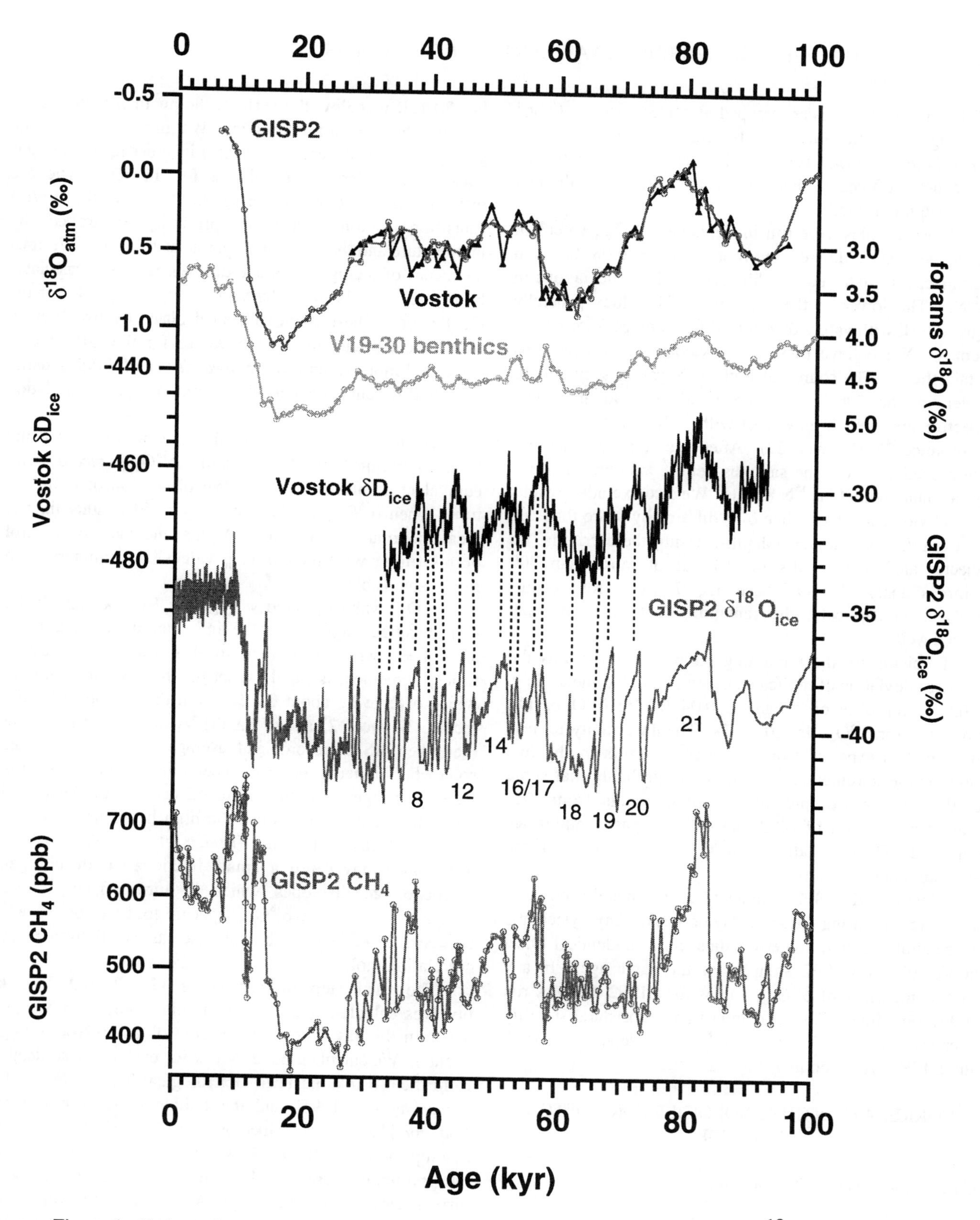

Figure 1. Various climate records plotted vs. age back to 100 ka. From top to bottom: (a) $\delta^{18}O_{atm}$ vs. age for Vostok and GISP2 (this paper). (b) $\delta^{18}O$ of Equatorial Pacific benthic forams from Shackleton and Pisias [1984]. (c) δD of the Vostok 5G ice core vs. age according to the timescale derived in this paper. (d) $\delta^{18}O$ of GISP2 ice vs. age according to the timescale of Meese et al. [1997]. (e) CH_4 concentration [Brook et al., 1996] vs. age.

of the firn where gases are trapped also fractionates ^{18}O and ^{15}N, again fractionating ^{18}O by (nearly) twice as much [*Severinghaus et al.*, 1998]. The standard "gravitational" correction suffices to correct for the effects of thermal fractionation as well.

Most samples were run in duplicate. $\delta^{18}O_{atm}$ values of some duplicates differed by more than ± 0.08 ‰. In certain cases, we traced the difference to large deviations in $\delta^{15}N$. The lower of the replicate $\delta^{15}N$ values generally agreed well with values determined for samples from nearby depths. We believe these latter values are reliable, and retain them. The source of high $\delta^{15}N$ values seems to be water in the "dry" air samples admitted to the mass spectrometer, which may react with $^{14}N_2$ in the ion source to produce $^{14}N_2H^+$ (mass 29). After this study we modified our procedure to dry the samples better, and eliminated the occasional anomalous ^{15}N results. We also excluded results for individual samples when $\delta^{15}N$ differed by more than 0.1 ‰ from values at adjacent depths. Using these criteria, we rejected analytical results for 13 out of more than 300 samples analyzed. Of these 13, 7 were from depths analyzed in triplicate; the remaining 2 from each depth agreed well.

Precision for the remaining samples, expressed as the standard deviation of replicate analyses from the mean, is ± 0.03 ‰ for $\delta^{15}N$ of N_2 and ± 0.04 ‰ for $\delta^{18}O$ of O_2. Precision for $\delta^{18}O_{atm}$ is ± 0.05 ‰. The uncertainty term is smaller than expected from propagating errors in the two isotopic measurements. The reason is probably that some of the error in isotopic measurements results from ~2:1 mass fractionation of $\delta^{18}O$ and $\delta^{15}N$ as gas samples are collected. The calculation of $\delta^{18}O_{atm}$ removes most of the aliasing due to this error.

Vostok and GISP2 samples of approximately the same age were run during the same day or week. Any systematic procedural or calibration errors thus cause an identical offset in Vostok and GISP2 samples of the same age. In any case, comparison of $\delta^{18}O_{atm}$ values for GISP2 samples run at different times (Fig. 2) demonstrates procedural stability over several years. Values for $\delta^{18}O_{atm}$ vs. depth for Vostok and GISP2 are given in Table 1 and Fig. 3.

CORRELATION CHRONOLOGIES FOR VOSTOK AND GISP2

Correlation of Gas and Ice Records

$\delta^{18}O_{atm}$ is plotted vs. gas age for GISP2 and Vostok in Fig. 1. The assumed curve of ice age vs. depth in GISP2 is identical to that in *Bender et al.* [1994]. Because air bubbles are trapped in the firn/ice transition region, which lies 70-120 m below the surface, the air in the bubbles is younger than the surrounding ice. We must correct for this gas age - ice age difference ("Δage") in order to use the gas correlation to derive chronologies for the ice. The age difference is very close to the age of the ice in the firn/ice transition region which is primarily controlled by temperature and the accumulation rate at the site. We used the record of ice age - gas age differences (Δage) calculated by *Brook et al.* [1996] to derive a gas age timescale for the GISP2 core. These Δage values calculated by Brook et al. were in close agreement with values for the GISP2 core calculated more recently by *Schwander et al.* [1997] using slightly different estimates of temperature and accumulation rate.

We derive the time scale for Vostok by correlating "control points" in the Vostok $\delta^{18}O_{atm}$ record with equivalent points in GISP2. We place control points at times when $\delta^{18}O_{atm}$ is changing very rapidly rather than at local maxima or minima. Depths and ages of control points, along with their $\delta^{18}O_{atm}$ values, are summarized in Fig. 3 and Table 2.

$\delta^{18}O_{atm}$ values plotted vs. age for Vostok and GISP2 (Fig. 1) agree very well for the time interval between 45 ka and 78 ka, which is critical to this study. For earlier times, agreement is good except for three Vostok samples which diverge by about 0.2 ‰ from the smoother GISP2 record. Between 27 and 45 ka, the Vostok record is noisier than the GISP2 record, and average $\delta^{18}O_{atm}$ values are somewhat higher. The gas concentration record of Vostok ice of this age is highly disturbed because of the partial formation of air hydrates and the high hydrostatic pressure of residual gas in bubbles [*Bender et al.*, 1995]. This disturbance may cause a small fractionation of nitrogen and/or oxygen isotopes which in turn offsets $\delta^{18}O_{atm}$ values. In any case $\delta^{18}O_{atm}$ appears to vary too slowly between 27-45 ka to be of use as a high-precision correlation tool.

The final step in correlating GISP2 and Vostok involves calculating a continuous chronology for Vostok ice from the gas ages of the 8 control points (or $\delta^{18}O_{atm}$ events). We calculate Δage values for each of these depths from the Vostok Extended Glaciological Timescale (EGT) (*Jouzel et al.*, 1996) and the bubble closeoff model of *Pimienta* [1988] as described by *Barnola et al* [1991]. Our assumption is that the EGT, which estimates accumulation rate and temperature directly from $\delta^{18}O_{ice}$, gives the best instantaneous estimate of Δage. Adding the Δage values to gas ages gives ice ages at each of the control points in Vostok. We calculate ages for other depths by assuming that, between each pair of control points, the relative accumulation rate varies as calculated in the EGT. We note

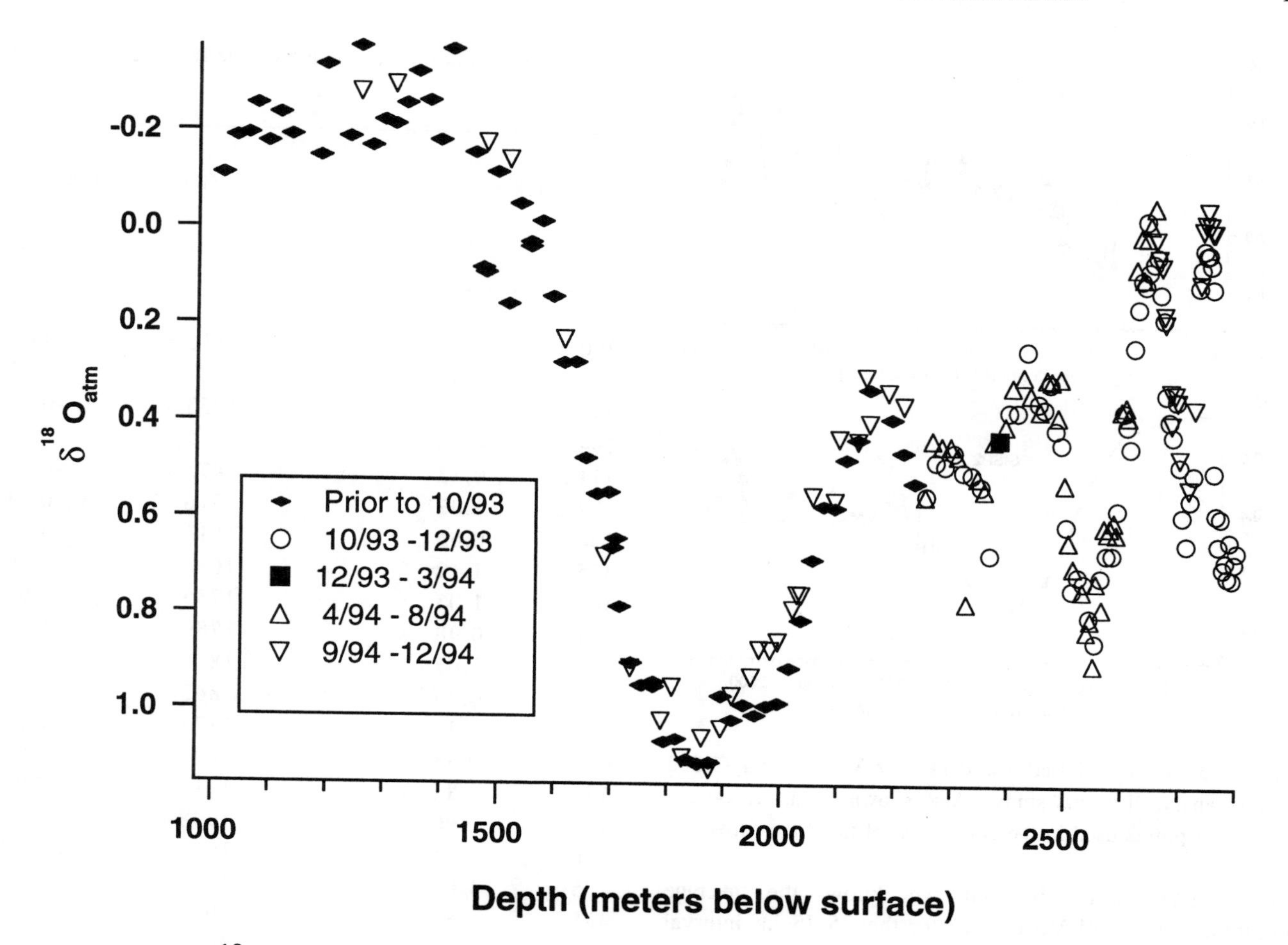

Figure 2. $\delta^{18}O_{atm}$ vs. depth for samples from the GISP2 ice core analyzed during different time periods. Systematic errors between data sets are 0.05 ‰ or less. Vostok and GISP2 samples used for the correlation in this study were run at the same time and are therefore free of systematic offsets with respect to each other.

that Δage values calculated in this way are slightly inconsistent with our age-depth curve, which differs from the EGT.

Relative ice ages for Vostok and GISP2 calculated as described above have 3 sources of error. The first is in relative gas ages of equivalent control points. At these points, $\delta^{18}O_{atm}$ is typically changing at the rate of 0.1 ‰/ka. An uncertainty of ± 0.05 ‰ in relative $\delta^{18}O_{atm}$ values translates to an age error of ± 0.5 kyr. The second source of error enters from interpolating ice ages between control points. Since there is an average of about 6 kyr of time between control points, interpolation errors within these intervals are unlikely to be more than a few hundred years. We assume an average uncertainty of ± 0.3 kyr. An important exception is the interval between the youngest 2 control points, with ice ages of 27 and 48 ka. Interpolation uncertainties are much greater because of the long duration of this interval. For this reason, we focus some aspects of our interpretation on the period between 45-75 ka.

The third, and largest, source of error is in the gas age-ice age difference (Δage values). Errors arise here because of uncertainties in both the closeoff depth and the accumulation rate, which are inferred by extrapolating their modern temperature dependence. It is very difficult to quantify the uncertainty largely because there is no modern analog for the glacial section of Vostok, with its very cold temperatures and low accumulation rates. However we must admit the possibility that uncertainties in Vostok Δage values lead to errors of ± 2 kyr or more in ice ages. We believe, therefore, that we cannot use ages of individual millennial timescale events to determine climate phasing between Greenland and East Antarctica.

We emphasize here that uncertainty in the average phase for all 15 interstadial events between 35-75 ka is much smaller than the age uncertainty of the individual

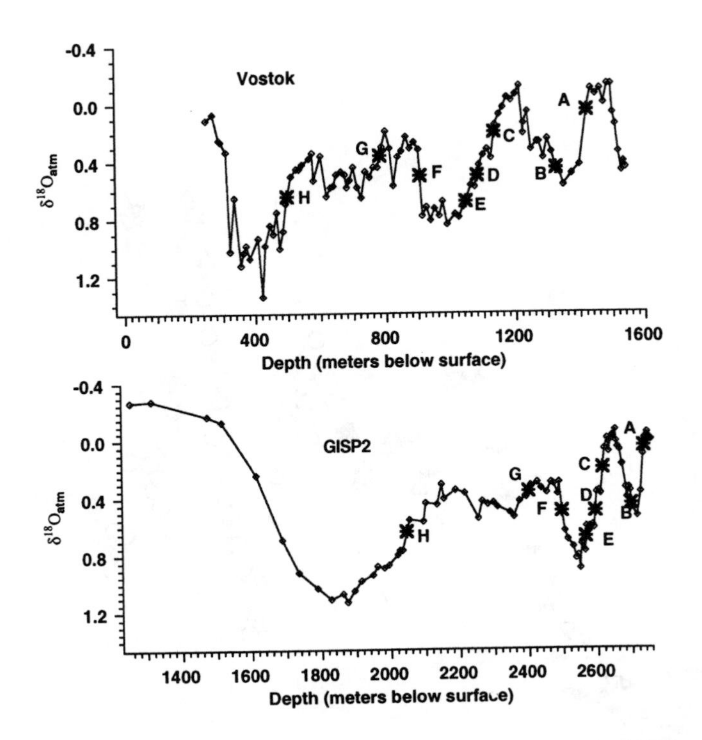

Figure 3. $\delta^{18}O_{atm}$ plotted vs. depth for Vostok and GISP2 samples analyzed in this study. Also shown (as asterisks) are the control points used in the correlation of the 2 ice cores.

Table 1. $\delta^{18}O_{atm}$ for Vostok and GISP2 samples analyzed in this study.

Vostok samples		GISP2 samples	
Depth	$\delta^{18}O_{atm}$	Depth	$\delta^{18}O_{atm}$
250.3	0.11	116.8	0.02
271	0.07	117.7	0.03
290	0.25	1253.9	-0.27
294	0.26	1315	-0.28
311	0.33	1475.1	-0.17
322.2	1.02	1516	-0.13
336	0.65	1615	0.24
355	1.12	1688	0.69
363	1.03	1735	0.92
371	0.98	1789	1.03
382	1.07	1828	1.11
409	0.93	1862	1.07
421	1.34	1875	1.13
430	0.98	1895	1.05
445	0.84	1915	0.98
455	0.90	1948	0.94
466	0.75	1962	0.88
475	1.00	1981.2	0.89
485	0.88	1995.1	0.87
494	0.69	2021	0.80
510	0.50	2028	0.77
526	0.45	2035	0.77
535	0.45	2055	0.56
545	0.42	2095	0.57
565	0.38	2102	0.44
575	0.34	2135	0.45
579.6	0.53	2148.1	0.31
600.9	0.36	2155.1	0.41
619	0.64	2188.1	0.35
631	0.58	2215.1	0.37
640	0.57	2254	0.55
649	0.49	2266	0.43
663	0.47	2282	0.45
675	0.49	2298	0.44
681	0.58	2310	0.47
690	0.53	2346	0.51
700	0.44	2358	0.54
713	0.58	2374	0.43
725	0.65	2394	0.40
738	0.47	2406	0.32
751	0.51	2426	0.30
763	0.44	2438	0.34

events. The reason is that we know the average accumulation rate of Vostok ice for the 35-75 ka interval much more accurately than we know the instantaneous accumulation rate at a given time within this interval. At the younger end of the interval, the age of interstadial 8 in the GISP2 chronology is in good agreement with the ^{14}C age of a correlated long interstadial event observed in North Atlantic deep sea sediments [*Bond and Lotti,* 1995]. At the older end, the age for a long warm event correlated with deep sea isotope stage 5a, 83 ka at Vostok and 84 ka at GISP2, is in excellent agreement with the age of coral reefs recording the high sea stand dated at about 82 ka (marine isotope stage 5a). Overall, we estimate that the uncertainty in the duration of the period dated between 35-75 ka is ± 4 ka (± 10%). Δage values between 35-75 ka average 5.3 kyr at Vostok. The systematic uncertainty from accumulation rate alone is thus ± 0.5 kyr (10% of 5.3 kyr).

The largest source of uncertainty in Δage comes from the uncertainty in the closeoff depths. Closeoff depths are themselves functions of accumulation rate and temperature. As described above, we calculate closeoff depths at Vostok following *Pimienta* [1988]. His equations express how temperature and accumulation rate influence the increase of density with depth as well as the density required to seal air in bubbles.

Table 1. (continued)

Vostok samples		GISP2 samples	
Depth	$\delta^{18}O_{atm}$	Depth	$\delta^{18}O_{atm}$
775	0.44	2454	0.37
788	0.30	2466	0.30
800	0.19	2474	0.31
813	0.31	2486	0.38
825	0.57	2490	0.30
838	0.37	2498	0.52
850	0.33	2506	0.64
863	0.23	2514	0.70
875	0.31	2530	0.75
888	0.27	2538	0.83
901	0.32	2544	0.81
913	0.78	2550	0.90
925	0.72	2554	0.73
938	0.81	2564	0.78
950	0.73	2568	0.61
964	0.78	2574	0.62
975	0.68	2578	0.61
988	0.84	2586	0.60
1013	0.77	2590	0.62
1025	0.79	2598	0.37
1038	0.72	2606	0.36
1049	0.66	2610	0.38
1062	0.57	2622	0.07
1075	0.58	2630	0.00
1088	0.43	2634	0.09
1100	0.36	2640	0.01
1113	0.32	2646	-0.02
1125	0.38	2654	-0.06
1138	0.15	2656	0.02
1150	0.08	2660	0.06
1163	0.03	2666	0.08
1175	-0.04	2672	0.18
1188	-0.02	2674	0.18
1201	-0.06	2684	0.34
1213	-0.12	2686	0.41
1225	0.21	2692	0.34
1226	0.14	2696	0.36
1238	0.06	2700	0.48
1250	0.32	2716	0.54
1268	0.27	2726	0.37
1275	0.27	2734	0.11
1288	0.38	2738	0.00
1300	0.25	2742	-0.01
1313	0.34	2746	-0.04

Table 1. (continued)

Vostok samples		GISP2 samples	
Depth	$\delta^{18}O_{atm}$	Depth	$\delta^{18}O_{atm}$
1325	0.44	2750	-0.01
1338	0.48	2754	0.01
1349	0.57	2758	0.01
1375	0.49		
1399	0.43		
1423	0.03		
1435	-0.10		
1449	-0.06		
1463	-0.10		
1475	0.00		
1486	-0.13		
1497	-0.13		
1502	0.07		
1510	0.15		
1519	0.34		
1529	0.47		
1535	0.41		
1541	0.45		

There are two reasons to believe that our estimates of Vostok closeoff depths may be in systematic error. First, $\delta^{15}N$ values of the trapped gas reflect the height of the stagnant column of air in the firn through which molecular diffusion leads to gravitational fractionation [*Craig et al., 1988; Sowers et al.*, 1989; *Schwander*, 1989; *Sowers et al.*, 1992]. Between 35-75 ka, $\delta^{15}N$ values from Vostok average 0.42 ± 0.03‰ and indicate that the average stagnant column height was 76 m. Assuming that the air in the top 12 m of Vostok firn was mixed by convection [*Bender et al.*, 1994a], the average closeoff depth would be 88 m, 30 m less than the mean depth of 118 m calculated from the densification model of *Pimienta* [1988]. Mean Δage values calculated from the Vostok $\delta^{15}N$ data are about 1.3 kyr less than values calculated from the densification model (5.3 ka). Applying Δage values from $\delta^{15}N$ would make the entire Vostok ice chronology younger by 1.3 ka.

Alternatively, closeoff depths would have been deeper, and true Δage values would have been greater, if Vostok temperatures between 35-75 ka were colder than estimated by *Jouzel et al.* [1996] from δD, because more time is required for firn to be metamorphosed into ice at colder temperatures. A preliminary indication for colder temperatures comes from the borehole modeling study of

Table 2. Control points. Δage is the gas age-ice age difference. Units: depth, meters; Δage, kyr, age, ka.

Point	Vostok depth	Vostok ice age	Vostok Δage	Vostok gas age	$\delta^{18}O_{atm}$	GISP2 depth	GISP2 ice age	GISP2 Δage	GISP2 gas age
A	1422	101.7	4.1	97.6	0.05	2736	98.1	0.5	97.6
B	1327	92.8	4.4	88.4	0.45	2699	89.0	0.6	88.4
C	1135	78.2	5.0	73.2	0.2	2617	73.8	0.6	73.2
D	1082	74.7	5.2	69.5	0.5	2594	70.3	0.8	69.5
E	1046	70.3	5.4	64.9	0.68	2566	66.3	1.4	64.9
F	906	61.2	4.5	56.7	0.5	2497	57.2	0.5	56.7
G	782	52.8	5.4	47.5	0.36	2400	48.2	0.7	47.5
H	498	32.7	6.4	26.3	0.64	2047	27.3	1.0	26.3

Salamatin et al. [1997], who concluded that the ice age temperature lowering at Vostok was about 1.5 times greater than estimated by *Jouzel et al.* [1996]. If Vostok temperatures during our study period were 2° C colder on average than estimated by *Jouzel et al.* [1996], Δage values would be greater by 0.8 kyr, and the ages of Vostok ice would be older by the same amount.

Similar considerations apply to the GISP2 core, but age uncertainties are much smaller because accumulation rates are about an order of magnitude higher. In comparing Vostok and GISP2 ages for 15 interstadial events between 35-75 ka, average errors in the correlation ages of the gas records are very small and can be neglected. We estimate the important systematic errors in mean correlation ages as follows: uncertainty in average GISP2 Δage values, ± 0.3 kyr; uncertainty in Vostok Δage values from accumulation rates, ± 0.6 kyr; uncertainty in Vostok Δage values from closeoff depths, -1.3 kyr to +0.8 kyr. Adding the errors quadratically gives an uncertainty in the mean relative age uncertainties for correlated events of - 1.4 to + 1.0 kyr. We average the uncertainty to ± 1.2 kyr. This number reflects our ability to measure mean leads and lags in the Vostok and GISP2 climate records between 35-75 ka.

Again, we acknowledge that uncertainties are considerably larger during the long interval between the two youngest control points at 27 and 48 ka gas age.

INTERSTADIAL CLIMATE CHANGE

Correlation of Interstadial Events Between Vostok and GISP2

In Fig. 1, we plot isotopic temperature records of GISP2 and Vostok vs. time for the last 100 ka. We plot Vostok prior to 35 ka only because its ice ages are not well constrained by $\delta^{18}O_{atm}$ correlations after this time.

Comparison of the two isotopic temperature records between 35-75 ka clearly confirms earlier conclusions [*Jouzel et al.,* 1994; *Bender et al.,* 1994b; Blunier et al., 1998] that interstadial events at Vostok occurred at about the same time as the long interstadial events observed in Greenland (events 8, 12, 14, 16/17, 19, and 20, each of which lasted longer than 2 kyr from baseline to baseline). We extend this earlier work by showing here that shorter interstadial events in GISP2 also appear to have their counterparts in Vostok. There are 3 δD maxima between interstadials 8 and 12 in Vostok which may correspond with interstadials 9-11 in GISP2. The largest event (greatest peak to baseline amplitude) in Vostok between interstadials 12 and 14 comes shortly before interstadial 12, as does interstadial event 13 in GISP2. Two short δD maxima occur in Vostok between interstadials 14 and 16; these may be linked to events 14a and 15 in GISP2. Finally, two prominent events between interstadials 17 and 19 in Vostok appear to coincide with events 18 and 18a in GISP2. This comparison accounts for the largest δ maxima between each pair of long interstadial events at GISP2 and Vostok. Only between interstadial events 12 and 14 is there an unaccounted peak in the Vostok record which is nearly as large in magnitude as the events we have linked to GISP2 interstadials.

Ages of both short and long interstadial events which we correlated between Vostok and GISP2 are in excellent agreement, without exception. In Table 3, we list the age of warmest temperature for each event between 45-75 ka. All events are dated to be synchronous to within 1.1 kyr or better. On average, the warmest time occurs 0.1 ± 0.5 ka (1σ, n=9) later at Vostok than at GISP2 (Table 3).

Table 3. Ages (ka) of maximum isotopic temperature at Vostok and GISP2 for each interstadial event. together with the length of time the age of maximum temperature at Vostok precedes the age at GISP2. Absolute age uncertainties are about ± 1.2 kyr before 50 ka, and larger for younger ages.

Interstadial No.	Vostok max	GISP2 max	Vostok lead
20	71.9	73.0	-1.1
19	68.1	68.7	-0.6
18a	66.3	66.0	0.3
18	62.2	62.1	0.1
17	58.1	57.8	0.3
16	57.0	56.5	0.5
15	54.0	53.8	0.2
14a	53.2	52.9	0.3
14	52.1	51.8	0.3
13		47.0	
12	43.8	45.4	-1.6
11	42.2	42.6	-0.4
10	40.2	41.2	-1
9	39.6	40.3	-0.7
8	37.9	38.3	-0.4
7	36.0	35.3	0.7
6	33.8	33.6	0.2

Agreement is much better than expected given our uncertainty estimates.

Leads and Lags Between Antarctica and Greenland

Our results thus demonstrate that interstadial events 14-20 are in phase, on average, between Greenland and Antarctica, to within about ± 1.3 kyr. We calculate this uncertainty by quadratically adding the uncertainty in relative ages (1.2 ka) and the standard error in these leads and lags (0.5 ka / $\sqrt{15}$); we then add the mean lead (0.1 kyr) calculated above.

At the outer limits of uncertainty, short interstadial events could be out of phase between Greenland and Antarctica. Long events (duration > 2 kyr) might be characterized by a lead in one of the polar regions; eventually the other region must catch up and climates come into phase for the remaining period.

Blunier et al. (1998) derive a common chronology for the Vostok, Byrd, and GRIP ice cores from 10-47 ka. Our Vostok-GISP2 ice age differences generally agree well with the Vostok-GRIP age differences of Blunier et al. within our stated uncertainty of ± 1.2 kyr. The largest differences between our ice correlations is at 2360 m depth in GISP2 and about 670 m depth in Vostok (about 45 ka). Here, the Vostok ice to which we correlate GISP2 (or GRIP) is about 1.5 kyr younger than in the correlation of *Blunier et al.* [1998]. In making this comparison, we correct for the offset between the age-depth curves of GISP2 and GRIP. Correlation depths agree to better than 1 kyr for ice ages of 47 ka and younger than 41 ka.

The exact values of the age differences, of course, have major implications for the climate dynamics. Warming at Vostok leads GRIP, at least for long interstadials, according to the Blunier chronology, whereas interstadials coincide according to our chronology. The Vostok chronology of *Blunier et al.* [1998] dates climate change at Vostok (central East Antarctic plateau) as coinciding with climate change at Byrd (West Antarctica). Byrd is robustly and accurately correlated with GRIP by the work of *Blunier et al.* [1998]. This accurate correlation is possible because both accumulate rapidly and have small (and accurately constrained) gas age-ice age differences. Thus *Blunier et al.* [1998] have established that long interstadial events occur in West Antarctica before they occur in Greenland. One could presume that interstadial events at Vostok coincide with Byrd, and the chronology of *Blunier et al.* [1998] supports this idea. However, our estimates of uncertainties in Δage indicate that the phasing of interstadial climate change between Vostok and Byrd cannot now be determined experimentally.

Extent of Interstadial Climate Change

There is compelling evidence that interstadial events were times of climate amelioration in temperate to polar regions of the northern hemisphere, and almost certainly in the tropics as well. Climate was wetter [*Chappellaz et al.*, 1993a; *Brook et al.*, 1996; *Porter and Zhisheng*, 1995; *Mayewski et al.*, 1994], winds were weaker [*Porter and Zhisheng*, 1995; *Mayewski et al.*, 1994], North Atlantic sea surface temperatures were warmer [*Bond et al.*, 1993], and continental ice sheets melted on the continents rather than terminating in tidewater glaciers calving into the oceans [*Bond and Lotti,* 1996]. Higher interstadial CH_4 concentrations [*Chappellaz et al.*, 1993a; *Brook et al.*, 1996] suggest that climate was wetter and perhaps warmer in the tropics, which is where most atmospheric CH_4 is produced [e. g., *Cicerone and Oremland*, 1989; *Aselman and Crutzen*, 1989; *Fung et al.*, 1991; *Hein et al.*, 1997]. Another indication for tropical climate change comes from the work of *Linsley* [1996], who correlated four minima in the $\delta^{18}O$ record of planktonic foraminifera from Sulu Sea deep sea sediment cores with long interstadial events in

GRIP. In a preliminary report, *Kipfstuhl et al.* [1996] suggest that accumulation of ice-rafted detritus was diminished in the Southern Ocean during interstadial events in Greenland, although the chronological link is not yet convincing. These results, along with work reported here, demonstrate the global reach of interstadial climate changes.

CLIMATE PHASING DURING INTERSTADIAL EVENTS AND TERMINATION I

Sowers and Bender [1995] and *Blunier et al.* [1998] correlated the gas records of ice cores from Byrd (West Antarctica) and GISP2 by matching their $\delta^{18}O_{atm}$ or CH_4 curves back to 30 ka (Fig. 4). They then transferred their gas chronology for Byrd to the ice by correcting for the gas age-ice age difference, and plotted isotopic temperature ($\delta^{18}O_{ice}$) vs. ice age for comparison to GISP2/GRIP. The most accurate part of their correlation is between ~8-15 ka, because $\delta^{18}O_{atm}$ and CH_4 are changing rapidly during most of this interval. The accumulation rate at Byrd is about an order of magnitude greater than at Vostok, and the uncertainty in the gas age-ice age difference is correspondingly reduced, as noted above. *Sowers and Bender* [1995] estimated a correlation uncertainty of about ± 0.6 ka between 8-15 ka, increasing to ± 1.5 ka at 22 ka. In this section we update information about climate phasing during the last termination in Antarctica. We then discuss the striking difference between the approximate synchroneity of interhemispheric climate change during interstadial events and the long Arctic lag during Termination 1.

The records of $\delta^{18}O_{atm}$ vs. depth, and the gas correlation in the depth domain of *Sowers and Bender* [1995] for GISP2 and Byrd stand as reported. However, their resulting record of $\delta^{18}O_{ice}$ vs. age for Byrd needs to be modified for several reasons. First, depths of $\delta^{18}O_{ice}$ samples in the Byrd core must be revised according to the recommendation of S. Johnsen [*Broecker*, 1997]. This change corresponds to age changes of a few hundred years at most. Second, the gas age - depth curve for GISP2 has changed because Δage (gas age - ice age difference) is temperature dependent. GISP2 Δage values were recalculated by *Brook et al.* [1996] using a smaller value for the dependence of $\delta^{18}O_{ice}$ on temperature [*Cuffey et al.*, 1995; *Johnsen et al.*, 1995]. The resulting glacial temperatures are colder, making glacial Δage values larger [*Schwander et al.*, 1997], and GISP2 gas ages are younger than those presented in *Sowers and Bender* [1995] by as much as 150 years during the coldest intervals. Correlated gas ages at Byrd are then also correspondingly younger, giving younger ages for ice at Byrd. Also, the recent recalibrations of the isotopic temperature scale at GRIP [*Johnsen et al.*, 1995] and GISP2 [*Cuffey et al.*, 1995] suggest that past temperature changes at Byrd may have been greater than the values estimated from the modern spatial dependence of $\delta^{18}O_{ice}$ on surface temperature, and *Boyle* [1997] explains why this may generally be true. In this case, Byrd Δage values would be up to a few hundred years greater than calculated by *Sowers and Bender* [1995], and correct Byrd ice ages would be older than we have estimated. We have no way to evaluate this possibility but note that lower values at Byrd would tend to increase the age of the ice, and accentuate the West Antarctic lead, during the glacial period and the beginning of the termination relative to the "spatial" value used here. Byrd ice ages plotted in Fig. 4 use the same Δage values as *Sowers and Bender* [1995]. The Byrd-GISP2 correlation for the last 20 ka used here is in good agreement with the correlation of *Blunier et al.* [1998] between Byrd and GRIP.

We can use the $\delta^{18}O_{atm}$ data of Vostok to correlate the Vostok gas record into the GISP2 gas chronology during Termination 1. However, we do not attempt to construct an accurate ice chronology for Vostok for this time. The reason is that Δage values for Vostok during the termination are very large (up to 7 kyr) and, we believe, uncertain. Such large uncertainties arise because the depth interval of the glacial termination in this very slowly accumulating core is comparable to the closeoff depth. In this interval, accumulation rate and surface temperature changed dramatically with time during the densification process. We do not have any empirical information which would allow us to evaluate how closeoff depths would change with time under such variable climate conditions.

Instead of trying to develop an accurate ice chronology for Vostok during Termination 1, we use a simple approach to constrain the minimum age for the start of the deglacial warming at Vostok. We begin by calculating the gas age of Vostok at 300 m depth. $\delta^{18}O_{atm}$ at this depth = 0.32 ‰ (Fig. 5), which corresponds to a gas age of 10.7 ka according to the chronology of GISP2 [Fig. 4]. According to the Extended Glaciological Timescale (EGT), Δage = 3.5 kyr at 300 m depth. Using this value, we calculate an ice age of 14.2 ka at this depth. This age is considerably older than the EGT ice age for this depth (12.6 ka), a result in agreement with *Jouzel et al.* [1995] who concluded that EGT overestimates the accumulation rate (and hence underestimates Δage) in this part of the core. Because of this overestimation, the EGT Δage value is a lower limit, and the ice age of 14.2 ka for 300 m depth is a lower limit to the true age. The deglacial warming at Vostok begins at 354 m depth (Fig. 5). According to the EGT, the age difference between 300 and 354 m depth is 3.5 kyr.

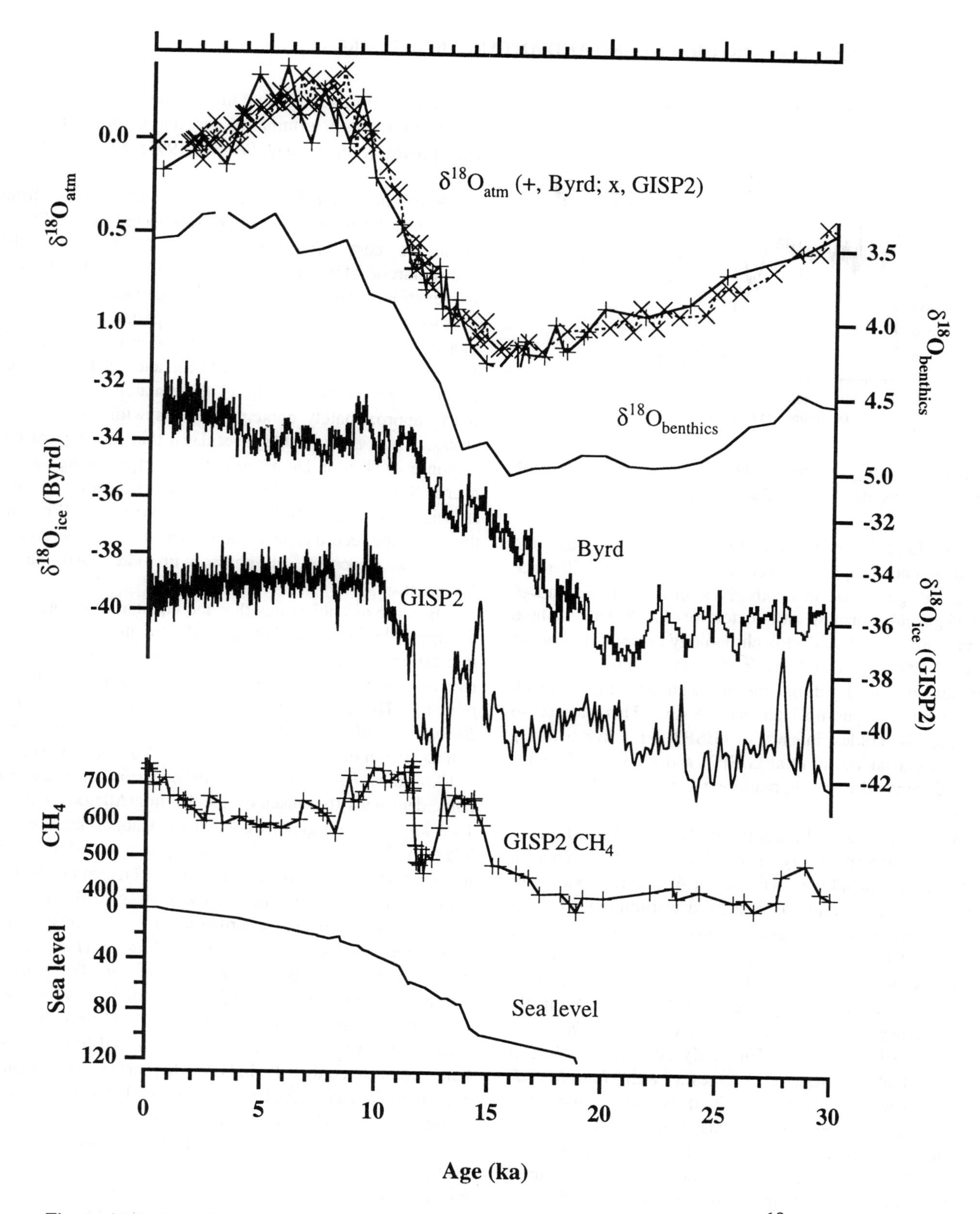

Figure 4. Various climate records plotted vs. age back to 30 ka. From top to bottom: (a) $\delta^{18}O_{atm}$ vs. gas age in the upper part of the Vostok core [*Malaize*, 1995] and the GISP2 core. (b) $\delta^{18}O$ of Equatorial Pacific benthic forams from *Shackleton* and *Pisias* [1984]. (c) $\delta^{18}O_{ice}$ vs. ice age for Byrd (timescale from *Sowers and Bender* [1995] as modified in this paper). (d) $\delta^{18}O_{ice}$ vs. ice age for GISP2 (timescale from *Meese et al.* [1997]). (e) GISP2 [CH_4]vs. age [*Brook et al.*, 1996]. (f) Sea level vs. age [*Fairbanks*, 1989; *Bard et al.*, 1990].

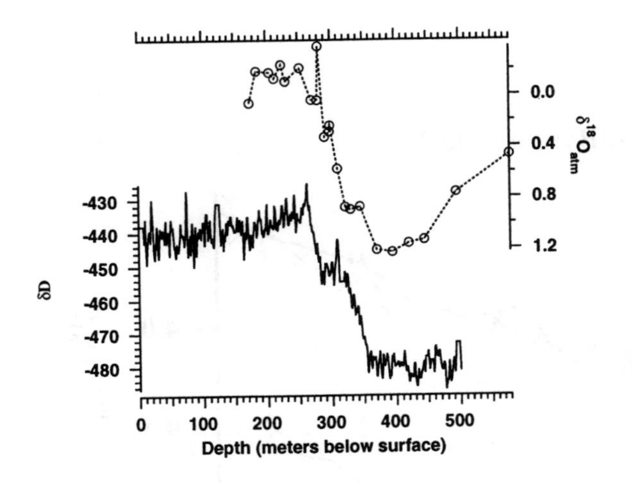

Figure 5. $\delta^{18}O_{atm}$ [Malaize, 1993] vs. depth in the Vostok core and δD vs. depth in the Vostok core to 580 m.

Because EGT is overestimating accumulation rates, this age difference is again a lower limit. Adding this difference to the ice age at 300 m depth gives an age for the start of the deglacial warming at Vostok of 17-18 ka. (These deductions about the Vostok chronology agree with *Blunier et al.* [1998].) The Vostok data thus agree with the conclusion from Byrd that the deglacial warming in much of Antarctica began no later than 18 ka. This date is ~6 kyr after the coldest interval of GISP2, ~1.5 kyr before a small deglacial warming at GISP2, and ~3 kyr before the sharp Greenland warming marking the onset of the Bolling period.

Greenland warmed abruptly at about 14.7 ka (Fig. 4), then cooled intermittently. The coldest part of the Younger Dryas in Greenland was between 11.6 and 12.9 ka. The Younger Dryas terminated in a rapid warming at 11.6 ka, followed by a slow warming into the relatively stable climates of the Holocene. Vostok and Byrd, in contrast, warmed slowly during the glacial termination after about 18 ka. The small "Antarctic cold reversal" recorded in the Byrd core began before the onset of the Younger Dryas and concluded before its end. This early Antarctic cooling is recorded in other Antarctic ice cores as well [*Jouzel et al.*, 1995]. The revision of the Byrd isotopic temperature record, discussed above, moves this event slightly closer in time to the Younger Dryas, but it remains early. At Vostok, the period of constant isotopic temperature during the termination begins long before the start of the Younger Dryas. As the preceding discussion of deglacial Vostok chronology shows, the Antarctic cold reversal is centered around a depth of 300 m, corresponding to an age of ≥ 14.2 ka. Hence a noteworthy feature of the polar climate records for the last termination is that the climate reversals as well as the onset of the termination are out of phase between hemispheres [*Sowers and Bender*, 1995; *Blunier et al.*, 1998].

A striking contradiction to this statement comes from the recently dated isotopic temperature record of the Taylor Dome ice core [*Steig et al.*, 1998], located in the Transantarctic Mountains at 77° 48' S, 158° 43' E, at 2374 m elevation. This record is poorly defined immediately prior to 15 ka, but indicates glacial temperatures. It records an abrupt warming at about 15 ka, roughly coincident with the Bolling/Allerod warming in Greenland. It then records a cooling which clearly follows the Antarctic cold reversal and is approximately contemporaneous with the Younger Dryas. *Steig et al.* [1998] speculated that the "Greenland" response of Taylor Dome reflected local warming in the nearby Southern Ocean linked to ocean circulation changes associated with the Bolling/Allerod warming. Whatever the cause, the record clearly shows that at least one area of Antarctica responded in phase with Greenland during Termination 1.

Because of conjectures that the equatorial region plays a large role in glacial/interglacial climate change, it is interesting to examine high resolution tropical climate records during the last termination. Six records are relevant. The CH_4 record [Fig. 4; *Chappellaz et al.*, 1993a; *Brook et al.*, 1996] shows that the atmospheric concentration of this gas rose by about 40 ppb between 17 - 17.5 ka. However the major deglacial increase was coincident with the onset of the Bolling/Allerod at about 14.7 ka. The terrigenous content in Arabian Sea core 74 KL [*Sirocko et al.*, 1996] serves as a proxy record of aridity in Africa and the Indian subcontinent. Terrigenous content (and precipitation) first fall below their glacial values at about 16 ka, then decrease further at 14.7 ka, corresponding to the onset of the Bolling/Allerod. *Hughen et al.* [1996] present gray scale indices of sediments in the Cariaco Trench (western equatorial Atlantic), which they interpret as proxy records of surface productivity. Their grey scale records can be closely aligned to the GISP2 isotopic temperature record. Periods of higher organic matter content in Cariaco Trench sediments correspond to times of colder climate in Greenland. They interpret their results as indicating stronger trade winds when Greenland was cold. As their records do not extend to calendar ages older than 15 ka, one cannot determine whether there was an early response in this oceanic region. Between 15 and 10 ka, however, the Cariaco Trench records align very closely with the GISP2/GRIP isotopic temperature records rather than those from Antarctica. *Linsley* [1996] presented a high-resolution record of the $\delta^{18}O$ of planktonic foraminifera from a Sulu Sea (western Pacific) core discussed earlier. The deglacial $\delta^{18}O$ decrease begins with a sharp change at about 15 ka. $\delta^{18}O$ reaches a minimum, then rises to a maximum during the Younger Dryas. $\delta^{18}O$ finally decreases very rapidly at the end of the Younger Dryas. Again the tropical response is in phase with Greenland. Recently, *Bard et al.* [1997] have shown that the UK'_{37} index of alkenone unsaturation rose abruptly from 0.88 to 0.92 at 15 ka in an Indian Ocean core from 20° S. The dependence of UK'_{37} on various physiological properties [*Epstein et al.*, 1998] makes it possible that some environmental change other than temperatures caused the change in this index. The results of *Bard et al.* [1997] nevertheless imply some change in oceanographic conditions in the southern tropical latitudes of the Indian Ocean at 15 ka. To summarize, an in - phase response is observed between Greenland and the equatorial latitudes of all oceans. Only the Sulu Sea, Arabian Sea, and south equatorial Indian Ocean records extend to 18 ka. None of these shows a deglacial response as early as 17.7 ka.

During Heinrich events, in-phase climate change with Greenland extended beyond the equatorial region to southern temperate latitudes. As noted earlier, ice advances in the Chilean Andes coincide with Heinrich event H1 [*Lowell et al.*, 1994]. With respect to the Younger Dryas, the situation is unclear. *Denton and Hendy* [1994] showed that an ice advance in the New Zealand Alps was nearly synchronous with the start of the Younger Dryas. The date for this ice advance also falls within the period of the Antarctic cold reversal. A moraine in the Cropp River Valley corresponds with the end of the Younger Dryas according to *Lowell et al.* [1995], who note that the existence of a Younger Dryas cooling in this region is controversial. Pollen records of *Lowell et al.* [1995] from northern Patagonia show no evidence for a Younger Dryas cooling.

While climate change in the tropics was largely delayed until the start of the Bolling/Allerod, the melting of northern hemisphere ice sheets apparently began at about 18 ka. The deglacial sea level rise [*Fairbanks,* 1989; *Bard et al.*, 1990] began by 18 ka, and reached about 20 m before the beginning of meltwater pulse 1a at the onset of the Bolling/Allerod. A rise of this magnitude presumably reflects melting of the northern hemisphere ice sheets, and northern glaciers did in fact retreat during the early part of the termination [*Jones and Keigwin*, 1988]. Thus parts of the northern hemisphere were warming, or at least deglaciating, coincidentally with Antarctica, but before the tropics, during the last termination.

In summary the history of deglacial warming differed between Greenland and Antarctica. Some tropical regions, southern hemisphere temperate regions, and at least one site in Antarctica followed Greenland, while some glaciated northern hemisphere areas followed Antarctica.

CONCLUSIONS

The main contribution of this study is to show that, between 35-75 ka, most Vostok isotopic temperature increases lasting more than about 0.5 kyr are associated with global climate change. There is compelling evidence that this is the case for long interstadial events. It is very likely to be true for the short warmings of smaller amplitude which we have linked with short interstadial events in Greenland. The same appears to be true for millennial-duration events which occurred during Termination I. It is well established that much of the lower frequency variability in the Vostok isotope temperature record is of orbital frequency [*Genthon et al.*, 1987]. This variance is also clearly linked with global climate change [*Jouzel et al.*, 1987].

Results presented here, together with work of *Imbrie et al.* [1992] and many others, indicate that we can characterize most variance in late Pleistocene climate records as being of two types. Variability associated with orbital changes is mostly in phase between the hemispheres [Fig. 1; *Imbrie et al.*, 1992; *Bender et al.*, 1994]. Orbital-timescale climate changes are relatively slow. They are associated with large changes in the Earth's radiation balance associated with variations in the concentration of the greenhouse gases CO_2, CH_4, and N_2O [e. g., *Barnola et al.*, 1991; *Chappellaz et al.*, 1990; 1993; *Leuenberger et al.*, 1992], variations in ice albedo from large continental ice sheets, and variations in dust albedo [*Yung*, 1996]. Variability associated with interstadial events is, at least occasionally, out of phase between the hemispheres [*Sowers and Bender*, 1995; *Yiou et al.*, 1997; *Blunier et al.*, 1998]; extremely rapid, particularly around the North Atlantic [*Alley et al.*, 1993; *Severinghaus et al.*, 1998]; associated with small changes in radiative forcing between the hemispheres [CO_2: *Stauffer et al.*, 1998; CH_4: *Chappellaz et al.*, 1993; *Brook et al.*, 1997]; and associated with abrupt changes in the thermohaline circulation [*Bond et al.*, 1993]. Both modes are global in extent.

Acknowledgements: Financial support was provided by the National Science Foundation, Office of Polar Programs. We appreciate stimulating discussions with Jean-Marc Barnola, Jerome Chappellaz, and Richard Alley.

REFERENCES

Adkins, J. F. and E. A. Boyle, Changing atmospheric $\Delta^{14}C$ and the record of deep water paleo-ventilation ages, *Science*, 280, 725-728, 1998.

Alley, R. B., D. A. Meese, C. A. Shuman, A. J. Gow, K. C. Taylor, P. M. Grootes, J. W. C. White, M. Ram, E. Waddington, P. A. Mayewski, and G. A. Zielinski, Abrupt increase in Greenland snow accumulation at the end of the Younger Dryas event, *Nature*, 362, 527-529, 1993.

Aselmann, I. and P. J. Crutzen, Global distribution of natural freshwater wetlands and rice paddies, their net primary productivity, seasonality and possible methane emissions, *J. Atm. Chem.*, 8, 307-358.

Bard, E. B., Hamelin, R., Fairbanks and A. Zindler, Calibration of the ^{14}C timescale over the past 30,000 years using mass spectrometric U-Th ages from Barbados corals, *Nature*, 345, 405-410, 1990.

Bard, E., F. Rostek and C. Sonzogni, Interhemispheric synchrony of the last deglaciation inferred from alkenone paleothermometry, *Nature*, 385, 707-710, 1997.

Barnola, J. M., P. Pimienta, D. Raynaud and Y.S. Korotkevich, CO_2-climate relationship as deduced from the Vostok ice core: A reexamination based on new measurements and on a reevaluation of the air dating, *Tellus*, 43B, 83-90, 1991.

Behl, R. J. and J. P. Kennett, Brief interstadial events in the Santa Barbara basin, NE Pacific during the past kyr, *Nature*, 379, 243-246, 1996.

Bender, M. L., T. Sowers, J. M. Barnola, and J. Chappellaz, Changes in the O_2/N_2 ratio of the atmosphere during recent decades reflected in the composition of air in the firn at Vostok Station, Antarctica. *Geophys. Res. Lett.*, 21, 189-192, 1994a.

Bender, M. L., T. Sowers, M. L. Dickson, J. Orchardo, P. Grootes, P. A. Mayewski, and D. A. Meese, Climate connection between Greenland and Antarctica during the last 100,000 years, *Nature*, 372, 663-666, 1994b.

Blunier, T., J. Chappellaz, J. Schwander, A. Dallenbach, B. Stauffer, T. F. Stocker, D. Raynaud, J. Jouzel, H. B. Claussen, C. U. Hammer, and S. J. Johnsen, Asynchrony of Antarctic and Greenland climate change during the last glacial period, *Nature*, 394, 739-743, 1998.

Bond, G., W. Broecker, S. Johnsen, J. McManus, L. Labeyrie, J. Jouzel, and G. Bonani, Correlations between climate records from North Atlantic sediments and Greenland ice, *Nature,* 365, 143-147, 1993.

Bond, G. C. and R. Lotti, Iceberg discharges into the North Atlantic on millennial time scales during the last glaciation, *Science*, 267, 1005-110, 1995.

Boyle, E. A., Cool tropical temperatures shift the global delta ^{18}O-T relationship: an explanation for the ice core delta ^{18}O-borehole thermometry conflict? *Geophys. Res. Lett.*, 24, 273-276, 1997.

Broecker, W., Glacial cooling: is water the villain? *EOS*, 77, S283, 1996 (abstract).

Broecker, W., Paleocean circulation during the last deglaciation: A bipolar seesaw? *Paleoceanography*, 13, 119-121, 1998.

Brook, E. J., T. Sowers and J. Orchardo, Rapid variations in atmospheric methane concentration during the past 110,000 years, *Science*, 273, 1087-1091.

Björck, S., B. Kromer, S. Johnsen, O. Bennike, D. Hammarlund, G. Lemdahl, G. Possnert, T. L. Rasmussen, B. Wohlfarth, C. U. Hammer, and M. Spurk, Synchronized terrestrial-atmospheric deglacial records around the North Atlantic, *Science,* 274, 1155-1160, 1996.

Chappellaz, J., J.-M. Barnola, D. Raynaud, Y. S. Korotkevich, and C. Lorius, Ice-core record of atmospheric methane over the past 160,000 years, *Nature*, 345, 127-131, 1990.

Chappellaz, J., T. Blunier, D. Raynaud, J. M. Barnola, J. Schwander, and B. Stauffer, Synchronous changes in atmospheric CH_4 and Greenland climate between 40 and 8 kyr B. P., *Nature*, 366, 443-445, 1993.

Charles, C. D., J. Lynch-Stieglitz, U. S. Ninnemann, and R. G. Fairbanks, Climate connections between the hemisphere revealed by deep sea sediment core/ice core correlations, *Earth Planet. Sci. Lett.*, 142, 19-27, 1996.

Cicerone, R. J., and R. S. Oremland, Biogeochemical aspects of atmospheric methane, *Global Biogeochem. Cycles*, 2, 299-327, 1988.

Craig, H., Y. Horibe, and T. A. Sowers, Gravitational separation of gases and isotopes in polar ice caps, *Science*, 242, 1675-1678, 1988.

Crowley, T. J., North Atlantic deep water cools the southern hemisphere, *Paleoceanogr.*, 7, 489-499, 1992.

Cuffey, K., G. Clow, R. Alley, M. Stuiver, E. Waddington, and R. Saltus, Large Arctic temperature change at the Wisconsin - Holocene glacial transition, *Science*, 270, 455-457, 1995.

Curry, W. B., and D. W. Oppo, Synchronous, high-frequency oscillations in tropical sea surface temperatures and North Atlantic deep water production during the last glacial cycle, *Paleoceanography*, 12, 1-14, 1997.

Denton, G. H., and C. H. Hendy, Younger Dryas Age advance of Franz Josef Glacier in the Southern Alps of New Zealand, *Science*, 264, 1434-1437, 1994.

Epstein, B. L., S. D'Hondt, J. G. Quinn, J. P. Zhang, and P. E. Hargraves, An effect of dissolved nutrient concentrations on alkenone-based temperature estimates, *Paleoceanogr.*, 13, 122-126, 1998.

Fairbanks, R., A 17,000 year glacio-eustatic sea level record: influence of glacial melting rates on the Younger Dryas event and deep ocean circulation, *Nature*, 342, 637-642, 1989.

Fung, I., J. John, J. Lerner, E. Matthews, M. Prather, L. P. Steele, and P. J. Fraser, Three-dimensional model synthesis of the global methane cycle, *J. Geophys. Res.* 96, 13,033-13,065, 1991.

Genthon, C., J.-M. Barnola, D. Raynaud, C. Lorius, J. Jouzel, N. I. Barkov, Y. S. Korotkevich, and V. M. Kotlyakov, Vostok ice core - climate response to CO_2 and orbital forcing changes over the last climatic cycle, *Nature*, 329, 414-418, 1987.

Greenland Ice Core Project (GRIP) Members, Climate instability during the last interglacial period recorded in the GRIP ice core, *Nature*, 364, 203-207, 1993.

Grootes, P. M., M. Stuvier, J. W. C. White, S. Johnsen, and J. Jouzel, Comparison of oxygen isotope records from the GISP2 and GRIP Greenland ice cores, *Nature*, 366, 552-555, 1993.

Guilderson, T., R. Fairbanks, and J. Rubenstone, Tropical temperature variations since 20,000 years ago: modulating interhemispheric climate change, *Science*, *263*, 663-665, 1994.

Hein, R., P. J. Crutzen, and M. Heimann, An inverse modeling approach to investigate the global atmospheric methane cycle, *Global Biogeochem. Cycles*, 11, 43-76, 1997.

Hughen, K. A., J. T. Overpeck, L. C. Peterson, and S. Trumbore, Rapid climate changes in the tropical Atlantic region during the last deglaciation, *Nature*, 380, 51-54, 1996.

Imbrie, J., E. Boyle, S. Clemens, A Duffy, W. Howard, G. Kukla, J. Kutzbach, D. Martinson, A. McIntyre, A. Mix, B. Molfino, J. Morley, L. Peterson, N. Pisias, W. Prell, M. Raymo, N. Shackleton and J. Toggweiler, On the structure and origin of major glaciation cycles: 1. Linear responses to Milankovitch forcing, *Paleoceanogr.*, 7, 701-738, 1992.

Johnsen, S., D. Dahl-Jensen, W. Dansgaard, and N. Gundestrup, Greenland paleotemperatures derived from GRIP bore hole temperature and ice core isotope profiles, *Tellus*, 47B, 624-629, 1995.

Jones, G. and L. D. Keigwin, Evidence from Fram Stait (78°N) for early deglaciation, *Nature,* 336, 56-59, 1988.

Jouzel, J, C. Lorius, S. J. Johnsen, and P. Grootes, Climate instabilities: Greenland and Antarctic records, *C.R. Acad. Sci. Paris,* 319 série II, 65-77, 1994.

Jouzel, J, C. Lorius, J. R. Petir, C. Genthon, N. I. Barkov, V. M. Kotlyakov, and V. M. Petrov, Vostok ice core - a continuous isotope temperature record over the last climate cycle (160,000 years), *Nature*, 329, 403-408, 1987.

Jouzel, J., R. Vaikmae, J. R. Petit, M. Martin, Y. Duclos, M. Stiévenard, C. Lorius, M. Toots, M. A. Mélières, L. H. Burckle, N. I. Barkov, V. M. Kotlyakov, The two-step shape and timing of the last deglaciation in Antarctica, *Clim. Dyn.,* 11, 151-161, 1995.

Jouzel, J., C. Waelbroeck, B. Malaize, M. Bender, J. R. Petit, M. Stievenard, N. I. Barkov, J. M. Barnola, T. King, V. M. Kotlyakov, V. Lipenkov, C. Lorius, D. Raynaud, C. Ritz, and T. Sowers, Climatic interpretation of the recently extended Vostok ice records, *Clim. Dyn.,* 12, 513-521, 1996.

Keigwin, L. D., and G. A. Jones, Western North Atlantic evidence for millennial-scale changes in ocean circulation and climate, J. *Geophys. Res.*, 99, 12,397-12,410, 1994.

Kennett, J. P., and B. L. Ingram, A 20,000-year record of ocean circulation and climate change from the Santa Barbara basin, *Nature*, 377, 510-514, 1995.

Kipfstuhl, J., A. Hofmann, Grobe, H. G. Kuhn, Synchronous calving events in the northern and southern hemispheres?, *EOS*, 77, F29 1997 (abstract).

Lehman, S. J., and D. W. Oppo, Changes in North Atlantic deep and intermediate water formation linked to millennial climate change in Greenland and Antarctica, submitted to *Science*, 1997.

Leuenberger, M. and U. Siegenthaler, Ice-age atmospheric concentration of nitrous-oxide from an Antarctic ice core, *Nature*, 360, 449-451, 1992.

Le Treut, H., Sulfate aerosol indirect effect and CO_2 greenhouse forcing: quilibrium response of the LMD GCM and associated cloud feedbacks, *J. Climate*, 7, 1673-1684.

Linsley, B. K., Oxygen-isotope record of sea level and climate variations in the Sulu Sea over the past 150,000 years, *Nature*, 380, 234-237, 1996.

Lowell, T. V., C. J. Heusser, B. G. Andersen, P. I. Moreno, A. Hauser, L. E. Heusser, C. Schlüchter, D. R. Marchant, and G. H. Denton, Interhemispheric correlation of Late Pleistocene glacial events, *Science*, 269, 1541-1549, 1995.

Manabe, S., and R. J. Stouffer, Simulation of abrubt climate change induced by freshwater input to the North Atlantic Ocean, *Nature*, 378, 165-167, 1995.

Manabe, S. and R. J. Stouffer, Coupled ocean-atmosphere model response to freshwater input: comparison to Younger Dryas event, *Paleoceanography*, 12, 321-336, 1997.

Maslin, M., N. Shackleton, and U. Pflaumann, Surface water temperature, salinity, and density changes in the northeast Atlantic during the last 45,000 years: Heinrich events, deep water formation, and climatic rebounds, *Paleoceanogr.*, 10, 527-544, 1995.

Mayewski, P., L. Meeker, S. Whitlow, M. Twickler, M. Morrison, P. Grootes, G. Bond, R. Alley, D. Meese, A.

Gow, K. Taylor, M. Ram, and M. Wumkes, Polar atmospheric cell and ocean ice cover variability over the North Atlantic region during the last 41,000 years, *Science*, 263, 1747-1751, 1994.

Pimienta, P. Etude du comportement mecanique des glaces polycristallines aux faibles contraintes: applications aux glaces des calottes polaires. Thesis, Universite Scientifique, Technique et Medicale de Grenoble, 163 pp.

Porter, S. C. and A. Zhisheng, Correlation between climate events in the North Atlantic and China during the last glaciation, *Nature*, 375, 305-308, 1995.

Rahmstorf, S., Bifurcations of the Atlantic thermohaline circulation in response to changes in the hydrological cycle, *Nature*, 378, 145-149, 1995.

Salamatin, A. N., V. Y. Lipenkov, N. I. Barkov, J. Jouzel, J.-R. Petit, and D. Raynaud, Ice core age dating and paleothermometer calibration based on isotope and temperature profiles from deep boreholes at Vostok Station (East Antarctica), *J. Geophys. Res.*, 103, 8693-8977, 1998.

Schwander, J. T. Sowers, J.-M. Barnola, B. Malaize, T. Blunier, A. Fuchs, and S. J. Johnsen, Age scale of the air in Summit ice - implication for glacial - interglacial temperature change, submitted to *J. Geophys. Res*, 102, 19483-19493, 1997.

Severinghaus, J. P., T. Sowers, E. Brook, R. Alley, and M. Bender, The timing of atmospheric temperature and methane-concentration changes at the end of the Younger Dryas period, *Nature*, 391, 141-146, 1997.

Sirocko, F., D. Garbe-Schönberg, A. McIntyre and B. Molfino, Teleconnections between the subtropical Monsoons and high-latitude climates during the last deglaciation, *Science*, 272, 526-529, 1996.

Sowers, T. and M. Bender, Climate records covering the last deglaciation, *Science*, 269, 210-214, 1995.

Sowers, T., M. L. Bender, and D. Raynaud, Elemental and isotopic composition of occluded O2 and N2 in polar ice, J. *Geophys. Res.*, 94, 5137-5150, 1989.

Sowers, T., M. Bender, D. Raynaud and Y. S. Korotkevich, The N of N2 in air trapped in polar ice: a tracer of gas transport in the firn and a possible constraint on ice age - gas age differences, J. *Geophys. Res.*, 97, 15,683-15,697, 1992.

Steig, E. J., E. J. Brook, J. W. C. White, C. M. Sucher, M. L. Bender, S. J. Lehman, D. L. Morse, E. D. Waddington, and G. D. Clow, Synchronous climate changes in Antarctica and the North Atlantic, *Science*, 282, 92-95, 1998.

Stocker, T. and D. Wright, Rapid changes in ocean circulation and atmospheric radiocarbon, *Paleoceanogr.*, 11, 773-795, 1996.

Stocker, T. F., D. G. Wright, and L. A. Mysak, A zonally averaged, coupled ocean-atmosphere model for paleoclimate studies, *J. Climate*, 5, 773-797, 1992a.

Stocker, T. F., D. G. Wright, and W. S. Broecker, The influence of high-latitude surface forcing on the global thermohaline circulation, *Paleoceanography*, 7, 529-541, 1992b.

Yiou, F., G. M. Raisbeck, S. Baumgartner, J. Beer, C. Hammer, S. Johnsen, J. Jouzel, P. W. Kubik, J. Lestringuez, M. Stievenard, M. Suter, and P. Yiou, Beryllium 10 in the Greenland Ice Core Project ice core at Summit, Greenland, *J. Geophys. Res.*, 102, 26,783-26.794, 1997.

Yung, Y., T. Lee, C. Wang and Y. Shieh, Dust: a diamnostic of the hydrologic cycle during the last glacial maximum, *Science*, 271, 962-963, 1996.

Michael Bender, Department of Geosciences, Princeton University, Princeton, NJ 08544

Jean Jouzel, Laboratoire des Sciences du Climat et l'Environnement, UMR CEA-CNRS 1572, 91191 Gif sur Yvette, France

Bruno Malaize, UFR des Science de la Terre et de la Mer, University of Bordeaux 1, 351 Cours de la Liberation, F 33405 Talence, France

Joseph Orchardo, Graduate School of Oceanography, University of Rhode Island, Kingston, RI 02881

Todd Sowers, Department of Geosciences, 447 Deike Bldg., Penn State University, University Park, PA 16802

Atmospheric Methane and Millennial-Scale Climate Change

Edward J. Brook and Susan Harder
Geology Department, Washington State University, Vancouver, WA

Jeff Severinghaus
Scripps Institution of Oceanography, University of California at San Diego, La Jolla, CA

Michael Bender
Department of Geosciences, Princeton University, Princeton, NJ

Studies of polar ice cores show that rapid changes in atmospheric methane were associated with millennial-scale climate oscillations during the last glacial period and deglaciation. These methane variations were most likely caused by changes in methane emissions from tropical and boreal wetlands. Methane emissions from modern wetlands are generally positively correlated with temperature and precipitation, suggesting that past methane variations reflect changes in these variables. Because northern hemisphere ice sheets covered at least part of the present boreal methane source areas, methane mixing ratio changes during the glacial period are thought to have been driven by changes in tropical climate. This inference is supported by measurements of the interpolar methane gradient, an indicator of the latitudinal distribution of methane sources and sinks. Interpolar gradient data for a number of time periods during the last 50,000 years suggest a dominance of tropical methane sources during both stadial and interstadial climates, although small shifts in the gradient indicate variations in the relative strength of boreal sources. During two climate oscillations where precise temporal phasing has been established (the end of the Younger Dryas and the onset of the Bølling warm interval), a rapid increase in temperature slightly lead an increase in methane mixing ratio, suggesting that the methane rise was a response to climate change, and that tropical and high-latitude northern hemisphere climate may be tightly coupled. An alternative hypothesis holds that large releases of methane from clathrates in marine sediments or permafrost were responsible for rapid methane changes, and perhaps were also significant forcing agents for millennial-scale climate change. However, the rate, timing, and magnitude of change observed in high resolution methane records from the GISP2 ice core do not appear to be consistent with this hypothesis.

Mechanisms of Global Climate Change at Millennial Time Scales
Geophysical Monograph 112

1. INTRODUCTION

Methane is a trace greenhouse gas with a variety of natural and anthropogenic sources, including natural wetlands, animals, rice paddies, fossil fuel extraction, and landfills [*Fung et al.*, 1991; *Hein et al.*, 1997]. The primary methane sink is destruction by the OH radical in the troposphere, and the current atmospheric lifetime is ~ 9 years [*Prinn et al.*, 1995]. Natural wetlands were the dominant methane source before the industrial revolution [*Chappellaz et al.* 1993a], and it is generally believed that past methane variations are indicative of changes in source strength, as changes in OH are thought to have been small [*Thompson*, 1992]. Ice core methane records, therefore, provide information about the terrestrial biosphere during changing climate conditions. In this paper we review ice core records of millennial-scale (and shorter) variability in atmospheric methane, summarize new high resolution results that illustrate the rate and magnitude of methane changes associated with millennial-scale climate change, and summarize new work on the interpolar methane gradient, which constrains the location of past methane sources.

Late Quaternary methane mixing ratios varied from 350 to 750 ppbV on a variety of time scales. Changes in radiative forcing due to these changes in methane can be approximated using the relation described by *Shine et al.* [1990]. For the maximum mixing ratio change in the ice core record (~400 ppbV), the change in forcing is ~0.2 W/m^2. Temperature changes produced by this change were probably small. For example, both *Raynaud et al.* [1988] and *Stauffer et al.* [1988] estimated that the 300 ppbV glacial-interglacial change in mixing ratio that they reported would have by itself caused an average temperature change of only ~0.1°C. *Raynaud et al.* [1988] *and Chappellaz et al.* [1990] further calculated that, when all feedbacks are included, this change might have caused an increase in global temperatures of ~0.5-0.7°C, out of the total of 4-5°C glacial-interglacial warming hypothesized at that time. Millennial-scale methane mixing ratio variations during the last glacial period were about 250 ppbV or less (Figure 1), and the total magnitude of global temperature changes during interstadial events, while uncertain, was probably at least several degrees. Therefore, increases in methane mixing ratio probably made only a minor contribution to warming during interstadial events. Nonetheless, methane records are of great interest as tracers of the influence of climate change on the terrestrial biosphere.

2. METHANE EMISSIONS FROM TERRESTRIAL ECOSYSTEMS

Methane is produced naturally by bacterial methanogenesis in anaerobic environments such as wetlands and the digestive system of ruminants and termites. There is little net methane production in the ocean [*Fung et al.*, 1991; *Bates et al.*, 1996], and methane mixing ratios in ice core air are generally viewed as tracers of terrestrial climatic and biological processes. Most estimates of the global methane budget indicate that wetlands are the major natural source, making up ~ 75% of the total natural budget for the preindustrial Holocene [*Chappellaz et al.*, 1993b]. The remaining natural sources are wild ruminants, termites, biomass burning, methane hydrates, and the oceans.

Much of what we know about the factors controlling methane emissions from wetlands comes from modern studies of wetlands on relatively small spatial scales. These studies indicate that a number of variables are correlated with methane emissions, including temperature, precipitation, water table height, and net ecosystem production (NEP = net primary production - soil and bacterial respiration) (e.g., *Whiting and Chanton* [1993]; *Bubier and Moore* [1994]; *Schlesinger*, [1996]). The general picture that emerges is that wet conditions (saturated soils) are necessary to maintain anoxic conditions, and given wet conditions, warmer temperatures enhance emissions.

Modern wetlands are concentrated in two latitudinal regions, the boreal belt between 50-70°N, and a tropical region between 30°N-30°S [*Bartlett and Harriss*, 1993]. These are regions of atmospheric convergence and high precipitation. An analogous belt of methane emissions is not present in the high latitude southern hemisphere due to the absence of large land areas.

The factors controlling methane emissions from other sources during past times, particularly during millennial-scale climate oscillations, are not as obvious. Changes in biome distribution driven by temperature changes probably influenced termite and ruminant populations [*Chappellaz et al.*, 1993b] and natural biomass burning, but available information is not sufficient to place quantitative constraints on such changes [*Chappellaz et al.*, 1993b]. Further studies examining changes in terrestrial vegetation on millennial time scales would help to address these issues, although these sources do not constitute a large fraction of the methane budget.

3. METHODS

Data discussed here come from the GISP2 and Taylor Dome ice cores. The GISP2 core was recovered in 1993 at 72°36'N 38°30'W in central Greenland. We have improved the resolution of our original GISP2 data set [*Brook et al.*, 1996a] by analyzing samples from 127 additional depths in the upper 2438 m, which represents the past 50 ka (Brook, E. J., S. Harder, J. Severinghaus, M. Bender, C. Sucher, and E. Steig, Atmospheric methane during the past 50,000 years: trends, interpolar gradient, and rate of change, in prep.; hereafter referred to as *Brook et al.*, in prep.). We focus here on abrupt transitions in the record, including the termination of the Younger Dryas period at ~11.8 ka, the beginning of the Bølling-Allerød period at ~14.5 ka, and the onsets of interstadials #8 and #12, rapid warming events that occurred at ~ 38.2 and 45.5 ka in the GISP2 isotope record. The GISP2 methane data set over the 0-50 ka interval now includes 287 sample depths.

The Taylor Dome ice core was recovered in 1994 at 77° 48'S 158° 43'E on Taylor Dome, a small ice dome in the Ross Sea sector of Antarctica. The Taylor Dome paleoclimate record extends to greater than 130 ka at a depth of 554 m [*Grootes and Steig*, 1994; *Steig*, 1996]. We focus here on methane measurements of 137 samples in the upper 445 m (0-50 ka), with high resolution measurements over the time intervals listed above.

Analytical methods are described in detail elsewhere [*Brook et al.*, 1996a; *Sowers et al.*, 1997; *Brook et al.*, in prep.]. Briefly, ice samples weighing ~35 g were placed in cubical stainless steel vacuum vessels. The vessels were sealed and evacuated at -25 to -30°C. The ice samples were then melted in vacuum and slowly refrozen at -80°C. This procedure expels the air trapped in the ice into the vessel headspace. The extracted gas was injected into a previously evacuated sample loop of a Hewlett-Packard Model 5890 Series II gas chromatograph equipped with a 6-port gas sampling valve, a 10 cm^3 sample loop, a Poropak Q column, and a flame ionization detector. The pressure in the sample loop was measured with a capacitance manometer. Methane mixing ratios were quantified by measuring peak areas using ELAB (OMS Tech, Miami, Florida) chromatographic software. Our working standard was a high pressure cylinder of synthetic air prepared by Scott-Marin Specialty Gases. This standard has a mixing ratio of 962 ± 6 ppbV (2-sigma) on the NOAA/CMDL mixing ratio scale. The procedural blank was 16 ± 4 ppbV (2-sigma, N=63) over the time period the analyses reported here were made. We analyzed at least two separate ice samples at each sample depth, and analyzed the headspace gas for each twice (averaging the results for the two injections). If analyses of duplicate ice samples disagreed by more than 5% a third replicate was analyzed. For 859 duplicate pairs (not all measured for this study) meeting the 5% criteria, analyzed between 1993 and 1997, the average deviation from the mean was 1.2%. Data are available from the authors.

The depth at which air is trapped in polar ice (the close-off depth) is a function of the temperature and accumulation rate, and is typically 50-100 m below the ice surface. Above that depth air is transported in the firn by convection and diffusion. The age of the gas is therefore younger than the age of the ice. This gas age-ice age difference (Δage) varies with time. It is not known *a priori* for past times, but can be calculated using a model of the snow densification process.

In the following discussions we present our results as a function of age for the Taylor Dome and GISP2 ice cores (Figures 1-4). We placed these data sets on a common gas-age time scale as follows. For GISP2 we used the gas-age time scale described by *Brook et al.* [1996a]. This time scale is based on the GISP2 layer counting ice-age time scale [*Meese et al.*, 1994]. Δage was calculated using the steady state *Herron and Langway* [1980] densification model [*Brook et al.*, 1996a]. Gas ages were calculated by subtracting Δage from the ice age at each sample depth. The GISP2 gas-age time scale used here agrees well (±100 years) with a later GISP2 gas-age time scale reported by *Schwander et al.* [1997], which incorporates gas age-ice age differences calculated from a dynamic densification model [*Barnola et al.*, 1991]. We estimate that the uncertainty of the GISP2 gas-age time scale, relative to the ice-age time scale, is ± 100 years [*Brook et al.*, 1996a]. This estimate does not include the uncertainty in the original ice-age scale on which the gas-age time scale is based. For the purposes of correlation between Greenland and Antarctica the latter is not relevant.

For Taylor Dome we created a correlation gas-age time scale by visually matching common inflection points in the GISP2 and Taylor Dome methane and $\delta^{18}O_{atm}$ (isotopic composition of atmospheric O_2) time series, each corresponding to a time when methane or $\delta^{18}O_{atm}$ was changing rapidly, and interpolating between those points [*Sucher*, 1998; *Steig et al.*, 1998]. Because the atmosphere is well mixed on time scales of greater than a year, inflections in the methane and $\delta^{18}O_{atm}$ records should occur at essentially the same time in both polar regions. Correlation using methane is complicated slightly by the existence of an interpolar methane gradient. We accounted for this complication by making the *a priori* assumption that the gradient was linearly proportional to the methane mixing ratio. We assumed that there was no gradient when methane mixing ratios were at their lowest levels and that the ratio of Greenland/Antarctic mixing ratios was 1.10 at the maximum mixing ratio observed. The latter figure is based on earlier work [*Rasmussen and Khalil*, 1984]. The assumption of 0 gradient at times of low mixing ratios is equivalent to assuming no boreal methane sources at those times. These assumptions were made for correlation purposes only, and we use the original data sets in calculations of the true interpolar gradient (described below).

The uncertainty of the resulting Taylor Dome gas-age time scale is a function of the resolution of the records and varies with time. For control points based on methane, we estimate the uncertainty in the correlation at the match points as ± 100 years between 10 and 15 ka, and ± 250 years at control points outside this range. Uncertainty introduced into the correlation due to our *a priori* assumption about the nature of the methane gradient is small. We estimate that a 50% uncertainty in the gradient causes a less than 20 year uncertainty in our correlation at times when methane levels were changing rapidly. For control points based on $\delta^{18}O_{atm}$, which varies more slowly than methane, we estimate the average uncertainty in the correlation at the control points as ± 450 years. Because accumulation rates probably varied between the match points, and time periods between control points are large for some intervals, uncertainties between the control points are harder to quantify. At this level the accuracy of the correlation between Taylor Dome and GISP2 is not central to the interpretation of our results, and is discussed further by *Steig et al.* [1998]. Note that we have little age control between 20 and 34 ka (Figure 4).

4. MILLENNIAL-SCALE METHANE VARIABILITY

4.1 General Trends

Ice core methane records show that methane mixing ratios varied with a variety of periods over the past 220,000 years [*Chappellaz et al.*, 1993a; *Jouzel et al.*, 1993; *Blunier et al.*, 1995; *Brook et al.*, 1996a]. Mixing ratios ranged between ~ 350 to 750 ppbV, compared to modern values in excess of

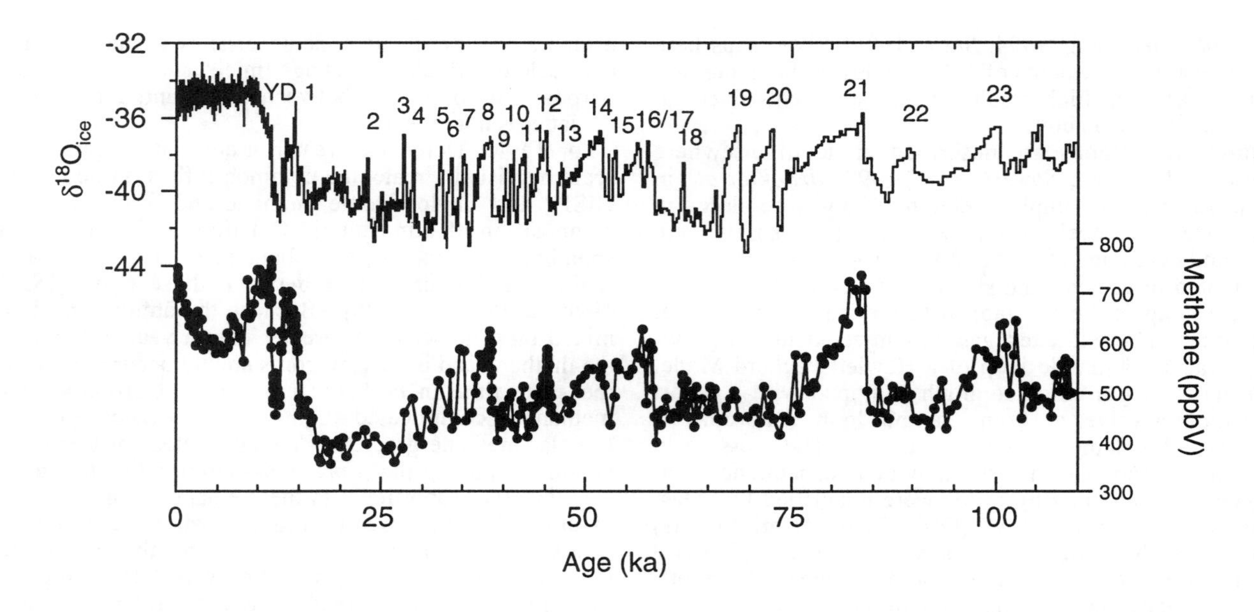

Figure 1. The methane (closed circles) and $\delta^{18}O_{ice}$ (thin solid line) records from the GISP2 ice core covering the past 110 ka. Data include published results of *Brook et al.* [1996a] and previously unpublished data from *Brook et al.* (in prep.). GISP2 isotope data are from *Grootes et al.* [1993]. Time scale for GISP2 methane record from *Brook et al.* [1996a], time scale for isotope record from *Meese et al.* [1994]. Numbers indicate interstadial events. YD = Younger Dryas.

1700-1800 ppbV. On orbital time scales methane variations appear to correspond with Northern Hemisphere insolation variations [*Chappellaz et al.*, 1990]. Orbitally forced changes in tropical monsoon strength have been proposed as a possible explanation - a more active monsoon and warmer temperatures could cause both greater wetland area and greater methane production in the tropics [*Chappellaz et al.*, 1990]. High resolution methane records from Greenland considerably expanded the picture of methane variability [*Chappellaz et al.*, 1993a; *Blunier et al.*, 1995; *Brook et al.*, 1996a]. These records confirm the long-term patterns [*Brook et al.*, 1996a] but also show higher frequency millennial-scale variability closely associated with temperature changes inferred from oxygen isotope records (Figure 1). This rapid variability has been confirmed from studies of several Antarctic ice cores [*Chappellaz et al.*, 1997; *Blunier et al.*, 1998; *Steig et al.*, 1998; *Brook et al.*, in prep].

All of the interstadial events in the GISP2 $\delta^{18}O_{ice}$ record, with the possible exception of interstadial #20, appear to be associated with increases in atmospheric methane (Figure 1). The amplitude of the response varies from < 50 ppbV (for example interstadial 19) to up to 300 ppbV for interstadials 21 and 1. The amplitude variation may reflect modulation of the methane response to interstadial warming by longer term, orbitally forced climate cycles [*Brook et al.*, 1996].

4.2 Rapid Methane Changes During Interstadial Events

Because the atmospheric methane lifetime is about 10 years, methane mixing ratios have the potential to change very quickly, over periods of a few decades. Detailed records of the rate, magnitude, and timing of methane changes in the ice core record therefore may enhance our understanding of the causes and implications of millennial-scale climate events. However, as described above, interpretation of the records is complicated by the nature of gas trapping in polar ice. Gases are trapped below the snow surface, typically between 50 and 100 m at polar sites. Above the "close-off depth" the atmosphere diffuses freely within the porous firn and transport is mainly by molecular diffusion, with the exception of a surface convection zone of up to ~10 m in thickness [*Sowers et al.*, 1991]. Density increases with depth as the firn transforms to ice, therefore diffusive mixing decreases as the firn-ice transition is approached [*Schwander et al.*, 1993]. Gas transport in the firn smoothes the record of very rapid changes in atmospheric composition. Also, bubble close-off in a given volume of the firn column occurs progressively as that firn passes through the firn-ice transition. This progressive trapping in a given volume may cause further smoothing of the record [*Schwander et al.*, 1993].

Precise determination of the temporal phasing of methane and temperature change in the ice core record can be problematic because traditional isotopic temperature proxies are recorded in the ice phase. This is a problem because uncertainties in the gas age - ice age difference can be several hundred years, even in high accumulation rate cores like GISP2 and GRIP. In Figure 2 we show sub-century resolution methane records from GISP2 (*Brook et al.*, in prep.) for four rapid climate shifts: the end of the Younger Dryas, the beginning of the Bølling-Allerød period, and the onsets of interstadials #8 and #12. In several cases, the methane mixing ratio apparently began to rise up to 300 years before

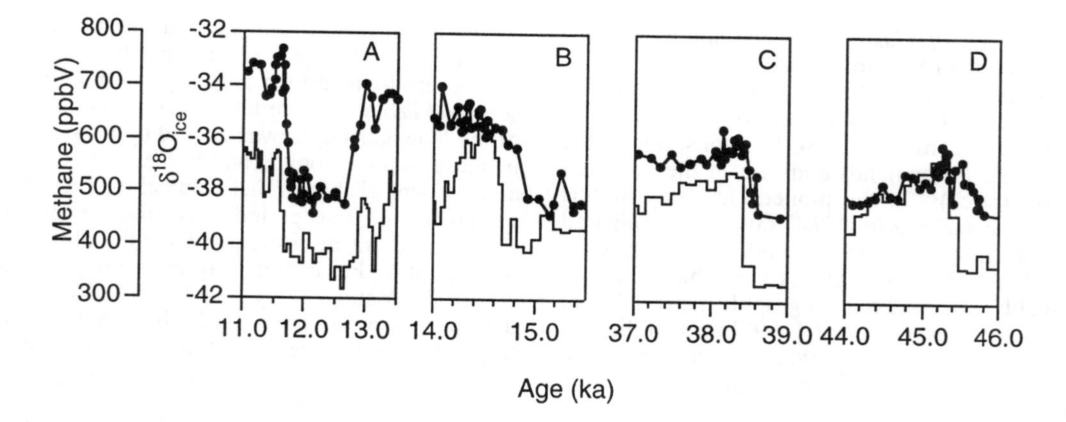

Figure 2. Methane (closed circles) records from GISP2 for the Younger Dyras (A), transition to the Bølling period (B), interstadial #8 (C), and interstadial #12 (D). Isotope data (thin solid line) from *Grootes et al.* [1993].

the associated temperature change, but this apparent lead is comparable to the uncertainty of the gas age-ice age difference.

The precise timing of methane versus temperature change at the end of the Younger Dryas was determined by *Severinghaus et al.* [1998] by employing anomalies in $\delta^{15}N$ of N_2 as gas phase markers of warming (Figure 3). By using an indicator of warming recorded in the trapped gases, this approach eliminates uncertainties associated with the gas age-ice age difference. *Severinghaus et al.* [1998] showed that methane change lagged temperature change at this time by 0-30 years. Similar results have been obtained for the transition to the Bølling-Allerød period [*Severinghaus and Brook*, 1998]. These results suggest that the change in methane was a response to, rather than a cause of, climate change. Further work on $\delta^{15}N$ of N_2 and methane at other transitions should allow us to establish temperature and methane phasing for most interstadials in the Greenland record.

The rate of change of methane mixing ratio at the beginning of interstadial events may also provide important information about the nature of these climate transitions. The records (Figure 2) suggest that these shifts of 200-300 ppbV took place over 100-300 years. In contrast, temperature changes that occurred over the same transitions are thought to have happened over several decades in some cases (e.g., *Alley et al.*, [1993]; *Stuiver et al.*, [1995]). The rate of change of methane mixing ratios is slow enough that it should not have been significantly affected by mixing in the firn, and therefore should closely approximate the true rate of change in the atmosphere (*Brook et al.*, in prep.).

High resolution data immediately following the rise in methane mixing ratios indicate maintenance of interstadial methane values for roughly the duration of the warming event. Mixing ratio trends seem to generally follow temperature trends as indicated by the $\delta^{18}O_{ice}$ record (Figure 2), with some exceptions (*Brook et al.*, in prep.). The large methane oscillation after the rise at the beginning of interstadial #12 is an example of such an exception. A similar feature appears in recent subtropical Atlantic sea surface temperature reconstructions [*Sachs and Lehman*, 1998] suggesting that the methane oscillation is a signature of a real climate event.

Chappellaz et al. [1993a] suggested that these rapid shifts in methane mixing ratio were indicative of significant climate changes in the tropics during interstadial events. Large tropical wetland areas exist today in South America, central and southern Africa, and south Asia, and presumably existed in some form in the late Quaternary. Changes in temperature and hydrologic balance in these regions are likely candidates for forcing millennial-scale oscillations in methane mixing ratio. However, large wetland areas also exist today in boreal regions. In addition, large-scale releases of methane from methane hydrates in marine sediments in polar and subpolar regions have been suggested as a significant source of atmospheric methane at the end of the last ice age [*Nisbet*, 1992]. Although high latitude northern hemisphere wetland regions were at least partly covered by expanded ice sheets prior to the beginning of the Holocene, the possibility that significant northern hemisphere methane sources existed during interstadial events can not be discounted *a priori* . Perhaps, for example, the large areas of methane producing wetlands that exist today in the global boreal belt simply moved further south during the last glacial period. Below we describe one approach to resolving this question, and follow that with further discussion of the clathrate hypothesis.

4.3 Interpolar Methane Gradient, 40-10 ka

The atmospheric methane mixing ratio varies with latitude and is controlled by the distribution of sources and sinks, and atmospheric transport. The methane sink (primarily tropospheric OH) is approximately hemispherically symmetrical. Methane removal is more rapid in the tropics where temperatures and OH levels are higher. Natural sources (primarily wetlands) are concentrated in the tropical (30°N-30°S) and boreal (50-70°N) regions. In the modern atmosphere, the source/sink distribution, combined with atmospheric transport patterns, results in mixing ratios in the high latitude northern hemisphere that are ~8% higher than in the high latitude southern hemisphere. This reflects the present dominance of northern hemisphere (including anthropogenic) sources. The N/S ratio can be calculated for past times using ice core data from Antarctica and Greenland [*Rassmussen and Khalil*, 1984; *Nakazawa et al.*, 1993;

Brook et al, 1996b; *Chappellaz et al*., 1997] and used to infer changes in the latitudinal source distribution.

Ideally, one would obtain methane mixing ratios from ice cores in a latitudinal transect for the entire globe. Practically we are presently confined to two locations, one in the northern hemisphere (Greenland) and one in the southern hemisphere (Antarctica). In the pioneering study of this problem, *Rasmussen and Khalil* [1984] favorably compared Holocene ice core methane data to results of a 4-box atmospheric model using estimates of methane source strength, removal by OH, and the geographic distribution of sources and sinks. *Chappellaz et al.* [1997] refined this approach for a study of the entire Holocene methane record and here we extend the analysis to the deglaciation and the latter part of the last glacial period.

In Figure 4 we plot the interpolar methane gradient (IPG), which we define as IPG = $[(C_G / C_A) - 1]$, for a number of time periods for which there are sufficient data and adequate chronology. (C_G = Greenland mixing ratio and C_A = Antarctic mixing ratio). This plot includes the Holocene data of *Chappellaz et al.* [1997] and our recent results (*Brook et al.*, in prep.). IPG's are calculated from time weighted means of methane mixing ratios for intervals when methane levels were relatively stable (Figure 4). Trends in IPG indicate a changing balance between northern and tropical methane sources. Lower values of the IPG characterize the glacial maximum, Bølling-Allerød, and Younger Dryas periods, while the IPG increased immediately after the end of the Younger Dryas. These variations suggest that tropical sources were primarily responsible for the increase in methane at the beginning of the Bølling-Allerød period, and that high latitude sources became more important after the Younger Dryas. The IPG during interstadial #8 is similar to that found that during the mid-late Holocene, suggesting significant northern hemisphere emissions. However, this value is constrained by only four Antarctic data points and final conclusions await further analysis. Significant IPG variations that occurred during the Holocene are described further by *Chappellaz et al.* [1997].

Quantitative modeling of changes in the IPG using the box model approach [*Rassmussen and Khalil*, 1984; *Chappellaz et al.*, 1997] yields changes in source distribution, although several assumptions are necessary for this calculation. We used the three box model of *Chappellaz et al.* [1997] to examine the record in Figure 4 for the period 40-10 ka (*Brook et al.*, in prep.). This model divides the atmosphere into a tropical box (30°S - 30°N), a northern box (30°-90°N), and a southern box (30°-90° S). We model inter-box mixing and methane sink strength as described by *Chappellaz et al.* [1997]. Ice core data from Greenland and Antarctica provide mixing ratios for the northern and southern boxes. The mixing ratio in the tropical box is unknown and is adjusted so that the southern hemisphere source is fixed at 15 Tg/yr (12 Tg/yr for the last glacial maximum). (All model parameters are identical to those used by *Chappellaz et al.* [1997] to allow comparison with their results). Uncertainties in the calculated source due to uncertainties in the specified sink strength and inter-box mixing are difficult to evaluate but probably do not grossly affect the results (*Brook et al.*, in prep.).

The model results indicate that the methane mixing ratio increase of ~ 250 ppbV at the transition from LGM to Bølling-Allerød was driven primarily by a doubling of tropical sources (Table 1). The later ~220 ppbV increase at the end of the Younger Dryas was caused by an approximately 50% increase in both tropical and boreal sources (Table 1). It is also evident (Table 1) that tropical methane sources are required for all time periods. Northern hemisphere sources alone could not be responsible for the observed changes in mixing ratio. For example, if the mixing ratio change at the end of the Younger Dryas and the Bølling transition were driven only by a change in high latitude emissions, while tropical emissions remained constant, the interpolar gradient would have reached 12 and 14%, respectively, significantly higher than our measurements indicate, and higher than at any time during the Holocene. Further studies of the interpolar gradient at earlier times, particularly during portions of Marine Isotope Stages 3 and 5, when northern hemisphere ice sheets may have been in intermediate configurations, may reveal further IPG variability.

4.4 Evaluation of the Clathrate Hypothesis

Large quantities of methane are stored as clathrates in marine sediments and permafrost [*Kvenvolden*, 1988]. The contribution to the modern atmospheric methane budget from these deposits is small [*Fung et al*., 1991]. However, *Nisbet* [1992], *Thorpe et al.* [1996], and others proposed that rapid release of this methane may have been responsible for some of the rapid changes in the ice core methane record, and that clathrate derived methane may have been an important climate forcing during the deglaciation. *Kennett et al.* [1996] have also proposed that instantaneous release of clathrate methane from marine sediments may have caused the interstadial methane shifts observed in the ice core record, and that the greenhouse effect of this methane may have driven or amplified global warming.

In a quantitative assessment of this type of hypothesis, *Thorpe et al.* [1996] considered the instantaneous release of 4000 Tg of methane (about 20x the annual preindustrial budget) in the high latitude northern hemisphere and predicted the atmospheric response with a 2-D chemical and transport model. They predicted that methane mixing ratios at high northern latitudes would reach in excess of 25,000 ppbV immediately after the event, and remain in excess of 2000 ppbV for over a year. (*Thorpe et al.* noted that although the quantity of methane released was large, the temperature change due to the radiative forcing of this hypothetical event would be less than 1°C). In their model these high mixing ratios mixed further in the atmosphere and were removed over the following several decades by reaction with atmospheric OH. Because interhemispheric mixing occurs on a one year time scale, the rapid mixing ratio increase would be observed in the southern hemisphere with a lag of about one year, smaller than might be resolvable in an ice core record.

Would such an event be recorded in an ice core? Mixing processes in the firn would cause such an extreme and rapid change to be smoothed, and the amplitude of the increase would be damped. We estimate [*Brook et al*., 1996b; *Brook et al*., in prep] that an instantaneous increase in atmospheric

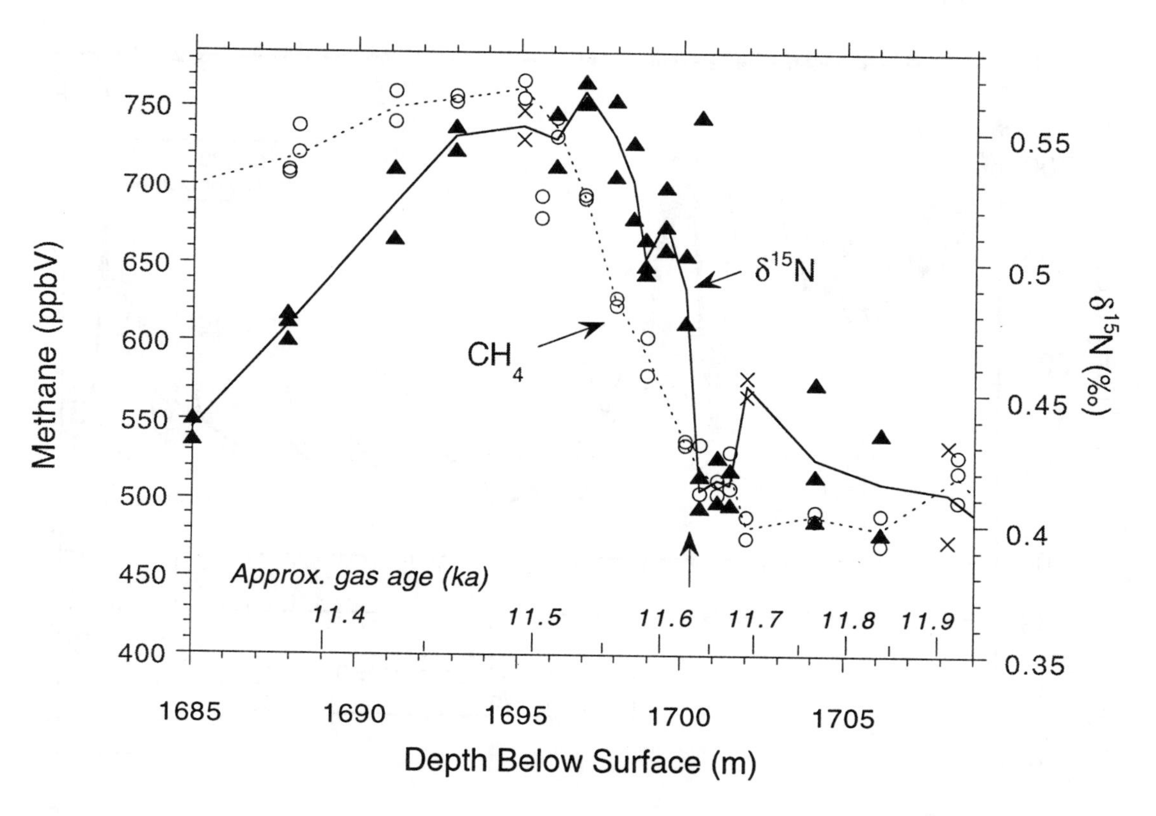

Figure 3. Methane (open circles) and $\delta^{15}N$ of N_2 (filled triangles) records for the Younger Dryas from GISP2 (from *Severinghaus et al.*, [1998]). The rise in $\delta^{15}N$ of N_2 (arrow) was caused by thermal diffusion fractionation due to the rapid warming of the surface of the firn at the end of the Younger Dryas. It precisely marks the time of the warming in the gas phase. The lag of the methane rise indicates a slight lag (0-30 years) in methane production relative to temperature increase in Greenland. This conclusion is independent of uncertainties in gas age-ice age difference. *See Severinghaus et al.* [1988] for further information.

mixing ratio would be recorded in the GISP2 record as an increase with an apparent duration of ~ 10 years, and that the *Thorpe et al.* [1996] scenario would produce maximum mixing ratios in GISP2 in excess of 1500 ppbV. To our knowledge, no published or unpublished ice core methane records illustrate effects like these. However, as *Thorpe et al.* suggested, if such an event were simply a one-time release of clathrate bound methane, elevated mixing ratios would be preserved over only 40-50 years of record, smaller than the sample spacing of most ice core methane records.

The data presented in Figure 2 are resolved to periods as short as 20-25 years preceding, during, and following rapid changes in methane, and do not reveal elevated mixing ratios or large mixing ratio changes on decadal time scales. In fact, the methane rise at the beginning of the events illustrated in Figure 2 took place over 100-300 years, and mixing ratios did not reach levels greater than about 750 ppbV. Modeling studies discussed previously indicate that mixing ratio changes on 100-300 year time scales, like those shown in Figure 2, would not be significantly smoothed by firn processes, and that faster changes, if they occurred, would be preserved (*Brook et al.*, in prep.).

There are at least two other arguments against the clathrate hypothesis and the role of clathrates in climate forcing. First, the IPG results do not indicate large sources in the high latitude northern hemisphere, as predicted for the clathrate release by *Nisbet* [1992]. The second argument is provided by *Severinghaus et al.* [1998] *and Severinghaus and Brook* [1998]. Their demonstration that methane change lagged temperature change at the end of the Younger Dryas and the beginning of the Bølling shows that methane was not forcing interstadial climate change at those times.

In summary, high resolution data show that the rate of change of methane mixing ratio at interstadial transitions was substantially slower than that which would accompany a rapid release of large quantities of methane from marine hydrates (the "clathrate gun" hypothesis of *Kennett et al.*, [1996]) and show no evidence of the elevated mixing ratios that would support this hypothesis. Of course, it is not possible to completely rule out smaller contributions from a clathrate source. A substantially slower, and smaller, clathrate release than stipulated by *Thorpe et al.* would be required to match the speed and magnitude of our data for the stadial-interstadial transitions. In addition, this source would have to either be maintained during the subsequent warm periods or fortuitously replaced by a wetland methane source immediately after the clathrate release. Maintenance of interstadial methane levels (for example during the

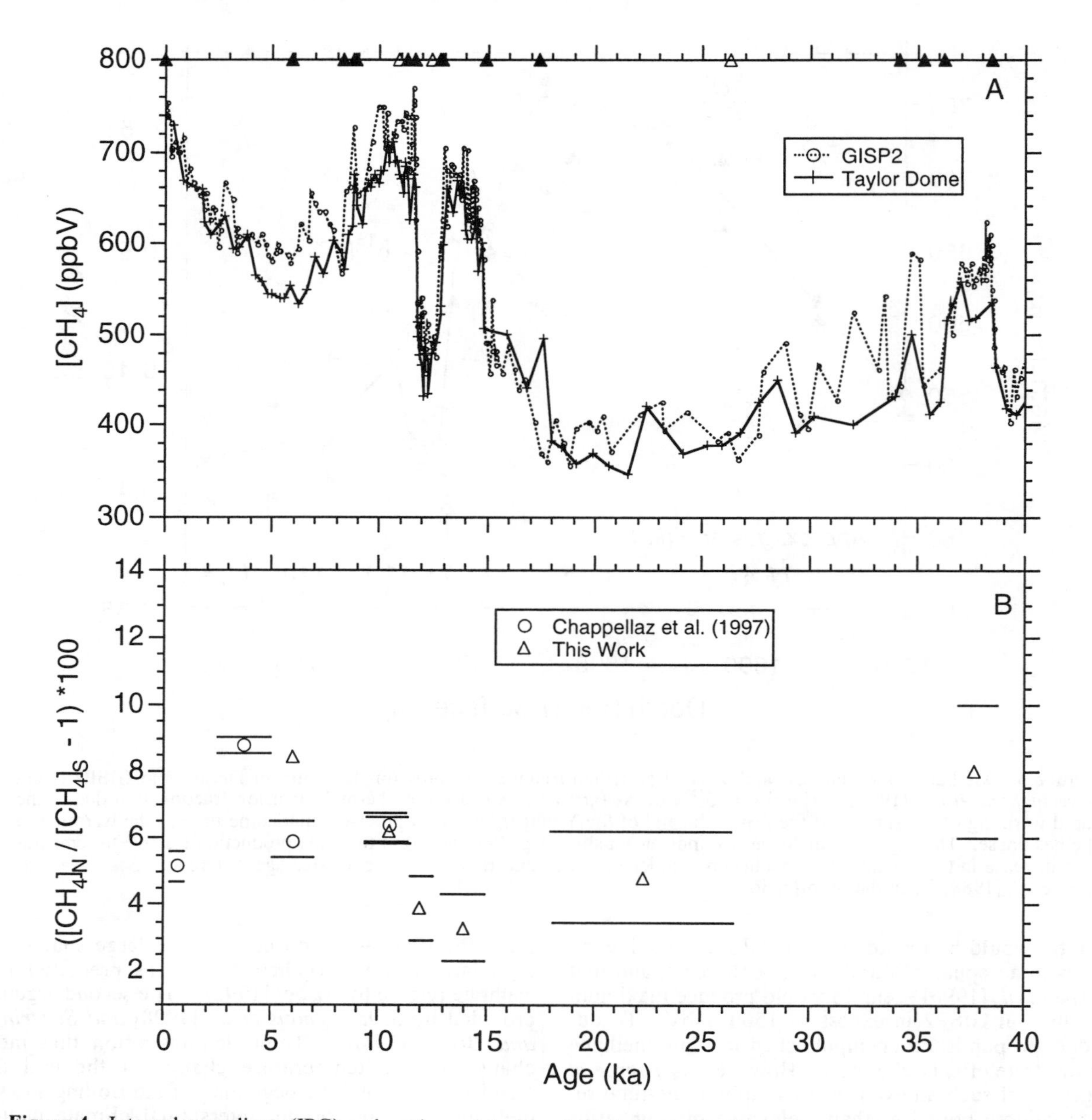

Figure 4. Interpolar gradient (IPG) and methane data for selected time periods from *Brook et al.* (in prep.) and *Chappellaz et al.* [1997]. The distance between the bars is the 95% confidence interval for the IPG. A) Methane data plotted are from the GISP2 (Greenland; see Figure 1) and Taylor Dome (Ross Sea region of Antarctica; *Brook et al.*, in prep) ice cores. Control points tying the GISP2 gas age chronology to the Taylor Dome gas age chronology are shown at the top of the figure. Open triangles = methane control points, filled triangles = $\delta^{18}O_{atm}$ control points. B) IPG results. IPG is defined as (C_G/C_A-1), where C is the time weighted average mixing ratio over the time period represented by the horizontal bars; G = Greenland, A = Antarctica. IPG data *from Chappellaz et al.* [1997] are based on data from the D47 and Byrd (Antarctica) and GRIP (Greenland) ice cores. IS#8 = Interstadial #8, LGM = last glacial maximum, B/A = Bølling-Allerød period, PB = Preboreal period, MH = mid Holocene. Data are insufficient to calculate uncertainties for MH results. Note that data and age control between 17 and 35 ka, and later than 6 ka, are sparse, and we do not attempt to calculate interpolar gradients for any portions of those time intervals.

Bølling/Allerød or preBoreal periods) by periodic large methane bursts can be ruled out because they would be detected in high resolution methane data [*Chappellaz et al.*, 1997]. Maintenance by a more constant source of methane from clathrate decomposition is still a possibility, but as discussed above, our interpolar gradient results still require dominance of tropical methane sources.

4.5 Implications of Millennial-Scale Changes in Methane Mixing Ratio

As discussed above, the interpolar gradient results (Figure 4; Table 1) indicate that changes in tropical methane sources are necessary to explain millennial-scale changes in the methane mixing ratio during the last glacial period and

Table 1. Methane emissions calculated from three box model using Taylor Dome and GISP2 measurements (Figure 4). Model is described in *Brook et al.* (in prep.) and *Chappellaz et al.* [1997].

Period	Time Interval (ka BP)	Methane Emissions[a] (Tg/yr) North	Tropics	South[b]	Total
Mid-Holocene	5-7	65±14	79±23	15	159
Preboreal	9.5-11.5	64±5	123±8	15	202
Younger-Dryas	11.5-12.5	39±6	86±11	15	141
Bølling-Allerød	13.5-14.8	43±7	127±15	15	185
LGM	18-26	34±6	65±9	12	111
Interstadial #8	37-39	63±7	80±11	15	158

[a]The error for the source estimate is the 95% confidence interval based on propagating errors in average mixing ratios. [b]Southern source is fixed in model (*Brook et al.*, in prep.).

the deglaciation. In all cases that we examined, tropical sources make up 50% or greater of the methane budget, and the rapid transitions always involved at least a 50% increase in the tropical methane source strength. This result verifies previous inferences about the close connection between millennial-scale climate change in Greenland and tropical climate variability [*Chappellaz et al.*, 1993; *Hughen et al.*, 1996; *Linsley et al.*, 1996; *Bard et al.*, 1997; *Schulz et al.*, 1998]. Variation in the strength of the tropical monsoon, inferred on millennial time scales from marine paleoclimate records in the Indian and Pacific Oceans (e.g., *Sirocko et al.*, [1996]; *Schulz et al.*, [1998]) is one likely contributor to the methane mixing ratio shifts, although we speculate that climate changes in the Amazon basin region may also have been important.

The relative sluggishness of the increase in methane mixing ratio at the initiation of interstadial events, compared to the speed of temperature changes inferred from the ice core record, suggests that methane production responded to climate change on a slower time scale. This delayed response may have been hydrologic - it may have taken one to two centuries for water tables in wetland regions to rise (and thereby create anoxic conditions) in response to changes in precipitation. Ecological factors may also have played a role - plant ecological shifts necessary to enhance or enlarge areas of methane emission may have required this longer time interval.

The inference that tropical methane emissions must have contributed significantly to methane changes on millennial time scales is relevant to proposed mechanisms for millennial-scale climate change. *Sowers and Bender* [1995], *Charles et al.* [1996], and *Blunier et al.* [1998] all suggested that millennial-scale climate oscillations in Antarctica and the southern ocean preceded similar events in Greenland by ~1000-2000 years. (We note, however, that *Steig et al.* [1998] suggested that the timing of Antarctic climate change may have been regionally variable, with some regions responding on a northern hemisphere time scale). *Charles et al.* speculated that tropical climate instability might have driven Antarctic and southern ocean climate variability, with the northern hemisphere responding later. The possible role of the tropics is also discussed further in this volume (see papers by Cane and Clement and Clement and Cane). However, the timing of changes in methane in the ice core record, combined with the interpolar gradient results described above, do not appear to support this hypothesis. Rather, these results indicate that methane variations appear to have a source in the tropics at all times, and that methane variations are closely coupled to processes with "northern hemisphere" timing [*Chappellaz et al.*, 1993; *Brook et al.*, 1996a; *Blunier et al.*, 1998; *Severinghaus et al.*, 1998; *Severinghaus and Brook*, 1998].

The lack of tropical methane response matching the time scale of Antarctic warming suggests that a different mechanism is necessary to explain the lag between Antarctic and Greenland climate change. Attention has been focused on oceanic mechanisms (for example: *Stocker et al.* [1992]; *Broecker* [1998]) that can explain the antiphase climate behavior between Greenland and Antarctica. Such mechanisms provide explanations for why cooling in Greenland and warming in Antarctica may have been coincident, but do not explain the link between tropical and high latitude northern hemisphere climate implied by the methane results. The methane results indicate that changes in tropical precipitation and temperature over land were coincident or nearly coincident with rapid temperature increases in Greenland, which are believed to be linked directly to changes in North Atlantic Deep Water formation. Furthermore, the constraints provided by $\delta^{15}N$ of N_2 [*Severinghaus et al.*, 1998; *Severinghaus and Brook*, 1998] show that, at least at the end of the Younger Dryas and start of the Bølling, the change in Greenland temperature preceded, by no more than 30 years, the associated rise in methane. This result suggests that a process directly linked to the high latitude northern hemisphere climate drove both the Greenland temperature change and the tropical climate change that contributed to increases in methane mixing ratios.

An adequate explanation for the tropical-high northern latitude connection most likely involves links between high latitude ocean circulation and aspects of the tropical hydrologic cycle, including tropospheric water vapor levels, the monsoon circulation, and the location of the Intertropical Convergence Zone. Such links are suggested by climate modeling studies that show large increases in water balance over tropical land areas in response to an increase in North Atlantic sea surface temperatures [*Fawcett et al.*, 1997; *Hostetler et al.*, in press]. Tropical surface temperatures in these and other models [*Manabe and Stouffer*, 1997] do not change substantially, but large precipitation changes alone could perhaps have led to increases in methane emissions. Forthcoming work on the phasing of temperature and methane change in the ice core record at other climate transitions, combined with more detailed models of climate and the factors influencing methane emissions, will be useful for understanding the nature of millennial-scale climate transitions.

Acknowledgements. We thank Melissa Swanson and Elizabeth Tuttle for expertly making some of the methane measurements presented here. Todd Sowers and Eric Steig contributed important data and comments on various aspects of this work. Richard Thorpe generously provided model calculations. The comments of two anonymous reviewers, and the hard work of the editors, strengthened the manuscript. This work was supported by grants from the Office of Polar Programs, National Science Foundation. Susan Harder and Jeff Severinghaus were also supported by NOAA Climate and Global Change Postdoctoral Fellowships.

REFERENCES

Alley, R. B., D. Meese, C. A. Shuman, A. J. Gow, K. Taylor, M. Ram, E. Waddington, J. W. C. White, and P. Mayewski, Abrupt accumulation increase at the Younger Dryas termination in the GISP2 ice core, *Nature*, *362*, 527-529, 1993.

Bard, E, F. Rostek, and C. Sonzongi, Interhemispheric synchrony of the last deglaciation inferred from alkenone palaeothermometry, *Nature*, *385*, 707-710, 1997.

Barnola, J-M., P. Pimienta, D. Raynaud and Y. S. Korotkevich, CO_2-climate relationship as deduced from the Vostok ice core: A reexamination based on new measurements and on a reevaluation of the air dating, *Tellus*, *B43*, 83-90, 1991.

Bartlett, K. B., and R. C. Harris, Review and assessment of methane emission from wetlands, *Chemosphere*, *26*, 261-320, 1993.

Bates, T. S., K. C. Kelly, J. E. Johnson, and R. H. Gammon, A reevaluation of the open ocean source of methane to the atmosphere, *J. Geophys. Res.*, *101*, 6,953-6,961, 1996.

Blunier, T., J. A. Chappellaz, J. Schwander, A. Dallenbach, B. Stauffer, T. F. Stocker, D. Raynaud, J. Jouzel, H. B. Clausen, C.U. Hammer, and S. J. Johnsen, Asynchrony of Antarctic and Greenland climate change during the last glacial period, *Nature*, *393*, 739-743, 1998.

Blunier, T., J. Chappellaz, J. Schwander, B. Stauffer, and D. Raynaud, Variations in atmospheric methane mixing ratio during the Holocene epoch, *Nature*, *374*, 46-49, 1995.

Broecker, W. S., Paleocean circulation during the last deglaciation: A bipolar seesaw?, *Paleoceanography*, *13*, 119-121, 1998.

Brook, E. J., J. P. Severinghaus, M. Swanson, and P. Grootes, Late glacial methane variations in Greenland and Antarctic ice cores, *EOS, Transactions of AGU*, *77*, S156, 1996b.

Brook, E. J., T. Sowers and J. Orchardo, Rapid variations in atmospheric methane concentration during the past 110,000 years, *Science*, *273*, 1087-1091, 1996a.

Bubier, J., and T. R. Moore, An ecological perspective on methane emissions from northern wetlands, *Trends in Ecology and Evolution*, *9*, 460-465, 1994.

Chappellaz, J. A, J-M. Barnola, D. Raynaud, Y. S. Korotkevich, and C. Lorius, Atmospheric methane record over the last climatic cycle revealed by the Vostok ice core, *Nature*, *345*, 127-131, 1990.

Chappellaz, J. A., I. Y. Fung and A. M. Thompson, The atmospheric methane increase since the last glacial maximum (1) Source estimates, *Tellus Series B*, *45b*, 228-241, 1993b.

Chappellaz, J. A., T. Blunier, D. Raynaud, J-M. Barnola, J. Schwander and B. Stauffer, Synchronous changes in atmospheric methane and Greenland climate between 40 and 8 kyr BP, *Nature*, *366*, 443-445, 1993a.

Chappellaz, J., T. Blunier, S. Kints, B. Stauffer, and D. Raynaud, Variations of the Greenland/Antarctic mixing ratio difference in atmospheric methane during the last 11,000 years, *J. Geophys. Res.*, *102*, 15,987-15,997, 1997.

Charles, C. D., J. Lynch-Steiglitz, U. S. Ninnemann, R. G. Fairbanks, Climate connections between the hemispheres revealed by deep sea sediment core/ice core correlations, *Earth Planet. Sci. Lett.*, *142*, 19-27, 1996.

Fawcett, P. J., A. M. Agustsdottir, R. B. Alley, and C. A. Shuman, 1997, The Younger Dryas termination and North Atlantic deep water formation: insights from climate model simulations and Greenland ice cores, *Paleoceanography*, *12*, 23-38, 1997.

Fung, I., J. John, J. Lerner, E. Matthews, M. Prather, L. P. Steele, and P. J. Fraser, Three-dimensional model synthesis of the global methane cycle, *J. Geophys. Res.*, *96*, 13,033-13,065, 1991.

Grootes, P. M. and E. J. Steig, Taylor Dome ice core study 1993/1994: An ice core to bedrock, *Antarctic J. U.S.*, *29*, 79-81, 1994.

Grootes, P. M., M. Stuiver, J. W. C. White, S. Johnsen, and J. Jouzel, Comparison of oxygen isotope records from the GISP2 and GRIP Greenland ice cores, *Nature*, *366*, 552-554, 1993.

Hein, R., P. J. Crutzen, and M. Heimann, An inverse modeling approach to investigate the global atmospheric methane cycle, *Global Biogeochem. Cycles*, *11*, 43-76, 1997.

Herron, M. M. and C. C. Langway, Firn densification: an empirical model, *J. Glaciology*, *25*, 373-385, 1980.

Hostetler, S. W., P. U. Clark, P. J. Bartlein, A. C. Mix, and N. G. Pisias, Atmospheric transmission of North Atlantic Heinrich events, *J. Geophys. Res.*, in press, 1999.

Hughen, K., J. Overpeck, L. Peterson, and S. Trumbore, Rapid climate changes in the tropical Atlantic region during the last deglaciation, *Nature*, *380*, 51-54, 1996.

Jouzel, J., N. I. Barkov, J-M. Barnola, M. Bender, J. Chappellaz, C. Genthon, V. M. Kotlyakov, V. Lipenkov, C. Lorius, J. R. Petit, D. Raynaud, G. Raisbeck, C. Ritz, T. Sowers, M. Stievenard, F. Yiou, and P. Yiou, Extending the Vostok ice-core record of paleoclimate to the penultimate glacial period, *Nature*, *364*, 407-412, 1993.

Kennett, J. P., L. L. Hendy, and R. J. Behl, Late Quaternary foraminiferal carbon isotope record in Santa Barbara Basin: implications for rapid climate change, *EOS, Transactions of AGU*, *77*, F294, 1996.

Kvenvolden, K. A., Methane hydrates-a major reserve of carbon in the shallow geosphere?, *Chemical Geology*, *7*, 41-51, 1988.

Linsley, B., Oxygen isotope record of sea level and climate variations in the Sulu Sea over the past 150,000 years, *Nature*, *380*, 234-237, 1996.

Manabe, S., and R. J. Stouffer, Coupled ocean-atmosphere model response to freshwater input: comparison to Younger Dryas event, *Paleoceanography*, *12*, 321-336, 1997.

Meese, D., R. Alley, T. Gow, P. Grootes, P. Mayewski, M. Ram, K. Taylor, E. Waddington, and G. Zielinski, Preliminary depth-age scale of the GISP2 ice core, *CRREL Spec. Rep. 94-1*, Cold Reg. Res. and Eng. Lab., Hanover, N. H., 1994.

Nakazawa, T., T. Machida, M. Tanaka, Y. Fujii, S. Aoki, and O. Watanabe, Differences of the atmospheric CH_4 concentration between the Arctic and the Antarctic regions in pre-industrial/pre-agricultural era, *Geophys. Res. Lett*, *20*, 943-946, 1993.

Nisbet, E. G., Sources of atmospheric methane in early postglacial time, *J. Geophys. Res*, *97*, 12,859-12,867, 1992.

Prinn, R. G., R. F. Weiss, B. R. Miller, J. Huang, F. N. Alyea, D. M. Cunnold, P. J. Fraser, D. E. Hartley, and P. G. Symonds, Atmospheric trends and lifetime of CH_3CCl_3 and global OH mixing ratios, *Science*, *269*, 187-192, 1995.

Rasmussen, R. A., and M. A. K. Khalil, Atmospheric methane in recent and ancient atmospheres: mixing ratios, trends, and interhemispheric gradient, *J. Geophys. Res.*, *89*, 11,599-11,605, 1984.

Raynaud, D., J. Chappellaz, J-M. Barnola, Y. S. Korotkevich, and C. Lorius, Climatic and methane cycle implications of glacial-interglacial methane change in the Vostok ice core, *Nature*, *333*, 655-657, 1988.

Sachs, J. P. and S. J. Lehman, Large century-millennial scale sea surface temperature fluctuations in the subtropical NW Atlantic during marine isotope stage 3, *EOS, Transactions of AGU*, *79*, F471.

Schlesinger, W.H., *Biogeochemistry: an analysis of global change*, Academic Press, 588 p., 1996.

Schulz, H., U. von Rad, and H. Erlenkeuser, Correlation between

Arabian Sea and Greenland climate oscillations of the past 110,000 years, *Nature, 393,* 54-57, 1998.

Schwander, J., J-M. Barnola, C. Andrié, M. Leunberger, A. Ludin, D. Raynaud, and B. Stauffer, The age of the air in the firn and the ice at Summit, Greenland, *J. Geophys. Res, 98*, 2,831-2,838, 1993.

Schwander, J., T. Sowers, J-M. Barnola, T. Blunier, A. Fuchs and B. Malaize, Age scale of the air in the summit ice: Implication for glacial-interglacial temperature change, *J. Geophys. Res. 102*, 19,483-19,4993, 1997

Severinghaus, J. and E. J. Brook, Simultaneous tropical-arctic abrupt climate change at the Bølling transition, *EOS, Transactions of AGU, 79*, F454, 1998.

Severinghaus, J. P., T. Sowers, E. J. Brook, R. B. Alley, and M. L. Bender, Timing of abrupt climate change at the end of the Younger Dryas interval from thermally fractionated gases in polar ice, *Nature, 391*, 141-146, 1998.

Shine, K. G., R. G. Derwent, D. J. Wuebbles, and J-J. Morcrette, Radiative forcing of climate change, in *Climate Change: The IPCC Scientific Assessment* , edited by J. T. Houghton, G.J. Jenkins, and J. J. Ephraums, pp. 41-68, Cambridge University Press, Cambridge, U.K., 1990.

Sirocko, F., D. Garbe-Schonberg, A. McIntyre, and B. Molfino, Teleconnections between the subtropical monsoons and high latitude climates during the last deglaciation, *Science, 272*, 526-529, 1996.

Sowers, T. A. and M. Bender, Climate records during the last deglaciation, *Science, 269,* 210-214, 1995.

Sowers, T. A., M. Bender, D. Raynaud, and Y. S. Korotkevich, $\delta^{15}N$ of N_2 in air trapped in polar ice cores: a tracer of gas transport in the firn and possible constraint on ice age-gas age differences, *J. Geophys. Res., 97*, 15,683-15,697, 1991.

Sowers, T., E. Brook, D. Etheridge, T. Blunier, A. Fuchs, M. Luenberger, J. Chappellaz, J.-M. Barnola, M. Whalen, B. Deck, and C. Weyhenmeyer, An interlaboratory comparison of techniques for extracting and analyzing trapped gases in ice cores, *J. Geophys. Res., 102*, 26,527-26,538, 1997.

Stauffer, B., E. Lochbronner, H. Oeschger, and J. Schwander, Methane mixing ratio in the glacial atmosphere was only half that of the preindustrial Holocene, *Nature, 332*, 812-814, 1988.

Steig, E. J., E. J. Brook, J. W. C. White, C. Sucher, S. J. Lehman, M. L. Bender, D. L. Morse, and E. D. Waddington, Synchronous climate changes in Antarctica and the North Atlantic during the last glacial-interglacial transition, *Science, 282,* 92-95, 1998.

Steig, E.J., Beryllium-10 in the Taylor Dome ice core: applications to Antarctic glaciology and paleoclimatology, Ph.D. thesis, University of Washington, 1996.

Stocker, T. F., D. G. Wright, and W. S. Broecker, The influence of high-latitude surface forcing on the global thermohaline circulation, *Paleoceanography, 7*, 529-541, 1992.

Stuiver, M., P. M. Grootes, and T. F. Braziunas, The GISP2 $\delta^{18}O$ climate record of the past 16,500 years and the role of the sun, ocean and volcanoes, *Quat. Res., 44*, 341-354, 1995.

Sucher, C., Atmospheric gases in the Taylor Dome Ice Core: implications for East Antarctic climate change, M.S. Thesis, University of Rhode Island, Narragansett, RI, 1997.

Thompson, A., The oxidizing capacity of the Earth's atmosphere: probable past and future changes, *Science, 256,* 1,157-1,165, 1992.

Thorpe, R. B., K. S. Law, S. Bekki, J. A. Pyle, and E. G. Nisbet, Is methane-driven deglaciation consistent with the ice core record?, *J. Geophys, Res. 101*, 28,627-28,635, 1996.

Whiting, G. J and J. P. Chanton, Primary production control of methane emission from wetlands, *Nature, 364*, 794-795, 1993.

Michael Bender, Department of Geosciences, Princeton University, Princeton NJ, 08544.

Edward J. Brook and Susan Harder, Geology Department, Washington State University, 14204 NE Salmon Creek Ave, Vancouver, WA 98686.

Jeff Severinghaus, Scripps Institution of Oceanography, University of California at San Diego, La Jolla, CA, 92093.

Freshwater Routing by the Laurentide Ice Sheet During the Last Deglaciation

Joseph M. Licciardi

Department of Geosciences, Oregon State University, Corvallis

James T. Teller

Department of Geological Sciences, University of Manitoba, Winnipeg, Canada

Peter U. Clark

Department of Geosciences, Oregon State University, Corvallis

Using new reconstructions of the Laurentide Ice Sheet and estimates of glacial-age precipitation suggested by a general circulation model, we calculated the freshwater fluxes derived from meltwater and precipitation runoff from North America to the North Atlantic and Arctic Oceans during the last deglaciation. Additional fluxes from iceberg-discharge events such as Heinrich events are not included, but would have supplemented fluxes calculated here. Meltwater plus precipitation runoff from each of 19 cryohydrological basins is routed to the oceans through one of five main injection sites: the Mississippi River to the Gulf of Mexico, the Hudson River to the North Atlantic, the St. Lawrence River to the North Atlantic, the Mackenzie and other Arctic rivers to the Arctic Ocean, and Hudson Strait to the Labrador Sea. We present new time series of freshwater fluxes to each of these injection sites for the last deglaciation. Our results indicate that large and rapid changes in freshwater flux to individual injection sites occurred as ice-margin fluctuations redirected runoff from one site to another. Although these abrupt rerouting events are not associated with a change in the total freshwater flux draining from the continent, they caused significant geographic changes in the flux entering the North Atlantic and Arctic Oceans during the last deglaciation which are of the same magnitude suggested by climate models to affect the rate of North Atlantic Deep Water formation, thus identifying routing events as a potentially important mechanism of abrupt climate change during the last deglaciation.

INTRODUCTION

Abrupt century- to millennial-scale climate change in the North Atlantic region during the last glaciation has been linked to changes in the formation of North Atlantic Deep Water (NADW) [*Boyle and Keigwin*, 1982, 1987;

Mechanisms of Global Climate Change at Millennial Time Scales
Geophysical Monograph 112

Broecker et al., 1985; *Keigwin et al.*, 1991; *Lehman and Keigwin*, 1992; *Oppo and Lehman*, 1995]. Similar climate changes are identified at sites far from the North Atlantic region [*Behl and Kennett*, 1996; *Benson et al.*, 1997; *Schulz et al.*, 1998], suggesting transmission of North Atlantic climate change elsewhere, a common forcing mechanism, or both [*Broecker*, 1994; *Bond and Lotti*, 1995; *Hostetler et al.*, 1999].

Changes in the freshwater flux to the North Atlantic may cause climate change on a global scale. Temperature and salinity determine the density of ocean water, and changes in these properties drive the thermohaline circulation (THC) responsible for the formation of NADW. In particular, ocean modeling indicates that the formation rate of NADW is sensitive to changes in the amount as well as the site of freshwater forcing, insofar as it affects the salinity structure of the ocean. An increase in the freshwater flux to the North Atlantic amplifies the sensitivity of NADW formation to changes in freshwater fluxes, and may result in a change in the site and depth of deep water formation, a complete shut down, or in inherently unstable behavior [*Manabe and Stouffer*, 1988, 1997; *Stocker and Wright*, 1991, 1996; *Rahmstorf*, 1994, 1995a; *Fanning and Weaver*, 1997; *Mikolajewicz et al.*, 1997; *Schiller et al.*, 1997; *Tziperman*, 1997]. Models also suggest a greater response of NADW formation to "surgical" or targeted injections of freshwater from high-latitude sources near sites of deep water formation in comparison to a more subdued response to low-latitude injections [*Maier-Reimer and Mikolajewicz*, 1989; *Rahmstorf*, 1995b; *Fanning and Weaver*, 1997; *Manabe and Stouffer*, 1997]. A decrease in the formation rate of NADW causes cooling in the North Atlantic region, and is transmitted as climate change elsewhere through the atmosphere [*Rind et al.*, 1986; *Fawcett et al.*, 1997; *Hostetler et al.*, 1999] and the ocean [*Manabe and Stouffer*, 1988, 1997; *Stocker and Wright*, 1991, 1996; *Rahmstorf*, 1995a; *Mikolajewicz et al.*, 1997].

During interglaciations, changes in the freshwater flux to the North Atlantic may be caused by changes in the hydrological cycle [*Manabe and Stouffer*, 1994; *Stocker and Schmittner*, 1997] or by export of Arctic Ocean sea ice [e.g., *Aagaard and Carmack*, 1989; *Mysak et al.*, 1990, *Häkkinen*, 1993; *Bond et al.*, 1997]. During glaciations, freshwater fluxes from circum-North Atlantic ice sheets varied on millennial timescales either by changes in melting or calving rates at an ice margin [e.g., *Bond and Lotti*, 1995] or by changes in the routing of freshwater to the North Atlantic and Arctic Oceans [*Johnson and McClure*, 1976; *Broecker et al.*, 1988, 1989; *Teller*, 1990a, 1990b].

In this paper, we reconstruct the history of freshwater routing from the North American continent to the North Atlantic and Arctic Oceans during the last deglaciation. Building on earlier work by *Teller* [1990a, 1990b], we quantify the freshwater runoff from 19 cryohydrological basins (CHBs) (Figure 1) derived from the melting of the Laurentide Ice Sheet (LIS) and from precipitation. Runoff is routed to the North Atlantic and Arctic Oceans through one of five drainage pathways or "injection sites" (Figure 1). Our results identify times and amounts of abrupt changes in the freshwater flux through individual injection sites as the fluctuating ice margin opened or closed basin outlets, rerouting freshwater from one or more CHBs to a different injection site.

METHODS

Defining Cryohydrological Basins

During glaciations, the presence of the LIS dramatically altered modern drainage basins in North America. "Glacial" drainage basins changed in size and configuration in response to the changing extent and surface morphology of the LIS, and runoff was redirected from one route to another as ice-margin fluctuations and glacial isostatic recovery opened or closed outlets.

In order to calculate the changing volume of meltwater and precipitation runoff, we defined a number of basins, which we refer to as cryohydrological basins (CHBs) (Figure 1, Table 1). At the last glacial maximum, all of these basins were covered or partly covered by ice. As the LIS melted, fewer and fewer of these CHBs contained ice, although most continued to have meltwater flow through them en route to the oceans. Because runoff from the expanding region beyond the edge of the LIS was mainly collected in drainage basins defined by modern topography, we used major modern drainage divides to define many of the boundaries of the CHBs (A to L, Figure 1); knowledge of the history of deglaciation was also used in establishing these basins, although practical considerations limited the actual number of basins defined.

Our definitions of CHBs A to L are listed in Table 1 and depicted in Figure 1. The presence of major ice divides resulted in most CHBs being further subdivided, each with a time-varying ice divide forming part of its boundary during at least part of its history, and resulting in a maximum of 19 CHBs at any one time. We denote these subdivisions by the subscripts "n" and "s" that indicate northern and southern sectors of a given CHB (e.g., H_n and H_s). Many of these CHBs are similar in outline to those

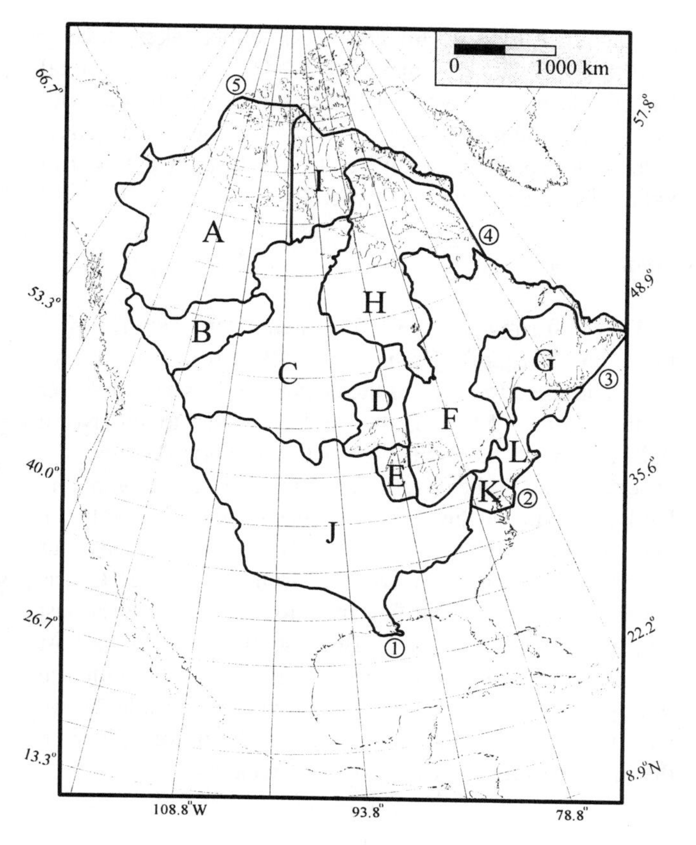

Figure 1. Major cryohydrological basins (A-L) used to partition runoff from the continent to one of five injections sites: (1) Mississippi River, (2) Hudson River, (3) St. Lawrence River, (4) Hudson Strait, and (5) the Arctic Ocean. Basins are subdivided through the last deglaciation depending on the location of the ice divides on the Laurentide Ice Sheet, resulting in up to 19 subbasins at any one time (Table 2).

defined by *Teller* [1990a, b] except that he used a fixed ice divide during deglaciation.

Estimating Runoff from Ice Melt

We used numerical reconstructions of the LIS during the last deglaciation ["maximum" reconstructions of *Licciardi et al.*, 1998] to calculate the contribution of ice volume loss to total runoff during the last deglaciation. This ice volume component, which we term "meltwater runoff," is thus equivalent to the sea level contribution from the LIS during the last deglaciation [*Licciardi et al.*, 1998] and includes ice volume loss from some unspecified combination of melting and calving. Additional fluxes from non-steady-state iceberg-discharge events such as Heinrich events are not included, but would have supplemented fluxes calculated here. We discuss the possible influence of Heinrich events as an additional source of freshwater in a later section.

Our reconstructions include the effect of a soft, deformable bed on ice flow, which results in less ice volume than in reconstructions modeled by *Hughes* [1987] and used by *Teller* [1990a, 1990b] for earlier calculations of meltwater runoff. Ice sheet reconstructions are for 18, 14, 13, 12, 11, 10, 9, 8.4, and 8 ^{14}C ka, equivalent to 21.4, 16.8, 15.6, 14.1, 13.0, 11.4, 10.2, 9.5, and 8.9 cal ka using the Calib 4.0 calibration program [*Stuiver and Reimer*, 1993; *Stuiver et al.*, 1998]. Meltwater runoff for the interval between two successive reconstructions is computed by subtracting the ice volume (in water equivalent) within each CHB for any given reconstruction from the ice volume within the same CHB for the next older reconstruction.

Estimating Runoff from Precipitation

Because runoff from part of the North American continent is not currently monitored, and because even the monitored basins had their runoff divided in varying ways between injection sites during deglaciation, we used modern precipitation as a proxy for modern runoff from the CHBs, with the exception of the Mississippi drainage. We assumed that infiltration would have been approximately balanced by groundwater discharge over long periods of time such as those studied here. Using annual precipitation as the volume of runoff from a basin does not take into account the loss by evaporation or transpiration, but subtracting evapotranspiration as *Teller* [1990b] did overestimates the loss of water from the land and glacier surface during late glacial time. Comparisons of monitored runoff with our calculated volume of modern precipitation falling over two large areas, the Mackenzie River basin and the Great Lakes-upper St. Lawrence Valley region, suggest that our calculated annual precipitation overestimates measured runoff in these basins by up to a factor of about 2. Comparisons of modern precipitation isohyets with evapotranspiration isohyets [*Fisheries and Environment Canada*, 1978] show that the loss by evapotranspiration ranges from 100% in the Canadian Prairies (i.e., no runoff) to ~30% in the wetter areas of eastern Canada.

We superposed annual precipitation isohyets, derived from the *WMO* [1979] climate atlas, onto maps of the CHBs. Because the isohyet contour interval published by *WMO* is at best 100 mm, we interpolated isohyets at 25 or 50 mm intervals in order to define more precisely

Table 1. Definition of Each CHB

CHB	Definition
A	Mackenzie River basin plus much of the high Arctic region to its east that drains mainly into the Arctic Ocean
B	headwaters of Mackenzie River
C	drainage into western Hudson Bay
D	Lake Superior Basin and adjacent area to the north that drains into Hudson Bay
E	Lake Michigan Basin
F	eastern Great Lakes, upper St. Lawrence Valley, drainage into eastern Hudson Bay, and adjacent drainage into Hudson Strait and Labrador Sea
G	lower St. Lawrence Valley and Gulf of St. Lawrence drainage
H	Hudson Bay and Foxe Basin
I	high Arctic region in the vicinity of the Gulf of Boothia and the northern slope of Baffin Island that drains into the Arctic Ocean and Baffin Bay
J	Mississippi River Basin
K	mid-Atlantic states' drainage from glaciated regions
L	Hudson River, Mohawk River, New England, and southern Nova Scotia drainage

precipitation in the CHBs. We then integrated the total annual volume of water falling as precipitation in each CHB.

Changes in (de)glacial-age boundary conditions (e.g., ice sheets, sea surface temperatures, vegetation, CO_2) would have modified precipitation patterns. Experiments with general circulation models (GCMs) simulate global climate under different boundary conditions, but most GCM experiments relevant to this study are only for the last glacial maximum (LGM = 21 cal ka). *Kutzbach et al.* [1998], however, conducted experiments with the Community Climate Model 1 (CCM1) GCM for the LGM as well as for several times during the last deglaciation (16, 14, 11, and 6 cal ka), thus providing a more complete temporal representation of the effects of changes in boundary conditions on global climate during the last deglaciation. In order to account for (de)glacial changes in precipitation, we calculated precipitation runoff (F_P) for each CHB at any particular time during the last deglaciation by using CCM1 anomalies in precipitation minus evaporation (P-E) to adjust our estimates of modern runoff as:

$$F_P = R_{MOD} + (CCM1_{GL} - CCM1_{CTRL})$$

where R_{MOD} is our proxy of modern runoff (i.e., modern precipitation) within each CHB, $CCM1_{GL}$ is the integrated glacial-age P-E value for each CHB either taken directly from the CCM1 experiments at 21, 16, 14, 11, and 6 cal ka, or interpolated between these times to coincide with the time intervals we use from our ice sheet reconstructions (see above), and $CCM1_{CTRL}$ is the control value of P-E integrated over each CHB.

Total Runoff

Total freshwater runoff (meltwater runoff plus precipitation runoff) from each CHB is routed to one of five main injection sites to the ocean: the Mississippi River to the Gulf of Mexico, the Hudson River to the North Atlantic, the St. Lawrence River to the North Atlantic, the Mackenzie and other Arctic rivers to the Arctic Ocean, and Hudson Strait to the Labrador Sea (Figure 1, Table 1). Our routing scheme (Table 2) is based on our understanding of how runoff from each CHB was allocated to each of the five injection sites during the last deglaciation (see below). Because the timing of most changes in routing does not always coincide with the timing of our ice sheet reconstructions (e.g., Table 2), we partitioned total runoff from each CHB to coincide with the time boundaries of each routing event as:

$$F_{T(t1)} = (X\%)(F_{T(t1)}) + (Y\%)(F_{T(t1)}) + ...$$

where $F_{T(t1)}$ is the total runoff from a CHB in a calendar-year time interval (t1) defined by our ice sheet reconstructions, X% is the percentage of time in that ice-sheet interval during which runoff from that CHB is routed to one drainage, Y% is the percentage of that interval during which runoff is routed to a different drainage, and so on. Runoff was calculated in $km^3\ yr^{-1}$ and then converted to sverdrups ($1\ Sv = 10^6\ m^3\ s^{-1}$) (Appendix A).

Our results indicate that the precipitation contribution to total runoff was greater than the meltwater contribution (Figure 2). CCM1 simulations suggest that precipitation was lower over the LIS during the last glacial maximum and throughout the last deglaciation [*Kutzbach et al.*, 1998]. Lowered temperatures over the ice sheet, however, caused a reduction in evaporation that dominated over the reduction in precipitation, resulting in small increases in P-E [*Kutzbach et al.*, 1998]. Precipitation runoff was thus greater early during deglaciation, but decreased as the size of the LIS, and its effect on evaporation, decreased (Figure 2). In contrast, our ice-sheet reconstructions [*Licciardi et al.*, 1998] suggest that an increase in meltwater runoff offset the decrease in precipitation runoff, resulting in a nearly constant freshwater flux from total runoff throughout much of the deglaciation until the final demise of the LIS (Figure 2). Two maxima in total runoff

Table 2. Routing History for Each CHB

Interval (^{14}C ka)	Mississippi River	St. Lawrence River	Hudson River	Hudson Strait	Arctic Ocean
18.0 - 16.5	$B_s+C_s+D_s+E+F_s+H_s+J$	$F_s'+G$	$K+L$	$A_n+B_n+C_n+D_n+F_n+H_n+I$	A_s
16.5 - 15.2	$B_s+C_s+D_s+E+J$	$F_s'+G$	F_s+H_s+K+L	$A_n+B_n+C_n+D_n+F_n+H_n+I$	A_s
15.2 - 14.0	$B_s+C_s+D_s+E+F_s+H_s+J$	$F_s'+G$	$K+L$	$A_n+B_n+C_n+D_n+F_n+H_n+I$	A_s
14.0 - 13.5	$B+C_s+D_s+E+F_s+H_s+J$	$F_s'+G$	$K+L$	$A_n+C_n+D_n+F_n+H_n+I$	A_s
13.5 - 13.0	$B+C_s+J$	$F_s'+G$	$D_s+E+F_s+H_s+K+L$	$A_n+C_n+D_n+F_n+H_n+I$	A_s
13.0 - 12.3	$B+C_s+D_s+E+F_s+H_s+J$	$F_s'+G$	$K+L$	$A_n+C_n+D_n+F_n+H_n+I$	A_s
12.3 - 12.0	$B+C_s+D_s+E+J$	$F_s'+G$	F_s+H_s+K+L	$A_n+C_n+D_n+F_n+H_n+I$	A_s
12.0 - 11.7	$B+C_s+J$	$F_s'+G$	$D+E+F_s+H_s+K+L$	$A_n+C_n+F_n+H_n+I_n$	A_s+I_s
11.7 - 11.5	$B+C_s+D+E+J$	$F_s'+G$	F_s+H_s+K+L	$A_n+C_n+F_n+H_n+I_n$	A_s+I_s
11.5 - 11.0	$B+C_s+D+E+J$	$F_s+F_s'+G+H_s$	$K+L$	$A_n+C_n+F_n+H_n+I_n$	A_s+I_s
11.0 - 10.0	J	$C_s+D+E+F_s+F_s'+G+H_s$	$K+L$	$A_n+C_n+F_n+H_n+I_n$	A_s+B+I_s
10.0 - 9.7	J	$D+E+F_s+G+H_s$	$K+L$	$C_n+F_n+H_n+I_n$	$A+B+C_s+I_s$
9.7 - 9.4	C_s+J	$D+E+F_s+G+H_s$	$K+L$	$C_n+F_n+H_n+I_n$	$A+B+I_s$
9.4 - 9.3	J	$C_s+D+E+F_s+G+H_s$	$K+L$	$C_n+F_n+H_n+I_n$	$A+B+I_s$
9.3 - 9.1	C_s+J	$D+E+F_s+G+H_s$	$K+L$	$C_n+F_n+H_n+I_n$	$A+B+I_s$
9.1 - 8.4	J	$C_s+D+E+F_s+G+H_s$	$K+L$	$C_n+F_n+H_n+I_n$	$A+B+I_s$
8.4 - 7.7	J	$C_s+D+E+F_s+G+H_s$	$K+L$	$C_n+F_n+H_n+I$	$A+B$
7.7 - 7.0	J	D_s+E+F_s+G	$K+L$	$C+D_n+F_n+H+I$	$A+B$
7.0 - 6.0	J	D_s+E+F_s+G	$K+L$	$C+D_n+F_n+H+I$	$A+B$
6.0 - 0.0	J	D_s+E+F_s+G	$K+L$	$C+D_n+F_n+H+I$	$A+B$

Note: subbasin F_S' is a small portion of subbasin F_S that was routed independently of that basin during times when an ice divide separated it from the main area of F_S.

occurred between 13 - 12 ^{14}C ka (15.6 - 14.1 cal ka) (0.31 Sv) and 10 - 8.4 ^{14}C ka (11.4 - 9.5 cal ka) (0.31 Sv) against a backdrop of a nearly constant freshwater flux of 0.28 - 0.29 Sv (Figure 2).

ROUTING HISTORY

Background

Our routing scheme (Table 2) is based on those times during the last deglaciation when runoff from any one of the CHBs was redirected from one injection site to another. Total runoff generated in any one CHB will flow to an injection site through the lowest-elevation drainage route available. Today, those routes are determined solely by continental topography, but during the last glaciation, the LIS covered many modern routes, thus forcing total runoff to flow through alternative, higher-elevation routes, which may have directed runoff to a different injection site. Because retreat of the LIS margin was mainly downslope, it uncovered successively lower elevation routes which, at certain times, resulted in rerouting to a different injection site. However, because the LIS margin readvanced several times during the last deglaciation, those lower routes were in some cases blocked again, causing runoff to be rerouted back toward the previous injection site. This complex history of routing changes associated with the fluctuating LIS margin resulted in century- to millennial-scale changes in the freshwater fluxes through each of the five injection sites.

Most of the routing events we identify (Figure 3) have long been recognized as significant events in the deglaciation of the LIS [e.g., *Prest*, 1970; *Dreimanis and Karrow*, 1972]. The deglacial history of the LIS has been studied for over a century, and is now relatively well known. Since the advent of radiocarbon dating in the 1950's, several hundred radiocarbon ages have been generated which constrain the timing of deglaciation events often to the resolution of the dating method [see reviews in *Fullerton*, 1980; *Mayewski et al.*, 1981; *Clayton and Moran*, 1982; *Mickelson et al.*, 1983; *Dyke and Prest*, 1987; *Teller*, 1995a]. These studies revealed that the deglaciation of many parts of the southern LIS ice margin were characterized by rapid and large (order of 100's of km) century- to millennial-scale oscillations of the ice margin superimposed on overall retreat [*Dreimanis*, 1977;

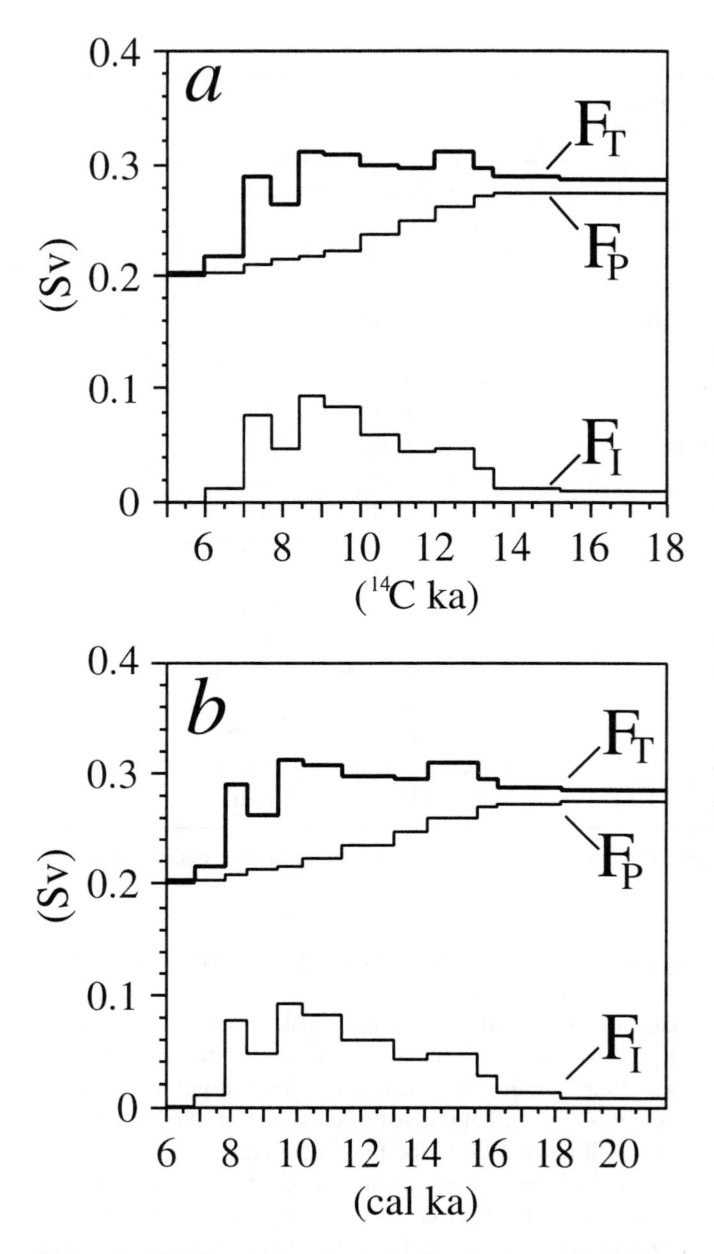

Figure 2. (A) Time series of our calculated values of freshwater fluxes (in Sv) during the last deglaciation on a radiocarbon timescale, including total runoff (F_T), precipitation runoff (F_P), and meltwater runoff (F_I). (B) Same as for (A), but on a calendar year timescale.

Clayton et al., 1985; *Hansel and Johnson*, 1992; *Clark*, 1994]. Many of these oscillations are identified regionally and have been formally defined as stades and interstades [*Dreimanis and Karrow*, 1972] or as phases [*Attig et al.*, 1985; *Johnson et al.*, 1997] of the last glaciation.

Our results suggest that the most significant variability in freshwater fluxes due to routing changes involved the southern (Mississippi River) and eastern (Hudson and St. Lawrence Rivers) injection sites associated with the large oscillations of the southern LIS margin. Retreat of the ice margin to open an eastern routing increased the freshwater flux through that outlet at the expense of the flux through the Mississippi River, whereas readvance of the ice margin had the opposite effect, resulting in an antiphase relationship between flux changes through eastern and southern injection sites. Significant routing changes affecting the more northerly injection sites (Mackenzie River, Hudson Strait) only occurred late in the deglaciation (<10 ^{14}C ka) associated with final collapse of the LIS over Hudson Bay.. Otherwise, neither geography nor ice-margin dynamics led to large and abrupt routing changes anywhere around the LIS except along its southern margin. Much of our discussion of routing history thus focuses on events along the southern LIS margin affecting the southern and eastern injection sites. An important implication of this result is that freshwater routing as a mechanism of abrupt climate change from North America was most common only during those times when the southern LIS margin had advanced south into the Great Lakes-St. Lawrence Valley region, where it had the potential to affect the interplay of routing through the Mississippi, Hudson, and St. Lawrence Rivers. The rerouting of Eurasian drainage during the last deglaciation, which represents an additional mechanism for changes in freshwater flux to the North Atlantic, is discussed in a later section.

The timing of routing events has also been interpreted from $\delta^{18}O$ records of planktonic foraminifera in marine cores recovered from sites that were located within the influence of freshwater fluxes entering the oceans through one of the injection sites (e.g., Gulf of Mexico, Gulf of St. Lawrence) [*Kennett and Shackleton*, 1975; *Leventer et al.*, 1982; *Broecker et al.*, 1988, 1989; *Keigwin et al.*, 1991; *Keigwin and Jones*, 1995; *de Vernal et al.*, 1996]. Although these records have the advantage over terrestrial records of being more complete, the interpretation of the $\delta^{18}O$ record as a signal of freshwater flux variations is far from straightforward because of the opposing influence of temperature on the $\delta^{18}O$ value. Without some independent measure of one of these two variables [e.g., *Duplessy et al.*, 1991; *Rostek et al.*, 1993; *Bard and Rostek*, 1998], it is thus impossible to uniquely constrain a freshwater flux history from a $\delta^{18}O$ record. The continental record, on the other hand, clearly identifies the timing of changes in freshwater fluxes to the oceans arising from routing events; when a lower-elevation outlet to a different injection site is uncovered by ice-margin retreat, the freshwater flux to that site will increase. Because the

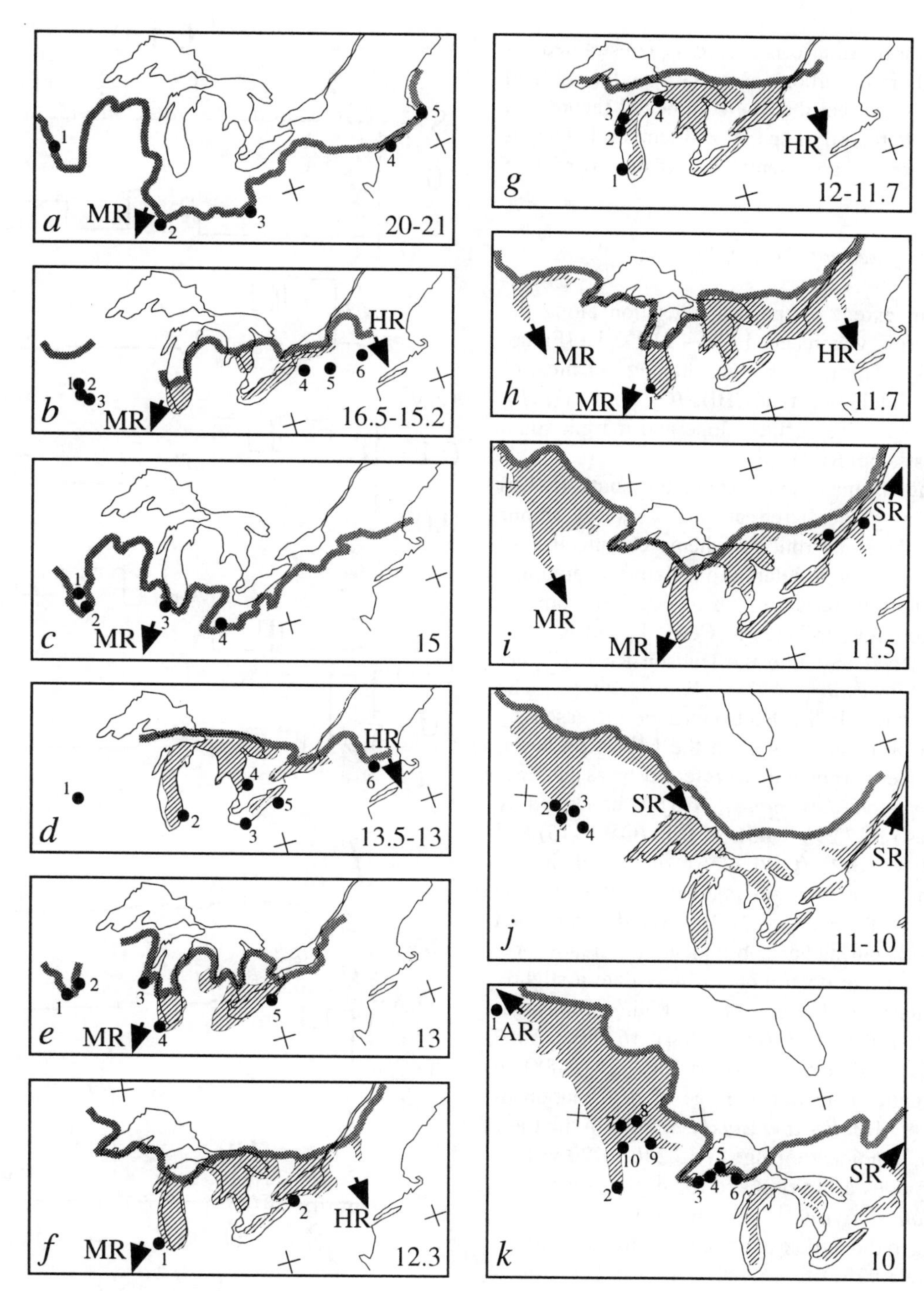

Figure 3. Paleogeographic maps showing changes in the position of the southern margin of the Laurentide Ice Sheet (heavy stippled line), proglacial and marine bodies (patterned regions), and directions of runoff to major injections sites during the last deglaciation (MR = Mississippi River, HR = Hudson River, SR = St. Lawrence River, and AR = Arctic). Ages in lower right corner of each map are in ^{14}C ka. Also shown are sites with radiocarbon ages which are used to constrain the timing of the deglaciation and associated rerouting events (keyed by site number to Appendix B). Information on ice-margin positions is from *Dreimanis* [1977], *Clayton and Moran* [1982], *Mickelson et al.* [1983], *Hansel et al.* [1985], *Teller* [1985], *Dyke and Prest* [1987], *Monaghan and Hansel* [1990], and *Ridge* [1997].

timing of the North American record is constrained by several hundred radiocarbon ages (Appendix B and references below), we feel that the continental record is a unique and reliable record of spatial and temporal changes in freshwater fluxes of continental runoff due to routing changes.

The Early Stages of Retreat (21 - 15.2 ^{14}C ka)

The maximum extent of the last glaciation along the southern LIS margin was reached ~20 - 21 ^{14}C ka (Figure 3A; Appendix B). During this time, drainage of most of the southern LIS, including five CHBs (C_s, D_s, E, F_s, H_s), was redirected by the ice surface slope and margin south through the Mississippi River.

After reaching its maximum extent, the southern LIS margin began a gradual retreat with several minor oscillations, but the same routing scheme continued until about 16.5 ^{14}C ka, when an interval of relatively rapid and widespread ice-margin retreat of up to 600 km from the LGM position rerouted drainage of CHBs F_s and H_s from the Mississippi River east into the Hudson River (Figure 3B; Table 2) [*Dreimanis*, 1977; *Ridge*, 1997]. The freshwater flux through the Hudson River increased by 0.04 Sv as a result (Figure 4A). In the Lake Erie basin, this interval of ice-margin retreat, referred to as the Erie Interstade [*Mörner and Dreimanis*, 1973], is bracketed by radiocarbon ages of 17,290 ± 436 yr B.P. (OWU-76) and 17,340 ± 390 yr B.P. (OWU-256) on wood in till deposited during an ice-margin readvance preceding the interstade, and 14,780 ± 192 yr B.P. (OWU-83) on wood in till deposited during a subsequent ice readvance, although this latter age determination may date a slightly younger readvance [*Fullerton*, 1980]. Radiocarbon ages of 16,380 +660/-710 yr B.P. (DIC-1884), 16,650 ± 1880 yr B.P. (BGS-86), and 16,650 ± 660 yr B.P. (GX-8488) on basal organic matter constrain the age of deglaciation of western New York [*Muller and Calkin*, 1993]. In the Lake Michigan basin, a radiocarbon age of 15,240 ± 120 yr B.P. (ISGS-465) provides a limiting age for deglaciation there [*Hansel and Johnson*, 1992]. To the west of the Lake Michigan basin, radiocarbon ages ranging from 15,140 ± 220 yr B.P. (Beta-9797) to 16,980 ± 180 yr B.P. (Beta-10525) suggest retreat of the Des Moines Lobe during this interval [*Bettis et al.*, 1996] (Appendix B).

Although the lithostratigraphy and radiocarbon ages identify widespread ice-margin retreat between ~17 and 15 ^{14}C ka, the actual duration of eastward drainage through the Hudson River is not well constrained. We suggest that the limiting radiocarbon ages from western New York, near the outlet through the Mohawk Valley to the Hudson River (Figure 3B; Appendix B), indicate that eastward

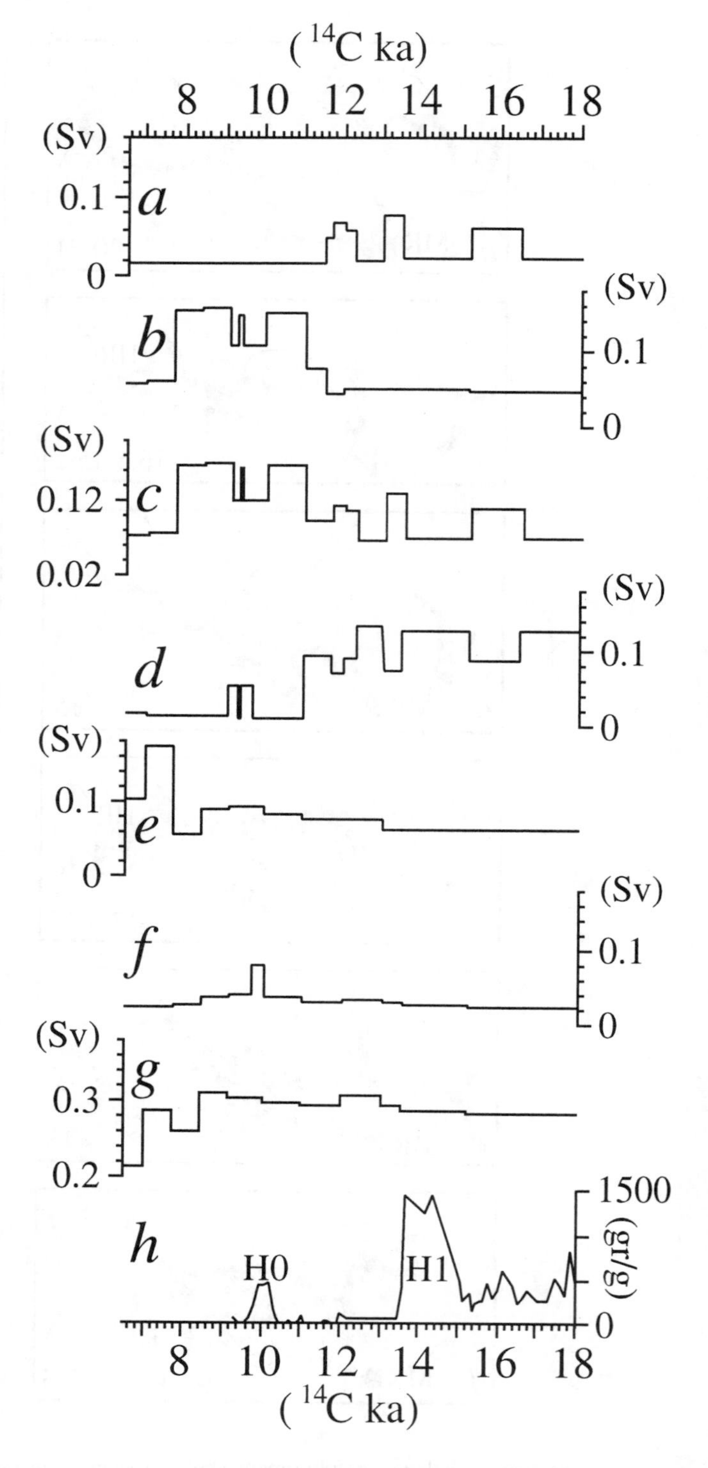

Figure 4. Time series of freshwater fluxes (in Sv) through major injections sites during the last deglaciation (18 - 7 ^{14}C ka). (A) Hudson River. (B) St. Lawrence River. (C) Combined Hudson and St. Lawrence Rivers. (D) Mississippi River. (E) Hudson Strait. (F) Arctic Ocean. (G) Total runoff. (H) Lithic record from marine core V23-81 showing episodes of increased iceberg flux out of Hudson Strait associated with H1 and H0 [*Bond and Lotti*, 1995].

drainage into the Hudson River began ~16.5 ^{14}C ka, and that radiocarbon ages constraining the age of subsequent ice-margin readvance (see below) suggest that rerouting from the Hudson River back south to the Mississippi River occurred at ~15.2 ^{14}C ka. Throughout the Erie Interstade, runoff from the region west of the Huron basin was directed south to the Gulf of Mexico (Figure 3B).

The Port Bruce Readvance

Subsequent readvance of the southern LIS margin covered the eastern outlet, forcing drainage of CHBs F_s and H_s back into the Mississippi River (Figure 3C, Table 2) and decreasing the freshwater flux through the Hudson River by 0.04 Sv (Figure 4A). In the eastern Great Lakes, the ice margin readvanced 500 to 600 km during the Port Bruce Stade nearly to its LGM position [*Dreimanis*, 1977; *Mickelson et al.*, 1983]. A radiocarbon age of 14,780 ± 192 yr B.P. (OWU-83) on wood in till indicates the ice reached northern Ohio shortly after 15 ^{14}C ka, although the sample may record the age of a younger advance [*Fullerton*, 1980]. Limiting ages on organic matter from depressions on or north of the ice limit suggest retreat following this advance began before 14,040 ± 75 yr B.P. (ISGS-348) to 14,500 ± 150 yr B.P. (ISGS-402) (Appendix B). The readvance of the Lake Michigan Lobe during the corresponding Crown Point Phase began after 15,240 ± 120 yr B.P. (ISGS-465) [*Hansel and Johnson*, 1992]. Limiting ages on wood in lake sediments which constrain subsequent ice-margin retreat range from 14,330 ± 250 yr B.P. (ISGS-1550) to 13,870 ± 170 yr B.P. (ISGS-1549) [*Hansel and Mickelson*, 1988] (Appendix B). Radiocarbon ages on organic matter underlying till suggest readvance of the Des Moines Lobe was underway sometime after 15,310 ± 180 yr B.P. (Beta-10838) to 15,140 ± 220 yr B.P. (Beta-9797) and reached its maximum extent between 13,680 ± 80 yr B.P. (ISGS-552) and 14,470 ± 400 yr B.P. (W-512) [*Clayton and Moran*, 1982; *Bettis et al.*, 1996] (Figure 3C; Appendix B).

The Mackinaw Interstade

Following the Port Bruce readvance, the southern LIS margin began renewed widespread retreat during the Mackinaw Interstade [*Dreimanis and Karrow*, 1972], redirecting drainage of four CHBs (D_s, E, F_s, H_s) from the Mississippi River east into the Hudson River beginning ~13.5 ^{14}C ka until 13 ^{14}C ka (Figures 3D, 5A, 5B; Table 2) when ice-margin readvance again closed the eastern routing to the North Atlantic. Freshwater flux to the Hudson River increased by 0.06 Sv during this interval (Figure 4A). A low lake level in the Lake Michigan basin, dated at 13,470 ± 130 yr B.P. (ISGS-1378), records the opening of an eastward drainage route across the Straits of Mackinaw into the Lake Huron basin during this interval [*Monaghan and Hansel*, 1990]. A radiocarbon age of 13,100 ± 110 yr B.P. (GSC-2213) on wood in till dates deglaciation of the Lake Huron basin during this time as well as subsequent ice-margin readvance [*Gravenor and Stupavsky*, 1976]. Radiocarbon ages of 13,360 ± 440 yr B.P. (BGS-929) [*Barnett*, 1985] and 12,920 ± 400 yr B.P. (W-430) [*Calkin and Feenstra*, 1985] constrain a low lake level in the Lake Erie basin, which corresponds to eastward drainage into the Hudson River.

The Port Huron Readvance

The southern LIS margin again began to readvance during the Port Huron Stade, reaching its maximum limit several hundred kilometers to the south by about 13 ^{14}C ka (Figure 3E) and closing all eastern routes again, forcing drainage of four CHBs (D_s, E, F_s, H_s) back into the Mississippi River (Table 2). The freshwater flux to the Hudson River correspondingly decreased by 0.06 Sv (Figure 4A). The radiocarbon age on wood in till from the Lake Huron basin discussed above (13,100 ± 110 yr B.P.; GSC-2213) provides a maximum age for readvance in this basin. High lake levels in the Lake Erie basin recording the Port Huron readvance are dated at 12,800 ± 250 yr B.P. (W-240) and 12,900 ± 200 yr B.P. (I-3175) [*Fullerton*, 1980]. Similarly, high lake levels in the Lake Michigan basin are dated at 12,650 ± 350 yr B.P. (W-140) and 12,660 ± 140 yr B.P. (ISGS-1190) [*Hansel et al.*, 1985]. Radiocarbon ages on basal organic matter from Devil's Lake, Wisconsin, suggest deglaciation of the neighboring Green Bay Lobe began after 12,880 ± 125 yr B.P. (WIS-1004) [*Maher and Mickelson*, 1996].

Establishment of Eastern Great Lakes Flow to the Hudson Valley

Following the Port Huron readvance, ice-margin retreat reopened drainage routes from the eastern Great Lakes to the North Atlantic that remained open for the remainder of the deglaciation. Variations in freshwater fluxes to the North Atlantic from routing changes continued, however, both as a result of the rerouting of runoff from the Hudson River to the St. Lawrence River as well as from ice-margin oscillations elsewhere along the southern LIS margin, which ultimately influenced the flux through these injections sites.

Retreat of the ice margin in the Lake Ontario basin allowed eastward drainage of two CHBs (F_s and H_s) through the Mohawk Valley into the Hudson River at

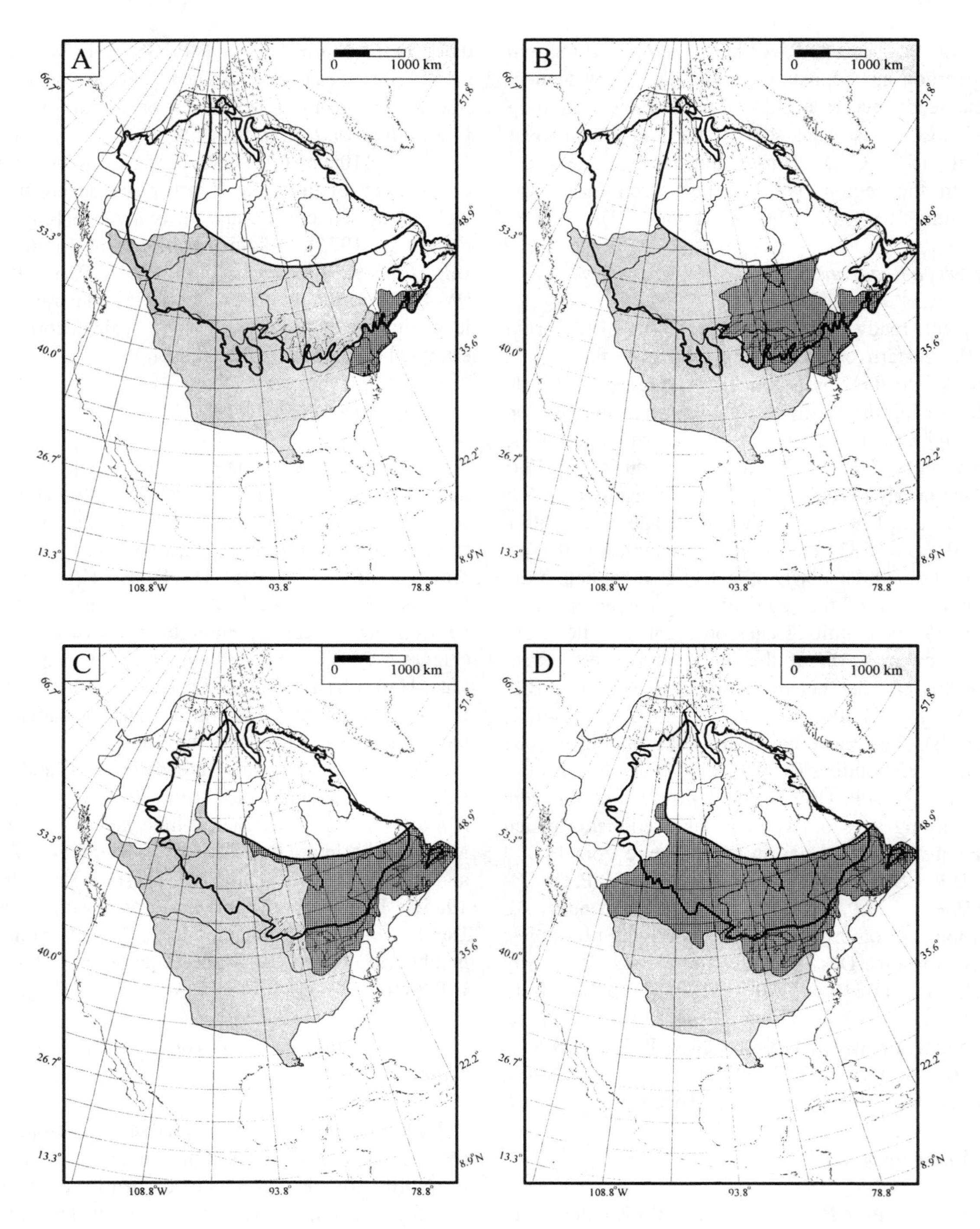

Figure 5. Paleogeographic maps illustrating partitioning of cryohydrological basins just before and after two major rerouting events. Heavy lines show the approximate location of the ice sheet margin and major ice divide at the time of the events. (A) Areas of North America that drained into the Mississippi River (area of gray shading) and the Hudson River (area of dark stippling) just before the start of the Mackinaw Interstade (13.5 ^{14}C ka). (B) Areas of North America that drained into the Mississippi River (area of gray shading) and the Hudson River (area of dark stippling) just after the start of the Mackinaw Interstade (13.5 ^{14}C ka). (C) Areas of North America that drained into the Mississippi River (area of gray shading) and the St. Lawrence River (area of dark stippling) just before 11 ^{14}C ka. (D) Areas of North America that drained into the Mississippi River (area of gray shading) and the St. Lawrence River (area of dark stippling) just after 11 ^{14}C ka.

~12.3 ^{14}C ka (Table 2), increasing the freshwater flux to the Hudson River by 0.04 Sv (Figure 4A). Initiation of eastward drainage at this time is recorded by the development of proglacial Lake Iroquois in the Ontario basin (Figure 3F). The age of Lake Iroquois is constrained by three radiocarbon ages on wood associated with beach deposits which average 12.3 ^{14}C ka: 12,100 ± 100 yr B.P. (I-838), and replicate analyses on the same sample of 12,080 ± 300 yr B.P. (W-883) and 12,660 ± 400 yr B.P. (W-861) [*Muller and Calkin*, 1993].

The Two Creeks Interstade

The ice margin in the Lake Michigan basin continued to block eastward drainage of two CHBs (D_s and E) at the time of renewed drainage through the Hudson River, as suggested by a radiocarbon age of 12,220 ± 350 yr B.P. (W-161) which dates a high lake level in the basin [*Hansel et al.*, 1985] (Figure 3F). Subsequent ice-margin retreat during the Two Creeks Interstade, however, opened an outlet across the Straits of Mackinaw, routing drainage of these two CHBs eastward between 12 to 11.7 ^{14}C ka (Figure 3G) and thus increasing the freshwater flux through the Hudson River by 0.01 Sv (Figure 4A). Organic material from the Two Creeks forest bed and correlative sites has been extensively dated [*Broecker and Farrand*, 1963; *Hansel et al.*, 1985; *Schneider*, 1990; *Larson et al.*, 1994] (Appendix B). *Kaiser* [1994] reported ten AMS radiocarbon ages on wood from the type Two Creeks site, ranging from 11,760 ± 90 yr B.P. (ETH-8271) to 12,035 ± 60 yr B.P. (ETH-8273) (Appendix B). The dendrochronology of Two Creeks wood suggests that the interstade lasted at least 250 yr [*Kaiser*, 1994].

The Greatlakean Readvance

The ice margin subsequently readvanced in the Lake Michigan basin at ~11.7 ^{14}C ka during the Greatlakean Phase (Figure 3H) [*Evenson et al.*, 1976], routing drainage of two CHBs (D_s and E) back into the Mississippi River (Table 2) and decreasing the freshwater flux through the Hudson River by 0.02 Sv (Figure 4A). The age of this readvance is constrained both by the youngest radiocarbon age on organics from sites recording the Two Creeks Interstade (e.g., 11,760 ± 90 yr B.P.) as well as by radiocarbon ages of 11,610 ± 70 yr B.P. (ISGS-985), 11,630 ± 120 yr B.P. (ISGS-1180), and 11,870 ± 100 yr B.P. (ISGS-1137) on wood in sediments recording a high proglacial lake level in the southern Lake Michigan basin (Appendix B) [*Hansel et al.*, 1985] that formed as a result of the ice margin blocking drainage eastward through the Straits of Mackinaw.

Establishment of the St. Lawrence Route to the North Atlantic

Ice-margin retreat from the St. Lawrence Valley at ~11.5 ^{14}C ka (Figure 3I) allowed drainage of the Great Lakes region eastward through the St. Lawrence River for the first time during the last deglaciation, and also permitted marine inundation by the Champlain Sea. The age of the ice retreat is based on AMS radiocarbon ages of 11,530 ± 90 yr B.P. (TO-703) and 11,460 ± 70 yr B.P. (TO-702) on marine shells from sediments of the Champlain Sea [*Rodrigues*, 1992]. Freshwater runoff derived from CHBs F_s and H_s was diverted from the Hudson River to the St. Lawrence River (Table 2), thus increasing the freshwater flux through the St. Lawrence River by 0.03 Sv (Figure 4B).

The ice margin in the Lake Michigan basin retreated north of the Straits of Mackinaw at ~11 ^{14}C ka, which is the age of the youngest wood in beach sediments (11,000 ± 80 yr B.P. (ISGS-1097)) (Appendix B) associated with the proglacial lake that existed while the Straits of Mackinaw were covered by ice [*Hansel et al.*, 1985; *Hansel and Mickelson*, 1988]. Retreat of the ice margin north of the Straits of Mackinaw rerouted drainage of CHBs D_s and E east into the St. Lawrence River (Figure 5C, 5D) (Table 2), increasing the freshwater flux through it by 0.02 Sv.

The Addition of Lake Agassiz Drainage to Outflow to the North Atlantic

At the same time as the western Great Lakes were routed east to the St. Lawrence (~11 ^{14}C ka), retreat of the LIS margin also opened an eastern outlet for proglacial Lake Agassiz (Figure 3J). Prior to this retreat, overflow from the Lake Agassiz basin had been directed south into the Mississippi River system (Figure 3H, 3I) [*Matsch and Wright*, 1967; *Clayton and Moran*, 1982; *Fenton et al.*, 1983; *Hobbs*, 1983]. Dates of 10,960 ± 300 yr B.P. (W-723) and 10,820 ± 190 yr B.P. (TAM-1) in North Dakota [*Clayton*, 1983] and 10,810 ± 120 yr B.P. (TO-1594) in northwestern Ontario [*Bajc*, 1991] on wood from shoreline sediment below the elevation of the southern outlet are the oldest to indicate that overflow had shifted to lower-elevation routes into the Great Lakes via the Lake Superior basin. The duration of this routing event is constrained by radiocarbon ages ranging from 10,960 ± 300 yr B.P. (W-723) to 10,000 ± 110 yr B.P. (TO-4286) (Appendix B), which come from the extensive surface that was subaerially exposed during the low-water (Moorhead) phase of Lake Agassiz [*Elson*, 1967; *Harris et al.*, 1974; *Fenton et al.*, 1983] and which was progressively drowned as differential isostatic rebound caused gradual re-flooding of that surface.

Overflow from the Lake Agassiz basin diverted drainage of CHB C_s to the Great Lakes (Figure 5C, 5D), which added an additional freshwater flux of 0.05 Sv to the St. Lawrence River, for a total increase of 0.07 Sv at 11 ^{14}C ka (Figure 4B). The freshwater flux was initially much higher as the level of Lake Agassiz dropped to the level of the new outlet. *Teller and Thorleifson* [1983] estimated the initial flood discharge through the eastern outlets of Lake Agassiz to be about 100,000 m^3 sec^{-1} (0.1 Sv). There were a number of these peak flood flows, as ice retreated from successively lower channels that linked the Agassiz and Nipigon-Superior basins, although the duration of each flood flow probably did not exceed more than a few years before baseline outflow was established [*Teller and Thorleifson*, 1983]; rapid isostatic rebound of these channels may actually have resulted in a decreasing baseline flow after each flood.

The Marquette Readvance

Eastward routing of Lake Agassiz waters ended ~10 ^{14}C ka when a combination of differential rebound and the Marquette readvance of the LIS margin in the Lake Superior basin closed the lake's eastern outlets, forcing drainage of CHB C_s northwest into the Arctic Ocean by way of the Clearwater outlet to the Mackenzie River (Figure 3K) [*Smith and Fisher*, 1993; *Fisher and Smith*, 1994] and decreasing the freshwater flux through the St. Lawrence River by 0.05 Sv (Figure 4B). The age of the Marquette advance into the Superior basin is based on wood dates, including rooted stumps in northern Michigan, that average about 10 ^{14}C ka [*Drexler et al.*, 1983; *Lowell et al.*, 1999] (Appendix B). Based on radiocarbon ages on wood from catastrophic flood and deltaic deposits in northern Saskatchewan, ranging from 10,310 ± 290 yr B.P. (GX-5301-II) to 9710 ± 130 yr B.P. (AECV-1183c) (Appendix B), *Smith and Fisher* [1993], *Fisher and Smith* [1994], and *Fisher and Souch* [1998] concluded that the rerouting of Lake Agassiz overflow associated with the closure of the eastern outlets was through a large channel in the extreme northwestern corner of the lake basin, the Clearwater-Athabasca Spillway, to the Mackenzie River and Arctic Ocean; *Elson* [1967] concluded the same. The higher level of Lake Agassiz that resulted (Emerson phase) is constrained by radiocarbon ages ranging from 9,700 ± 140 yr B.P. (GSC-797) to 10,000 ± 280 yr B.P. (GSC-1428) (Appendix B).

As differential isostatic uplift raised the northwestern (Clearwater) outlet, overflow returned to the Mississippi River at ~9.7 ^{14}C ka, rerouting drainage of CHB C_s into the Mississippi River (Table 2, Figure 4D). This event is marked by the establishment of the extensive Upper Campbell strandline of Lake Agassiz [*Elson*, 1967; *Fenton et al.*, 1983; *Teller*, 1985; *Thorleifson*, 1996]. Radiocarbon ages indicate that relatively high lake levels of the Emerson phase and, therefore, overflow through the southern outlet, continued until 9.5 - 9.4 ^{14}C ka [*Elson*, 1967; *Teller and Thorleifson*, 1983; *Thorleifson*, 1996; *Teller et al.*, 1999] (Appendix B).

The evidence described above for routing changes to the Mississippi River is contrary to the conclusion of *Marchitto and Wei* [1995] that major meltwater flow was permanently diverted away from the Gulf of Mexico at ~11.3 ^{14}C ka. *Marchitto and Wei* [1995] used the concentration of reworked calcareous nannofossils in Orca Basin sediments as a proxy for sediment erosion and hence meltwater flow into the Gulf of Mexico, and argued that low concentrations of nannofossils after the Younger Dryas indicated cessation of major meltwater flow into the Gulf. We suggest that the dearth of nannofossils may alternatively be explained by changes in sediment transport caused by rising sea levels, changes in ocean circulation, and other processes unrelated to changes in meltwater flux. Furthermore, we believe that the weight of positive evidence from the continental record for diversion of meltwater through the Mississippi River during this time interval is more convincing than the lack of evidence from the marine record for such a diversion.

Reopening of the Eastern Outlets of Lake Agassiz

By 9.4 ^{14}C ka, ice-margin retreat in the Lake Superior basin reopened drainage routes to the east, increasing the freshwater flux to the St. Lawrence River by 0.04 Sv (Figure 4B). This routing event is suggested by beaches formed below the Campbell level, which are dated at 9340 ± 80 yr B.P. (GSC-4950) and 9400 ± 100 yr B.P. (GSC-4933) and represent a phase of Lake Agassiz too low to drain through the southern outlet [*Teller et al.*, 1999]. Additional evidence for the reopening of eastern outlets at this time comes from Great Lakes' basins to the east. In the Lake Superior basin, radiocarbon ages of 9380 ± 150 yr B.P. (GSC-287) and 9345 ± 240 yr B.P. (GX-4883) on a proglacial lake (Minong) indicate that the LIS ice margin had retreated from that basin, allowing eastward drainage [*Zoltai*, 1965; *Drexler et al.*, 1983; *Teller and Thorleifson*, 1983]. In the Lake Huron basin, *Lewis et al.* [1994]

interpreted a rise in lake level and an increase in $\delta^{18}O$ values measured on ostracodes as reflecting the increased inflows from the Agassiz basin between 9.6 to 9.3 ^{14}C ka (Appendix B).

Thorleifson and Kristjansson [1993] and *Thorleifson* [1996] argue that overflow from Lake Agassiz returned to the Gulf of Mexico between about 9.3 and 9.1 ^{14}C ka, as ice readvanced into the Nipigon basin north of Lake Superior and covered the eastern outlets of Lake Agassiz. This diversion is reflected in the Lake Huron basin, where *Lewis et al.* [1994] concluded that the reduced inflow resulted in a lower lake level and a decrease in $\delta^{18}O$ values measured on ostracodes between 9.3 and 9.1 ^{14}C ka (Appendix B). The freshwater flux to the St. Lawrence River was reduced by 0.04 Sv during this interval (Figure 4B).

Final opening of eastern outlets of Lake Agassiz occurred at ~9.1 ^{14}C ka, increasing the freshwater flux to the St. Lawrence River by 0.05 Sv. This rerouting again caused an increase in flow into the Lake Huron basin, raising the lake level there and increasing the $\delta^{18}O$ values measured on ostracodes starting at ~9.1 ^{14}C ka [*Lewis et al.*, 1994] (Appendix B). This event is also registered in the $\delta^{18}O$ record measured on ostracodes in the Lake Michigan basin [*Colman et al.*, 1994] (Appendix B).

Opening of Hudson Bay

The last major late-glacial routing change of consequence occurred when final deglaciation of the LIS from Hudson Bay established modern drainage routes for most major rivers. Until this time, northward runoff toward Hudson Bay had flowed into proglacial Lakes Agassiz and Barlow-Ojibway, which drained east through outlets to the St. Lawrence River [e.g., *Dyke and Prest*, 1987; *Veillette*, 1994]. Final deglaciation of Hudson Bay allowed rapid drainage of these proglacial lakes. The age of final deglaciation of Hudson Bay has long been estimated at ~8 - 8.2 ^{14}C ka based on radiocarbon ages on marine shells which date this event. *Klassen* [1983] and *Veillette* [1994], however, suggested that the deglaciation may be younger, and that Lakes Agassiz and Barlow-Ojibway thus drained a few hundred years later. *D.C. Barber et al.* [Forcing of the cold event 8200 years ago by outburst drainage of Laurentide proglacial lakes, submitted to Nature] find support for this argument based on a new estimate of the reservoir age for marine shells in Hudson Bay, which suggests deglaciation occurred at ~7.7 ^{14}C ka. At this time, they estimate that the rapid (10 - 1 yr) drainage of proglacial Lakes Agassiz and Barlow-Ojibway released 0.6 - 6 Sv through Hudson Strait into the Labrador Sea.

OTHER FRESHWATER FLUXES TO THE NORTH ATLANTIC

Our calculated freshwater fluxes to the North Atlantic represent only those from precipitation and from "meltwater," or ice-volume changes computed from results of steady-state ice-sheet reconstructions [*Licciardi et al.*, 1998]. Although our calculations of ice volume change during the last deglaciation thus implicitly include some ice volume loss due to calving, we do not account for non-steady-state changes arising from the abrupt iceberg-discharge events referred to as Heinrich events. The evidence from deep-sea cores, however, indicates that significant increases in the freshwater flux to the North Atlantic occurred during Heinrich events [*Andrews and Tedesco*, 1992; *Broecker et al.*, 1992; *Bond and Lotti*, 1995; *McManus et al.*, 1998], and that these would have supplemented our calculated fluxes. The timing of these iceberg-discharge events is well known, but the freshwater flux associated with them remains poorly constrained. Ice-sheet modeling of a Heinrich event suggests a freshwater flux ranging from values similar to those we calculate from routing changes (0.03 Sv) [*Marshall and Clarke*, 1997] to much higher values (0.3 Sv) [*MacAyeal*, 1993; *Alley and MacAyeal*, 1994]. Planktonic foraminiferal $\delta^{18}O$ records indicate that freshening of North Atlantic surface waters was greatest during Heinrich events [*Bond et al.*, 1993; *Labeyrie et al.*, 1995; *Maslin et al.*, 1995; *Cortijo et al.*, 1997], and $\delta^{13}C$ records measured on benthic foraminifera suggest that the greatest reduction in the rate of formation of NADW occurred at these times [*Sarnthein et al.*, 1994; *Keigwin and Lehman*, 1994].

During the last deglaciation, Heinrich ice-rafting events H1 (~14.5 - 13 ^{14}C ka) and H0 (~11 - 10 ^{14}C ka) [*Andrews et al.*, 1995; *Bond and Lotti*, 1995; *Jennings et al.*, 1996] would have supplemented the freshwater fluxes we calculate from Hudson Strait (Figure 4). H1 was preceded by an increase in freshwater flux to the North Atlantic (Figure 4A) through the Hudson River during the Erie Interstade (16.5 - 15.2 ^{14}C ka). The freshwater flux through the Hudson River decreased at 15.2 ^{14}C ka as the southern LIS margin readvanced across the eastern outlet during the Port Bruce Stade (Figure 4C), ending several hundred years before the start of H1 (Figure 4H) [*Bond and Lotti*, 1995; *Jennings et al.*, 1996]. H1 ice rafting ended at about the same time as the start of the Mackinaw Interstade (13.5 ^{14}C ka) (Figure 4H), when the southern

LIS margin again retreated far enough to open eastward drainage, increasing the freshwater flux through the Hudson River for the next ~500 yr (Figure 4A). These relations suggest that there was a prolonged interval of increased freshwater flux to the North Atlantic from 16.5 to 13 ^{14}C ka which resulted from a complex series of events involving fluctuations of the southern LIS margin and surging of ice through Hudson Strait. H0, on the other hand, occurred at the same time as the major routing event through the St. Lawrence River between 11 and 10 ^{14}C ka (Figure 4B, 4H), thus augmenting the freshwater flux associated with this routing event. *Miller and Kaufman* [1990] estimated the freshwater flux through Hudson Strait associated with H0 at 0.010 - 0.076 Sv.

The Baltic Ice Lake developed as the southern margin of the Scandinavian Ice Sheet retreated from the Baltic Sea basin during the last deglaciation. *Björck* [1995] argued that the lake level dropped rapidly at ~11 - 11.2 ^{14}C ka and again at ~10.4 ^{14}C ka. These two rapid drainage events would have provided significant, but brief (order of 10 yr), increases in freshwater flux to the North Atlantic [*Björck*, 1995] at the start of and during the Younger Dryas. *Lehman et al.* [1991] and *Lehman and Keigwin* [1992] proposed that evidence of these two lake-discharge events is recorded as low δ^{18}O values measured on foraminifera in a core off the coast of Norway. *Boden et al.* [1997] also found evidence of these discharge events in shallow marine cores recovered near the Baltic Ice Lake outlet. A δ^{18}O "meltwater spike" suggests the first lake-discharge event occurred ~10,900 ^{14}C ka and lasted on the order of 10 yr. *Boden et al.* [1997] argued that the second lake-discharge event occurred in two closely spaced steps. Radiocarbon ages suggest a younger date (~ 10 ^{14}C ka) than estimated by earlier workers, and biostratigraphic studies suggest the second event occurred during the early Preboreal warm interval.

Asia's main northward-flowing rivers (Kotuy, Lena, Ob, Yenisei) to the Arctic Ocean were impacted by the growth of ice sheets across Siberia, although the timing and extent of ice cover is much debated [cf. *Rutter*, 1995]. If modern runoff is used as an approximation of the volume of water that would have been diverted as a result of glacial damming, the change in freshwater flux from the Arctic to the mid-latitude North Atlantic by way of the Mediterranean would have been ~ 2,000 km^3 yr^{-1} (~0.06 Sv), but the timing and amount of any routing changes to this flux remains unknown. Some [e.g. *Grosswald*, 1980, 1988, 1998; *Denton and Hughes*, 1981] advocate an expansive ice cover, where the Kara and Barents ice sheets of western Siberia joined together and merged with the Scandinavian Ice Sheet. In these reconstructions, northward-flowing rivers were dammed and their waters rerouted south and west through the Aral, Caspian, Black, and Mediterranean Seas to the mid-Atlantic Ocean at Gibraltar. Many workers question this ice-sheet extent during the Late Weichselian [*Velichko et al.*, 1984; *Astakhov*, 1992; *Velichko*, 1995], but find support for earlier widespread glaciation (e.g. Early Weichselian) which may have brought about a major reorganization of Asia's drainage, and a rerouting of its freshwaters to the mid-Atlantic Ocean [e.g. *Arkhipov et al.*, 1995; *Arkhipov*, 1998; *Astakhov*, 1998]. Given the magnitude of the freshwater flux from Asia, determining its routing history remains an important question in understanding how those changes may have impacted NADW formation [*Aagaard and Carmack*, 1989; *Rahmstorf*, 1994].

DISCUSSION AND CONCLUSIONS

Research on the deglacial history of the LIS has long identified a series of millennial-scale fluctuations of the southern margin superimposed on overall orbital-scale retreat [e.g., *Prest*, 1970; *Willman and Frye*, 1970; *Dreimanis and Karrow*, 1972; *Fullerton*, 1980; *Clayton and Moran*, 1982; *Mickelson et al.*, 1983]. *Teller* [1987, 1990a, 1990b, 1995b] identified the implications of these ice-margin fluctuations in controlling the routing history of freshwater to the North Atlantic, and devised a strategy to quantitatively evaluate the changes in freshwater fluxes to the North Atlantic Ocean that resulted from this routing history. We have elaborated on this analysis, combining new ice-sheet reconstructions to compute meltwater runoff [*Licciardi et al.*, 1998] with modern precipitation estimates adjusted by paleoclimate GCM results [*Kutzbach et al.*, 1998] to compute precipitation runoff for the last deglaciation. Our routing history (Table 2, Figures 3 and 4) is based on more than 150 ^{14}C ages (Appendix B) which constrain the timing of well-known deglacial routing events often to the resolution of the dating method.

Our results show that the total freshwater flux due to meltwater and precipitation runoff did not change appreciably during the last deglaciation, delivering on order of 0.3 Sv to the North Atlantic and Arctic Oceans (Figure 2). The most significant change in total runoff occurred upon final deglaciation of the LIS when the contribution from meltwater ended and the total runoff decreased ~35% (0.11 Sv) between 8.4 and 6 ^{14}C ka (Figure 2). Insofar as some ocean models have found that preconditioning of the North Atlantic by freshwater makes the thermohaline circulation (THC) more unstable [*Tziperman*, 1997] or more responsive to a "pulsed" freshwater forcing [*Fanning and Weaver*, 1997], overall meltwater runoff to the North Atlantic may have increased the sensitivity of the THC to freshwater forcing events during glaciations, such as the routing events we describe, whereas the loss of this meltwater contribution may account for the relative stability of the THC since final deglaciation of the LIS [*Bond et al.*, 1997].

Although only relatively small changes in total runoff from North America to the ocean were reconstructed for the last deglaciation, we find that significant flux variations occurred through each of the five injection sites as a result of routing changes. Most major routing events resulting in spatial changes in freshwater fluxes to the ocean occurred along the southern margin of the LIS and involved changes in the routing of flow to the Mississippi River versus to the Hudson or St. Lawrence Rivers. As long as the ice margin extended far enough south to block drainage to the east, runoff from much of the southern sector of the LIS was routed south through the Mississippi River drainage. Subsequent recession of the ice margin eventually uncovered lower outlets to the east, thus rerouting some fraction of water from the Mississippi River to the Hudson or St. Lawrence rivers. The interplay between these eastern and southern injection sites produced an "anti-phasing" in the time series of freshwater fluxes, whereby an increase in the flux through an eastern outlet occurred at the expense of the flux through the southern outlet, and vice versa (Figure 4). The total flux from the southern margin, however, remained nearly constant.

Ocean and coupled ocean-atmosphere models predict that the site as well as the flux of freshwater injection to the North Atlantic is important in affecting NADW formation, with a freshwater forcing from the St. Lawrence River having a greater effect on the rate of formation of NADW than an equivalent or greater forcing from the Mississippi River [*Maier-Reimer and Mikolajewicz*, 1989; *Rahmstorf*, 1995b; *Manabe and Stouffer*, 1997]. Furthermore, *Rahmstorf* [1994, 1995a] finds that the rate of formation of NADW is sensitive to changes in freshwater fluxes of the same order as we reconstruct (0.015 - 0.06 Sv). In addition, *Fanning and Weaver* [1997] show that preconditioning of sites of NADW formation by freshwater from the Mississippi River increases the sensitivity of the THC to a subsequent freshwater forcing from the St. Lawrence River, such as would accompany one of the rerouting events described here. Although not specifically tested yet by ocean models, we suggest that because the Hudson River injection site also delivers freshwater closer to sites of NADW formation than does the Mississippi River, a similar sensitivity difference between these two injection sites may apply as well.

These climate model results suggest that the routing events described in this study, which influence the freshwater flux to eastern versus southern outlets, may have had a significant effect on rates of NADW formation, and thus caused abrupt climate change during the last deglaciation. Supplementing these spatially and temporally varying shifts in freshwater delivery to the North Atlantic were episodic influxes of icebergs through Hudson Strait (Heinrich events). *Broecker et al.* [1988, 1989] proposed that the routing event involving diversion of Lake Agassiz drainage between 11 and 10 ^{14}C ka from the Mississippi River to the St. Lawrence River (and back again) (Figure 4B, 4D, 4F, 5C, 5D) was responsible for the Younger Dryas. Iceberg discharge through Hudson Strait during the Younger Dryas (H0) would have augmented the freshwater flux to the North Atlantic associated with this rerouting event. Our results suggest that there were several similar routing events both before and after this particular Lake Agassiz event which, combined with earlier Heinrich events, may have similarly influenced climate. The cumulative freshwater forcing associated with the Erie Interstade, H1, and the Mackinaw Interstade at about 16.5 - 13 ^{14}C ka (Figure 4), for example, spans the Oldest Dryas cold period which, like the Younger Dryas, interrupts warming during the last deglaciation [*Wohlfarth*, 1996; *McCabe and Clark*, 1998].

The mechanism of ice-margin controlled switching of freshwater runoff between eastern and southern outlets would have occurred whenever the southern LIS margin advanced into the Great Lakes-St. Lawrence River region [*Teller*, 1995b]. Because the advance of the LIS to its LGM position (Figure 3A) eroded much of the record of previous routing changes, it is not possible to reconstruct a similar time series of freshwater fluxes as for the last deglaciation (Figure 4). However, our time series are extremely important for modeling studies which evaluate the sensitivity of NADW formation to spatial and temporal variations in freshwater fluxes [e.g., *Fanning and Weaver*, 1997; *Manabe and Stouffer*, 1997].

Only late in deglaciation did major flux variations by routing changes occur through Hudson Strait and to the Arctic Ocean. Heinrich events, however, greatly supplemented the episodic fluxes through Hudson Strait. These iceberg-discharge events, in combination with routing events along the southern LIS margin, resulted in significant spatial and temporal variability in the freshwater flux to the North Atlantic during the last glaciation.

APPENDIX A. PRECIPITATION, ICE MELT, AND TOTAL RUNOFF VALUES

Radiocarbon ages are calibrated using the Calib 4.0 calibration program [*Stuiver and Reimer*, 1993; *Stuiver et al.*, 1998]. Freshwater volumes are converted to sverdrups using calendar ages of time intervals.

Total Runoff (10^5 km^3)

(^{14}C ka)	(cal ka)	Mississippi	St. Lawrence	Hudson River	Hudson Strait	Arctic	Total
18.0 - 16.5	21.39 - 19.67	67.66	25.75	12.41	33.94	13.70	153.45
16.5 - 15.2	19.67 - 18.17	41.12	22.45	28.71	29.60	11.95	133.83
15.2 - 13.5	18.17 - 16.21	78.08	29.62	13.85	38.78	16.52	176.85
13.5 - 13.0	16.21 - 15.63	13.31	8.96	14.38	11.55	5.53	53.73
13.0 - 12.3	15.63 - 14.29	55.78	20.39	7.49	31.74	14.86	130.27
12.3 - 12.0	14.29 - 14.07	6.28	3.35	4.11	5.21	2.44	21.39
12.0 - 11.7	14.07 - 13.68	8.62	5.46	8.47	9.49	3.94	35.99
11.7 - 11.5	13.68 - 13.46	6.35	3.08	3.29	5.36	2.22	20.30
11.5 - 11.0	13.46 - 13.00	13.28	11.06	2.26	11.20	4.65	42.45
11.0 - 10.0	13.00 - 11.44	4.46	74.32	7.19	40.93	18.79	145.70
10.0 - 9.7	11.44 - 11.17	0.86	9.02	1.20	7.96	7.00	26.04
9.7 - 9.4	11.17 - 10.62	9.00	18.36	2.45	16.21	7.02	53.04
9.4 - 9.3	10.62 - 10.46	0.51	7.45	0.71	4.71	2.04	15.43
9.3 - 9.1	10.46 - 10.24	3.60	7.35	0.98	6.48	2.81	21.22
9.1 - 8.4	10.24 - 9.45	2.76	38.86	3.57	22.13	9.83	77.15
8.4 - 7.7	9.45 - 8.44	3.78	48.56	4.66	17.59	8.72	83.31
7.7 - 7.0	8.44 - 7.81	2.57	12.10	2.94	34.13	5.25	56.99
7.0 - 6.0	7.81 - 6.81	4.63	17.59	4.76	32.81	7.86	67.64
6.0 - 0.0	6.81 - 0.00	37.46	117.77	33.76	186.30	55.69	430.97

Total Runoff (Sv)

(^{14}C ka)	(cal ka)	Mississippi	St. Lawrence	Hudson River	Hudson Strait	Arctic	Total
18.0 - 16.5	21.39 - 19.67	0.1247	0.0475	0.0229	0.0626	0.0253	0.2829
16.5 - 15.2	19.67 - 18.17	0.0869	0.0475	0.0607	0.0626	0.0253	0.2829
15.2 - 13.5	18.17 - 16.21	0.1263	0.0479	0.0224	0.0627	0.0267	0.2861
13.5 - 13.0	16.21 - 15.63	0.0728	0.0490	0.0786	0.0631	0.0302	0.2937
13.0 - 12.3	15.63 - 14.29	0.1320	0.0482	0.0177	0.0751	0.0352	0.3083
12.3 - 12.0	14.29 - 14.07	0.0905	0.0482	0.0592	0.0751	0.0352	0.3083
12.0 - 11.7	14.07 - 13.68	0.0701	0.0444	0.0689	0.0772	0.0321	0.2926
11.7 - 11.5	13.68 - 13.46	0.0915	0.0444	0.0475	0.0772	0.0321	0.2926
11.5 - 11.0	13.46 - 13.00	0.0915	0.0762	0.0156	0.0772	0.0321	0.2926
11.0 - 10.0	13.00 - 11.44	0.0091	0.1511	0.0146	0.0832	0.0382	0.2962
10.0 - 9.7	11.44 - 11.17	0.0101	0.1059	0.0141	0.0934	0.0822	0.3058
9.7 - 9.4	11.17 - 10.62	0.0519	0.1059	0.0141	0.0934	0.0405	0.3058
9.4 - 9.3	10.62 - 10.46	0.0101	0.1476	0.0141	0.0934	0.0405	0.3058
9.3 - 9.1	10.46 - 10.24	0.0519	0.1059	0.0141	0.0934	0.0405	0.3058
9.1 - 8.4	10.24 - 9.45	0.0111	0.1560	0.0143	0.0888	0.0395	0.3097
8.4 - 7.7	9.45 - 8.44	0.0119	0.1524	0.0146	0.0552	0.0274	0.2615
7.7 - 7.0	8.44 - 7.81	0.0130	0.0609	0.0148	0.1718	0.0264	0.2869
7.0 - 6.0	7.81 - 6.81	0.0147	0.0558	0.0151	0.1040	0.0249	0.2145
6.0 - 0.0	6.81 - 0.00	0.0174	0.0548	0.0157	0.0867	0.0259	0.2007

Ice Melt Runoff (10^5 km^3)

(^{14}C ka)	(cal ka)	Mississippi	St. Lawrence	Hudson River	Hudson Strait	Arctic	Total
18.0 - 16.5	21.39 - 19.67	1.51	0.76	0.82	1.52	0.08	4.70
16.5 - 15.2	19.67 - 18.17	0.16	0.67	1.87	1.32	0.07	4.10
15.2 - 13.5	18.17 - 16.21	3.03	0.97	1.04	2.14	0.74	7.93
13.5 - 13.0	16.21 - 15.63	1.32	0.36	0.88	0.92	0.68	4.16
13.0 - 12.3	15.63 - 14.29	6.91	1.49	0.29	8.09	3.38	20.16
12.3 - 12.0	14.29 - 14.07	0.79	0.25	0.39	1.33	0.55	3.31
12.0 - 11.7	14.07 - 13.68	1.63	0.23	-0.22	2.98	0.71	5.33
11.7 - 11.5	13.68 - 13.46	1.21	0.13	-0.42	1.68	0.40	3.01
11.5 - 11.0	13.46 - 13.00	2.54	-0.64	0.03	3.51	0.84	6.28
11.0 - 10.0	13.00 - 11.44	0.00	12.47	0.00	14.49	2.65	29.61
10.0 - 9.7	11.44 - 11.17	0.00	1.83	0.00	3.44	1.83	7.10
9.7 - 9.4	11.17 - 10.62	1.77	3.72	0.00	7.01	1.96	14.46
9.4 - 9.3	10.62 - 10.46	0.00	1.60	0.00	2.04	0.57	4.21
9.3 - 9.1	10.46 - 10.24	0.71	1.49	0.00	2.80	0.79	5.78
9.1 - 8.4	10.24 - 9.45	0.00	9.95	0.00	10.30	2.83	23.08
8.4 - 7.7	9.45 - 8.44	0.00	15.54	0.00	-0.59	0.00	14.96
7.7 - 7.0	8.44 - 7.81	0.00	1.17	0.00	14.14	0.00	15.31
7.0 - 6.0	7.81 - 6.81	0.00	0.05	0.00	3.56	0.00	3.61
6.0 - 0.0	6.81 - 0.00	0.00	0.00	0.00	0.00	0.00	0.00

Ice Melt Runoff (Sv)

(^{14}C ka)	(cal ka)	Mississippi	St. Lawrence	Hudson River	Hudson Strait	Arctic	Total
18.0 - 16.5	21.39 - 19.67	0.0028	0.0014	0.0015	0.0028	0.0002	0.0087
16.5 - 15.2	19.67 - 18.17	0.0003	0.0014	0.0040	0.0028	0.0002	0.0087
15.2 - 13.5	18.17 - 16.21	0.0049	0.0016	0.0017	0.0035	0.0012	0.0128
13.5 - 13.0	16.21 - 15.63	0.0072	0.0019	0.0048	0.0050	0.0037	0.0228
13.0 - 12.3	15.63 - 14.29	0.0164	0.0035	0.0007	0.0191	0.0080	0.0477
12.3 - 12.0	14.29 - 14.07	0.0114	0.0035	0.0056	0.0191	0.0080	0.0477
12.0 - 11.7	14.07 - 13.68	0.0133	0.0019	-0.0018	0.0242	0.0058	0.0433
11.7 - 11.5	13.68 - 13.46	0.0175	0.0019	-0.0060	0.0242	0.0058	0.0433
11.5 - 11.0	13.46 - 13.00	0.0175	-0.0044	0.0002	0.0242	0.0058	0.0433
11.0 - 10.0	13.00 - 11.44	0.0000	0.0253	0.0000	0.0295	0.0054	0.0602
10.0 - 9.7	11.44 - 11.17	0.0000	0.0215	0.0000	0.0404	0.0215	0.0834
9.7 - 9.4	11.17 - 10.62	0.0102	0.0215	0.0000	0.0404	0.0113	0.0834
9.4 - 9.3	10.62 - 10.46	0.0000	0.0317	0.0000	0.0404	0.0113	0.0834
9.3 - 9.1	10.46 - 10.24	0.0102	0.0215	0.0000	0.0404	0.0113	0.0834
9.1 - 8.4	10.24 - 9.45	0.0000	0.0399	0.0000	0.0413	0.0114	0.0926
8.4 - 7.7	9.45 - 8.44	0.0000	0.0488	0.0000	-0.0018	0.0000	0.0470
7.7 - 7.0	8.44 - 7.81	0.0000	0.0059	0.0000	0.0712	0.0000	0.0770
7.0 - 6.0	7.81 - 6.81	0.0000	0.0002	0.0000	0.0113	0.0000	0.0114
6.0 - 0.0	6.81 - 0.00	0.0000	0.0000	0.0000	0.0000	0.0000	0.0000

Precipitation Runoff (10^5 km^3)

(^{14}C ka)	(cal ka)	Mississippi	St. Lawrence	Hudson River	Hudson Strait	Arctic	Total
18.0 - 16.5	21.39 - 19.67	66.15	24.98	11.59	32.42	13.62	148.76
16.5 - 15.2	19.67 - 18.17	40.96	21.79	26.84	28.27	11.88	129.73
15.2 - 13.5	18.17 - 16.21	75.05	28.65	12.81	36.63	15.78	168.92
13.5 - 13.0	16.21 - 15.63	11.99	8.61	13.49	10.62	4.85	49.57
13.0 - 12.3	15.63 - 14.29	48.87	18.89	7.20	23.66	11.48	110.11
12.3 - 12.0	14.29 - 14.07	5.49	3.10	3.72	3.88	1.89	18.08
12.0 - 11.7	14.07 - 13.68	6.99	5.23	8.70	6.52	3.23	30.66
11.7 - 11.5	13.68 - 13.46	5.14	2.95	3.71	3.68	1.82	17.30
11.5 - 11.0	13.46 - 13.00	10.74	11.70	2.23	7.69	3.81	36.17
11.0 - 10.0	13.00 - 11.44	4.46	61.86	7.19	26.44	16.14	116.09
10.0 - 9.7	11.44 - 11.17	0.86	7.19	1.20	4.52	5.17	18.94
9.7 - 9.4	11.17 - 10.62	7.23	14.64	2.45	9.20	5.06	38.58
9.4 - 9.3	10.62 - 10.46	0.51	5.85	0.71	2.68	1.47	11.22
9.3 - 9.1	10.46 - 10.24	2.89	5.86	0.98	3.68	2.02	15.43
9.1 - 8.4	10.24 - 9.45	2.76	28.91	3.57	11.83	6.99	54.07
8.4 - 7.7	9.45 - 8.44	3.78	33.01	4.66	18.18	8.72	68.35
7.7 - 7.0	8.44 - 7.81	2.57	10.94	2.94	19.99	5.25	41.69
7.0 - 6.0	7.81 - 6.81	4.63	17.54	4.76	29.25	7.86	64.03
6.0 - 0.0	6.81 - 0.00	37.46	117.77	33.76	186.30	55.69	430.97

Precipitation Runoff (Sv)

(^{14}C ka)	(cal ka)	Mississippi	St. Lawrence	Hudson River	Hudson Strait	Arctic	Total
18.0 - 16.5	21.39 - 19.67	0.1220	0.0461	0.0214	0.0598	0.0251	0.2742
16.5 - 15.2	19.67 - 18.17	0.0866	0.0461	0.0567	0.0598	0.0251	0.2743
15.2 - 13.5	18.17 - 16.21	0.1214	0.0464	0.0207	0.0593	0.0255	0.2733
13.5 - 13.0	16.21 - 15.63	0.0656	0.0471	0.0738	0.0581	0.0265	0.2710
13.0 - 12.3	15.63 - 14.29	0.1157	0.0447	0.0170	0.0560	0.0272	0.2606
12.3 - 12.0	14.29 - 14.07	0.0791	0.0447	0.0536	0.0560	0.0272	0.2606
12.0 - 11.7	14.07 - 13.68	0.0568	0.0425	0.0707	0.0530	0.0263	0.2493
11.7 - 11.5	13.68 - 13.46	0.0740	0.0425	0.0535	0.0530	0.0263	0.2493
11.5 - 11.0	13.46 - 13.00	0.0740	0.0806	0.0154	0.0530	0.0263	0.2493
11.0 - 10.0	13.00 - 11.44	0.0091	0.1257	0.0146	0.0537	0.0328	0.2360
10.0 - 9.7	11.44 - 11.17	0.0101	0.0844	0.0141	0.0530	0.0607	0.2224
9.7 - 9.4	11.17 - 10.62	0.0417	0.0844	0.0141	0.0530	0.0292	0.2224
9.4 - 9.3	10.62 - 10.46	0.0101	0.1160	0.0141	0.0530	0.0292	0.2224
9.3 - 9.1	10.46 - 10.24	0.0417	0.0844	0.0141	0.0530	0.0292	0.2224
9.1 - 8.4	10.24 - 9.45	0.0111	0.1160	0.0143	0.0475	0.0281	0.2170
8.4 - 7.7	9.45 - 8.44	0.0119	0.1036	0.0146	0.0571	0.0274	0.2146
7.7 - 7.0	8.44 - 7.81	0.0130	0.0550	0.0148	0.1006	0.0264	0.2098
7.0 - 6.0	7.81 - 6.81	0.0147	0.0556	0.0151	0.0927	0.0249	0.2030
6.0 - 0.0	6.81 - 0.00	0.0174	0.0548	0.0157	0.0867	0.0259	0.2007

APPENDIX B. RADIOCARBON AGES USED TO CONSTRAIN THE AGES OF ROUTING EVENTS

21 - 20 ^{14}C ka (LGM) (Figure 3A)

(1) 20,000 ± 800 (O-1325), 20,500 ± 800 (I-1864-A), 25,190 ± 280 (Beta-2763) [*Hallberg and Kemmis*, 1986]. (2) 21,460 ± 470 (ISGS-1486), 19,680 ± 460 (ISGS-532) [*Hansel and Johnson*, 1996]. (3) 20,910 ± 240 (ISGS-44) [*Mickelson et al.*, 1983]; 20,200 ± 140 (PITT-0507), 19,200 ± 140 (PITT-0508) [*Lowell et al.*, 1990]. (4) 21,750 ± 750 (SI-1590) [*Sirkin and Stuckenrath*, 1980]. (5) 20,700 ± 2000 (I-751) [*Mickelson et al.*, 1983].

16.5 - 15.2 ^{14}C ka (Erie Interstade) (Figure 3B)

(1) 15,775 ± 145 (Beta-10836) [*Bettis et al.*, 1996]. (2) 16,100 ± 500 (I-1024), 15,290 ± 110 (Beta-10836) [*Bettis et al.*, 1996]. (3) 16,980 ± 180 (Beta-10525) [*Bettis et al.*, 1996]. (4) 16,380 +660/-710 (DIC-1884) [*Muller and Calkin*, 1993]. (5) 16,650 ± 1880 (BGS-86) [*Muller and Calkin*, 1993]. (6) 16,650 ± 660 (GX-8488) [*Muller and Calkin*, 1993].

15.2 - 13.4 ^{14}C ka (Port Bruce Stade) (Figure 3C)

(1) 15,140 ± 220 (Beta-9797), 15,310 ± 180 (Beta-10838) [*Bettis et al.*, 1996]. (2) 14,470 ± 400 (W-512), 14,380 ± 180 (W-9765), 14,200 ± 500 (I-1402), 13,910 ± 400 (W-517), 13,900 ± 400 (I-1268), 13,820 ± 400 (W-513), 13,680 ± 80 (ISGS-552) [*Clayton and Moran*, 1982]. (3) 15,240 ± 120 (ISGS-465), 14,330 ± 250 (ISGS-1550) (these two ages bracket age of maximum ice-margin advance in Lake Michigan basin during this event) [*Hansel and Johnson*, 1992, 1996], 14,100 ± 640 (ISGS-1570), 13,890 ± 120 (ISGS-1649), 13,870 ± 170 (ISGS-1549) (these three ages bracket recessional phase of ice margin while it was still at southern end of the Lake Michigan basin) [*Hansel and Johnson*, 1992, 1996]. (4) 14,780 ± 192 (OWU-83) (wood in till), 14,500 ± 150 (ISGS-402), 14,300 ± 450 (W-198), 14,290 ± 130 (ISGS-72), 14,050 ± 75 (ISGS-348) (from depressions on or north of ice-margin position marking limit of advance) [*Fullerton*, 1980; *Mickelson et al.*, 1983].

13.5 - 13 ^{14}C ka (Mackinaw Interstade) (Figure 3D)

(1) 13,030 ± 250 (W-625) [*Ruhe*, 1969]. (2) 13,470 ± 130 (ISGS-1378) [*Monaghan and Hansel*, 1990]. (3) 12,920 ± 400 (W-430) [*Calkin and Feenstra*, 1985]. (4) 13,360 ± 440 (BGS-929) [*Barnett*, 1985]. (5) 13,100 ± 110 (GSC-2213) [*Gravenor and Stupavsky*, 1976]. (6) 12,915 ± 175 (GX-14781) [*Ridge and Larson*, 1990].

13 - 12.3 ^{14}C ka (Port Huron Stade) (Figure 3E)

(1) 12,000 ± 105 (DIC-1362), 12,020 ± 170 (I-8768) [*Clayton and Moran*, 1982]. (2) 12,920 ± 250 (WIS-626) [*Bettis et al.*, 1996]. (3) 12,880 ± 125 (WIS-1004), 12,260 ± 115 (WIS-1073), 12,520 ± 160 (WIS-1075) (basal ages that date deglaciation of Green Bay lobe) [*Maher and Mickelson*, 1996]. (4) 12,650 ± 350 (W-140), 12,660 ± 140 (ISGS-1190) [*Hansel et al.*, 1985]. (5) 12,800 ± 250 (W-240) [*Goldthwait*, 1958], 12,900 ± 200 (I-3175) [*Fullerton*, 1980].

12.3 ^{14}C ka (Lake Iroquois) (Figure 3F)

(1) 12,220 ± 350 (W-161) [*Hansel et al.*, 1985]. (2) 12,080 ± 300 (W-883), 12,660 ± 400 (W-861), 12,100 ± 100 (I-838) [*Muller and Calkin*, 1993].

12 - 11.7 ^{14}C ka (Two Creeks Interstade) (Figure 3G)

(1) 11,810 ± 90 (ISGS-1159), 11,870 ± 100 (ISGS-1137) [*Hansel et al.*, 1985]. (2) 11,760 ± 90 (ETH-8271), 11,805 ± 95 (ETH-8610), 11,865 ± 65 (ETH-8270), 11,885 ± 100 (ETH-8274), 11,890 ± 95 (ETH-8609), 11,915 ± 100 (ETH-8272), 11,965 ± 95 (ETH-8608), 11,980 ± 95 (ETH-8611), 12,015 ± 90 (ETH-8612), 12,035 ± 60 (ETH-8273) [*Kaiser*, 1994]. (3) 11,650 ± 170 (ISGS-1234), 11,700 ± 110 (ISGS-1061) [*Schneider*, 1990]. (4) 12,050 ± 80 (ETH-9241), 12,100 ± 100 (Beta-50967) [*Larson et al.*, 1994].

11.7 - 11 ^{14}C ka (Greatlakean Stade) (Figure 3H)

(1) 11,000 ± 80 (ISGS-1097), 11,180 ± 160 (ISGS-1121), 11,250 ± 180 (ISGS-1165), 11,610 ± 70 (ISGS-985), 11,630 ± 120 (ISGS-1180), 11,740 ± 100 (ISGS-1147), 11,740 ± 270 (ISGS-1117), 11,810 ± 90 (ISGS-1159), 11,815 ± 640 (IU-67), 11,850 ± 150 (ISGS-1454), 11,870 ± 100 (ISGS-1137), 11,960 ± 130 (GX-11934) [*Hansel et al.*, 1985; *Hansel and Mickelson*, 1988].

11.5 ^{14}C ka (Opening of St. Lawrence) (Figure 3I)

(1) 11,530 ± 90 (TO-703) [*Rodrigues*, 1992]. (2) 11,460 ± 70 (TO-702) [*Rodrigues*, 1992].

11 - 10 ^{14}C ka (Moorhead phase of Lake Agassiz) (Figure 3J)

(1) 10,050 ± 300 (W-1005), 10,080 ± 280 (W-900), 10,340 ± 170 (I-5213), 10,820 ± 190 (TAM-1), 10,960 ± 300 (W-723) [*Moran et al.*, 1973]. (2) 10,000 ± 150 (GSC-870), 10,200 ± 80 (GSC-1909), 10,550 ± 200 (Y-411) [*Teller*, 1980]. (3) 10,000 ± 110 (TO-4286), 10,040 ± 70 (TO-4871), 10,140 ± 80 (TO-4872), 10,340 ± 100 (TO-4285) [*Teller et al.*, 1999]. (4) 10,020 ± 120 (BGS-1303), 10,050 ± 180 (WAT-1935), 10,080 ± 160 (BGS-1320), 10,100 ± 180 (WAT-1936), 10,100 ± 200 (WAT-1689),), 10,400 ± 160 (WAT-1749) [*Bajc*, 1991].

10.3 - 9.7 ^{14}C ka (Clearwater Outlet of Lake Agassiz) (Figure 3K)

(1) 9,710 ± 130 (AECV-1183c), 9,860 ± 230 (GS-5031), 9,910 ± 190 (GSC-4302), 10,015 ± 320 (GX-5306), 10,310 ± 290 (GX-5301-II) [*Smith and Fisher*, 1993].

10 - 9.7 ^{14}C ka (Marquette Readvance) (Figure 3K)

(3) 9,730 ± 140 (I-5082), 10,100 ± 100 (WIS-409) [*Drexler et al.*, 1983]. (4) 10,100 ± 250 (W-1540), 10,250 ± 250 (W-1541) [*Peterson*, 1982]; 10,230 ± 500 (W-1414) [*Drexler et al.*, 1983]. (5) 9,500 ± 350 (W-1150), 9,600 ± 350 (W-965) [*Hack*, 1965]; 10,230 ± 280 (W-964), [*Levin et al.*, 1965]. (6) 9,545 ± 225 (DAL-340), 9,780 ± 250 (W-3904), 9,850 ± 300 (W-3866), 10,220 ± 215 (DAL-338), 10,320 ± 300 (W-3896) [*Drexler et al.*, 1983]; 9,895 ± 55 (A-7878), 9,910 ± 55 (A-7876), 9,965 ± 55 (A-7877), 10,040 ± 65 (A-7879), 10,050 ± 55 (A-7883), 10,075 +95/-90 (A-7880), 10,155 ± 65 (A-7882), 10,200 ± 55 (A-7875) [*Lowell et al.*, 1999].

10 - 9.4 ^{14}C ka (Emerson phase of Lake Agassiz) (Figure 3K)

(2) 9,810 ± 300 (W-1360), 9,820 ± 300 (W-1361), 9,890 ± 300 (I-4853), 9,900 ± 400 (W-993), 9,930 ± 280 (W-388), 9,940 ± 160 (I-3880) [*Clayton and Moran*, 1982]. (7) 9,700 ± 140 (GSC-797), 9,990 ± 160 (GSC-391), 10,000 ± 280 (GSC-1428) [*Teller*, 1980], and 9,510 ± 90 (GSC-4490), 9,600 ± 70 (TO-534) [*Teller*, 1989]. (8) 9,330 ± 80 (TO-4856), 9,340 ± 90 (TO-4855), 9,380 ± 90 (TO-4873), 9,460 ± 90 (TO-4284), 9,950 ± 90 (TO-4874) [*Teller et al.*, 1999]. (9) 9,530 ± 140 (WAT-1934) (Bajc, 1991). (10) 9,350 ± 100 (WIS-1324), 10,050 ± 100 (WIS-1325) [*Björck and Keister*, 1983].

9.4 - 9.3 ^{14}C ka

Ages of 9,340 ± 80 (GSC-4950) and 9,400 ± 100 (GSC-4933) on wood in a beach below the Campbell level of Lake Agassiz indicate eastward drainage [*Teller et al.*, 1999]. Ages of 9,345 ± 240 (GX-4883) and 9,380 ± 150 (GSC-287) on the Minong beach in the Lake Superior basin also indicate ice retreat from the Lake Superior basin to allow eastward drainage [*Zoltai*, 1965; *Drexler et al.*, 1983; *Teller and Thorleifson*, 1983]. An age of 9,560 ± 160 (GSC-1360) associated with a higher lake level in the Lake Huron basin ("early" Mattawa) due to increased discharge from Lake Agassiz [*Lewis et al.*, 1994].

9.3 - 9.1 ^{14}C ka

Ages on a low lake level in the Lake Huron basin ("mid" Stanley) due to reduced discharge from Lake Agassiz [*Lewis et al.*, 1994]: 8,985 ± 100 (AA-8770), 9,130 ± 140 (I-4036), 9,235 ± 100 (AA-8771), 9,260 ± 290 (GSC-1971), 9,350 ± 95 (AA-95).

9.1 - 7.8 ^{14}C ka

Ages which constrain a rise in lake level in the Lake Huron basin ("main" Mattawa) suggest increased inflow from Lake Agassiz after 9.1 ^{14}C ka [*Lewis et al.*, 1994]: 8,310 ± 130 (GSC-1979), 8,475 ± 90 (AA-8774), 8,760 ± 250 (GSC-514), 8,785 ± 145 (I-7857). Ages which constrain influx of Lake Agassiz waters into the Lake Michigan basin [*Colman et al.*, 1994]: 7,900 ± 70 (AA-5912), 8,040 ± 70 (AA-4978), 8,350 ± 80 (AA-5385), 8,590 ± 70 (AA-6868), 8,890 ± 90 (AA-4615), 8,910 ± 70 (AA-5897), 9,180 ± 80 (AA-4616).

Acknowledgments. This work was funded by grants from the National Science Foundation (P.U.C.) and the Natural Sciences and Engineering Research Council of Canada (J.T.T.). We thank Dave Leverington and Grant Penn for their help in data processing, Steve Hostetler for helpful discussions, and J.T. Andrews, A.S. Dyke, and R.S. Webb for reviews.

REFERENCES

Aagaard, K., and E.C. Carmack, The role of sea ice and other fresh water in the Arctic circulation, *Journal of Geophysical Research*, *94*, 14,485-14,498, 1989.

Alley, R.B., and D.R. MacAyeal, Ice-rafted debris associated with binge/purge oscillations of the Laurentide Ice Sheet, *Paleoceanography*, *9*, 503-511, 1994.

Andrews, J.T., and K. Tedesco, Detrital carbonate-rich

sediments, northwestern Labrador Sea: Implications for ice-sheet dynamics and iceberg rafting (Heinrich) events in the North Atlantic, *Geology*, *20*, 1087-1090, 1992.

Andrews, J.T., A.E. Jennings, M. Kerwin, W. Manley, G.H. Miller, G. Bond, and B. MacLean, A Heinrich-like event, H-0 (DC-0): Source(s) for detrital carbonate in the North Atlantic during the Younger Dryas chronozone, *Paleoceanography*, *10*, 943-952, 1995.

Arkhipov, S.A., Stratigraphy and paleoecology of the Sartan Glaciation in West Siberia, *Quaternary International*, *45/46*, 29-42, 1998.

Arkhipov, S.A., J. Ehlers, R.G. Johnson, and H.E. Wright, Jr., Glacial drainage towards the Mediterranean during the Middle and Late Pleistocene, *Boreas*, *24*, 196-206, 1995.

Astakhov, V., Last glaciation of West Siberia, *Sveriges Geologiska Undersokning, Serie Ca*, *81*, 21-30, 1992.

Astakhov, V., The last ice sheet of the Kara Sea; terrestrial constraints on its age, *Quaternary International*, *45/46*, 19-28, 1998.

Attig, J.W., L. Clayton, and D.M. Mickelson, Correlation of late Wisconsin glacial phases in the western Great Lakes area, *Geological Society of America Bulletin*, *96*, 1585-1593, 1985.

Bajc, A.F., Glacial and glaciolacustrine history of the Fort Francis-Rainy River area, Ontario, Canada, Ph.D. thesis, University of Waterloo, Waterloo, 1991.

Bard, E., and F. Rostek, Abrupt climatic changes during the last deglaciation in the north-west Pacific, paper presented at Chapman Conference on Mechanisms of Millennial-Scale Global Climate Change, Snowbird, Utah, 1998.

Barnett, P.J., Glacial retreat and lake levels, north central Lake Erie basin, Ontario, in *Quaternary Evolution of the Great Lakes*, edited by P.F. Karrow, and P.E. Calkin, pp. 185-194, Geological Association of Canada Special Paper 30, St. John's, 1985.

Behl, R.J., and J.P. Kennett, Brief interstadial events in the Santa Barbara basin, NE Pacific, during the past 60 kyr, *Nature*, *379*, 243-246, 1996.

Benson, L.V., J.W. Burdett, S.P. Lund, M. Kashgarian, and S. Mensing, Nearly synchronous Northern Hemispheric climate change during the last glacial termination?, *Nature*, *388*, 263-265, 1997.

Bettis, E.A., Jr., D.J. Quade, and T.J. Kemmis, Hogs, bogs, and logs: Quaternary deposits and environmental geology of the Des Moines Lobe, in *Geological Survey Bureau Guidebook 18*, pp. 1-170, Iowa Department of Natural Resources, 1996.

Björck, S.V., A review of the history of the Baltic Sea, 13.0 - 8.0 ka BP, *Quaternary International*, *27*, 19-40, 1995.

Björck, S.V., and C.M. Keister, The Emerson Phase of Lake Agassiz, independently registered in northwestern Minnesota and northwestern Ontario, *Canadian Journal of Earth Sciences*, *20*, 1536-1542, 1983.

Boden, P., R.G. Fairbanks, J.D. Wright, and L.H. Burckle, High-resolution stable isotope records from southwest Sweden: The drainage of the Baltic Ice Lake and Younger Dryas ice margin oscillations, *Paleoceanography*, *12*, 39-49, 1997.

Bond, G., W.S. Broecker, S. Johnsen, J. McManus, L. Labeyrie, J. Jouzel, and G. Bonani, Correlations between climate records from North Atlantic sediments and Greenland ice, *Nature*, *365*, 143-147, 1993.

Bond, G., and R. Lotti, Iceberg discharges into the North Atlantic on millennial time scales during the last deglaciation, *Science*, *267*, 1005-1010, 1995.

Bond, G., W. Showers, M. Cheseby, R. Lotti, P. Almasi, P. deMenocal, P. Priore, H. Cullen, I. Hajdas, and G. Bonani, A pervasive millennial-scale cycle in North Atlantic Holocene and glacial climates, *Science*, *278*, 1257-1266, 1997.

Boyle, E.A., and L.D. Keigwin, North Atlantic circulation during the last 20,000 years linked to high-latitude surface temperature, *Nature*, *330*, 35-40, 1982.

Boyle, E.A., and L.D. Keigwin, North Atlantic thermohaline circulation during the past 20,000 years linked to high-latitude surface temperature, *Nature*, *330*, 35-40, 1987.

Broecker, W.S., Massive iceberg discharges as triggers for global climate change, *Nature*, *372*, 421-424, 1994.

Broecker, W.S., and W.R. Farrand, Radiocarbon age of the Two Creeks forest bed, Wisconsin, *Geological Society of America Bulletin*, *74*, 795-802, 1963.

Broecker, W.S., D.M. Peteet, and D. Rind, Does the ocean-atmosphere system have more than one stable mode of operation?, *Nature*, *315*, 21-26, 1985.

Broecker, W.S., M. Andree, W. Wolfli, H. Oeschger, G. Bonani, J. Kennett, and D. Peteet, The chronology of the last deglaciation: Implications of the cause of the Younger Dryas event, *Paleoceanography*, *3*, 1-19, 1988.

Broecker, W.S., J.P. Kennett, B.P. Flower, J.T. Teller, S. Trumbore, G. Bonani, and W. Wolfli, Routing of meltwater from the Laurentide ice sheet during the Younger Dryas cold episode, *Nature*, *341*, 318-321, 1989.

Broecker, W.S., G. Bond, M. Klas, E. Clark, and J. McManus, Origin of the northern Atlantic's Heinrich events, *Climate Dynamics*, *6*, 265-273, 1992.

Calkin, P.E., and B.H. Feenstra, Evolution of the Erie-Basin Great Lakes, in *Quaternary Evolution of the Great Lakes*, edited by P.F. Karrow and P.E. Calkin, pp. 149-170, Geological Association of Canada Special Paper 30, St. John's, 1985.

Clark, P.U., Unstable behavior of the Laurentide Ice Sheet over deforming sediment and its implications for climate change, *Quaternary Research*, *41*, 19-25, 1994.

Clayton, L., Chronology of Lake Agassiz drainage to Lake Superior, in *Glacial Lake Agassiz*, edited by J.T. Teller and L. Clayton, pp. 291-307, Geological Association of Canada Special Paper 26, St. John's, 1983.

Clayton, L., and S.R. Moran, Chronology of late Wisconsinan glaciation in middle North America, *Quaternary Science Reviews*, *1*, 55-82, 1982.

Clayton, L., J.T. Teller, and J.W. Attig, Surging of the southwestern part of the Laurentide Ice Sheet, *Boreas*, *14*, 235-241, 1985.

Colman, S.M., L.D. Keigwin, and R.M. Forester, Two episodes

of meltwater influx from glacial Lake Agassiz into the Lake Michigan basin and their climatic contrasts, *Geology*, *22*, 547-550, 1994.

Cortijo, E., L. Labeyrie, L. Vidal, M. Vautravers, M. Chapman, J.-C. Duplessy, M. Elliot, M. Arnold, J.L. Turon, and G. Auffret, Changes in sea surface hydrology associated with Heinrich event 4 in the North Atlantic Ocean between 40° and 60°N, *Earth and Planetary Science Letters*, *146*, 29-45, 1997.

de Vernal, A., C. Hillaire-Marcel, and G. Bilodeau, Reduced meltwater outflow from the Laurentide ice margin during the Younger Dryas, *Nature*, *381*, 774-777, 1996.

Denton, G.H., and T.J. Hughes, The Arctic ice sheet: an outrageous hypothesis, in *The Last Great Ice Sheets*, edited by G.H. Denton and T.J. Hughes, pp. 440-467, Wiley-Interscience, New York, 1981.

Dreimanis, A., and P.F. Karrow, Glacial history of the Great Lakes-St. Lawrence region, the classification of the Wisconsin(an) stage, and its correlatives, in *24th International Geological Congress*, pp. 5-15, Montreal, 1972.

Dreimanis, A., Late Wisconsin glacial retreat in the Great Lakes region, North America, *New York Academy of Science Annals*, *288*, 70-89, 1977.

Drexler, C.W., W.R. Farrand, and J.D. Hughes, Correlations of glacial lakes in the Superior basin with eastward discharge events from Lake Agassiz, in *Glacial Lake Agassiz*, edited by J.T. Teller and L. Clayton, pp. 309-329, Geological Association of Canada Special Paper 26, St. John's, 1983.

Duplessy, J.-C., L. Labeyrie, A. Juillet-Leclerc, F. Maitre, J. Duprat, and M. Sarnthein, Surface salinity reconstruction of the North Atlantic Ocean during the last glacial maximum, *Oceanologica Acta*, *14*, 311-324, 1991.

Dyke, A.S., and V.K. Prest, Late Wisconsinan and Holocene history of the Laurentide Ice Sheet, *Géographie Physique et Quaternaire*, *41*, 237-263, 1987.

Elson, J.A., Geology of Glacial Lake Agassiz, in *Life, Land and Water*, edited by W.J. Mayer-Oakes, pp. 37-95, University of Manitoba Press, Winnipeg, 1967.

Evenson, E.B., W.R. Farrand, D.F. Eschman, D.M. Mickelson, and L.J. Maher, Greatlakean substage- A replacement for Valderan substage in the Lake Michigan basin, *Quaternary Research*, *6*, 411-424, 1976.

Fanning, A.F., and A.J. Weaver, Temporal-geographical meltwater influences on the North Atlantic conveyor: Implications for the Younger Dryas, *Paleoceanography*, *12*, 307-320, 1997.

Fawcett, P.J., A.M. Águstsdóttir, R.B. Alley, and C.A. Shuman, The Younger Dryas termination and North Atlantic Deep Water formation: Insights from climate model simulations and Greenland ice cores, *Paleoceanography*, *12*, 23-38, 1997.

Fenton, M.M., S.R. Moran, J.T. Teller, and L. Clayton, Quaternary stratigraphy and history in the southern part of the Lake Agassiz basin, in *Glacial Lake Agassiz*, edited by J.T. Teller and L. Clayton, pp. 49-74, Geological Association of Canada Special Paper 26, St. John's, 1983.

Fisher, T.G., and D.G. Smith, Glacial Lake Agassiz: its northwest maximum extent and outlet in Saskatchewan (Emerson Phase), *Quaternary Science Reviews*, *13*, 845-888, 1994.

Fisher, T.G., and C. Souch, Northwest outlet channels of Lake Agassiz, isostatic tilting and a migrating continental drainage divide, Saskatchewan, Canada, *Geomorphology*, *25*, 57-73, 1998.

Fisheries and Environment Canada, *Hydrological Atlas of Canada*, Department of Energy, Mines, and Resources of Canada, 1978.

Fullerton, D.S., Preliminary correlation of post-Erie interstadial events (16,000-10,000 radiocarbon years before present), central and eastern Great Lakes region, and Hudson, Champlain, and St. Lawrence lowlands, United States and Canada, *Geological Survey Professional Paper 1089*, 52 pp., Washington, 1980.

Goldthwait, R.P., Wisconsin age forests in western Ohio: I. Age and glacial events, *Ohio Journal of Science*, *58*, 209-219, 1958.

Gravenor, C.P., and M. Stupavsky, Magnetic, physical, and lithologic properties and age of till exposed along the east coast of Lake Huron, Ontario, *Canadian Journal of Earth Sciences*, *13*, 1655-1666, 1976.

Grosswald, M.G., Late Weichselian ice sheets of northern Eurasia, *Quaternary Research*, *13*, 1-32, 1980.

Grosswald, M.G., An Antarctic-style ice sheet in the Northern Hemisphere: toward a new global glacial theory, *Polar Geography and Geology*, *12*, 239-267, 1988.

Grosswald, M.G., Late Weichselian ice sheets in Arctic and Pacific Siberia, *Quaternary International*, *45/46*, 3-18, 1998.

Hack, J.T., Postglacial drainage evolution and stream geometry in the Ontonagon area, Michigan, *U.S. Geological Survey Professional Paper 504-B*, 1-40, Washington, 1965.

Häkkinen, S., An Arctic source for the Great Salinity Anomaly: a simulation of the Arctic ice-ocean system for 1955-1975, *Journal of Geophysical Research*, *98*, 16397-16410, 1993.

Hallberg, G.R., and T.J. Kemmis, Stratigraphy and correlation of the glacial deposits of the Des Moines and James Lobes and adjacent areas in North Dakota, South Dakota, Minnesota, and Iowa, *Quaternary Science Reviews*, *5*, 11-16, 1986.

Hansel, A.K., D.M. Mickelson, A.F. Schneider, and C.E. Larsen, Late Wisconsinan and Holocene history of the Michigan basin, in *Quaternary Evolution of the Great Lakes*, edited by P.F. Karrow and P.E. Calkin, pp. 39-53, Geological Association of Canada Special Paper 30, St. John's, 1985.

Hansel, A.K., and D.M. Mickelson, A reevaluation of the timing and causes of high lake phases in the Lake Michigan basin, *Quaternary Research*, *29*, 113-128, 1988.

Hansel, A.K., and W.H. Johnson, Fluctuations of the Lake Michigan lobe during the late Wisconsin subepisode, *Sveriges Geologiska Undersokning*, *81*, 133-144, 1992.

Hansel, A.K., and W.H. Johnson, Wedron and Mason Groups: Lithostratigraphic reclassification of deposits of the Wisconsin Episode, Lake Michigan Lobe area, *Illinois State Geological Survey Bulletin 104*, 116 pp., 1996.

Harris, K.L., S.R. Moran, and L. Clayton, Late Quaternary stratigraphic nomenclature, Red River Valley, North Dakota and Minnesota, *North Dakota Geological Survey Miscellaneous Series 52*, 47 pp., 1974.

Hobbs, H.C., Drainage relationships of Glacial Lakes Aitkin and Upham and early Lake Agassiz in northeastern Minnesota, in *Glacial Lake Agassiz*, edited by J.T. Teller and L. Clayton, pp. 245-259, Geological Association of Canada Special Paper 26, St. John's, 1983.

Hostetler, S.W., P.U. Clark, P.J. Bartlein, A.C. Mix, and N.G. Pisias, Atmospheric transmission of North Atlantic Heinrich events, *Journal of Geophysical Research, 104*, 3947-3952, 1999.

Hughes, T., Ice dynamics and deglaciation models when ice sheets collapsed, in *North America and adjacent oceans during the last deglaciation*, edited by W.F. Ruddiman, and H.E. Wright, Jr., pp. 183-220, Geological Society of America, Boulder, 1987.

Jennings, A.E., K.A. Tedesco, J.T. Andrews, and M.E. Kirby, Shelf erosion and glacial ice proximity in the Labrador Sea during and after Heinrich events (H-3 or 4 to H-0) as shown by foraminifera, in *Late Quaternary Paleoceanography of the North Atlantic Margins*, edited by J.T. Andrews, W.E.N. Austin, H. Bergsten, and A.E. Jennings, pp. 29-49, Geological Society of London Special Publication, 1996.

Johnson, R.G., and B.T. McClure, A model for northern hemisphere continental ice sheet variation, *Quaternary Research, 6*, 325-353, 1976.

Johnson, W.H., A.K. Hansel, E.A. Bettis, Jr., P.F. Karrow, G.J. Larson, T.V. Lowell, and A.F. Schneider, Late Quaternary temporal and event classifications, Great Lakes region, North America, *Quaternary Research, 47*, 1-12, 1997.

Kaiser, K.F., Two Creeks Interstade dated through dendrochronology and AMS, *Quaternary Research, 42*, 288-298, 1994.

Keigwin, L.D., G.A. Jones, S.J. Lehman, and E.A. Boyle, Deglacial meltwater discharge, North Atlantic deep circulation, and abrupt climate change, *Journal of Geophysical Research, 96*, 16811-16826, 1991.

Keigwin, L.D., and S.J. Lehman, Deep circulation change linked to Heinrich event 1 and Younger Dryas in a middepth North Atlantic core, *Paleoceanography, 9*, 185-194, 1994.

Keigwin, L.D., and G.A. Jones, The marine record of deglaciation from the continental margin off Nova Scotia, *Paleoceanography, 7*, 499-520, 1995.

Kennett, J.P., and N.J. Shackleton, Laurentide ice sheet meltwater recorded in Gulf of Mexico deep-sea cores, *Science, 188*, 147-150, 1975.

Klassen, R.W., Lake Agassiz and the late glacial history of northern Manitoba, in *Glacial Lake Agassiz*, edited by J.T. Teller and L. Clayton, pp. 97-115, Geological Association of Canada Special Paper 26, St. John's, 1983.

Kutzbach, J., R. Gallimore, S. Harrison, P. Behling, R. Selin, and F. Laarif, Climate and biome simulations for the past 21,000 years, *Quaternary Science Reviews, 17*, 473-506, 1998.

Labeyrie, L., L. Vidal, E. Cortijo, M. Paterne, M. Arnold, J.C. Duplessy, M. Vautravers, M. Labracherie, J. Duprat, J.L. Turon, F. Grousset, and T. van Weering, Surface and deep hydrology of the northern Atlantic Ocean during the past 150,000 years, *Philosophical Transactions Royal Society of London B, 348*, 255-264, 1995.

Larson, G.J., T.V. Lowell, and N.E. Ostrom, Evidence for the Two Creeks interstade in the Lake Huron basin, *Canadian Journal of Earth Sciences, 31*, 793-797, 1994.

Lehman, S.J., G.A. Jones, L.D. Keigwin, E.S. Andersen, G. Butenko, and S.-R. Østmo, Initiation of Fennoscandian ice-sheet retreat during the last deglaciation, *Nature, 349*, 513-516, 1991.

Lehman, S.J., and L.D. Keigwin, Sudden changes in North Atlantic circulation during the last deglaciation, *Nature, 356*, 757-762, 1992.

Leventer, A., D.F. Williams, and J.P. Kennett, Dynamics of the Laurentide ice sheet during the last deglaciation: Evidence from the Gulf of Mexico, *Earth and Planetary Science Letters, 59*, 11-17, 1982.

Levin, B., P.C. Ives, C.L. Oman, and M. Rubin, U.S. Geological Survey radiocarbon dates, *Radiocarbon, 7*, 376, 1965.

Lewis, C.F.M., T.C. Moore, Jr., D.K. Rea, D.L. Dettman, A.M. Smith, and L.A. Mayer, Lakes of the Huron basin: Their record of runoff from the Laurentide Ice Sheet, *Quaternary Science Reviews, 13*, 891-922, 1994.

Licciardi, J.M., P.U. Clark, J.W. Jenson, and D.R. MacAyeal, Deglaciation of a soft-bedded Laurentide Ice Sheet, *Quaternary Science Reviews, 17*, 427-448, 1998.

Lowell, T.V., K.M. Savage, C.S. Brockman, and R. Stuckenrath, Radiocarbon analyses from Cincinnati, Ohio, and their implications for glacial stratigraphic interpretations, *Quaternary Research, 34*, 1-11, 1990.

Lowell, T.V., G.J. Larson, J.D. Hughes, and G.H. Denton, Age verification of the Marquette buried forest and the Younger Dryas advance of the Laurentide Ice Sheet, *Canadian Journal of Earth Sciences*, in press, 1999.

MacAyeal, D.R., Growth/purge oscillations of the Laurentide ice sheet as a cause of the North Atlantic's Heinrich events, *Paleoceanography, 8*, 775-784, 1993.

Maher, L.J., Jr., and D.M. Mickelson, Palynological and radiocarbon evidence for deglacial events in the Green Bay Lobe, Wisconsin, *Quaternary Research, 46*, 251-259, 1996.

Maier-Reimer, E., and U. Mikolajewicz, Experiments with an OGCM on the cause of the Younger Dryas, in *Oceanography, 1988*, edited by A. Ayala-Castanares, W. Wooster, and A. Yanez-Arancibia, pp. 87-100, Universidad Nacional Autonoma de Mexico Press, Mexico City, 1989.

Manabe, S., and R.J. Stouffer, Two stable equilibria of a coupled ocean-atmosphere model, *Journal of Climate, 1*, 841-866, 1988.

Manabe, S., and R.J. Stouffer, Multiple-century response of a coupled ocean-atmosphere model to an increase of atmospheric carbon dioxide, *Journal of Climate, 7*, 5-23, 1994.

Manabe, S., and R.J. Stouffer, Coupled ocean-atmosphere model response to freshwater input: Comparison to Younger Dryas event, *Paleoceanography, 12*, 321-336, 1997.

Marchitto, T.M., and K. Wei, History of Laurentide meltwater flow to the Gulf of Mexico during the last deglaciation, as revealed by reworked calcareous nannofossils, *Geology, 23*, 779-782, 1995.

Marshall, S.J., and G.K.C. Clarke, A continuum mixture model of ice stream thermomechanics in the Laurentide Ice Sheet. 2. Application to the Hudson Strait Ice Stream, *Journal of Geophysical Research, 102*, 20615-20637, 1997.

Maslin, M.A., N.J. Shackleton, and U. Pflaumann, Surface water temperature, salinity, and density changes in the northeast Atlantic during the last 45,000 years: Heinrich events, deep water formation, and climatic rebounds, *Paleoceanography, 10*, 527-544, 1995.

Matsch, C.L., and H.E. Wright, Jr., The southern outlet of Lake Agassiz, in *Life, Land, and Water*, edited by W.J. Mayer-Oakes, pp. 121-140, University of Manitoba Press, Winnipeg, 1967.

Mayewski, P.A., G.H. Denton, and T.J. Hughes, Late Wisconsin ice sheets of North America, in *The Last Great Ice Sheets*, edited by G.H. Denton and T.J. Hughes, pp. 67-178, John Wiley and Sons, New York, 1981.

McCabe, A.M., and P.U. Clark, Ice-sheet variability around the North Atlantic Ocean during the last deglaciation, *Nature, 392*, 373-377, 1998.

McManus, J.F., R.F. Ardison, W.S. Broecker, M.Q. Fleisher, and S.M. Higgins, Radiometrically determined sedimentary fluxes in the sub-polar North America during the last 140,000 years, *Earth and Planetary Science Letters, 155*, 29-43, 1998.

Mickelson, D.M., L. Clayton, D.S. Fullerton, and H.W. Borns, Jr., The late Wisconsin glacial record of the Laurentide ice sheet in the United States, in *Late Quaternary Environments of the Western United States. Vol. 1. The Late Pleistocene*, edited by S.C. Porter, pp. 3-37, University of Minnesota Press, Minneapolis, 1983.

Mikolajewicz, U., T.J. Crowley, A. Schiller, and R. Voss, Modelling teleconnections between the North Atlantic and North Pacific during the Younger Dryas, *Nature, 387*, 384-387, 1997.

Miller, G.H., and D.S. Kaufman, Rapid fluctuations of the Laurentide Ice Sheet at the mouth of Hudson Strait: New evidence for ocean/ice-sheet interactions as a control on the Younger Dryas, *Paleoceanography, 5*, 907-919, 1990.

Monaghan, G.W., and A.K. Hansel, Evidence for the intra-Glenwood (Mackinaw) low-water phase of glacial Lake Chicago, *Canadian Journal of Earth Sciences, 27*, 1236-1241, 1990.

Moran, S.R., L. Clayton, M. Scott, and J. Brophy, Catalog of North Dakota radiocarbon dates, *North Dakota Geological Survey Miscellaneous Series, 53*, 1-51, 1973.

Mörner, N.-A., and A. Dreimanis, The Erie interstade, in *The Wisconsinan Stage: Geological Society of America Memoir 136*, edited by R.F. Black, R.P. Goldthwait, and H.B. Willman, pp. 107-134, 1973.

Muller, E.H., and P.E. Calkin, Timing of Pleistocene glacial events in New York State, *Canadian Journal of Earth Sciences, 30*, 1829-1845, 1993.

Mysak, L.A., D.K. Manak, and R.F. Marsden, Sea-ice anomalies observed in Greenland and Labrador seas during 1901-1984 and their relation to an interdecadal Arctic climate cycle, *Climate Dynamics, 5*, 111-133, 1990.

Oppo, D.W., and S.J. Lehman, Suborbital timescale variability of North Atlantic Deep Water during the past 200,000 years, *Paleoceanography, 10*, 901-910, 1995.

Peterson, W.L., Preliminary surficial geologic map of the Iron River 1 x 2 quadrangle, Michigan and Wisconsin, *U.S. Geological Survey Open-File Report 82-301*, 1-18, 1982.

Prest, V.K., Quaternary geology of Canada, in *Geology and Economic Minerals of Canada*, edited by R.W. Douglas, pp. 676-764, Geological Survey of Canada, 1970.

Rahmstorf, S., Rapid climate transitions in a coupled ocean-atmosphere model, *Nature, 372*, 82-85, 1994.

Rahmstorf, S., Bifurcations of the Atlantic thermohaline circulation in response to changes in the hydrological cycle, *Nature, 378*, 145-149, 1995a.

Rahmstorf, S., Multiple convection patterns and thermohaline flow in an idealized OGCM, *Journal of Climate, 8*, 3028-3039, 1995b.

Ridge, J.C., and F.D. Larsen, Reevaluation of Antevs' New England varve chronology and new radiocarbon dates of sediments from glacial Lake Hitchcock, *Geological Society of America Bulletin, 102*, 889-899, 1990.

Ridge, J.C., Shed Brook discontinuity and Little Falls gravel: Evidence for the Erie interstade in central New York, *Geological Society of America Bulletin, 109*, 652-665, 1997.

Rind, D., D. Peteet, W.S. Broecker, A. McIntyre, and W.F. Ruddiman, The impact of cold North Atlantic sea-surface temperatures on climate: Implications for the Younger Dryas cooling (11-10 k), *Climate Dynamics, 1*, 3-33, 1986.

Rodrigues, C.G., Succession of invertebrate microfossils and the late Quaternary deglaciation of the central St. Lawrence lowland, Canada and United States, *Quaternary Science Reviews, 11*, 503-534, 1992.

Rostek, F., G. Ruhland, F.C. Bassinot, P.J. Müller, L.D. Labeyrie, Y. Lancelot, and E. Bard, Reconstructing sea surface temperature and salinity using $\delta^{18}O$ and alkenone records, *Nature, 364*, 319-321, 1993.

Ruhe, R.V., *Quaternary Landscapes of Iowa*, 255 pp., Iowa State University Press, Ames, 1969.

Rutter, N., Problematic ice sheets, *Quaternary International, 28*, 19-37, 1995.

Sarnthein, M., K. Winn, S. Jung, J.-C. Duplessy, L. Labeyrie, H. Erlenkeuser, and G. Ganssen, Changes in east Atlantic deepwater circulation over the last 30,000 years - an eight time slice reconstruction, *Paleoceanography, 9*, 209-268, 1994.

Schiller, A., U. Mikolajewicz, and R. Voss, The stability of the thermohaline circulation in a coupled ocean-atmosphere general circulation model, *Climate Dynamics, 13*, 325-348, 1997.

Schneider, A.F., Radiocarbon confirmation of the Greatlakean age of the type Two Rivers till of eastern Wisconsin, in *Late Quaternary History of the Lake Michigan Basin*, edited by A.F. Schneider and G.S. Fraser, pp. 51-56, Geological Society of America, Boulder, 1990.

Schulz, H., U. von Rad, and H. Erlenkeuser, Correlation between

Arabian Sea and Greenland climate oscillations of the past 110,000 years, *Nature, 393*, 54-57, 1998.

Sirkin, L., and R. Stuckenrath, The Portwashingtonian warm interval in the northern Atlantic coastal plain, *Geological Society of America Bulletin, 91*, 332-336, 1980.

Smith, D.G., and T.G. Fisher, Glacial Lake Agassiz: The northwestern outlet and paleoflood, *Geology, 21*, 9-12, 1993.

Stocker, T.F., and D.G. Wright, Rapid transitions of the ocean's deep circulation induced by changes in surface water fluxes, *Nature, 351*, 729-732, 1991.

Stocker, T.F., and D.G. Wright, Rapid changes in ocean circulation and atmospheric radiocarbon, *Paleoceanography, 11*, 773-795, 1996.

Stocker, T.F., and A. Schmittner, Influence of CO_2 emission rates on the stability of the thermohaline circulation, *Nature, 388*, 862-865, 1997.

Stuiver, M., and P.J. Reimer, Extended ^{14}C data base and revised Calib 3.0 ^{14}C age calibration program, *Radiocarbon, 35*, 215-230, 1993.

Stuiver, M., P.J. Reimer, E. Bard, J.W. Beck, G.S. Burr, K.A. Hughen, B. Kromer, G. McCormac, J. van der Plicht, and M. Spurk, INTCAL98 radiocarbon age calibration, 24,000 - 0 cal B.P., *Radiocarbon, 40*, 1041-1084, 1998.

Teller, J.T., Radiocarbon dates in Manitoba, *Manitoba Mineral Resources Division, Geological Report GR80-4*, 1-61, 1980.

Teller, J.T., Glacial Lake Agassiz and its influence on the Great Lakes, in *Quaternary Evolution of the Great Lakes*, edited by P.F. Karrow and P.E. Calkin, pp. 1-16, Geological Association of Canada Special Paper 30, St. John's, 1985.

Teller, J.T., Proglacial lakes and the southern margins of the Laurentide ice sheet, in *North America and Adjacent Oceans during the last deglaciation*, edited by W.F. Ruddiman, and H.E. Wright, Jr., pp. 39-69, Geological Society of America, Boulder, 1987.

Teller, J.T., Importance of the Rossendale site in establishing a deglacial chronology along the southwestern margin of the Laurentide Ice Sheet, *Quaternary Research, 32*, 12-23, 1989.

Teller, J.T., Volume and routing of late-glacial runoff from the southern Laurentide Ice Sheet, *Quaternary Research, 34*, 12-23, 1990a.

Teller, J.T., Meltwater and precipitation runoff to the North Atlantic, Arctic, and Gulf of Mexico from the Laurentide Ice Sheet and adjacent regions during the Younger Dryas, *Paleoceanography, 5*, 897-905, 1990b.

Teller, J.T., History and drainage of large ice-dammed lakes along the Laurentide Ice Sheet, *Quaternary International, 28*, 83-92, 1995a.

Teller, J.T., The impact of large ice sheets on continental paleohydrology, in *Global Continental Paleohydrology*, edited by K.J. Gregory, L. Starkel, and V.R. Baker, pp. 109-129, John Wiley & Sons, New York, 1995b.

Teller, J.T., and L.H. Thorleifson, The Lake Agassiz-Lake Superior connection, in *Glacial Lake Agassiz*, edited by J.T. Teller and L. Clayton, pp. 261-290, Geological Association of Canada Special Paper 26, St. John's, 1983.

Teller, J.T., J. Risberg, G. Matile, and S. Zoltai, Postglacial history and paleoecology of Wampum, Manitoba, a former lagoon in the Lake Agassiz basin, *Geological Society of America*, in press, 1999.

Thorleifson, L.H., Review of Lake Agassiz history, in *Sedimentology, geomorphology, and history of the central Lake Agassiz basin*, edited by J.T. Teller, L.H. Thorleifson, G. Matile, and W.C. Brisbin, pp. 55-84, Geological Association of Canada, Winnipeg, 1996.

Thorleifson, L.H., and F.J. Kristjansson, Quaternary geology and drift prospecting, Beardmore-Geraldton area, Ontario, *Geological Survey of Canada Memoir 435*, 146 pp., 1993.

Tziperman, E., Inherently unstable climate behaviour due to weak thermohaline ocean circulation, *Nature, 386*, 592-595, 1997.

Veillette, J.J., Evolution and paleohydrology of glacial Lakes Barlow and Ojibway, *Quaternary Science Reviews, 13* (945-971), 1994.

Velichko, A.A., The Pleistocene termination in northern Eurasia, *Quaternary International, 28*, 105-111, 1995.

Velichko, A.A., V. Makeyev, G. Matishov, and G. Faustova, Late Pleistocene glaciation of the Arctic shelf, and the reconstruction of Eurasian ice sheets, in *Late Quaternary Environments of the Soviet Union*, edited by A.A. Velichko, H.E. Wright, Jr., and C.W. Barnosky, pp. 35-44, University of Minnesota Press, Minneapolis, 1984.

Willman, H.B., and J.C. Frye, Pleistocene stratigraphy of Illinois, *Illinois State Geological Survey Bulletin 94*, 1-204, 1970.

WMO, *Climate atlas of North and Central America, Vol. I*, World Meteorological Organization, Geneva, 1979.

Wohlfarth, B., The chronology of the last termination: a review of radiocarbon-dated, high-resolution terrestrial stratigraphies, *Quaternary Science Reviews, 15*, 267-284, 1996.

Zoltai, S.C., Glacial features of the Quetico-Nipigon area, Ontario, *Canadian Journal of Earth Sciences, 2*, 247-269, 1965.

P.U. Clark and J.M. Licciardi, Department of Geosciences, Oregon State University, Corvallis, OR 97331. (e-mail: clarkp@geo.orst.edu; licciarj@geo.orst.edu)

J.T. Teller, Department of Geological Sciences, University of Manitoba, Winnipeg, Manitoba R3T 2N2, Canada. (e-mail: jt_teller@umanitoba.ca)

Records of Millennial-Scale Climate Change From the Great Basin of the Western United States

Larry Benson

U. S. Geological Survey, Boulder, CO

High-resolution (decadal) records of climate change from the Owens, Mono, and Pyramid Lake basins of California and Nevada indicate that millennial-scale oscillations in climate of the Great Basin occurred between 52.6 and 9.2 ^{14}C ka. Climate records from the Owens and Pyramid Lake basins indicate that most, but not all, glacier advances (stades) between 52.6 and ~15.0 ^{14}C ka occurred during relatively dry times. During the last alpine glacial period (~60.0 to ~14.0 ^{14}C ka), stadial/interstadial oscillations were recorded in Owens and Pyramid Lake sediments by the negative response of phytoplankton productivity to the influx of glacially derived silicates. During glacier advances, rock flour diluted the TOC fraction of lake sediments and introduction of glacially derived suspended sediment also increased the turbidity of lake water, decreasing light penetration and photosynthetic production of organic carbon. It is not possible to correlate objectively peaks in the Owens and Pyramid Lake TOC records (interstades) with Dansgaard-Oeschger interstades in the GISP2 ice-core δ^{18}O record given uncertainties in age control and difference in the shapes of the OL90, PLC92 and GISP2 records. In the North Atlantic region, some climate records have clearly defined variability/cyclicity with periodicities of 10^2 to 10^3 yr; these records are correlatable over several thousand km. In the Great Basin, climate proxies also have clearly defined variability with similar time constants, but the distance over which this variability can be correlated remains unknown. Globally, there may be minimal spatial scales (domains) within which climate varies coherently on centennial and millennial scales, but it is likely that the sizes of these domains vary with geographic setting and time. A more comprehensive understanding of the mechanisms of climate forcing and the physical linkages between climate forcing and system response is needed in order to predict the spatial scale(s) over which climate varies coherently.

INTRODUCTION

Background

Studies of ice cores and marine sediments from the North Atlantic region indicate that millennial-scale climate

Mechanisms of Global Climate Change at Millennial Time Scales
Geophysical Monograph 112

variability was common during the last glacial cycle [Dansgaard *et al.*, 1971; Johnsen *et al.*, 1972; Dansgaard *et al.*, 1984; Bond *et al.*, 1992; Johnsen *et al.*, 1992; Dansgaard *et al.*, 1993; Grootes *et al.*, 1993; Bond *et al.*, 1993; Taylor *et al.*, 1993; Mayewski *et al.*, 1994; Bond and Lotti, 1995; Brook *et al.*, 1996; Bond *et al.*, 1997]. Millennial-scale oscillations in proxies of climate change have also been documented throughout other regions of the Northern Hemisphere [Allen and Anderson, 1993; Phillips *et al.*, 1994; Porter and Zhisheng, 1995; Behl and Kennett, 1996; Benson *et al.*, 1996a; Benson *et al.*, 1997; Benson *et al.*, 1998a; Hughen *et al.*, 1996; Oviatt, 1997; Lund and Mix, 1998; and Lin *et al.*, 1998]. Whether these oscillations were in phase, out of phase, or completely unrelated remains a fundamental question which, if answered, will lead to a more thorough understanding of the fundamental nature of the causes of global climate variability.

The Great Basin of the western United States provides an opportunity for obtaining high-resolution records of climate change from a region located outside the direct influence of the Atlantic Ocean. Lakes in the Great Basin have deep-water sedimentation rates as high as 2 mm yr^{-1} [Benson *et al.*, 1991], allowing recovery of less-than-decadal-scale records of climate variability. In addition, chemical and isotopic proxies of change in the size of lakes and glaciers have been developed and applied in high-resolution studies of California's Owens and Mono Lakes [Benson *et al.*, 1996a; Benson *et al.*, 1997; Benson *et al.*, 1998a,b]. Studies of Great Basin climate change, however, have limitations not encountered in the marine environment. The Great Basin is tectonically active [e.g., Huber, 1981; Beanland and Clark, 1994; Dixon *et al.*, 1995], causing basin shapes to change and rivers to divert [Benson and Peterman, 1995]. These changes in hydrologic setting impose restrictions on interpretations of climate change based on chemical and physical indicators (proxies).

In the first part of this paper, I have attempted to present a concise summary of the results of high-resolution climate studies published by myself and my colleagues between 1996 and 1999. For details on laboratory procedures, age control, and the application of proxy indicators of climate, the reader is referred to the original publications [Benson *et al.*, 1996a; Benson *et al.*, 1997; Benson *et al.*, 1998a,b]. In the second part of this paper, the lake based records of climate change are compared with climate records of the North Atlantic region and, in particular, with the timing of Heinrich events and stadial/interstadial oscillations documented in the Greenland GISP2 ice core.

Hydrology and Climate of the Northern Great Basin

Most of the precipitation falling in Sierra Nevada catchment areas that supply the Owens, Mono, and Pyramid lake basins is in the form of winter snow. The progression of maximum precipitation along the western flank of the Sierra Nevada is associated with southward movement of the mean position of the polar jet stream (PJS) [Horn and Bryson, 1960; Pyke, 1972; Riehl *et al.*, 1954]. During summer, the westerlies weaken and Pacific storm tracks move north and the lake basins receive only minor amounts of moisture, originating primarily in the Gulf of California [Tang and Reiter, 1984; Adams and Comrie, 1997 and references therein]. Precipitation and streamflow in this region is uncorrelated with the El Nino/Southern Oscillation [Ropelewski and Halpert, 1986; Redmond and Koch, 1991].

During the late Pleistocene Tioga glaciation, Sierran ice fields extended from 36.4 to 39.7°N [Wahrhaftig and Birman, 1956] and valley glaciers formed between 36.2 and 40.2°N [Clark, 1995]. Pyramid, Owens, and Mono lakes occupied basins east of the Sierras between 36.2 and 41.8°N (Figure 1). During the past 50 ka, Pyramid Lake was one of several bodies of water that combined to form Lake Lahontan, a lake that covered an area of ~23,000 km^2 [Russell, 1885]. During the same period, Owens Lake intermittently overflowed [Gale, 1914] but Mono Lake remained hydrologically closed, falling ~33 m short of its spill point [Lajoie, 1968].

Objectives

The ability to correlate climate oscillations across the Northern Hemisphere depends on several factors, including: (1) the ability of climate indicators (proxies) to distinguish between the fundamental elements of climate (temperature, precipitation, humidity, clouds, and wind), (2) age control and temporal resolution of climate records, (3) the spatial and temporal intensity of the climate signal, and (4) the rates of responses of climate proxies to climate forcing.

The objectives of this paper are to: (1) briefly discuss chemical parameters that have been used to indicate oscillations in alpine glaciation and hydrologic balances of Great Basin lakes, (2) review the strengths and limitations of methods of age control of lacustrine sediments, (3) present and discuss high-resolution records of climate change from the Owens, Pyramid, and Mono Lake basins, (4) discuss the correlation of these records with the GISP2 record of climate change, and (5) to examine the concept of spatial and temporal uniformity of climate change.

LACUSTRINE PROXIES OF CLIMATE CHANGE

Proxies of Alpine Glaciation

Alpine glaciers greatly increase the sediment load of streams that emanate from them [Hallet *et al.*, 1996]. Sediment yields from glaciated basins generally depend on the extent of ice cover; however, other processes such as meltwater production, sliding speed and ice flux also affect yields [Hallet *et al.*, 1996]. Nearly all indicators of alpine glaciation reflect the production of glacial rock flour and its input to a lake basin. Three proxies of glacier activity have recently been used in studies of Owens Lake sediments: magnetic susceptibility (MS), chemical composition of the carbonate-free clay-size fraction, and total organic carbon (TOC) [Benson et al., 1996a, 1998b; Bischoff et al., 1977a].

Magnetic susceptibility. Benson *et al.*, [1996a] used variations in MS as a proxy for oscillations in the abundance of glacial rock flour. Assumptions implicit in this application are: (1) the fraction of bedrock-derived rock flour found in lake sediment is greater during times of glacial activity, and (2) that the magnetic moment/susceptibility of a sediment is proportional to the amount of glacial material in that sediment. During interglacial periods, magnetite concentrations can be diluted by autochthonous silicates and carbonates and by allochthonous detrital silicates [Bischoff et al., 1977a]. During interstades, magnetite concentrations can be diluted by organic carbon and amorphous silica derived from diatoms [Benson et al., 1998b].

Chemical composition of the carbonate-free clay-size fraction. Weathering of granite rocks in alpine and subalpine environments produces a grain size distribution similar to the parent material. Few rock fragments produced by weathering are in the clay-size fraction (<2 μm); glacial abrasion of granitic bedrock, however, produces clay-size particles. Bischoff *et al.*, [1997a] used this concept, showing that the rock flour component of lake sediment can be detected using elements that are relatively common to rock flour but rare in nonglacial materials (e.g., Na_2O and TiO_2).

Total organic carbon. Input of glacial rock flour to a lake dilutes the TOC fraction of lake sediment. The glacially-derived suspended sediment also increases the turbidity of lake water, decreasing photosynthetic production of organic carbon [Syvitski *et al.*, 1990; Benson *et al.* 1996a]. Benson *et al.*, [1996a] have suggested that seasonal ice cover and decreased water temperature further decrease productivity. Thus a number of physical and biological processes associated with glacier activity tend to reduce the amount of TOC deposited in lake sediment. In lakes in which diatoms are highly productive, their soft parts tend to increase the TOC content of lake sediment and their hard parts tend to increase the SiO_2 content of lake sediment [Benson *et al.*, 1998b].

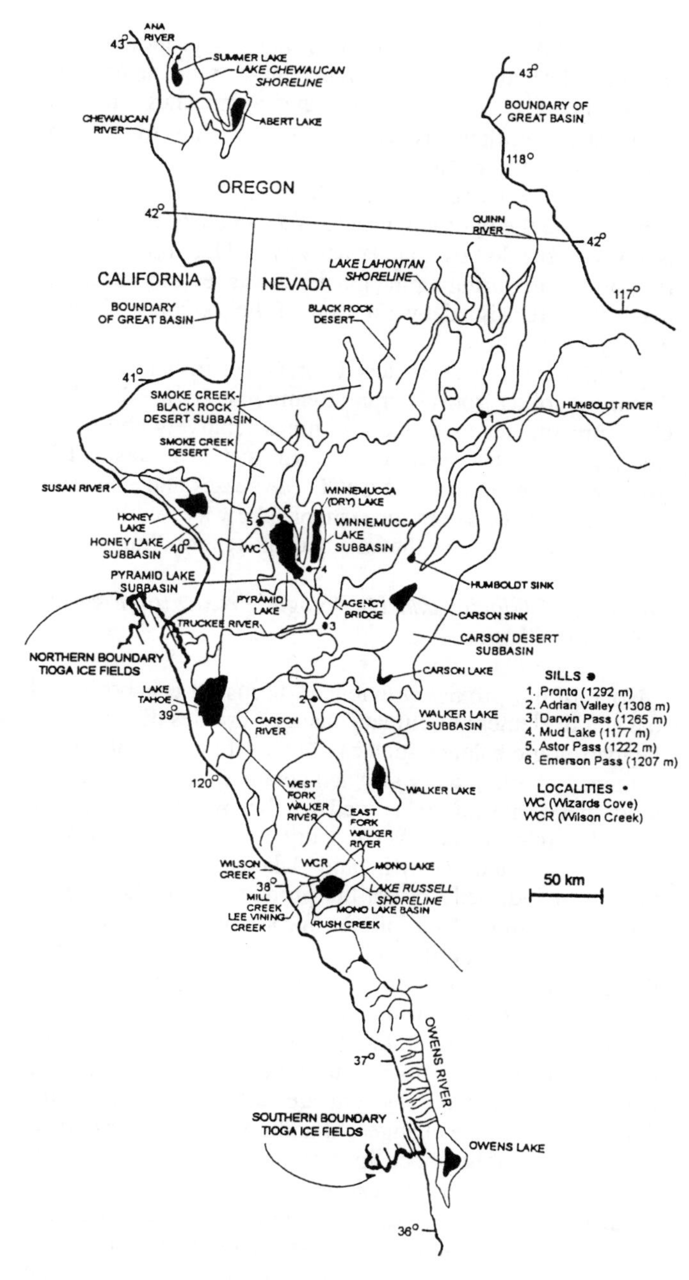

Figure 1. Map from Benson *et al.* [1998b] showing location of Pyramid Lake, Mono Lake and Owens Lake basins of California and Nevada. Latitudinal extent of Tioga glaciation (ice fields) is shown schematically [Wahrhaftig and Birman, 1956]. Highstand lake-surface areas at ~13.0 ^{14}C ka are also indicated.

Proxies of Change in the Hydrologic Balance

Oxygen-18 (^{18}O). The application of $\delta^{18}O$ to the hydrologic balances of Great Basin lakes has been detailed in Benson [1994], Benson et al., [1996a,b] and references therein. The $\delta^{18}O$ value of a steady-state hydrologically closed lake in the Great Basin will be ~ 0 $^o/oo$. When input of isotopically light surface water exceeds the amount of water exported as evaporation, lake level rises and $\delta^{18}O$ decreases and vice versa. The greater the rate of change in volume of a lake, the greater the excursion in $\delta^{18}O$. The $\delta^{18}O$ value of lake water is path dependent; i.e., a specific lake volume cannot be assigned to a unique values of $\delta^{18}O$.

When a lake overflows, its $\delta^{18}O$ value is proportional to the spill/discharge ratio. When that ratio approaches unity, the $\delta^{18}O$ value of lake water approaches the $\delta^{18}O$ value of streamflow discharge (-13 ± 2 °/oo for streams headed in the Sierra Nevada).

Total inorganic carbon (TIC). TIC can be used to approximate the hydrologic state of a paleolake if its signal is not confounded by glacier activity. The input of rock flour to a lake during alpine glacier advances can effectively mask the TIC signal by dilution of the carbonate fraction with silicate debris.

When hydrologically-closed Great Basin lakes receive dissolved solids from Sierran streams, nearly all dissolved Ca^{2+} precipitates as $CaCO_3$ within a relatively short time (months to a few years). When a Great Basin lake overflows, precipitation of $CaCO_3$ decreases and may even cease because the saturation state of lake water with respect to $CaCO_3$ decreases as the lake becomes more dilute.

Linkages of Glacial and Hydrologic Proxies to Elements of Climate Change

None of the proxies previously discussed can be linked to a single element of climate change. Proxies of change in the hydrologic balance such as $\delta^{18}O$ are functions of several elements of climate (water temperature, air temperature, wind speed, humidity) as well as the hydrologic state of the lake (closed or overflowing). Proxies of glacier oscillations such as MS and lake-sediment chemistry directly or indirectly reflect the production of glacial rock flour and its input to a lake basin, a complex process that is a function of multiple elements of climate that control glacier advances and retreats and the transport of glacial sediment to the lake basin. The fraction of TOC reaching a lake's sediment-water interface can be linked to glacier activity; however, biological processes which recycle carbon to the aqueous system are difficult to link to the elements of climate change. In addition, the TOC proxy is not sensitive to the upper limit of glacier extent; i.e., TOC can potentially achieve near zero values before rock-flour induced turbidity reaches a maximum. If lake overflow occurs during glaciation, clay-size material may leave the basin in the outflow, causing the intensity of glacial erosion to be underestimated.

PROBLEMS OF AGE CONTROL

Radiocarbon age controls for both lake and marine records are clouded by questions regarding the magnitudes of reservoir effects. A study by Broecker and Kaufman [1959] indicated modern reservoir effects of ~150, ~600, and ~1700 yr in Walker, Pyramid, and Mono Lakes. Radiocarbon ages of Great Basin lake sediments have usually been obtained on the TOC fraction. Only small amounts of detrital carbon reach Great Basin lakes [Galat *et al.*, 1981; Benson *et al.*, 1991]; therefore, the TOC based dates largely reflect the $^{14}C/C$ ratio of algae which obtain their carbon from dissolved inorganic carbon (DIC) in a lake.

Little is known of the magnitudes of reservoir effects in Great Basin lakes during the past 50 ka. Thus, uncertainties in ^{14}C-based age models of lacustrine and marine sediment cores are often of the same magnitude as the frequencies of the climate oscillations being compared, making it difficult to determine if the records are synchronous, asynchronous, or unrelated.

Studies of late Pleistocene lake sediments in the Mono Lake basin have historically relied on ^{14}C ages of subaerially exposed carbonates (tufas). Porous tufas, typical of the Mono Basin, do not usually remain closed systems with respect to carbon. Such carbonates acquire ^{14}C in the subaerial environment by a dissolution-reprecipitation process that occurs when low-pH rain comes in contact with the carbonate. In the Pyramid Lake basin, this process has been shown to decrease the apparent ^{14}C age of a porous carbonate by more than 1000 years [Thompson *et al.*, 1986].

The use of paleomagnetic secular variation for correlation of lake records across the western United States has proven useful [Liddicoat and Coe, 1979; Negrini *et al.*, 1984; Lund *et al.*, 1988; and Liddicoat, 1992]. Lund [1996] in a comparison of nine Holocene PSV records from North America, has shown that distinctive field features in inclination and declination can often be traced more than 4000 km without significant change in pattern. The spatial extent of synchronous magnetic change remains a subject of debate and continued investigation [Thompson, 1983; Lund, 1993].

METHODS

Detailed coring procedures, sampling, and methods of chemical and physical analysis are described in Benson *et al.* [1996a, 1997, 1998a, 1998b] and Bischoff et al. [1997a]. In general, the $\delta^{18}O$ records discussed in this paper were created using $\delta^{18}O$ values of the TIC fraction contained in bulk sediment samples. Sample reproducibility was ≤0.3 °/oo. TIC and total carbon (TC) measurements were made with a commercial CO_2 coulometer and TOC determined by difference. TIC and TOC values are generally reproducible to at least two significant figures.

HIGH-RESOLUTION RECORDS OF CLIMATE CHANGE FROM THE GREAT BASIN

Owens Lake

Core OL92 (150 to 15 ka). To place the high-resolution Owens Lake data in perspective, medium-resolution (~1500 yr) $\delta^{18}O$, TIC, and Na_2O data records from core OL92 (Figure 2) for the period 150 to 15 ka

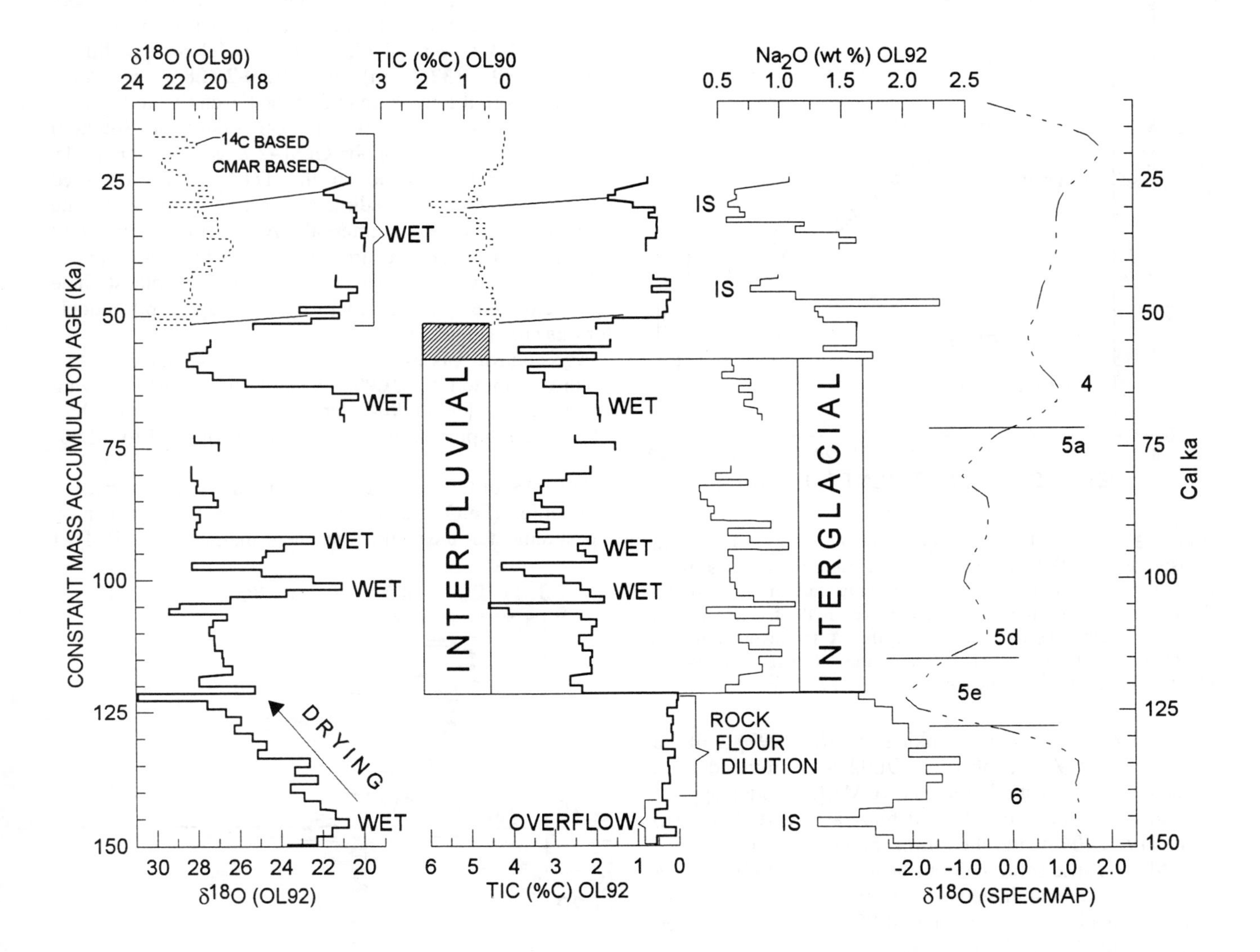

Figure 2. Oxygen-18 ($\delta^{18}O$), total inorganic carbon (TIC), and Na_2O values of the clay-size fraction in sediment samples from Owens Lake cores OL90-1, OL90-2, and OL92. Dotted lines are plots of 1000-yr averages of data from OL90-1 and -2. Solid lines are plots of ~1500-yr channel samples from OL92. Age control for OL92 data (solid lines) is based on constant mass accumulation rate (CMAR) model. CMAR model for OL92 yields ages ~2000 yr younger than the ^{14}C based age model applied to OL90-1 and -2 at 50 ka [Benson *et al.*, 1996a]. The CMAR model was generated using a 3rd degree polynomial fit to the data in Table 4 of Bischoff *et al.* [1997b]. The interpluvial period between 121 and 51 ka is defined by generally low values of $\delta^{18}O$ and TIC; the interglacial period is defined by low values of Na_2O in the carbonate-free clay-size fraction. Note that wet-dry oscillations occurred during the alpine interglacial period and that glacier oscillations occurred during the last alpine glacial period (IS refers to interstades). Note also that interpluvial and alpine interglacial periods were not synchronous. The SPECMAP $\delta^{18}O$ time series [Imbrie *et al.*, 1984] with the boundary between isotope stages 4, 5a, 5d, 5e, and 6 [Shackleton, 1969] are shown for comparison.

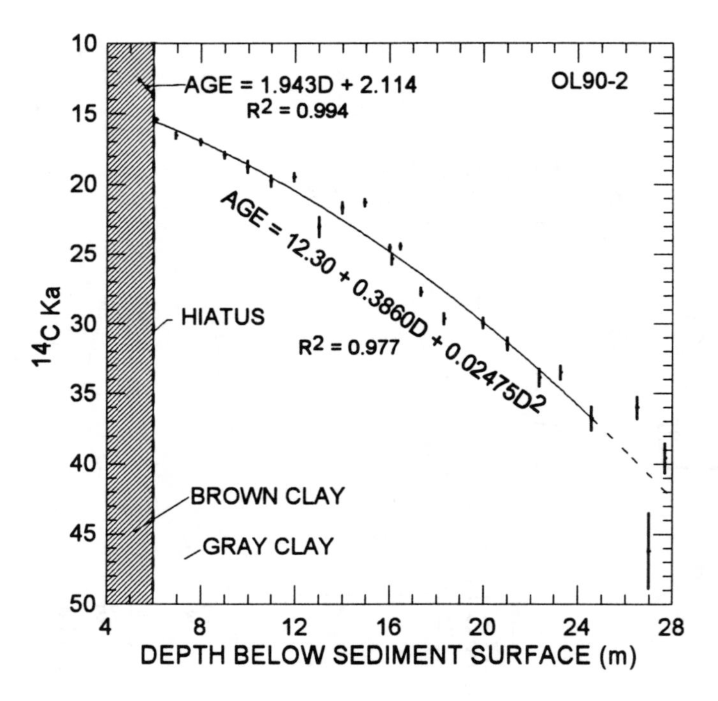

Figure 3. Radiocarbon (^{14}C) age-depth plot for OL90-2 from Benson *et al.* [1996a]. Prior to analysis all samples were pretreated with HCL to remove inorganic carbon. The oldest three samples were not used in the polynomial fit. 'D' in this and subsequent figures refers to depth. A sediment hiatus at ~5.94 m is indicated by an offset in the ^{14}C ages and by an abrupt change in sediment color.

[Bischoff *et al.*, 1997a; Menking *et al.*, 1997] will be examined first. Age control for OL92 was provided by a constant mass accumulation rate (CMAR) model applied to sediments between the Bishop ash bed (759 ka) boundary at 304 m and the top of the core [Bischoff *et al.*, 1997b]. Because some sections near the top of OL92 were fluidized and contaminated during coring, 1000-yr averages of $\delta^{18}O$ and TIC data from core OL90 were plotted (dotted line in Figure 2) beside the OL92 data.

The various proxies discussed above can be used to separate the 150,000 to 15,000 yr period into intervals that are dominantly glacial/interglacial or pluvial/interpluvial (wet/dry). A case can be made that alpine glacial intervals were generally wet and interglacials were generally dry; however, none of the these intervals were climatically monotonic. The generally dry interpluvial was punctuated by three very wet periods of overflow, each of which lasted ~5000 yr, and three interstades occurred during dominantly glacial intervals (Figure 2).

Pluvial/interpluvial and glacial/interglacial boundaries are not always coincident and the approach to those boundaries can be gradual or abrupt. The Na_2O record indicates that the interstade centered on 145 ka was followed by a strong glacial oscillation that peaked ~135 ka; and at 122 ka, the climate shifted abruptly into an interglacial mode. The $\delta^{18}O$ record, on the other hand, shows a gradual wet-to-dry transition between 145 and 122 ka. Overflow followed by input of glacial rock flour combined to suppress TIC values between 150 and 122 ka.

Cores OL90-1 and -2 (52.6 to 12.5 ^{14}C ka). Sediment cores, OL90-1 (32.75 m), and OL90-2 (28.20 m), were obtained from the southern end of the Owens Lake basin in 1990 by Steve Lund of the University of Southern California. Age control for OL90-2 was based on AMS ^{14}C determinations made on the TOC fraction of cored sediment. A hiatus in sedimentation at 5.94 m spans the ≤15.47 to 13.66 ^{14}C ka interval. A 1st-degree polynomial was fit to the youngest three ages above the hiatus and a 2nd-degree polynomial was fit to all but the oldest three ages below the hiatus in order to provide age-depth models for this core (Figure 3).

Age control was extended to OL90-1 by matching MS features common to OL90-1 and -2 (Figure 4). Having derived equivalent OL90-2 depths for OL90-1, the OL90-2 ^{14}C age-depth polynomial (Figure 5) was applied to OL90-1.

The MS of Owens Lake sediments derives from detrital magnetite (Fe_3O_4) and greigite (Fe_3S_4). Sequential measurement of the natural remanent magnetism (NRM) of

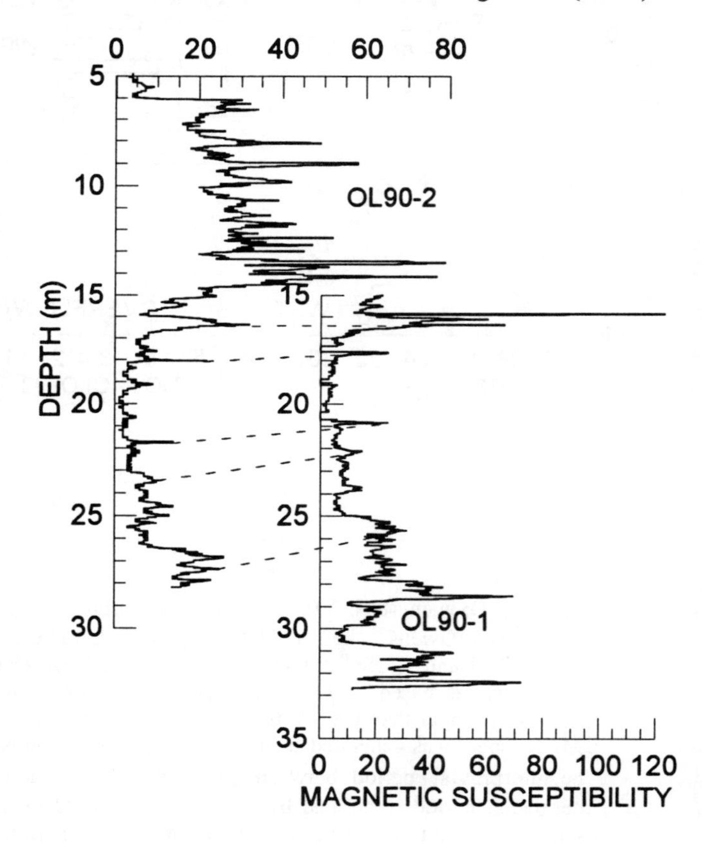

Figure 4. From Benson *et al.* [1998b]; Correlation of the magnetic susceptibility (MS) records between parts of OL90-1 and -2. Not all 30 correlations used in Figure 5 are shown.

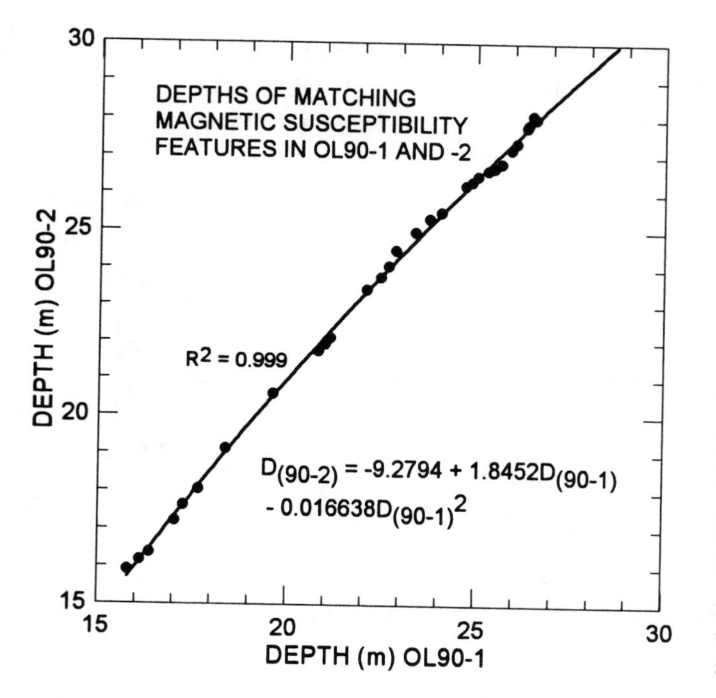

Figure 5. From Benson *et al.* [1998b]; 2nd-degree polynomial fit to MS features in OL90-1 and -2.

samples from OL90-2, indicates that greigite is the most common magnetic mineral deposited between 23.2 and 15.5 ^{14}C ka. This conclusion is supported by thermomagnetic measurements of magnetic mineral blocking temperatures (unpublished data of S. Lund). Sequential measurements of NRM values of samples deposited between 13.66 and 12.5 ^{14}C ka and between 52.6 and 35.0 ^{14}C ka indicate that magnetite makes up the greater part of the magnetic mineral assemblage in these intervals (unpublished data of S. Lund).

As discussed in Benson et al. [1998b], high MS values in core OL90 distinguish two intervals (52.6 to 40.0 ^{14}C ka and 25.5 to 15.47 ^{14}C ka, Figure 6). Except for the low MS interval between 34.5 and 29.0 ^{14}C ka, maxima in MS can be associated with minima in TOC. Each of the TOC minima and MS maxima are interpreted to result from a glacier advance (stade), during which detrital silicates (containing magnetic minerals) scoured from Sierran bedrock were dumped into Owens Lake, diluting the TOC fraction, reducing the photosynthetic production of organic carbon.

To further test for the presence of glacial oscillations, Benson et al. [1988b] Cs-exchanged the carbonate-free clay-size fraction of 22 samples taken from peaks and troughs in the TOC record of OL90-1 and –2. These samples were then analyzed by inductively coupled plasma-atomic emission spectrometry (ICP-AES) and by inductively coupled plasma-mass spectrometry (ICP-MS). In general, all major rock forming elements excepting SiO_2 were negatively correlated with TOC (see e.g., Figure 6). TiO_2, a proxy for biotite, hornblende, and sphene, achieved its highest values during the Tioga glaciation when MS and TOC values were high and its lowest values at ~30 ^{14}C ka when MS and TOC values were low. In addition, the highest MS and lowest TOC values occur during the Tioga 2 and 3 glacier advances whose moraines mark the furthest advances of glaciers during the past 30 ^{14}C ka. These observations confirm that TOC and MS values are, at least, semiquantitative indicators of glacial extent.

Chemical analysis of the clay-size fraction indicated that 25 of 48 elements were negatively correlated (R^2>0.5) with TOC, including Cs_2O, indicating that this oxide was associated with primary minerals such as feldspar and biotite [Hinkley, 1974] or biotite's alteration product vermiculite and not with secondary smectite clays [Bischoff *et al.*, 1997a]. Eight clay-size samples left over from the ICP analyses were Mg exchanged, glycolated, and X-rayed. Their X-ray patterns showed that four of the samples did not contain detectable smectite and that mixed-layer clay was present in only minor amounts in the other four samples (Figure 6).

Optical inspection of the clay-size fraction of Owens Lake sediments indicated that much of the SiO_2 determined in the ICP-AES analyses was associated with diatom fragments (amorphous SiO_2). X-ray diffraction studies also revealed the presence of a broad peak centered at ~22° 2Θ in the high-SiO_2 samples, adding further support to the concept that diatom frustules were the source of much of the SiO_2. These observations suggest that, during the last alpine glacial period, stadial/interstadial oscillations were recorded in Owens Lake sediments by the response of phytoplankton productivity to the influx of glacially derived silicates.

Millennial-scale oscillations in δ^{18}O and TIC also occurred between 52.0 and 13.68 ^{14}C ka (Figure 7). Assuming that calcite precipitation occurred at ~10°C, the δ^{18}O value of Owens Lake water between 52.6 and 13.68 ^{14}C ka was often ≤ -10 o/oo, indicating numerous periods of overflow. Thus, the last alpine glacial period was generally a time of extreme wetness.

A series of pronounced maxima in δ^{18}O between 52.6 and ~40.0 ^{14}C ka occur at about the same time as minima in TOC (Figures 7 and 8). The simplest explanation of the data is that the δ^{18}O maxima represent closed-basin conditions (C-1...C-6). If oscillations in calcite δ^{18}O values were caused by the effect of decreased water temperature on water-calcite isotopic fractionation, temperature would have had to oscillate between 13 and 23°C within a few hundred years. These ranges seem extreme given previous estimates of the temperature difference (3 to 7°C) between the glacial maximum and the Holocene time periods [Van Devender, 1973; Porter *et al.*, 1983; Spaulding, 1983; Dohrenwend, 1984; Phillips *et al.*, 1986; Benson and Klieforth, 1989; Stute *et al.*, 1992].

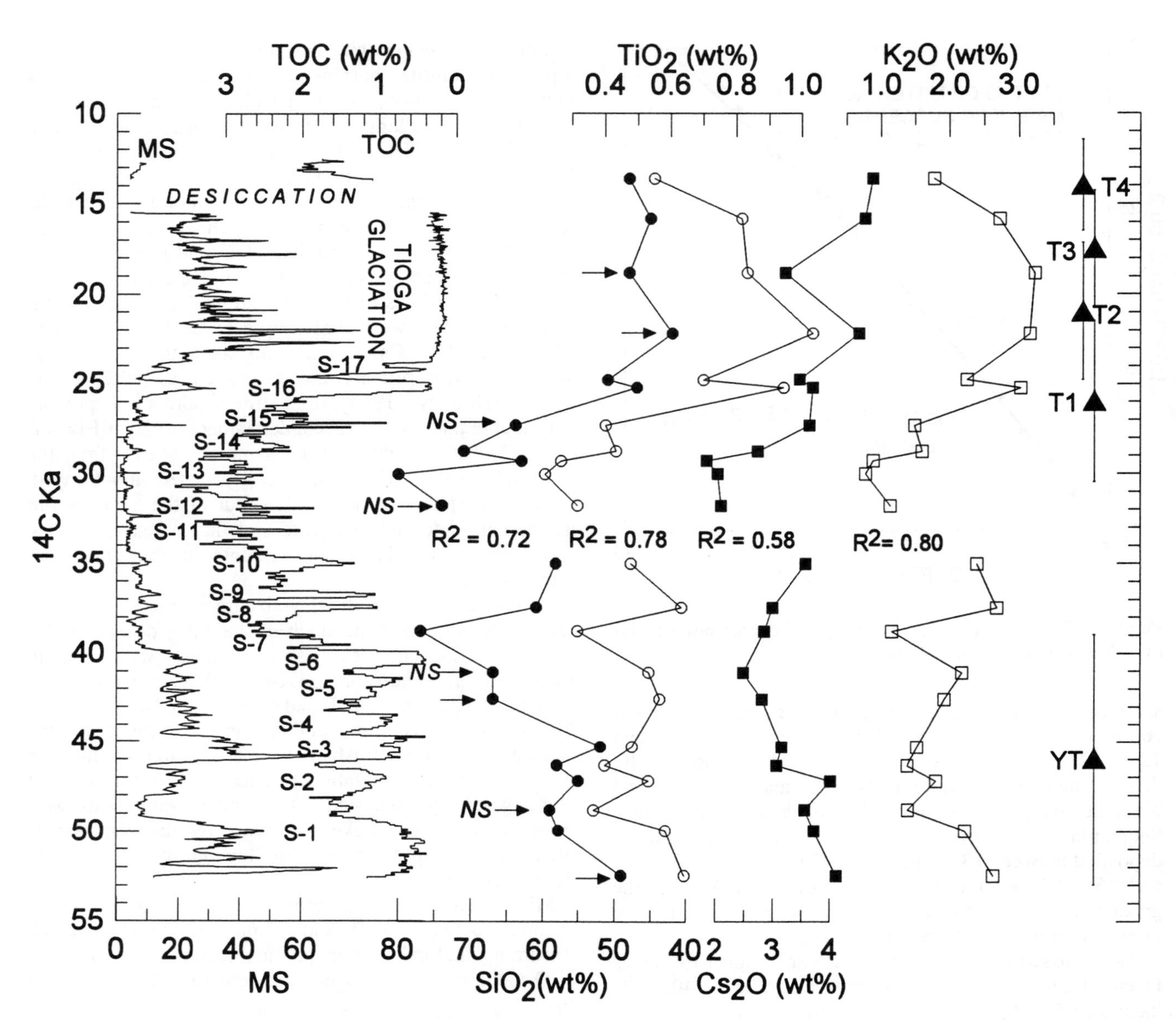

Figure 6. Plot of MS, TOC, SiO_2, TIO_2, Cs_2O, and K_2O in the carbonate-free clay-size fraction of OL90. Note that SiO_2 and TOC values are reversed relative to the other parameter values. 'T' and 'YT' indicate ^{36}Cl based ages of late Pleistocene Tioga and Younger Tahoe Sierran moraines [Phillips *et al.*, 1996]. The error bar associated with each moraine age includes maximum estimates of both systematic and random uncertainties [Phillips *et al.*, 1997]. 'S' refers to alpine glacial stades. The R^2 value refers to correlation of the metal oxide with TOC. The arrows point to eight samples that were Mg saturated, glycolated and X-rayed. The X-ray diffraction patterns showed that smectite was absent from the four samples marked 'NS' and present in very small amounts in the other samples.

In addition, a 13 to 23°C change in condensation air temperature would have decreased the $\delta^{18}O$ of precipitation by 7 to 13 ‰, more than offsetting any increase in calcite $\delta^{18}O$ caused by decreased water temperatures.

Modern precipitation in the Lake Tahoe area of California has a volume-weighted $\delta^{18}O$ value -14.6 ‰ [Benson, 1994] and $\delta^{18}O$ values of the West Fork of the Carson River, which drains the area immediately east of Lake Tahoe, average -14.3 ‰ (unpublished data of L. Benson). Modern ground water from the Carson and Eagle Valleys located to the east and southeast of Lake Tahoe has $\delta^{18}O$ values ranging from -12.7 to -15.0 ‰ [Welch, 1994].

Data from the Honey Lake basin (Figure 1) that borders the eastern flank of the Sierra Nevada near Susanville, California, indicate that ground water with uncorrected ^{14}C

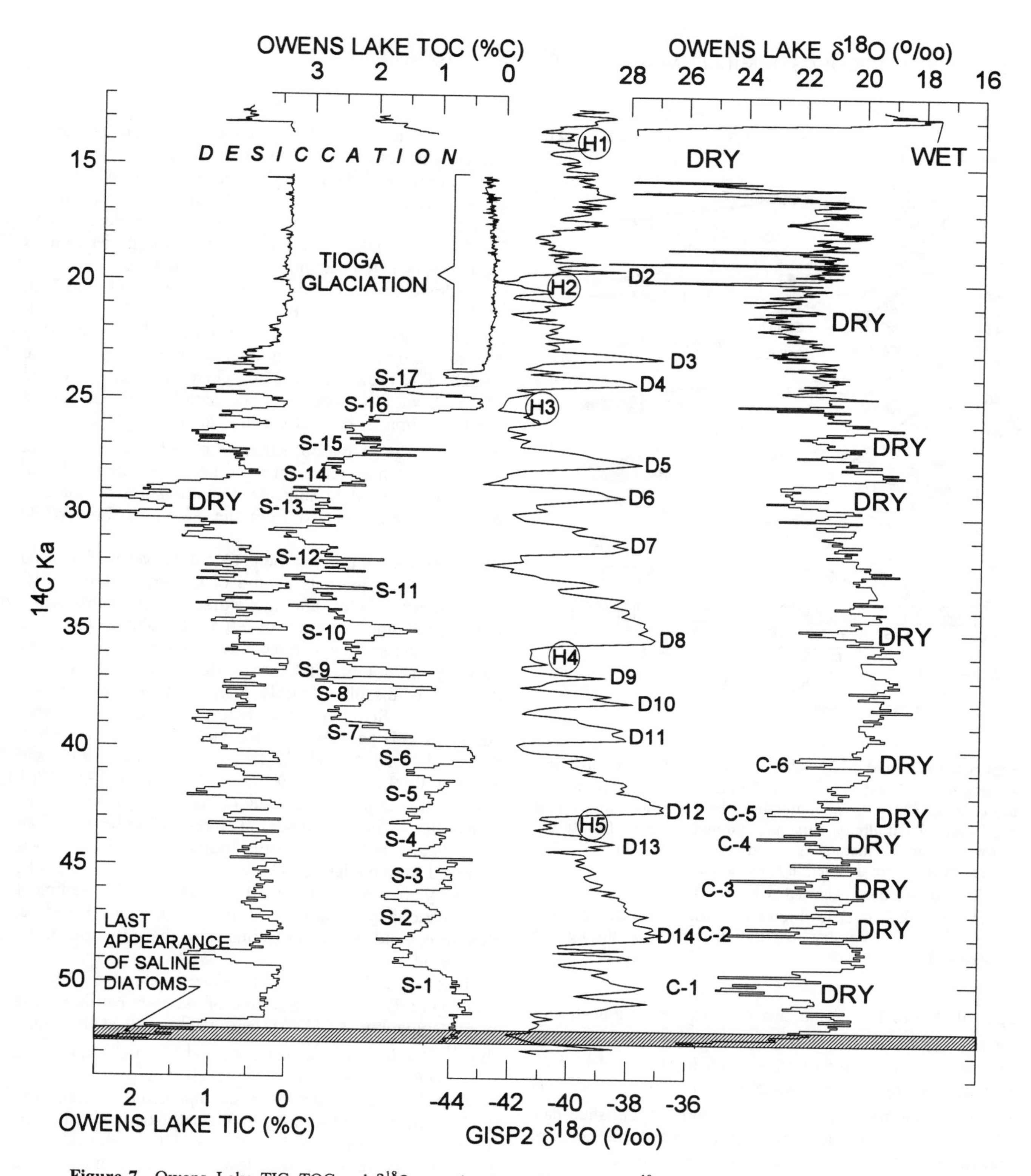

Figure 7. Owens Lake TIC, TOC and $\delta^{18}O$ records compared with the $\delta^{18}O$ record from GISP2 [Grootes *et al.*, 1993]. The GISP2 data have been stretched to place Heinrich (H) events at 14.0 (H1), 20.0 (H2), 25.0 (H3), 36.0 (H4), and 43.0 (H5) ^{14}C ka The ^{14}C dates for Heinrich events H2 through H5 were obtained using the correlation made by Bond *et al.* [1993] between $\delta^{18}O$ minima in North Atlantic and Greenland ice cores (GISP2). The GISP2 calendar age for each event was then transformed to a ^{14}C age using the data of Bard *et al.* [1993], [1996], [1990], and Kitagawa and van der Plicht [1998]. 'S' refers to Sierran stades, circled 'H' refers to North Atlantic Heinrich events, 'D' refers to Dansgaard-Oeschger stades, and 'C' refers to prominent times of hydrologic closure of Owens Lake before 35.0 ^{14}C ka

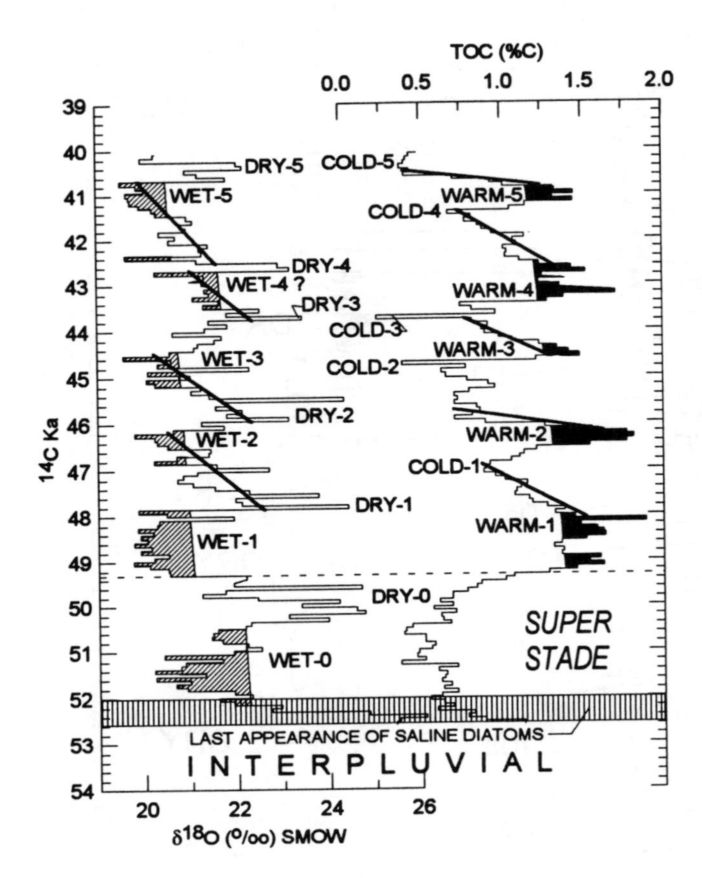

Figure 8. TOC and $\delta^{18}O$ values of sediments from OL90-1 and -2 between 52.6 and 40.0 ^{14}C ka Saline diatoms are gone from the core by 52.0 ^{14}C ka indicating the transition to pluvial conditions (J. Bradbury personal communication). Relatively warm glacial interstades indicated by high TOC concentrations are denoted by the word 'WARM'. Each of the five interstades occurred when $\delta^{18}O$ values were low, indicating 'WET' overflowing conditions. Between 52.3 and 49.3 ^{14}C ka, a wet-dry oscillation occurred during an exceptionally long stade.

ages of 19.0 to 14.4 ^{14}C ka has $\delta^{18}O$ values ranging from -14.1 to -15.2 ‰ [Rose *et al.*, 1997]. Ground water with an uncorrected ^{14}C age of 15.4 ± 0.8 ka from a drill hole located south of Walker Lake, Nevada (Figure 1), has $\delta^{18}O$ values ranging from -15.4 to -15.7 ‰ (unpublished data of L. Benson); and ground water from beneath the northern and eastern sides of the Owens Lake Playa with uncorrected ^{14}C ages of 28.7 ± 0.9 ka and 33.5 ± 1.2 ka has $\delta^{18}O$ values of -16.8 and -16.9 ‰ [Font, 1995]. These data suggest that late Pleistocene precipitation was only ~2 ‰ more negative than modern-day precipitation, implying that condensation air temperatures were only a few degrees colder in the late Pleistocene than today. Therefore, air and water temperature changes were not sufficiently large to account for observed changes in $\delta^{18}O$ between 52.6 and 40.0 ^{14}C ka further supporting the argument that alpine glacial stades were characterized by *relatively* dry climates during a very wet period.

Variations in $\delta^{18}O$ and TOC between 48.0 and 40.6 ^{14}C ka are asymmetric and warm-cold and dry-wet transitions are offset (Figure 8). Some glacier advances (decreases in TOC) tend to take ~1000 yr, whereas all glacier retreats (increases in TOC) take only a few hundred years.

The lowest values of $\delta^{18}O$ occur between 40.0 and 32.0 ^{14}C ka and at 28.2 and 26.2 ^{14}C ka (Figure 7). The 40.0 to 32.0 ^{14}C ka interval exhibits much less $\delta^{18}O$ variability than the 52.6 to ~40.0 ^{14}C ka interval, containing only one notable $\delta^{18}O$ oscillation at 35.0 ^{14}C ka. Between 26.0 and ~15.5 ^{14}C ka, $\delta^{18}O$ variability is generally small although a broad maxima in $\delta^{18}O$ is centered at ~21.5 ^{14}C ka. The five very large (>26 ‰) values in $\delta^{18}O$ between 20.0 and ~15.5 ^{14}C ka may represent rapid drops in lake level; but more probably, they indicate contamination of samples with carbonates derived from the Sierra or Inyo-White Mountains.

The hiatus in the sediment record between 15.47 and 13.68 ^{14}C ka represents a time of extreme aridity. TIC measurements, before and after extraction of soluble salts, demonstrate the presence of ~0.5 % soluble inorganic carbon in the sediment immediately below the hiatus; and decrease in soluble TIC with depth indicates vertical diffusion of dissolved solids from a shallow saline lake (Figure 9). Sediments that bracket this hiatus in other cores (OL84B and OL97) have yielded dates of ~14.8 ± 0.10 and 13.38 ± 0.07 ^{14}C ka [Benson *et al.*, 1997] and 16.38 ± 0.09 and 13.31 ± 0.06 ^{14}C ka (CAMS# 41701 and 41702; unpublished data of M. Kashgarian and L. Benson). The difference in ages of sediment from immediately below the hiatus indicate that material was removed from the lake bed by water erosion (when the lake was very shallow) or by wind erosion (after the sediment was subaerially exposed). Dates on the lower surface, therefore, yield maximum estimates of the initiation of the dry period.

Owens Lake $\delta^{18}O$ reached its lowest values at 13.0 ^{14}C ka (Figure 7), indicating a time of extreme overflow. If a water temperature of 15°C is assumed to exist at this time, the $\delta^{18}O$ value of lake water was -13 ‰, implying a overflow/discharge ratio that approached unity.

Most (14 of 17) stades were accompanied by decreases in TIC, indicating dilution with glacially derived silicates. This process neutralizes the ability of TIC to function as a proxy for change in the hydrologic balance for much of the OL90 record. An exception to this generalization occurs between 30.5 and 28.5 ^{14}C ka when TIC and $\delta^{18}O$ values were both high, indicating aridity. Between 24.0 and 15.5 ^{14}C ka, transport of materials of glacial origin to Owens Lake was sufficient to almost completely mask the TIC fraction (Figure 7).

Core OL84B (13.4 to 9.2 ^{14}C ka). Samples from OL84B were used to obtain $\delta^{18}O$, TIC, and TOC records

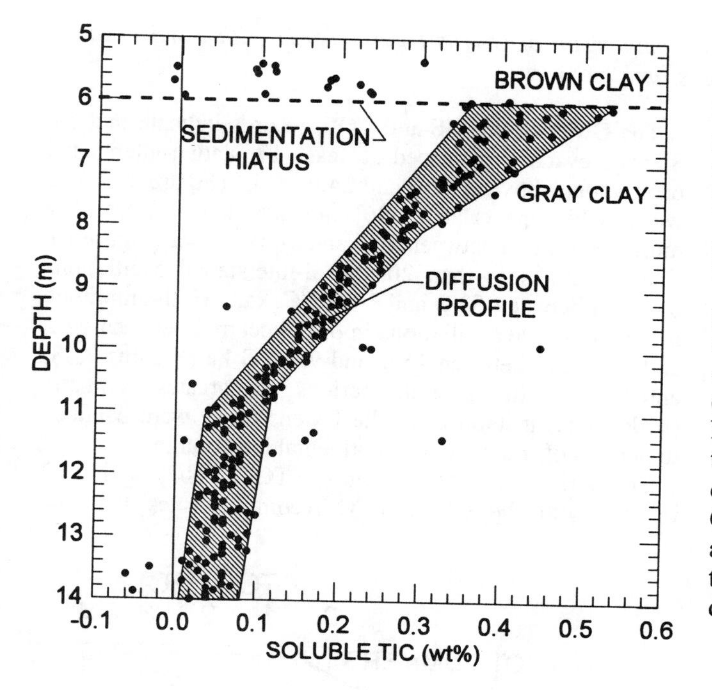

Figure 9. Weight % soluble TIC as a function of depth in OL90-2. A hiatus in sedimentation occurs at 6.0 m. Below that depth, the amount of soluble carbon decreases, indicating that a high-carbonate brine occupied Owens Lake basin prior to the desiccation of Owens Lake.

for the period 13.6 to 9.2 ^{14}C ka (Figure 10). Age control for OL84B is based on 11 AMS ^{14}C determinations made on the total organic carbon (TOC) fraction of cored sediment (Figure 11). The ^{14}C data indicate hiatuses in the sedimentary record at 2.25 and 9.20 m (6.1 to 4.3 and 14.8 to 13.38 ^{14}C ka). The 9.20-m hiatus in OL84B is, therefore, equivalent to the hiatus at 5.94 m in OL90-2. Judging from the OL84B and OL90-2 records, the W_{1a} wet event appears to have begun later in OL84B, probably reflecting a 300-yr offset in the age models applied to the two cores.

In OL84B, the $\delta^{18}O$ and TIC records (Figure 10) indicate a series of abrupt and extreme oscillations between 13.0 and 9.5 ^{14}C ka. TIC closely parallels variation in $\delta^{18}O$ after 12.85 ^{14}C ka This parallelism, and the magnitude of variability in $\delta^{18}O$ (10 ‰), suggest that $\delta^{18}O$ maxima represent closed-basin conditions and $\delta^{18}O$ minima represent times of overflow. Four dry (D_n) intervals occur during this interval, including D_1 (the late-Wisconsin desiccation) and D_4 which marks the beginning of the Holocene. If a water temperature of 25°C is assumed for the dry (D_n) intervals, a calculation indicates that the $\delta^{18}O$ value of lake water was 0 ± 1 ‰, a value consistent with hydrologically closed conditions.

In detail, abrupt increases in TIC in core OL84B tend to occur ~1000 yr prior to major increases in $\delta^{18}O$, probably indicating the effect of residence time on the amount of Ca^{2+} and CO_3^{2-} dissolved in lake water. When the overflow/discharge ratio approached unity, little $CaCO_3$ precipitated from lake water; but when overflow slowed, and the overflow/discharge ratio decreased, $CaCO_3$ began to precipitate in quantity.

The TOC data for OL84B do not exhibit the oscillatory behavior documented in OL90-1 and -2. From 13.2 to 9.2 ^{14}C ka, TOC gradually increases with no indication of productivity decreases that signal glacier activity. This is consistent with the work of Clark and his colleagues [Clark *et al.*, 1995; Clark, 1997; Clark and Gillespie, 1997] who demonstrated that all but the highest cirques in the Sierra Nevada were deglaciated by 13.1 ± 0.07 ^{14}C ka and that the last Pleistocene Sierran glacier advance (Recess Peak) occurred between 12.2 ± 0.06 and 11.19 ± 0.07 ^{14}C ka. Clark and Gillespie's [1997] reconstruction of glaciers along the crest of the Sierra Nevada indicated that during the Recess Peak advance, equilibrium-line altitudes (ELA) dropped by only ~20% of the maximum late-Pleistocene

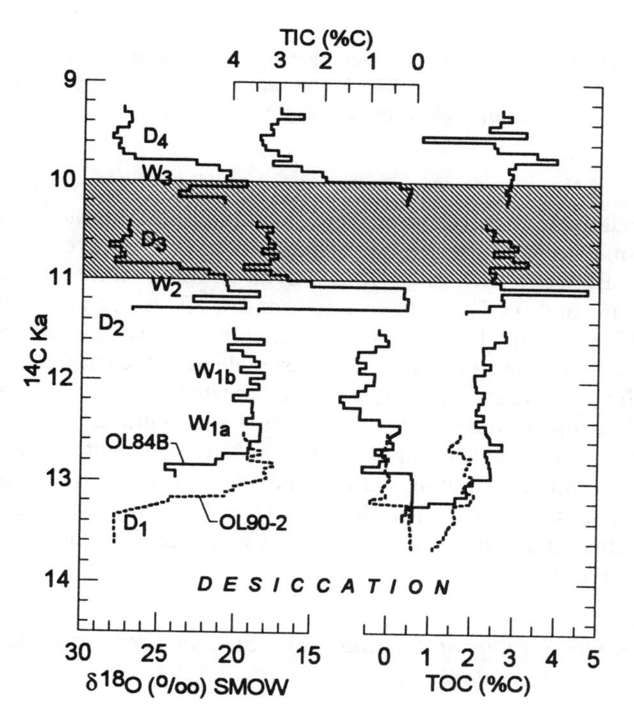

Figure 10. TIC, TOC, and $\delta^{18}O$ values of sediments from OL84B between 13.6 and 9.2 ^{14}C ka (solid lines) plotted using age control of Figure 11. 'D' indicates dry and 'W' indicates wet periods. Dashed line indicates same data for OL90-2 subsequent to desiccation of Owens Lake. An abrupt oscillation in wetness occurred during the Younger Dryas chronozone (shaded rectangle). During wet-dry transitions, TIC increases before decreases in $\delta^{18}O$. This may be due to the rapid response of carbonate precipitation during hydrologic closure [Benson *et al.*, 1996a].

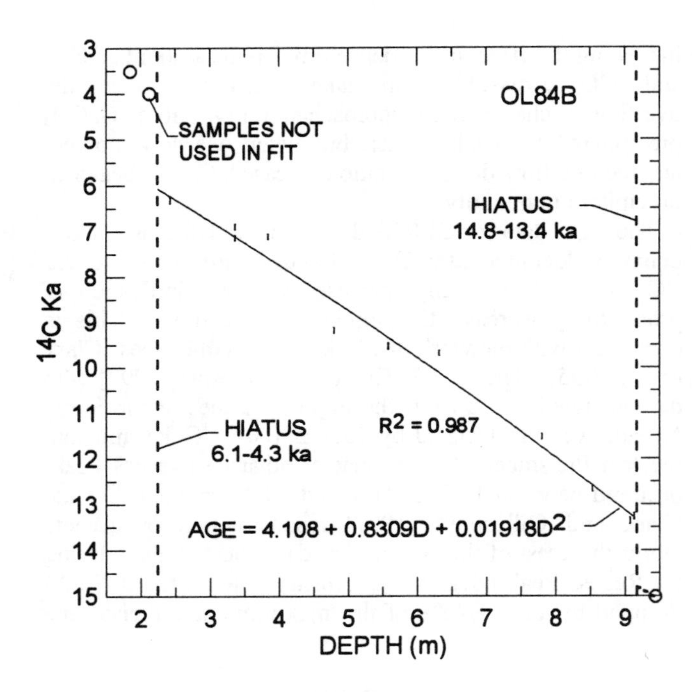

Figure 11. Radiocarbon age-depth plot for samples between 1.85 and 9.43 m in OL84B. The samples indicated by empty circles were not used in the 2nd-degree polynomial fit.

ELA change. The production of rock flour by this relatively small and perhaps brief advance appears too small to have affected the productivity of Owens Lake.

Between 13.7 and ~13.2 ^{14}C ka (OL90-2 time scale), TIC and TOC values are low and δ^{18}O values are high (Figure 10). TIC values of six samples in this interval are near zero and only two samples yielded sufficient $CaCO_3$ for isotopic analysis. These data indicate that after the late-Wisconsin desiccation of Owens Lake, climate became very wet and glacially derived detrital silicates were remobilized, diluting the TIC and TOC fractions in Owens Lake sediment. The high δ^{18}O values may represent contamination of the TIC fraction with pre-Wisconsin-age carbonates.

Solar Forcing of Wisconsin-age Sierran Glaciers in the Owens Lake Basin

Low-frequency changes in the size of Sierran glaciers appear related to summer solar insolation (Figure 12). The period of most extensive alpine glaciation, the Tioga, occurred between 24.0 and ~15.0 ^{14}C ka, when summer insolation was very low. Glacier advances between 34.0 and 26.0 ^{14}C ka appear to have been blunted by high values of summer insolation when summer ablation rates increased relative to winter accumulation rates.

Comparison of Owens Lake Records With North Atlantic Climate Events

The Owens Lake MS and TOC records indicate that the Sierra Nevada experienced at least 17 stadial-interstadial oscillations between 52.0 and 24.0 ^{14}C ka (Figure 7). The work of Phillips et al. [1996] indicates that at least three moraines formed between 24.0 and 14.0 ^{14}C ka (Figure 6). Therefore, no less than 20 stadial-interstadial oscillations occurred between 52.0 and 14.0 ^{14}C ka. High-amplitude millennial-scale oscillations in δ^{18}O occurred between 52.5 and 40.0 and between 15.0 and 9.5 ^{14}C ka (Figures 7, 8, and 10). In general, periods of greatest climatic (hydrologic) instability in the Owens basin were confined to periods of intermediate continental ice volume.

Comparison of OL90-1 and -2 TOC, TIC, and δ^{18}O records with the GISP2 δ^{18}O record indicates that the

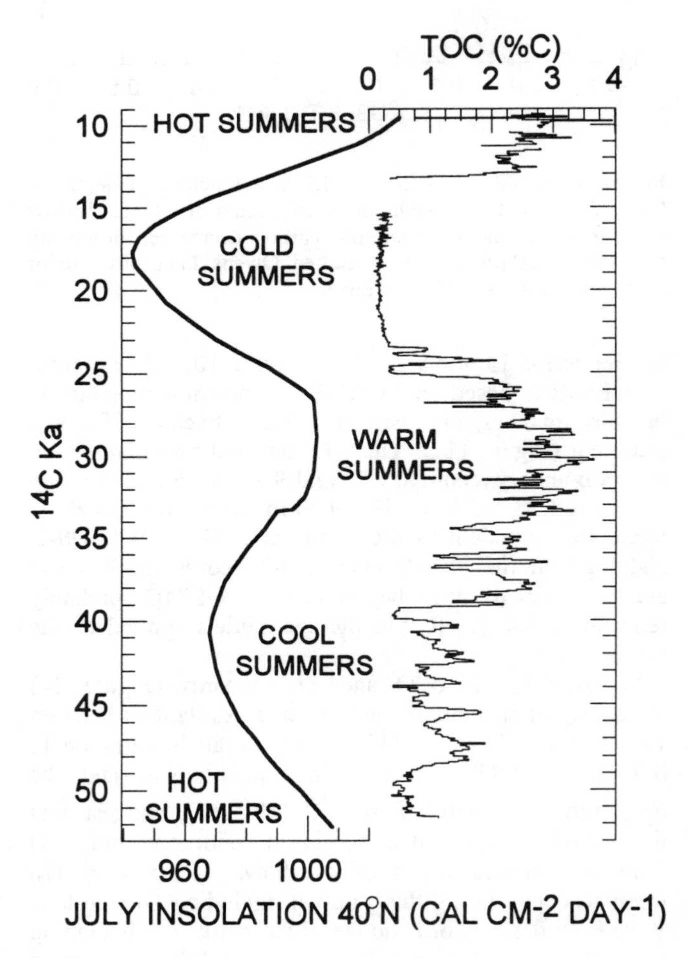

Figure 12. Plot comparing the TOC proxy for Sierran glaciation with July solar insolation at 40°N. Note that summer temperatures appear to have modulated the size of alpine glaciers.

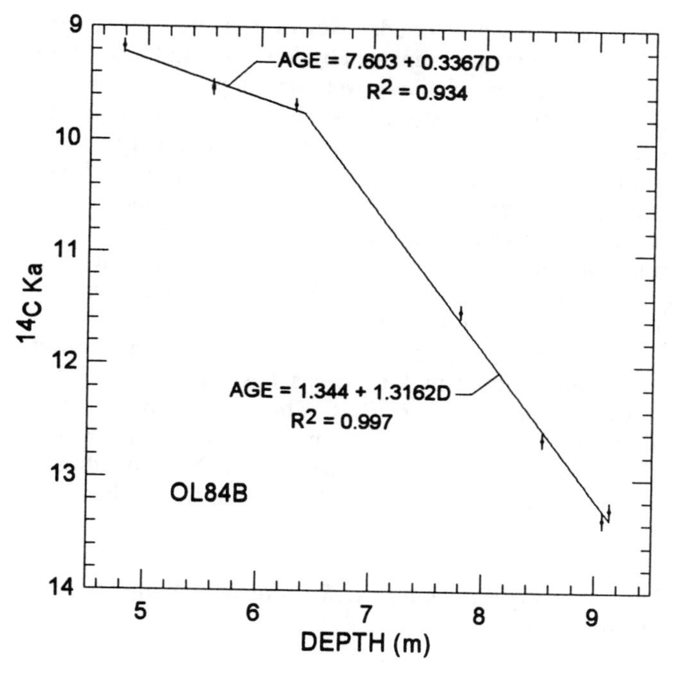

Figure 13. Linear-segment age model for OL84B between 4.79 and 9.13 m.

climate of the Owens Lake basin was *relatively* dry during Heinrich events H1 and H2 (Figure 7). The Tioga glaciation occurred during an interval when Greenland's climate was very cold and relatively stable; i.e., oscillations in $\delta^{18}O$ were relatively minor. It is not possible to objectively correlate peaks in the Owens Lake TOC record (interstades) with Dansgaard-Oeschger interstades in the GISP2 $\delta^{18}O$ record given uncertainties in age control and the difference in the shapes of the OL90 and GISP2 records. For example, the multimillennial asymmetric decreases in $\delta^{18}O$ that follow Dansgaard-Oeschger events D8, D12, and D14 have no counterpart in the Owens Lake records.

The $\delta^{18}O$ data obtained in the study of OL84B indicate four wet-dry oscillations between ~13.6 and 9.0 ^{14}C ka (Figure 10). During the 11.0 to 10.0 ^{14}C ka Younger Dryas chronozone, the Owens basin appears to have experienced a wet-dry-wet (W_2-D_3-W_3) oscillation in climate. The age model on which this conclusion is based does not provide a good fit to the age-depth data between 10.0 and 9.0 ^{14}C ka (Figure 11). To provide a different and perhaps more accurate age model, linear segments were fitted to the age-depth data between 13.4 and 9.2 ^{14}C ka (Figure 13). Although the new age model results in a 200-yr shift in the timing of the wet-dry oscillations, a climate oscillation (W_2-D_3) still characterizes the Younger Dryas chronozone (Figure 14).

Pyramid Lake

Core PLC92B (40.8 to 12.5 ^{14}C ka). In 1992, a 17.35-m sediment core was taken from the Wizards Cove area of Pyramid Lake, Nevada (Figure 1). An age model for core PLC92B was obtained by fitting a 3rd-degree polynomial (Figure 15) to 12 of 14 samples from the ^{14}C age-depth data set.

A change from organic-rich to organic-poor sediments occurs at ~24.5 ^{14}C ka, marking the beginning of the Tioga glaciation (Figure 16). Prior to ~24.5 ^{14}C ka, millennial-scale oscillations in TOC are evident. The TOC oscillations indicate productivity changes that accompanied the advance and retreat of Sierran glaciers. It is difficult to correlate TOC oscillations recorded in the Pyramid and Owens lakes basins (Figure 17). Both records, however,

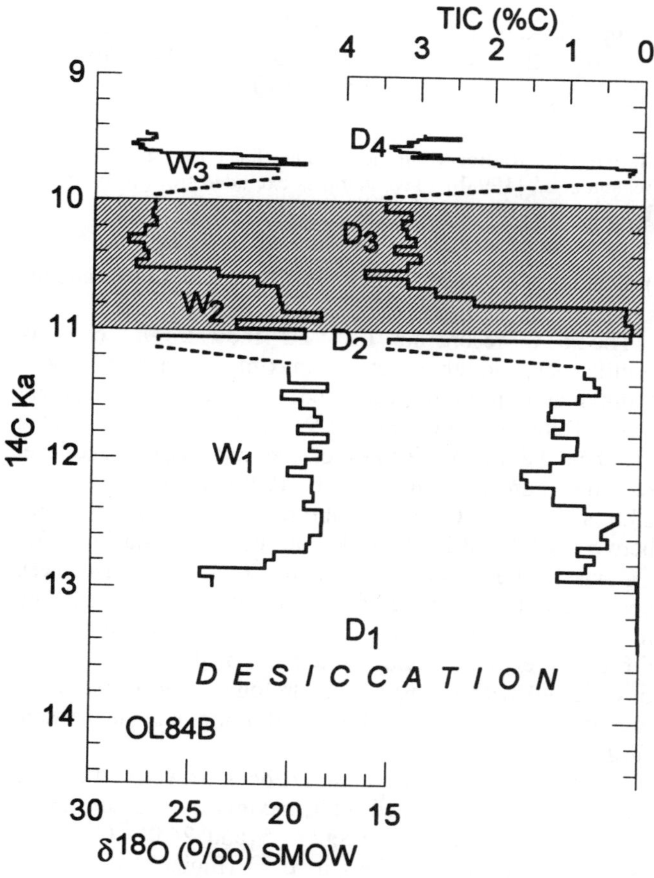

Figure 14. TIC and $\delta^{18}O$ values of sediments from OL84B between 13.5 and 9.4 ^{14}C ka plotted using age control of Figure 13. This age model indicates that an abrupt wet-dry oscillation occurred during the Younger Dryas chronozone (shaded rectangle).

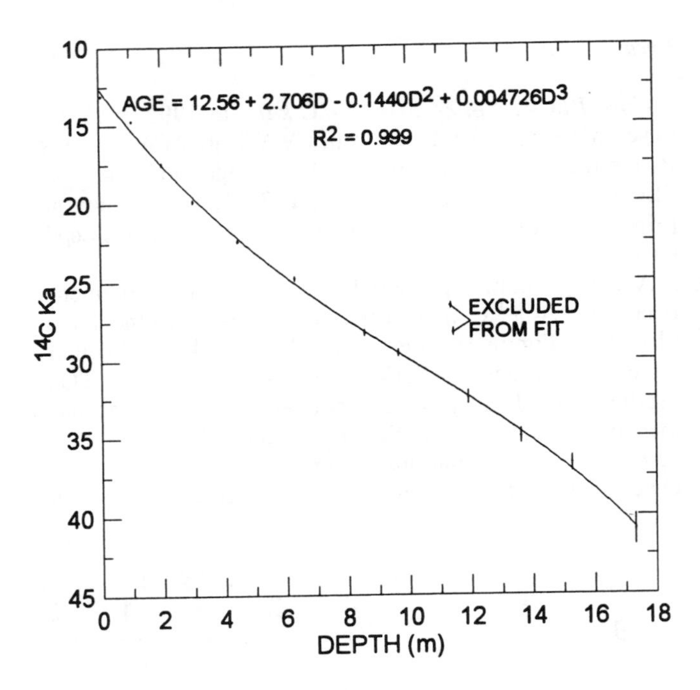

Figure 15. Radiocarbon age model for core PLC92B from Benson *et al.* [1998b]. Two samples excluded from polynomial fit may have organic carbon derived from roots.

indicate that between 10 and 13 TOC oscillations occurred between 40.0 and ~24.5 ^{14}C ka.

The δ^{18}O record for PLC92B indicates a series of oscillations, including several maxima (lowstands) and a prominent minimum (highstand) at 13.6 ^{14}C ka (Figure 16). Between 25.0 and 12.5 ^{14}C ka, the δ^{18}O record matches the major features of the tufa based lake-level record (Figure 18, Benson et al., 1995), with lowstands at 23.0 and 14.5 ^{14}C ka and abrupt increases in lake size between 22.0 and 20.0 ^{14}C ka and between 14.5 and 13.6 ^{14}C ka. The lowstand at 27.0 ^{14}C ka has also been observed in the tufa based δ^{18}O record of figure 9 published in Benson *et al.* [1996b]. In general, however, it is difficult to explicitly associate features of the Pyramid Lake δ^{18}O record with changes in hydrologic balance given the topographic complexity (three spill points) of the system (Figure 1).

Most of the TOC minima (stades) between 40.6 and 24.0 ^{14}C ka are associated with maxima in TIC and many of the TOC minima between 37.5 and 24.0 ^{14}C ka are associated with high values in δ^{18}O (Figure 16). These relationships suggest that glacial introduction of silicate materials was not sufficient to mask the TIC signal in the Pyramid Lake basin and further implies that many of the alpine stades occurred during relatively dry times. The latter part of this conclusion is consistent with data from the Owens Lake basin between 52.6 and 40.0 ^{14}C ka, reinforcing the concept that alpine glacier advances occurred during relatively dry periods. However, some stades, e.g., S-8 and S-9 were accompanied by low TIC and low δ^{18}O values suggesting relatively wet conditions. The fact that the PLC92B TIC record was not masked by glacially derived silicates prior to the Tioga glaciation may be due to one of more of the following processes: (1) glacier activity was weaker in the Truckee River catchments than in the Owens valley catchments, (2) substantial amounts of glacial debris were trapped in the Lake Tahoe basin, and (3) the location of PLC92B was far from the input source of silicate rock flour to Pyramid Lake.

It was previously noted (see above) that a dry-wet oscillation occurred in the Owens Lake basin during the Tioga glaciation (Figure 7). This oscillation is also evident in the δ^{18}O and TIC records from PLC92B (Figure 16). In particular the TIC record of PLC92B exhibits a maximum between 24.4 and 21.3 ^{14}C ka that indicates a long shallow-lake interval.

Comparison of Pyramid Lake Records with North Atlantic Climate Events

Comparison of the PLC92B δ^{18}O records with the GISP2 δ^{18}O record indicates that the climate of the Pyramid Lake basin was relatively dry at about the times of Heinrich events H1 and H2 (Figure 16). It is, however, not possible to correlate objectively peaks in the Pyramid Lake TOC record (interstades) with Dansgaard-Oeschger interstades in the GISP2 δ^{18}O record, given uncertainties in age control and the difference in the shapes and amplitudes of the PLC92B and GISP2 records.

Mono Lake

Wilson Creek Formation (35.4 to 12.86 ^{14}C ka). Mono Lake (Pleistocene Lake Russell) is a hydrologically closed lake located directly north of the Owens River drainage (Figure 1). Three streams (Lee Vining, Rush, and Mill Creeks) contribute most of the surface-water discharge to the lake. Mono Lake appears to have remained closed throughout the past 100 ka [Lajoie, 1968].

The Wilson Creek Formation contains laminated muds and silts separated by 19 tephra layers (ashes) [Lajoie, 1968]. At the type section, sediments sandwiched between Ash 4 and 5 were eroded and reworked during a lowstand [Benson *et al.*, 1998a]. At its South Shore site, the Wilson Creek Formation is thicker than at its type section and 17 m lower in elevation. At this site, sediments between Ash 4 and 5 remain unreworked. Nearly continuous sets of 2-cm-thick samples were taken from outcrop near the type section and from between Ashes 4 and 5 at the South Shore site for δ^{18}O and TIC analyses.

An age model for sediments at the Wilson Creek type section was constructed using ^{14}C ages of carbonate

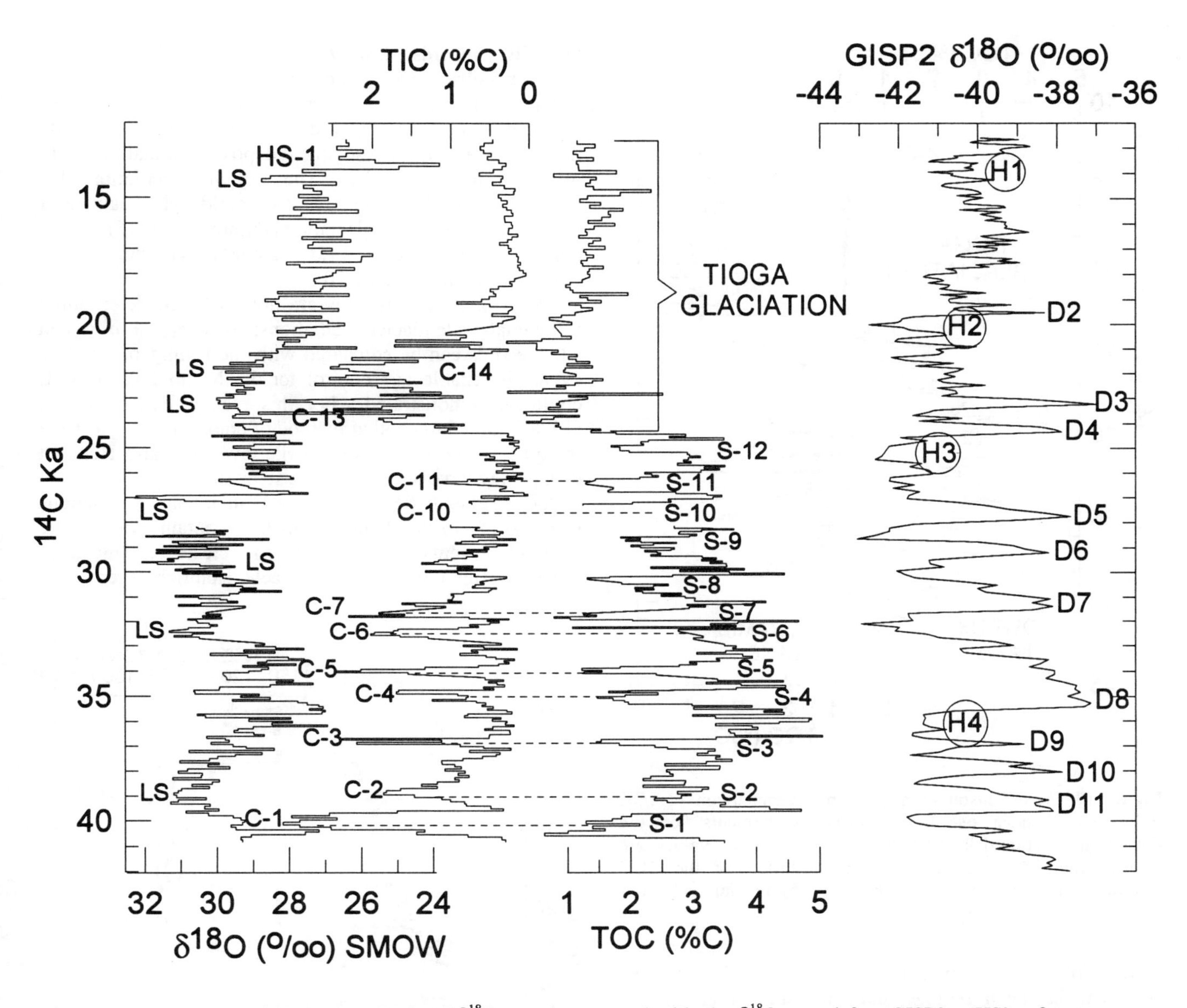

Figure 16. Pyramid Lake TIC, TOC and δ^{18}O records compared with the δ^{18}O record from GISP2. 'HS' refers to Lahontan highstand; LS refers to lowstands, 'S' refers to Sierran stades. 'H' refers to North Atlantic Heinrich events, and 'D' refers to Dansgaard-Oeschger interstades. Dotted lines indicate when Sierran stades occurred during periods of closure of Owens Lake; i.e., when the climate was cold and relatively dry .

samples [Table IV in Benson *et al.*, 1990] collected from several localities along the Wilson Creek drainage by Ken Lajoie of the U.S. Geological Survey. To construct the age model, it was assumed that the location of a carbonate sample relative to its confining ash layers at the type section was the same as at the locality from which the sample was taken. Linear regressions of data between Ashes 1 and 4 and between Ashes 5 and 19 (Figure 19a) were used to estimate the ^{14}C ages of individual tephra layers [Table 1 in Benson *et al.*, 1998a]. The estimated ash ages and two AMS ^{14}C determinations on ostracode shells from the isotope section were subsequently used to derive an age-depth relationship for the Wilson Creek isotope sampling site (Figure 19b). A linear age-depth relationship was applied to samples taken from the South Shore site.

Given the fit of the linear age-depth models and the questionable reliability of the porous carbonates that were dated, the uncertainties of the models are considered no better than a thousand years and the age of the older sediments may be underestimated by a few to several thousand years.

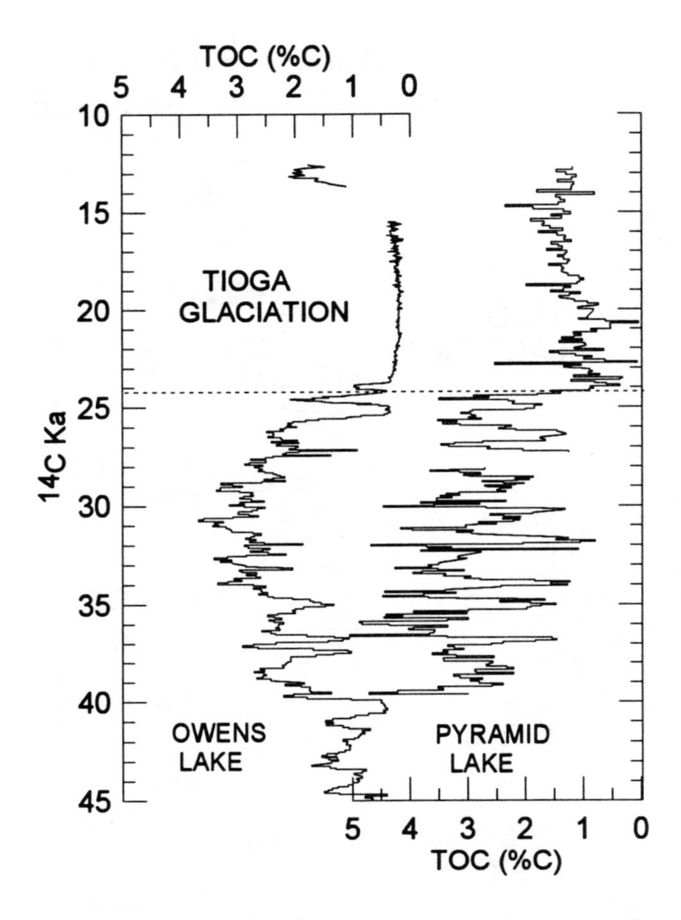

Figure 17. From Benson *et al.* [1998b]. Comparison of TOC indicators of glacial oscillations in cored sediments from the Owens and Pyramid lake basins. Note that TOC values are smaller after 24.5 ^{14}C ka, indicating dilution of the TOC fraction and suppression of productivity by glacial rock flour from the Tioga glaciation.

Most of the TOC in subaerially exposed Mono Lake sediments has been oxidized and is not suitable as a proxy of glacier activity. Low values of TIC between 26.0 and 14.0 ^{14}C ka probably reflect dilution of the TIC fraction by rock flour from the Tioga glaciation (Figure 20).

The δ^{18}O record displays high-amplitude fluctuations with four principal maxima centered at 34, 27, 21 and 15 ^{14}C ka (Figure 20). The simplest interpretation of these maxima is that they represent lowstands. However, the maxima may also indicate approaches to closed-basin isotopic equilibrium that did not involve lake-volume decreases. Two major δ^{18}O minima (highstands) occur at 18.0 and 13.0 ^{14}C ka

The TIC and δ^{18}O records parallel each other between 35.5 and 26.0 ^{14}C ka, suggesting that glacier activity wasn't sufficient to mask the usefulness of the TIC record as a hydrologic proxy. The parallelism of both records supports the interpretation of δ^{18}O maxima as lowstand indicators. Lowstands L3 and L4 are clearly expressed in the TIC records; however, glacially derived (Tioga) sediments obscure lowstands L1 and L2.

The 31 to 26 ^{14}C ka period was a time of increased carbonate deposition. Lowstand L4 was an exceptionally dry event; ~60% of the sediment deposited at that time was in the form of $CaCO_3$. The dry-wet oscillation that occurred in the Owens Lake and Pyramid Lake basins after the onset of the Tioga glaciation (Figures 7 and 16) is also evident in the δ^{18}O and TIC records from the Wilson Creek Formation (Figure 20).

The Mono Lake records are not useful in determining millennial-scale glacier oscillations; however, the δ^{18}O and TIC records can be compared with the timing of Heinrich events if reliable age control for the Mono Lake records could be demonstrated. For this reason, Benson *et al.* [1998a] turned to another method of comparison to test the hypothesis that some Mono Lake lowstands and Heinrich events may have occurred at the same times.

Paleomagnetic field directional and intensity variations recorded in North Atlantic cores that contain evidence for Heinrich events were used to create a magnetic chronostratigraphy in which each Heinrich event was

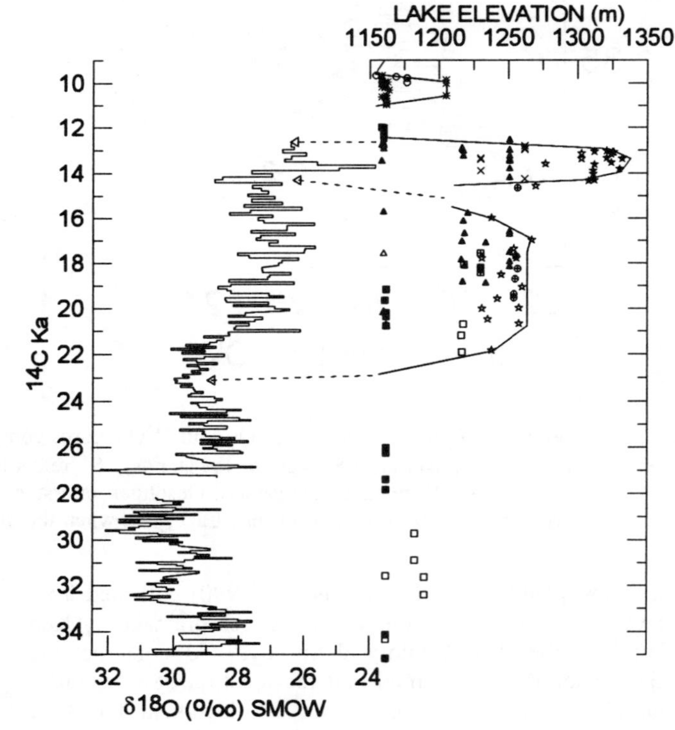

Figure 18. Comparison of Pyramid Lake δ^{18}O record with carbonate based lake-level envelope for Lake Lahontan between 35.0 and 13.0 ^{14}C ka. Different symbols indicate differing styles of carbonate deposits discussed in Benson *et al.* [1995]. Dashed lines connect times of low lake levels observed in both records.

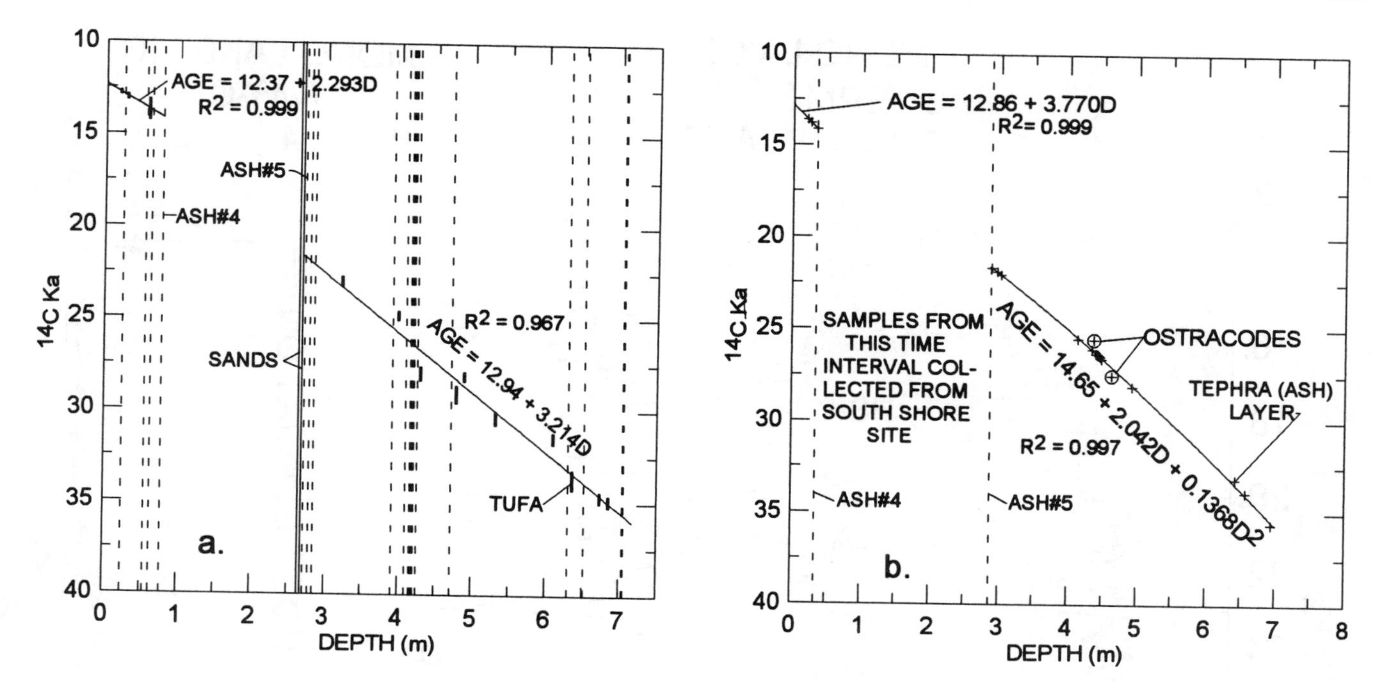

Figure 19. (a) From Benson *et al.* [1998a]. Radiocarbon ages and depths of carbonates in the Wilson Creek area used to estimate the ages of 19 tephra layers (ashes) exposed in the Wilson Creek Formation along Wilson Creek. Ashes are depicted as vertical dashed lines. Data used in this figure were taken from Wilson Creek type section listing in Table IV of Benson *et al.* [1990]. In the regression equations, D refers to depth in meters (m). In constructing the two regression lines, some samples from Table IV were rejected. Sample USGS-1435 came from the reworked Ash 4-5 interval. Radiocarbon ages of samples USGS-1436 and L-1167C were splits from the same collection of ostracode valves whose radiocarbon ages are anomalously young (~23,000 instead of ~26,000 ^{14}C yr). We expect that modern carbon was added to many of the thin-walled ostracode valves. The radiocarbon age of sample USGS-362, a nodular tufa with carbonate coatings from the base of the Wilson Creek Formation, was also clearly too young (28,600 instead of ~36,000 ^{14}C yr). Sample USGS -276 was rejected because its ^{14}C age was infinite (39,600 ^{14}C yr). (b) From Benson *et al.* [1998a]. Radiocarbon age control for Wilson Creek isotope section. Age control is based on estimated ages of 18 tephra layers [Table 1 in Benson *et al.*, 1998a]. The two ostracode AMS ^{14}C ages provide a check on the reliability of the tephra-derived age model for the interval 27.5 to 25.0 ^{14}C ka

associated with directional features of the magnetic records [Benson et al., 1998a]. North Atlantic intensity and directional records (both inclination and declination) were then correlated with Wilson Creek paleomagnetic secular variation (PSV) features to determine what Mono Lake δ^{18}O features were associated with Heinrich events. From the PSV data, it appears that Heinrich events H1, H2, and H4 and Mono Lake lowstands L1, L2, and L4 overlap in time (Figures 20 and 21). Heinrich event H3, however, does not overlap any Mono Lake lowstand and there are other lowstands such as the one at 30.3 ka that bear no relation to Heinrich events.

SUMMARY AND CONCLUSIONS

During the past 150,000 yr, the Great Basin witnessed intervals that were dominantly glacial/interglacial or pluvial/interpluvial (wet/dry); however, none of the intervals were climatically monotonic on the millennial scale. 5000± yr wet periods occurred during the last interpluvial and 5000± yr interstades occurred during glacial intervals (Figure 2). A comparison of the Owens Lake medium-resolution records [Bischoff *et al.*, 1997a; Menking *et al.*, 1997] with the SPECMAP δ^{18}O proxy for continental ice volume (Figure 2) indicates that the last alpine interglacial (121 to 62 ka) did not occur at the same time as the last continental interglacial (stage 5e, 128 to 115 ka) nor did it occur at the same time as stage 5 (128 to 71 ka) when continental ice volumes were relatively small. In addition, the Owens Lake δ^{18}O, TIC, and Na_2O records have rectilinear shapes and bear little similarity to the sawtooth shaped marine δ^{18}O record.

Studies of Greenland ice cores and North Atlantic sediments indicate that centennial-to-millennial-scale climate variability occurred often during the last ice age. Hydrologic-balance records from each of the three lake

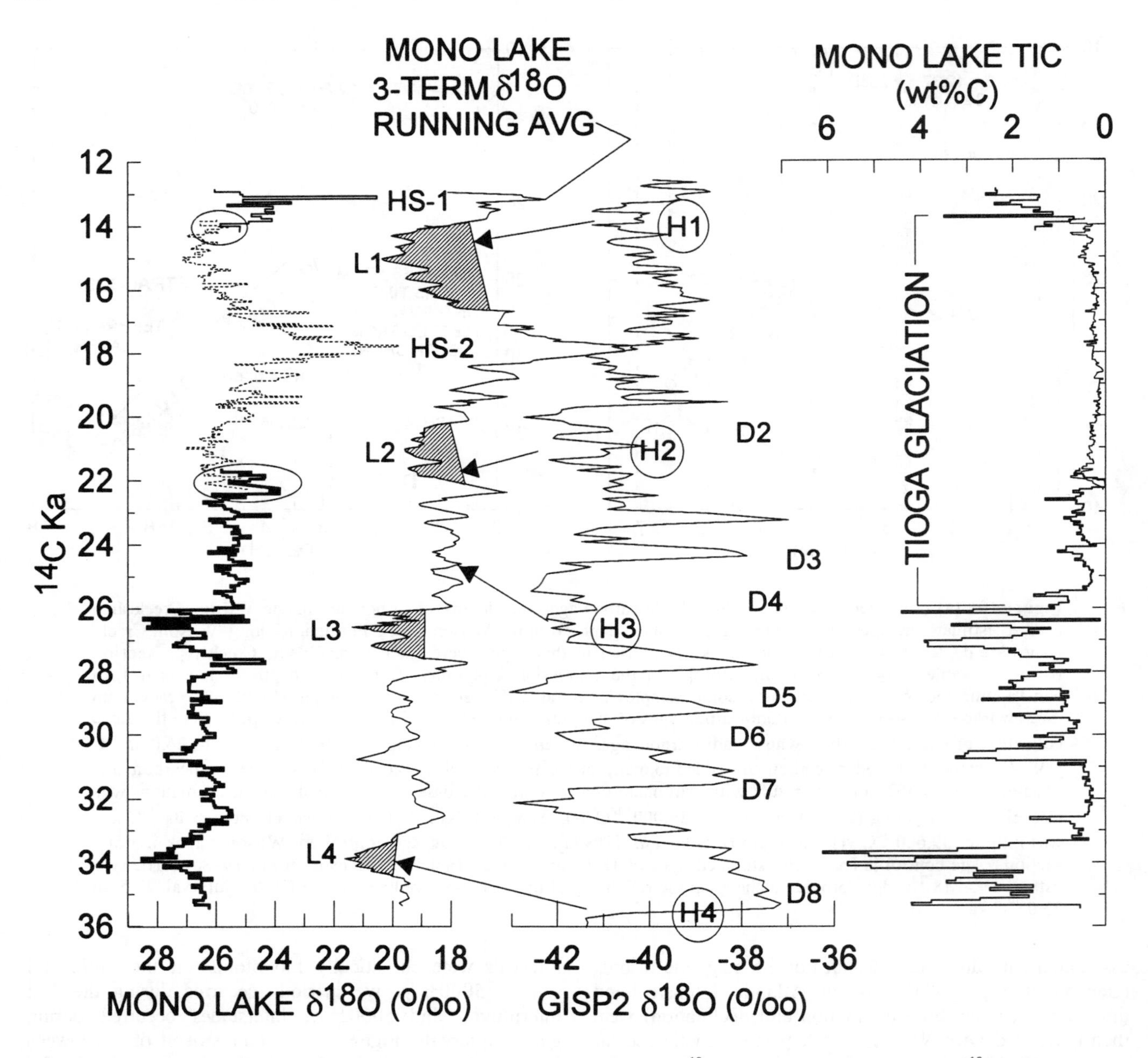

Figure 20. Modified from Benson *et al.* [1998a]. Mono Lake TIC and $\delta^{18}O$ records compared with the $\delta^{18}O$ record from GISP2. 'L' refers to lowstands of Mono Lake; circled 'H' refers to North Atlantic Heinrich events; 'D' refers to Dansgaard-Oeschger stades and 'HS' refers to highstands of Mono Lake at 18.0 and 13.0 ^{14}C ka Arrows point to the location of Heinrich events in the Mono Lake records based on comparison of paleomagnetic secular variation waveforms in North Atlantic cores and Mono Lake (Wilson Creek) sediments (see Figure 21). Low TIC values between 25.0 and 14.0 ^{14}C Ka indicate dilution of TIC fraction by Tioga-age glaciers.

basins and alpine-glacial records from Owens and Pyramid Lake exhibit millennial-scale variability (Figures 6, 7, 10, 16, 17, 20). In the Owens Lake basin, the highest amplitude oscillations in hydrologic balance occurred near the beginning and end of the last Sierran alpine glacial period when major changes in the large-scale pattern of atmospheric circulation were occurring in response to rapid growth and destruction of the Laurentide Ice Sheet. Climate records from the Owens and Pyramid Lake basins indicate that most, but not all, glacier advances (stades) occurred during relatively dry times (Figures 7, 16).

In the North Atlantic region, some climate records have clearly defined variability/cyclicity (e.g., Dansgaard-Oeschger and Heinrich events) with periodicities of 10^2 to

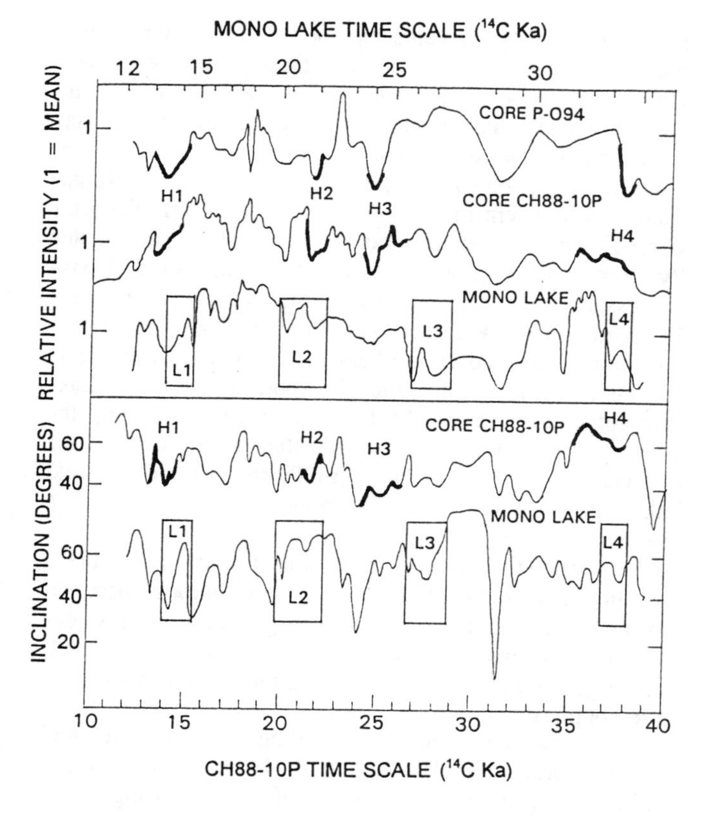

Figure 21. Modified from Benson *et al.* [1998a]. Comparison of Heinrich event and Mono Lake lowstand ages using paleomagnetic field variability (intensity and inclination) for chronostratigraphic control. Methodology for placement of the Heinrich events on the paleomagnetic waveforms is discussed in Benson *et al.*, [1998a]. The stratigraphic locations of Heinrich events in North Atlantic cores CH88-10P, and P-094 are indicated by thick black lines labeled H1 through H4. Mono Lake lowstands are shown as rectangles labeled L1 through L4. The paleomagnetic data indicate that Heinrich events H1, H2, and H4 occurred during Mono low-lake intervals, whereas Heinrich event H3 did not. Note that the durations of the lowstands and accompanying Heinrich events differ. The ^{14}C chronology for CH88-10P [Keigwin and Jones, 1989; Keigwin 1994] was used to develop a common time scale for the marine records). Radiocarbon ages of Mono Lake sediments are shown in the upper abscissa and radiocarbon ages of marine sediments are shown in the lower abscissa. Declination data (not shown) were also used in the construction of this diagram.

10^3 yr and these records are correlatable over several thousand km. In the Great Basin, climate proxies also have clearly defined variability with similar (and even smaller) periodicities, but the distance over which this variability can be correlated remains unknown. Globally, there may be minimal spatial scales (domains) within which climate varies coherently on centennial and millennial scales. The sizes of those domains change with location and the time constant of climate forcing. A more thorough understanding of the mechanisms of climate forcing and the physical linkages between climate forcing and system response is needed in order to predict the spatial scale(s) over which climate varies coherently.

A question remains as to whether oscillations in the sizes of Great Basin glaciers and lakes can or should be linked to climatic oscillations documented in the North Atlantic region. Some data suggests that the Owens Lake basin was relatively dry during Heinrich events H1, H2, and H4 (Figure 20). In addition, each of the lakes discussed in this paper experienced relatively low levels at the times of H1 and H2 (Figure 22). The presence of lake-level minima in all three lake basins at ~21 and ~14 ka as well as lake-level maxima at ~18 and 13 ka may, however, have resulted from an oscillation in the mean position of

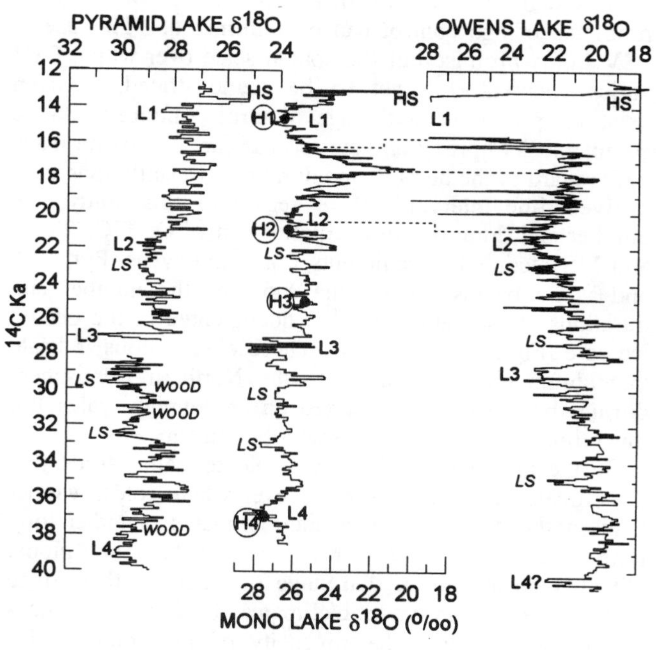

Figure 22. Comparison of Pyramid, Mono and Owens Lake δ^{18}O data for the period 40.0 to ~12.5 ^{14}C ka 'L' refers to prominent lowstands in the three basins; '*LS*' refers to other unnumbered lowstands; 'H' and solid circles (•) denote the paleomagnetic secular variation (PSV) based positions of Heinrich events in the Mono Lake record; and 'HS' indicates the last major highstand/overflow event in each of the three basins. 'WOOD' indicates times when tufa coated roots were found at low elevations around the Pyramid Lake basin [Benson *et al.*, 1995]. The location of Heinrich events in the Mono Lake section are based on a PSV match with CH88-10P and P-094 (see Figures 20 and 21]. The chronology of the Wilson Creek section has been stretched to match the ^{14}C dates for Heinrich events H1 through H4 in CH88-10P [Keigwin, 1989, 1994]. The dotted lines indicate PSV correlations between Mono Lake and Owens Lake sediments (unpublished data of S. Lund).

the PJS that occurred in response to the gradual change in size of the Laurentide Ice Sheet [Benson et al., 1995].

Alpine glaciers and surface-water systems do not come to instantaneous equilibrium with a new climate state. The more abrupt the change in climate and the larger its departure from the existing climate state, the greater the time the system will take to come to equilibrium. If abrupt changes in climate occur frequently, glacier and surface-water systems never achieve complete equilibrium with the climate system. The scale of these lag times must be determined if accurate comparisons of proxy records from marine, lacustrine and alpine glacier systems are to be made. This can best be done using numerical models that link changes in lake and glacier sizes to changes in the elements of climate.

Intrahemispheric correlation of marine and continental records will remain more of an art than a science until problems of age control can be minimized. The use of PSV holds promise but the spatial scale over which PSV varies coherently is yet to be demonstrated. Given existing problems with age control, intrahemispheric comparisons of proxy climate records can at least be made with regard to the number and duration of oscillations over a given time interval. With regard to this study, the number and durations of oscillations in δ^{18}O, TIC, TOC, and MS recorded in sediments from the Owens, Pyramid, and Mono basins are not the same as the number and durations of oscillations in δ^{18}O documented in the GISP2 ice core (Figures 7, 16, 20). These observations indicate that abrupt climate changes in the Northern Hemisphere during the Wisconsin occurred on millennial scales but that climate change was not spatially uniform.

On the global scale, it is possible to demonstrate that the climate of one time slice is everywhere colder/warmer than another time slice if average climates of rather long time slices are compared. Most, and perhaps all, global climate records indicate that Stage 2 was colder than Stage 1. However, as the synoptic climates of thinner time slices are compared, spatial heterogeneity of temperature fields soon become apparent; i.e., the spatial heterogeneity within each time slice begins to approach the weather endmember of the climate spectrum as the time slice increases in resolution.

One climate teleconnection that has been invoked in explaining the rise and fall of Great Basin lakes is the PJS. The PJS can be thought of as a teleconnection that links hemispheric weather anomalies along a moderately narrow band. In general, the precipitation field is not only inhomogeneous it is also discontinuous in nature. Today, the width of the precipitation anomaly associated with the PJS spans ~2000 km [Starrett, 1949]. Climate is colder north and warmer south of the PJS core and wetter under the core, implying that regions of uniform and synchronous climate change in the Northern Hemisphere associated with the statistical position of PJS may take the form of irregularly shaped bands. If this conjecture is correct, Dansgaard-Oeschger events cannot be linked with alpine glacier and lake-size fluctuations throughout the western United States because the trajectory and intensity of the climate signal that connects Greenland to the western United States varied in space and time.

Evidence for even regional nonuniformity of climate change is apparent in hydrologic balance records from the Great Basin. Oxygen-18 records from Owens, Mono, and Pyramid Lake basins indicate that all three basins witnessed a wet-dry-wet oscillation in climate between 18.0 and 13.0 ^{14}C ka. However, there appears to have been a profound north-south gradient in the effective wetness of the lake basins during the dry period. Owens Lake completely desiccated ~14.0 ^{14}C ka (Figure 7); but the surface of Mono Lake stood higher than at any time in the historical period (above the South Shore site) and Pyramid Lake data show only a moderate increase in δ^{18}O (Figure 16).

δ^{18}O and TOC records from the Owens Lake basin indicate that substantial wet-dry oscillations occurred during what may be interpreted as generally cold stades (52.0 to 49.0 and 24.0 to 14.0 ^{14}C ka)(Figures 7 and 8). This suggests that precipitation and temperature variations were not phase locked.

One of the more interesting findings of this study and those that preceded it [Benson *et al.*, 1996a, Benson *et al.*, 1997] is that different frequencies of climate change have different polarity combinations; e.g., a climate frequency which varies from cold-wet to warm-dry versus a frequency that varies from cold-dry to warm-wet. The medium-resolution Owens Lake records of Bischoff *et al.* [1997a] and Menking *et al.* [1997](Figure 2) show quite clearly that cold-wet alpine glaciations alternated with warm-dry interglacials on Milankovitch time scales. The high-resolution Owens Lake records (Figures 7 and 8) of Benson *et al.* [1996a] show that during the last alpine glacier interval, cold-dry stades alternated with warm-wet interstades on millennial (Dansgaard-Oeschger) time scales.

Acknowledgments. The author is grateful for reviews of this manuscript by John Andrews, Jim Dungan Smith, Jack Oviatt, and John Flager.

REFERENCES

Adams, D.K., and A.C. Comrie, The North American Monsoon, *Bulletin of the American Meteorological Society*, *78*, 2197-2213, 1997.

Allen, B.D., and R.Y. Anderson, Evidence from western North America for rapid shifts in climate during the last Glacial Maximum, *Science*, *260*, 1920-1923, 1993.

Bard, E., M. Arnold, R.G. Fairbanks, and B. Hamelin, ^{230}Th-^{234}U and ^{14}C ages obtained by mass spectrometry on corals, *Radiocarbon*, *35* (1), 191-199, 1993.

Bard, E., B. Hamelin, M. Arnold, L. Montaggioni, G. Cabioch, G. Faure, and F. Rougerie, Deglacial sea-level record from

Tahiti corals and the timing of global meltwater discharge, *Nature*, *382*, 241-244, 1996.

Bard, E., B. Hamelin, R.G. Fairbanks, and A. Zindler, Calibration of the ^{14}C time scale over the past 30,000 years using mass spectrometric U-Th ages from Barbados corals, *Nature*, *345*, 405-410, 1990.

Beanland, S., and M. Clark, The Owens Valley fault zone, eastern California, and surface rupture associated with the 1872 earthquake, in *U.S. Geological Survey Bulletin 1982*, pp. 29, 1994.

Behl, R.J., and J.P. Kennett, Brief interstadial events in the Santa Barbara Basin, Northeastern Pacific, during the past 60 kyr, *Nature*, *379*, 243-246, 1996.

Benson, L.V., Stable isotopes of oxygen and hydrogen in the Truckee River-Pyramid Lake surface-water system. 1. Data analysis and extraction of paleoclimatic information, *Limnology Oceanography*, *39*, 344-355, 1994.

Benson, L.V., J. Burdett, S. Lund, M. Kashgarian, and S. Mensing, Nearly synchronous climate change in the Northern hemisphere during the last glacial termination, *Nature*, *388*, 263-265, 1997.

Benson, L.V., J.W. Burdett, M. Kashgarian, S.P. Lund, F.M. Phillips, and R.O. Rye, Climatic and Hydrologic oscillations in the Owens Lake Basin and adjacent Sierra Nevada, California, *Science*, *274*, 746-749, 1996a.

Benson, L.V., D.R. Currey, R.I. Dorn, K.R. Lajoie, C.G. Oviatt, S.W. Robinson, G.I. Smith, and S. Scott, Chronology of expansion and contraction of four Great Basin lake systems during the past 35,000 years, *Palaeogeography, Palaeoclimatology, Palaeoecology*, *78*, 241-286, 1990.

Benson, L.V., M. Kashgarian, and M. Rubin, Carbonate deposition, Pyramid Lake subbasin, Nevada. 2. Lake levels and polar jet stream positions reconstructed from radiocarbon ages and elevations of carbonates (tufas) deposited in the Lahontan Basin, *Palaeogeography, Palaeoclimatology, Palaeoecology*, *117*, 1-30, 1995.

Benson, L.V., and H. Klieforth, Stable isotopes in precipitation and ground water in the Yucca Mountain region, southern Nevada: Paleoclimatic implications, in *Aspects of Climate Variability in the Pacific and Western Americas*, edited by D.H. Peterson, pp. 41-59, 1989.

Benson, L.V., S.P. Lund, J.W. Burdett, M. Kashgarian, T.P. Rose, J.P. Smoot, and M. Schwartz, Correlation of Late-Pleistocene lake-level oscillations in Mono Lake, California, with North Atlantic climate events, *Quaternary Research*, *49*, I-10, 1998a.

Benson, L.V., H.M. May, R.C. Antweiler, T.I. Brinton, M. Kashgarian, J.P. Smoot, and S. P. Lund, Continuous lake-sediment records of glaciation in the Sierra Nevada between 52,600 and 12,500 ^{14}C yr B.P., *Quaternary Research, 50*, 113-127, 1998b.

Benson, L.V., P.A. Meyers, and R.J. Spencer, Change in the size of Walker Lake during the past 5000 years, *Palaeogeography, Palaeoclimatology, Palaeoecology*, *81*, 189-214, 1991.

Benson, L.V., and Z. Peterman, Carbonate deposition, Pyramid Lake subbasin, Nevada: 3. The use of ^{87}Sr values in carbonate deposits (tufas) to determine the hydrologic state of paleolake systems, *Palaeogeography, Palaeoclimatology, Palaeoecology*, *119*, 201-213, 1995.

Benson, L.V., L.D. White, and R. Rye, Carbonate deposition, Pyramid Lake Subbasin, Nevada: 4. Comparison of the stable isotope values of carbonate deposits (tufas) and the Lahontan lake-level record, *Palaeogeography, Palaeoclimatology, Palaeoecology*, *122*, 45-76, 1996b.

Bischoff, J.L., K.M. Menking, J.P. Fitts, and J.A. Fitzpatrick, Climatic oscillations 10,000-155,000 yr B.P. at Owens Lake, California, reflected in glacial rock flour abundance and lake salinity in Core OL-92, *Quaternary Research*, *48*, 313-325, 1997a.

Bischoff, J.L., T.W. Stafford, and M. Rubin, A time-depth scale for Owens Lake sediments of core OL-92: Radiocarbon dates and constant mass-accumulation rate, in *An 800,000-year paleoclimatic record from Core OL-92, Owens Lake, Southeast California*, edited by G.I. Smith, and J.L. Bischoff, pp. 91-98, 1997b.

Bond, G., W. Broecker, S. Johnsen, J. McManus, L. Labeyrie, J. Jouzel, and G. Bonani, Correlations between climate records from North Atlantic sediments and Greenland ice, *Nature*, *365*, 143-147, 1993.

Bond, G., H. Heinrich, W. Broecker, L. Labeyrie, J. McManus, J. Andrews, S. Houn, R. Jantschik, S. Clasen, C. Simet, K. Tedesco, M. Kias, G. Bonani, and S. Ivy, Evidence for massive discharges of icebergs into the North Atlantic ocean during the last glacial period, *Nature*, *360*, 245-249, 1992.

Bond, G., W. Showers, M. Cheseby, R. Lotti, P. Almasi, P. deMenocal, P. Priore, H. Cullen, I. Hajdas, and G. Bonani, A pervasive millennial-scale cycle in North Atlantic Holocene and glacial climates, *Science*, *278*, 1257-1266, 1997.

Bond, G.C., and R. Lotti, Iceberg discharges into the North Atlantic on millennial time scales during the last glaciation, *Science*, *267*, 1005-1010, 1995.

Broecker, W.S., and A. Kaufman, The geochemistry of ^{14}C in freshwater systems, *Geochemistry, 16*, 15-38, 1959.

Brook, E.J., T. Sowers, and J. Orchanrdo, Rapid variations in Atmospheric methane concentration during the past 110,000 Years, *Science*, *273*, 1087-1091, 1996.

Clark, D.H., Extent timing and climatic significance of latest Pleistocene and Holocene glaciation in the Sierra Nevada, California, unpublished doctoral thesis, University of Washington, Washington, 1995.

Clark, D.H., A new alpine lacustrine sedimentary record from the Sierra Nevada: Implications for late-Pleistocene paleoclimate reconstructions and cosmogenic isotope production rates, 1997.

Clark, D.H., P.R. Bierman, and P. Larsen, Improving *in situ* cosmogenic chronometers, *Quaternary Research*, *44*, 367-377, 1995.

Clark, D.H., and A.R. Gillespie, Timing and significance of late-glacial and Holocene cirque glaciation in the Sierra Nevada, California, *Quaternary International*, *38/39*, 21-38, 1997.

Dansgaard, W., S.J. Johnsen, H.B. Clausen, D. Dahl-Jensen, N. Gundestrup, C.U. Hammer, and H. Oeschger, North Atlantic climatic oscillations revealed by deep Greenland ice cores, in *Climate Processes and Climate Sensitivity*, edited by J.E. Hansen, and T. Takahashi, pp. 288-298, AGU, Washington, D.C., 1984.

Dansgaard, W., S.J. Johnsen, H.B. Clausen, D. Dahl-Jensen, N.S. Gundestrup, C.U. Hammer, C.S. Hvidberg, J.P.

Steffensen, A.E. Sveinbjornsdottir, J. Jouzel, and G. Bond, Evidence for general instability of past climate from a 250-kyr ice-core record, *Nature*, *364*, 218-364, 1993.

Dansgaard, W., S.J. Johnsen, H.B. Clausen, and C.C. Langway Jr., Climate record revealed by the Camp Century ice core, in *The Late Cenozoic Glacial Ages*, edited by K.K. Turekian, pp. 37-56, Yale University Press, New Haven, 1971.

Dixon, T.H., J.L. Robaudo, J. Lee, and M.C. Reheis, Constraints on present-day basin and range deformation from space geodesy, *Tectonics*, *14*, 755-772, 1995.

Dohrenwend, J.C., Nivation landforms in the western Great Basin and their paleoclimatic significance, *Quaternary Research*, *22*, 275-288, 1984.

Font, K.R., Geochemical and Isotopic Evidence for Hydrologic Processes at Owens Lake, California, Master thesis, University of Nevada, Reno, 1995.

Galat, D.L., E.L. Lider, S. Vigg, and S.R. Robertson, Limnology of a large, deep, North American terminal lake, Pyramid Lake, Nevada, USA, *Hydrobiologia*, *82*, 281-317, 1981.

Gale, H.S., Notes on the Quaternary lakes of the Great Basin with special reference to the deposition of potash and other salines, *U.S. Geological Survey Bulletin 540-N*, 399-406, 1914.

Grootes, P.M., M. Stuiver, J.W.C. White, S. Johnsen, and J. Jouzel, Comparison of oxygen isotope records from the GISP2 and GRIP Greenland ice cores, *Nature*, *366*, 552-554, 1993.

Hallet, B., L. Hunter, and J. Bogen, Rates of erosion and sediment evacuation by glaciers: A review of field data and their implications, *Global and Planetary Change*, *12*, 213 - 235, 1996.

Hinkley, T., Alkali and alkaline earth metals: Distribution and loss in a High Sierra Nevada watershed, *Geological Society of America Bulletin*, *85*, 1333 - 1338, 1974.

Horn, L.H., and R.A. Bryson, Harmonic analysis of the annual march of precipitation over the United States, *Annals of the Association of American Geographers*, *50*, 157-171, 1960.

Huber, N.K., Amount and time of late Cenozoic uplift and tilt of central Sierra Nevada, California, *U.S. Geological Survey Professional Paper* 1197, pp. 1-28, 1981.

Hughen, K.A., J.T. Overpeck, L.C. Peterson, and S. Trumbore, Rapid climate changes in the tropical Atlantic region during the last deglaciation, *Nature*, *380*, 51-54, 1996.

Imbrie, J., J.D. Hays, D.G. Martinson, A. McIntyre, A.C. Mix, J.J. Morley, N.G. Pisias, W.L. Prell, and N.J. Shackleton, The orbital theory of Pleistocene climate: Support from a revised chronology of the marine $d^{18}O$ record, *Milankovitch and Climate, Part 1 (NATO ASI Series C)*, *126*, 269 -305, 1984.

Johnsen, S.J., H.B. Clausen, W. Dansgaard, K. Fuhrer, N. Gundestrup, C.U. Hammer, P. Iversen, J. Jouzel, B. Stauffer, and J.P. Steffensen, Irregular glacial interstadials recorded in a new Greenland ice core, *Nature*, *359*, 311-313, 1992.

Johnsen, S.J., W. Dansgaard, H.B. Clausen, and J. Langway, C.C., Oxygen isotope profiles through the Antarctic and Greenland ice sheets, *Nature*, *235*, 429-434, 1972.

Keigwin, L.D., and G.A. Jones, Glacial-Holocene stratigraphy, chronology, and paleoceanographic observations on some north Atlantic sediment drifts, *Deep Sea Research*, *36*, 845-867, 1989.

Keigwin, L.D., and G.A. Jones, Western north Atlantic evidence for millennial-scale changes in ocean circulation and climate, *Journal of Geophysical Research*, *99*, 12397-12410, 1994.

Kitagawa, H., and J. van der Plicht, Atmospheric radiocarbon calibration to 45,000 yr B.P.: late glacial fluctuations and cosmogenic isotope production, *Science*, *279*, 1187-1190, 1998.

Lajoie, K.R., Late quaternary stratigraphy and geologic history of Mono Basin Eastern California, Doctoral thesis, University of California, Berkeley, 1968.

Liddicoat, J.C., Mono Lake excursion in Mono Basin, California, and at Carson Sink and Pyramid Lake, Nevada, *Geophysical Journal, International*, *108*, 442-452, 1992.

Liddicoat, J.D., and R.S. Coe, Mono Lake Geomagnetic Excursion, *Journal of Geophysical Research*, *84*, 261-271, 1979.

Lin, J.C., W.S. Broecker, S.R. Hemming, I. Hajdas, R.F. Anderson, G.I. Smith, M. Kelley, and G. Bonani, A reassessment of U-Th and ^{14}C ages for late-glacial high-frequency hydrological events at Searles Lake, California, *Quaternary Research*, *49*, 11-23, 1998.

Lund, D.C., and A.C. Mix, Millennial-scale deep water oscillations: Reflection of the North Atlantic in the deep Pacific from 10 to 60 ka, *Paleoceanography*, *13*, 10-19, 1998.

Lund, S.P., Paleomagnetic secular variation, in *Trends in Geophysical Research*, pp. 423-438, Council of Scientific Research Integration, Trivandrum, India, 1993.

Lund, S.P., A comparison of Holocene paleomagnetic secular variation records from North America, *Journal of Geophysical Research*, *101*, 8007-8024, 1996.

Lund, S.P., J.C. Liddicoat, K.R. Lajoie, T.L. Henyey, and S.W. Robinson, Paleomagnetic evidence for long-term (10^4 year) memory and periodic behavior in the Earth's core dynamo process, *Geophysical Research Letters*, *15*, 1101-1104, 1988.

Mayewski, P.A., L.D. Meeker, S. Whitlow, M.S. Twickler, M.C. Morrison, P. Bloomfield, G.C. Bond, R.B. Alley, A.J. Gow, P.M. Grootes, D.A. Meese, M. Ram, K.C. Taylor, and W. Wumkes, Changes in atmospheric circulation and ocean ice cover over the North Atlantic during the last 41,000 years, *Science*, *263*, 1747-1751, 1994.

Menking, D.M., J.L. Bischoff, and J.A. Fitzpatrick, Climatic/Hydrologic Oscillations since 155,000 yr B.P. at Owens lake, California, reflected in abundance and stable isotope composition of sediment carbonate, *Quaternary Research*, *48*, 58-68, 1997.

Negrini, R.M., J.O. David, and K.L. Verosub, Mono Lake geomagnetic excursion found at Summer Lake, Oregon, *Geology*, *12* (643-646), 1984.

Oviatt, C.G., Lake Bonneville fluctuations and global climate change, *Geology*, *25*, 155-158, 1997.

Phillips, F.M., A.R. Campbell, G.I. Smith, and J.L. Bischoff, Interstadial climatic cycles: A link between western North American and Greenland?, *Geology*, *22*, 1115-1118, 1994.

Phillips, F.M., L.A. Peeters, and M.K. Tansey, Paleoclimatic inferences from an isotopic investigation of groundwater in the Central San Juan Basin, New Mexico, *Quaternary Research*, *26*, 179-193, 1986.

Phillips, F.M., M.G. Zreda, L.V. Benson, M.A. Plummer, D.

Elmore, and P. Sharma, Chronology for fluctuations in Late Pleistocene Sierra Nevada glaciers and lakes, *Science, 274* (749-751), 1996.

Phillips, F.M., M.G. Zreda, J.C. Gosse, J. Klein, E.B. Evenson, R.D. Hall, O.A. Chadwick, and P. Sharma, Cosmogenic ^{36}Cl and ^{10}Be ages of Quaternary glacial and fluvial deposits of the Wind River Range, Wyoming, *Geological Society of America Bulletin 109*, 1453-1463, 1997.

Porter, S.C., K.L. Pierce, and T.D. Hamilton, Late Wisconsin mountain glaciation in the western United States, in *Late-Quaternary Environments of the United States*, edited by J. Wright, H. E., pp. 407, University of Minnesota Press, Minneapolis, 1983.

Porter, S.C., and A. Zhisheng, Correlation between climate events in the North Atlantic and China during the last glaciation, *Nature, 375*, 305-308, 1995.

Pyke, C.B., Some meteorological aspects of the seasonal distribution of precipitation in the Western United States and Baja, California, in *University of California Water Resources Center Contribution*, pp. 139, 1972.

Redmond, K.T., and R.W. Koch, Surface climate and streamflow variability in the western United States and their relationship to large-scale circulation indices, *Water Resources Research, 27*, 2381-2399, 1991.

Riehl, H., M.A. Alaka, C.L. Jordan, and R.J. Renard, The jet stream, *Meteorology and Monograph, 2*, 23-47, 1954.

Ropelewski, C.F., and M.S. Halpert, North American precipitation and temperature patterns associated with the El Nino/Southern Oscillation (ENSO), *Monthly Weather Review, 114*, 2352-2362, 1986.

Rose, T.P., M.L. Davisson, G.B. Hudson, and A.R. Varian, Environmental Isotope Investigation of Groundwater Flow in the Honey Lake Basin, California and Nevada, pp. 42, Lawrence Livermore National Laboratory, UCRL-ID-127978, Livermore, CA, 1997.

Russell, I.C., Geological History of Lake Lahontan, a Quaternary lake of Northwestern Nevada, in *U.S. Geological Survey Monograph 11*, pp. 288, 1885.

Shackleton, N.J., The last interglacial in the marine and terrestrial records, *Proceedings of the Royal Society of London, B174*, 135-154, 1969.

Spaulding, W.G., Vegetation and climates of the last 45,000 years in the vicinity of the Nevada Test Site, south-central Nevada, in *U.S. Geological Survey Open-File Report 83-535*, pp. 205, 1983.

Starrett, L.G., The relation of precipitation patterns in North America to certain types of jet streams at the 300-millibar level, *Journal of Meteorology, 6*, 347-352, 1949.

Stute, M., P. Schlosser, J.F. Clark, and W.S. Broecker, Paleotemperatures in the southwestern United States derived from noble gases in ground water, *Science, 256*, 1000-1003, 1992.

Syvitski, J.P., LeBlanc, K.W., and R.E. Cranston, The flux and preservation of organic carbon in Baffin Island fjord,in *Glaciomarine Environments: Processes and Sediments*, edited by J.A. Dowdeswell, and J.D. Scourse, pp. 217-239, Geological Society of London Special Publication 53, 1990.

Tang, M., and E.R. Reiter, Plateau Monsoons of the Northern Hemisphere: A comparison between North America and Tibet, *Monthly Weather Review, 112*, 617-637, 1984.

Taylor, K.C., G.W. Lamorey, G.A. Doyle, R.B. Alley, P.M. Grootes, P.A. Mayewski, J.W.C. White, and L.K. Barlow, The "flickering switch" of Late Pleistocene climate change, *Nature, 361*, 432-436, 1993.

Thompson, R., Global Holocene magnetostratigraphy, *Hydrobiologia, 103*, 45-51, 1983.

Thompson, R.S., L.V. Benson, and E.M. Hattori, A revised chronology for the last Pleistocene lake cycle in the central Lahontan basin, *Quaternary Research, 25*, 1-9, 1986.

Van Devender, T.R., Late Pleistocene Plants and Animals of the Sonoran Desert, a Survey of Ancient Packrat Middens in Southwestern Arizona, Doctorate thesis, University of Arizona, Tucson, 1973.

Wahrhaftig, C., and J.H. Birman, The Quaternary of the Pacific Mountain System in California. In *Means of Correlation of Quaternary Successions*, Vol. 8: Proceedings VII Congress International Association for Quaternary Research edited by R. Morrison, and H. E. Wright, pp. 293, 1956.

Welch, A.H., Ground-Water Quality and Geochemistry in Carson and Eagle Valleys, Western Nevada and Eastern California, in *U.S. Geological Open-File Report 93-33*, pp. 99, 1994.

Larry Benson, U.S. Geological Survey, 3215 Marine St., Boulder, CO 80303

Paleoecological Evidence of Milankovitch and Sub-Milankovitch Climate Variations in the Western U.S. During the Late Quaternary

Cathy Whitlock and Laurie D. Grigg

Department of Geography, University of Oregon, Eugene, Oregon

Vegetation history provides a record of responses to paleoenvironmental variations occurring across a number of spatial and temporal scales. Effects at particular scales cannot be identified easily from a single site, and a single pollen spectrum cannot be interpreted without understanding the higher and lower frequencies of variation. The vegetation history of the western U.S. shows a strong response to Milankovitch forcing, including the effects that variations in insolation and ice sheet size have had on atmospheric circulation, the position of the jet stream, and subregional precipitation regimes. The response to millennial-scale changes is contained in the residual signal. While the registration of Dansgaard-Oeschger (D-O) and Heinrich events is clear in terrestrial records from regions surrounding the North Atlantic, the signal is greatly attenuated in other parts of the world. In the western U.S., many sites show millennial-scale variability, but few exhibit a spatially coherent pattern that can be ascribed to specific D-O cycles or Heinrich events. Well-dated sites in the Pacific Northwest with close-interval sampling, however, indicate climatic oscillations associated with H1, H2, and H3. Based on these records, the vegetational response may have been greatest when the Laurentide Ice Sheet was at its maximum size, possibly because the climate changes in the North Atlantic were amplified regionally by the presence of the glacial anticyclone. The Younger Dryas cool event (12.9-11.6 Cal ka) is evident in coastal sites and high-elevation sites inland. In the Pacific Northwest the cooling event began later at 12.4 Cal ka and lasted until 11 Cal ka. Millennial-scale climate changes are not evident in most Holocene pollen records from the western U.S., except in those from treeline locations. Charcoal records of past fire occurrence may provide a more sensitive climate proxy during interglacial periods.

Mechanisms of Global Climate Change at Millennial Time Scales
Geophysical Monograph 112

SPATIAL AND TEMPORAL CONTROLS OF VEGETATION CHANGE

Our understanding of the response of vegetation to large-scale changes in the climate system is based primarily on an assessment of fossil pollen and plant-macrofossil records. The quality of the information varies greatly depending on the analytical precision and chronologic control. As the number of sites increases, it becomes possible to examine environmental change across a spatial network and from that pattern determine the nature of regional climate change. Such a network is essential to separate local causes of vegetation change from those that might be regional or global in nature. The focus of this paper is on the causes of vegetation change in the western U.S. during the late Quaternary, and especially the response of vegetation to millennial-scale climate changes. The analysis of well-dated pollen and packrat-midden records helps to clarify the extent and nature of the change, as well as the mechanism of transmission. At present, numerous paleoecologic records span the Holocene, whereas comparatively few records are available for marine oxygen-isotope stage (MIS) 2 and older (See Fig. 1, for sites discussed in text).

The controls on late-Quaternary vegetation change are hierarchical in nature ranging from Milankovitch time scales to century time scales and global-to-local spatial scales. Sub-Milankovitch climate changes are best known from marine and ice core records from the North Atlantic region. Climatic instability within MIS 2 and 3 is apparent in oxygen-isotope profiles from the Greenland ice cores (Dansgaard et al., 1993; Grootes et al., 1993). Twenty or so episodes of cold-warm cycles, so-called Dansgaard-Oeschger or D-O cycles, are identified, with each cycle lasting about 1500 years (Alley, 1998). The cold stages of D-O cycles are associated with iceberg surging that probably led to a weakening or shutdown of North Atlantic Deep Water formation and are followed by abrupt warmings or interstadials. Progressive cooling is evident in several sets of D-O cycles that constitute the larger bundles (Bond cycles). The coldest interval of a Bond cycle is marked by an influx of ice-rafted debris in the North Atlantic; these are termed Heinrich events and are spaced 5000-17,000 years apart. Six Heinrich events are identified: H0 (Younger Dryas start)=12.9 Cal ka; H1=17.1 Cal ka; H2/IS2=24.5 Cal ka; H3/IS3=29.5 Cal ka; H4/IS8= 38 Cal ka; H5/IS8=50 Cal ka; H6/IS6=66.5 Cal ka. (These age assignments, with the exception of H0, are based on a mean calendar age of the Heinrich events reported in Bond et al., 1992, 1993; Grimm et al., 1993; Dansgaard et al., 1993; Broecker, 1994; Bender et al., 1994; Jouzel, 1994; Bond and Lotti, 1995; Porter and An, 1995; Behl and Kennett, 1996. Radiocarbon years were converted to calendar years [Cal ka] by use of the CALIB3 program for the last 20 ^{14}C ka [Stuiver and Reimer, 1993], the U-Th calibration for ages between 20 and 30 ^{14}C ka [Bard et al., 1990] and the geomagnetic record for the period from 30 to 50 ^{14}C ka [Mazaud et al., 1991]). H0, the Younger Dryas event (11.6-12.9 Cal ka; Alley et al., 1993), is registered clearly in ice core and ocean records in the North Atlantic and pollen

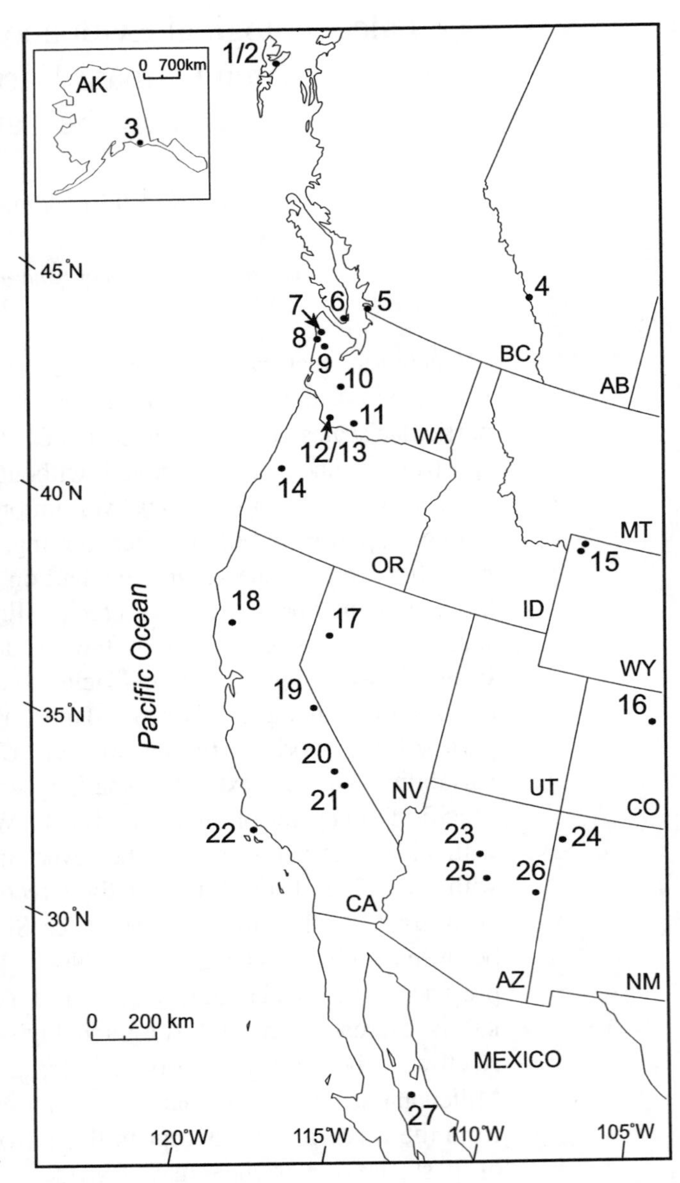

Figure 1. Paleoclimatic records referred to in the text. 1=Pilot Mill, 2=Cape Ball, 3=Pleasant Island, 4=Crowfoot Lake, 5=Marion Lake, 6=Dashwood, 7=Bogachiel River, 8=Kalaloch Sea cliffs, 9=Humptulips Bog, 10=Davis Lake, 11=Carp Lake, 12=Battle Ground Lake, 13=Fargher Lake, 14=Little Lake, 15=Yellowstone lakes, 16=Sky Pond, 17=Pyramid Lake, 18=Clear Lake, 19=Mono Lake, 20= Owens Lake, 21=Searles Lake, 22=Santa Barbara Basin, 23=Walker Lake, 24=Deadman Lake, 25=Potato Lake, 26=Hay Lake, 27= Guaymas Basin.

data from northwest Europe, and it is also described for other parts of the world. Evidence of cold, dry conditions corresponding with H1, H2, H3, and in some cases H4 is reported in southern South America (Lowell et al., 1995), New Zealand (Denton and Hendy, 1994), the loess plateau of China (Porter and An, 1995), and the pluvial lakes of the American Southwest (Benson et al., 1996, 1998). Although first described in MIS 2 and 3 (Bond and Lotti, 1995), millennial-scale variations in climate have recently been recorded in the Holocene and early Pleistocene (Bond et al., 1997; Oppo et al. 1998; Raymo et al., 1998).

Several reviews describe the vegetation response to climate variations occurring on different temporal and spatial scales in the Quaternary (Bartlein, 1988; Prentice, 1992; Webb and Bartlein, 1992; Whitlock, 1993). At the largest scale are changes in the seasonal cycle of insolation, in continental ice sheet size, and in atmospheric composition that arise from orbital variations on time scales of 10^4 to 10^5 years. The residuals in the pollen record, after Milankovitch-scale variations are removed, represent vegetation changes on shorter time scales. This high-frequency signal, however, must be separated from subregional variations in climate, such as a coastal-versus-inland location or the influence of topography. For example, the differences between full-glacial vegetation on opposite sides of the Cascade Range suggest that few storms penetrated east of the mountain range. This orographic effect probably explains why moisture increased earlier on the west side of the range than on the east side in late-glacial time. Similarly, the climate history of northern Yellowstone is different from that in southern Yellowstone as a result of trade-offs in precipitation regimes in a topographically complex region. Thus, subregional variations may lead to spatially discontinuous registration of a globally transmitted climate event.

At the scale of a single site, the effects of climate on vegetation cannot always be disentangled from those caused by other site conditions, including substrate, biotic interactions, and disturbance regime. For example, the prominence of *Pinus contorta* (lodgepole pine) in central Yellowstone was established at the beginning of the Holocene and has been maintained ever since by nutrient-poor soils (Whitlock, 1993). Even when the vegetation changed elsewhere in the region during the Holocene, the forests of central Yellowstone remained dominated by *Pinus contorta*. Such an area would be unlikely to record millennial-scale climate changes in the pollen record. In addition, biotic interactions, like plant succession following disturbance, account for some of the pollen signal at the local scale.

VEGETATION RESPONSE ON MILANKOVITCH TIME SCALES

An example of the importance of Milankovitch variations is provided by a pollen record of the last 130 Cal ka from Carp Lake, Washington (Whitlock and Bartlein, 1997) (Fig. 2). The record shows gradual shifts between periods dominated by high-elevation taxa and those dominated by low-elevation species. Most of the variation in the record is explained by changes in summer insolation and ice sheet size, as evidenced by the close similarity of the July insolation anomaly at 45°N and the marine oxygen-isotope stratigraphy over the last 130 Cal ka. Correlation of the vegetation changes with the insolation and isotopic record suggests negligible lags between large-scale climatic forcing and local vegetation response. Apparently, species kept pace with Milankovitch climate changes by expanding their ranges a short distance from discontinuous refugia. However, it is important to note that, despite the continuity of the changes in large-scale controls, the abundance of particular taxa at Carp Lake changes abruptly. The response of individual species is likely to depend on its sensitivity to the specific climate change and the possibility of natural disturbance to accelerate the change.

Other long records from western North America show the strong influence of Milankovitch variations on vegetation change. For example, the pollen record at Clear Lake, California has comparable resolution for the early Wisconsin to that at Carp Lake (Fig. 2). Adam (1988) identifies multiple warm periods with high *Quercus* (oak) pollen that are generally correlated with summer insolation maxima in MIS 4 and 5 and cool periods with low *Quercus* percentages during insolation minima. The oscillations are similar to those recorded by the pollen ratios at Carp Lake. Heusser (1995, 1998) makes a similar comparison of *Quercus* pollen percentages and the oxygen-isotope record from ODP Core 893 from the Santa Barbara Basin. Heusser and Heusser (1990) use the profile of *Tsuga heterophylla* (western hemlock) as an indicator of warm periods on the Olympic Peninsula, Washington and again note the correlation with the oxygen-isotope record.

A network of paleoecologic sites discloses the geographic extent of vegetation response to Milankovitch variations. For example, Pacific Northwest records of the last glacial maximum indicate that subalpine parkland and open forest expanded into areas that today support temperate rain forest. Similarly, present-day temperate steppe was replaced by communities of cold dry steppe and scattered *Picea* (spruce). Changes in the geographic ranges of particular taxa provide further evidence of cold dry

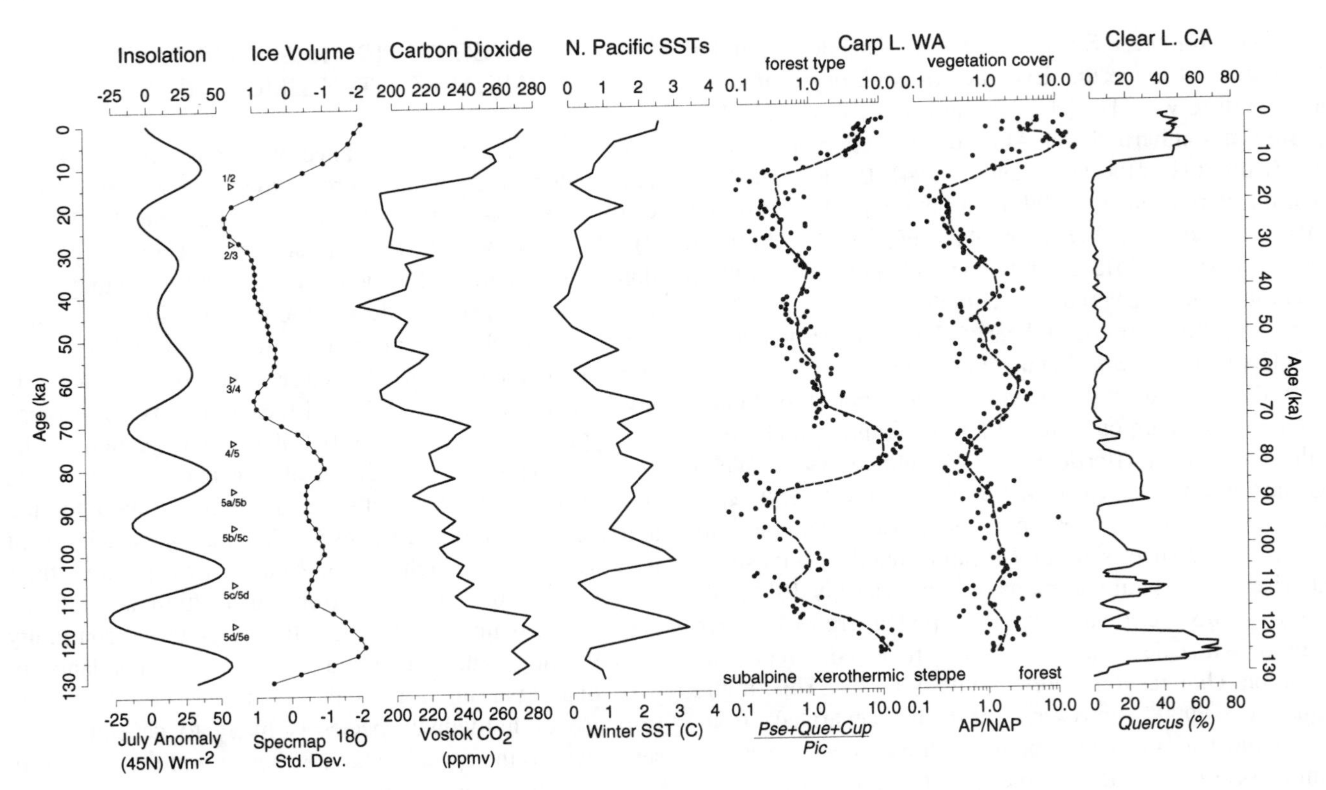

Figure 2. Comparison of Milankovitch-time scale controls of climate change (July insolation anomalies at latitude 45° N and global ice volume) and pollen data from Carp Lake, Washington (Whitlock and Bartlein, 1997), Clear Lake, California (Adam, 1988), and Grande Pile, France (Woillard and Mook, 1982) over the last 130 kyr. The Carp Lake data are a smooth record of the ratio of *Pseudotsuga* + *Quercus* + Cupressaceae pollen to *Picea* pollen to indicate warm and cold vegetation types. High percentages of *Quercus* at Clear Lake indicate warm dry conditions, and at Grande Pile high values of arboreal to nonarboreal pollen (AP/NAP) indicate periods of forest.

conditions during full-glacial time. *Picea engelmannii* (Engelmann spruce), a conifer confined today to the Cascade Range and Rocky Mountain region, extended its range westward into the Coast Range, implying a more continental climate than at present. The dominant controls on full-glacial climate (reduced summer insolation, large ice sheet, and lowered atmospheric CO_2) apparently affected the climate of the Pacific Northwest in three ways (Thompson et al., 1993): (1) a steeper-than-present latitudinal temperature gradient caused average annual temperatures to be between 5 and 10°C colder; (2) a southward shift of the jet stream robbed the Pacific Northwest and northern Rockies of winter precipitation; and (3) the glacial anticyclone strengthened surface easterlies, which increased aridity in the region. The results were cold dry conditions and a steepened west-to-east precipitation gradient within the Pacific Northwest, in contrast to the pluvial conditions of the American Southwest.

Greater-than-present summer insolation in western North America between 12 and 7 Cal ka led to contrasts between early- and late-Holocene climate and vegetation at different spatial scales. In the Pacific Northwest and northern Rockies, increased summer insolation directly increased summer temperatures and decreased effective moisture, and indirectly strengthened the northeast Pacific subtropical high pressure system, which further intensified drought. In response, vegetation across the Pacific Northwest and the northern Rockies show an expansion of xerophytic taxa (although the specific species varied from region to region). Upper treeline shifted to higher elevations than at present in response to warmer summer temperatures, and lower treeline shifted to higher elevations and extended as much as 100 km north of its present location because of greater summer drought. As the seasonal cycle of insolation attenuated in the middle and late Holocene, the importance of xerophytic species decreased in most regions.

EVIDENCE FOR MILLENNIAL-SCALE VARIATIONS IN MIS 2 AND 3

The interesting questions concerning millennial-scale fluctuations from the perspective of terrestrial paleoecology are: How much of the vegetation response can be explained by sub-Milankovitch climate variations? Do vegetation changes coincide with the timing of D-O cycles and Heinrich events in the North Atlantic? Does the sensitivity of vegetation to millennial-scale variations change as a result of changes in large-scale controls? Does the spatial variability of vegetation change suggest the influence of particular components of the climate system (i.e., oceanic or atmospheric circulation) in triggering or transmitting sub-Milankovitch climate change?

Closely sampled, well-dated pollen records from regions near the North Atlantic disclose variations on millennial time scales that appear to be associated with D-O cycles and Heinrich events in MIS 2 and 3. (The Younger Dryas event is discussed later.) The clear registration of high-frequency variations in the vicinity of the North Atlantic suggests that it was the source area of these variations. A detailed pollen record from Lake Tulane in central Florida, for example, displays clear oscillations in *Pinus* percentages that are as sharp as variations recorded in faunal, isotopic, and lithologic records from the adjacent North Atlantic and Greenland (Grimm et al., 1993). These peaks in *Pinus* percentages at Lake Tulane are attributed to wet periods associated with H1 through H5. The strong response of *Pinus* closely matches the GRIP record and the percentages of *Neogloboquadrina pachyderma* in Atlantic core V23-81 (Bond et al., 1993). Similarly, cooling during Younger Dryas time is best registered in pollen records from northwestern Europe and northeastern North America (Peteet, 1995), close to the region of greatest change in ocean circulation.

Evidence of millennial-scale variations is clearly shown in the alternations of arboreal and nonarboreal taxa at Grand Pile, France over the last 130 Cal ka (Woillard and Mook, 1982; Fig. 2). It is also expressed in a detailed pollen record from Lago di Monticchio, a crater lake from southern Italy (Watts et al., in press). The vegetational diversity of that region makes the response more complex than at Lake Tulane. Three interstadial pollen diagrams at Monticchio, dated by varve chronology and tephrochronology, are correlated with interstadial events 20 (72.1 Cal ka), 14 (50.9 Cal ka), and 12 (43.1 Cal ka) of the GRIP record. The longest interstadial (IS 12) of MIS 3 shows a sequence of warming in which *Betula* (birch) and *Quercus* expanded first, followed by mesic deciduous trees, including *Ulmus* (elm), *Acer* (maple), *Fraxinus* (ash), *Tilia* (lime), and *Fagus* (beech). The sequence ended with an expansion of *Abies* (fir), implying a return to cool conditions. This and the two other warm pollen zones began and ended abruptly and do not show the progressive cooling found in ocean and ice core records.

At greater distances both down the global ocean conveyor belt and upwind of the North Atlantic, the vegetational response to millennial events becomes attenuated. For example, the Lake District of Chile shows clear fluctuations in glacier position associated with H1, H2, and H3. Ice-proximal pollen records, in contrast, indicate the presence of *Nothofagus* (southern beech) parkland and Magellanic moorland, and show little temporal change during MIS 2 (Lowell et al., 1995). The relative complacency of pollen records suggests that the vegetation was not responding to local climate changes, nor was it greatly influenced by the changing position of glacial ice. More likely, the vegetation was responding to changes in the large-scale controls, including the position of the polar jet during MIS 2 (Kutzbach et al., 1998). Similar conclusions have been reached in New Zealand, where the lack of response in pollen data during YD time (McGlone, 1995) contrasts with evidence of glacial fluctuations (Denton and Hendy, 1994).

Ten sites between the Queen Charlotte Islands of British Columbia and the Oregon Coast Range have close-interval sampling of the period from 50 to 15 Cal ka (Fig. 1). Clear Lake (Adam, 1988) and Santa Barbara Basin ODP 893 (Heusser, 1995) offer suitable records from California. Three sites from Arizona--Walker Lake, Hay Lake, and Potato Lake--extend between 50 and 35 Cal ka, and diatom and ostracod data from Owens Lake in the Great Basin (Bradbury and Forester, in press) extend back to MIS 3 and 4 (Fig. 1). Radiocarbon chronologies, however, at these sites are of variable quality, and tephrochronology or paleomagnetic variations are often used to extend the chronology beyond the limit of radiocarbon dating. Carp Lake and Little Lake in the Pacific Northwest and ODP Core 893 from the Santa Barbara Basin are the best dated of these records. Several of the Pacific Northwest sites show a climatic oscillation at the time of H2 (ca. 24.5 Cal ka). Carp Lake and Little Lake also record an oscillation at the time of H1 and possibly H3. None of the sites offer convincing evidence of an oscillation associated with earlier Heinrich events. One explanation of the strong response during H1, H2 and H3 may be the fact that the prevailing vegetation was subalpine parkland or tundra during MIS 2. During earlier periods, the vegetation was forested and the response of the vegetation may have been less dramatic.

A closer look at the Carp Lake record provides an example of the response of vegetation to short-term events (Fig. 3). The low ratio of arboreal to nonarboreal (AP/NAP) pollen during H1, H2, and H3 times is consistent with a period of cold steppe. During MIS 3 and 4, the site supported a mixed forest (with higher AP values) and the climate was warmer. *Picea* pollen is a good proxy of summer temperature, although its specific significance depends on the particular climate conditions. Increases in *Picea* pollen during MIS 2 would imply warming, which allowed *P. engelmannii* to establish during a cold period. In contrast, an increase in *Picea* pollen during a period dominated by pine forest or temperate steppe would indicate cool conditions. Thus, a single species, *Picea engelmannii*, has different interpretations depending on the overall climate and vegetation conditions.

Changes in *Picea* percentages at Carp Lake suggest a shift from cold to warm conditions associated with H0 (although there seems to be a lag), H1, H2, and possibly H3 (Fig. 3). H4 time is not closely sampled at Carp Lake, and intervals correlating to H5 and H6 are not associated with variations in *Picea* pollen. Abrupt increases in *Picea* percentages following increased *Artemisia* (sagebrush) values coincide with the timing of H1 and H2. A gradual decline in *Picea* and an increase in *Artemisia* pollen led to the succeeding Heinrich Event. The changes during H3 time are less easy to interpret, because the overall vegetation was changing between ca. 33 and 27 Cal ka from *Pinus* forest to cold steppe and *Picea* parkland. Increased *Picea* percentages and moderate percentages of *Pinus* during H3 time suggest cooler conditions than before, but warmer than full-glacial conditions. *Picea* gradually replaced *Pinus* over the next 4000 years as the climate continued to cool.

The implied warming at Carp Lake associated with H1 and H2 time may be related to shifts in the size of the Laurentide Ice Sheet and associated changes in the position of the jet stream and the strength of the glacial anticyclone. During MIS 3 and 4, the glacial anticyclone and attendant easterly winds were not a strong influence on Pacific Northwest climate. The presence of a large ice sheet thus may have amplified the regional climate response to millennial-scale events in the North Atlantic.

The pollen record from the last glacial period at Little Lake in the Oregon Coast Range reveals a series of sub-Milankovitch climate changes (L. Grigg, unpublished data). The chronology is constrained by eight AMS dates between ca. 43 and 16 Cal ka. A general pattern of alternating peaks in pollen from cold- and warm-adapted taxa is seen throughout MIS 3, but it is less evident in MIS 2. MIS 3 at Little Lake is generally characterized by a mixed forest of *Pinus*, *Abies*, and *Tsuga heterophylla* and suggests cool wet conditions. However, a series of significant peaks in warm-adapted taxa, such as *T. heterophylla*, *Pseudotsuga menziesii* (Douglas-fir), and *Thuja plicata*-type (western red cedar), are repeatedly preceded by smaller increases in pollen of cold-adapted taxa, such as *Picea* and *Tsuga mertensiana* (mountain hemlock) or *Pinus monticola*-type (western white pine). This sequence of pollen changes suggests a series of climatic oscillations, each lasting ca. 2000 years. The warm intervals within this climatic cycle correspond in time with seven of the ten interstadials identified in the Santa Barbara record (Behl and Kennett, 1996). Included within this pattern is a climatic oscillation at 29.5 Cal ka that correlates with H3. Evidence for a cool period associated with H4 is less clear. Increases in *Pinus*

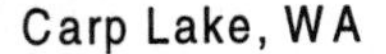

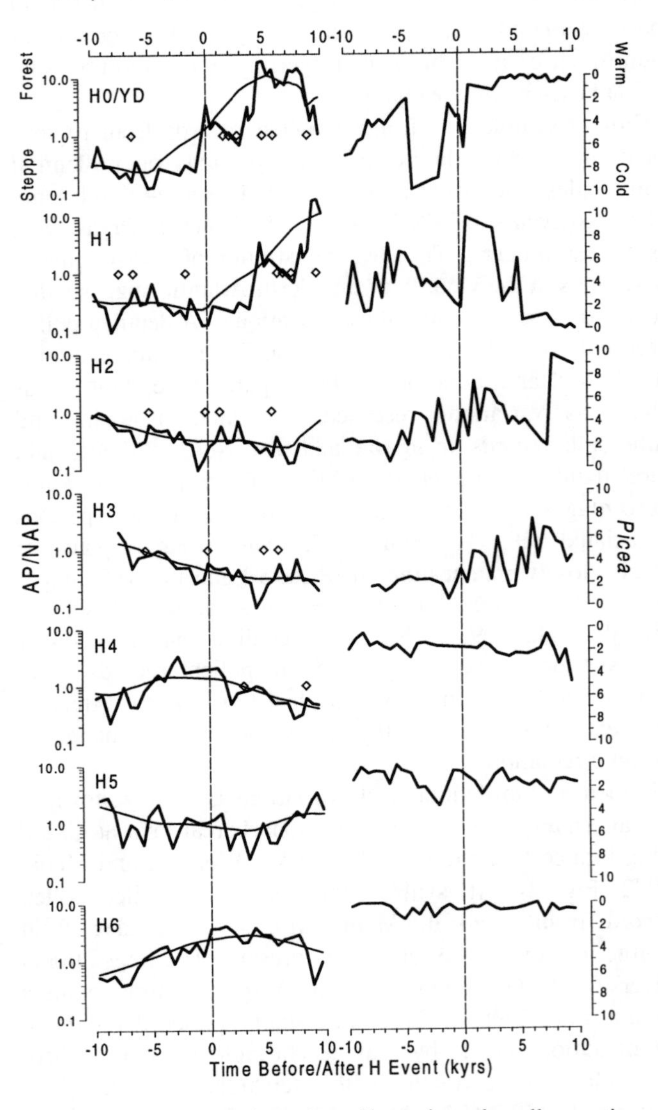

Figure 3. Changes in arboreal/nonarboreal pollen ratios (AP/NAP) and *Picea* percentage data at Carp Lake, Washington, at the time of Heinrich events. The graphs show the pollen values in the 5000 years before (negative values) and after (positive values) H0 through H6. Diamonds denote the position of calibrated radiocarbon dates (Whitlock and Bartlein, 1997). High AP/NAP values indicate periods of greater forest cover during H4, H5, and H6, whereas steppe vegetation prevailed at the time of H0, H1, H2, and possibly H3. The smoothed curves are drawn with a lowess procedure (Cleveland, 1993). *Picea* percentages are plotted to show warm periods at the top of each graph. Warm conditions (based on changes in *Picea* percentages) immediately follow H1, H2, and possibly H3 (see text for discussion).

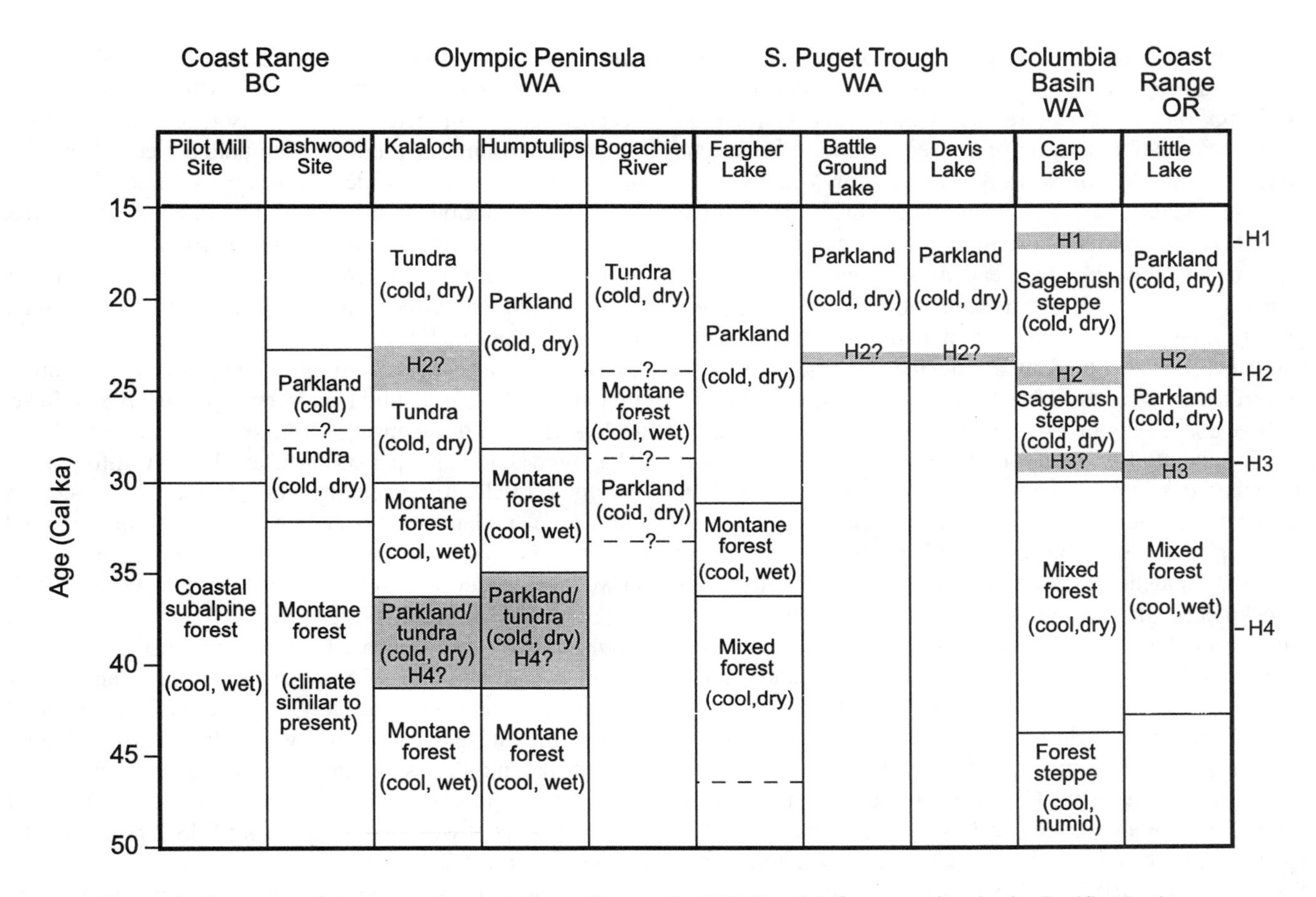

Figure 4. Summary of the vegetation and climate history for MIS 2 and 3 from ten sites in the Pacific Northwest. Shading shows intervals of sub-Milankovitch climate change that coincide with the timing of Heinrich events in the North Atlantic (shown on right side of figure).

pollen and a small increase in *Tsuga heterophylla* pollen during MIS 2 suggest a shift towards slightly warmer conditions between 24 and 21 Cal ka. This warm interval was preceded by a cool interval, indicated by *Picea* and *Tsuga mertensiana* pollen, which coincides with H2. Closer examination of the record during H1 has yet to be undertaken.

Other paleoecological records from the Pacific Northwest, although not as well dated as Carp and Little lakes, also show sub-Milankovitch climate change. Several Pacific Northwest sites indicate warm conditions following the time of H2 (ca. 24.5 Cal ka.) (Fig. 4). A peat section from the Kalaloch sea cliffs on the Olympic Peninsula shows an increase in arboreal pollen between 25 and 22.6 Cal ka. (Heusser, 1972). This warm interval is also found in two sites in the Fraser Lowland based on beetle and plant-macrofossil evidence (Mathewes, 1991). Two sites in the southern Puget Trough (Battle Ground Lake and Davis Lake) show an increase in *Tsuga heterophylla* pollen at ca. 23 Cal ka that also suggests warmer conditions (Barnosky, 1985, 1981). A climatic oscillation between 41 and 36 Cal ka recorded in two sites from the Olympic Peninsula (Kalaloch and Humptulips) overlap with the timing of H4 (38 Cal ka) (Heusser, 1972; Heusser and Heusser, 1990). In addition, a series of glacial advances from the Olympic Peninsula has been tentatively correlated with H4, H3, and possibly H2 (Thackray, 1998). Other paleoenvironmental records, including those from Pilot Mill and Dashwood in coastal British Columbia (Warner, 1984; Alley, 1979), Bogachiel River on the Olympic Peninsula (Heusser, 1978), and Fargher Lake in the southern Puget Trough (Heusser and Heusser, 1980), do not show any evidence of climate change during Heinrich events (Fig. 4). Widespread expression of a climatic oscillation during H2 suggests a teleconnection between the Pacific Northwest and the North Atlantic region during MIS 2. Shifts in the size of the ice sheet and its influence on the position of the jet stream and the glacial anticyclone may explain the linkage between H2 and climatic fluctuations in the Pacific Northwest. Warm periods, for example, would occur during times of reduced ice volume, which would decrease the intensity of easterly winds and allow for northward penetration of storm tracks otherwise shifted south. The correlation between warm intervals during MIS 3 at Little Lake and interstadials in the Santa Barbara Basin record suggests that changes in oceanic circulation may have caused climatic oscillations along the west coast of North America during MIS 3.

In the Santa Barbara Basin core ODP 893, high percentages of conifer pollen dominate glacial periods and imply a 5°C lowering of mean annual temperature from present (Heusser, 1995). Fluctuations in *Pinus*, *Juniperus*-type (juniper), and *Quercus* pollen and in charcoal abundance are also evident during the last 20 Cal ka and interpreted as a series of brief climatic events, each lasting <200 years. Peaks in *Pinus* pollen, in particular, occurred at 1550- and 1070-yr intervals and are attributed to episodes of increased winter precipitation and associated runoff in the Sierra Nevada. They seem to correlate with pluvial events in the Great Basin and New Mexico, and their periodicity matches monsoon episodes in the Indian Ocean (Sirocko et al., 1993). Heusser and Sirocko (1997) suggest that a trans-Pacific, ENSO-like atmospheric forcing may have caused wet periods in the western U.S. and Indian Ocean. The periodicity of the peaks also matches variations in the GISP2 record during the last deglaciation (Heusser and Sirocko, 1997), implying a global teleconnection.

Pollen records from the Colorado Plateau that span MIS 2 and 3 include Potato Lake (Anderson, 1993), Walker Lake, (Hevly, 1985), Hay Lake (Jacobs, 1985), and Deadman Lake (Wright et al., 1973). There are also several packrat midden localities in the Southwest that encompass this period (Betancourt et al., 1990). Packrat midden records represent a discontinuous time series and are not well suited for an examination of short-term changes. The pollen records show intriguing variations in *Artemisia*, *Picea*, *Pinus*, and Poaceae (grass) percentages during MIS 2 and 3, but the relatively wide sampling and poor dating of these intervals make it difficult to resolve millennial-scale variability.

Owens Lake, Mono Lake, Pyramid Lake, and Searles Lake in the Great Basin are sampled at high-enough resolution to document millennial-scale climate changes (Benson et al., 1996, 1998; Lin et al., 1998). Variations in lithology, oxygen isotopes, sediment magnetism, and geochemistry in cores taken from these sites are used to reconstruct the hydrologic state of the lakes and, at Owens Lake, the snowpack fluctuations in the Sierra Nevada (Benson et al., 1996). The variations in hydrology are correlated with climate changes in the GISP2 and North Atlantic records. The global transmission of Heinrich events and D-O cycles are attributed to shifts in the strength and position of winter storm tracks. It should be noted, however, that at Owens Lake, stratigraphic changes in freshwater and saline diatom and ostracod taxa between 50 and 25 Cal ka (Bradbury and Forester, in press) do not align well with the hydrologic reconstruction implied by oxygen-isotope and lithologic records (Benson et al., 1996). The pollen record from this site also shows little millennial-scale variation (Litwin et al., 1997), although the sampling interval is coarse. Apparently, different climate proxies from a single site show a non-uniform response to climate change, just as different pluvial systems in the Great Basin vary in their response (Benson, 1998).

Bradbury and Forester (in press) propose a mechanism that might link millennial-scale changes at Owens Lake, California with those during MIS 3 off the California Coast. They note that warm periods at Clear Lake, California are recorded by high percentages of *Quercus* and *Sequoia* (redwood) pollen. Coastal fog, required for redwood growth, is today associated with times of increased upwelling off the California coast. Gardner et al. (1997) suggest that upwelling is related to the strength and persistence of the northeast Pacific subtropical high in the summer, and the trade-off between dry summers and winter storms from the Aleutian low. Upwelling was reduced during the full-glacial period, when the subtropical high was weak and the Aleutian low was strong and located south of its present position (Gardner et al., 1997). At Clear Lake, cool wet conditions during MIS 2 promoted the development of *Pinus* forest.

During MIS 3, the Owens Lake record shows fluctuations in summer-season planktic diatoms *Aulacoseira granulata* and *A. ambigua* that may represent times of increased freshwater input, presumably associated with higher snowpack in the Sierra Nevada and wetter periods at Clear Lake. Times of high snowpack are also evidenced by intervals of heavy minerals in ocean cores (due to increased runoff) (Gardner et al., 1997). Such wet periods are probably caused by a strengthened Aleutian low. Bradbury and Forester (in press) further suggest that the freshwater assemblages at Owens Lake should be out-of-phase with the record of the planktic foraminfera *Globigerina bulloides* in the Santa Barbara Basin. *G. bulloides* is high during times of increased upwelling and presumably reduced winter precipitation. Although direct correlation of the Owens Lake and Santa Barbara Basin records has not been possible given uncertainties in the chronologies, trying to understand the regional linkages between climatic, terrestrial, and marine records within a region before seeking global correlations seems like a good approach.

Thus, while millennial-scale variability is clear in and around the North Atlantic, the signal in the western U.S. is temporally discontinuous and spatially highly variable. No paleoecologic record shows variations that match directly

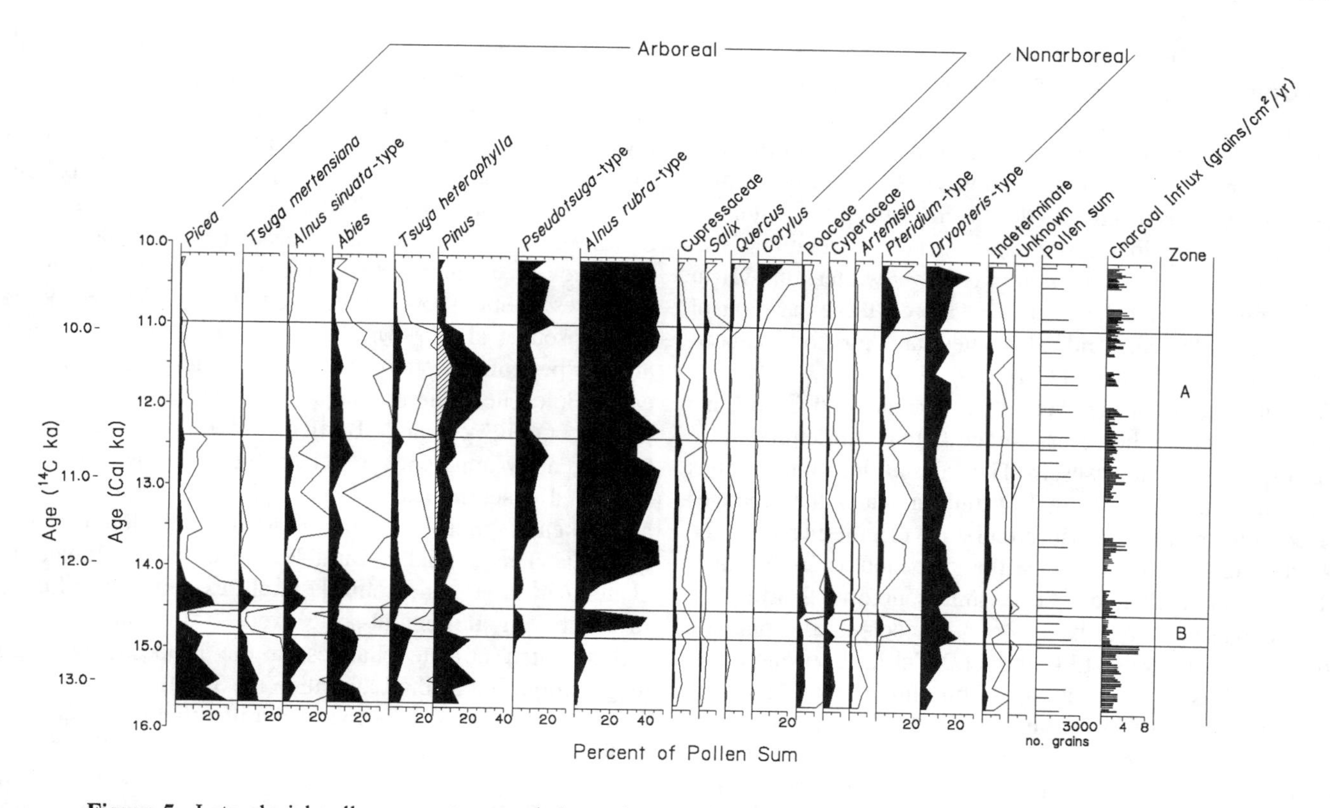

Figure 5. Late-glacial pollen percentages and charcoal accumulation rates from Little Lake, OR (based on Grigg and Whitlock, 1998). Open curves are a five-fold exaggeration of black curves. In *Pinus* curve hatched lines show abundance of *Pinus monticola*-type. Interval A is the increase in *P. monticola* associated with YD cooling. Interval B shows the establishment of *Pseudotsuga* after a peak in charcoal accumulation. The radiocarbon age scale was derived using the CALIB 3.0 calibration program (Stuiver and Reimer, 1993).

with D-O cycles, and these cycles may be best expressed in marine and coastal sites. Similarly, only a few sites show vegetation changes that correspond with specific Heinrich events. Exceptions may be evidence of a climatic oscillation during H2 in several records from the Pacific Northwest and during H1, H2, and possibly H3 at Carp Lake, and H2 and H3 at Little Lake. Further examination of the Carp Lake and Little Lake cores around the time of H1 and H4 is still needed. Additional uncertainties are also apparent in the different responses of paleoclimatic proxy from Owens Lake. Millennial-scale variations are noted in other sites, but it seems premature to assume that they have global significance until their climatic interpretation is better understood at a regional scale.

THE YOUNGER DRYAS IN THE PACIFIC NORTHWEST

The complex registration of millennial-scale climate variations is illustrated by an examination of the Younger Dryas (YD/H0). Recent studies identify a YD event not only in the North Atlantic, but also in the Pacific, South America, New Zealand, and western North America (see review by Peteet, 1995). Several alpine glaciers in the western U.S. fluctuated in late-glacial time, however, most do not show an advance during YD time. Clear signals of a vegetation change during YD time come from the Pacific Northwest, where several coastal sites in British Columbia (Cape Ball and Marion Lake) show an increase in *Tsuga mertensiana* or herbaceous pollen between 12.6 and 11 Cal ka (Mathewes, 1993). This change implies a shift towards cooler wetter conditions. An increase in herbaceous taxa between 12.7 and 10.9 Cal ka is also noted in southeastern Alaska (Pleasant Island) (Engstrom et al., 1990; Hansen and Engstrom, 1996). Farther inland, high-elevation sites from the Colorado and Canadian Rockies show subtle shifts in treeline during YD time (Menounos and Reasoner, 1997; Reasonser and Huber, in press).

Little Lake records a cooling event that began 500 years after the YD event in the North Atlantic (Grigg and Whitlock, 1998). The evidence is an increase in *Pinus monticola*-type pollen, between 12.4 to 11 Cal ka (Fig. 5; Interval A). A similar "pine" period is noted at other sites from western Washington and Oregon, although none are well dated. Carp Lake shows a decrease in *Picea* pollen approximately at YD time (Fig. 3). One explanation for the cool period at all these sites invokes gradual changes in the Milankovitch controls of regional climate: the northward shift of the jet stream as a source of increased winter precipitation occurred at the same time that summer insolation was increasing and winter insolation was decreasing. This combination of cool wet winters and warm dry summers would have briefly favored the expansion of *Pinus* at Little Lake and other sites, and possibly reduced the abundance of *Picea* at Carp Lake.

The abruptness of the event at Little Lake and Carp Lake and its timing, however, suggest that millennial-scale variations were the cause. The coastal location of sites featuring a cooling during YD time indicates that changes in oceanic or atmospheric circulation originating over the North Pacific was probably the proximal cause of late-glacial climate change. An oceanic source is also implied by increased bioturbation and cool-water planktic foraminifera between 13.0 and 11.2 Cal ka from the Santa Barbara Basin (Kennett and Ingram, 1995) and the Guaymas Basin (Keigwin and Jones, 1990). Records from the Pacific Northwest indicate that the late-glacial cooling episode was warmer than full-glacial conditions and that it lasted 500-600 years beyond the end of the YD event in the North Atlantic. This discrepancy requires either a lag in the transmission of the YD event to the Northeast Pacific or a delayed response on the part of the vegetation. Lund and Mix (1998) propose that ventilation of the Northeast Pacific lagged behind that in the Santa Barbara Basin by 500 years (where a YD signal is noted by a shift to unlaminated sediments), perhaps reflecting lags in the adjustment of global thermohaline circulation. Recent paleoclimate model simulations suggest that cooling of the North Pacific decreased sea-surface temperatures and increased sea-ice, which caused further changes to atmospheric circulation (Mikolajewicz et al., 1997). The model also implies that the greatest decrease in sea-surface temperatures occurred north of 50°N as a result of changes in upwelling, which may explain a stronger YD signal in coastal British Columbia and Alaska.

EVIDENCE OF HOLOCENE VARIABILITY

Milankovitch-type controls constitute the dominant forcing on Holocene vegetation changes in the western U.S. The strongest signal is the amplification of the seasonal cycle of insolation in the early Holocene and its attenuation in the late-Holocene. Because the western U.S. lay "upwind" of the continental ice sheet, there was little lingering effect of the ice sheet on regional climate after 14 Cal ka. Seasonal summer drought in the early Holocene associated with greater-than-present summer insolation resulted in higher-than-present treeline, a northward and upward shift in the lower forest/grassland ecotone (compared with present), an expansion of xerophytic conifer species, and greater-than-present fire frequencies (Whitlock, 1992). In the American Southwest, there is some paleoecological evidence for expansion of summer-wet conditions as a result of stronger-than-present monsoonal circulation in the early Holocene (Thompson et al., 1993). In the Rocky Mountains, some areas experienced wetter-than-present conditions during the early Holocene, whereas adjacent areas were drier than present (Whitlock and Bartlein, 1993). This spatial heterogeneity apparently reflected the juxtaposition of the expanded subtropical high pressure system and strengthened monsoons in a topographically complex region. The spatially variable response of climate to changes in large-scale controls might explain the difficulty in identifying millennial-scale events. In addition, sub-Milankovitch climate changes in the Holocene were no longer amplified by the Laurentide Ice Sheet.

Within the late Holocene, submillennial-scale climate variations are suggested by glacial advances in the Rocky Mountain region (e.g., Luckman, 1996; Davis, 1988; Carrara, 1987), Cascade Range (Burbank, 1981), and Coast Mountains (Clague and Mathewes, 1996), but the timing of these advances is asynchronous. The Little Ice Age (LIA) (1350-1850 AD; Bradley and Jones, 1993) offers an example of one climate oscillation. Glacial advances occurred in the 13th and 14th centuries in Canada, but not until the 16th century in Wyoming and the late 17th century in Colorado (Leonard, 1986; Luckman, 1986; Smith et al., 1995; Richmond, 1986; Benedict, 1981). Dendrochronologic data also suggest LIA climate oscillations. In the Canadian Rockies, these records indicate modest cooling in the late 14th century, followed by warming, and then a more-severe cooling from 1690-1705 AD. The remainder of the 18th century was cool, with the most severe stage of the LIA beginning about 1800-1838 AD (Luckman et al., 1997). Tree-ring records from the central and southern Rockies exhibit similar cool phases within the LIA, and many of these indicate that the last phase was the most severe. In the Cascade Range, cooling was centered on the period from 1590 to 1900 AD (Graumlich and Brubaker, 1986).

Perhaps a more sensitive proxy of millennial-scale climate change than pollen or glacial data comes from high-resolution charcoal data that disclose variations in fire frequency. Only a few such records are available for the western U.S. (Millspaugh, 1997; Long et al., 1998; Mohr, 1997). In many areas where pollen data show only gradual changes in vegetation, the fire frequency data indicate short-term variations that imply periods of drought. Moreover, shifts in fire frequency that accompany climate change may have been the proximal trigger of vegetation change. For example, at Little Lake, an abrupt shift from *Picea* forest to *Pseudotsuga* forest occurred at about 14.5 Cal ka (Fig. 5; Interval B). The pollen record of this event is associated with a large peak of sedimentary charcoal, suggesting that one or more fires destroyed stands of *Picea*, and allowed replacement by fire-adapted *Pseudotsuga*. The warming responsible for this shift lasted about 350 years, after which *Picea* again became the forest dominant (Grigg and Whitlock, 1998).

High-resolution records from Yellowstone National Park show that during some 500-year periods the fire frequency was synchronized across the region (Millspaugh, 1997). For example, fire occurrence was widespread between 6.24-5.95, 3.90-3.15, 2.75-1.6 and 1.00-0.50 Cal ka (Fig. 6). The period of high fire incidence between 1.00-0.50 Cal ka overlaps with the Medieval Warm Period (Lamb, 1977) and also with high fire-related debris flows (Meyer et al., 1995) and an increase in xerophytic mammals (Hadly, 1996). The close correspondence suggests that multi-centennial periods of drought and increased convective summer storms affected the region.

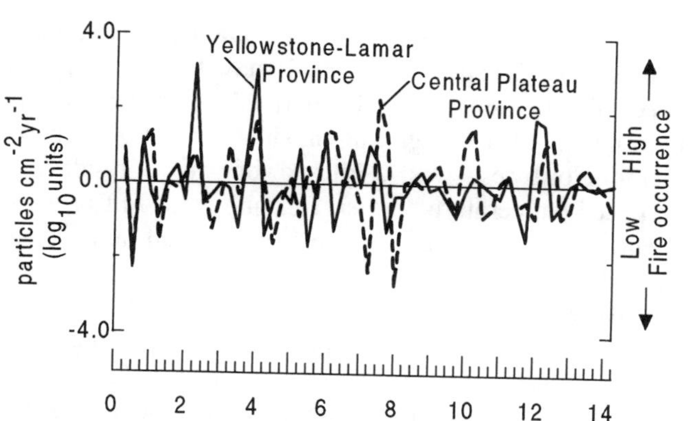

Figure 6. Fire occurrence during 500-year periods in Yellowstone National Park, based on high-resolution charcoal data from two lakes in different regions. The charcoal accumulation rates were transformed to standard z scores and plotted at the center of 500 year periods. Each 500-year period is offset by 250 years (Millspaugh, 1997). Note that there are several periods when fires occurred in both regions co-occur suggesting a large-scale climate control.

CONCLUSIONS

Millennial-scale fluctuations, such as Heinrich events and D-O cycles, suggest rapid reorganizations of the climate system at the global scale. Understanding how the terrestrial biosphere responded to rapid environmental change presents a challenge to terrestrial paleoecologists, because of the necessity to obtain long records, develop good chronologies, and sample at close intervals. This type of high-resolution analysis has generally not been undertaken, in part, because attention has focused on vegetation responses to Milankovitch variations. As a result, few records are of adequate resolution to evaluate the vegetation response to short events. A coordinated effort, similar to that mounted by the marine scientific community over the last few decades, is needed to obtain long continental records and bring the analysis of these records to a common standard. Such an effort is now underway in the IGBP PAGES program.

Few sites in the western U.S. meet the criteria necessary to evaluate millennial-scale climate changes. Moreover, western records are composed of a few taxa, such as *Pinus*, *Artemisia*, and Poaceae, that produce large amounts of widely-dispersed pollen. Small fluctuations in the abundance of these pollen taxa are not necessarily evidence of a climate change, and one often must turn to minor pollen types, which represent <10% of a pollen sum to find a meaningful signal. In this regard, the rain forest regions of the Pacific Northwest are well suited for studies of millennial-scale change. *Pinus*, *Artemisia*, and Poaceae are not the dominant taxa at present nor have they been in the past, and variations in the abundance of minor taxa provide a sensitive climate record.

Beyond the limitations of the database is the nature of the vegetation response itself. The western flora is relatively depauperate, having survived major climatic cooling and drought throughout the late Cenozoic (Barnosky, 1987). The survivors of these long-term climatic trends are species with a large ecological amplitude, which were able to survive glacial and interglacial variations with little change in range. The response to shorter less-extreme climatic events may have been, in fact, no change in range or abundance at all. For these reasons, changes in fire frequency may be better proxy of short-term climate change during interglacials. Fire occurrence requires suitable fuel conditions, including low fuel moisture, an ignition source, and weather conditions to carry the fire. These are directly tied to climate without appreciable lag.

Another important consideration is that the response of the regional climate depends on the combination of large-scale controls operating on Milankovitch time scales. Transmission of warm/cool intervals, such as D-O events, via changes in ocean circulation may not significantly perturb the climate of the western U.S., especially far inland.

Heinrich events represent potentially a larger jolt to the climate system, and their signal may be transmitted to North Pacific surface and intermediate waters by the atmosphere. During glacial maxima, the climate of the Pacific Northwest was cold and dry as a result of a steepened temperature gradient, a southward shift of the jet stream, and a stronger-than-present glacial anticyclone. These large-scale controls allowed H1, H2, and H3 to be strongly expressed, whereas H5 and H6 are weak signals.

The YD event is best expressed in the western U.S. along the coast and at high elevations. Few glacial advances in the western U.S. are truly YD in age (12.9-11.5 Cal ka), although Reasoner and Huber (in press) and Menounos and Reasoner (1997) offer notable exceptions. The pollen records show a weak signal, except in Alaska and coastal British Columbia. In Oregon, the signal comes 500 years later and is found only in a coastal setting.

Millennial-scale events have not been studied adequately in Holocene pollen records, although forested sites may not be sensitive enough to record short-term climate changes. For example, few records, with the exception of those at treeline, register a response to changes in climate during the Little Ice Age. In western North America, fire records hold more promise for providing a sensitive climate proxy during the Holocene. When charcoal records are examined at high resolution, they show variations in fire frequency that suggest short periods of drought.

Acknowledgements. We thank P.J. Bartlein for discussion of the topic and W.A. Watts for permission to cite unpublished material. Eric Grimm and Vera Markgraf provided helpful reviews. The research was supported by NSF grant ATM-9615822.

REFERENCES

Alley, N.F., Middle Wisconsin stratigraphy and climatic reconstruction, southern Vancouver Island, British Columbia. *Quat. Res., 11*, 213-237, 1979.

Alley, R.B., Meese, D.A., Shuman, C.A., Gow, A.J., Taylor, K.C., Grootes, P.M., White, J.W.C., Ram, M., Waddington, E.D., Mayewski, P.A., and Zielinski, G.A., Abrupt increase in Greenland snow accumulation at the end of the Younger Dryas event. *Nature, 262*, 527-529, 1993.

Alley, R.B., Icing the North Atlantic. *Nature 392*, 335-337, 1998.

Adam, D. P., Palynology of two Upper Quaternary cores from Clear Lake, Lake County, California, in US Geological Survey Professional Paper, 1363, pp. 86, 1988.

Anderson, R.S. A 35,000 year vegetation and climate history from Potato Lake, Mogollon Rim, Arizona. *Quat. Res., 40*, 351-359, 1993.

Bard, E., Hamelin, B., Fairbanks, R.G., and Zindler, A., Calibration of the ^{14}C timescale over the past 30,000 years using mass spectrometric U-Th ages from Barbados corals. *Nature, 345*, 405-410, 1990.

Barnosky, C.W., A record of late Quaternary vegetation from Davis Lake, southern Puget Lowland, Washington. *Quat. Res., 16*, 221-239, 1981.

Barnosky, C.W., Late Quaternary vegetation near Battle Ground Lake, southern Puget Trough, Washington. *Geol. Soc. Am., 96*, 263-271, 1985.

Barnosky, C.W., Response of vegetation to climatic changes of different duration in the late Neogene. *Trends in Ecol. and Evol. 2,* 247-250, 1987.

Bartlein, P.J., Late-Tertiary and Quaternary Palaeoenvironments, in Vegetation History, edited by B. Huntley and T. Webb III, pp. 113-152, Kluwer Academic Publishers, Dordrecht, 1988.

Behl, R.J., and Kennett, J.P., Brief interstadial events in the Santa Barbara basin, NE Pacific, during the past 60 kyr. *Nature 379*, 243-246, 1996.

Bender, M., Sowers, T., Dickson, M.L., Orchardo, J., Grootes P., Mayewski, P.A., and Meese, D.A., Climate correlations between Greenland and Antarctica during the last 100,000 years. *Nature 372*, 663-666, 1994.

Benedict, J.B., Arapaho Pass: glacial geology and archeology at the crest of the Colorado Front Range. *Center for Mountain Archeology, Res. Rept, 3*, Ward, Colorado, 139 pp, 1985.

Benson, L.V., The nonuniform nature of climate change: some examples from the Great Basin [abstract], p. 13-14, in Mechanisms of millennial-scale global climate change, AGU Chapman Conference, June 14-16, Snowbird, UT, 1998.

Benson, L.V., Burdett, J.W., Kashgarian, M., Lund, S.P., Phillips, F.M., and Rye, R.O., Climatic and hydrologic oscillations in the Owens Lake Basin and adjacent Sierra Nevada, California. *Science, 274*, 746-749, 1996.

Benson, L.V., Lund, S.P., Burdett, J.W., Kashgarian, M., Rose, T.P., Smoot, J.P., and Schwartz, M., Correlation of late-Pleistocene lake-level oscillations in Mono Lake, California, with North Atlantic climate events. *Quat. Res. 49*, 1-10, 1998.

Betancourt, J.L., Van Devender, T., and Martin, P.S., Packrat middens, the Last 40,000 Years of Biotic Change, Univ. Arizona Press, Tucson, 1990.

Bond, G.C., Heinrich, H., Broecker, W.S., Labeyrine, L., McManus, J., Andrews, J., Huon, S., Jantschik, R., Clasen, S., Simet, C., Tedesco, K., Klas, M., Bonani, G., and Ivy, S., Evidence for massive discharges of icebergs into the North Atlantic ocean during the last glacial period. *Nature 360, 245-249*, 1992.

Bond, G.C., Broecker, W.S., Johnsen, S.J., McManus, L., Labeyrie, J., Jouzel, J., Bonani, G., Correlations between climate records from North Atlantic sediments and Greenland ice. *Nature, 365*, 143-147, 1993.

Bond, G.C., and Lotti, R., Iceberg discharges into the North Atlantic on millennial time scales during the last glaciation. *Science, 267*, 1005-1010, 1995.

Bond, G., Shower, W., Cheseby, M., Lotti, R., Almasi, P., deMenocal, P., Priore, P., Cullen, H., Hajdas, I., and Bonani, G., A pervasive millennial-scale cycle in North Atlantic

Holocene and Glacial Climates. *Science, 278*, 1257-1266, 1997.

Bradbury, J.P., and Forester, R.M., A paleolimnologic record of climate change from Owens Lake, California for the past 50 kyr. Smithsonian Institution, in press.

Bradley, R.S., and Jones, P.D., 'Little Ice Age' summer temperature variations: their nature and relevance to recent global warming trends. *The Holocene*, 3, 367-376, 1993.

Broecker, W.S., Massive icebergs discharges as triggers for global change. *Nature, 372*, 421-424, 1994.

Burbank, D.W., A chronology of late Holocene glacier fluctuations on Mount Rainier, Washington. *Arctic Alpine Res., 13*, 369-386, 1981.

Carrara, P.E., Holocene and latest Pleistocene glacial chronology, Glacier National Park, Montana. *Can. J. Earth Sci., 24*, 387-395, 1987.

Clague, J.J., and Mathewes, R.W., Neoglaciation, glacier-dammed lakes, and vegetation change in northwestern British Columbia, Canada. *Arctic and Alpine Res.*, 28, 10-24, 1996.

Cleveland, W.S., Visualizing Data, Hobard, Summit, 1993.

Dansgaard, W., Johnsen, S.J., Clausen, H.B., Dahl-Jensen, D., Gundestrup, N.S., Hammer, C.U., Hvidberg, C.S., Steffensen, J.P., Sveinbjörnsdottir, A.E., Jouzel. J., and Bond, G., Evidence for general instability of past climate from a 250-kyr ice-core record. *Nature, 364*, 218-220, 1993.

Denton, G.H., and Hendy, C.H., Younger Dryas age advance of Franz Josef Glacier in the Southern Alps of New Zealand. *Science, 264*, 1434-1437, 1994.

Davis, P.T., Holocene glacier fluctuations in the American Cordillera. *Quat. Sci. Rev., 7*, 129-157, 1988.

Engstrom, D.R., Hansen, B.C.S. and Wright, H.E. Jr., A possible Younger Dryas record in southwestern Alaska. *Science, 250*, 1383-1385, 1990.

Gardner, J.V., Dean, W.E., and Partnell, P., Biogenic sedimentation beneath the California current system for the past 30kyr and its paleoceanographic significance, *Paleoceanography, 12*, 207-225, 1997.

Graumlich, L.J., and Brubaker, L.B., Reconstruction of annual temperature (1590-1979) for Longmire Washington, derived from tree-rings. *Quat. Res., 25*, 223-234, 1986.

Grigg, L.D., and Whitlock, C., Late-glacial climate and vegetation changes in western Oregon. *Quat. Res., 49*, 287-298, 1998.

Grimm, E.C., Jacobson, G.L., Watts, W.A., Hansen, B.C.S., and Maasch, K.A., A 50,000-year record of climate oscillations from Florida and its temporal correlation with Heinrich events. *Science, 261*, 198-200, 1993.

Grootes, P.M., Stuiver, M., White, J.W.C., Johnsen, S., and Jouzel, J., Comparison of oxygen isotope records from the GISP2 and GRIP Greenland ice cores. *Nature, 366*, 552-554, 1993.

Hadly, E.A. Influence of late-Holocene climate on northern Rocky Mountain mammals. *Quat. Res., 46*, 298-310, 1996.

Hansen, B.C.S, and Engstrom, D.R., Vegetation history of Pleasant Island, southeastern Alaska since 13,000 yr B.P. *Quat. Res., 46*, 161-175, 1996.

Heusser, C.J., Palynology and phytogeographical significance of a late-Pleistocene refugium near Kalaloch, Washington. *Quat. Res., 2*, 189-201, 1972.

Heusser, C.J., Palynology of Quaternary deposits of the lower Bogachiel River area, Olympic Peninsula, Washington. *Can. J. of Earth Sci., 15*, 1568-1578, 1978.

Heusser, C.J., and Heusser, L.E., Sequence of pumiceous tephra layers and the late Quaternary environmental record near Mt. St. Helens. *Science, 210*, 1007-1009, 1980.

Heusser, C.J., and Heusser, L.E. Long continental pollen sequence from Washington state (U.S.A.): correlation of upper levels with marine pollen-oxygen isotope stratigraphy through substage 5e. *Palaeogeog., Palaeoclim., Palaeoecol., 79*, 63-71, 1990.

Heusser, L.E., Pollen stratigraphy and paleoecologic interpretation of the 160-k.y. record from Santa Barbara basin, hole 893A, in Proceedings of the Ocean Drilling Program. Scientific Results, 146, edited by J.P. Kennett, J.G. Baldauf, and M. Lyle, pp. 265-279, 1995.

Heusser, L.E., Direct correlation of millennial-scale changes in western North American vegetation and climate with changes in the California Current system over the past 60 kyr. *Paleoceanography, 13*, 252-262, 1998.

Heusser, L.E., and Sirocko, F., Millennial pulsing of environmental change in southern California from the past 24 k.y.: A record of Indo-Pacific ENSO events? *Geology, 25*, 243-246, 1997.

Hevly, R.H., A 50,000 year record of Quaternary environments in southern New Mexico, in Late Quaternary Vegetation and Climates of the American Southwest, edited by B.F. Jacobs, P.L. Fall, and O.K. Davis, pp. 141-154, Amer. Assoc. Stratigraphic Palynologists, Contrib. 15, AASP, Dallas, 1985.

Jacobs, B.F., A middle Wisconsin pollen record from Hay Lake, Arizona. *Quat. Res., 24*, 121-130, 1985.

Jouzel, J., Ice cores north and south. *Nature, 372*, 612-613, 1994.

Keigwin, L.D., and Jones, G.A., Deglacial climatic oscillations in the Gulf of California. *Paleoceanography., 5*, 1009-1023, 1990.

Kennett, J.P., and Ingram, B.L., A 20,000-year record of ocean circulation and climate change from the Santa Barbara basin. *Nature, 377*, 305-308, 1995.

Kutzbach, J., Gallimore, R., Harrison, S., Behling, P., Selin, R., and Laarif, R., Climate and biome simulations for the past 21,000 years. *Quat. Sci. Rev., 17*, 473-506, 1998.

Lamb, H.H., Climate: present, past, and future, in Climate history and the future, v.2. Methuen, London, 1977

Leonard, E.M., Use of lacustrine sedimentary sequences as indicators of Holocene glacial history, Banff National Park, Alberta, Canada. *Quat. Res., 26*, 218-231, 1986.

Lin, J.O., Broecker, W.S., Hemming, S.R., Hajdas, I, Anderson, R.F., Smith, G.I., Kelley, M., and Bonani, G., A reassessment of U-Th and 14C ages for late-glacial high-frequency hydrological events at Searles Lake, California, *Quat. Res., 49*, 11-23, 1998.

Litwin, R.J., Adam, D.P., Frederiksen, N.O., and Woolfenden, W.B., A palynomorph record of Owens Lake sediments of core

OL-92: preliminary analyses, in An 800,000-year paleoclimatic record from core OL-92, Owens Lake, southeast California, edited by G.I. Smith and J.L. Bischoff, pp. 127-142, *Geol. Soc. Amer. Spec. Pap.*, 1997.

Long, C.J., Whitlock, C., Bartlein, P.J., Millspaugh, S.H., A 9000-year fire history from the Oregon Coast Range, based on a high-resolution charcoal study. *Can. J. For. Res.*, *28*, 774-787, 1998.

Lowell, T.V., Heusser, C.J., Anderson, B.G., Moreno, P.I., Hauser, A., Heusser, L.E., Schlüchter, C., Marchant, D.R., Denton, G.H., Interhemispheric correlation of late Pleistocene glacial events. *Science*, *269*, 1541-1549, 1995.

Luckman, B.H., Reconstruction of 'Little Ice Age' events in the Canadian Rockies. *Geogr. Phys. et Quat.*, *XL*, 17-28, 1986.

Luckman, B.H., Reconciling the glacial and dendrochronological records for the last millennium in the Canadian Rockies, in Climatic variations and forcing mechanisms of the last 2000 years, 85-108, edited by R.S. Bradley, P.D. Jones, and J. Jouzel, Springer-Verlag, Berlin, 1996.

Luckman, B.H., Briffa, K.R., Jones, J.D., and Schweingruber, F.H., Tree-ring based reconstruction of summer temperatures at the Columbia Ice field, Alberta, Canada, AD 1073-1983. *The Holocene*, *7*, 375-389, 1997.

Lund, D.C., and Mix, A.C., Millennial-scale deep water oscillations: reflections of the North Atlantic in the deep Pacific from 10 to 60 ka. *Paleoceanography*, *13*, 10-19, 1998.

Mathewes, R.W., Evidence for Younger Dryas-age cooling on the North Pacific coast of America. *Quat. Sci. Rev.*, *12*, 321-331, 1993.

Mathewes, R.W., Climatic conditions in the western and northern Cordillera during the last glaciation: Paleoecological evidence. *Géographie Physique et Quaternaire*, *45*, 333-339, 1991.

Mazaud, A., Laj, C., Bard, E., Arnold, M., and Tric, E., Geomagnetic field control of ^{14}C production over the last 80 ky: Implications for the radiocarbon time-scale. *Geophys. Res. Lett.*, *18*, 1885-1888, 1991.

McGlone, M.S., Late glacial landscape and vegetation change and the Younger Dryas climatic oscillation in New Zealand, *Quat. Sci. Rev.*, *14*, 867-881, 1995.

Menounos, B., and Reasoner, M.A., Evidence for cirque glaciation in the Colorado Front Range during the Younger Dryas chronozone. *Quat. Res.*, *48*, 38-47, 1997.

Meyer, G.A., Wells, S.G., Jull, A.J.T., Fire and alluvial chronology in Yellowstone National Park: climatic and intrinsic controls on Holocene geomorphic processes. *Geol. Soc. Am. Bull.*, *107*, 1211-1230, 1995.

Mikolajewicz, U., Crowley, T.J., Schiller, A., and Voss, R., Modelling teleconnections between the North Atlantic and Pacific during the Younger Dryas. *Nature*, *387*, 384-387, 1997.

Millspaugh, S.H., Late-glacial and Holocene variations in fire frequency in the Central Plateau and Yellowstone-Lamar Provinces of Yellowstone National Park. Ph.D. dissert., Univ. Oregon, Eugene, 1997.

Mohr, J.A., Postglacial vegetation and fire history near Bluff Lake, Klamath Mountains, California, M.A. thesis, Dept. Geography, Univ. Oregon, Eugene, 1997.

Oppo, D.W., McManus, J.F., Cullen, J.L., Abrupt climate events 500,000 to 340,000 years ago: evidence from subpolar North Atlantic sediments. *Science*, *279*, 1335-1338, 1998.

Peteet, D., Global Younger Dryas? *Quat. Inter.*, *28*, 93-104, 1995.

Porter, S.C., and An, Z., Correlation between climate events in the North Atlantic and China during the last glaciation. *Nature*, *375*, 305-308, 1995.

Prentice, C., Climate change and long-term vegetation dynamics, in Plant Succession: Theory and Prediction, pp. 293-339, edited by D.C. Glenn-Lewin, R.K. Peet, T.T. Veblen, Chapman & Hall, London, 1992.

Raymo, M.E., Ganley, K., Carter, S., Oppo, D.W., and McManus, J., Millennial-scale climate instability during the early Pleistocene epoch. *Nature*, *392*, 699-702, 1998.

Reasoner, M.A., and Huber, U.M., Postglacial palaeoenvironments of the upper Bow Valley, Banff National Park, Alberta, Canada. *Quat. Sci. Rev.*, in press.

Reasoner, M.A., Osborn, G., and Rutter, N.W., Age of the Crowfoot advance in the Rocky Mountains: a glacial event coeval with the Younger Dryas oscillation. *Geology*, *22*, 439-442, 1994.

Richmond, G.M., Stratigraphy and correlation of glacial deposits of the Rocky Mountains, the Colorado Plateau, and the ranges of the Great Basin. *Quat. Sci. Rev.*, *5*, 99-127, 1986.

Sirocko, F., Sarnthein, M., Erlenkeuser, H., Lange, H., Arnold, M., and Duplessey, J.-C., Century-scale events in monsoonal climate over the past 24,000 years, *Nature*, *264*, 322-324, 1993.

Smith, D.J., McCarthy, D.P., and Colenutt, M.E., Little Ice Age glacial activity in Peter Lougheed and Elk Lakes provincial parks, Canadian Rocky Mountains. *Can. J. Earth Sci.*, *32*, 579-589, 1995.

Stuiver, M. and Reimer, P.J., Extended ^{14}C data base and revised CALIB 3.0 ^{14}C age calibration program. *Radiocarbon*, *35*, 215-230, 1993.

Thackray, G.D., Glaciation and neotectonic deformation on the western Olympic Peninsula, Washington. Ph.D. Dissertation, University of Washington, 1996.

Thompson, R.S., Whitlock, C., Bartlein, P.J., Harrison, S., and Spaulding, W.G. Climatic changes in the western United States since 18,000 yr B.P., in Global climates since the Last Glacial Maximum, pp. 468-513, edited by H.E. Wright, Jr., J.E. Kutzbach, T. Webb III, W.F. Ruddiman, F.A. Street-Perrott, and P.J. Bartlein, Univ. Minn. Press, Minneapolis, 1993.

Warner, B.G., Clague, J.J., and Mathewes, R.W., Geology and paleoecology of a Mid-Wisconsin peat from the Queen Charlotte Islands, British Columbia, Canada. *Quat. Res.*, *21*, 337-350, 1984.

Watts, W.A., Allen, J.R.M., and Huntley, B., Palaeoecology and geochronology of three interstadial events during oxygen-isotope Stages 3 and 4: a lacustrine record from Lago Grande di Monticchio, southern Italy. *Palaeogeog., Palaeoclim., and Palaeoecol.*, in review.

Webb III, T., and Bartlein, P.J., Global changes during the last 3 million years: climatic controls and biotic responses. *Ann. Rev. of Ecology and Systematics*, *23*, 141-173, 1992.

Whitlock, C., Vegetational and climatic history of the Pacific Northwest during the last 20,000 years: Implications for understanding present-day biodiversity. *Northwest Environ. J., 8*, 5-28, 1992.

Whitlock, C., Postglacial vegetation and climate of Grand Teton and southern Yellowstone National Parks. *Ecol. Monogr., 63*, 173-198, 1993.

Whitlock, C. and Bartlein, P.J., Spatial variations of Holocene climatic change in the Yellowstone region. *Quat. Res. 39*, 231-238, 1993.

Whitlock, C., and Bartlein, P.J., Vegetation and climate change in northwest America during the past 125 kyr. *Nature, 388*, 57-61, 1997.

Woillard, G.M.., and Mook, W.G., Carbon-14 dates at Grande Pile: Correlation of land and sea chronologies, *Science, 215*, 159-161.

Wright, H.E., Bent, A.M., Hansen, B.S., and Maher, L.J., Jr., Present and past vegetation of the Chuska Mountains, northwestern New Mexico, *Geol. Soc. Amer. Bull., 84*, 1150-1180, 1982.

Grigg, Laurie D., Department of Geography, University of Oregon, Eugene, OR 97403

Whitlock, Cathy, Department of Geography, University of Oregon, Eugene, OR 97403; whitlock@oregon.uoregon.edu

A Glaciological Perspective on Heinrich Events

Garry K. C. Clarke and Shawn J. Marshall

Department of Earth and Ocean Sciences, University of British Columbia, Vancouver, Canada

Claude Hillaire-Marcel, Guy Bilodeau, and Christine Veiga-Pires

GEOTOP, Université du Québec à Montréal, Montréal, Québec, Canada

Heinrich events, the massive episodic disgorgement of sediment-laden ice from the Laurentide Ice Sheet to the North Atlantic Ocean, are a puzzling instability of the Ice-Age climate system. Although there is broad agreement on the defining characteristics of Heinrich events, the glaciological mechanisms remain controversial. Paleoceanographic records show that Heinrich events tend to occur at the culmination of a cooling cycle, termed the Bond cycle, and this has invited the interpretation that the events are a fast response of the Laurentide Ice Sheet to external atmospheric changes. A vexing issue for glaciologists is how a fast and timely response to an external forcing can possibly be reconciled with the known physics of glaciers and ice sheets. Fast changes in glacier behavior can only occur if some flow instability is excited. Thus glaciologists tend to favor the idea that the climate change occurring at the culmination of a Bond cycle is an atmospheric response to ice sheet instability. However, a free-running cyclic flow instability, such as that exhibited by surging glaciers, could not satisfy the timing requirements. Using computer modeling we explore ways to resolve these conflicts.

INTRODUCTION

The record of ice-rafted sedimentation in the North Atlantic ocean attests to the growth and decay of the Laurentide and Fennoscandian ice sheets during the last glacial cycle [*Ruddiman,* 1977]. A surprising feature of the marine sedimentary record is the evidence for synchronous and areally-extensive sedimentation events [*Heinrich,* 1988]. The provenance of these ice-rafted sediments can be traced to the Hudson Bay region of Canada [*Gwiazda et al.,* 1993a,b] and the accepted interpretation is that "Heinrich events" are associated with massive episodic disgorgement of sediment-laden icebergs from the Laurentide Ice Sheet [*Broecker et al.,* 1992]. An excellent recent review has been published by *Andrews* [1998].

Different scientific communities view Heinrich events from different perspectives—indeed from within different belief systems. To many glaciologists, Heinrich events fulfill the expectation that known glacier-scale flow instabilities are also expressed at the scale of ice streams and ice sheets. For several decades, glacial geologists have found this assumption useful [e.g., *Wright,* 1973; *Clayton, et al.,* 1985], although this hardly constitutes an observational proof. Among the important

Mechanisms of Global Climate Change at Millennial Time Scales
Geophysical Monograph 112

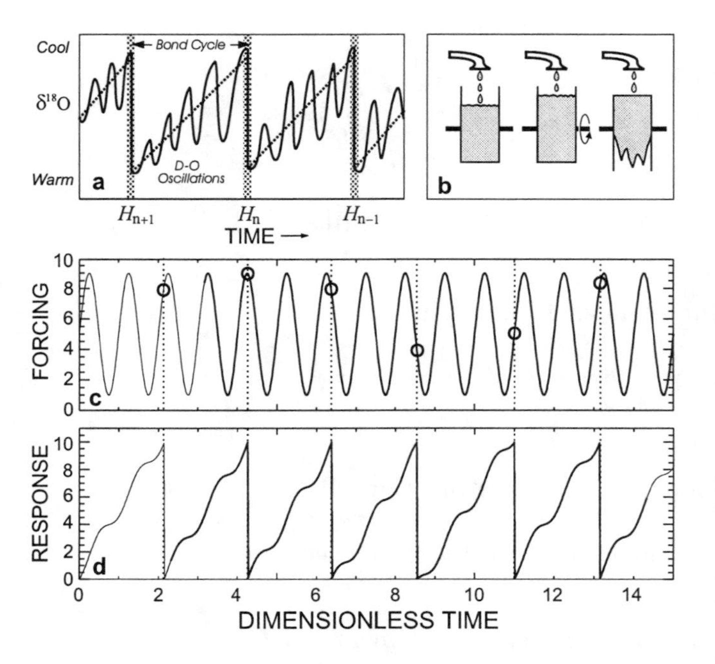

Figure 1. (a) Heinrich events bracket the Bond cycle. At the start of a Bond cycle climate is warm and at the end it is cold. Superimposed on this trend are millenial-scale climate oscillations known as Dansgaard-Oeschger oscillations. (b) MacAyeal's simple binge/purge oscillator is an example of a free-running oscillator. The timing of purge events is not synchronized with the external forcing. (c) Input forcing to a simple binge-purge simulation model. The forcing is water flux into the tipping bucket. A sinusoidal variation in forcing is assumed. (d) Response of simple binge-purge simulation model. The response is taken to be the water level in the tipping bucket. Purge events correspond to instantaneous drops in water level. Note that timing of purge events is independent of whether the filling rate is increasing or decreasing with time.

glaciological issues raised by Heinrich events, the following seem central:

(1) What was the order of events and what were the cause–effect relationships between continental ice dynamics and other components of the global climate system?

(2) Accepting that Heinrich events are evidence for large-scale flow instability of continental ice sheets, exactly what was the Laurentide Ice Sheet doing during Heinrich events? Which of several possible instability mechanisms were involved? How much ice was transferred from land to oceans?

(3) Can a physically consistent narrative be developed that does not violate present understanding of ice stream and ice sheet physics?

What follows is an overview of the glaciological issues that must be confronted, a commentary on inadmissible or unlikely possibilities, and an attempt to test some of these ideas using computer models.

ORDER OF EVENTS AND ISSUES OF TIMING

The marine and ice core paleoclimatic records agree on the principal features of Heinrich events and their relationship to other climate oscillations [*Bond et al.,* 1992, 1993; *Broecker,* 1994] and expose the problem of timing. Figure 1a sketches several Bond cycles; these are long-term cooling trends that are terminated by Heinrich events and upon which Dansgaard-Oeschger oscillations are superimposed. The coincidence of the onset of a Heinrich event with a pronounced climate switch has invited conflicting interpretations.

MacAyeal [1993a,b] has led the scientific attack by glaciologists and his "simple kitchen-built binge/purge oscillator" provides a good starting point for discussion of timing issues. Suppose the tipping bucket illustrated in Figure 1b is subjected to a sinusoidally-varying input flux (Figure 1c). Water level in the bucket rises at a variable rate until a critical threshold is reached and the tipping instability is triggered (Figure 1d). The instant at which the unstable response occurs is not synchronized with the forcing cycle (alignment of Figures 1c and 1d is indicated by vertical time lines; circle annotations on Figure 1c mark individual purge events). One cannot claim that the forcing and response are uncoupled but they lack the phase-lock that appears to characterize the ice-climate coupling associated with Heinrich events.

Anatomy of Heinrich Events

For complex phase-locked systems the question of distinguishing forcings from responses is murky. In principle the marine sedimentary record can resolve this difficulty but, for Heinrich events, coring sites distant from Hudson Strait [e.g., *Heinrich,* 1988] present a partial picture whereas those very close are cluttered by local effects [e.g., *Andrews et al.,* 1994]. The Orphan Knoll site (50°12.26N; 45°41.14W; 3448 m) appears to possess the advantages of proximity without the attendant disadvantages. A location map (Figure 2) shows the coring site and geographical features relevant to this paper. The site was first cored during a 1991 cruise of the CSS-Hudson (91-045-094, henceforth P-094; *Hillaire-Marcel et al.,* 1994), and again during the first North Atlantic campaign of the Marion-Dufresne II (MD-2024; *Stoner et al.,* 1998). The 11 m core P-094 spans isotopic stages 5a-1, whereas the 26 m core MD-2024 spans isotopic stages 5d-1. In spite of their large difference in length, both cores cover roughly the same time interval, the last Ice Age. Core MD-2024 yielded "stretched" records due to coring artifacts. Unfortunately, these also resulted in smoothing effects. Sedimentological, magnetic, and geochemical studies at 10 cm intervals (P-094) or 5 cm intervals (MD-2024), i.e., having a time resolution of a ~1000 yr and 250 yr, respectively, have already been published, with special attention to Heinrich layers [*Stoner et al.,* 1996]. In the present paper, we will refer to a new data set from P-094. It is based on samples collected at 1 cm intervals, i.e., with a theoretical resolution of 100 yr, but for the smoothing effect of bioturbations [see *Wu and Hillaire-Marcel,* 1994a]. The choice of core P-094, in preference to MD-2024, is based on the consideration that smoothing effects linked to coring artifacts, make MD-2024 records less reliable than those of P-094 for documenting the effects of short-term climate instability on sedimentation.

Figure 3 shows physical and sedimentological properties of the upper 6 m sequence in core P-094, a sequence that spans the time interval 0–45 kyr. We restrict the present study to the upper 6 m section of the core, because of its much better time control. The conversion of depth to calendar years is controlled by AMS ^{14}C age determinations (triangular annotations to the graph of $\delta^{18}O$ stratigraphy) and on high-resolution paleomagnetic correlations with other deep sea records [*Stoner et al.,* 1998; *Roberts et al.,* 1997]. Variations in $\delta^{18}O$ values in *Neogloboquadrina pachyderma* (s.) are taken as a proxy for salinity and, to a lesser extent, temperature changes in the upper water layer (i.e., a water layer between ~100 m and 500 m deep; see *Wu and Hillaire-Marcel* [1994b]). At the Orphan Knoll site, this proxy mainly indicates the dilution by ice meltwater of a water layer corresponding today to the upper Labrador Sea water mass [*Lucotte and Hillaire-Marcel,* 1994]. The ratio of anhysteretic remanent magnetization susceptibility K_{ARM} to low-frequency susceptibility K decreases with increasing magnetic grain size when the magnetic mineralogy is dominantly magnetite [*Stoner et al.,* 1996]. This indicator is taken as a proxy for the settling of suspended sediment produced by turbiditic flow down the Northwest Atlantic Mid-Oceanic Channel (NAMOC in Figure 2). The coarse fraction (>125 μm weight per cent) is viewed as an indicator of ice-rafted debris (IRD) [*Hillaire-Marcel et al.,* 1994a] as well as of coarse debris-flow sedimentation channeled by the NAMOC [e.g., *Hesse and Rakofsky,* 1992]. In the fast-deposited units, such as Heinrich layers, the weight per cent $CaCO_3$ is associated with silt-size suspended sediment transport and/or with overspilled turbidites from the NAMOC, rather than with ice-rafting. This material originates from glacial erosion of Paleozoic carbonates in Hudson Strait area [e.g., *Andrews et al.,* 1994]. The ratio $^{230}Th_0/^{232}Th$ (Figure 4) is an indicator of sedimentation rate. ($^{230}Th_0$ denotes the ^{230}Th activity corrected for radioactive decay since deposition, i.e., the ^{230}Th activity of the sediment when it settled. ^{232}Th labels terrigenous fluxes, whereas ^{230}Th includes a terrigenous fraction, a diagenetic fraction due to the decay of the diagenetically uptaken U in the sediment, and most importantly for this indicator, a fraction produced by the decay of U dissolved in the overlying water column and a fraction scavenged by the settling particles [see *Hillaire-Marcel et al.,* 1994b].)

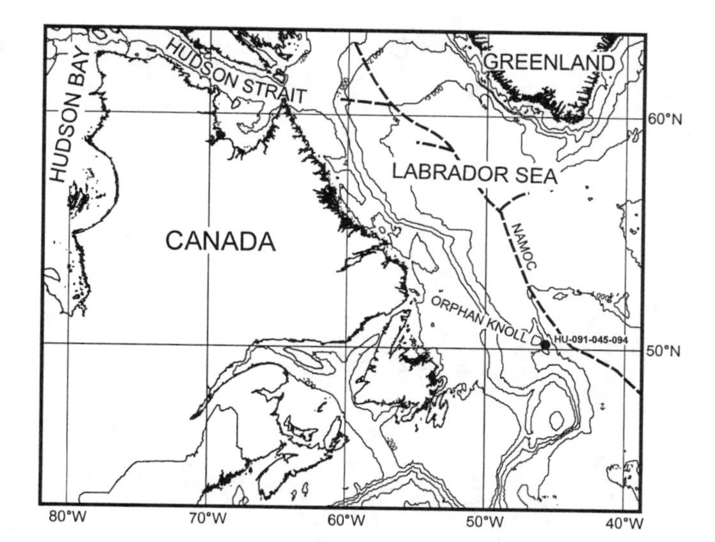

Figure 2. Map of Labrador Sea region of Eastern Canada showing bathymetric contours and the location the drilling site east of Orphan Knoll from which core HU-91-045-094P was recovered. The Northwest Atlantic Mid-Oceanic Channel (NAMOC) is indicated by a dashed line.

Scrutiny of the $\delta^{18}O$ variations shown in Figure 3 reveals that all the Heinrich events started while the ocean state was cold and that onset of a Heinrich event is signalled by a sharp increase in K_{ARM}/K, indicating a rapid increase in sediment flow along NAMOC which guides submarine suspended sediment from Hudson Strait to the core site. Over the five Heinrich events, this increase tends to be well correlated with increases in coarse fraction and $CaCO_3$. The timing of maximum dilution by iceberg melting (indicated by local minima on the $\delta^{18}O$ plot and a second peak in coarse fraction

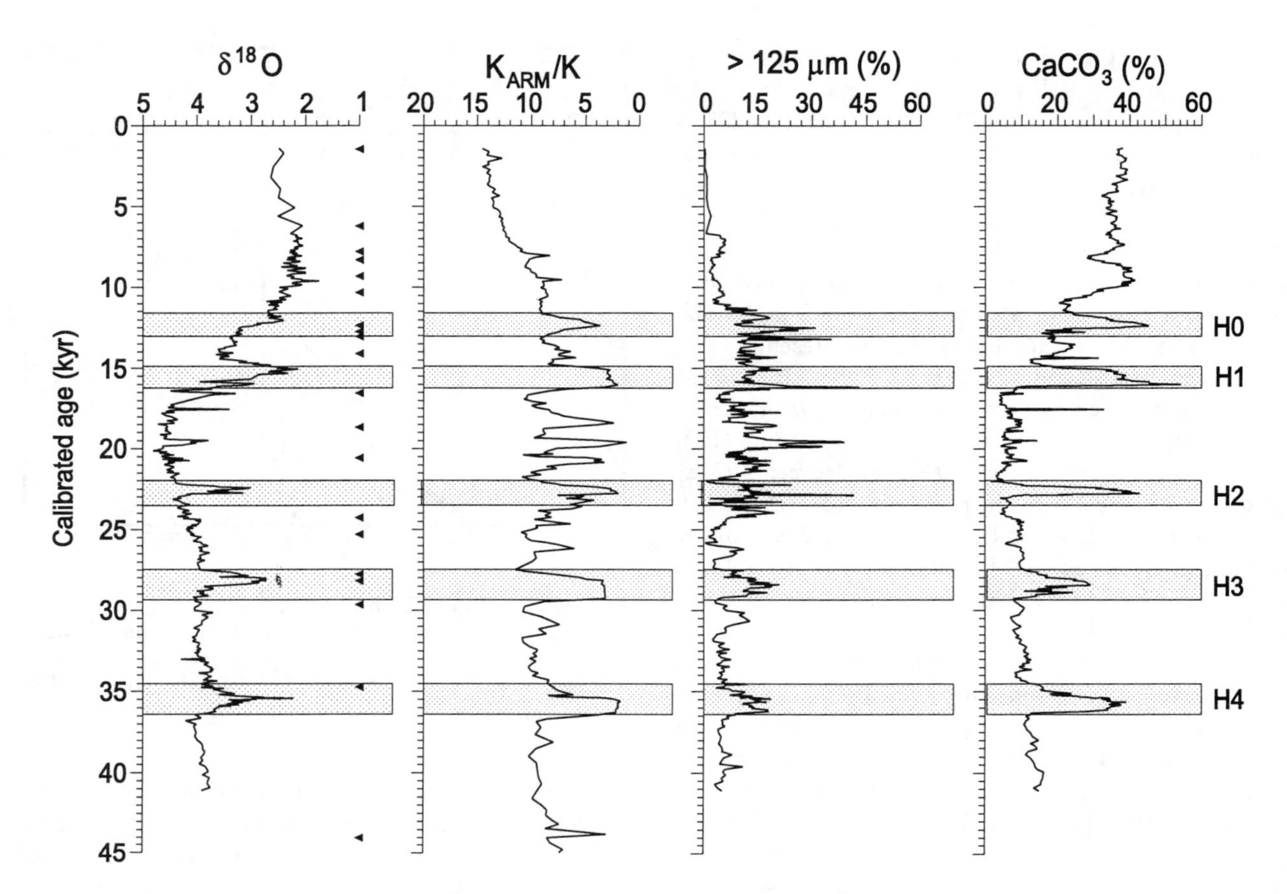

Figure 3. Properties of the studied sequence in core HU-91-045-094P. Heinrich events H0 to H4 are indicated. (K_{ARM}/K values are from *Stoner et al.* [1996].)

content) is delayed relative to these sedimentation signals. This interpretation is entirely based on analysis of the HU-91-045-094P core, which yields the most detailed available record of the composite nature of Heinrich layers.

Figure 4 provides a closer look at the H1 and H0 events and includes results of $^{230}Th_0/^{232}Th$ measurements from *Veiga-Pires and Hillaire-Marcel* [under revision] which help to document the relative changes in sedimentation rates during these two events. We shall use the calendar age axis as an aid to cross-referencing between the graphs but it is important to recognize that these values are subject to all the ambiguities associated with the interpolation of ages between precisely dated levels. Our references to calendar ages are intended to facilitate discussion of the sequence of events, not to fix their absolute timing or duration. At 16.4 kyr the $\delta^{18}O$ plot shows a cold peak immediately preceding the onset of H1. From 16.4 kyr–15.1 kyr there is a decreasing trend in $\delta^{18}O$ that is interpreted as a progressive isotopic lightening of near-surface waters by iceberg melt, 15.1 kyr corresponding to maximum dilution. The major peak in coarse fraction occurs at 16.2 kyr and is accompanied by a pronounced increase in K_{ARM}/K. The absence of a correspondingly strong local minimum in $\delta^{18}O$ suggests to us that this increase in coarse fraction deposition is not primarily an effect of enhanced IRD, because a larger flux of icebergs should result in diluted low $\delta^{18}O$ surface waters. Instead, we believe that the increase in coarse fraction is an indication of active debris-flows channeled along NAMOC, possibly signalling advancing ice and accompanying sediment disturbance in the Hudson Strait area. From 16.1 kyr–16.0 kyr there is a striking increase in $CaCO_3$ accompanied by a rapid increase in sedimentation rate (decrease in $^{230}Th_0/^{232}Th$). The $CaCO_3$ increase is not a result of IRD and the grain size is suggestive of glacial flour; this could be associated with suspended sediment released by the subglacial water system of an actively-flowing ice stream. The interval from 16.1 kyr–15.1 kyr is characterized by fast deposition (low $^{230}Th_0/^{232}Th$), predominantly of fine-grained $CaCO_3$. From 15.2 kyr–15.1 kyr

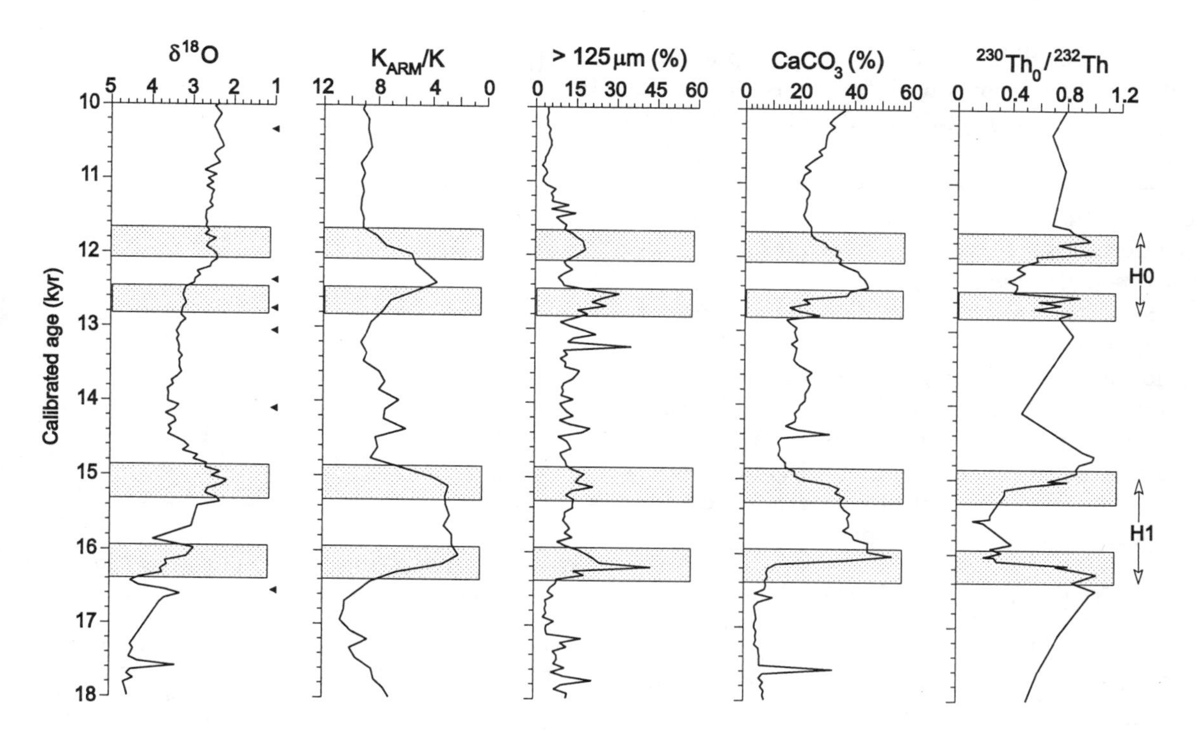

Figure 4. Details of H1 and H0 in core HU-91-045-094P. (K_{ARM}/K values are from *Stoner et al.* [1996] and $^{230}Th_0/^{232}Th$ values are from *Veiga-Pires and Hillaire-Marcel* [under revision].)

a minor peak in coarse fraction is very well correlated with the $\delta^{18}O$ minimum at maximum dilution. This is interpreted as the principal IRD peak for H1.

A similar close examination of H0 (Figure 4) leads to the following interpretation. Around 12.8 kyr the first indications of H0 appear as a progressive increase in coarse fraction (12.8 kyr–12.5 kyr) leading to a major coarse fraction peak at 12.5 kyr. This interval is followed by a rapid increase in fine-grained $CaCO_3$ beginning around 12.5 kyr and coinciding with a fast-deposition interval (12.5 kyr–12.0 kyr as indicated by $^{230}Th_0/^{232}Th$). Significantly, the $\delta^{18}O$ signal is unaffected by the initial major coarse fraction peak suggesting that sedimentation over the interval 12.8 kyr–12.5 kyr is associated with debris-flows along NAMOC and not with IRD, as also seen in the H1 layer. From 12.5 kyr–12.0 kyr there is a progressive decrease in $\delta^{18}O$ and maximum dilution occurs at 12.0 kyr, followed at 11.9 kyr by a minor peak in coarse fraction which we interpret as the IRD peak for H0.

In the near-field, both H1 and H0 reveal a two-part structure highlighted by their two coarse fraction peaks. This twin-peak structure is not a calculation artifact linked to the dilution of the total sediment by the fine-detrital carbonates. The two peaks remain clearly defined when coarse fraction contents are evaluated for $CaCO_3$-free sediment [*Veiga-Pires and Hillaire-Marcel,* under revision] and they constitute a typical feature of most Heinrich layers in the Orphan Knoll area. At this site, the first sedimentary signal of Heinrich events is associated with enhanced debris-flow deposition presumed to be linked to advancing ice in the vicinity of Hudson Strait, i.e., at the head of the principal NAMOC tributaries. The major IRD peak is delayed relative to the onset of these Heinrich events and is associated with maximum dilution by iceberg melting. Additional evidence for the involvement of two distinct depositional mechanisms for the coarse fraction peaks of Heinrich layers is provided by the fact that at least two of these layers show erosional surfaces at the very base of the bottom peak (notably H3 and H6; see *Hillaire-Marcel et al.* [1994a]). Furthermore, lithic fragment counts in the H1-H0 sequence, which we looked at in greater detail, indicate that the bottom coarse fraction peak has a proportion of carbonate debris typically exceeding 50%, likely due to their restricted source in the carbonate-rich area of Hudson Strait, whereas the top peak shows a substantially reduced proportion of such carbonate fragments (~20%) and more silicate grains, suggesting IRD by icebergs from more scattered sources. A modeling study of IRD during Heinrich events [Alley and MacAyeal, 1994] predicts a two-peak structure in IRD sedimentation but differs in detail from our interpretation. The essential point of difference is that in the modeling study the two peaks result from a single depositional mechanism whereas evidence from core HU-91-045-094P points to the sequential operation of two distinct mechanisms.

It is informative to consider the question: "How would the signals that are preserved in the Orphan Knoll core be transmitted to mid-Atlantic sites?" In particular, if the major coarse fraction peak at Orphan Knoll is indeed associated with turbiditic transport along NAMOC and not with IRD, then this signal would not be found in cores at sites distant from Hudson Strait. At these sites the paleoceanographic record of Heinrich events would be that of a progressive decrease in $\delta^{18}O$ culminating in an IRD episode.

PHYSICAL CONSTRAINTS ON SURFACE-TO-BED COUPLING

The Orphan Knoll data show no evidence of a premonitory climate forcing causing a cryospheric response. Most glaciologists favor the view that Heinrich events are not climate-triggered and computer models have been developed to demonstrate that Hudson Strait Ice Stream can plausibly behave as a free-running oscillator [*MacAyeal,* 1993a,b; *Marshall and Clarke,* 1997a,b]. The problem with this conclusion is that a free-running oscillator, such as the simple kitchen-built oscillator, would lack the phase-locking between climate state and ice volume that seems to be a feature of the ice–climate system before the onset of a Heinrich event. The occurrence of Heinrich events during the cold phase of a Dansgaard-Oeschger (D-O) oscillation needs explanation. One might suggest that, like Heinrich events, the D-O oscillations are a consequence of cryospheric forcing of climate but the marine sedimentary record leaves this question undecided [*Bond et al.,* 1997]. It appears that during the long onset period, ice–climate coupling is an expression of climate forcing of a cryospheric response. How this can be effected is a challenging question for glaciologists.

It is important to appreciate that the flow of glaciers and ice sheets is largely dictated by processes that are active at the glacier bed whereas atmospheric forcing operates at the upper surface. For any surface process to produce a flow response there must be transmission of this influence from the surface to the bed. The list of surface-to-bed couplings is short and several processes can be dismissed as untenable.

Extreme and rapid changes in surface temperature are propagated to the bed with substantial attenuation

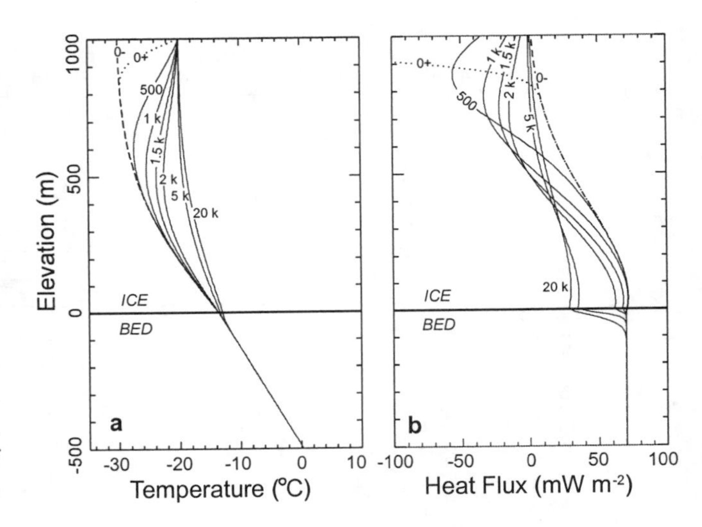

Figure 5. Response of a 1000 m thick ice slab to a 10°C increase in surface temperature. Ice rests upon bedrock. The ice–bed interface is taken as the elevation datum. The time slices are for 0^- (immediately before the temperature increase), 0^+ (immediately after), 500 yr, 1 kyr, 1.5 kyr, 2 kyr, 5 kyr, and 20 kyr. (a) Temperature as a function of elevation. (b) Upward heat flux as function of elevation.

and delay. For conductively-dominated heat transport the characteristic time constant for diffusion is $t_C = \rho_I c_I H_I^2/K_I$, where ρ_I is ice density, c_I is the specific heat capacity for ice, H_I is ice thickness, and K_I is the thermal conductivity of ice. For advectively-dominated heat transport the time constant is $t_A = H_I/b_I$ where b_I is the ice-equivalent surface accumulation rate. Taking $H_I = 1000$ m, $b_I = 0.25$ m yr^{-1}, and physical constants consistent with *Paterson* [1994] gives $t_C \sim 28$ kyr and $t_A \sim 4$ kyr. Figure 5 presents simulation results that illustrate the combined effects of diffusion and advection in a 1000 m thick ice slab subjected to an instantaneous 10°C increase in surface temperature. Even after 2 kyr there is no appreciable increase in bottom temperature. All paths point to the same conclusion: surface temperature variations, regardless of how large they are and how rapidly they occur, cannot be effectively transmitted to the beds of ice streams and ice sheets.

Hydrological coupling, an alternative to thermal coupling, can be both fast and strong. If surface melting accompanies an increase in surface temperature and meltwater can percolate from the ice sheet surface to the bed, a fast flow response might be coupled to some surface forcing, for example an increase in surface temperature or a decrease in surface albedo (e.g., from dust deposition). Modern mountain glaciers demonstrate this coupling in the form of strong diurnal and seasonal cycles in flow rate. It is extremely unlikely that this is a significant process in modern continental ice sheets and it is unlikely that the Laurentide Ice Sheet behaved differently except during the late stages of its disintegration. As evidenced in Figure 5, summer meltwater would have to percolate through great thicknesses of extremely cold ice before a surface-to-bed coupling could be activated. Although hydrological coupling between the ice surface and the bed can occur rapidly and produce a large ice flow response, we do not consider this a likely possibility throughout most of the lifetime of the Laurentide Ice Sheet.

The transmission of stress from the surface of an ice sheet to its bed occurs virtually instantaneously. Any increase in surface loading, for example from increased accumulation rate, would produce an instantaneous mechanical response at the glacier bed. A problem with mechanical coupling between the surface and bed is that it is difficult to load the surface rapidly; a greatly increased surface accumulation rate cannot produce a rapid thickness increase. In summary, this form of surface-to-bed coupling is very strong, but it is difficult to justify a surface forcing of substantial amplitude. Nevertheless mechanical coupling is the only viable surface-to-bed coupling of those identified. We shall revisit this discussion in a subsequent section.

MECHANISMS OF FAST FLOW AND FLOW INSTABILITY

Ice creep is a slow but ubiquitous process. Fast flow, such as that associated with ice stream motion and the active state of surging glaciers, is caused by basal flow processes—sliding and deformation of subglacial sediment. The relative importance of these fast-flow processes remains controversial [*Alley*, 1989; *Kamb*, 1991; *Iverson et al.*, 1995] and is likely to vary both spatially and temporally. Whereas the creep process can operate in cold ice, sliding and bed deformation are only effective if the bed is at the ice melting temperature. Although this is a necessary condition for fast flow, it is by no means a sufficient condition. Fast flow is enabled by thermal conditions but activated by hydrological conditions. There are numerous examples of valley glaciers that are permanently at their melting temperature at the bed but which slide negligibly unless subglacial water pressure is high [e.g., *Iken and Bindschadler*, 1986]. Because the thermal and mechanical physics of glaciers is well known but the processes controlling subglacial hydrology, sliding, and bed deformation are comparatively poorly known, most current ice sheet models ignore hydrology entirely.

Ice Stream Surging

MacAyeal [1993a,b] was the first to use computer models to explore the possibility that Heinrich events are associated with episodic surging of Hudson Strait Ice Stream. His low-order model has similar physics to a thermal model developed to describe cyclic surging of Trapridge Glacier, Yukon Territory, Canada [*Clarke*, 1976] and shares the major shortcoming of that model—onset and termination of fast flow are exclusively controlled by thermal conditions at the bed. Field studies of Variegated Glacier, Alaska, during its 1982-83 surge [*Kamb et al.*, 1985; *Raymond*, 1987] demonstrate that surging is predominantly controlled by subglacial hydrological processes rather than thermal processes and that the cause of fast flow is elevated subglacial water pressure.

Generalized characteristics of a surge cycle for a glacier that, like Hudson Strait Ice Stream, discharges ice into the ocean are presented in Figure 6. Modern examples of such glaciers can be found in Svalbard [*Schytt*, 1969] and East Greenland [*Reeh et al.*, 1994] and it is apparent that individual behaviors will be influenced by factors such as water depth and bed geometry. At the onset of a surge, ice flow rate increases dramatically and the terminus advances (Figure 6a). As the advance proceeds, the calving rate increases (Figure 6b) and the ice stream thins. The stopping mechanism for ice stream surges is not known. For mountain glaciers it appears that the subglacial water system switches from an areally-distributed sheet-like configuration to a spatially-localized conduit configuration and this transformation leads to a rapid drop in subglacial water pressure [*Kamb et al.*, 1985]. A similar explanation might apply to ice streams but it is also possible that the combined effects of reduced ice thickness and surface slope eventually lead to a reduction in bottom stress and that the accompanying decrease in subglacial meltwater production terminates the surge (Figure 6c). Following the surge, the ice–bed contact might refreeze but this is not an essential feature of the surge mechanism.

Ice Shelf Breakup

The recent breakup of Larsen Ice Shelf in the Antarctic Peninsula [*Vaughn and Doake*, 1996; *Doake et al.*, 1998] reminds glaciologists that large iceberg fluxes are not necessarily associated with fast-flowing ice. In the case of ice shelf breakup, the ice flow rate need not vary markedly and the most conspicuous manifestation of breakup is upflow migration of the calving front. Figure

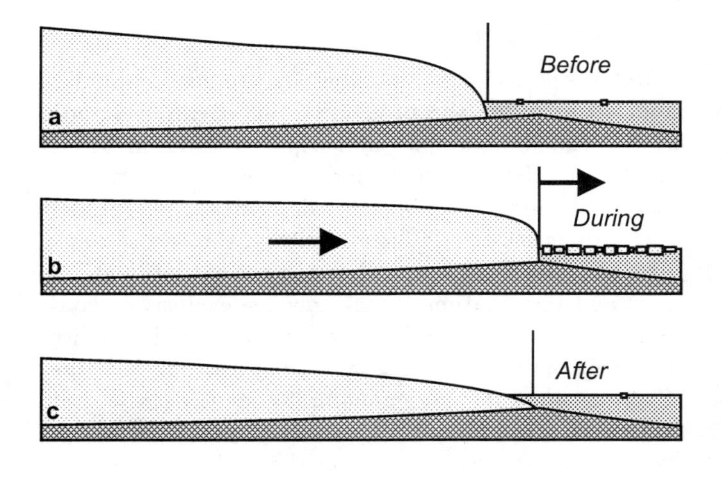

Figure 6. Ice stream surging mechanism. (a) Pre-surge state. (b) Active phase. (c) Post-surge state.

7 summarizes the sequence of events. Prior to breakup the shelf extends well beyond its grounding line (Figure 7a); lack of nourishment, or some other trigger, causes breakup to initiate (Figure 7b) and the calving front moves inland; breakup ceases when the shelf disappears (Figure 7c). *Hulbe* [1997] has successfully modeled this sequence of events.

Tidewater Instability

Like the ice shelf breakup mechanism, tidewater instability is associated with retreating, rather than advancing ice. The dramatic retreat of Columbia Glacier, Alaska, provides a well-documented modern example of this instability in operation [*Meier and Post,* 1987; *Meier et al.,* 1994; *Kamb et al.,* 1994]. Ice calving lies at the heart of this instability and, unfortunately, this process is poorly understood. Adopting ideas from *Reeh* [1968], *Brown et al.* [1983], and *Meier and Post* [1987], *Marshall et al.* [submitted] have expressed the calving discharge per unit width Q_c (in $\mathrm{m^2\,yr^{-1}}$) as

$$Q_c = k_c H_W H_I \tag{1}$$

where H_W is water depth, H_I is ice thickness, and k_c is a rate parameter which can include the effects of calving geometry, ice temperature, and longitudinal stress. The key feature of (1) and most empirical calving laws is that calving rate is proportional to water depth. This introduces the possibility of unstable behavior. If dH_W/ds (the downflow slope of the sea floor) is negative then water depth decreases as ice flows toward the calving margin. During advance of a tidewater glacier, ice volume is increasing. The further ice advances, the lower the water depth and, by (1), the lower the calving rate. Ice input exceeds output and thickness increases until the calving rate and inflow are in balance (Figure 8a). If ice supply then decreases, the calving rate would exceed the inflow and the calving margin would retreat from its stable position. In retreat, the system cannot produce a stable response. As the calving front recedes, water depth increases as does the calving rate (Figure 8b). This positive feedback continues until the calving front reaches a region where a new equilibrium can be established (Figure 8c).

For glaciers it is known that surging is a cyclic internal flow instability and external factors such as climate change have only a secondary influence on the cycle. In contrast, tidewater instability is ultimately a response to trends that affect glacier mass balance; thus this instability is the more amenable to a climate trigger. An alternative trigger, suggested by *Paterson* [1998], draws on *Alley's* [1991] suggestion that a tidewater retreat can be initiated by the advance of a tidewater glacier beyond its stable grounding position.

Other Possibilities

Although there are no modern examples, compound instabilities that involve combinations of the foregoing mechanisms cannot be ruled out. Possible examples include: (i) ice-shelf breakup triggering a surge, (ii) ice shelf breakup triggering a catastrophic tidewater retreat, and (iii) a surge advance followed by tidewater instability.

It has recently been suggested that Heinrich events are triggered by earthquakes that are induced by ice

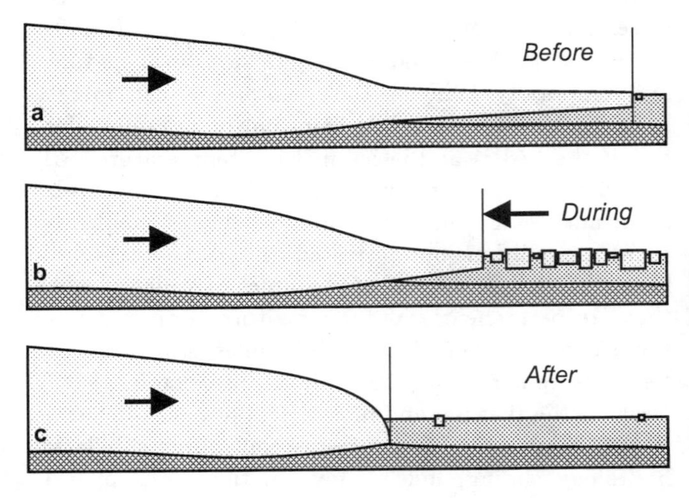

Figure 7. Ice shelf breakup mechanism. (a) Before onset of breakup instability. (b) During active breakup phase. (c) Post-breakup state.

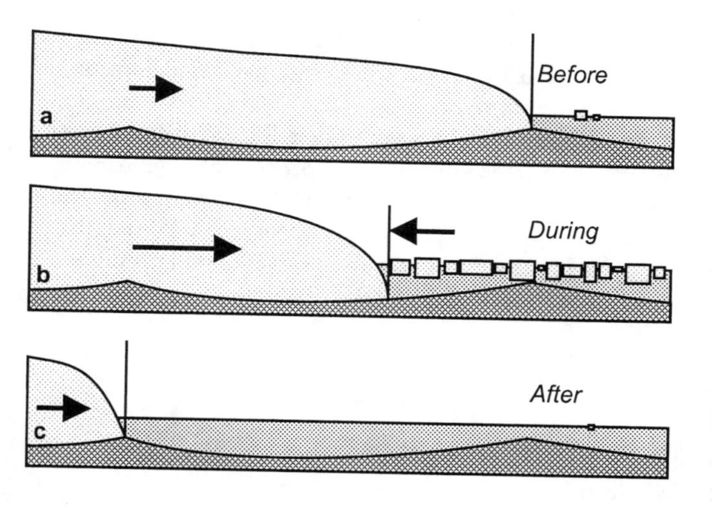

Figure 8. Tidewater instability mechanism. (a) Before onset of calving instability. (b) During catastrophic retreat. (c) Following termination of calving instability.

loading in the vicinity of Hudson Strait [*Hunt and Malin*, 1998]. Apart from the loose claim that Hudson Strait Ice Stream might behave as an "ice slide", the proposal has no basis in glacier physics and, indeed ignores the entire relevant literature on the subject. The ice slide hypothesis can be traced to *Saussure* [1779-96] and was superceded by the field studies of *Forbes* [1842]. The earthquake-advance theory is associated with *Tarr and Martin* [1914] who happened to be studying Alaskan glaciers shortly after the occurrence of a major earthquake. By an unremarkable coincidence one of Alaska's many surge-type glaciers was observed to be actively surging and this led to their plausible inference. *Post* [1965] formally tested the earthquake-advance theory and concluded that it was unsupported by observations. While wrong, *Tarr and Martin's* [1914] theory is at least sensible; Alaskan glaciers subjected to large earthquakes can receive an increased surface load contributed by earthquake-triggered avalanches. For Hudson Strait Ice Stream, avalanche loading is not a possibility and earthquake-triggered decoupling of the ice stream from its bed is completely inconsistent with modern understanding of ice stream dynamics.

Problems of sediment entrainment and retention

Another challenging problem associated with Heinrich events is to explain the vast quantities of sediment exported to the North Atlantic Ocean from the Hudson Bay region of Canada. Vast amounts of sediment imply vast amounts of ice [*Alley and MacAyeal,* 1994; *Dowdeswell et al.,* 1995]. This is not the end to the difficulty; even if large ice volumes are accepted, it seems necessary to invoke a highly-efficient sediment entrainment mechanism.

A further problem relating to sediment is explaining sediment retention. Entrainment processes operate most effectively near the ice–bed contact and the foreseeable result is accretion of a sub-horizontal layer of debris-rich ice. Bottom melting works against sediment entrainment and retention and, because fast flow processes are launched and sustained by geothermal and frictional meltwater production at the ice–bed contact, the requirements of fast ice flow and efficient sediment entrainment appear to conflict. The rate of bottom melting $\dot{m}$ (in $\mathrm{m\,yr^{-1}}$) is given by the expression

$$\dot{m} = \frac{1}{n_I \rho_I L}\left(q_F + q^- - q^+\right) \tag{2}$$

where n_I is the volume fraction of ice, L is the latent heat of melting for ice, $q_F = v_S \tau_B$ is the heat flux generated by sliding friction (where v_S is the sliding rate and τ_B is the basal shear stress), $q^- = -K^-(\partial T/\partial z)^-$ is the heat flux immediately below the ice–bed contact, and $q^+ = -K^+(\partial T/\partial z)^+$ is the heat flux immediately above it. Heat flux q can be discontinuous across this boundary as can the thermal conductivity K and temperature gradient $\partial T/\partial z$; superscripts "+" and "−" respectively denote the regions immediately above and below the contact.

Ice that calves from the Laurentide Ice Sheet is subjected to melting by sea water. Melting rates depend on temperature and salinity [*Russell-Head,* 1980] and, for bergs that do not roll over, bottom-accreted sediment is the first casualty of melting. These considerations suggest that a substantial thickness of sediment-rich ice is required and that entrainment mechanisms that inject sediment beyond the ice–bed contact favor long-term sediment retention.

Figure 9 illustrates five known mechanisms for sediment entrainment. Thermal freeze-down (Figure 9a) is the classic mechanism [*Weertman*, 1961] and the mechanism associated with *MacAyeal's* [1993a,b] binge/purge model. A shortcoming of the freeze-down mechanism is that it requires downward migration of the freezing plane and, typically, this occurs very slowly beneath thick ice. If $Z(t)$ is the elevation of the freezing plane then

$$\frac{dZ}{dt} = \frac{1}{n_E \rho_I L}\left(q^- - q^+\right) \tag{3}$$

where n_E is the porosity of unfrozen water-saturated bed material, and q^+ and q^- are as previously defined.

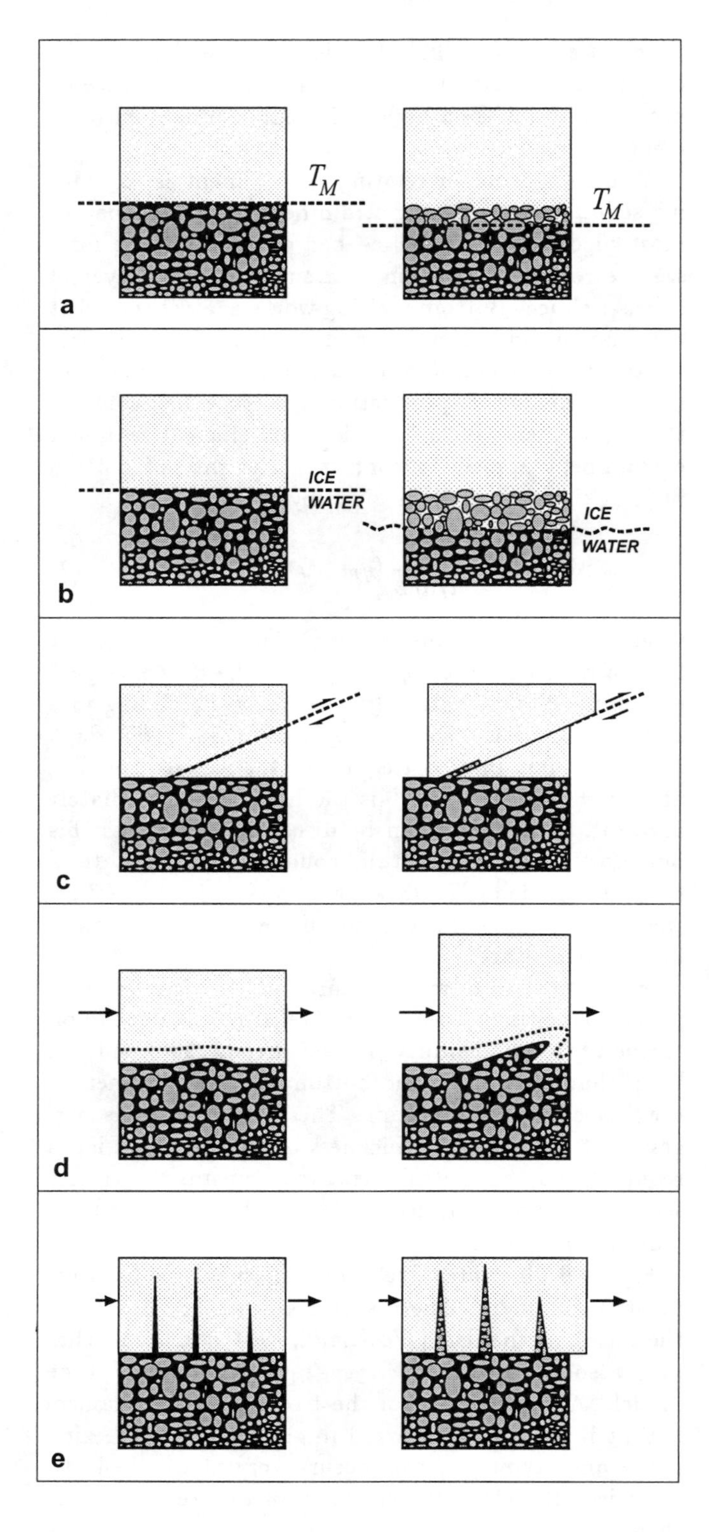

Figure 9. Sediment entrainment mechanisms. (a) Thermal freeze-down. (b) Regelation infiltration. (c) Thrusting. (d) Folding. (e) Bottom crevassing.

From Equation (3) it is evident that the large latent heat of fusion for water works against rapid motion of the freezing plane.

An entrainment process that is likely to be more effective than freeze-down is the process of regelation infiltration (Figure 9b) [*Iverson,* 1993; *Iverson and Semmens,* 1995]. According to the theory of *Phillip* [1980], ice at its melting temperature infiltrates underlying sediment at a rate

$$\frac{dZ}{dt} = -k(p_I - p_W) \tag{4}$$

where $p_I = \rho_I g H_I$ is the ice overburden pressure and p_W is the water pressure beneath the ice–bed contact. The rate factor k decreases as the thickness of ice-entrained sediment increases. The infiltration process underlying (4) is regelation; pressure melting occurs at the contact with a resisting clast and refreezing follows as the clast is passed. If subglacial water pressure is low, the process can proceed rapidly. Like the thermal freeze-down mechanism, there is a moving phase boundary but the migration rate is not controlled by macroscopic heat fluxes.

The entrainment processes of thrusting [*Clarke and Blake,* 1991], folding, and bottom crevassing [*Sharp,* 1985] are tectonic (Figures 9c–9e) and difficult to quantify. Tectonic processes share the attractive feature that sediment is stored at some distance from the ice–bed contact and this style of entrainment would delay the timing of sediment release by berg melting.

EVALUTION OF CANDIDATE MECHANISMS

The foregoing discussion sets the stage for an evaluation of the candidate instability mechanisms. Although each of the candidates, surging, ice shelf breakup, and tidewater instability, seems capable of explaining the large iceberg fluxes associated with Heinrich events, their sediment entrainment and sediment preservation characteristics differ.

Cyclic surging of mountain glaciers is associated with pronounced cyclic variations of stress [*Raymond et al.,* 1987] and subglacial water pressure [*Kamb et al.,* 1985]; for sub-polar glaciers there is also a cyclic variation in thermal structure [*Clarke and Blake,* 1991]. All of these factors are highly favorable to the sediment entrainment mechanisms depicted in Figure 9. As discussed above, surging is a hydrological instability and, for ice streams, bottom melting is necessary to activate fast flow. A possible shortcoming of surging as an explanation of Heinrich events is that bottom melting erodes ice-entrained sediment and, during fast flow episodes, frictional bottom melting could be very high. It is possible that this problem is less serious than might be imagined. Fast flowing ice streams can become largely decoupled from their beds, restraint being provided by the ice stream margins rather than bed friction. In this way, fast flow and modest bottom melting rates might be reconciled.

The ice shelf breakup mechanism does not invoke fast-flow processes and therefore does not require or provoke large cyclic variations in basal stress, temperature, and water pressure. The absence of fast flow is advantageous for sediment retention and, as suggested by *Hulbe* [1997], ice accretion beneath a floating ice shelf would defend entrained sediment from the initial consequences of berg melting. The problem with the ice shelf breakup mechanism is that it relies entirely on normal rates of sediment entrainment and it is our conviction that normal rates cannot explain the volume of sediment delivered to the North Atlantic during Heinrich events. Finally, Heinrich event H0, though it may not be representative, is known to have been associated with an abrupt advance, not a retreat, of Hudson Strait Ice Stream [*Kaufman et al.*, 1993]. This behavior is consistent with the surging mechanism and inconsistent with ice shelf breakup.

Tidewater instability, as manifested in the recent catastrophic retreat of Columbia Glacier, involves similar stress and hydrological cycles to those associated with the surge instability. For tidewater disintegration of an ice stream, thermal cycling similar to that experienced by a surging ice stream, would also occur. In our opinion, both surging and tidewater instability are equally viable in terms of the criteria presented above.

A POSSIBLE MECHANISM FOR FAST SURFACE-TO-BED COUPLING

If we accept that Heinrich events are initiated by fast motion of the Hudson Strait Ice Stream and that fast flow must result from the triggering of an intrinsic instability, several problems remain. For an instability to be triggered, the system must be near an instability threshold prior to receiving the *coup de grâce.* Why does the instability tend to be triggered during cooling excursions of Dansgaard-Oeschger oscillations? Such behavior suggests that the ice stream is "sensitive", already primed for instability, and argues for the existence of a mechanism for fast coupling between the ice stream surface and its bed. As suggested in a previous section, the only viable fast coupling mechanisms are mechanical. An increase in ice thickness or surface slope will be immediately expressed as an increase in basal shear stress $\tau_B = \rho_I g H_I \sin\theta$ where $g = 9.80\,\mathrm{m\,s^{-2}}$ is the gravitational acceleration and θ is the surface slope. The nonlinear flow law of ice implies that for a simple inclined ice slab the discharge Q_I resulting from ice creep varies strongly with ice thickness. For an $n = 3$ flow law [see *Paterson*, 1994, pp. 95-98] $Q_I \propto H_I^5$, but even vigorous activation of the creep process is unlikely to produce a substantial flow response. For initiation of fast flow, a hydrological response of the bed is far more effective than a purely mechanical response.

One avenue that has not been explored is the possible role of strain heating and creep instability [*Clarke et al.*, 1976; *Yuen and Schubert*, 1979] in mediating a fast coupling between the surface and bed. Before delving into this matter we review some elementary glaciology. The geophysical classification of glaciers, introduced by *Ahlmann* [1935], categorizes glaciers according to their thermal structure. Polar glaciers, the coldest, are everywhere below their melting temperature (Figure 10a); sub-polar glaciers reach the melting temperature at the bed but are otherwise cold (Figure 10b); temperate glaciers are at the melting temperature throughout their entire thickness (Figure 10d). For ice, the melting temperature decreases with increasing pressure so that $T_M = -c_t \rho_I g H_I$ where $c_t = 7.42 \times 10^{-8}\,\mathrm{K\,Pa^{-1}}$ is the pressure melting coefficient. Thus, for example, beneath a 1000 m ice column the melting temperature is $T_M = -0.65°\mathrm{C}$. Years after Ahlmann's system was established, applied mathematicians [*Fowler and Larsen*, 1978; *Hutter*, 1982] pointed to the logical necessity of an intermediate category between sub-polar and temperate glaciers; such glaciers are referred to as polythermal (Figure 10c) and their existence has been observationally confirmed [*Classen*, 1977; *Blatter and Kappenberger*, 1988].

Traditionally those interested in thermal instability of ice sheets have concentrated on triggering a transition from polar to sub-polar thermal regimes [e.g., *Clarke*, 1976; *Clarke et al.*, 1977; *MacAyeal*, 1993a,b]. This fascination is a legacy of the belief that the onset of sliding necessarily accompanies the transition from a cold to melting bed. The transition from sub-polar to polythermal conditions has no effect on bottom temperature, but a closer examination will show that this transition can have an important effect on subglacial hydrological

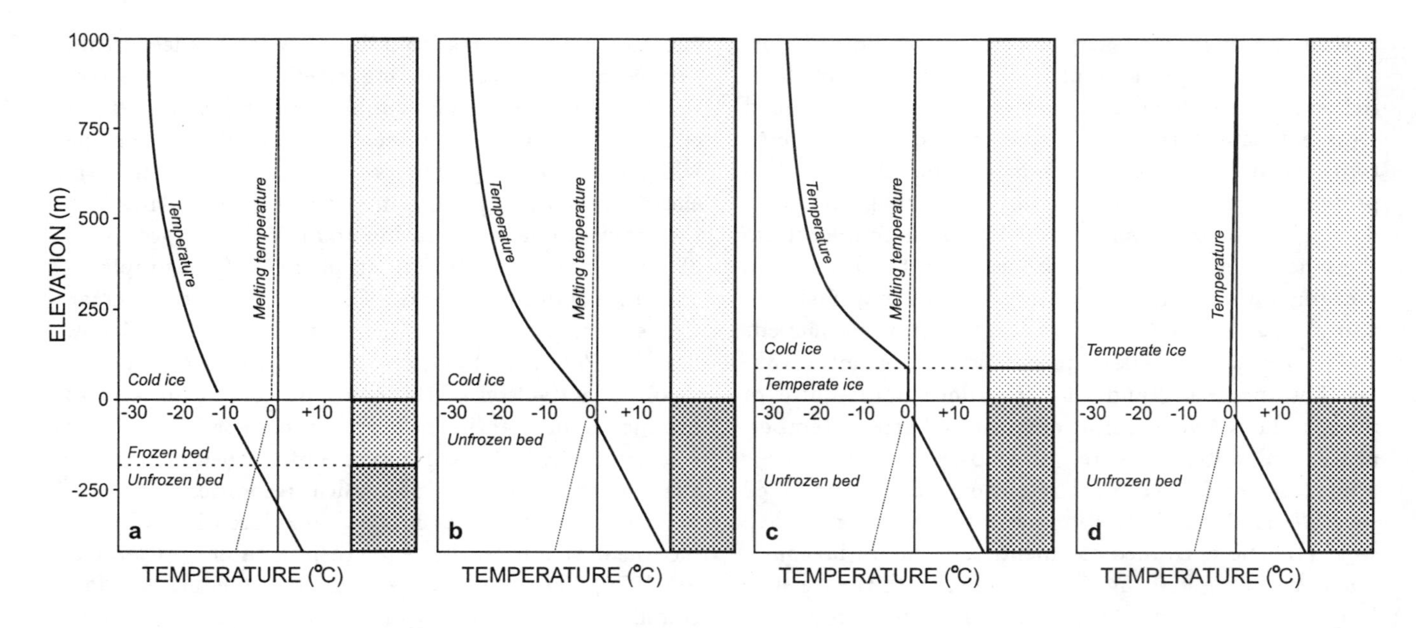

Figure 10. Thermal classification of glaciers and ice sheets. (a) Polar. (b) Sub-polar. (c) Polythermal. (d) Temperate.

conditions and thus could be the agent promoting the transition from slow to fast flow.

For one-dimensional heat transport in an ice column overlying a non-deforming bed, the governing equations for temperature $T = T(z,t)$ are

$$\rho_I c_I \frac{\partial T}{\partial t} = K_I \frac{\partial^2 T}{\partial z^2} - \rho_I c_I v \frac{\partial T}{\partial z} + \Phi \qquad (5a)$$

$$\rho_E c_E \frac{\partial T}{\partial t} = K_E \frac{\partial^2 T}{\partial z^2} \qquad (5b)$$

where Equation (5a) applies in ice and (5b) in the bed, c_I is the specific heat capacity of ice, ρ_E and c_E are the density and specific heat capacity of the bed, v is the bed-normal component of ice velocity, and Φ is the frictional dissipation of the deforming ice. We make standard simplifying assumptions and, taking $z = 0$ to correspond to the ice–bed contact, write $v = -v_0 z/H_I$, $\tau_B = \rho_I g H_I \sin\theta$, and

$$\Phi(z,t) = 2B_0 \exp(-Q/RT)\tau_B^{n+1}[1 - z/H_I(t)]^{n+1} \qquad (6)$$

where B_0 is a constant coefficient, Q is the creep activation energy for ice and $n \sim 3$ is the exponent in Glen's flow law. For appropriate values of B_0 and Q the reader is referred to *Paterson* [1994, p. 97]. The assumed form for τ_B greatly simplifies the mathematical treatment but eliminates interesting and realistic possibilities involving mechanical decoupling of the bed and the accompanying transfer of longitudinal stress. Intriguing features of (6) are the presence of a positive feedback (the greater the temperature the greater the strain heating), the concentration of frictional dissipation near the ice–bed contact, and the strong dependence of strain heating on ice thickness.

We shall assume that the thickness of the ice slab varies with time in response to changes in the ice-equivalent surface accumulation rate $b(t)$ and take

$$b(t) = b_0 + \frac{1}{2}\Delta b \sin(2\pi t/\Delta t). \qquad (7)$$

If v_0 is assumed to be constant then

$$\frac{dH_I}{dt} = -v_0 + b(t). \qquad (8)$$

With the foregoing assumption, the ice column can thicken or thin depending on the relative magnitudes of v_0 and b; as thickness changes so does the magnitude of the strain heating term. The dimensionless Brinkman number, here defined as

$$\mathrm{Br}(t) = \frac{H_I(t)\Phi(0,t)}{K_I T(0,t)}, \qquad (9)$$

with T expressed as absolute temperature, is a useful indicator of the significance of strain heating. In our work, $\mathrm{Br} = 1$ roughly delineates the threshold between low and high Brinkman numbers.

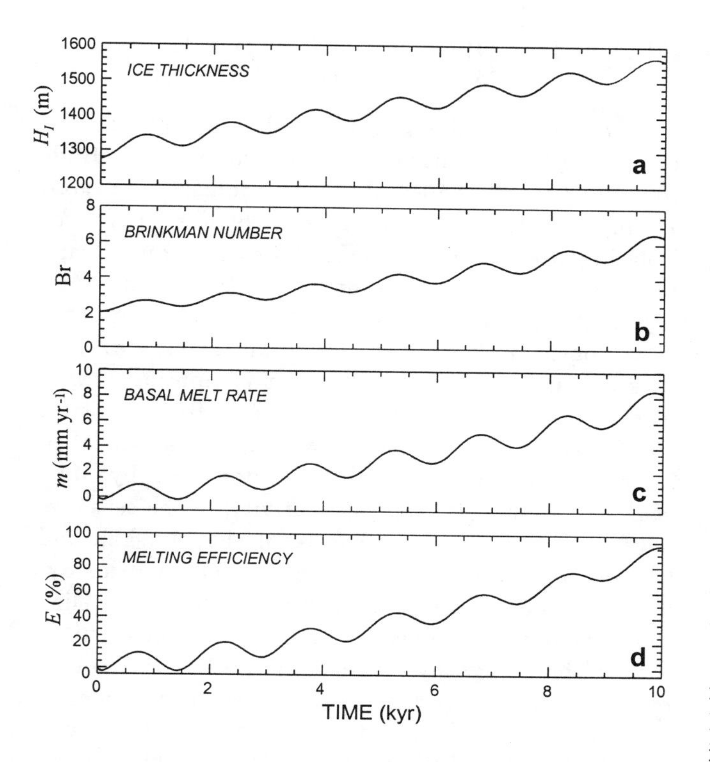

Figure 11. Time evolution of one-dimensional thermal model. (a) Ice thickness. (b) Brinkman number. (c) Basal melting rate. (d) Melting efficiency parameter.

Equations (5a) and (5b) are subject to boundary conditions and initial conditions. For simplicity we assume that at $t = 0$ the system is in a steady-state. As boundary conditions we take $\partial T/\partial z = -q_G/K_E$ at the lower boundary where q_G is the geothermal flux and $T(H_I, t) = T_S(t)$ at the upper boundary, where $T_S(t)$ is a specified surface temperature variation. At the ice–bed contact

$$-K_E \frac{\partial}{\partial z} T(0^-, t) + K_I \frac{\partial}{\partial z} T(0^+, t) = 0 \quad (10)$$

if the bed temperature is below the melting temperature, and $T(0, t) = -c_t \rho_I g H_I(t)$ if the bed is at the melting temperature. If the bed is at the melting temperature and sliding friction is neglected then, with $n_I = 1$, Equation (2) gives the basal melting rate as

$$\dot{m} = \frac{1}{\rho_I L} \left\{ -K_E \frac{\partial}{\partial z} T(0^-, t) + K_I \frac{\partial}{\partial z} T(0^+, t) \right\}. \quad (11)$$

Figure 11 shows a computer simulation of the effects of strain heating on melt generation in an ice column subjected to a time-varying surface accumulation rate. Because we have taken $b_0 > v_0$ the long-term trend is for ice thickness to increase with time. A sinusoidal fluctuation in b, of period 1 kyr, accounts for the sinusoidal fluctuation in ice thickness (Figure 11a). Figure 11b shows, not surprisingly, that sinusoidal fluctuations in Brinkman number are synchronized with thickness variations. Furthermore, time variation in strain heating gives rise to a synchronous variation in the basal melt rate (Figure 11c). This effect can be expressed in terms of time variations in a melting efficiency parameter defined as $E = 1 - q^+/q^-$ where $q^+ = -K_I \partial T(0^+, t)/\partial z$ and $q^- = -K_E \partial T(0^-, t)/\partial z$. The inequality $0 < E < 1$ defines the existence region for sub-polar glaciers. At lower levels of strain heating than those consistent with $0 < E < 1$ the thermal regime is polar; at higher levels of strain heating than those consistent with this inequality, the thermal regime is polythermal or temperate. Polar, polythermal, and temperate glacier types can be viewed as insensitive because subglacial melt production is unaffected by ice thickness fluctuations.

Figure 12 illustrates how sensitivity is achieved. The results are extracted from the simulation that yielded Figure 11 and focuses attention on conditions near the ice–bed contact. The temperature vs. depth curves (Figure 12a) are for times $t = 0$ kyr (solid line), $t = 6.5$ kyr (dot-dashed line), and $t = 10$ kyr (dotted line). The melting efficiencies at these times are roughly 0%, 50%, and 100%. Figure 12b shows the vertical distribution of heat flux at the same time snapshots. The solid line ($E \approx 0\%$) shows a continuous variation in heat flux as the ice–bed contact is traversed, implying that no portion of the inflowing heat flux is available for

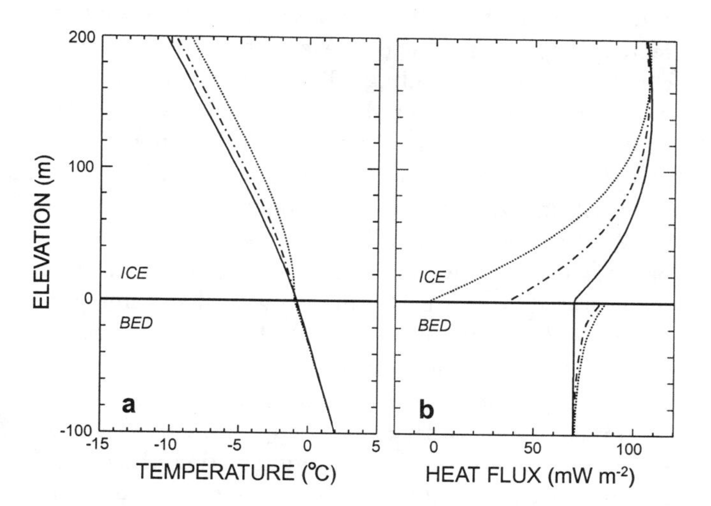

Figure 12. Results of one-dimensional thermal model showing switching from sub-polar to polythermal regime. (a) Temperature distribution. (b) Heat flux distribution.

meltwater production. The dot-dashed line ($E \approx 50\%$) shows a sharp discontinuity in heat flux across the ice–bed boundary, and the dotted line ($E \approx 100\%$) shows an even more pronounced discontinuity. Temporal and spatial variation in the subglacial heat flux are a consequence of the change in melting temperature associated with a thickening ice column.

Critical State Conjecture and Its Attractions

We have demonstrated that a sub-polar ice mass having a high Brinkman number is in a sensitive state. Changes in ice thickness or surface slope are immediately communicated to the bed as changes in basal stress τ_B and, through (6), these are transformed to strain heating at the bed and increased basal meltwater production. If the Hudson Strait Ice Stream was sensitive in this way, atmospheric forcing (e.g., increased accumulation rate leading to increased ice thickness) and a flow response could become phase-locked and an extreme forcing could, plausibly, launch one of several possible flow instabilities. Sensitivity alone seems incapable of explaining the vigorous ice discharge associated with Heinrich events; an instability must be triggered.

Rather than dismiss sensitivity as an unique feature of Hudson Strait Ice Stream, we propose that under certain circumstances it could be a commonplace one. As cold glaciers and ice sheets thicken, they change from a predominantly polar thermal regime (Figure 10a) to a predominantly sub-polar thermal regime (Figure 10b). By itself this transition is not likely to be accompanied by a dramatic change in flow rate, thus thickening might continue until strain heating became strong enough to render the ice mass sensitive and thus responsive to changes in surface loading. This line of thought is reminiscent of self-organized criticality [*Bak et al.,* 1987, 1988] and it is our conjecture that, during epochs when global ice volume is increasing, a substantial fraction of the existing glaciers and ice sheets evolve to a sensitive state. If this speculation is correct then the seemingly paradoxical evidence for a quasi-synchronous climate-orchestrated response of unlinked glacial systems [e.g., *Elliot et al.,* 1998; *Grousset et al.,* 1998; *van Kreveld et al.,* 1998] might be explained.

We offer no proof of these assertions but believe they are amenable to testing using appropriately designed computer models. Indeed, self-organization, though not self-organized criticality, has been demonstrated by the three-dimensional thermomechanical ice sheet model of *Payne and Dongelmans* [1997].

MODELING OF SURGING AND TIDEWATER INSTABILITY MECHANISMS

Computer modeling has provided a valuable tool for examining the viability of glaciological models of Hudson Strait ice dynamics. Reduced models [*MacAyeal,* 1993a,b; *Verbitsky and Saltzman,* 1995] are useful for isolating process interactions and identifying promising research directions. Flow-line models [*Greve and MacAyeal,* 1996] and non-thermodynamic two-dimensional flow models [*Pfeffer et al.,* 1997] add realism without making unreasonable demands on computing power. The first attempt to embed surges of Hudson Strait Ice Stream within a full thermomechanical model of Laurentide Ice Sheet dynamics is described by *Marshall and Clarke* [1997a,b]. Sensitivity, in the sense we have discussed, is not a feature of the Marshall and Clarke (MC) model nor of other conventional ice dynamics models. Indeed sensitivity presents a modeling challenge because the phenomenon is associated with a thermal boundary layer and these are poorly resolved by existing coarse-gridded thermomechanical models.

Details of the MC ice stream surge model are fully described in *Marshall and Clarke* [1997b] and here we simply review the main results. Figure 13 shows the ice surface topography and barycentric ice surface velocity predicted by the MC reference model 1 kyr after the surge onset. The fast flow channel through Hudson Strait is clearly apparent from the velocity vectors (Figure 13b) as is the deflection of flow from surrounding sheet ice toward the ice stream. For this model the calculated surge duration was 750 yr and the duration of the quiescent period was found to be 4500 yr, values that are broadly consistent with those extracted from the paleoceanographic record. The maximum outlet velocity was found to be $3320\,\mathrm{m\,yr^{-1}}$, considerably less the $18000\,\mathrm{m\,yr^{-1}}$ predicted by *MacAyeal's* [1993b] binge/purge model. Similarly the predicted freshwater flux (icebergs plus meltwater) for the MC reference model is $102\,\mathrm{km^3\,yr^{-1}}$ ($0.0032\,\mathrm{Sv}$) in contrast to MacAyeal's value of $2479\,\mathrm{km^3\,yr^{-1}}$ ($0.079\,\mathrm{Sv}$). Even the MC maximum model [*Marshall and Clarke,* 1997b], aimed at placing physically reasonable upper limits on flow velocity ($6680\,\mathrm{m\,yr^{-1}}$) and freshwater discharge ($278\,\mathrm{km^3\,yr^{-1}}$ or $0.0088\,\mathrm{Sv}$) falls far short of the binge/purge predictions. Other points of comparison are provided by freshwater discharge estimates based on ocean sedimentation measurements and with the results of ocean dynamics modeling experiments. The net freshwater flux of the MC maximum model compares very

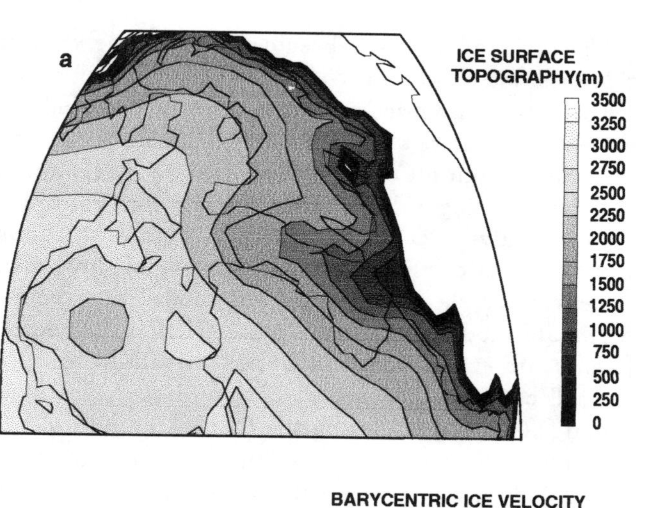

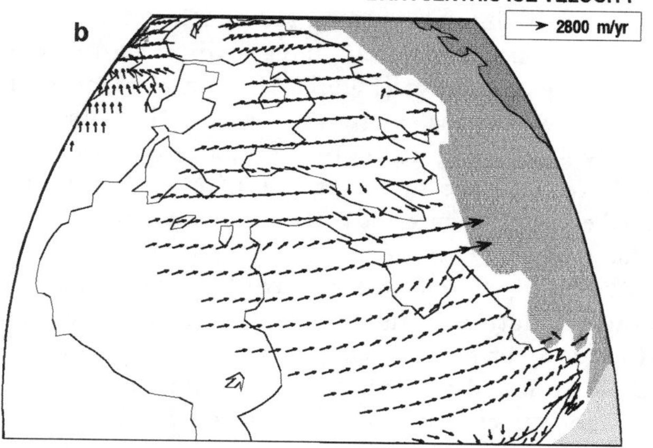

Figure 13. Model of Hudson Strait Ice Stream surge. (a) Ice surface topography 1 kyr after surge onset. (b) Barycentric (a weighted average) ice surface velocity 1 kyr after surge onset.

favorably to that estimated by *Dowdeswell et al.* [1995] based on measurements of the thickness of ice-rafted sediment for events H1 and H2. Several ocean dynamics modeling studies have explored the effect on thermohaline circulation (THC) of increased freshwater input to the North Atlantic. *Rahmstorf* [1995] examined bifurcations in the equilibrium response rather than sensitivity to event-like disturbances; he found that an increase of 0.075 Sv was sufficient to switch the THC from its present value of ~17 Sv to essentially full shutdown. *Manabe and Stouffer* [1995] assumed a 1 Sv increase in freshwater input, sustained for 10 yr, and triggered an abrupt weakening in THC followed by a complicated response. The study of *Weaver* [this volume] is most directly relevant to Heinrich events and is based on the MC estimates of freshwater flux. For the MC reference model the THC decreases from an LGM rate of 11 Sv to ~5 Sv; for the maximum model the rate drops to ~3.5 Sv.

The MC reference model predicts cyclic oscillations in geometry, temperature, and flow rate (Figure 14) for a constant climate. Figure 14a shows the simulated ice velocity at Hudson Strait outlet. At the surge onset, velocity rapidly increases from near zero to 3320 m yr^{-1} then slowly decreases to around 1800 m yr^{-1} before the surge abruptly shuts down. Basal temperature also

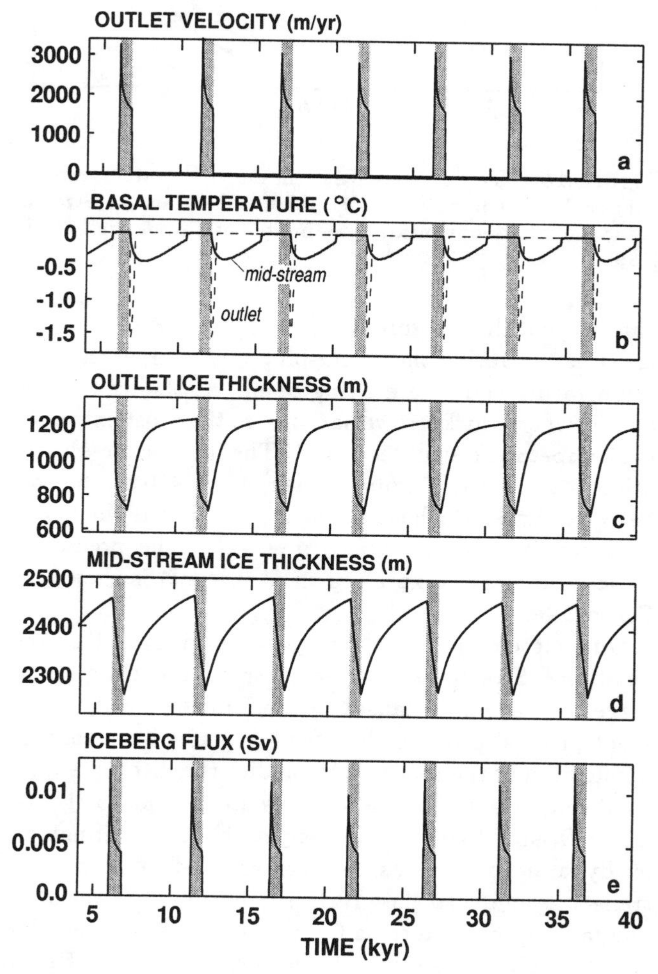

Figure 14. Time series results of Hudson Strait Ice Stream surge model. (a) Ice surface velocity at the ice stream outlet. (b) Basal temperature at the ice stream outlet and at a site halfway between the outlet and the head of the ice stream. (c) Ice thickness at the ice stream outlet. (d) Ice thickness at a site halfway between the outlet and the head of the ice stream. (e) Iceberg flux from ice stream.

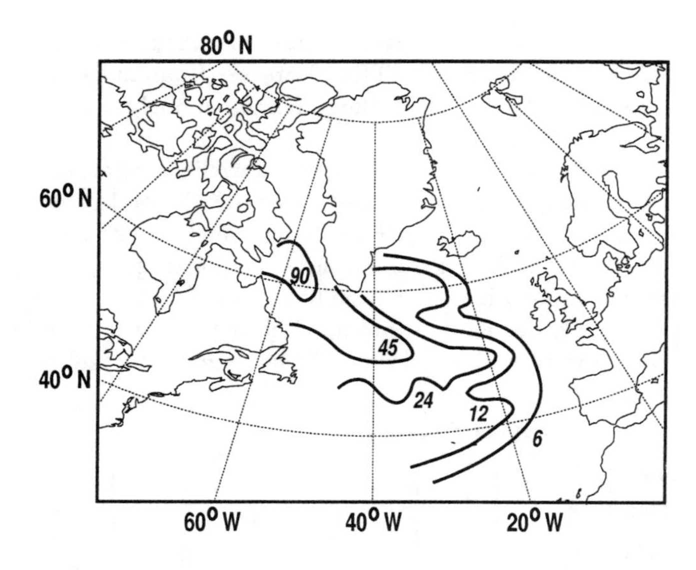

Figure 15. Iceberg melt rates (in cm yr^{-1} water equivalent) based on computer model of Hudson Strait Ice Stream surge and the sedimentation estimates of *Dowdeswell et al.* [1995].

varies cyclically (Figure 14b) along the length of the ice stream. During the active phase of a surge, the bed temperature is at the melting point. At the ice stream outlet, where thickness variations are the most extreme, the temperature swing is 1.5°C. The ice thickness variation at the outlet (Figure 14) is roughly 450 m whereas halfway toward the head of the ice stream the thickness variation is only 200 m (Figure 14d). The predicted iceberg discharge (expressed in Sverdrups) is shown in Figure 14e.

Our success in modeling cyclic surging of Hudson Strait Ice Stream lends credibility to the claim that surges are a viable causative mechanism for Heinrich events, as first proposed by *MacAyeal* [1993b]. A second application of the model is to predict the discharge and total volume of icebergs and freshwater to the North Atlantic Ocean (Figure 14e). As yet the MC continental ice dynamics model has not been coupled to an iceberg transport model so the areal distribution of freshwater flux, an important forcing for ocean circulation models, is not predicted. As an interim measure we use Heinrich sedimentation estimates [*Dowdeswell et al.,* 1995] as a proxy for iceberg melt, and areally distribute the iceberg water flux predicted by the MC model in proportion to the IRD sedimentation estimates. Figure 15 shows the results of this calculation. The reader is referred to *Weaver* [this volume] for the results of computer simulations of the ocean circulation response to this freshwater forcing.

The MC model has also been used to examine whether tidewater instability is a viable glaciological mechanism for explaining Heinrich events. Models of this kind are extremely sensitive to the form of the calving law as well as to the magnitude of the rate constant k_c in Equation (1); by choosing a large value for the rate constant, vigorous calving can be promoted. Figure 16 shows how vigorous calving affects the growth of the North American ice sheet. Here the MC model is forced by a climate model controlled by the GRIP $\delta^{18}O$ record [*Dansgaard et al.,* 1993] and GCM temperature and $P-E$ fields computed for present day and last glacial maximum (LGM) climate using the CCCv2.0 model of the Canadian Climate Centre for Modelling and Analysis. Details are given in *Marshall et al.* [submitted]. The ice dynamics simulation is started at 120 kyr BP from a fully-deglaciated state and integrated forward to the present day. At 50 kyr (Figures 16a and 16b) much of Hudson Bay remains ice-free and open to the Labrador Sea. By 40 kyr (Figure 16c) this embayment has been greatly reduced and by 20 kyr (the LGM) it has disappeared entirely (Figure 16d). One immediate conclusion of this modeling exercise is that it would be difficult to justify a more vigorous calving law than that employed for this simulation because more vigorous calving would further delay or prevent ice coverage of Hudson Bay.

Close examination of these simulation results reveals that minor fluctuations of the Hudson Strait calving margin occur as the large-scale trends of infilling and formation of an ice dome proceed. Figure 17 shows the changes in surface topography, ice thickness, surface velocity, and the sum of basal and internal melt rates that accompany one such fluctuation. The values presented in this figure are obtained by averaging modeling results over the latitude range 60–64°N (six cell widths). The calving front retreat of 2° longitude (~110 km) occurs over the time interval 20 kyr to 16 kyr. In fact the active retreat occurs over a much smaller time interval, from 18.00 kyr to 17.95 kyr. The rapid 200 km retreat of a tidewater glacier would, by modern measures, seem impressive. Relative to the ice volume requirements of Heinrich events, the simulated tidewater retreat is completely inadequate.

Our modeling results raise serious questions about the viability of tidewater disintegration as an explanation of large episodic iceberg fluxes from Hudson Strait. A more vigorous calving law might promote a more impressive retreat of the calving margin but would also delay or entirely prevent the formation of an ice dome covering Hudson Bay. One could escape this difficulty by switching between a weak and strong calving law

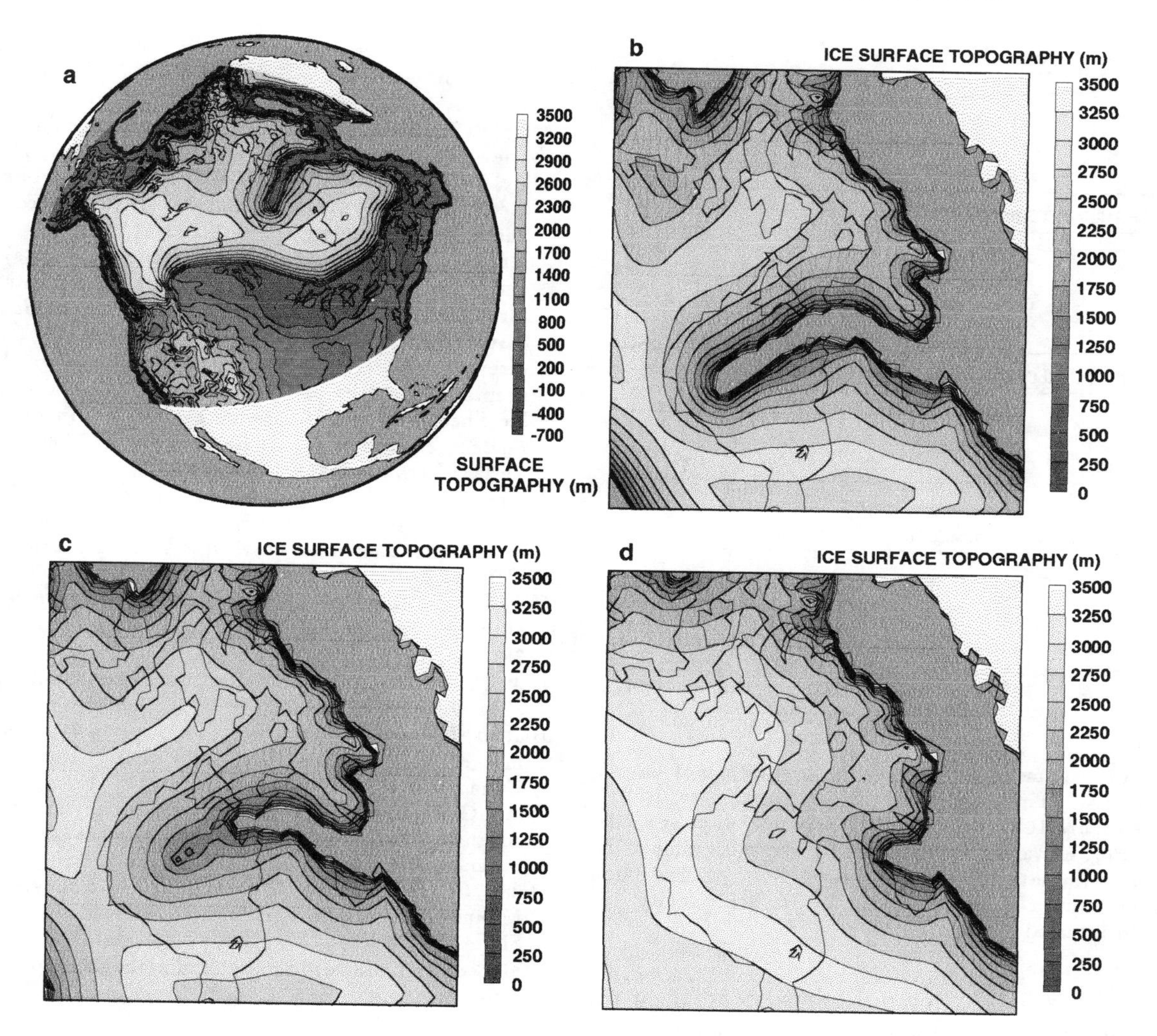

Figure 16. Evolution of North American Ice Sheet if vigorous calving is assumed. (a) Map of North American Ice Sheet at 50 kyr BP. (b) Detail of ice distribution in Hudson Strait region at 50 kyr BP. (c) Detail of ice distribution in Hudson Strait region at 40 kyr BP. (d) Detail of ice distribution in Hudson Strait region at 20 kyr BP.

but this "hand-of-God" expedient has no clear physical justification. Given the state of knowledge of calving physics and the subgrid nature of the processes governing calving and grounding line retreat, it is premature to dismiss the tidewater instability mechanism.

CONCLUSIONS

The paleoceanographic record at Orphan Knoll supports the assertion that there is no premonitory climate signal associated with Heinrich events. The first expression of Heinrich event onset is increased submarine sediment transport along the Northwest Atlantic Mid-Oceanic Channel that links the ice source at Hudson Strait with the Orphan Knoll core site. A delayed Heinrich event signal is the increased sedimentation of ice-rafted debris and dilution of near-surface waters by iceberg melt. "Normal" ice sheet and ice stream flow rates are believed to be insufficient for explaining the large volumes of ice and sediment exported by Heinrich events. It is therefore likely that the onset of a Heinrich event occurs when a glacier flow instability is triggered. Based on our assessment of the merits and shortcomings of candidate instability mechanisms as well as attempts to model the surging and tidewater instability mechanisms, the preferred explanation of increased ice flux through Hudson Strait is that Hudson Strait Ice Stream surges episodically. Although small-scale tidewater instability events can be simulated, large-scale instability is difficult or impossible to excite unless we manipulate

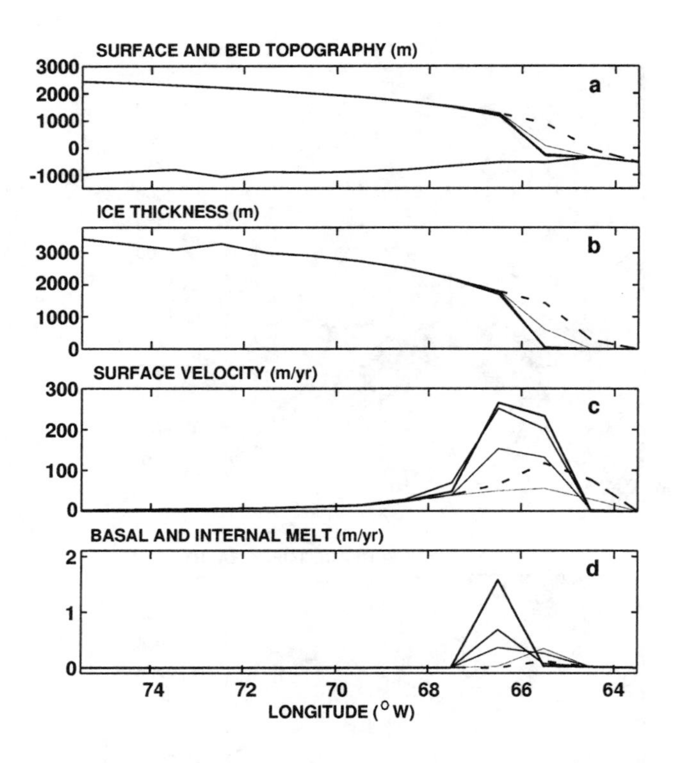

Figure 17. Time slices showing the occurrence of tidewater instability. The simulation run is the same as used to generate Figure 16 and assumes an extremely vigorous calving law. Plotted values are obtained by averaging simulation results over the latitude interval 60–64°N. (a) Surface and bed topography. (b) Ice thickness. (c) Surface velocity. (d) Basal and internal melt rates.

the ice calving law. Incomplete understanding of the calving process, rather than non-viability of the tidewater instability mechanism, may lie at the heart of this difficulty. Finally, we speculate that during the build-up to a Heinrich event, phase-locking between an atmospheric forcing applied to the ice surface and subglacial meltwater production, a necessary process for activating fast flow, can be achieved if the ice mass is in sensitive condition, i.e., if the ice–bed contact is at the melting temperature and strain heating is appreciable.

Acknowledgments. This paper is a contribution to Phase 2 of the Climate System History and Dynamics Program (CSHD) that is jointly sponsored by the Natural Sciences and Engineering Research Council of Canada (NSERC) and the Atmospheric Environment Service of Canada. We greatly appreciate the helpful criticism of Tad Pfeffer and an anonymous reviewer whose comments have substantially improved this submission.

REFERENCES

Ahlmann, H. W., Contribution to the physics of glaciers, *Geogr. J., 86,* 97-113, 1935.

Alley, R. B., Water-pressure coupling of sliding and bed deformation: II. Velocity-depth profiles, *J. Glaciol., 35*(119), 119-129, 1989.

Alley, R. B., Sedimentary processes may cause fluctations of tidewater glaciers, *Ann. Glaciol., 15,* 119-124, 1991.

Alley, R. B., and D. R. MacAyeal, Ice-rafted debris associated with binge/purge oscillations of the Laurentide Ice Sheet, *Paleoceanography, 9*(4), 503-511, 1994.

Andrews, J. T., Abrupt changes (Heinrich events) in late Quaternary North Atlantic marine environments: a history and review of data and concepts, *J. Quat. Sci., 13*(1), 3-16, 1998.

Andrews, J. T., K. Tedesco, and A. E. Jennings, Heinrich events: chronology and processes, east-central Laurentide Ice Sheet and NW Labrador Sea, in *Ice in the Climate System,* edited by W. R. Peltier, pp. 167-186, Springer-Verlag, Berlin, 1993.

Andrews, J. T., K. Tedesco, W. M. Briggs, and L. W. Evans, Sediments, sedimentation rates, and environments, southeast Baffin Shelf and northwest Labrador Sea, 8–26 ka, *Can. J. Earth Sci., 31,* 90-103, 1994.

Bak, P., C. Tang, and K. Wiesenfeld, Self-organized criticality: An explanation of $1/f$ noise, *Phys. Rev. Lett., 59,* 381-384, 1987.

Bak, P., C. Tang, and K. Wiesenfeld, Self-organized criticality. *Phys. Rev. A, 38,* 364-374, 1988.

Blatter, H., and Kappenberger, G., Mass balance and thermal regime of Laika Ice Cap, Coburg Island, N.W.T., Canada, *J. Glaciol., 34,* 102-110, 1988.

Bond, G. C., and R. Lotti, Iceberg discharges into the North Atlantic on millenial time scales during the last glaciation, *Science, 267,* 1005-1010, 1995.

Bond, G., H. Heinrich, W. Broecker, L. Labeyrie, J. McManus, J. Andrews, S. Huon, R. Jantschik, S. Clasen, C. Simet, K. Tedesco, M. Klas, G. Bonani, and S. Ivy, Evidence for massive discharges of icebergs into the North Atlantic ocean during the last glacial period, *Nature, 360,* 245- 249, 1992.

Bond, G., W. Broecker, S. Johnsen, J. McManus, L. Labeyrie, J. Jouzel, and G. Bonani, Correlations between climate records from North Atlantic sediments and Greenland ice, *Nature, 365,* 143-147, 1993.

Bond, G., W. Showers, M. Cheseby, R. Lotti, P. Almansi, P. deMenocal, P. Priore, H. Cullen, I. Hajdas, and G. Bonani, A pervasive millenial-scale cycle in North Atlantic Holocene and glacial climates, *Science, 278,* 1257-1266, 1997.

Broecker, W. S., Massive iceberg discharges as triggers for global climate change, *Nature, 372,* 421-424, 1994.

Broecker, W., G. Bond, M. Klas, E. Clark, and J. McManus, Origin of the northern Atlantic's Heinrich events, *Clim. Dyn., 6,* 265-273, 1992.

Brown, C. S., W. G. Sikonia, A. Post, L. A. Rasmussen, and M. F. Meier, Two calving laws from grounded iceberg-calving glaciers, *Ann. Glaciol., 4,* 295, 1983.

Clarke, G. K. C., Thermal regulation of glacier surging, *J. Glaciol., 16*(74), 231-250, 1976.

Clarke, G. K. C., and E. W. Blake, Geometric and thermal evolution of a surge-type glacier in its quiescent state: Trapridge Glacier 1969–89, *J. Glaciol., 37*(125), 158-169, 1991.

Clarke, G. K. C., U. Nitsan, and W. S. B. Paterson, Strain

heating and creep instability in glaciers and ice sheets, *Rev. Geophys., 15*(2), 235-247, 1977.

Classen, D. F., Temperature profiles for the Barnes Ice Cap surge zone, *J. Glaciol., 18,* 391-405, 1977.

Clayton, L., J. T. Teller, and J. W. Attig, Surging of the southwestern part of the Laurentide ice sheet, *Boreas, 14,* 235-242. 1985.

Dansgaard, W., et al., Evidence for general instability of past climate from a 250-kyr ice-core record, *Nature, 364,* 218-220, 1993.

Doake, C. S. M., H. F. J. Corr, H. Rott, P. Skvarca, and N. W. Young, Breakup and conditions for stability of the northern Larsen Ice Shelf, Antarctica, *Nature, 391*, 778-780, 1998.

Dowdeswell, J. A., M. A. Maslin, J. T. Andrews, and I. N. McCave, Iceberg production, debris rafting, and the extent and thickness of Heinrich layers (H-1, H-2) in North Atlantic sediments, *Geology, 23*(4), 301-304, 1995.

Elliot, M., L. Labeyrie, G. Bond, E. Cortijo, J.-L. Turon, N. Tisherat, and J.-C. Duplessy, Millenial-scale iceberg discharges in the Irminger Basin during the last glacial period: Relationship with the Heinrich events and environmental settings, *Paleoceanography, 13,* 433-446, 1998.

Fowler, A. C., and D. A. Larson, On the flow of polythermal glaciers I. Model and preliminary analysis. *Proc. R. Soc. London A, 363*, 217-242, 1978.

Greve, R., and D. R. MacAyeal, Dynamic/thermodynamic simulations of Laurentide ice-sheet instability, *Ann. Glaciol., 23,* 328-335, 1996.

Grousset, F. E., H. Snoeckx, and M. Revel, Did the European Ice Sheet surges trigger the North Atlantic Heinrich events? Poster presentation at AGU Chapman Conference on "Mechanisms of Millenial-Scale Global Climate Change", Snowbird, Utah, June 14-18, 1998.

Gwiazda, R. H., S. R. Hemming, and W. S. Broecker, Tracking the sources of icebergs with lead isotopes: The provenance of ice-rafted debris in Heinrich layer 2, *Paleoceanography, 11,* 77-93, 1993a.

Gwiazda, R. H., S. R. Hemming, and W. S. Broecker, The provenance of icebergs during in Heinrich event 3 and the contrast to their sources during other Heinrich episodes, *Paleoceanography, 11,* 371-378, 1993b.

Heinrich, H., Origin and consequences of cyclic ice rafting in the Northeast Atlantic Ocean during the past 130,000 years, *Quat. Res., 29,* 142-152, 1988.

Hesse, R., and A. Rakofsky, Deep-sea channel/submarine-Yazoo system of the Labrador Sea: a new deep-water facies model, *Amer. Assoc. Petrol. Geol. Bull., 76,* 680-707, 1992.

Hillaire-Marcel, C., A. de Vernal, G. Bilodeau, and G. Wu, Isotope stratigraphy, sedimentation rates, deep circulation, and carbonate events in the Labrador Sea during the last ~200 ka, *Can. J. Earth Sci., 31,* 63-89, 1994a.

Hillaire-Marcel, C., A. de Vernal, M. Lucotte, A. Mucci, G. Bilodeau, A. Rochon, S. Vallières, and G.-P. Wu, Productivité et flux de carbone dans la mer du Labrador au cours des derniers 40 000 ans, *Can J. Earth Sci., 31,* 139-158, 1994b.

Hulbe, C. L., An ice shelf mechanism for Heinrich layer production, *Paleoceanography, 12*(5), 711-717, 1997.

Hunt, A. G., and P. E. Malin, Possible triggering of Heinrich events by ice-load-induced earthquakes, *Nature, 393,* 155-158, 1998.

Hutter, K., A mathematical model of polythermal glaciers and ice sheets. *Geophys. Astrophys. Fluid Dynamics, 21,* 201-224, 1982.

Iken, A., and R. A. Bindschadler, Combined measurements of subglacial water pressure and surface velocity of Findelengletscher, Switzerland: conclusions about drainage system and sliding mechanism, *J. Glaciol., 32,* 101-119, 1986.

Iverson, N. R., Regelation of ice through debris at glacier beds: Implications for sediment transport, *Geology, 21,* 559-562, 1993.

Iverson, N. R., and D. J. Semmens, Intrusion of ice into porous media by regelation: A mechanism of sediment entrainment by glaciers, *J. Geophys. Res., 100*(B7), 10,219-10,230, 1995.

Iverson, N. R., B. Hanson, R. LeB. Hooke, and P. Jansson, Flow mechanism of glaciers on soft beds, *Science, 267,* 80-81, 1995.

Kamb, B., Rheological nonlinearity and flow instability in the deforming bed mechanism of ice stream motion, *J. Geophys. Res., 96*(B10), 16,585–16,595, 1991.

Kamb, B., C. F. Raymond, W. D. Harrison, H. Engelhardt, K. A. Echelmeyer, N. Humphrey, M. M. Brugman, and T. Pfeffer, Glacier surge mechanism: 1982-1983 surge of Variegated Glacier, Alaska, *Science, 227*(4686), 469-479, 1985.

Kamb, B., H. Engelhardt, M. Fahnestock, N. Humphrey, M. Meier, and D. Stone, Mechanical and hydrologic basis for the rapid motion of a large tidewater glacier 2. Interpretation, *J. Geophys. Res., 99*(B8), 15,231-15,244, 1994.

Kaufman, D. S., G. H. Miller, J. A. Stravers, and J. T. Andrews, Abrupt early Holocene (9.9–9.6 ka) ice-stream advance at the mouth of Hudson Strait, Arctic Canada, *Nature, 21,* 1063-1066, 1993.

Lucotte, M., and C. Hillaire-Marcel, Identification et distribution des grandes masses d'eau dans les mers du Labrador et d'Irminger, *Can. J. Earth Sci., 31,* 5-13, 1994.

MacAyeal, D. R., A low-order model of the Heinrich event cycle, *Paleoceanography, 8*(6), 767-773, 1993a.

MacAyeal, D. R., Binge/purge oscillations of the Laurentide ice sheet as a cause of the North Atlantic's Heinrich events, *Paleoceanography, 8*(6), 775-784, 1993b.

Manabe, S., and R. J. Stouffer, Simulation of abrupt climate change induced by freshwater input to the North Atlantic Ocean, *Nature, 378,* 165-167, 1995.

Marshall, S. J., and G. K. C. Clarke, A continuum mixture model of ice stream thermomechanics in the Laurentide Ice Sheet 1. Theory, *J. Geophys. Res., 102*(B9), 20,599-20,613, 1997a.

Marshall, S. J., and G. K. C. Clarke, A continuum mixture model of ice stream thermomechanics in the Laurentide Ice Sheet 2. Application to the Hudson Strait Ice Stream, *J. Geophys. Res., 102*(B9), 20,615-20,637, 1997b.

Marshall, S. J., L. Tarasov, G. K. C. Clarke, and W. R. Peltier, Glaciology of Ice Age cycles: Physical processes and modelling challenges, *Can. J. Earth Sci.,* submitted.

Meier, M. F., and A. Post, Fast tidewater glaciers, *J. Geophys. Res., 92*(B9), 9051-9058, 1987.

Meier, M., S. Lundstrom, D. Stone, B. Kamb, H. Engelhardt, N. Humphrey, W. W. Dunlap, M. Fahnestock, R.

M. Krimmel, and R. Walters, Mechanical and hydological basis for the rapid motion of a large tidewater glacier 1. Observations, *J. Geophys. Res., 99*(B9), 15,219-15,229, 1994.

Paterson, W. S. B., *The Physics of Glaciers*, Pergamon, Tarrytown, N. Y., 1994.

Paterson, W. S. B., Some aspects of the physics of glaciers, in *Ice Physics in the Natural Environment*, edited by J. S. Wettlaufer, J. G. Dash, and N. Untersteiner, pp. 69-88, NATO Advanced Study Institute, *56, Ser. I*, Springer-Verlag, Berlin, 1998.

Payne, A. J., and P. W. Dongelmans, Self-organization in the thermomechanical flow of ice sheets, *J. Geophys. Res., 102*(B6), 12,219-12,233, 1997.

Pfeffer, W. T., M. Dyurgerov, M. Kaplan, J. Dwyer, C. Sassolas, A. Jennings, B. Raup, and W. Manley, Numerical modeling of late Glacial Laurentide advance of ice across Hudson Strait: Insights into terrestrial and marine geology, mass balance, and calving flux, *Paleoceanography, 12*(1), 97-110, 1997.

Phillip, J. R., Thermal field during regelation, *Cold Regions Sci. Technol., 3*, 193-203, 1980.

Post, A. S., Alaskan glaciers: recent observations in respect to the earthquake-advance theory, *Science, 148*, 366-368, 1965.

Rahmstorf, S., Bifurcations of the Atlantic thermohaline circulation in response to changes in the hydrological cycle, *Nature, 378*, 145-149, 1995.

Raymond, C. F., How do glaciers surge? A review, *J. Geophys. Res., 92*(B9), 9121-9134, 1987.

Raymond, C., T. Johannesson, T. Pfeffer, and M. Sharp, Propagation of a glacier surge into stagnant ice, *J. Geophys. Res., 92*(B9), 9037-9049, 1987.

Reeh, N., On the calving of ice from floating glaciers and ice shelves, *J. Glaciol., 7*, 215-232, 1968.

Reeh, N., C. E. Bøggild, and H. Oerter, Surge of Storstrømmen, a large outlet glacier from the Inland Ice of North-East Greenland, *Rapp. Grønlands geol. Unders., 162*, 201-209, 1994.

Roberts, A. P., B. Lehman, R. J. Weeks, K. L. Verosub, and C. Laj, Relative paleointensity of the geomagnetic-field over the last 200,000 years from ODP sites 883 and 884, North Pacific Ocean. *Earth Planet. Sci. Lett., 152*, 11-23, 1997.

Ruddiman, W. F., Late Quaternary deposition of ice-rafted sand in the subpolar North Atlantic (lat 40° to 65°N), *Geol. Soc. Am. Bull., 88*, 1813-1827, 1977.

Russell-Head, D. S., The melting of free-drifting icebergs, *Ann. Glaciol., 1*, 119-127, 1980.

Saussure, H.-B. de, *Voyages dan les Alpes, précédés d'un essai sur l'histoire naturelle des environs de Geneve* (4 vols.), S. Fauche, Neuchâtel, Switzerland, 1779-96.

Schytt, V., Some comments on glacier surges in eastern Svalbard, *Can. J. Earth Sci., 6*, 867-871, 1969.

Sharp, M. J., 'Crevasse-fill' ridges: a landform type characteristic of surging glaciers? *Geografiska Annaler, 67A*, 213-220, 1985.

Stoner, J. S., J. E. T. Channell, C. Hillaire-Marcel, The magnetic signature of rapidly deposited detrital layers from the deep Labrador Sea: Relationship to North Atlantic Heinrich layers, *Paleoceanography, 11*(3), 309-325, 1996.

Stoner, J. S., J. E. T. Channell, and C. Hillaire-Marcel, A 200 ka geomagnetic chronostratigraphy for the Labrador Sea: Indirect correlation of the sediment record to SPECMAP, *Earth Planet. Sci. Let.*, 165-181, 1998.

Tarr, R. S., and L. Martin, *Alaska Glacier Studies*, National Geographic Society, Washington, D.C., 1914.

van Kreveld, S., U. Pflaumann, M. Sarnthein, P. Grootes, and M. J. Nadeau, Correlation between temperature records from Reykjanes Ridge sediment and GISP2 ice-core, Poster presentation at AGU Chapman Conference on "Mechanisms of Millenial-Scale Global Climate Change", Snowbird, Utah, June 14-18, 1998.

Veiga-Pires, C., and C. Hillaire-Marcel, U-Th constraints on the duration of Heinrich events H0 to H4 in southeastern Labrador Sea, *Paleoceanography,* under revision.

Vaughn, D. G., and C. S. M. Doake, Recent atmospheric warming and retreat of ice shelves on the Antarctic Peninsula, *Nature, 379*, 328-331, 1996.

Verbitsky. M., and B. Saltzman, A diagnostic analysis of Heinrich glacial surge events, *Paleoceanography, 10*, 59-66, 1995.

Weaver, A. J., Millenial timescale variability in ocean/climate models, this volume.

Weertman, J., Mechanism for the formation of inner moraines found near the edge of cold ice caps and ice sheets, *J. Glaciol., 3*(30), 965-978, 1961.

Wright, H. E., Tunnel values, glacial surges and subglacial hydrology of the Superior Lobe, Minnesota, in *The Wisconsinan Stage*, edited by R. F. Black, R. P. Goldthwait, and H. B. Williams, Geol. Soc. Am. Memoir 36, 251-276, 1973.

Wu, G.-P., and C. Hillaire-Marcel, AMS radiocarbon stratigraphies in deep Labrador Sea cores: paleoceanographic implications, *Can. J. Earth Sci. 31*, 38-47, 1994a.

Wu, G.-P., and C. Hillaire-Marcel, Oxygen isotope compositions of sinistral Neogloboquadrina pachyderma tests in surface sediments: North Atlantic Ocean, *Geochim. Cosmochim. Acta, 58*, 1303-1312, 1994b.

Yuen, D. A., and G. Schubert, The role of shear heating in the dynamics of large ice masses, *J. Glaciol. 24*(90), 195-212, 1979.

G. K. C. Clarke and S. J. Marshall, Department of Earth and Ocean Sciences, 2219 Main Mall, University of British Columbia, Vancouver, B.C., V6T 1Z4, Canada. (e-mail: clarke@eos.ubc.ca; marshall@eos.ubc.ca)

C. Hillaire-Marcel, G. Bilodeau, C. Veiga-Pires, GEOTOP, Université du Québec à Montréal, Montréal, Québec, H3C 3P8, Canada. (e-mail: chm@uqam.ca)

Physical and Biogeochemical Responses to Freshwater-Induced Thermohaline Variability in a Zonally Averaged Ocean Model

Olivier Marchal, Thomas F. Stocker, and Fortunat Joos

Climate and Environmental Physics, Physics Institute, University of Bern

Freshwater perturbation experiments are conducted with a latitude-depth, circulation-biogeochemistry ocean model coupled to an energy balance model of the atmosphere. The aim is to identify potential effects of different changes of the Atlantic thermohaline circulation (THC). Strong THC reductions (> 50%) lead to cooling at high northern latitudes and warming in the southern hemisphere. For moderate reductions, however, cooling in the north is not accompanied by temperature changes in the south. These results are discussed in relation with a recent synchronization of isotopic records from Greenland and Antarctic ice cores based on methane, which documents north-south thermal antiphasing during the largest Greenland $\delta^{18}O$ oscillations and no clear Antarctic counterparts during the other, shorter oscillations of the last glacial period. Simulations show that strong THC reductions result in PO_4 enrichment and $\delta^{13}C$ depletion below 1 km in the North Atlantic reaching, on average, about 0.5 $mmol\,m^{-3}$ and –0.3‰ for a complete THC collapse. These chemical and isotopic changes are due to an imbalance between organic matter oxidation and import of nutrient-poor waters from the northern North Atlantic. The THC reductions also lead to a drop in $\delta^{13}C$ air-sea disequilibrium in the Atlantic where the surface waters stay longer in contact with the atmosphere. Thus, in the upper kilometer, cold waters in the northern North Atlantic become isotopically heavier (by more than 1‰), whereas warm waters further south become slightly lighter (~ –0.2‰). The simulated chemical and isotopic shifts are much smaller below 1 km in the South Atlantic and Southern Ocean. These results indicate that the same circulation change could produce completely different PO_4 and $\delta^{13}C$ anomalies at different locations and depths in the Atlantic and Southern Ocean. This might have strong implications for the interpretation of marine Cd/Ca and $\delta^{13}C$ sediment records obtained from different oceanic regions.

1. INTRODUCTION

A major area of current climate research is the study of the large, millennial-scale variability observed in many paleoclimate records. The prominent and abrupt $\delta^{18}O$ oscillations observed in Greenland ice cores during the

Mechanisms of Global Climate Change at Millennial Time Scales
Geophysical Monograph 112

last glacial period [*Dansgaard et al.*, 1982; *Oeschger et al.*, 1984; *Johnsen et al.*, 1992] prompted a search for such variability in records from various parts of the climate system. When translated into temperature (T) changes from the modern spatial relationship between the annual mean snow $\delta^{18}O$ and T in polar regions, these "Dansgaard-Oeschger" (D-O) events correspond to warming-cooling cycles with an amplitude of $\sim$ 7°C [*Johnsen et al.*, 1992]. The estimated temperature changes are even larger if the relationship is calibrated by borehole temperature measurements [*Johnsen et al.*, 1995; *Cuffey and Clow*, 1997]. The best investigated abrupt oscillation in Greenland $\delta^{18}O$ records is the Younger Dryas cold event (YD), dated by annual layer counting to between 12,700±100 yr BP–11,550±70 yr BP in the GRIP ice core [*Johnsen et al.*, 1992] and 12,940±260 yr BP–11,640±250 yr BP in the GISP2 ice core [*Alley et al.*, 1993]. The fractionation of nitrogen and argon isotopes at the transition from the YD to the Preboreal in the GISP2 core indicates that Central Greenland was 15±3°C colder during the Younger Dryas than today [*Severinghaus et al.*, 1998].

There has been a long controversy about whether the Younger Dryas has been a global event or not. Continental pollen sequences reveal that the YD affected western Europe and eastern North America, but do not provide definitive evidence of the event in other regions [*Peteet*, 1995]. Gas measurements on polar ice cores, on the other hand, document that the atmospheric CH_4 concentration followed Greenland $\delta^{18}O$ shifts of the last glaciation, including the YD [*Chappellaz et al.*, 1993; *Brook et al.*, 1996]. These measurements suggest that the D-O events were at least hemispheric in extent, as a main source of atmospheric CH_4 during the last glacial period was tropical wetlands [*Chappellaz et al.*, 1993]. Climatic records from glacier stands in New Zealand [*Denton and Hendy*, 1994] and Southern Chile [*Lowell et al.*, 1995] document notable glacier advances during the last deglaciation. If glacier stands are faithful recorders of temperature only, then the advances come close in time to the Younger Dryas. *Denton and Hendy* [1994] concluded thus, that Younger Dryas must be a cold event of global extent, possibly with an enhanced signal in the north. However, uncertainties in the time scale and interpretation of wood remains in moraines [*Mabin*, 1996] as well as recent analyses of pollen assemblages [*Singer et al.*, 1998] do not support this conclusion.

Measurements of the $^{18}O/^{16}O$ ratio of O_2 and of the CH_4 content in air trapped in ice permits the synchronization of climate records from both polar regions. *Bender et al.* [1994a] placed the δD record from Vostok (East Antarctica) and the GISP2 $\delta^{18}O$ record on a common chronology based on $^{18}O/^{16}O$ measurements on O_2. They showed that the last glacial Greenland interstadials longer than $\sim$ 2,000 yr have counterparts in Antarctica. The same approach was adopted by *Sowers and Bender* [1995] who demonstrated that warming in the Byrd $\delta^{18}O$ record (West Antarctica) began approximately 3,000 yr before the onset of the warm Bølling epoch in the GISP2 $\delta^{18}O$ record. Past changes in the $^{18}O/^{16}O$ ratio of atmospheric oxygen, however, were relatively small, which prevents the determination of the phase relationships between Greenland and Antarctica during the abrupt climate changes of the last glacial period [*Bender et al.*, 1994b]. By contrast, fast CH_4 variations during this period permits a fit of ice core CH_4 records from both polar regions to ± 200 yr [*Blunier et al.*, 1998]. *Blunier et al.* [1997] and *Blunier et al.* [1998] used CH_4 measured in GRIP, Byrd, and Vostok ice cores to put the isotopic records from these different sites on the same timescale. They found a conspicuous antiphase relationship between the high northern and southern latitudes during the prominent Greenland $\delta^{18}O$ events 8 and 12 [*Dansgaard et al.*, 1993] and the Bølling/Allerød/YD sequence. However, this relationship does not hold for all of the 14 D-O cycles of the last 50,000 years, as the other, shorter Greenland $\delta^{18}O$ events do not have clear counterparts in Antarctica. The possible wide geographic expression of the YD and other D-O events has profound implications for our understanding of climate dynamics, as it provides insight into the propagation of anomalies caused by abrupt changes in the atmosphere-ocean-ice system.

A possible cause of the millennial-scale, warming-cooling cycles of the last glacial period is changes in the Atlantic thermohaline circulation (THC) [*Broecker et al.*, 1985]. These changes may have been triggered, in turn, by the discharge of low-density meltwater from the big ice sheets which covered the northern hemisphere during this period. A major support to the THC hypothesis comes from various model simulations showing that a reduction of the THC causes a cooling in the North Atlantic region [*Wright and Stocker*, 1993; *Manabe and Stouffer*, 1995; *Manabe and Stouffer*, 1997; *Fanning and Weaver*, 1997; *Mikolajewicz et al.*, 1997; *Schiller et al.*, 1997]. Furthermore, consistent with polar isotopic records, climate models point to an antiphase relationship, whereby the south exhibits a warming when the North cools abruptly and vice versa (for a review see *Stocker*, 1999). The mechanism is simple: an active THC in the Atlantic draws heat from the Southern Ocean into the Atlantic basin [*Crowley*, 1992; *Stocker et al.*, 1992a]. If it is switched off, the excess heat tends to warm subsurface and surface waters in the South Atlantic and Southern Ocean. A sudden initiation of the THC, on the other hand, should lead to a cooling in the south. If the mechanism is indeed operating, this behaviour should be particularly present during the transitions into and out of the YD and other, similarly strong events.

Deep sea records of fossil benthic foraminiferal Cd/Ca and $\delta^{13}C$, however, have led to contradictory conclusions regarding the state of the THC during the Younger Dryas event [*Boyle and Keigwin*, 1987; *Keigwin et al.*, 1991; *Keigwin and Lehman*, 1994; *Jansen and Veum*, 1990; *Veum et al.*, 1992; *Charles and Fairbanks*, 1992; *Bard et al.*, 1994]. Data of the difference between benthic and planktonic foraminifera ^{14}C ages are still too sparse to constrain ventilation changes during this event [*Adkins and Boyle*, 1997]. The apparent inconsistency between various sediment records could be due to the coarse temporal resolution and sampling frequency in some of the deep sea cores [*Boyle*, 1995]. Other possible causes for such inconsistencies are that these records come from different locations and depths and that different paleocirculation proxies, such as the foraminiferal Cd/Ca, $\delta^{13}C$, and $\Delta^{14}C$, are influenced by distinct oceanic processes [*Jansen and Veum*, 1990].

In this paper we examine the effect of large and small freshwater-induced changes in the Atlantic thermohaline circulation on the climatic coupling between the two hemispheres and on the distribution of $\Delta^{14}C$, PO_4, and $\delta^{13}C$ in the deep sea. We conduct various freshwater perturbation experiments with a latitude-depth, circulation-biogeochemistry ocean model [*Marchal et al.*, 1998a] coupled to an energy balance model of the atmosphere. The ocean model accounts for the main features of the thermohaline circulation and biological cycling which govern the large-scale distribution of major chemical and isotopic tracers. It allows us to perfom extensive and millennial-scale numerical integrations.

2. MODEL DESCRIPTION

The model includes physical and biogeochemical components. These components have been detailed in previous publications and we give only a brief description here. The ocean physical component is the zonally averaged circulation model of *Wright and Stocker* [1992] with the dynamical closure of *Wright et al.* [1998]. The Atlantic, Indian, and Pacific basins are represented individually and connected by a well-mixed Southern Ocean (for model grid see *Stocker and Wright*, 1996). The set of circulation parameters adopted here produces a reasonable agreement with the basin mean vertical profiles of temperature, salinity, and $\Delta^{14}C$ of dissolved inorganic carbon observed in the modern oceans [*Marchal et al.*, 1998a]. The ocean component is coupled to an energy balance model of the atmosphere [*Stocker et al.*, 1992a] and to a thermodynamic sea ice model (described in *Wright and Stocker*, 1993) in order to permit transitions between different climatic states (for parameters see *Wright and Stocker*, 1993).

The ocean biogeochemical component is a description of the cycles of organic carbon and $CaCO_3$ [*Marchal et al.*, 1998a]. Nine tracers are considered: phosphate (taken as the biolimiting nutrient), dissolved inorganic carbon (DIC), alkalinity (ALK), labile dissolved organic carbon (DOC_l), dissolved oxygen, and ^{13}C and ^{14}C in DIC and in DOC_l. River input and sediment burial are omitted, i.e. all the organic carbon and $CaCO_3$ produced in the euphotic zone (top 100 m) are entirely recycled in the water column below. The production of organic carbon exported from the euphotic zone (J_{org}) depends on the local PO_4 availability via Michaelis-Menten kinetics:

$$J_{org} = J_{pot} \cdot \frac{PO_4}{K_{PO_4} + PO_4}, \quad (1)$$

where J_{pot} is a potential export production diagnosed from the spin-up and K_{PO_4} is a half-saturation constant for PO_4 uptake [*Marchal et al.*, 1998b]. The export production is partitioned between fast sinking particulate organic carbon (POC) and DOC_l, both are recycled below the euphotic zone. POC is remineralized without delay according to a spatially uniform vertical profile consistent with sediment trap data [*Bishop*, 1989], whereas DOC_l is oxidized assuming first-order kinetics. The air-sea fluxes of the carbon isotopes are calculated with a constant CO_2 transfer coefficient (Appendix C). The biogeochemical parameters were constrained so as to produce a reasonable fit to the distributions of PO_4, apparent oxygen utilization, DIC, ALK, and $\delta^{13}C$ (of DIC) observed in the modern oceans [*Marchal et al.*, 1998a].

The land biogeochemical component is a 4-pool land biosphere model [*Siegenthaler and Oeschger*, 1987]. We use this model for ^{13}C and ^{14}C perturbations in order to account for the relatively large dilution effect caused by the land biosphere. The biospheric fluxes of organic carbon and the CO_2 fluxes between the land biosphere and the atmosphere are kept constant. Thus we assume no changes in carbon storage and productivity on land [*Marchal et al.*, 1998b]. The ocean and the land biosphere exchange $^{12}CO_2$, $^{13}CO_2$, and $^{14}CO_2$ via the atmosphere which is considered well mixed with respect to these isotopes. Throughout the paper the term anomaly is used for any variable to denote the difference compared to the initial model steady state.

3. RESULTS

3.1. Physical Changes during Abrupt Events

The initial model steady state is characterized by a maximum rate of thermohaline overturning in the Atlantic of about 24 Sv (1 Sv = 10^6 m^3 s^{-1}; left panel in Fig. 1a). We simulate water masses [*Stocker et al.*, 1992b], conservative "colors" with specified origin of formation, in order to help the interpretation of Δ^{14}C, PO_4, and δ^{13}C in the transient experiments. These water masses are designated North Atlantic Deep Water (formed between 55°N–80°N), Central Water (47.5°S–32.5°N), and Southern Ocean Water (70°S–47.5°S). In the initial steady state NADW dominates the deep Atlantic (Fig. 1b). NADW is sandwiched between two components of SOW: bottom waters formed along the Antarctic perimeter (70°S–62.5°S) and intermediate waters formed between 62.5°S–47.5°S (Fig. 1c). Central Water is essentially confined between 50°S–40°N in the Atlantic, with the 80% contour located above 250 m (Fig. 1d).

Different strengths of the THC are obtained by applying a freshwater flux anomaly (FFA) at the surface between 32.5°N–45°N in the Atlantic. We assume that the FFA follows a linear increase and then a linear decrease at the same rate r (Sv kyr^{-1}). The total volume of freshwater released, $V = 6 \cdot 10^6$ km^3, is the same in all the experiments and corresponds to a sea level rise of $\sim$ 17 m. Thus, the experiments differ only by the duration D of the FFA, or equivalently, by the rate of the FFA $r = 4V/D^2$.

A series of experiments demonstrates the sensitivity of the THC and water mass distribution in the Atlantic to the rapidity of the freshwater perturbation. A slow FFA has a relatively minor impact on the THC as it does not permit the development of a low-density cap in the North Atlantic region (Fig. 2). However, a threshold exists (corresponding to $r \sim 0.4$ Sv kyr^{-1} in our model) beyond which the FFA is sufficiently rapid to reduce drastically deep water formation and the thermohaline overturning rate in this region. NADW retreats, whereas SOW becomes more abundant in the deep Atlantic for a greater reduction of the THC (Fig. 1b–c). The reduced influence of NADW and the advance of waters of southern origin are due to the altered buoyancy contrast between high latitude surface waters when a low-density cap develops in the North Atlantic [*Stocker et al.*, 1992b]. Similarly, the Central Water remains confined in the upper water column in the North Atlantic, but reaches deeper levels south of the equator (Fig. 1d).

A hypothesis to explain the variable climate coupling between the two hemispheres [*Blunier et al.*, 1998] is that strong reductions of the THC occurred during prominent D-O events and that only partial shutdowns of the THC occurred during shorter D-O events [*Stocker*, 1999]. A partial shut-down is either a reduction in thermohaline overturning strength or a more southward change in convection location, both with the net result of a decreased heat supply by the THC to the North Atlantic region.

We test this hypothesis using our simplified, coupled ocean-atmosphere model. For illustration, we consider two cases: a partial reduction of the THC (experiment E2 where the THC drops from 24 Sv to 12 Sv) and a complete shut-down (E4 with THC dropping to 3 Sv). In the two cases heat accumulates in the South Atlantic which leads to a subsurface warming (Fig. 3a–b), confirming the concept explained by *Crowley* [1992]. A warming also occurs in the Indian and Pacific as deep water formation in the North Atlantic is the main mechanism adding cold water to the interior to counter the effect of vertical mixing. The partial reduction of the THC has a maximum subsurface warming in the top kilometer at $\sim$ 10°N in the Atlantic (Fig. 3a) while for the complete shut-down, the maximum warming is located in the same depth interval, but around 40°S (Fig. 3b). The former is in remarkable agreement with the temperature anomalies in a freshwater experiment with a coupled atmosphere-ocean, three-dimensional circulation model showing a partial shut-down [*Manabe and Stouffer*, 1997]. We note that the distribution of temperature anomaly in experiments E1 and E4 is qualitatively similar to that in experiments E2 and E4, respectively (maximum subsurface warming in E1, however, is in the North Atlantic).

The present model also simulates the changes in the zonally averaged surface air temperature, T_{atm}. It is clear that temperature changes over the northern North Atlantic are locally underestimated, because the model is zonally averaged. The two experiments show that the qualitative response of T_{atm} is also strongly dependent on the intensity of the THC reduction. A partial reduction leads to a cooling in the northern hemisphere only

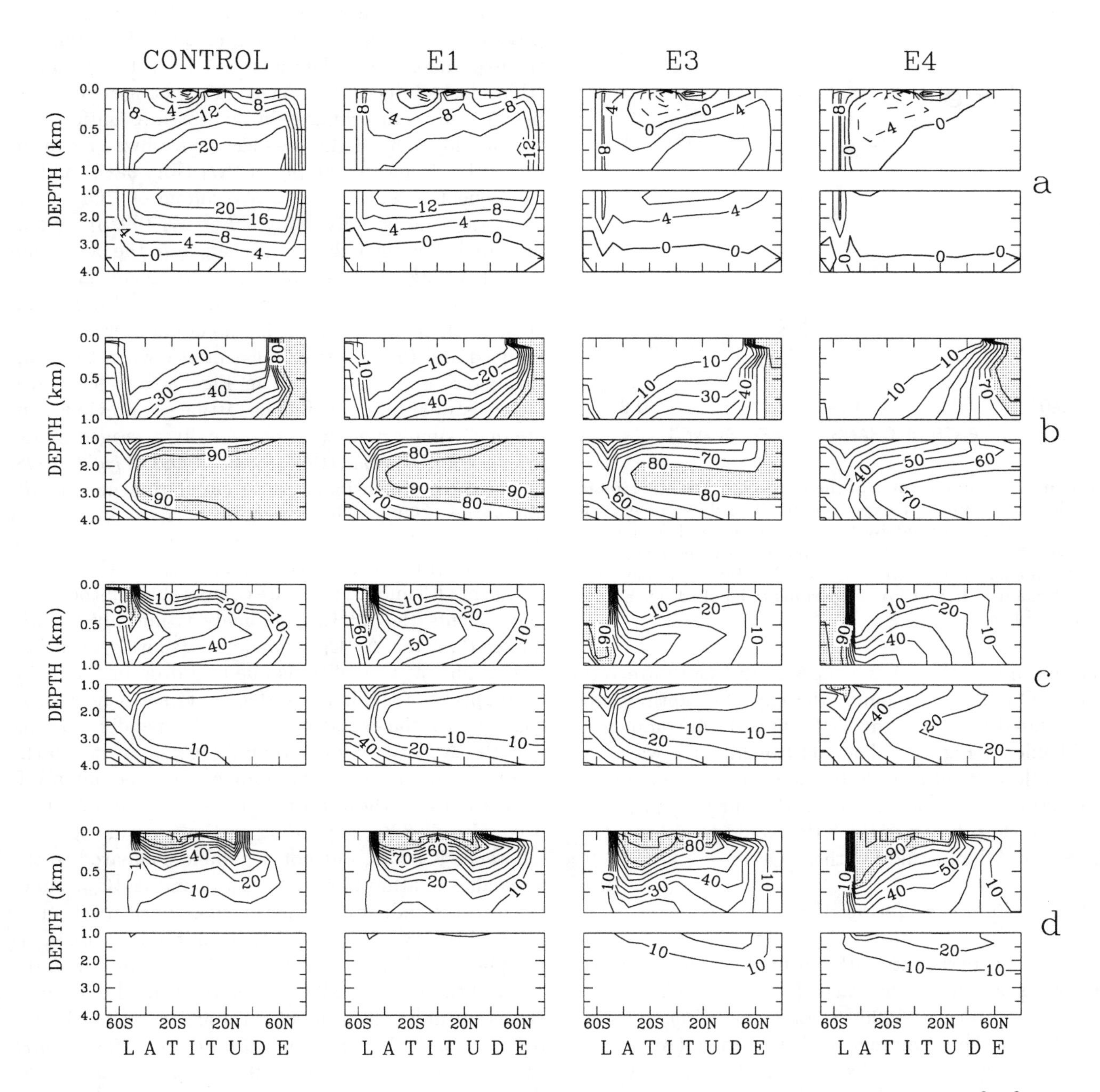

Figure 1. Latitude-depth distribution in the Atlantic of (a) the stream function (in Sv, 1 Sv = 10^6 m^3 s^{-1}) and (b–d) major water masses (%) at the initial model steady state (1st column) and in transient experiments E1 (2nd col.), E3 (3rd col.), and E4 (4th col.). The circulation response to freshwater flux anomaly (FFA) in E2 is similar to that in E3 (Fig. 2). The water masses are (b) North Atlantic Deep Water (formed between 55°N–80°N), (c) Southern Ocean Water (70°S–47.5°S), and (d) Central Water (47.5°S–32.5°N). The duration of the FFA is 2500 yr, 1350 yr, and 1000 yr in experiments E1, E3, and E4, respectively. The transient distributions, corresponding to 1.3 kyr (E1) and 1.0 kyr (E3 and E4) after the start of the FFA, are those predicted at the time of minimum thermohaline overturning in the Atlantic. Contour intervals are 4 Sv and 10%. Regions where the water mass abundance is greater than 80% are shaded.

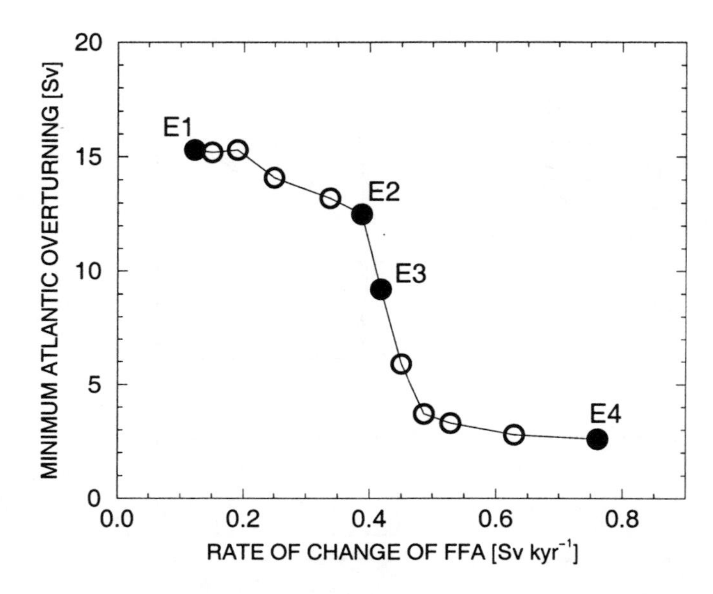

Figure 2. Minimum value of the maximum Atlantic stream function in 12 different experiments. The different experiments are characterized by the same volume of freshwater released ($6 \cdot 10^{15}$ m^3) and geographic application of the FFA (between 32.5°N–45°N in Atlantic), but by a different time rate of change of the FFA. Experiments labelled E1–E4 are examined here.

(dashed line in Fig. 4). Changes in the south are very small. On the other hand, the expected antiphasing is simulated whenever the THC shuts down completely (solid line in Fig. 4). The south exhibits a warming that is close in magnitude to the cooling in the north. Interestingly, our model simulations suggest that only complete shut-downs of the THC would be recorded in the South Atlantic and Southern Ocean.

3.2. $\Delta^{14}C$ Changes during Abrupt Events

Previous experiments with our model included radiocarbon as an inorganic tracer [*Stocker et al.*, 1992b; *Stocker and Wright*, 1996; *Stocker and Wright*, 1998]. We introduce here a more realistic treatment (Appendix) where ^{14}C is affected also by the ocean biological cycling (organic matter and $CaCO_3$ cycle). In the following we use the indexes "inorg" and "org" for $\Delta^{14}C$ to refer to the cases when ^{14}C is included as an inorganic and organic tracer, respectively.

We first consider the changes in global ocean export production in the three experiments examined below (E1, E3, and E4). These experiments are representative of, respectively, a small, intermediate, and strong reduction of the circulation in the North Atlantic (Fig. 2). We assume no isotope fractionation during $CaCO_3$ formation, so that $CaCO_3$ cycling does not directly affect the distribution of tracers here. The global ocean export production first decreases (by 5 to 20%, depending on the experiment) and then increases in each experiment (thin, dashed line in Fig. 5a–c). The changes in export production occur primarily in the North Atlantic where the development of the low-density freshwater cap inhibits the deepwater supply of PO_4 (Fig. 6a–b) and its subsequent erosion leads to a recharge of PO_4 in the euphotic zone (see also *Marchal et al.*, 1998b). We illustrate below the influence of changes in the ocean biological cycling on the atmospheric and oceanic $\Delta^{14}C_{org}$.

In each experiment the atmospheric $\Delta^{14}C_{org}$ first increases when the Atlantic thermohaline overturning is reduced and then decreases, essentially when the overturning resumes (solid line in Fig. 5a–c). This evolution is qualitatively similar to that in previous "inorganic" simulations with our zonally averaged model [*Stocker and Wright*, 1996; *Stocker and Wright*, 1998] and a three-dimensional ocean circulation model [*Mikolajewicz*, 1996]. The maximum positive anomaly in atmospheric ^{14}C activity is higher for a more drastic reduction of the THC, although $\Delta^{14}C_{org}$ reached in experiments E3 and E4 are quite comparable. In these two experiments the $\Delta^{14}C_{org}$ anomaly is initially slightly higher than the $\Delta^{14}C_{inorg}$ anomaly (short dashed line in Fig. 5b–c). This offset is due to a drop in the biological uptake of ^{14}C in the surface waters which varies in concert with the export production (Appendix). Subsequently, $\Delta^{14}C_{org}$ is lower than $\Delta^{14}C_{inorg}$ owing to the increasing ocean biological uptake. The maximum difference between the atmospheric $\Delta^{14}C_{inorg}$ and $\Delta^{14}C_{org}$ is equal to 7–8‰, depending on the experiment. This difference is small, but not neglibible compared to the anomalies of atmospheric $\Delta^{14}C_{inorg}$ and $\Delta^{14}C_{org}$ simulated in the individual experiments, which reach maxima of 20–35‰. We note that the initial increase in atmospheric $\Delta^{14}C_{inorg}$ and $\Delta^{14}C_{org}$ is consistent with the rise in atmospheric $\Delta^{14}C$ at the onset of the Younger Dryas documented in fossil coral and varved sediment records [*Edwards et al.*, 1993; *Goslar et al.*, 1995; *Björck et al.*, 1996; *Hughen et al.*, 1998; *Kitagawa and van der Plicht*, 1998; *Goslar et al.*, 1999]. Our simulations, however, do not exhibit a decrease in atmospheric $\Delta^{14}C$ during the model cold phase, in contrast to the gradual decline in atmospheric $\Delta^{14}C$ (corrected for changes in geomagnetic ^{14}C production) during the YD documented in the same records (see *Marchal et al.*, 1999). This decline remains a major problem to be solved in the future [*Goslar et al.*, 1999].

We now examine the distribution of $\Delta^{14}C_{org}$ in the three oceanic basins at the time of minimum Atlantic thermohaline overturning in experiments E1, E3, and

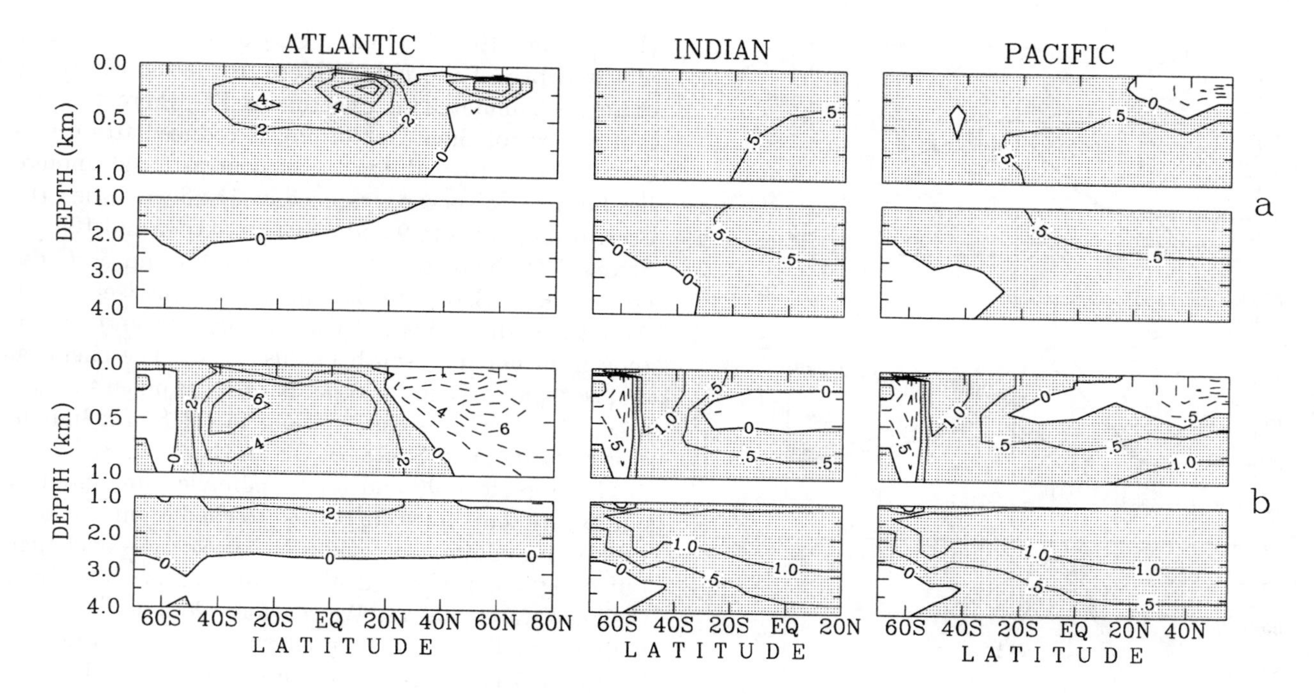

Figure 3. Latitude-depth distribution of the temperature anomaly (°C) simulated at the time of minimum Atlantic thermohaline overturning in experiments (a) E2, partial THC shut-down (0.8 kyr after start of FFA; THC drop from 24 Sv to 12 Sv) and (b) E4, complete shut-down (1.0 kyr after start of FFA; THC drop to 3 Sv). The contour interval is 2°C in the Atlantic and 0.5°C in the other basins. Regions with positive values are shaded.

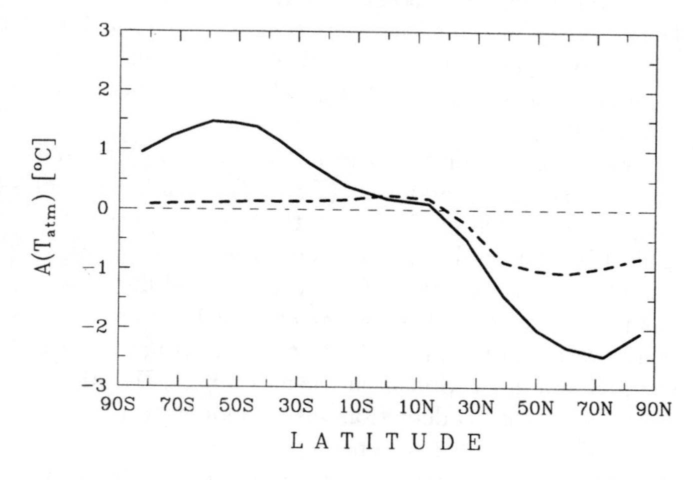

Figure 4. Meridional profile of the atmospheric temperature anomaly at the time of minimum THC in experiments E2 (– – –) and E4 (————).

E4 (Fig. 7b–d). $\Delta^{14}C_{org}$ exhibits an increase generally above ~ 2000 m in each basin, and a decrease at depth, essentially in the North Atlantic (Fig. 8a–c). Again, these results are in line with previous inorganic simulations with our model and a 3-d model [*Mikolajewicz*, 1996]. The maximum difference between the anomalies of oceanic $\Delta^{14}C_{inorg}$ and $\Delta^{14}C_{org}$ ranges from –9 to +7‰, depending on the ocean basin and the experiment. More than 99% of the variance of $\Delta^{14}C_{org}$ is explained by a linear regression against $\Delta^{14}C_{inorg}$ in E1, E3, and E4. Thus, changes in biological cycling have a minor effect on the oceanic $\Delta^{14}C_{org}$ anomalies simulated by our model and these anomalies can safely be interpreted as reflecting a local imbalance between air-sea gas exchange, oceanic transport, and radioactive decay.

The largest negative $\Delta^{14}C_{org}$ anomalies, amounting to about –100‰ in each experiment, are predicted in the deep North Atlantic (Fig. 8a–c). These are clearly related to the retreat of freshly formed North Atlantic Deep Water which is rich in ^{14}C (Fig. 1b and Fig. 7a). The advance of the deep component of Southern Ocean Water (Fig. 1c) is insufficient to compensate the effect of this retreat, because this component contains much less ^{14}C than the freshly formed NADW as it reaches the deep North Atlantic. On the other hand, the largest positive $\Delta^{14}C_{org}$ anomaly, greater than 60‰, is simulated in the upper kilometer of the South Atlantic in experiment E4 (Fig. 8c). Interestingly, the anomaly is coincident with the prominent subsurface warming predicted in this experiment (Fig. 3b). We infer that the ^{14}C enrichment and temperature increase are associated with the partial replacement of cold SOW by the

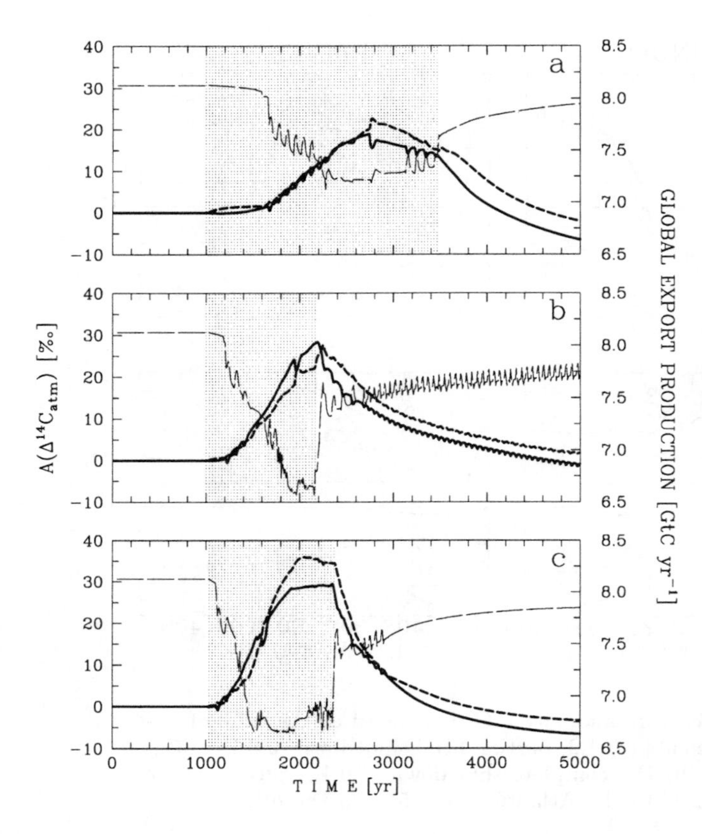

Figure 5. Atmospheric $\Delta^{14}C$ anomaly in experiments (a) E1, (b) E3, and (c) E4. The $\Delta^{14}C$ anomaly simulated when ^{14}C is included in the model as an inorganic tracer (— — —) and as an organic tracer (————) are represented. The global ocean export production (thin, dashed line) and the period during which the THC is altered (shaded area) are reported in each panel.

warmer Central Water (Fig. 1c–d), in a region where the convective activity and hence ^{14}C transport to depth is enhanced. Finally, we note that the positive $\Delta^{14}C_{org}$ anomalies simulated at intermediate depths (< 2000 m) in the Southern Ocean, Indian, and Pacific (Fig. 8a–c) cannot be due to $\Delta^{14}C_{org}$ changes occurring below since the former are negative. These anomalies evidence a far-field effect, whereby the reduction of deepwater formation in the North Atlantic produces an increase in the atmospheric ^{14}C activity which is then transmitted to the other basins through the sea surface by gas exchange.

3.3. Oceanic PO_4 and $\delta^{13}C$ Changes during Abrupt Events

The reduction of the Atlantic thermohaline overturning in the different experiments has a variable influence on the distribution of PO_4 and $\delta^{13}C$ in the ocean (Figs. 9 and 10). Negligible PO_4 and $\delta^{13}C$ anomalies at the depth of the NADW core in the model steady state are simulated in experiment E1 where the Atlantic thermohaline overturning is reduced to only ~ 15 Sv (Fig. 11a and 12a). By contrast, the anomalies become paleoceanographically significant when the overturning drops to 9 and 3 Sv (Fig. 11b–c and 12b–c). Strong PO_4 increases (> 0.3 mmol m^{-3}) and $\delta^{13}C$ decreases (< –0.3 ‰) are then present in the deep North Atlantic, with a general decline of their absolute amplitude from north to south in this basin. The chemical and isotopic shifts in the other basins, are much smaller, except for $\delta^{13}C$ at some locations in the South Indian, South Pacific, and Southern Ocean (Fig. 12b–c).

The oceanic PO_4 and $\delta^{13}C$ anomalies simulated in our experiments must be related to changes in the local rate of organic matter oxidation, air-sea gas exchange (affecting the preformed $\delta^{13}C$), and/or proportion of the deep water masses. We consider plots of $\delta^{13}C$ anomaly versus PO_4 anomaly predicted in the Atlantic and Southern Ocean in order to identify the most influential processes (Fig. 13). The general trend expected if these anomalies were caused only by organic matter cycling, is indicated by a solid line in each plot. This Redfield line has a negative slope of –1.2 ‰ (mmol m^{-3})$^{-1}$ based on model stoichiometry [*Marchal et al.*, 1998a]. Departures from this line, represented by dashed lines labeled with numbers expressed in ‰, reflect the effects of the air-sea gas exchange which influences $\delta^{13}C$, but not PO_4.

Two major features appear in the plots of $\delta^{13}C$ anomaly versus PO_4 anomaly (Fig. 13). First, waters deeper than 1000 m in the North Atlantic (ocean domain I below) exhibit strong PO_4 enrichment and $\delta^{13}C$ depletion. The departure from the Redfield line is less than +0.5 ‰ in each experiment (filled circles). Second, waters in the upper 1000 m in the Atlantic south of 65°N and in the Southern Ocean (domain II) exhibit a very strong $\delta^{13}C$ depletion with a moderate PO_4 decrease ("+" in Fig. 13). The departures from the Redfield line are larger for a greater reduction of the THC and reach more than 1 ‰ at some locations in experiments E3 and E4 (Fig. 13b–c). We note that $\delta^{13}C$ of waters in the upper 1000 m north of 65°N in the Atlantic (domain III) rises by more than 1 ‰ in experiment E4 (× in Fig. 13c). On the other hand, waters between 1–4 km in the South Atlantic and Southern Ocean (domain IV; open circles in Fig. 13) exhibit much lower chemical and isotopic shifts than in the same depth interval in the North Atlantic.

The first major feature in the $\delta^{13}C$ anomaly–PO_4 anomaly plots is examined with more detail. The small

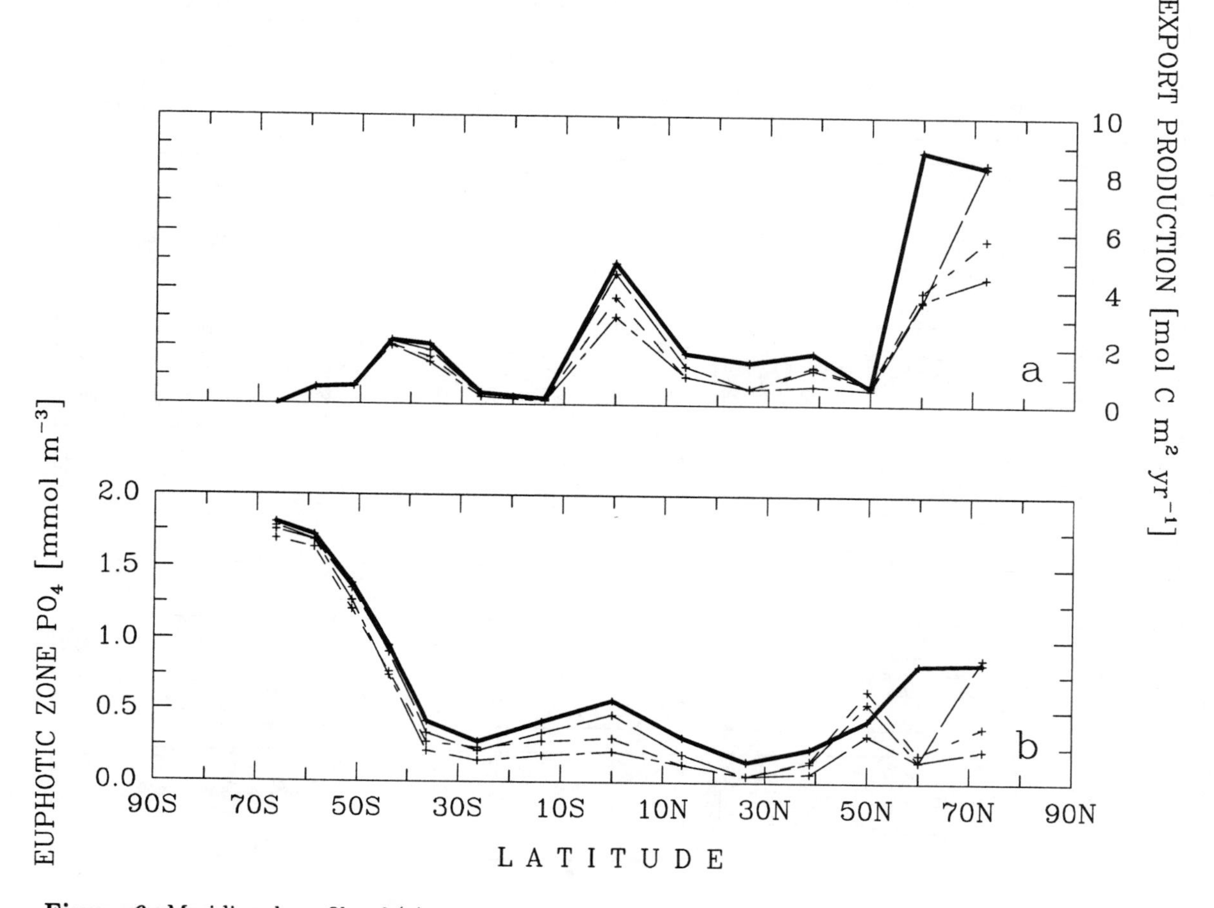

Figure 6. Meridional profile of (a) export production and (b) mean PO_4 concentration in the euphotic zone (top 100 m) in the Atlantic. The different curves correspond to the initial model steady state (————) and to the time of minimum THC in experiments E1 (—— —— ——), E3 (—— - ——), and E4 (– – –).

departure from the Redfield line indicates that the air-sea gas exchange has a relatively minor influence on the $\delta^{13}C$ anomalies and points to organic matter cycling as an important process. We identify whether the strong PO_4 and $\delta^{13}C$ anomalies simulated at the depth of the NADW core are related to an increase in the local rate of organic matter oxidation, Δ_{org}. This increase is calculated as:

$$\Delta_{org} = \int_{t_0}^{t_0+\Delta t} J_{org}(t)dt - J_{org}(t_0) \cdot \Delta t, \qquad (1)$$

where J_{org} is here the local rate of DIC production through the oxidation of POC and DOC_l, t_0 is the time of the initial steady state, and $t_0 + \Delta t$ is the time corresponding to the PO_4 and $\delta^{13}C$ anomalies, i.e. $\Delta t =$ 1.3 kyr in experiment E1 and 1.0 kyr in experiments E3 and E4. In experiments E3 and E4, Δ_{org} is negative in the North Atlantic (non shaded areas in Fig. 11b–c and 12b–c). This shift is due to the lowered injection of surface, DOC_l-rich waters to depth in the northern North Atlantic and, to a lesser extent, to the drop in the oxidation rate of POC associated with the decline in export production. Thus, the prominent chemical and isotopic changes predicted in the NADW core cannot result from an increase in the local rate of organic matter oxidation. These changes are related, instead, to a drop in the ventilation by PO_4-poor and $\delta^{13}C$-rich waters which becomes insufficient to balance the effects of organic matter oxidation at depth.

The second major feature in the $\delta^{13}C$ anomaly–PO_4 anomaly plots is the atypical $\delta^{13}C$ depletion in the upper 1000 m in the Atlantic south of 65°N and in the Southern Ocean. A possible effect here is a decrease in surface $\delta^{13}C$, associated with the decrease in export

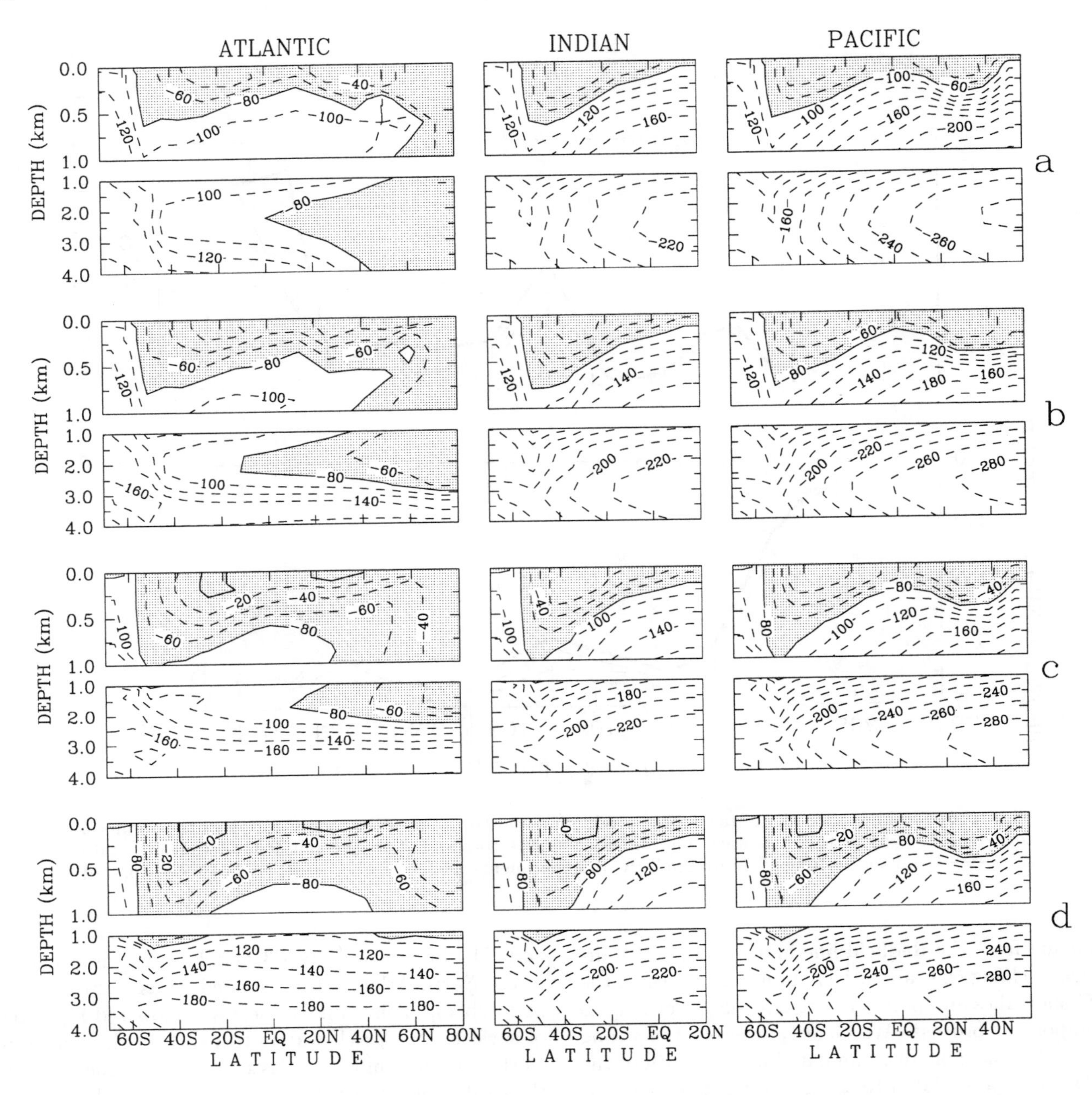

Figure 7. Latitude-depth distribution of $\Delta^{14}C$ (‰) simulated when ^{14}C is included as an organic tracer at the initial steady state (a) and at the time of minimum THC in experiments E1 (b), E3 (c), and E4 (d). The contour interval is 20‰. Regions where $\Delta^{14}C > -80$‰ are shaded.

production, which would occur in the absence of phosphate. Whereas the export production generally declines (Fig. 6a), the PO_4 level in the euphotic zone does not go to depletion in the Atlantic and Southern Ocean (Fig. 6b). This leads us to conclude that $\delta^{13}C$ of domain II-waters are markedly influenced by the effects of air-sea gas exchange. We now inspect the changes in the surface $\delta^{13}C$, $\delta^{13}C_{surf}$, and in the surface $\delta^{13}C$ expected from isotopic equilibrium with the atmosphere, $\delta^{13}C_{eq}$ (Fig. 14a–b). In the three experiments (E1, E3, and E4) the surface waters south of 30°N in the Atlantic become isotopically lighter. By contrast, the surface waters further north in the same basin exhibit generally an isotopic enrichment. $\delta^{13}C_{eq}$, on the other

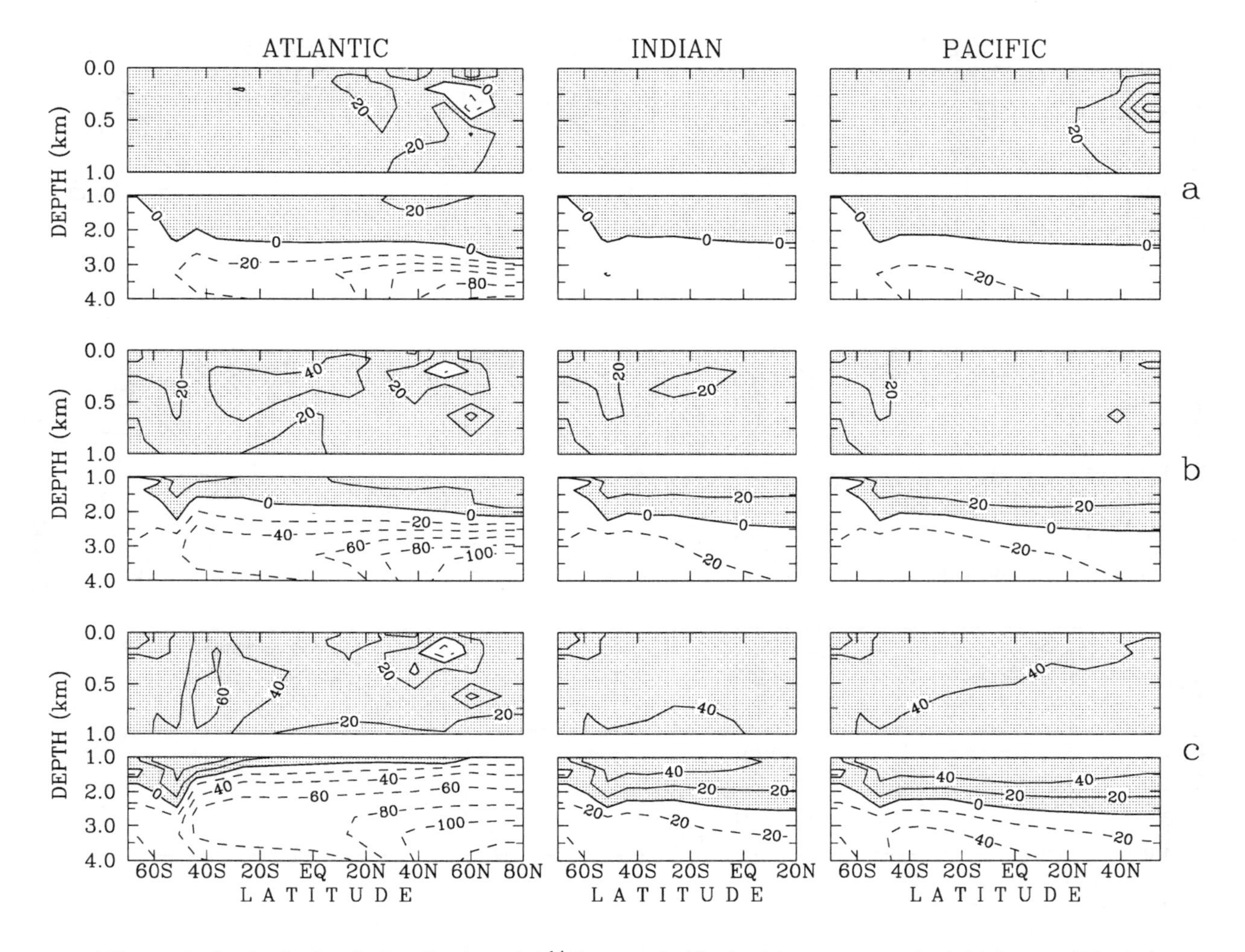

Figure 8. Latitude-depth distribution of $\Delta^{14}C$ anomaly (‰) with respect to the initial state (Fig. 7a) simulated when ^{14}C is included as an organic tracer at the time of minimum THC in experiments (a) E1, (b) E3, and (c) E4. The contour interval is 20‰. Regions with positive anomalies are shaded.

hand, is slightly depressed south of 30°N and enhanced north of this latitude. The fact that $\delta^{13}C_{surf}$ follow the $\delta^{13}C_{eq}$ changes suggests that the air-sea gas exchange is influential in producing the surface isotopic anomalies. More important, the absolute value of the isotopic disequilibrium, $\delta^{13}C_{surf} - \delta^{13}C_{eq}$, decreases generally at all latitudes in the Atlantic (Fig. 14c). A major contributor to this is likely the increasing residence time of waters at the surface in the Atlantic when the THC is partially (experiments E1 and E3) or completely collapsed (E4), which permits a better isotopic equilibration with the atmosphere. Interestingly, the isotopic anomalies are propagated to depth once they are generated at the surface (Fig. 9c–d and 10c–d). In the South Atlantic, this occurs through the deepening of the Central Water (Fig. 1d) which contains a small amount of PO_4 but is strongly depleted in $\delta^{13}C$. Thus, we infer that a combination between changes in the air-sea gas exchange and oceanic transport is responsible for the PO_4 and $\delta^{13}C$ anomalies simulated in the upper 1000 m of the Atlantic (Fig. 13b–c).

4. SUMMARY AND CONCLUSIONS

4.1. Physical Changes

It was initially proposed that variations in the flux of the relatively warm NADW would lead to a temperature evolution in phase in the two hemispheres [*Weyl*, 1968]. *Imbrie et al.* [1992] argued that this "NADW–Antarctic connection" contributed to the progression of climatic anomalies at Milankovitch frequencies from the Arctic to the other regions. *Bender et al.* [1994a] observed that

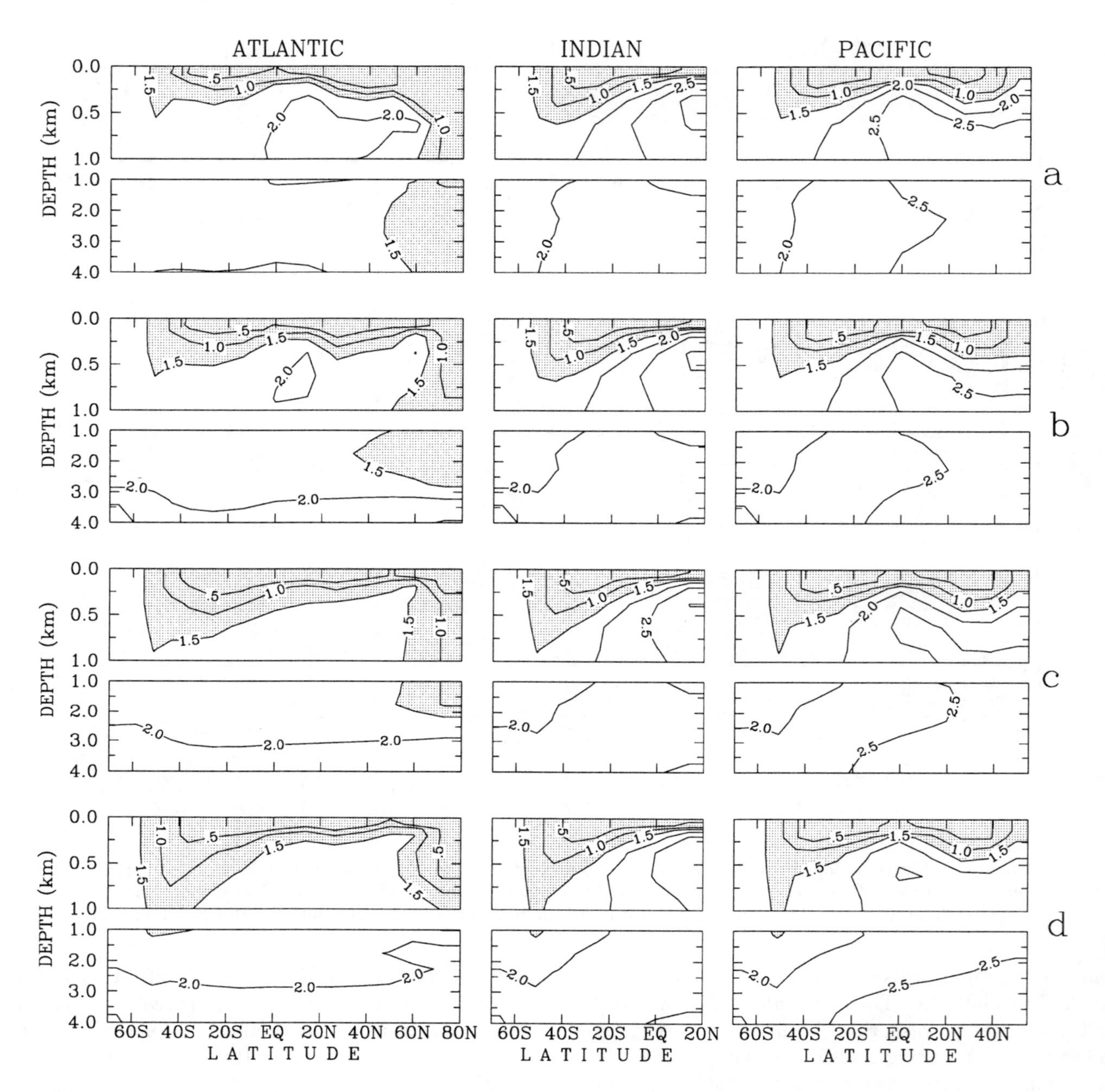

Figure 9. Latitude-depth distribution of PO_4 (mmol m^{-3}) at the initial steady state (a) and at the time of minimum THC in experiments E1 (b), E3 (c), and E4 (d). The contour interval is 0.5 mmol m^{-3}. Regions where PO_4 < 1.5 mmol m^{-3} are shaded.

isotopic events are more rapid and more numerous in ice records from Greenland than from Antarctica. They inferred that long interstadials first originated in the northern hemisphere and were then transmitted to the other hemisphere. This would have occurred through partial deglaciation and changes in ocean circulation [*Bender et al.*, 1994a]. These authors cautioned, however, that their inference cannot be proven (past changes in the $^{18}O/^{16}O$ ratio of atmospheric O_2 are too small to permit a sufficently accurate synchronization of the ice cores).

The occurrence of a NADW–Antarctic connection at millennial time scale is not supported by the following lines of evidence. First, the CH_4-synchronization of ice core records from both polar regions implies that Greenland interstadials which occurred between 47–23 kyr BP lagged their isotopic counterparts at Byrd and Vostok by 1–2.5 kyr on average [*Blunier et al.*, 1998].

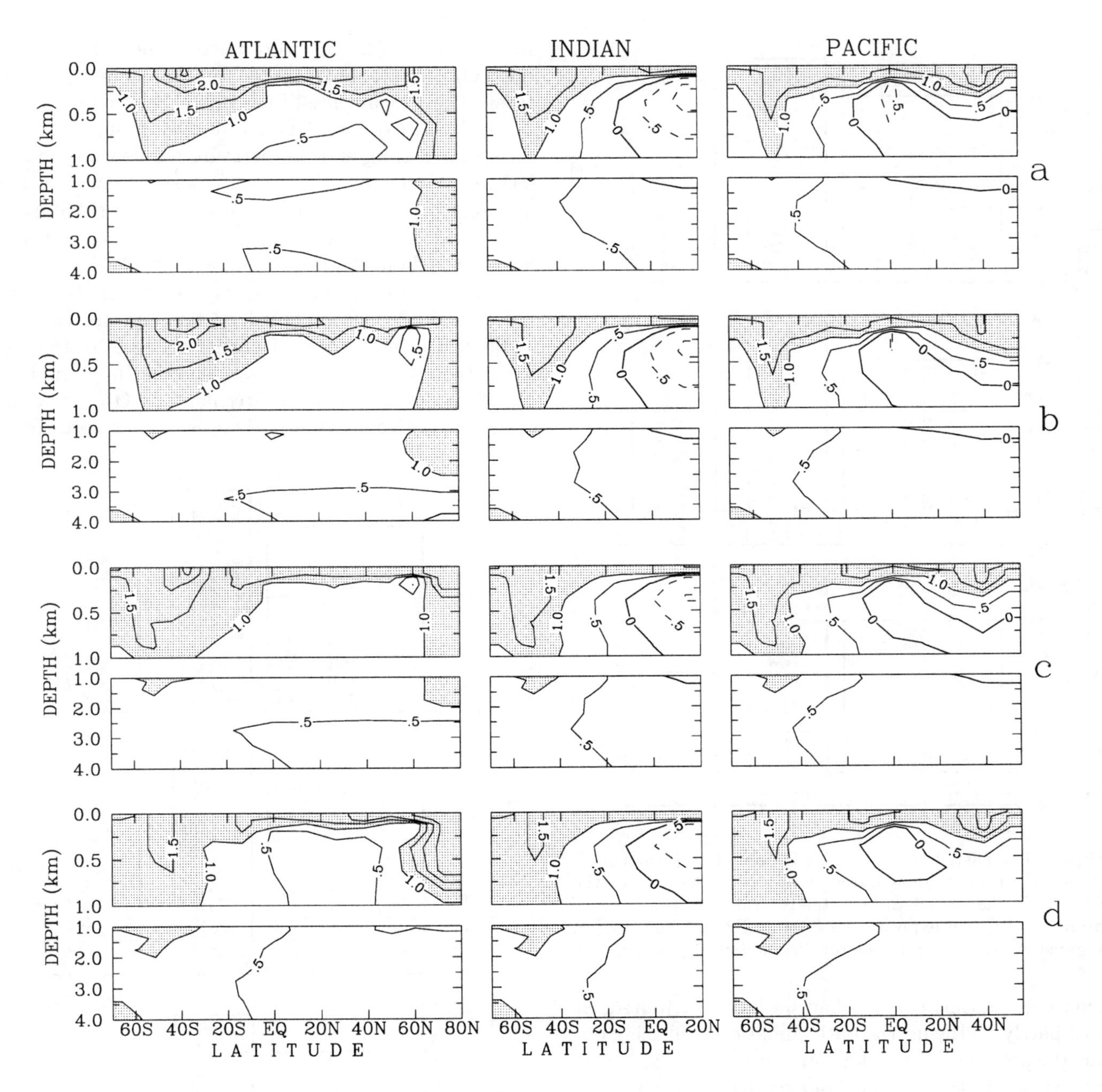

Figure 10. Latitude-depth distribution of $\delta^{13}C$ (‰) at the initial steady state (a) and at the time of minimum THC in experiments E1 (b), E3 (c), and E4 (d). The contour interval is 0.5 ‰. Regions where $\delta^{13}C > 1$ ‰ are shaded.

Second, in a high-resolution sediment core raised from the Southern Ocean, the planktonic $\delta^{18}O$ anomalies (a proxy of local SST) lead the benthic $\delta^{13}C$ anomalies (a proxy of deepwater production in the northern North Atlantic) by ~ 1,500 yr [*Charles et al.*, 1996]. Finally, the NADW–Antarctic connection is inconsistent with various model simulations which exhibit pronounced climate antiphasing between high northern and southern latitudes during periods where the THC is substantially altered [*Stocker*, 1999].

In our simplified model, freshwater-induced reductions of the Atlantic thermohaline circulation produce cooling at high northern latitudes and warming at high southern latitudes. Conversely, the resumption of the THC from a state of total collapse promotes warming in the north and cooling in the south. An interesting

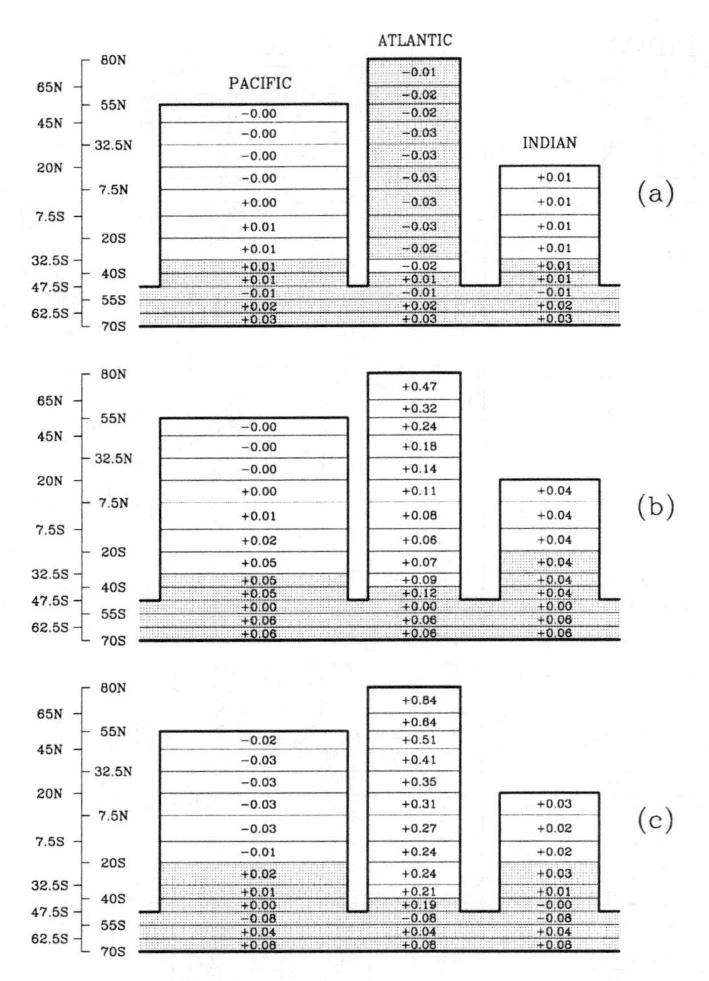

Figure 11. Distribution of the PO_4 anomaly (mmol m^{-3}) at the depth of 2250 m (corresponding to the core of the NADW in the initial model steady state) at the time of minimum THC in experiments E1 (a), E3 (b), and E4 (c). Regions where $\Delta_{org} > 0$ (see text for definition) are shaded.

hypothesis is that this mechanism has contributed, at least partly, to the north-south climatic antiphasing during the prominent Greenland interstadials 12, 8, 1 [*Blunier et al.*, 1998] and the Younger Dryas termination [*Blunier et al.*, 1997]. When these northern hemisphere warming events initiated (within several decades) the gradual warming that occurred previously in Antarctica was interrupted and followed either by a strong cooling, such as after the Greenland interstadials 12 and 8 [*Blunier et al.*, 1998], or a moderate cooling or a plateau such as the Antarctic Cold Reversal after interstadial 1 (Bølling/Allerød) and after the YD termination [*Jouzel et al.*, 1995; *Sowers and Bender*, 1995; *Blunier et al.*, 1997]. Recently, *Steig et al.* [1998] synchronized the δD record from Taylor Dome, a near-coastal East Antarctic site, to the GISP2 δD record based on measurements of the $^{18}O/^{16}O$ ratio of O_2 and of the CH_4 content in both cores. They documented that, unlike Byrd and Vostok, Taylor Dome experienced an abrupt warming at the onset of the Bølling and temperature minima during the YD in the Greenland record. A possible cause of the apparent inconsistency between the different Antarctic records is that short-term climate changes were not uniform throughout Antarctica.

Moreover, in our experiments, cooling in the north is less and not accompanied by temperature changes in the south in the case of partial collapses of the Atlantic thermohaline circulation. Another hypothesis is therefore that the shorter Greenland $\delta^{18}O$ oscillations do not have clear counterparts in isotopic records from

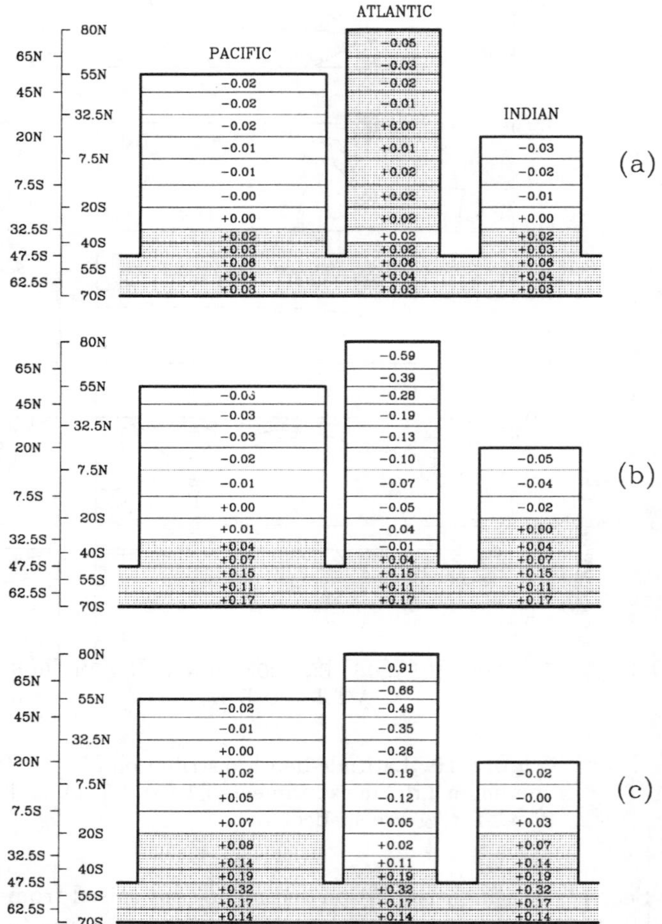

Figure 12. Distribution of the $\delta^{13}C$ anomaly (‰) at the depth of 2250 m (corresponding to the core of the NADW in the initial model steady state) at the time of minimum THC in experiments E1 (a), E3 (b), and E4 (c). Regions where $\Delta_{org} > 0$ (see text for definition) are shaded.

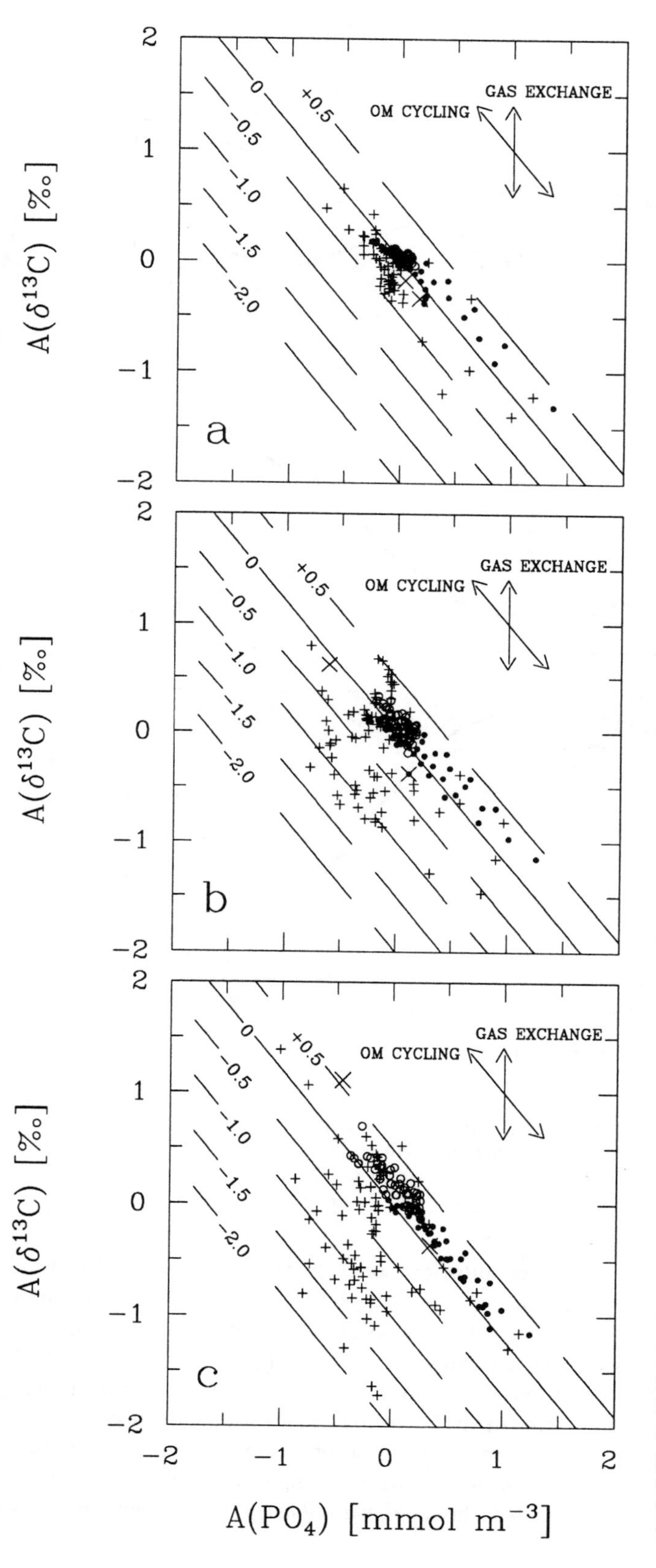

Byrd and Vostok [*Blunier et al.*, 1998] because the two hemispheres are decoupled in this case. Clearly, major questions remain regarding the millennial-scale climate changes of the last glacial period [*Stocker*, 1998; *Cane*, 1998].

First, we stress that we have examined only one potential mechanism with a particular model. On the one hand, a clear distinction must be made between the freshwater perturbation experiments done here and the more general issue of reducing the ocean thermohaline circulation. It is obvious that factors other than meltwater discharges in the North Atlantic basin and not included in our experiments could also alter the THC. For instance, simulations with a 3-d ocean circulation model illustrate that southern hemisphere climate could impact the northward inflow of warm waters into the North Atlantic and the southward ouflow from this basin through the westerly wind stress in the circumpolar region [*Toggweiler and Samuels*, 1993]. In addition, many processes which could have contributed to the fast climatic changes of the last glacial period are not represented. These processes include, for example, the feedback between the northward heat flux by the THC and glacial meltwater discharge at high northern latitudes, sea level changes affecting the stability of remote ice shelves, and variations in the atmospheric transport of water vapor. On the other hand, the

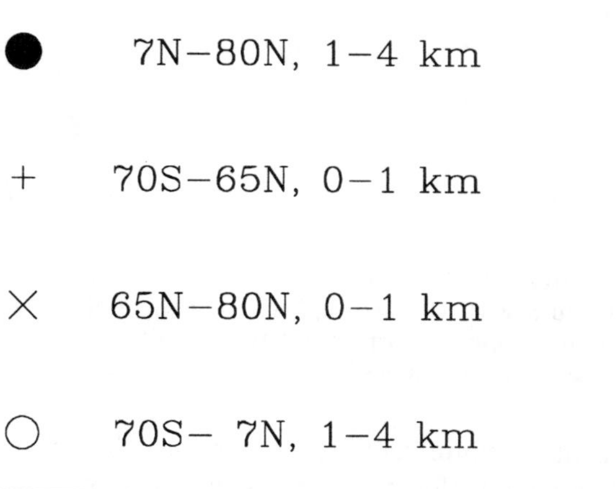

Figure 13. δ^{13}C anomaly versus PO_4 anomaly in four different domains in the Atlantic at the time of minimum THC in experiments E1 (a), E3 (b), and E4 (c). The solid line has a slope of -1.2 ‰ (mmol m^{-3})$^{-1}$ and illustrates the composition change expected from the effect of organic matter cycling. The dashed lines represent various departures from this line and illustrate composition changes expected from the effect of air-sea gas exchange.

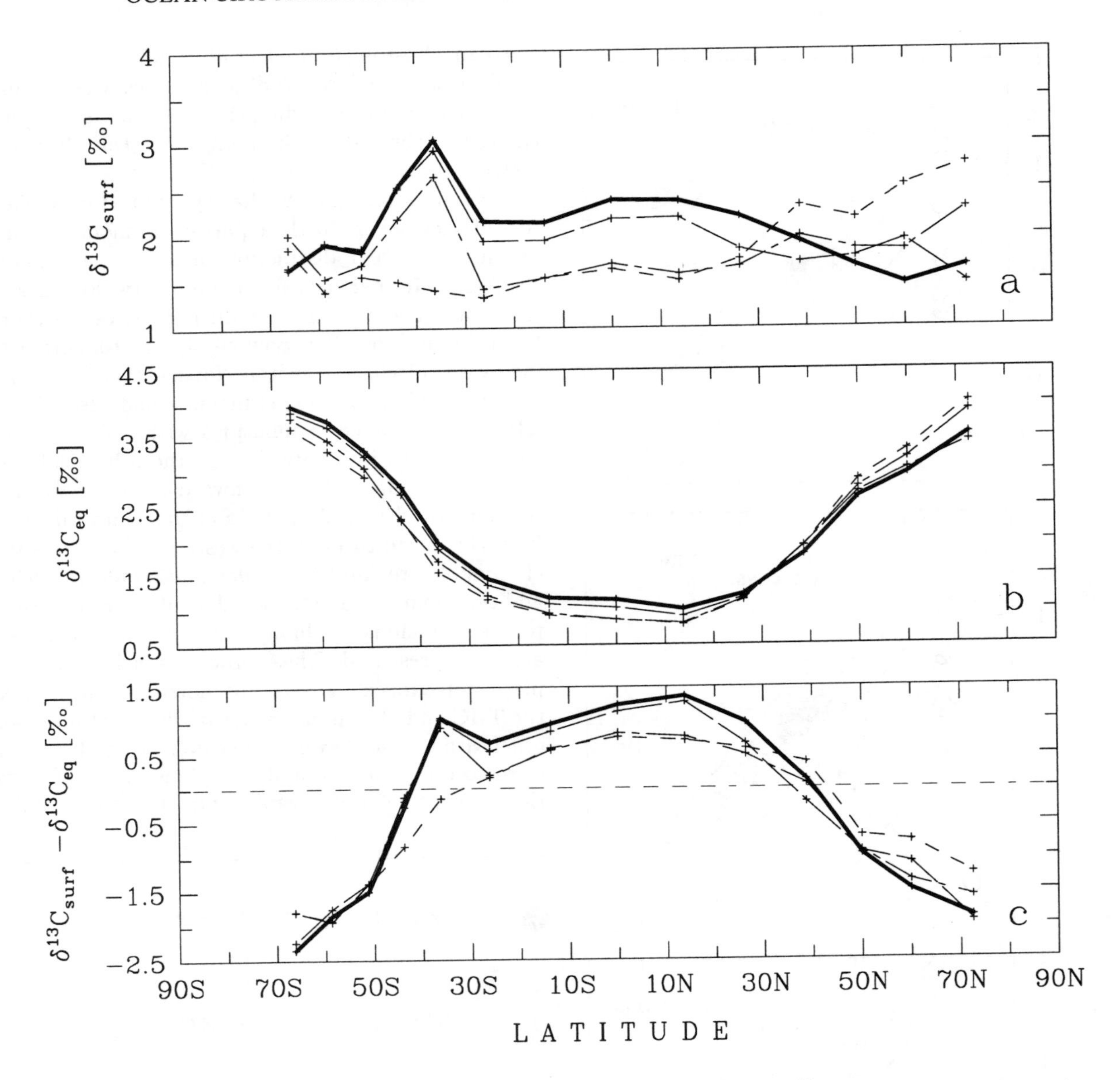

Figure 14. Meridional profile of (a) surface $\delta^{13}C$, $\delta^{13}C_{surf}$, (b) surface $\delta^{13}C$ expected from the equilibrium with the atmosphere, $\delta^{13}C_{eq}$, and (b) air-sea isotopic disequilibrium $\delta^{13}C_{surf} - \delta^{13}C_{eq}$. The different curves correspond to the initial model steady state (———) and to the time of minimum THC in experiments E1 (—— —— ——), E3 (—— - ——), and E4 (– – –).

limitations of our simplified model must be acknowledged. A major limitation comes from the zonal average representation of the ocean and atmosphere (for a discussion see *Wright and Stocker*, 1993). It is thus imperative that the scenario illustrated here be confirmed by more complete models in order to obtain a more detailed understanding of the north-south thermal antiphasing documented for abrupt changes during the last glacial period.

Another question concerns the geographical location and nature of the trigger(s) of the climatic sequences documented in paleoarchives from the northern and southern hemispheres. According to *Imbrie et al.* [1992], both theory and observation show that the initial response to orbital forcing must occur at high northern latitudes. Recent observations would suggest that millennial-scale climate changes originated rather in the southern hemisphere [*Charles et al.*, 1996; *Blunier*

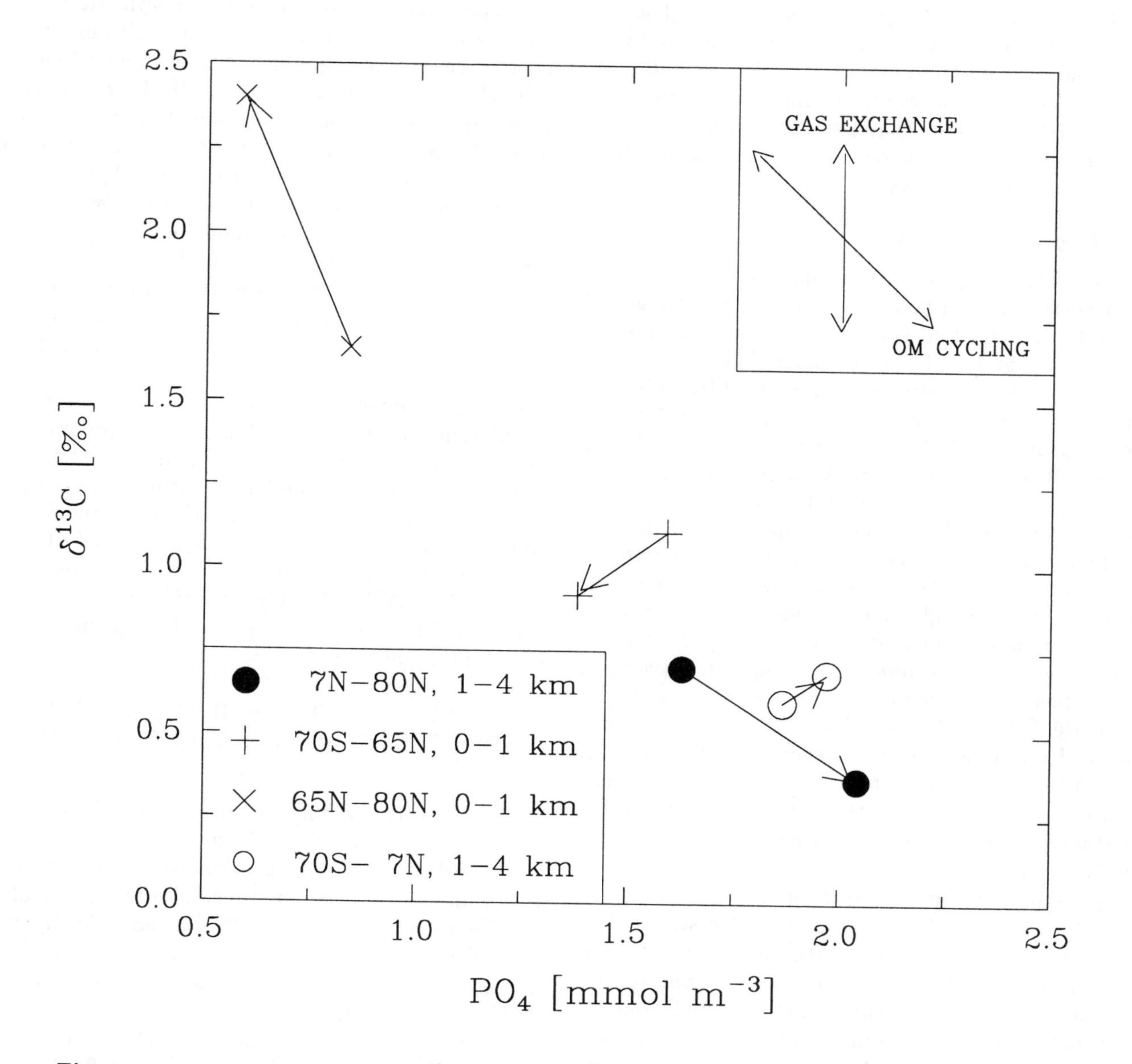

Figure 15. Volume-weighted mean $\delta^{13}C$ versus volume-weighted mean PO_4 in four different domains comprising the whole Atlantic basin (70°S–80°N) in experiment E4. The composition changes from the initial steady state to the time where the THC drops to a minimum of 3 Sv are illustrated by arrows. The composition changes expected from the effects of organic matter cycling and air-sea gas exchange are represented in the top right inset.

et al., 1998]. On the other hand, the role of the tropics and their interaction with middle and high latitudes need to be considered [*Chappellaz et al.*, 1993; *Guilderson et al.*, 1994; *McIntyre and Molfino*, 1996; *Bard et al.*, 1997; *Thompson et al.*, 1997]. The nature of interstadial forcing and response may well vary from one event to the next, as speculated by *Bender et al.* [1994a]. The search for cause-effect relations for millennial-scale climate changes is difficult much like for a mechanical system of two coupled oscillators in which the coupling constant can change in time.

4.2. Biogeochemical Changes

In addition to producing climate antiphasing between north and south, freshwater-induced changes of the THC also lead to spatially variable anomalies of $\Delta^{14}C$, PO_4, and $\delta^{13}C$ in our model. The simulated tracer anomalies are strongest when the THC collapses completely (experiment E4). In experiment E4, the average prominent PO_4 enrichment and $\delta^{13}C$ depletion between 1000 m in the North Atlantic (domain I) cannot only be due to mixing with a nutrient-rich end member, because the average composition of domain I-waters becomes extreme in the $\delta^{13}C$-PO_4 plot (filled circles in Fig. 15). The only mechanism to achieve this composition is through an imbalance between the remineralization of organic matter and the import of PO_4-poor and $\delta^{13}C$-rich waters. Interestingly, the averaged $\delta^{13}C$ depletion in domain I is less than that predicted by the Redfield line (see also Fig. 13c, which shows that domain I-waters are most generally shifted above this line). A likely contributor to this is the $\delta^{13}C$ enrichment of surface waters in the northern North Atlantic which are still transported to depth during the early phases of the THC collapse.

Waters in the upper 1000 m in the Atlantic south of 65°N and in the Southern Ocean (domain II; "+" in Fig. 15) experience, on average, moderate PO_4 and $\delta^{13}C$ depletions. The change in average composition is approximately orthogonal to that expected from organic matter cycling. This indicates that a combination between the air-sea gas exchange and oceanic transport is here influential. This is in line with our previous interpretation of the shift of domain II-waters to the bottom left quadrant in Fig. 13c.

Waters in the upper 1000 m north of 65°N in the Atlantic (domain III; × in Fig. 15) exhibit generally moderate PO_4 depletion but very strong $\delta^{13}C$ enrichment. Again, the change in average composition cannot be due to mixing with a PO_4-poor and $\delta^{13}C$-rich end member because the composition becomes extreme in the $\delta^{13}C$-PO_4 plot. The only possibility is therefore through a combination between surface gas exchange and organic matter cycling.

Finally, waters below 1000 m in South Atlantic and Southern Ocean (domain IV; open circles in Fig. 15) experience, on average, much smaller chemical and isotopic shifts than those in domains I-III. Here, the change in average composition is upward and to the right in the $\delta^{13}C$-PO_4 plot, which allows us to rule out a dominant effect from organic matter cycling. The slight PO_4 enrichment must be associated with mixing with waters of the deep North Atlantic (domain I) which become strongly nutrient-rich. The small $\delta^{13}C$ increase, on the other hand, must be due to mixing with waters above 1000 m (domain II) whose average $\delta^{13}C$, though it has decreased, remains higher than that in domain IV below.

Our model experiments suggest that the same ocean circulation change in the Atlantic can produce very distinct anomalies of $\Delta^{14}C$, PO_4, and $\delta^{13}C$ between different depths, latitudes, and basins. Some regions exhibit prominent, but opposite chemical and isotopic shifts, whereas others are weakly sensitive to THC changes. These results suggest that deep sea records of foraminiferal $\Delta^{14}C$, Cd/Ca, and $\delta^{13}C$ need not necessarily exhibit a uniform response throughout the deep ocean during abrupt climatic changes.

APPENDIX A: BIOGEOCHEMICAL PROCESSES IN THE EUPHOTIC ZONE ($z < z_{eup}$)

The biological cycling and air-sea gas exchange of ^{14}C are based on the theoretical expectation that, for photosynthesis and surface gas exchange, the fractionation factor for the ^{14}C–^{12}C pair should be the square of the fractionation factor for the ^{13}C–^{12}C pair [*Craig*, 1954]. We consider separately the formulation of biogeochemical processes in the euphotic zone, in the aphotic zone, and at the sea surface.

Dissolved inorganic radiocarbon, $DI^{14}C$, is biologically consumed in the euphotic zone (top 100 m) through the formation of organic matter and carbonate particles. The volumetric rates of $DI^{14}C$ removal through the formation of organic matter, $J_{org}^{DI^{14}C}$, and carbonate particles, $J_{car}^{DI^{14}C}$, are expressed as:

$$J_{org}^{DI^{14}C} = R_w \alpha_{org}^2 J_{org}, \quad (A1)$$

$$J_{car}^{DI^{14}C} = R_w \alpha_{car}^2 r_p J_{org}, \quad (A2)$$

where R_w is the $DI^{14}C/DIC$ ratio, α_{org} and α_{car} are the fractionation factors for the pair ^{13}C–^{12}C for photosynthesis and calcification, respectively, J_{org} is the rate of DIC removal through the formation of organic matter, and r_p is the production ratio, i.e. the ratio between the production of $CaCO_3$ to the production of organic carbon in the euphotic zone. In our model, α_{org} depends on the concentration of aqueous CO_2 [*Rau et al.*, 1989], $\alpha_{car} = 1$ [*Mook*, 1986], J_{org} is described as a function of the concentration of PO_4 through Michaelis-Menten kinetics, and r_p is related to temperature [*Drange*, 1994].

The labile dissolved organic radiocarbon, $DO^{14}C_l$, is biologically produced in the euphotic zone owing to the formation of organic matter:

$$J_{org}^{DO^{14}C_l} = -\sigma \cdot J_{org}^{DI^{14}C} = -\sigma \cdot R_w \alpha_{org}^2 J_{org}, \quad (A3)$$

$$J_{car}^{DO^{14}C_l} = 0, \quad (A4)$$

where $\sigma = 0.5$ is the fraction of organic carbon sequestred into DOC_l [*Marchal et al.*, 1998a].

APPENDIX B: BIOGEOCHEMICAL PROCESSES IN THE APHOTIC ZONE $(z > z_{eup})$

$DI^{14}C$ is produced in the aphotic zone (below 100 m) through the remineralization of organic matter and the dissolution of carbonate particles. We denote by $J_{pom}^{DI^{14}C}$ the recycling of fast sinking particulate organic matter, and by $J_{dom}^{DI^{14}C}$ the recycling of labile dissolved organic matter. Thus,

$$J_{org}^{DI^{14}C} = J_{pom}^{DI^{14}C} + J_{dom}^{DI^{14}C} = -\frac{\partial F_{pom}^{DI^{14}C}}{\partial z} + \kappa\, DO^{14}C_l, \quad (B1)$$

$$J_{car}^{DI^{14}C} = -\frac{\partial F_{car}^{DI^{14}C}}{\partial z}, \quad (B2)$$

where $F_{pom}^{DI^{14}C}$ and $F_{car}^{DI^{14}C}$ are the fluxes of $DI^{14}C$ associated with fast-sinking POM and carbonate particles at a given depth in the water column, and κ is a first-order decay rate calculated so that the ocean inventory of DOC_l remains constant in the simulations [*Najjar et al.*, 1992].

$F_{pom}^{DI^{14}C}$ and $F_{car}^{DI^{14}C}$ depend on the fluxes of fast-sinking POM and $CaCO_3$ at the base of the euphotic zone:

$$F_{pom}^{DI^{14}C}(z) = F_{pom}^{DI^{14}C}(z_{eup}) \cdot \left(\frac{z}{z_{eup}}\right)^{-\epsilon} = -(1-\sigma)\int_0^{z_{eup}} J_{org}^{DI^{14}C}\, dz \cdot \left(\frac{z}{z_{eup}}\right)^{-\epsilon}, \quad (B3)$$

$$F_{car}^{DI^{14}C}(z) = F_{car}^{DI^{14}C}(z_{eup}) \cdot \left(e^{-(z-z_{eup})/L_{dis}}\right) = -\int_0^{z_{eup}} J_{car}^{DI^{14}C}\, dz \cdot \left(e^{-(z-z_{eup})/L_{dis}}\right), \quad (B4)$$

where $\epsilon = 0.858$ is the exponent in the fast sinking POM remineralization profile [*Bishop*, 1989] and $L_{dis} = 3000$ m is the length scale of the $CaCO_3$ dissolution profile [*Marchal et al.*, 1998a]. The fluxes $F_{pom}^{DI^{14}C}$ and $F_{car}^{DI^{14}C}$ at the ocean bottom are recycled in the deepest model layer as our model does not include sediment burial.

Finally, $DO^{14}C_l$ is oxidized in the aphotic zone according to

$$J_{org}^{DO^{14}C_l} = -\kappa\, DO^{14}C_l, \quad \text{and} \quad (B5)$$

$$J_{car}^{DO^{14}C_l} = 0. \quad (B6)$$

APPENDIX C: GAS EXCHANGE

The net flux of $^{14}CO_2$ from the ocean to the atmosphere, $F_{wa,n}^{^{14}CO_2}$, is expressed as:

$$F_{wa,n}^{^{14}CO_2} = R_w \alpha_{wa}^2 F_{wa}^{CO_2} - R_a \alpha_{aw}^2 F_{aw}^{CO_2}, \quad (C1)$$

where $F_{wa}^{CO_2}$ and $F_{aw}^{CO_2}$ are the gross fluxes of CO_2 from the ocean to the atmosphere and from the atmosphere to the ocean, R_w and R_a are the $^{14}C/(^{12}C + ^{13}C)$ ratios of surface DIC and atmospheric CO_2, and α_{wa} and α_{aw} are the fractionation factors for the pair ^{13}C–^{12}C for the air-sea gas exchange. In our model, $F_{wa}^{CO_2}$ and $F_{aw}^{CO_2}$ are related to the air-sea difference of the partial pressure of CO_2 via a constant transfer coefficient $\mu = 0.067$ mol m^{-2} yr^{-1} μatm^{-1}, whereas α_{aw} and α_{wa} depend on temperature and DIC speciation in the surface water [*Marchal et al.*, 1998a].

We introduce appropriate scalings in the formulation of $F_{wa,n}^{^{14}CO_2}$ in the case where ^{14}C is included as an inorganic tracer, in order to compare consistently $\Delta^{14}C_{org}$ with $\Delta^{14}C_{inorg}$. The formulation of *Stocker and Wright* [1996] is used:

$$F_{wa,n}^{^{14}CO_2} = g\alpha_{wa}^2 \xi [DI^{14}C] - g\alpha_{aw}^2 [^{14}CO_2]_a^* \quad (C2)$$

where g is the gas transfer velocity for CO_2, ξ is the buffer factor for $^{14}CO_2$ in seawater, and $[^{14}CO_2]^*_a$ is the concentration of atmospheric $^{14}CO_2$ in units of oceanic concentration. g is related to the CO_2 transfer coefficient μ through $g = \mu \cdot pCO^\circ_{2,w}/[DIC]^\circ$, where $pCO^\circ_{2,w}$ and $[DIC]^\circ$ are reference values of the partial pressure of CO_2 and DIC concentration in surface seawater, respectively. With $\mu = 0.067$ mol m^{-2} yr^{-1} μatm^{-1}, $pCO^\circ_{2,w}$ = 280 μatm, and $[DIC]^\circ = 2.052$ mol m^{-3}, we obtain $g = 9.1$ m yr^{-1}. We choose $\xi = 1$, which is a good approximation for $^{14}CO_2$ in seawater. Finally, $[^{14}CO_2]^*_a$ is calculated as $[^{14}CO_2]^*_a = [^{14}CO_2]_a \cdot \overline{[DIC]}^\circ / [CO_2]^\circ_a$, where $[^{14}CO_2]_a$ is the concentration of $^{14}CO_2$ in the atmosphere (mol m^{-3} of air), $\overline{[DIC]}^\circ = 2.250$ mol m^{-3} is a reference, ocean mean concentration of DIC, and $[CO_2]^\circ_a = 1.184 \cdot 10^{-2}$ mol m^{-3} is a reference concentration of atmospheric CO_2.

The concentrations of radiocarbon in the ocean and in the atmosphere are expressed in conventional $\Delta^{14}C$ units. In the organic case, $\Delta^{14}C$ is calculated as

$$\Delta^{14}C = (\frac{r_N}{r_{St}} - 1) \cdot 1000. \quad (C3)$$

$r_{St} = 1.176 \cdot 10^{-12}$ is the standard $^{14}C/^{12}C$ ratio and r_N is the ^{13}C-normalized activity given by

$$r_N = r\left(1 - \frac{2(\delta^{13}C + 25‰)}{1000‰}\right), \quad (C4)$$

where r is the $^{14}C/^{12}C$ ratio and $\delta^{13}C$ denotes the reduced isotopic ratio referenced to the PDB standard [*Craig*, 1957].

In the inorganic case, we omit isotopic fractionation during the gas exchange, i.e. $\alpha^2_{wa} = \alpha^2_{aw} = 1$. Thus, $\Delta^{14}C$ is calculated without the correction for isotopic fractionation. For the atmosphere:

$$\Delta^{14}C = \left(\frac{[^{14}CO_2]/[CO_2]^\circ_a}{r_{St}} - 1\right) \cdot 1000‰. \quad (C5)$$

For the ocean:

$$\Delta^{14}C = \left(\frac{[DI^{14}C]/\overline{[DIC]}^\circ}{r_{St}} - 1\right) \cdot 1000‰. \quad (C6)$$

Acknowledgments. J. R. Toggweiler and an anonymous reviewer provided useful comments. TS would like to thank P. Clark and R. Webb for a very stimulating Chapman Conference and the editorial efforts in completing this volume. This study was made possible by the Swiss National Science Foundation and the Swiss Federal Office of Science and Education through the European Projects ENV4-CT95-0131 "Variability of the Glacial and Interglacial Climates and Abrupt Climatic Changes" and ENV4-CT95-0130 "North-South Climatic Connection and Carbon Cycle over the last 250 kyr".

REFERENCES

Adkins, J. F., and E. A. Boyle, Changing atmospheric $\Delta^{14}C$ and the record of deep water paleoventilation ages, *Paleoceanography, 12*, 337–344, 1997.

Alley, R. B., D. A. Meese, C. A. Shuman, A. J. Gow, K. C. Taylor, P. M. Grootes, J. W. C. White, M. Ram, E. D. Waddington, P. A. Mayewski, and G. A. Zielinski, Abrupt increase in Greenland snow accumulation at the end of the Younger Dryas event, *Nature, 362*, 527–529, 1993.

Bard, E., M. Arnold, J. Mangerud, M. Paterne, L. Labeyrie, J. Duprat, M.-A. Mélières, E. Sønstegaard, and J.-C. Duplessy, The North Atlantic atmosphere–sea surface ^{14}C gradient during the Younger Dryas climatic event, *Earth Planet. Sci. Lett., 126*, 275–287, 1994.

Bard, E., F. Rostek, and C. Sonzogni, Interhemispheric synchrony of the last deglaciation inferred from alkenone paleothermometry, *Nature, 385*, 707–710, 1997.

Bender, M., T. Sowers, M.-L. Dickson, J. Orchardo, P. Grootes, P. A. Mayewski, and D. A. Meese, Climate correlations between Greenland and Antarctica during the past 100'000 years, *Nature, 372*, 663–666, 1994a.

Bender, M., T. Sowers, and L. Labeyrie, The Dole effect and its variations during the last 130,000 years as measured in the Vostok ice core, *Global. Biogeochem. Cycles, 8*, 363–376, 1994b.

Bishop, J. K. B., Regional extremes in particulate matter composition and flux: effects on the chemistry of the ocean interior, in *Productivity of the Ocean: Present and Past*, edited by W. H. Berger, V. S. Smetacek, and G. Wefer, pp. 117–137, John Wiley, New-York, 1989.

Björck, S., B. Kromer, S. Johnsen, O. Bennike, D. Hammarlund, G. Lemdahl, G. Possnert, T. L. Rasmussen, B. Wohlfahrt, C. U. Hammer, and M. Spurk, Synchronised terrestrial-atmospheric deglacial records around the North Atlantic, *Science, 274*, 1155–1160, 1996.

Blunier, T., J. Chappellaz, J. Schwander, A. Dällenbach, B. Stauffer, T. F. Stocker, D. Raynaud, J. Jouzel, H. B. Clausen, C. U. Hammer, and S. J. Johnsen, Asynchrony of Antarctic and Greenland climate change during the last glacial period, *Nature, 394*, 739-743, 1998.

Blunier, T., J. Schwander, B. Stauffer, T. Stocker, A. Dällenbach, A. Indermühle, J. Tschumi, J. Chappellaz, D. Raynaud, and J.-M. Barnola, Timing of temperature variations during the last deglaciation in Antarctica and the atmospheric CO_2 increase with respect to the Younger Dryas event, *Geophys. Res. Let., 24*, 2683–2686, 1997.

Boyle, E. A., Last-Glacial-Maximum North Atlantic Deep Water: on, off or somewhere in-between ?, *Phil. Trans. Roy. Soc., London, 348*, 243–253, 1995.

Boyle, E. A., and L. D. Keigwin, North Atlantic thermohaline circulation during the past 20,000 years linked to high-latitude surface temperature, *Nature, 330*, 35–40, 1987.

Broecker, W. S., D. Peteet, and D. Rind, Does the ocean-atmosphere system have more than one stable mode of operation ?, *Nature, 315*, 21–25, 1985.

Brook, E. J., T. Sowers, and J. Orchardo, Rapid variations in atmospheric methane concentration during the past 110,000 years, *Science, 273*, 1087–1091, 1996.

Cane, M. A., A role of the tropical Pacific, *Science, 282*, 59–61, 1998.

Chappellaz, J., T. Blunier, D. Raynaud, J. M. Barnola, J. Schwander, and B. Stauffer, Synchronous changes in atmospheric CH_4 and Greenland climate between 40 and 8 kyr BP, *Nature, 366*, 443–445, 1993.

Charles, C., and R. G. Fairbanks, Evidence from Southern Ocean sediments for the effect of North Atlantic deep-water flux on climate, *Nature, 355*, 416–419, 1992.

Charles, C. D., J. Lynch-Stieglitz, U. S. Ninnemann, and R. G. Fairbanks, Climate connections between the hemisphere revealed by deep sea sediment core/ice core correlations, *Earth Planet. Sci. Lett., 142*, 19–27, 1996.

Craig, H., Carbon 13 in plants and the relationship between carbon 13 and carbon 14 variations in nature, *J. Geol. 62*, 115–149, 1954.

Craig, H., Isotopic standards for carbon and oxygen and correction factors for mass-spectrometric analysis of carbon dioxide, *Geochim. Cosmochim. Acta, 12*, 133–149, 1957.

Crowley, T. J., North Atlantic deep water cools the southern hemisphere, *Paleoceanography, 7*, 489–497, 1992.

Cuffey, K. M., and G. D. Clow, Temperature, accumulation, and ice sheet elevation in central Greenland through the last deglacial transition, *J. Geophys. Res., 102*, 26383–26396, 1997.

Dansgaard, W., H. B. Clausen, N. Gundestrup, C. U. Hammer, S. F. Johnsen, P. M. Kristinsdottir, and N. Reeh, A new Greenland deep ice core, *Science, 218*, 1273–1277, 1982.

Dansgaard, W., S. J. Johnsen, H. B. Clausen, D. Dahl-Jensen, N. S. Gundestrup, C. U. Hammer, C. S. Hvidberg, J. P. Steffensen, A. E. Sveinbjornsdottir, J. Jouzel, and G. Bond, Evidence for general instability of past climate from a 250-kyr ice-core record, *Nature, 364*, 218–220, 1993.

Denton, G. H., and C. H. Hendy, Younger Dryas age advance of Franz Josef glacier in the southern alps of New Zealand, *Science, 264*, 1434–1437, 1994.

Drange, H., An isopycnic coordinate carbon cycle model for the North Atlantic; and the possibility of disposing of fossil fuel CO_2 in the ocean, Ph. D. thesis, 286 pp., Department of Mathematics, University of Bergen, Norway, 1994.

Edwards, R. L., W. J. Beck, G. S. Burr, D. J. Donahue, J. M. A. Chappell, A. L. Bloom, E. R. M. Druffel, and F. W. Taylor, A large drop in atmospheric $^{14}C/^{12}C$ and reduced melting in the Younger Dryas, documented with ^{230}Th ages of corals, *Science, 260*, 962–968, 1993.

Fanning, A. F., and A. J. Weaver, Temporal-geographical meltwater influences on the North Atlantic conveyor: Implications for the Younger Dryas, *Paleoceanography, 12*, 307–320, 1997.

Goslar, T., M. Arnold, E. Bard, T. Kuc, M. F. Pazdur, M. Ralska-Jasiewiczowa, K. Różański, N. Tisnerat, A. Walanus, B. Wicik, and K. Więckowski, High concentration of atmospheric ^{14}C during the Younger Dryas cold episode, *Nature, 377*, 414–417, 1995.

Goslar, T., B. Wohlfarth, S. Björck, G. Possnert, and J. Björck, Variations of atmospheric ^{14}C concentrations over the Allerød-Younger Dryas transition, *Clim. Dyn., 15*, 29–42, 1999.

Guilderson, T. P., R. G. Fairbanks, and J. L. Rubenstone, Tropical temperature variations since 20,000 years ago: Modulating interhemispheric climate change, *Science, 263*, 663–665, 1994.

Hughen, K. A., J. T. Overpeck, S. J. Lehman, M. Kashgarian, J. Southon, L. C. Peterson, R. Alley, and D. M. Sigman, Deglacial changes in ocean circulation from an extended radiocarbon calibration, *Nature, 391*, 65–68, 1998.

Imbrie, J., et al., On the structure and origin of major glaciation cycles, 1. Linear responses to Milankovitch forcing, *Paleoceanography, 7*, 701–738, 1992.

Jansen, E., and T. Veum, Evidence for two-step deglaciation and its impact on North Atlantic deep-water circulation, *Nature, 343*, 612–616, 1990.

Johnsen, S. J., H. B. Clausen, W. Dansgaard, K. Fuhrer, N. Gundestrup, C. U. Hammer, P. Iversen, J. Jouzel, B. Stauffer, and J. P. Steffensen, Irregular glacial interstadials recorded in a new Greenland ice core, *Nature, 359*, 311–313, 1992.

Johnsen, S. J., D. Dahl-Jensen, W. Dansgaard, and N. Gundestrup, Greenland palaeotemperatures derived from GRIP bore hole temperature and ice core isotope profiles, *Tellus, Ser. B, 47*, 624–629, 1995.

Jouzel, J., R. Vaikmae, J. R. Petit, M. Martin, Y. Duclos, M. Stievenard, C. Lorius, M. Toots, M. A. Mélières, L. H. Burckle, N. I. Barkov, and V. M. Kotlyakov, The two-step shape and timing of the last deglaciation in Antarctica, *Clim. Dyn., 11*, 151–161, 1995.

Keigwin, L. D., G. A. Jones, and S. J. Lehman, Deglacial meltwater discharge, North Atlantic deep circulation and abrupt climate change, *J. Geophys. Res., 96*, 16811–16826, 1991.

Keigwin, L. D., and S. J. Lehman, Deep circulation change linked to Heinrich event 1 and Younger Dryas in a mid-depth North Atlantic core, *Paleoceanography, 9*, 185–194, 1994.

Kitagawa, H., and J. van der Plicht, Atmospheric radiocarbon calibration to 45,000 yr BP: Late glacial fluctuations and cosmogenic isotope production, *Science, 279*, 1187–1190, 1998.

Lowell, T. V., C. J. Heusser, B. G. Anderson, P. I. Moreno, A. Hauser, L. E. Heusser, C. Schlüchter, D. R. Marchant, and G. H. Denton, Interhemispheric correlation of Late Pleistocene glacial events, *Science, 269*, 1541–1549, 1995.

Mabin, M. C. G., The age of the Waiho Loop glacial event, *Science, 271*, 668, 1996.

Manabe, S., and R. J. Stouffer, Simulation of abrupt climate change induced by freshwater input to the North Atlantic Ocean, *Nature, 378*, 165–167, 1995.

Manabe, S., and R. J. Stouffer, Coupled ocean-atmosphere model response to freshwater input: Comparison to

Younger Dryas event, *Paleoceanography, 12*, 321–336, 1997.

Marchal, O., T. F. Stocker, and F. Joos, A latitude-depth, circulation-biogeochemical ocean model for paleoclimate studies. Development and sensitivities, *Tellus, Ser. B, 50*, 290–316, 1998a.

Marchal, O., T. F. Stocker, and F. Joos, Impact of oceanic reorganizations on the ocean carbon cycle and atmospheric carbon dioxide content, *Paleoceanography, 13*, 225–244, 1998b.

Marchal, O., T. F. Stocker, F. Joos, A. Indermühle, T. Blunier, and T. Tschumi, Modelling the concentration of atmospheric CO_2 during the Younger Dryas climate event, *Clim. Dyn., 15*, 341–354, 1999.

McIntyre, A., and B. Molfino, Forcing of Atlantic Equatorial and Subpolar millennial cycles by precession, *Science, 274*, 1867–1870, 1996.

Mikolajewicz, U., A meltwater induced collapse of the 'conveyor belt' - Thermohaline circulation and its influence on the distribution of $\Delta^{14}C$ and $\delta^{18}O$, *Tech. Rep. 189*, 25 pp., Max-Planck-Inst. für Meteorol., Hamburg, Germany, 1996.

Mikolajewicz, U., T. J. Crowley, A. Schiller, and R. Voss, Modelling teleconnections between the North Atlantic and North Pacific during the Younger Dryas, *Nature, 387*, 384–387, 1997.

Mook, W. G., ^{13}C in atmospheric CO_2, *Netherlands J. Sea. Res., 20*, 211–223, 1986.

Najjar, R. G., J. L. Sarmiento, and J. R. Toggweiler, Downward transport and fate of organic matter in the ocean: Simulations with a general circulation model, *Global. Biogeochem. Cycles, 6*, 45–76, 1992.

Oeschger, H., J. Beer, U. Siegenthaler, B. Stauffer, W. Dansgaard, and C. C. Langway, Late glacial climate history from ice cores, in *Climate Processes and Climate Sensitivity, Geophys. Monogr. Ser., vol. 29*, edited by J. E. Hansen and T. Takahashi, pp. 299–306, AGU, Washington, D. C., 1984.

Peteet, D., Global Younger Dryas?, *Quat. Int., 28*, 93–104, 1995.

Rau, G. H., T. Takahashi, and D. J. D. Marais, Latitudinal variations in plankton $\delta^{13}C$: Implications for CO_2 and productivity in past oceans, *Nature, 341*, 516–518, 1989.

Schiller, A., U. Mikolajewicz, and R. Voss, The stability of the North Atlantic thermohaline circulation in a coupled ocean-atmosphere general circulation model, *Clim. Dyn., 13*, 325–347, 1997.

Severinghaus, J. P., T. Sowers, E. J. Brook, R. B. Alley, and M. L. Bender, Timing of abrupt climate change at the end of the Younger Dryas interval from thermally fractionated gases in polar ice, *Nature, 391*, 141–146, 1998.

Siegenthaler, U., and H. Oeschger, Biospheric CO_2 emissions during the past 200 years reconstructed by convolution of ice core data, *Tellus, Ser. B, 39*, 140–154, 1987.

Singer, C., J. Shulmeister, and B. McLea, Evidence against a significant Younger Dryas cooling event in New Zealand, *Science, 281*, 812–814, 1998.

Sowers, T., and M. Bender, Climate records covering the last deglaciation, *Science, 269*, 210–213, 1995.

Steig, E. J., E. J. Brook, J. W. C. White, C. M. Sucher, M. L. Bender, S. J. Lehman, D. L. Morse, E. D. Waddington, and G. D. Clow, Synchronous climate changes in Antarctica and the North Atlantic, *Science, 282*, 92–95, 1998.

Stocker, T. F., The seesaw effect, *Science, 282*, 61–62, 1998.

Stocker, T. F., Past and future reorganisations in the climate system, *Quat. Sci. Rev.*, 1999 (in press).

Stocker, T. F., and D. G. Wright, Rapid changes in ocean circulation and atmospheric radiocarbon, *Paleoceanography, 11*, 773–796, 1996.

Stocker, T. F., and D. G. Wright, The effect of a succession of ocean ventilation changes on radiocarbon, *Radiocarbon, 40*, 359–366, 1998.

Stocker, T. F., D. G. Wright, and L. A. Mysak, A zonally averaged, coupled ocean-atmosphere model for paleoclimate studies, *J. Clim., 5*, 773–797, 1992a.

Stocker, T. F., D. G. Wright, and W. S. Broecker, The influence of high-latitude surface forcing on the global thermohaline circulation, *Paleoceanography, 7*, 529–541, 1992b.

Thompson, L. G., T. Yao, M. E. Davis, K. A. Henderson, E. Mosley-Thompson, P.-N. Lin, J. Beer, H.-A. Synal, J. Cole-Dai, and J. F. Bolzan, Tropical climate instability: The last glacial cycle from a Qinghai-Tibetan ice core, *Science, 276*, 1821–1825, 1997.

Toggweiler, J., and B. Samuels, Is the magnitude of the deep outflow from the Atlantic Ocean actually governed by southern hemisphere winds ?, in *The Global Carbon Cycle*, edited by M. Heimann, pp. 303–331, NATO ASI Ser., Ser. I, 15, Springer Verlag, 1993.

Veum, T., E. Jansen, M. Arnold, I. Beyer, and J.-C. Duplessy, Water mass exchange between the North Atlantic and the Norwegian Sea during the past 28,000 years, *Nature, 356*, 783–785, 1992.

Weyl, P., The role of the ocean in climatic change: a Theory of the ice ages, *Meteorol. Monogr. 8*, 37–62, 1968.

Wright, D. G., and T. F. Stocker, Sensitivities of a zonally averaged global ocean circulation model, *J. Geophys. Res., 97*, 12,707–12,730, 1992.

Wright, D. G., and T. F. Stocker, Younger Dryas experiments, in *Ice in the Climate System*, edited by W. R. Peltier, pp. 395–416, NATO ASI Ser., Ser. I, 12, Springer Verlag, 1993.

Wright, D. G., T. F. Stocker, and D. Mercer, Closures used in zonally averaged ocean models, *J. Phys. Oceanogr., 28*, 791–804, 1998.

O. Marchal, T. F. Stocker, and F. Joos, Climate and Environmental Physics, Physics Institute, University of Bern, 5 Sidlerstraße, CH-3012 Bern, Switzerland. (e-mail: marchal@climate.unibe.ch; stocker@climate.unibe.ch; joos@climate.unibe.ch)

Millennial Timescale Variability in Ocean/Climate Models

Andrew J. Weaver

School of Earth & Ocean Sciences, University of Victoria, Victoria, B.C., Canada

A review of mechanisms for millennial timescale variability from a hierarchy of ocean models is presented together with a comparison with observations from the last glaciation (Dansgaard-Oeschger oscillations, Heinrich events and Bond Cycles). Special attention is given to a review of modeling efforts aimed at unraveling the causes and consequences of the Younger Dryas cooling event (12,700 – 11,650 years BP) as well as potential mechanisms for interglacial millennial timescale variability. Finally, some recent experiments are included which examine the influence of Heinrich event runoff (obtained from a continental ice sheet model) on the global ocean circulation in a coupled atmosphere-ocean-sea ice model of intermediate complexity.

1. INTRODUCTION

Early ice core records for the last glaciation have revealed intense variability on the millennial timescale characterized by abrupt warming events (interstadials) lasting from several hundred to several thousand years. These oscillations (Fig. 1), known as Dansgaard/Oeschger (D-O) oscillations (after the pioneering work of *Oeschger et al.* [1984] and *Dansgaard et al.* [1984]) are also apparent in North Atlantic sediment records [*Bond et al.*, 1993], suggesting a role or response of the ocean. The last such event, known as the Younger Dryas event, took place between 12,700 and 11,650 years BP [*Dansgaard et al.*, 1989; *Taylor et al.*, 1993; *Stocker,* 1998] and terminated abruptly within a few decades [*Dansgaard et al.*, 1989; *Alley et al.*, 1993]. Evidence from the Santa Barbara basin [*Kennett and Ingram,* 1995; *Behl and Kennett,* 1996] and the Northeast Pacific [*Lund and Mix,* 1998] suggests that a signature of these D-O oscillations is also present in the Pacific, while further recent sediment analyses suggest they may be an inherent part of late [*Oppo et al.*, 1998] and early [*Raymo et al.*, 1998] Pleistocene climate.

Mechanisms of Global Climate Change at Millennial Time Scales
Geophysical Monograph 112

In an attempt to provide a mechanism for the observed D-O variability, *Broecker et al.* [1990] proposed that during glacial times, when the northern end of the Atlantic Ocean was surrounded by ice sheets, a stable mode of operation of the *conveyor belt* for North Atlantic Deep Water (NADW) was not possible. They further suggested that when the NADW conveyor was weakened or shut down and there were growing ice sheets, there would be little oceanic salt export from the Atlantic to the other world basins. Assuming the North Atlantic surface freshwater balance in this weakened or shut down conveyor state remained with evaporation dominating over precipitation, the North Atlantic salinity would continually increase with the moisture being deposited on land as snow, thereby growing ice sheets. Upon reaching a critical salinity, deep convection and subsequently the conveyor would turn on, transporting and releasing heat to the North Atlantic and thereby melting back the ice sheets. The flux of fresh water into the North Atlantic from the melting ice sheets (or enhanced ice berg calving as reconstructed by *Bond and Lotti* [1995]) eventually would reduce or shut off the conveyor and the process would begin anew. The results from a simple model previously developed by *Birchfield and Broecker* [1990] provided quantitative support for this mechanism.

Heinrich [1988], in analyzing marine sediments in three cores from the North Atlantic, noted the presence of six anomalous concentrations of lithic fragments over the last glaciation. Since the source for these fragments was the

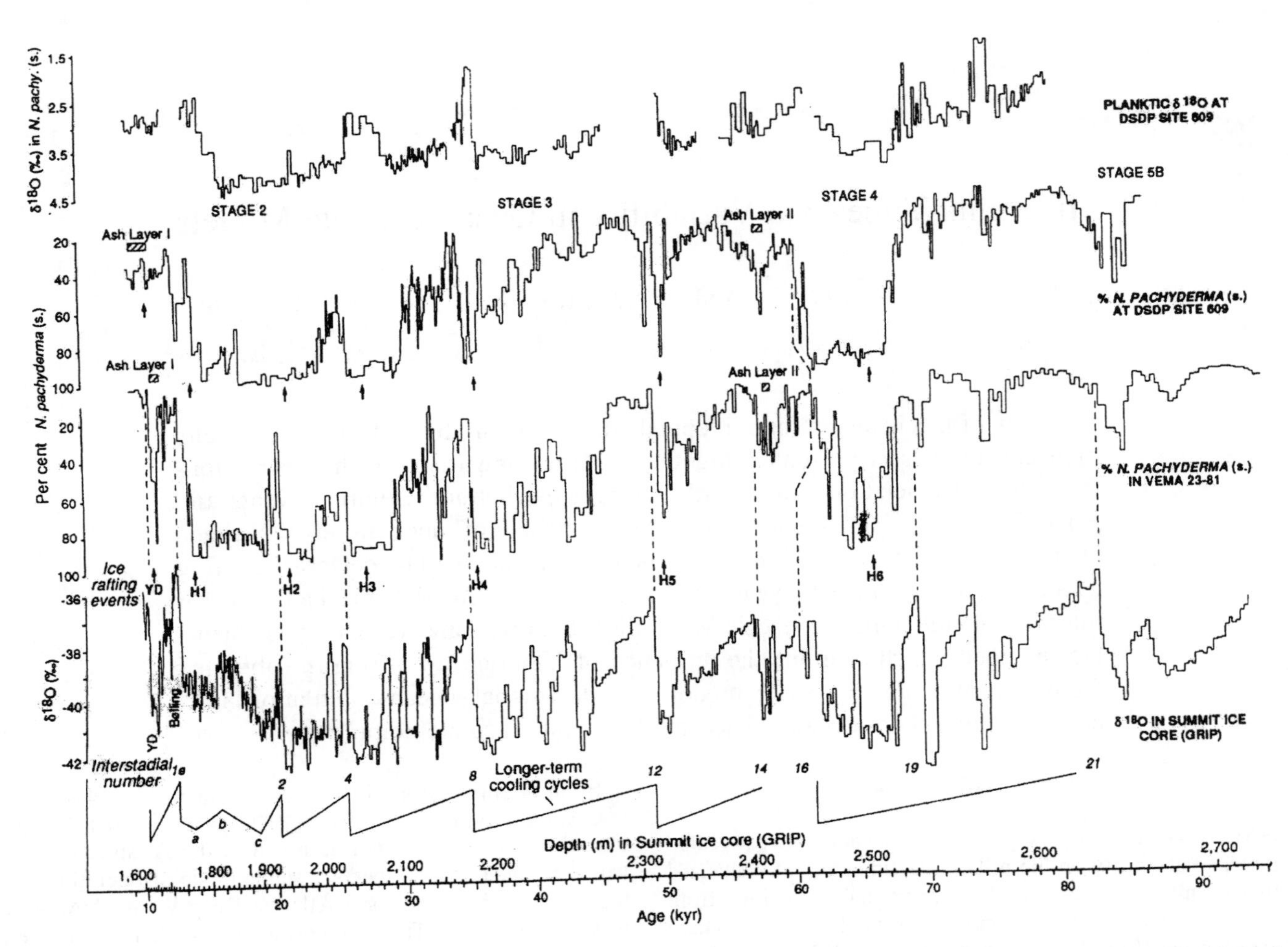

Figure 1. Sediment foraminiferal records from Bond et al. [1993] and ice core oxygen isotope records from Dansgaard et al. [1993] and GRIP [1993] as correlated by Bond et al. [1993]. The Younger Dryas (YD) and six Heinrich evens (H1–6) are indicated. Also shown at the bottom is the saw tooth Bond Cycle pattern of successively weaker interstadials associated with a sequence of D-O oscillations following a Heinrich event (see Bond et al., 1993 for a more detailed discussion — taken from Bond et al. [1993]).

land (and in particular from Canada — *Bond et al.* [1992]), he argued that this provided evidence for six anomalous surges of icebergs into the North Atlantic (Fig. 1). *Broecker et al.* [1992] noted that the so called Heinrich events were even more striking when expressed in terms of the ratio of lithic fragments to the sum of lithic fragments and foraminifera shells, due to the low foraminifera counts in the Heinrich sediment layers. *MacAyeal* [1993] developed a simple model to illustrate the mechanism for Laurentide ice-sheet instability which ultimately would give rise to an ice-sheet surge and a Heinrich event. He argued that the Heinrich cycle consisted of two phases. In the growth phase the Laurentide ice sheet would grow through snow accumulation while remaining attached to the bedrock. In the purge phase, he suggested that the high pressures under the deep ice sheet would cause thawing near its base, thereby allowing it to surge seaward through Hudson Strait. He pointed out that the resulting freshwater discharge into the North Atlantic would be of the order of 0.16 Sv (1 Sv ≡ $10^6\,m^3\,s^{-1}$) over a period as short as 250–500 years. Recent evidence [*Hewitt et al.*, 1997] suggests that Heinrich events (and associated ice rafted debris) also have a signature in the northeast Pacific Ocean.

As noted by *Bond et al.* [1993] and *Broecker* [1994] the Heinrich events, appearing about every 10,000 years, occur at the end of a sequence of D-O cycles during a prolonged cold period. *Bond et al.* [1993] further noted that the sequences of D-O oscillations tended to follow a saw-tooth cycle (now termed a Bond Cycle) with successive D-O oscillations involving progressively cooler interstadials (Fig. 1). They argued that this Bond Cycle was terminated by a Heinrich event, after which a rapid warming occurred and the process began anew. One possible interpretation of these results is that rather than terminating a Bond Cycle, the Heinrich event starts the Bond Cycle with the ocean subsequently responding like a damped oscillator to the large initial perturbation.

In this article a review is presented detailing previous modeling efforts at attempting to understand millennial timescale climate variability (section 2). In addition, experiments are presented (section 3) in which a coupled atmosphere ocean model [*Fanning and Weaver*, 1996] is used to examine the climatic response to meltwater discharge for two typical Heinrich events as obtained in the continental ice model of *Marshall and Clarke* [1997a,b].

2. MILLENNIAL TIMESCALE VARIABILITY IN OCEAN/CLIMATE MODELS

In this section previous modeling efforts aimed at understanding millennial timescale climate variability are reviewed. As many of these modeling efforts used ocean-only models, a brief discussion of the boundary conditions often used in these models is initially presented (section 2.1). A more general discussion of the mechanisms for sustained internal millennial timescale variability in ocean models is then given (section 2.2), moving to two specific examples involving the Younger Dryas event (section 2.3) and the Eemian interglacial period which occurred between 115,000–135,000 years BP (section 2.4).

2.1. Mixed Boundary Conditions

The heat and freshwater flux coupling between the ocean and the atmosphere occur on different timescales and involve different physical processes. The lag of sea surface temperature (SST) behind the seasonal cycle of insolation is on the order of six weeks [*Bretherton*, 1982], while the longwave emission, sensible heating and atmospheric saturation specific humidity (hence latent heat fluxes) depend on temperature, allowing for the use of a simple, linear, restoring boundary condition. The upper layer of the ocean (or the reservoir representing it in a box, or more complicated, model), is restored to an appropriate reference temperature on a fast timescale of typically 1-2 months [*Haney*, 1971]. The boundary condition therefore takes the form of a variable flux (in W m^{-2}; positive Q_T means heat out of the ocean):

$$Q_T = \frac{\rho_0 C_p \Delta z_1}{\tau_R}\Big(T_1(\lambda,\phi) - T_a(\lambda,\phi)\Big) \tag{1}$$

where $T_1(\lambda,\phi)$ is the temperature of the upper ocean box (with thickness Δz_1) at longitude λ and latitude ϕ. $T_a(\lambda,\phi)$ is the atmospheric reference temperature, C_p is the specific heat at constant pressure (~4000 J kg^{-1} °C^{-1}), ρ_0 is a reference density (~1000 kg m^{-3}), and τ_R is a restoring timescale [*Haney*, 1971].

In ocean models it is appropriate to represent freshwater fluxes at the ocean surface (due to evaporation, precipitation, river runoff or ice formation/melting) as a surface boundary condition on salinity (but see *Huang* [1993]). However, evaporation is mainly a function of the air-sea temperature difference while the distribution of precipitation depends upon complicated small and large scale atmospheric processes. The use of a restoring boundary condition on salinity (2 — units are g salt m^{-2}s^{-1})

$$Q_S = \frac{\rho_0 \Delta z_1}{\tau_R}\Big(S_1(\lambda,\phi) - S_a(\lambda,\phi)\Big), \tag{2}$$

is therefore not possible to justify physically as it implies a definite timescale (τ_R) for the removal of sea surface salinity (SSS) anomalies, which is not observed. Furthermore, (2) implies that the amount of precipitation or evaporation at any given place depends on the local SSS $S_1(\lambda,\phi)$, which is clearly incorrect. To resolve this problem in ocean-only models, one prefers to impose either *specified* salinity fluxes Q_S or a salinity flux which depends weakly on the atmosphere-ocean temperature difference. The salinity fluxes Q_S may then be converted to implied freshwater water fluxes (P–E) by

$$P - E\,(m/yr) = -\frac{syr \times Q_S}{\rho_0 S_0}, \tag{3}$$

where syr is the number of seconds in a year and S_0 is a constant reference salinity (~34.7 psu). A constant reference salinity is used in (3) instead of the local salinity $S_1(\lambda,\phi)$ so that when (3) is integrated over the surface of the ocean, zero net P-E corresponds to zero net Q_S.

In ocean modeling, the term *mixed boundary conditions* has been given to surface boundary conditions which involve a restoring condition on temperature and a specified flux on salinity (in a strict mathematical sense, mixed boundary conditions should refer to a mixed Dirichlet/ Neumann boundary condition on one variable). While these boundary conditions are admittedly crude, they do reflect the different nature of the observed SSS and SST coupling between the ocean and the atmosphere. In the uncoupled models discussed below these boundary conditions will usually be used.

Due to the lack of open ocean observations needed to determine evaporation through bulk formulae and precipitation directly, it was common to obtain a surface freshwater flux for use in uncoupled ocean models by spinning up a model to equilibrium under restoring boundary conditions on both T and S and then diagnosing the salt flux at the steady state. That is, (1) and (2) are used in the initial spin up and then at steady state the right hand side of (2) is diagnosed at each grid box to yield a two dimensional salt flux field (this can then be converted to an implied freshwater flux using Eq. 3). The rationale for this approach is that by spinning up the model using some specified climatological surface restoring fields, one obtains an equilibrium in which the surface fields of T and S are close to climatology. The diagnosed P-E field is then that field which, in theory, should yield the climatological SSS field. Furthermore, the equilibrium under restoring boundary conditions is also an equilibrium under the diagnosed mixed boundary conditions. Paradoxically, however, if the model simulates the SSS field exactly under restoring boundary conditions then there is zero P-E. Others have chosen to avoid this paradox by simply specifying a prescribed, often idealized, P-E field together with a restoring boundary condition on SST.

2.2. Millennial Timescale Variability

Marotzke [1989] spun up a single-hemisphere ocean general circulation model (OGCM) under restoring boundary conditions on temperature and salinity with no wind forcing. Upon switching to the diagnosed mixed boundary conditions and adding a small fresh perturbation to the high latitude salinity budget at equilibrium, a polar halocline catastrophe [*Bryan,* 1986] set in. Several thousand years later the system evolved into a quasi-steady state with weak equatorial downwelling (a weak inverse circulation). This state was not stable since low latitude diffusion acted to make the deep waters warm and saline with horizontal diffusion acting to homogenize these waters laterally. Eventually, at high latitudes the deep waters became sufficiently warm such that the water column became statically unstable and rapid convection set in. As in his zonally-averaged model [*Marotzke et al.*, 1988], the result was a *flush* in which a violent overturning (up to 200 Sv) occurred whereby the ocean lost all the heat it had taken thousands of years to store in a matter of a few decades. At the end of the flush the system continued to oscillate for a few decades until the circulation once more collapsed. In the presence of wind forcing, *Marotzke* [1990] found that no flush existed. The inevitability of the occurrence of a flush under a purely buoyancy forced, diffusive regime was elegantly illustrated in an analytic model developed by *Wright and Stocker* [1991].

Weaver and Sarachik [1991] undertook experiments of similar design to those of *Marotzke* [1989, 1990]. Contrary to the findings of *Marotzke* [1989, 1990] they observed the occurrence of flushes (Fig. 2a) even when wind forcing was included. Once more, in the collapsed state (Fig. 3a) low latitude diffusion and the subsequent horizontal homogenization of these waters tended to warm the deep waters (Fig. 2b) until static instability was detected at high latitudes. The result was the onset of a violent flush (Fig. 3b) which, as in *Marotzke* [1989, 1990], released all the heat stored over hundreds of years in a matter of a few decades. In the two-hemisphere experiments of *Weaver and Sarachik* [1991] they found that flushes still occurred although they were slightly weaker and of one-cell (pole-to-pole) structure. *Winton and Sarachik* [1993], *Winton* [1993] and *Winton* [1995] also found flushes (see also *Huang* [1994]; *Chen and Ghil* [1995]; *Sakai and Peltier* [1995, 1996]) in their frictional planetary geostrophic model under mixed boundary conditions but now termed them *deep decoupling oscillations.* In the coupled phase (where the deep and surface ocean were strongly coupled) they showed that advective coupling (Fig. 3b) acted to cool the global ocean whereas in the decoupled phase (Fig. 3a), diffusion of heat from the surface to the deep ocean was the dominant process linking the surface and deep layers (as in *Weaver and Sarachik* [1991] and *Wright and Stocker* [1991]).

The apparent discrepancy between the work of *Marotzke* [1989, 1990] and *Weaver and Sarachik* [1991] regarding the occurrence of flushes was resolved by *Weaver et al.* [1993]. They showed that the existence of flushes is linked to both the relative importance of freshwater flux over thermal forcing, and the strength of the wind forcing compared to

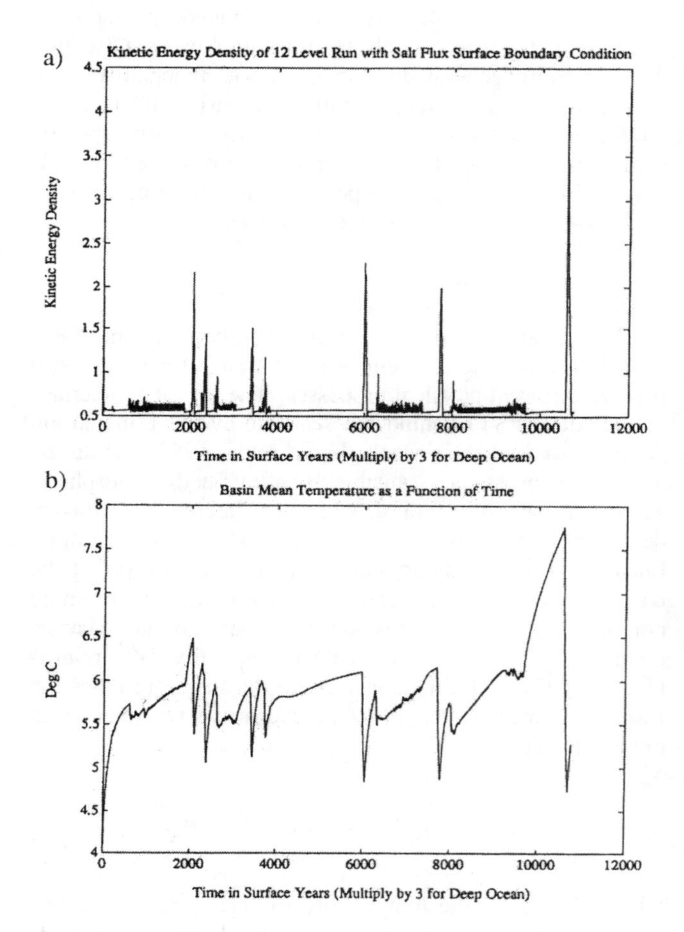

Figure 2. a)— Kinetic energy density (10^{-1} kg m^{-1} s^{-2}); b)— basin mean temperature (°C) throughout one of the single-hemisphere integrations of Weaver and Sarachik [1991]. The sharp peaks in a) and b) represent the occurrence of flushes whereas the more rapid oscillations indicate decadal variability (taken from Weaver and Sarachik [1991]).

the high latitude freshening. The latter balance was also investigated in detail in *Marotzke* [1990]. Comparing the equilibria obtained under restoring boundary conditions with and without wind, he found that at middle and high latitudes the thermohaline circulation provided the dominant transport mechanism for the meridional fluxes of heat and salt. Moreover, the strength of the meridional overturning was little influenced by the wind field, except for the Ekman transport in the top layer and its return flow which takes place in the 200 meters below the top layer. Because the stratification is nearly homogeneous in near-surface layers at high latitudes, the *Ekman cells* (i.e. Ekman transport plus return flow) contribute very little to the meridional transports. The situation changes, however, after the polar halocline catastrophe has occurred in, for example, the northern hemisphere: The surface layer is very fresh, compared to the layers below, and the southward Ekman transport of very fresh water is compensated by a northward transport of more saline water, resulting in a net northward transport of salt. Moreover, the northward salt transport due to the horizontal subtropical gyre increases substantially during the collapsed phase of the thermohaline overturning.

When the thermohaline circulation has collapsed, the wind-driven northward transport of salt amounts to about half the value of the total transport of the spin-up steady state. Thus, the high-latitude surface freshening is counteracted by the wind-driven salt transport, which in the case of *Marotzke* [1989, 1990] was strong enough to make the high-latitude surface waters sufficiently saline again, so that deep convection resumed and the thermohaline circulation reestablished itself. Strong surface freshening (as in *Weaver and Sarachik* [1991]) cannot be compensated by the wind-driven salt transport, and the thermohaline circulation remains in the collapsed state until a flush sets in.

Weaver et al. [1993] further examined the robustness of these flushes to the inclusion of a stochastic term added to the imposed surface freshwater flux field. In particular, they showed that as the magnitude of the stochastic term increased, the frequency of the flushing events increased while their intensity decreased. When there was no stochastic forcing, deep ocean temperatures warmed up to order 9°C in their model before static instability was detected at high latitudes, thereby inducing convection and a flush. With the inclusion of a stochastic term to the freshwater forcing field, flushes tended to occur earlier, before the ocean had warmed as much, and even earlier still as the magnitude of the stochastic forcing was increased. With increasing magnitude of the stochastic freshwater flux forcing, the probability of a sufficiently large evaporation anomaly increases, and the flushes occur more often. The basin mean temperature has not warmed as much so that the ocean loses less heat during the (thus weaker) flushing event. A similar result regarding the frequency and intensity of flushes was found when a seasonal cycle was imposed on the freshwater flux field [*Myers and Weaver,* 1992].

Zhang et al. [1993] argued, through use of a planetary geostrophic ocean model coupled to a *Schopf* [1983] zero-heat capacity atmosphere, that the occurrence of the polar halocline catastrophe and subsequent flush/collapse oscillation was less likely when the atmosphere was allowed to have a finite heat capacity (a restoring boundary condition on temperature assumes an infinite heat capacity atmosphere). They found a critical timescale of anomaly damping (~200 days) greater than which the polar halocline did not form in their model. Other more recent studies with atmospheres of finite heat capacity (*Mikolajewicz and Maier Reimer* [1994]; *Tziperman et al.* [1994]; *Power et al.* [1994]; *Pierce et al.* [1996]; *Rahmstorf and Willebrand,* [1995] — see also discussion in *Chen and Ghil* [1996]) have also suggested that the ocean may have a lesser degree of internal thermohaline variability than portrayed in mixed boundary condition experiments when a fast restoring

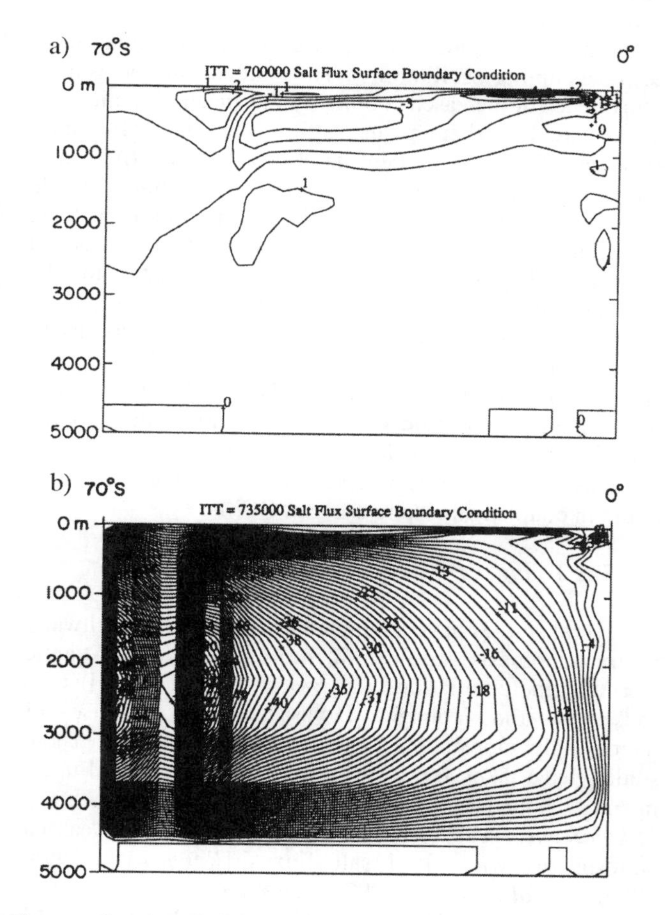

Figure 3. a)— Collapsed thermohaline state corresponding to year 1971 (refer to Fig. 2) immediately before the flush shown in b) at year 2067. All contours are in Sv (taken from Weaver and Sarachik [1991]).

timescale is used on SST. Those studies that included salinity in the equation of state still retained a fixed salt flux at the surface for salinity. *Hughes and Weaver* [1996] increased the realism of the fixed salt flux component of mixed boundary conditions by allowing for a feedback between SST and local evaporation. They found that this feedback both reduced the magnitude and the period of the flushing events although they were still possible to realize. The increased period and reduced intensity was due to an upper ocean advective timescale which dominated over the slower deep ocean diffusive timescale, discussed above, as the collapsed stage of the thermohaline circulation was never fully realized.

Using an energy balance atmosphere model coupled to a meridional-vertical plane ocean model, once more with a fixed salt flux used as a boundary condition on salinity, *Winton* [1997] showed that the flush/collapse or deep decoupling oscillation could indeed exist (see also *Sakai and Peltier* [1997]). He showed that the stability of the thermohaline circulation strongly depends on the mean climatic state itself with colder climates allowing polar halocline catastrophes (and subsequent flush/collapse or deep decoupling oscillations). He pointed out the general equivalence of increased cooling of the mean climatic state to increased freshening in setting the stage for their existence. Thus, even when the atmosphere has finite heat capacity, these oscillations can exist if the high latitude freshening is strong enough or the mean climatic state is sufficiently cool. It is therefore apparent that the actual thermal and freshwater forcing is fundamental to the existence of internal ocean variability on the millennial timescale whereas the existence of a finite heat capacity in the atmosphere is not. As pointed out by *McWilliams* [1996], however, the issue as to whether this mechanism for millennial timescale variability has application to the real climate system is far from resolved. To date such sustained millennial timescale variability has not been found in coupled atmosphere-ocean GCMs.

2.3. The Younger Dryas Event

Several studies have attempted to examine the meltwater influx hypothesis and its influence on the North Atlantic thermohaline circulation (e.g. *Broecker et al.* [1985]) as a cause for the Younger Dryas through the use of increasingly more sophisticated numerical models. These studies have taken the form of an ocean-only model under mixed boundary conditions [*Maier-Reimer and Mikolajewicz*, 1989], ocean models coupled to idealized atmospheres with fixed salt fluxes [*Rahmstorf*, 1994, 1995; *Mikolajewicz*, 1997], idealized coupled ocean-atmosphere models [*Stocker and Wright*, 1991], an OGCM coupled to an energy/moisture balance atmospheric model [*Fanning and Weaver*, 1997] and complex coupled ocean-atmosphere GCMs [*Manabe and Stouffer*, 1995, 1997; *Schiller et al.*, 1997]. Most of these studies have employed

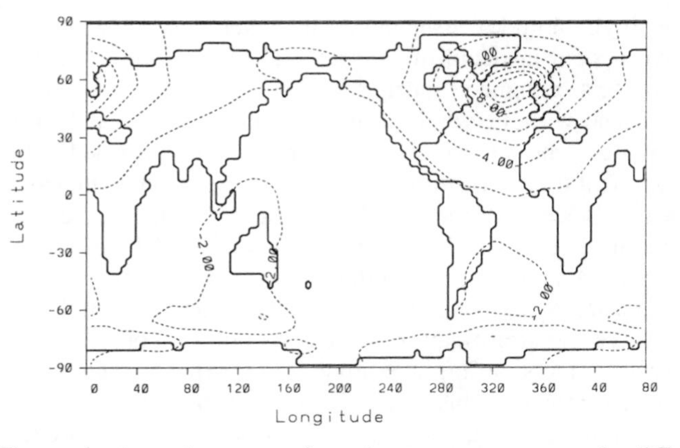

Figure 4. Annual mean surface air temperature anomaly (°C) from the present day equilibrium climatology in the middle of the Younger Dryas experiment. The contour interval is 1°C (taken from Fanning and Weaver [1997]).

highly idealized freshwater perturbations generally neglecting either temporal or geographical variations. *Fanning and Weaver* [1997] on the other hand investigated the temporal and geographical significance of meltwater pulses emanating from the Laurentide Ice Sheet using meltwater diversion estimates from *Teller* [1990]. From their equilibrium present-day climatology they conducted long integrations subject to three 500 year meltwater pulse episodes.

Through the analysis of a number of sensitivity experiments, *Fanning and Weaver* [1997] suggested that prior to the Younger Dryas, preconditioning by meltwater discharge to the Mississippi (in their first 500 year interval) was important in pushing NADW beyond the limit of its sustainability. They then argued that the diversion of meltwater to the St. Lawrence (second 500 year interval) merely served to completely inhibit NADW production. It is on the results from this work that attention is now focused.

The global mean 'Younger Dryas' surface air temperature (SAT) from *Fanning and Weaver* [1997] shown in Fig. 4 reveals a global pattern and regional magnitudes which generally agree with the changes seen in paleoclimatic reconstructions (see *Fanning and Weaver* [1997] for a comparison). The strong local response in the region surrounding the North Atlantic is intimately linked to a reduction in NADW formation there (Fig. 5). Interhemispheric teleconnection of the signal to the Southern Ocean arises from changes in the thermohaline circulation while changes in the atmospheric heat transport (in response to a global redistribution of oceanic heat transport) also provide a mechanism for interbasin teleconnection. Reestablishment of NADW occurs once the freshwater pulse is removed (after the third 500 year interval of Younger Dryas runoff from *Teller* [1990]) due to wind stress/speed feedbacks. As also found by *Schiller et al.* [1997] and *Mikolajewicz* [1997] when the thermohaline circulation has collapsed and a fresh halocline exists, the wind-driven salt transport (Ekman plus subtropical gyre) is an efficient mechanism for removal of the fresh anomaly so that deep convection can resume and NADW formation reestablish (see section 2.2).

The timing for the collapse of NADW in *Fanning and Weaver* [1997] from initial freshwater perturbation to subsequent reestablishment is about 2500 years, slightly longer than in observations. This is potentially misleading as the cold phase (collapsed state) of the integration is only about 1000 years (Fig. 5) surrounded by a relatively rapid (500 year) weakening and slow (1000 year) reestablishment. While the mechanism for collapse and reestablishment is the same as other coupled modeling studies [*Schiller et al.*, 1997; *Mikolajewicz*, 1997; *Manabe and Stouffer*, 1997], the timescale in *Fanning and Weaver* [1997] is longer. The reason for the discrepancy is unclear but may be associated with the use of fixed salt flux fields applied in previous studies [*Mikolajewicz*, 1997], or the duration, strength, and geographical location of the imposed and idealized meltwater applied [*Schiller et al.*, 1997; *Mikolajewicz*, 1997; *Manabe and Stouffer*, 1997]. Alternatively, as *Fanning and Weaver* [1997] applied the *Teller* [1990] runoff estimates as an external perturbation to their internal hydrological cycle (and hence runoff), their meltwater runoff data may be overestimated.

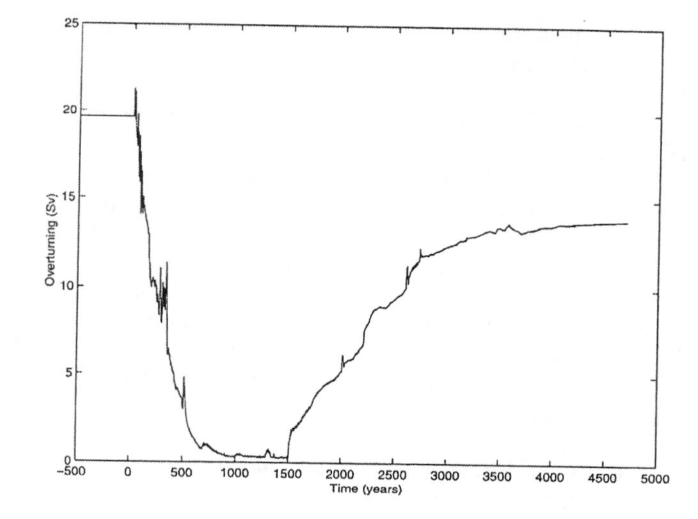

Figure 5. Time evolution of the maximum meridional overturning streamfunction (Sv) in the North Atlantic basin (strength of the conveyor) for the coupled model Younger Dryas meltwater experiment [Fanning and Weaver, 1997]. The Teller [1990] meltwater pulse data was used thereby defining three 500-year characteristic meltwater periods (via diversions between the Mississippi, St. Lawrence, Davis Strait and Arctic). The wind stress feedback was turned on at year 1500.

2.4. The Eemian

Prior to remarkable findings from the Greenland ice core project [*GRIP*, 1993; *Dansgaard et al.*, 1993], the existing coarse temporal resolution paleoclimatic data suggested that interglacial periods were relatively stable and free of intense millennial timescale climate variability. The oxygen isotope records from GRIP, however, suggested that the climate of the last interglacial (referred to as the Eemian—115,000 to 135,000 years BP) may have been characterized by three climatic states. A state very much like today's climate and states significantly warmer and colder than today's climate. It was also found that the transition between these states occurred very rapidly and that the climate system never remained in one state for more than about 2,000 years. Nevertheless, subsequent findings from the US Greenland Ice-Sheet Project 2 (GISP2) threw caution into the interpretation of the GRIP results for the Eemian [*Taylor et al.*, 1993; *Grootes et al.*, 1993]. *Grootes et al.* [1993] presented GISP2 oxygen isotope analyses from a core drilled 28km to the west of the GRIP core. While the GISP2 and GRIP cores correlated extremely well over the top 90% of the record, the correlation in the rest of the core, including the Eemian, was poor. Several reasons (most probably related to flow deformation at one or both of the core sites) for this poor correlation were discussed by *Taylor et al.* [1993] and *Grootes et al.* [1993]. Nevertheless, *Dansgaard et al.* [1993] and *Grootes et al.* [1993] reported that over the first 2847 m (~129,000 years) of the GRIP core, which includes most of the Eemian (~115,000–135,000 years ago), there was no significant disturbance of the ice layering. Disturbances in the ice layering were first observed at 2,678 m in the GISP2 core (well above the component of the core associated with the Eemian).

Since 1993 numerous observational studies have presented conflicting evidence either in support of the existence of Eemian climate instability, consistent with the GRIP findings, or in support of a more stable Eemian period. Sediment records in the North Atlantic are inconsistent with some [*Cortijo et al.*, 1994; *Seidenkrantz et al.*, 1995, *Sejrup et al.*, 1995; *Fronval and Jansen*, 1996, 1997] providing evidence in support of an unstable Eemian climate and others [e.g. *Keigwin et al.*, 1994; *McManus et al.*, 1994; *Oppo et al.*, 1997] finding no such evidence. Additional support for the potential existence of climate instability during the Eemian is provided by European pollen records [*Tzedakis*, 1994; *Thouveny et al.*, 1994; *Field et al.*, 1994] as well as magnetic susceptibility and organic carbon records from maar lake deposits [*Thouveny et al., 1994*]. On the other hand, the comparison of methane records between Greenland and Vostok ice cores (which should correlate highly as atmospheric trace gases are well mixed globally) has suggested that the deep GRIP ice core record (including the Eemian) was disturbed [*Chappellaz et al.*, 1997]. *Johnsen et al.* [1997] systematically addressed numerous potential issues with respect to the reliability of the GRIP Eemian ice core record. The argued that the last major cooling event seen in the earlier *GRIP* [1993] and *Dansgaard et al.* [1993] analyses was a real climate event (based on correlation with the Camp Century core collected earlier in northwest Greenland), they noted the apparent correlation over much of the period with the *Frontval and Jansen* [1996] Nordic Seas sediment data, but also noted the lack of correlation with numerous other proxy indicators and sediment data. At the same time they were unable to prove that the GRIP ice core was disturbed in the Eemian component and hence concluded: "since we are not able to reconcile the evidence, pertinent to the question of a disturbed GRIP Eemian, we have to admit to serious flaws in our understanding related to the basic behavior of some of the ice-core parameters discussed above".

Clearly the debate is far from concluded as to whether or not the last interglacial was a period of relative stability (like the Holocene) or whether it witnessed climate instability unparalleled in the Holocene. Below, modeling evidence is presented that would support and provide a physical mechanism for enhanced variability in a climate warmer than the Holocene (the Eemian was on average about 2°C warmer than the Holocene — *White* [1993]).

Hughes and Weaver [1994] used an idealized two-basin global ocean (under mixed boundary conditions) to investigate the potential existence of multiple equilibria under 'present-day' forcing. Through an exhaustive procedure of conducting numerous integrations starting from a variety of initial conditions they found three possible realizations of the conveyor. The first realization, which was also the most frequently obtained, corresponded to the present-day situation with NADW sinking to a depth of several thousand metres and overriding inflowing Antarctic Bottom Water (AABW) as it flowed out to the Southern Ocean. The second realization corresponded to a case in which there was no NADW production (and subsequent enhanced AABW intrusion into the North Atlantic) while the third realization corresponded to a *super-conveyor* (with much stronger NADW production). *Mikolajewicz et al.* [1993] noted three similar states in the Hamburg large-scale geostrophic ocean model. *Weaver and Hughes* [1994] noted the potential analogy of their three states with the three states found in the GRIP ice core data noted above. They also noted that the Eemian was on average slightly warmer than the present, and the response of coupled atmosphere-ocean GCM experiments (see reviews in *Mitchell et al.* [1990]; *Gates et al.* [1992]; *Kattenberg et al.* [1996]) to increasing atmospheric CO_2 involved an enhancement of the hydrological cycle (see also *Tang and Weaver* [1995]). They therefore parameterized the effects of a warmer mean climate as an increase in the magnitude of stochastic component of the freshwater flux forcing (i.e. enhanced variability of the freshwater flux forcing relative to the present).

Weaver and Hughes [1994] showed that when only a weak stochastic forcing was used, with a standard deviation

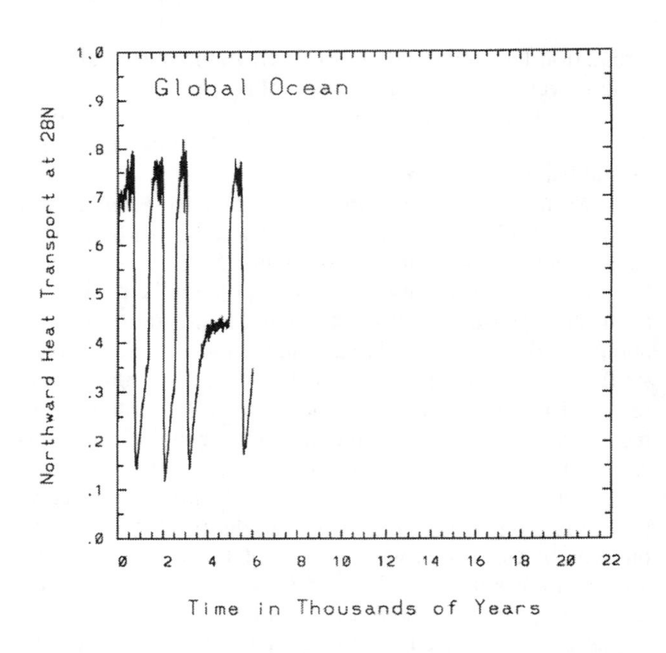

Figure 6. Poleward heat transport at 24°N over a 6000 year integration analogous to Fig. 4b of Weaver and Hughes [1994] but with a 100 day restoring timescale used on SST.

of 16 mm/month (i.e. 20% of the globally-averaged annual mean precipitation, *Baumgartner and Reichel* [1975], as used by *Mikolajewicz and Maier Reimer* [1990] for present conditions), the present day overturning remained largely unchanged, with weak internal variability at both decadal and century timescale linked to horizontal and overturning advective timescales, respectively (see *Weaver et al.* [1993]). When the standard deviation of stochastic forcing was increased to 48 mm/month the system oscillated internally between states involving weak and strong overturnings in the North Atlantic, passing through a period of normal (similar to the present) overturning along the way. The partitioning of time in each state was similar to that observed in the Eemian with the transition between the states happening very rapidly (over several decades) and the system remaining in a particular state for periods up to a thousand years or so. Once more, superimposed on the millennial-timescale variability was variability on both the decadal and century timescale. Figure 6 gives an example from a very similar experiment to that conducted by *Weaver and Hughes* [1994] but using a less-restrictive restoring timescale of 100 days instead of 50 days (which they used). The three states between which the system oscillated are readily visible.

In summary, *Weaver and Hughes* [1994] showed that rapid transitions between three modes of NADW formation in their ocean model could be excited through the addition of a simple random forcing to the mean freshwater flux forcing field. Their model results suggest that a source for the controversial Eemian variability may well lie in the dynamics of the ocean's thermohaline circulation which responds to an enhanced hydrological cycle associated with the warmer mean Eemian climate.

3. THE OCEAN RESPONSE TO HEINRICH EVENT FRESHWATER FORCING

In this section the response of the climate system to meltwater discharge typical of a Heinrich event is examined. To this end, the freshwater flux for two such events, as captured by a continental ice sheet model, is used as external forcing in a coupled atmosphere-ocean-sea ice model. The freshwater fluxes were obtained from a three-dimensional, thermomechanical ice sheet model which employs continuum mixture theory to incorporate ice streams (*Marshall and Clarke* [1997a] — see also *Clarke and Marshall* [1998]). *Clarke and Marshall* [1997a] spun up a Last Glacial Maximum (LGM) ice sheet to obtain initial ice thickness and temperature fields for the Laurentide Ice Sheet. From this initial model, they allowed the possibility of Laurentide ice stream activity whenever ice was at the melting point at the bed. This thermal switch gave free internal oscillations of ice stream flow in their model, with a periodicity of 5-10 kyr as observed in Heinrich events.

Two different surge scenarios of the Hudson Strait Ice Stream are used. The first consists of a brief, high-flux surge event, which corresponds to full flotation of the lower reaches of the ice stream (i.e. water pressure in the subglacial water system exceeds the ice pressure). This scenario corresponds to a 105 yr surge duration (Fig. 7a) with 43,767 km^3 of icebergs being released yielding an average flux of 0.013 Sv over this time. The second, longer surge event is the result of a case with partial flotation (linearly increasing water pressure along the Hudson Strait transect with full flotation at the outlet only). This scenario corresponds to a 510 yr surge duration with 85,144 km^3 of icebergs being released yielding an average flux of 0.005 Sv over this time. Both cases are described in *Marshall and Clarke* [1997b] in more detail.

The freshwater fluxes were derived (by S. Marshall) by taking the iceberg flux time series (Fig. 7a) and distributing this ice volume (areally-weighted and conserving the total ice volume of the surge) over the North Atlantic region which witnessed Heinrich events. The spatial distribution (Fig. 7b) was based on the observational record of Heinrich layer thickness in H1 and H2 [*Dowdeswell et al.*, 1995] assuming that sediment content in the melting icebergs is uniform throughout the trans-Atlantic voyage. In reality, the initial stages of melt will likely be *dirtier* (basal ice), so one might expect that the distributed iceberg meltwater fluxes overestimate the melting in the Labrador Sea area and underestimate the melting to the east.

The coupled atmosphere-ocean model of intermediate complexity used in this analysis is that described by *Fanning and Weaver* [1996]. The atmospheric component

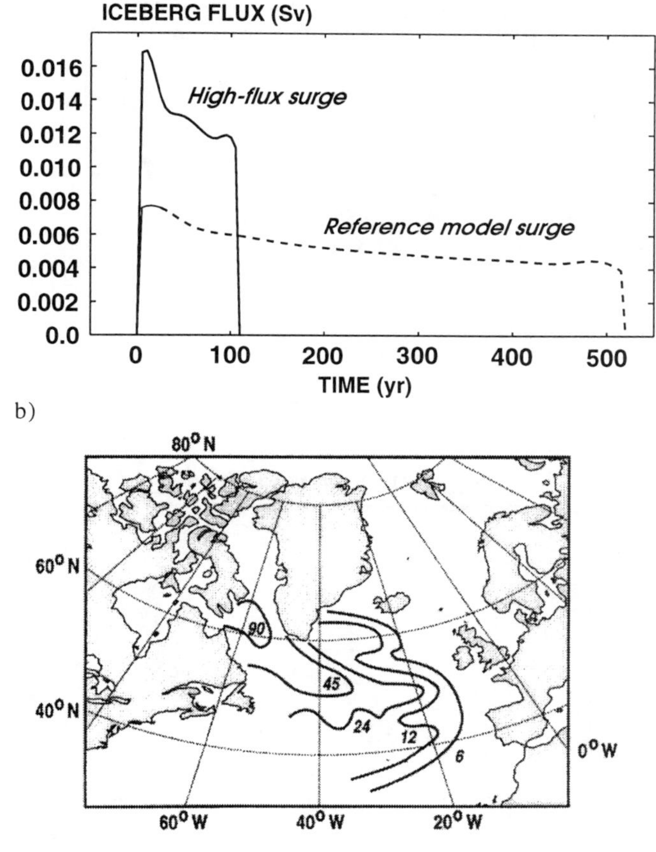

Figure 7. a)— Time series of iceberg flux (Sv) in the purge phase of two surge scenarios in Hudson Strait. Solid line: full flotation in the lower reaches of the ice stream. Dashed line: partial flotation upstream of the outlet. The two surge scenarios are: 1) 105 yr surge duration; 43,767 km^3 of icebergs released with an average flux of 0.013 Sv; 2) 510 yr surge duration; 85,144 k m^3 of icebergs released with an average flux of 0.005 Sv. b)— Average freshwater delivery to the North Atlantic from iceberg melt during the high-flux (105 year) surge (cm/yr). The horizontal distribution for the 510-yr surge looks identical and so is not plotted. To obtain correct magnitudes for this scenario the numbers in b) should be scaled by 0.005 Sv/0.013 Sv for 36 cm/yr in the Labrador Sea (instead of 90 cm/yr), for example. Note that in this case though, the input lasts 5 times longer, with twice the total volume of icebergs (figures from, and more details in, Marshall and Clarke [1997a,b]).

of the coupled model is an energy-moisture balance model while the ocean component is a three-dimensional OGCM based on the GFDL code (as described in *Weaver and Hughes* [1996]). The model configuration differs slightly from that of *Fanning and Weaver* [1996] and *Fanning and Weaver* [1997] and these differences are discussed in *Weaver et al.* [1998]. The equilibrium climatology of this coupled model yields 16 Sv of overturning in the North Atlantic

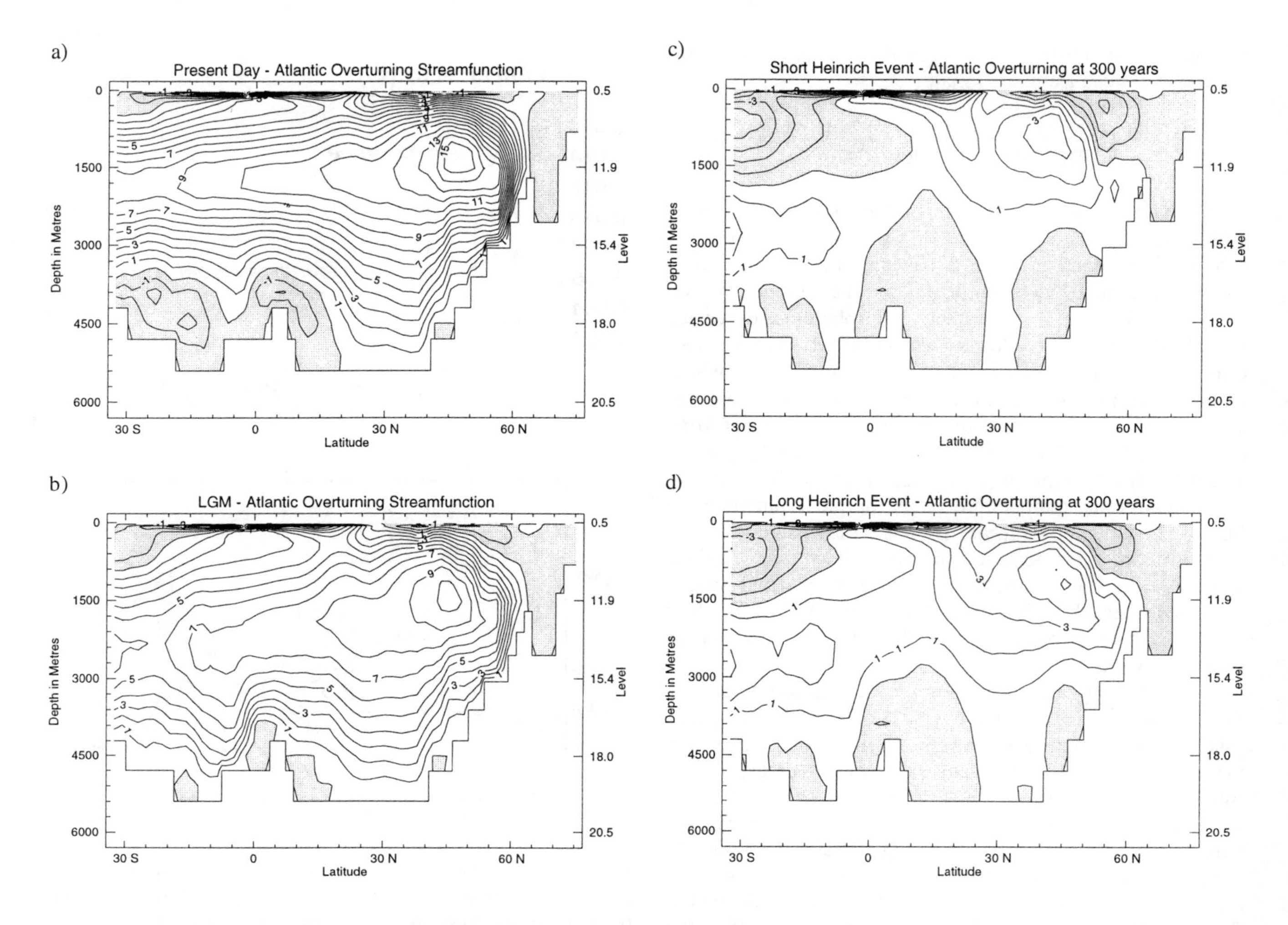

Figure 8. Meridional overturning streamfunction in the Atlantic Ocean (Sv) for a)— The present day equilibrium; b)— The LGM equilibrium; c)— 300 years into the short Heinrich event integration; d)— 300 years into the long Heinrich event integration. Shaded regions indicate counterclockwise circulation.

(Fig. 8a), reducing to 10Sv when LGM orbital forcing, atmospheric CO_2 levels, and continental ice sheets (specified via increased albedo over land — see Weaver et al., 1998) are included (Fig. 8b). The LGM equilibrium is used as the initial condition from which the meltwater perturbations are applied.

The time series of the maximum overturning in the North Atlantic (Fig. 9) through the 2500 years of integration of the coupled model indicates that in both the long and short Heinrich runs, the actual duration of the overturning weakening is nearly the same (~800 years), suggesting the duration is set by internal ocean advective and diffusive processes rather than by the forcing itself. The short Heinrich event run, with a high-flux surge (Fig. 7), however reveals a more substantial drop in the North Atlantic overturning (down to ~3.5 Sv — see Fig. 8c and 9) than the longer, yet weaker, surge event (down to ~5 Sv — see Fig. 8d and 9). In both cases the thermohaline circulation never entirely collapses and eventually reestablishes to a slightly stronger state (~11 Sv) than in its initial LGM equilibrium (Fig. 9).

The change in SST and SAT 300 years into the integrations (Figs. 10, 11) reveals a local maximum of cooling in the North Atlantic due to a reduction in the conveyor there (3.25°C [2.74°C] and 3.84°C [2.77°C] for the SST and SAT, respectively in the short [long] Heinrich surge scenario). The South Atlantic also reveals a local maximum in surface warming (0.56°C [0.48°C] and 0.28°C [0.24°C] for the SST and SAT, respectively in the short [long] Heinrich surge scenario), associated with an enhancement of Antarctic Intermediate Water production (Figs. 8c,d) and subsequent southward near-surface advection of warmer, subtropical, surface waters. Globally- and annually-averaged, the short Heinrich event case dropped 0.08°C and 0.12°C for SST and SAT, respectively, while the long Heinrich event case dropped 0.07°C and 0.10°C for SST and SAT, respectively. These differences of course largely disappear once the thermohaline circulation

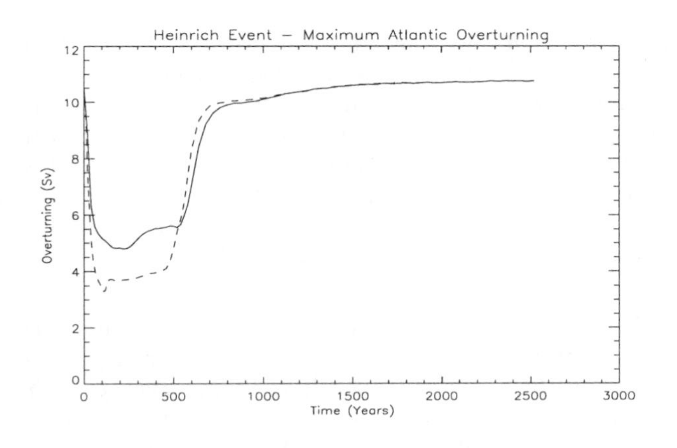

Figure 9. Maximum value of the overturning streamfunction (Sv) as a function of time for the short (dashed) and long (solid) Heinrich event integrations.

has reestablished. Combined with the LGM cooling relative to the present in this coupled model of 2.2°C for SST and 3.2°C for SAT [*Weaver et al.*, 1988], the total cooling (relative to the present) 300 years into the integrations is 2.3°C (SST) and 3.3°C (SAT) for both Heinrich event scenarios (Figs. 10, 11).

The results from these two Heinrich Event freshwater flux experiments differ, at first glance, substantially from the results from the Younger Dryas experiment analysed in section 2.3. First, in the Younger Dryas case the collapse and reestablishment timescale of the conveyor was about 2500 years (with the overturning completely collapsed for about 700 years), whereas in the present case it was only about 800 years. In addition, the reestablishment in the Younger Dryas case was to a slightly weaker overturning (Fig. 5), whereas in the present case reestablishment occurred to a slightly stronger overturning than in the initial state prior to the freshwater discharge. Finally, the SAT in the middle of the Younger Dryas experiment is negative everywhere (Fig. 4) whereas in the present case, it is more reminiscent of the *Broecker* [1998] bipolar seesaw, with slight warming in the Pacific and South Atlantic coincident with the strong cooling in the region around the North Atlantic (Fig.10).

On closer analysis, these results can easily be understood by examining the different forcing employed in the two cases. *Fanning and Weaver* [1997] undertook a sensitivity analysis to the geographical location of the freshwater discharge. They noted that without preconditioning by pre-Younger Dryas meltwater runoff (with a strong signal entering the Gulf of Mexico), the application of a Younger Dryas meltwater perturbation into the North Atlantic (via Davis Strait and the St. Lawrence River) merely caused the overturning to weaken and then reestablish (as in the Heinrich Event cases). This followed since the freshwater discharge acted to curtail Labrador Sea convection with a lesser impact on convection in the Nordic Seas. *Fanning and Weaver* [1997] also applied their freshwater perturbation (from *Teller* [1990]) over a substantially longer period of time than in the Heinrich Event cases. As such, the whole collapse/reestablishment timescale is longer in the Younger Dryas case (Fig. 5) than in the Heinrich Event cases (Fig. 9).

It is not surprising that both sets of experiments reequilibrate to a state with a conveyor strength that has a slightly different strength from its initial condition. As pointed out by *Lenderink and Haarsma* [1994], potentially convective regions, once triggered, can be self-sustaining leading to the possibility of multiple equilibria. In addition, the application of a freshwater perturbation over an extended period of time substantially alters the mean ocean salinity and hence mean climatic state.

The different SAT response between the Younger Dryas case (Fig. 4) and the Heinrich Event cases (Fig. 10) can be understood by realising that in the Younger Dryas case, the conveyor had completely shut down and the climate system had sufficient time to equilibrate. In the Heinrich Event cases, the conveyor was in a continuous state of change (never shutting down). In addition, the initial condition from which the Heinrich perturbation was applied was a

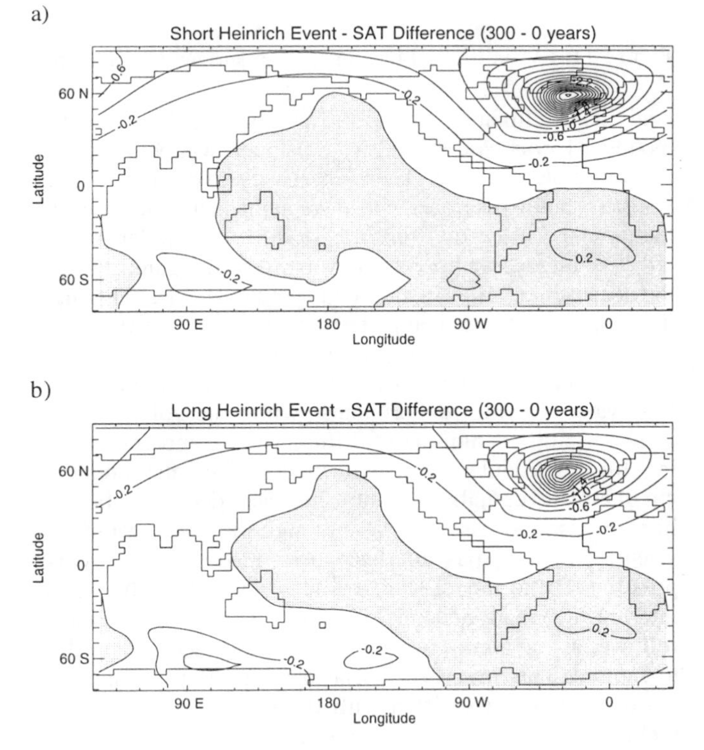

Figure 10. Surface air temperature difference (°C) between Heinrich event integrations (300 years into the integration — Fig. 9) and the Last Glacial Maximum initial condition a)— short Heinrich event experiment; b)— Long Heinrich event experiment. The contour interval is 0.02°C and shaded areas indicate positive temperature differences.

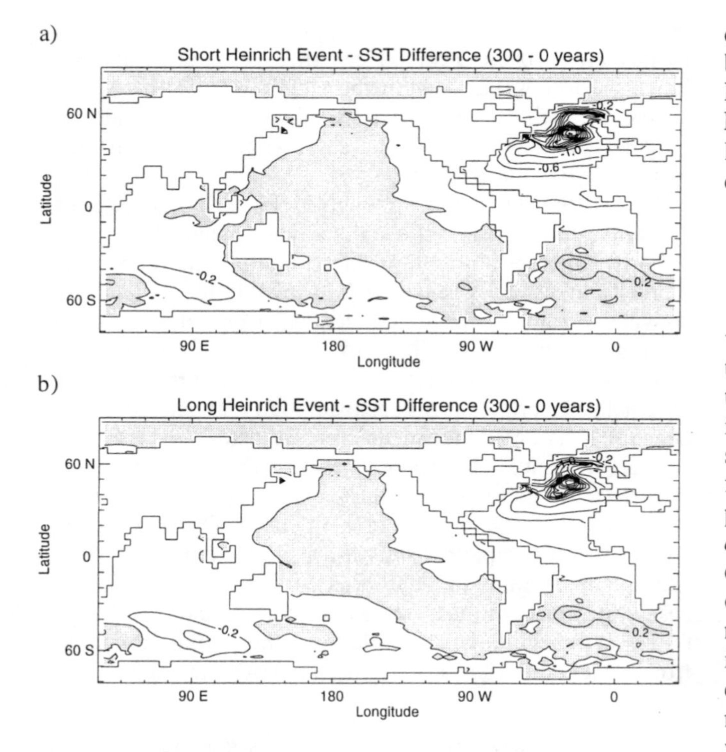

Figure 11. As in Fig. 10 but for the sea surface temperature.

cold LGM climate whereas the complete collapse in the Younger Dryas case started from a much warmer 'Present-day' climate. In the Heinrich Event experiments the southern hemisphere responded in a manner analogous to the bipolar seesaw of *Broecker* [1998]. The bipolar seesaw effect is larger swamped by the dramatic reorganization of the ocean circulation in the Younger Dryas case, although the actually geographical pattern of change (if not the sign) is similar between both sets of experiments.

In comparing the results from the Heinrich Event freshwater forcing experiments with the observations shown in Fig. 1 a number of differences are apparent. First, in this particular coupled model the thermohaline circulation eventually reestablishes and does not behave like a damped oscillator, although a full parameter sensitivity study has not been performed and it may be possible for the model to reveal less stability with weaker internal mixing. Second, Heinrich events are usually followed by an abrupt warming which the model is unable to capture. This may be associated with abrupt changes in albedo over land associated with the surges and subsequent melting of ice sheets (a process which is not included here), or increasing atmospheric CO_2 levels through changes in ocean ventilation. Finally, Heinrich events do not appear to create significant regional cooling in the observational record, but rather occur in a coldest phase of the Bond Cycle. This might suggest that the North Atlantic conveyor is completely shut down prior to the Heinrich event so that a subsequent further freshening has little effect. It is therefore apparent that we are very far from being able to capture the essence and climate response of Heinrich events in coupled atmosphere ocean models. Further advances in our understanding of these events will likely come from the incorporation of ice sheet and carbon cycle models into these models.

4. DISCUSSION

In this article the mechanisms for millennial timescale variability, as found in coarse resolution ocean models of the thermohaline circulation, have been reviewed. The timescale for such variability ultimately arises from the internal diffusive timescale of the ocean although it can be shortened by increasing atmospheric and ice sheet freshwater flux forcing variability or internal advective processes. The glacial salt oscillator theory of *Broecker et al.* [1990] has stood the test of time for the most comprehensive explanation for the existence of D-O oscillations in the last glaciation, although it has yet to be realized in coupled atmosphere-ocean-sea ice GCMs including interactive ice sheets and carbon cycle as no such comprehensive climate model exists. Combining this mechanism with a potential mechanism for variability in a warmer climate, such as the Eemian, through an enhancement of the hydrological cycle it is tempting to speculate that the present climate is stable as it is neither too cold (last glaciation) nor too warm (Eemian), but rather has an intermediate mean temperature.

Bond et al. [1997] have added a new twist to the debate concerning the existence of millennial timescale climate variability in interglacial periods. Through analyses of North Atlantic deep sea cores they provide compelling evidence for such variability even in the Holocene, albeit an order of magnitude weaker than the D-O oscillations in the last glaciation (see also *Oppo* [1997]). Similar evidence has also recently been provided by *Mayewski et al.* [1997] through analysis of GISP2 Greenland ice core records. When combined with the observations from the last glaciation (e.g. *Oeschger et al.* [1984] and *Dansgaard et al.* [1984]), early Pleistocene records [*Raymo et al.*, 1998], late Pleistocene records [*Oppo et al.*, 1998] records, and evidence from the Eemian period (section 2.4), these results suggest that millennial timescale variability is an inherent component of the Pleistocene climate, albeit with variable magnitude. Nevertheless, it is quite clear that a satisfactory quantitative analysis is still lacking. Further understanding will likely come about as fully interactive and non flux adjusted coupled atmosphere-ocean-sea ice-ice sheet-carbon cycle GCMs are developed, thereby allowing for a detailed understanding of climate variability as a function of the mean climatic state itself. The capturing of millennial timescale variability and its packaging into Bond Cycles in cold climates, its association with Heinrich events, and its dependence on the mean climatic state remains one of the greatest challenges for paleoclimate modelers.

Acknowledgments. This work was supported by NSERC operating and Steacie grants, AES/CICS, the NSERC/CICS CSHD project and an IBM SUR grant. Numerical computations were conducted on a suite of IBM RS6000's including two SP2s. I am indebted to S. Marshall for providing me with the freshwater flux output used in section 3 (as well as Fig. 7 and a detailed description of his model and its output). I am also grateful to M. Eby for conducting the numerical integrations used in section 3, to A. Fanning for providing me with Figs. 4 and 5, and to two anonymous reviewers and R. Webb for their constructive comments. Figure 1 was reprinted with permission from Nature [*Bond et al.*, 1993] © 1993, Macmillan Magazines Ltd.

REFERENCES

Alley, R, D.A. Meese, C.A. Shuman, A.J. Gow, K.C. Taylor, P.M. Grootes, J.W.C. White, M. Ram, E.D. Waddington, P.A. Mayewski, and G.A. Zielinski, Abrupt increase in Greenland snow accumulation at the end of the Younger Dryas event, *Nature*, *362*, 527–529, 1993.

Baumgartner, A., and E. Reichel, 1975: *The World Water Balance*, Elsevier, 179pp, 1975.

Behl, R.J., and J.P. Kennett, Brief interstadial events in the Santa Barbara basin, NE Pacific, during the past 60 kyr, *Nature*, *379*, 243–246, 1996.

Birchfield, G.E., and W. Broecker, A salt oscillator in the glacial Atlantic?, 2, A "scale analysis" model, *Paleoceanogr.*, *5*, 835–843, 1990.

Bond, G.C., and R. Lotti, Iceberg discharges into the North Atlantic on millennial timescales during the last glaciation, *Science*, *267*, 1005–1010, 1995.

Bond, G., W. Broecker, S. Johnsen, J. McManus, L. Labeyrie, J. Jouzel, and G. Bonani, Correlations between climate records from North Atlantic sediments and Greenland ice, *Nature*, *365*, 143–147, 1993.

Bond, G., H. Heinrich, S. Huon, W. Broecker, L. Labeyrie, J. Andrews, J. McManus, S. Clasen, K. Tedesco, R. Jantschik, C. Simet, and M. Klas, Evidence for massive discharges of icebergs into the North Atlantic ocean during the last glaciation, *Nature*, *360*, 245–249, 1992.

Bond, G., W. Showers, M. Cheseby, R. Lotti, P. Almasi, P. deMenocal, P. Priore, H. Cullen, I. Hajdas, and G. Bonani, A pervasive millennial-scale cycle in North Atlantic Holocene and glacial cycles, *Science*, *278*, 1257–1266, 1997.

Bretherton, F.P., Ocean climate modelling, *Prog. Oceanogr.*, *11*, 93–129, 1982

Broecker, W.S., Massive iceberg discharges as triggers for global climate change, *Nature*, *372*, 421–424, 1994.

Broecker, W.S., Paleocean circulation during the last deglaciation: A bipolar seesaw? *Paleoceanogr.*, *13*, 119–121, 1998.

Broecker, W., G. Bond, M. Klas, G. Bonani, and W. Wolfli, A salt oscillator in the glacial Atlantic?, 1, The concept, *Paleoceanogr.*, *5*, 469–477, 1990.

Broecker, W., G. Bond, M. Klas, E. Clark, J. McManus, Origin of the northern Atlantic's Heinrich events, *Clim. Dyn.*, *6*, 265–273, 1992.

Broecker, W.S., D. Peteet, D. Rind, Does the ocean-atmosphere have more than one stable mode of operation? *Nature*, *315*, 21–25, 1985.

Bryan, F., High-latitude salinity effects and interhemispheric thermohaline circulations, *Nature*, *323*, 301–304, 1986.

Chappellaz, J. , E. Brook, T. Blunier, and B. Malaizé, CH_4 and $\delta^{18}O$ of O_2 records from Antarctic and Greenland ice: A clue for stratigraphic disturbance in the bottom part of the Greenland Ice Core Project and the Greenland Ice Sheet Project 2 ice cores, *J. Geophys. Res.*, *102*, 26547–26557, 1997.

Chen, F., and M. Ghil, Interdecadal variability of the thermohaline circulation and high-latitude surface fluxes, *J. Phys. Oceanogr.*, *25*, 2547–2568, 1995.

Chen, F., and M. Ghil, Interdecadal variability in a hybrid coupled ocean-atmosphere model, *J. Phys. Oceanogr.*, *26*, 1561–1578, 1996.

Clarke, G.K.C., and S.J. Marshall, A glaciological perspective of Heinrich events, in *The Role of High and Low Latitudes in Millennial-Scale Global Climate Change, Geophys. Mon. Ser.*, edited by R.S. Webb, P.U. Clark, and L.D. Keigwin, this issue, AGU, Washington, D.C., 1998.

Cortijo, E., J.C. Duplessy, L. Labeyrie, H. Leclaire, J. Duprat, and T.C.E. van Weering, Eemian cooling in the Norwegian Sea and the North Atlantic Ocean preceding continental ice-sheet growth, *Nature*, *372*, 446–449, 1994.

Dansgaard, W., S.J. Johnsen, H.B. Clausen, D. Dahl-Jensen, N. Gundestrup, C.U. Hammer, and H. Oeschger, 1984: North Atlantic climate oscillations revealed by deep Greenland ice cores, in *Climate Processes and Climate Sensitivity*, *Geophys. Mon. Ser.*, vol. 29, edited by J.E. Hansen, and T. Takahashi, pp. 288–298, AGU, Washington, D.C., 1984.

Dansgaard, W., S.J. Johnsen, H.B. Clausen, D. Dahl-Jensen, N.S. Gundestrup, C.U. Hammer, C.S. Hvidberg, J.P. Steffensen, A.E. Sveinbjörnsdottir, J. Jouzel, and G. Bond, Evidence for general instability of past climate from a 250-kyr ice-core record, *Nature*, *364*, 218–220, 1993.

Dansgaard, W., J.W.C. White, and S.J. Johnsen, The abrupt termination of the Younger Dryas climate event, *Nature*, *339*, 532–534, 1989.

Dowdeswell, J.A., M.A. Maslin, J.T. Andrews, and I.N. McCave, Iceberg production, debris rafting, and the extent and thickness of "Heinrich layers" (H-1, H-2) in North Atlantic sediments, *Geology*, *23*, 301–304, 1995.

Fanning, A.F., and A.J. Weaver, An atmospheric energy moisture-balance model: climatology, interpentadal climate change and coupling to an OGCM, *J. Geophys. Res.*, *101*, 15111–15128, 1996.

Fanning, A.F., and A.J. Weaver, Temporal-geographical meltwater influences on the North Atlantic conveyor: Implications for the Younger Dryas, *Paleoceanogr.*, *12*, 307–320, 1997.

Field, M.H., B. Huntley, and H. Muller, Eemian climate fluctuations observed in a European pollen record, *Nature*, *371*, 779–783, 1994.

Fronval, T., and E. Jansen, Rapid changes in ocean circulation and heat flux in the Nordic Seas during the last interglacial period, *Nature*, *383*, 806–810, 1996.

Fronval, T., and E. Jansen, Eemian and early Weichselian (140–60 ka) paleoceanography and paleoclimate in the Nordic seas with comparisons to Holocene conditions, *Paleoceanogr.*, *12*, 443–462, 1997.

Gates, W.L., J.F.B. Mitchell, G.J. Boer, U. Cubasch U, and V.P. Meleshko, 1992, Climate modelling, climate prediction and model validation, in *Climate Change 1992*, *the Supplementary Report to the IPCC Scientific Assessment*, edited by J.T. Houghton, B.A. Callander, and S.K. Varney, pp. 97–134, Cambridge Univ. Press., 1992.

GRIP Project Members, Climate instability during the last interglacial period recorded in the GRIP ice core, *Nature*, *364*, 203–207, 1993.

Grootes, P.M., M. Stuiver, J.W.C. White, S. Johnsen, and J. Jouzel, Comparison of oxygen isotope records from GISP2 and GRIP Greenland ice cores, *Nature*, *366*, 552–554, 1993.

Haney, R.L., Surface thermal boundary condition for ocean circulation models, *J. Phys. Oceanogr.*, *1*, 241–248, 1971.

Heinrich, H., Origin and consequences of cyclic ice rafting in the northeast Atlantic Ocean during the past 130,000 years, *Quat. Res.*, *29*, 143–152, 1988.

Hewitt, A.T., D. McDonald, and B.D. Bornhold,: Ice-rafted debris in the North Pacific and correlation to North Atlantic climatic events, *Geophys. Res. Let.*, *24*, 3261–3264, 1997.

Huang, R.X., Real freshwater flux as the upper boundary condition for the salinity balance and thermohaline circulation forced by evaporation and precipitation, *J. Phys. Oceanogr.*, *23*, 2428–2446, 1993.

Huang, R.X., Thermohaline circulation: Energetics and variability in a single-hemisphere basin model, *J. Geophys. Res.*, *99*, 12471–12485, 1994.

Hughes, T.M.C., and A.J. Weaver, Multiple equilibria of an asymmetric two-basin ocean model, *J. Phys. Oceanogr.*, *24*, 619–637, 1994.

Hughes, T.M.C., and A.J. Weaver, Sea surface temperature–evaporation feedback and the ocean's thermohaline circulation, *J. Phys. Oceanogr.*, *26*, 644–654, 1996.

Johnsen, S.J., H.B. Clausen, W. Dansgaard, N.S. Gundestrup, C.U. Hammer, U. Andersen, K.K. Andersen, C.S. Hvidberg, D. Dahl-Jensen, J.P. Steffensen, H. Shoji, A.E. Sveinbjörnsdóttir, J. White, J. Jouzel, and D. Fisher, The $\delta^{18}O$ record along the Greenland Ice Core Project deep ice core and the problem of possible Eemian climatic instability, *J. Geophys. Res.*, *102*, 26397–26410, 1997.

Kattenberg, A., F. Giorgi, H. Grassl, G.A. Meehl, J.F.B. Mitchell, R.J. Stouffer, T. Tokioka, A.J. Weaver, and T.M.L. Wigley, 1996, Climate models—projections of future climate, in *Climate Change 1995, the Science of Climate Change*, edited by J.T. Houghton, L.G. Meira Filho, B.A. Callander, N. Harris, A. Kattenberg, and K. Maskell, pp. 285–357, Cambridge Univ. Press., 1996.

Keigwin, L.D., W.B. Curry, S.J. Lehman, and S. Johnsen, The role of the deep ocean in North Atlantic climate change between 70 and 130 kyr ago, *Nature*, *371*, 323-326, 1994.

Kennett, J.P., and B.L. Ingram, A 20,000 year record of ocean circulation and climate change from the Santa Barbara basin, *Nature*, *377*, 510–514, 1995.

Lenderink, G., and R.J. Haarsma, Variability and multiple equilibria of the thermohaline circulation associated with deep water formation, *J. Phys. Oceanogr.*, *24*, 1480-1493, 1994.

Lund, D.C., and A.C. Mix, Millennial-scale deep water oscillations: Reflections of the North Atlantic in the deep Pacific from 10 to 60 ka, *Paleoceanogr.*, *13*, 10–19, 1998.

MacAyeal, D.R., Binge/purge oscillations of the Laurentide ice sheet as a cause of the North Atlantic's Heinrich events, *Paleoceanogr.* *8*, 775–784, 1993.

Maier-Reimer, E., and U. Mikolajewicz, Experiments with an OGCM on the cause of the Younger Dryas, in *Oceanography*, edited by A. Ayala-Castanares, W. Wooster, and A. Yanez-Arancibia, pp. 87–99, UNAM Press, Mexico, 1989.

Manabe, S., and R.J. Stouffer, Simulation of abrupt climate change induced by freshwater input to the North Atlantic ocean, *Nature*, 378, 165-167, 1995.

Manabe, S., and R.J. Stouffer, Coupled ocean-atmosphere model response to freshwater input: Comparison to Younger Dryas event, *Paleoceanogr.*, *12*, 321–336, 1997.

Marotzke, J., Instabilities and multiple steady states of the thermohaline circulation, in *Oceanic Circulation Models: Combining Data and Dynamics.*, edited by D.L.T. Anderson and J. Willebrand, pp. 501–511, NATO ASI series, Kluwer, 1989.

Marotzke, J., *Instabilities and multiple equilibria of the thermohaline circulation*, Ph.D. thesis, Ber. Inst. Meeresk. Kiel, 194, 126pp, 1990.

Marotzke, J., P. Welander, and J. Willebrand, Instability and multiple steady states in a meridional-plane model of the thermohaline circulation, *Tellus, 40A*, 162–172, 1988.

Marshall, S.J., and G.K.C. Clarke, A continuum mixture model of ice stream thermomechanics in the Laurentide Ice Sheet, I, Theory, *J. Geophys. Res.*, *102*, 20599–20614, 1997a.

Marshall, S.J., and G.K.C. Clarke, A continuum mixture model of ice stream thermomechanics in the Laurentide Ice Sheet, II, Application to the Hudson Strait Ice Stream, *J. Geophys. Res.*, *102*, 20615–20638, 1997b.

Mayewski, P.A., L.D. Meeker, M.S. Twickler, S. Whitlow, Q. Yang, W.B. Lyons, and M. Prentice, Major features and forcing of high-latitude northern hemisphere atmospheric circulation using a 110,000-year-long glaciochemical series, *J. Geophys. Res.*, *102*, 26345–26366, 1997.

McManus, J.F., G.C. Bond, W.S. Broecker, S. Johnson, L. Labeyrie, and S. Higgins, High-resolution climate records from the North Atlantic during the last interglacial, *Nature*, *371*, 326-329, 1994.

McWilliams, J.C., Modeling the oceanic general circulation, *Ann. Rev. Fluid Mech,* 28, 215–248, 1996.

Mikolajewicz, U., A meltwater-induced collapse of the 'conveyor belt' thermohaline circulation and its influence on the distribution of $\Delta^{14}C$ and $\delta^{18}O$ in the oceans, in *Tracer Oceanography*, Proceedings of the Maurice Ewing Symposium on Applications of Trace Substance Measurements to Oceanographic Problems, October 1995, in press, 1997.

Mikolajewicz, U., and E. Maier-Reimer, Mixed boundary conditions in ocean general circulation models and their influence on the stability of the model's conveyor belt, *J. Geophys. Res.*, *99*, 22633–22644, 1990.

Mikolajewicz, U., and E. Maier-Reimer, Internal secular variability in an ocean general circulation model, *Clim. Dyn.*, *4*, 145–156, 1994.

Mikolajewicz, U., E. Maier-Reimer, T.J. Crowley, and K.Y. Kim, Effect of Drake and Panamanian gateways on the circulation of an ocean model, *Paleoceanogr.*, *8*, 409–426, 1993.

Mitchell, J.F.B., S. Manabe, V. Meleshko, and T. Tokioka, Equilibrium climate change — and its implications for the future, in *Climate Change The IPCC Scientific Assessment*, edited by J.T. Houghton, G.J. Jenkins, and J.J. Ephraums, pp. 131–172, Cambridge Univ. Press., 1990.

Myers, P.G., and A.J. Weaver AJ, Low-frequency internal oceanic variability under seasonal forcing, *J. Geophys. Res., 97*, 9541–9563, 1992.

Oeschger, H., J. Beer, U. Siegenthaler, B. Stauffer, W. Dansgaard, and C.C. Langway, Late glacial climate history from ice cores, in *Climate Processes and Climate Sensitivity, Geophys. Mon. Ser.*, vol. 29, edited by J.E. Hansen, and T. Takahashi, pp. 299–306, AGU, Washington, D.C., 1984.

Oppo, D., Millennial climate oscillations, *Science, 278*, 1244–1246, 1997.

Oppo, D.W., M. Horowitz, and S.J. Lehman, Marine core evidence for reduced deep water production during Termination II followed by a relatively stable substage 5e (Eemian), *Paleoceanogr., 12*, 51–63, 1997.

Oppo, D.W., J.F. McManus, and J.L. Cullen, Abrupt climate events 500,000 to 340,000 years ago: Evidence from subpolar North Atlantic sediments, *Science, 279*, 1335–1338, 1998.

Pierce, D.W., K.Y. Kim, and T.P. Barnett, Variability of the thermohaline circulation in an ocean general circulation model coupled to an atmospheric energy balance model, *J. Phys. Oceanogr., 26*, 725–738, 1996.

Power, S.B., A.M. Moore, D.A. Post, N.R. Smith, and R. Kleeman, Stability of North Atlantic deep water formation in a global ocean general circulation model, *J. Phys. Oceanogr., 24*, 906–916, 1994.

Rahmstorf, S., Rapid climate transitions in a coupled ocean-atmosphere model, *Nature, 372*, 82-85, 1994.

Rahmstorf, S., Bifurcations of the Atlantic thermohaline circulation in response to changes in the hydrological cycle, *Nature, 378*, 145-149, 1995.

Rahmstorf, S., and J. Willebrand, The role of temperature feedback in stabilizing the thermohaline circulation, *J. Phys. Oceanogr., 25*, 787–805, 1995.

Raymo, M.E., K. Ganley, S. Carter, D.W. Oppo, and J. McManus, Millennial-scale climate instability during the early Pleistocene epoch, *Nature, 392*, 699–702, 1998.

Sakai, K., and W.R. Peltier, A simple model of the Atlantic thermohaline circulation: Internal and forced variability with paleoclimatological implications, *J. Geophys. Res., 100*, 13455–13479, 1995.

Sakai, K., and W.R. Peltier, A multibasin reduced model of the global thermohaline circulation: Paleoceanographic analyses of the origins of ice-age variability, *J. Geophys. Res., 101*, 22535–22562, 1996.

Sakai, K., and W.R. Peltier, Dansgaard-Oeschger oscillations in a coupled atmosphere-ocean climate model, *J. Climate, 10*, 949–970, 1997.

Schiller, A., U. Mikolajewicz, and R. Voss, The stability of the thermohaline circulation in a coupled ocean-atmosphere general circulation model, *Clim. Dyn., 13*, 325–347, 1997.

Schopf, P.S., On equatorial Kelvin waves and El Niño. II: Effects of air-sea coupling, *J. Phys. Oceanogr., 13*, 1878–1893, 1983.

Seidenkrantz, M.-S., P. Kristensen, and K.L. Knudsen, Marine evidence for climatic instability during the last interglacial in shelf records from northwest Europe, *J. Quat. Sci., 10*, 77–82, 1995.

Sejrup, H.P., H. Haflidason, A.D. Kristensen, and S.J. Johnsen, Last interglacial and Holocene climatic development in the Norwegian Sea region: Ocean front movements and ice-core data, *J. Quat. Sci. 10*, 385–390, 1995.

Stocker, T.F., The role of the deep ocean circulation for past and future climate change, *Nature*, submitted, 1998.

Stocker, T.F., and D.G. Wright, Rapid transitions of the ocean's deep circulation induced by changes in surface water fluxes, *Nature, 351*, 729-732, 1991.

Tang, B., and A.J. Weaver AJ, Climate stability as deduced from an idealized coupled atmosphere-ocean model, *Clim. Dyn., 11*, 141–150, 1995.

Taylor, K.C., C.U. Hammer, R.B. Alley, H.B. Clausen, D. Dahl-Jensen, A.J. Gow, N.S. Gundestrup, J. Kipfstuhl, J.C. Moore, and E.D. Waddington, Electrical conductivity measurements from the GISP2 and GRIP ice cores, *Nature, 366*, 549–552, 1993.

Taylor, K.C., G.W. Lamorey, G.A. Doyle, R.B. Alley, P.M. Grootes, P.A. Mayewski, J.W.C. White, and L.K. Barlow, The 'flickering switch' of late Pleistocene climate change, *Nature, 361*, 432–436, 1993.

Teller, J.T., Meltwater and precipitation runoff to the North Atlantic, Arctic, and Gulf of Mexico from the Laurentide Ice sheet and adjacent regions during the Younger Dryas, *Paleoceanogr., 5*, 897-905, 1990.

Thouveny, N., J.-L. de Beaulieu, E. Bonifay, K.M. Creer, J. Guiot, M. Icole, S. Johnsen, J. Jouzel, M. Reille, T. Williams, and D. Williamson, Climate variations in Europe over the past 140 kyr deduced from rock magnetism, *Nature, 371*, 503-506, 1994.

Tzedakis, P.C., K.D. Bennett, and D. Magrl, Climate and pollen record, *Nature, 370*, 513, 1994.

Tziperman, E., J.R. Toggweiler, Y. Feliks, and K. Bryan, Instability of the thermohaline circulation with respect to mixed boundary conditions: Is it really a problem for realistic models? *J. Phys. Oceanogr., 24*, 217–232, 1994.

Weaver, A.J., and T.M.C. Hughes, Rapid interglacial climate fluctuations driven by North Atlantic ocean circulation, *Nature, 367*, 447–450, 1994.

Weaver, A.J., and T.M.C. Hughes, On the incompatibility of ocean and atmosphere models and the need for flux adjustments, *Clim. Dyn., 12*, 141–170, 1996.

Weaver, A.J. and E.S., Sarachik, The role of mixed boundary conditions in numerical models of the ocean's climate, *J. Phys. Oceanogr., 21*, 1470–1493, 1991.

Weaver, A.J., M. Eby, A.F. Fanning, and E.C. Wiebe, Simulated influence of CO_2, orbital forcing and ice sheets on the climate of the last glacial maximum, *Nature, 394*, 847–853, 1998.

Weaver, A.J., J. Marotzke, P.F. Cummins, and E.S. Sarachik, Stability and variability of the thermohaline circulation, *J. Phys. Oceanogr., 23*, 39–60, 1993.

White, J.W.C., Don't touch that dial, *Nature, 364*, 186, 1993.

Winton, M., Deep decoupling oscillations of the oceanic thermohaline circulation, in *Ice in the Climate System., NATO ASI series,* Vol. 12, edited by W.R. Peltier, pp. 417–432, Springer–Verlag, 1993.

Winton, M., Energetics of deep-decoupling oscillations, *J. Phys. Oceanogr., 25*, 420–427, 1995.

Winton, M., The effect of cold climate upon North Atlantic

deep water formation in a simple ocean-atmosphere model, *J. Climate.*, *10*, 37–51, 1997.

Winton, M., and E.S. Sarachik, Thermohaline oscillations induced by strong steady salinity forcing of ocean general circulation models, *J. Phys. Oceanogr.*, *23*, 1389–1410, 1993.

Wright, D.G., and T.F. Stocker, A zonally averaged ocean model for the thermohaline circulation. Part 1: Model development and flow dynamics, *J. Phys. Oceanogr.*, *21*, 1713–1724, 1991.

Zhang, S., R.J. Greatbatch, and C.A. Lin, A reexamination of the polar halocline catastrophe and implications for coupled ocean-atmosphere modelling, *J. Phys. Oceanogr.*, *23*, 287–299, 1993.

A. J. Weaver, School of Earth and Ocean Sciences, University of Victoria, PO Box 3055, Victoria, BC, V8W 3P6, Canada. (e-mail: weaver@uvic.ca)

Ice-Core Evidence of Late-Holocene Reduction in North Atlantic Ocean Heat Transport

R. B. Alley and A. M. Ágústsdóttir

Earth System Science Center and Department of Geosciences, The Pennsylvania State University, University Park, PA

P. J. Fawcett

Department of Earth & Planetary Sciences, University of New Mexico, Albuquerque, NM

The several-millennial cooling in central Greenland since the middle Holocene probably was caused by a trend to reduced oceanic heat transport as well as by orbital forcing, based on several lines of evidence. The late-Holocene trend is similar in many ways to the older coolings associated with abrupt, millennial climate changes. Climate records from Greenland ice cores indicate that both abrupt and gradual coolings involved: (i) greater temperature decrease in winter than in summer; (ii) greater accumulation-rate decrease in winter than in summer causing normally calibrated ice-isotopic changes to underestimate temperature changes; and (iii) increasing interannual climate variability. Paleoclimatic data and model results show that abrupt coolings have been linked to reduction in North Atlantic oceanic heat transport; we suggest that decreasing North Atlantic oceanic heat transport has also contributed to the Holocene trend. Gradual reductions in ocean heat transport precede abrupt reductions in both paleoclimatic records and model results, so it is likely that the natural trend has been toward the threshold for abrupt change in the late Holocene.

INTRODUCTION

Rapid climate change in the North Atlantic would have significant impact on its coastal peoples, and perhaps on people much farther away [e.g., *Rahmstorf, 1995; Stocker and Schmittner, 1997*]. Paleoclimatic studies show that climatic variations have occurred on time scales ranging from sub-annual to orbital and longer. Abrupt changes are especially linked to millennial oscillations [e.g., *McCabe and Clark, 1998; Alley, 1998*].

An emerging paradigm includes the following elements [*e.g., Broecker et al., 1990; Lehman and Keigwin, 1992; Bond et al., 1993; Keigwin and Lehman, 1994; Rahmstorf, 1995; Fawcett et al., 1997; Stocker and Schmittner, 1998; Alley and Clark, 1999*]:

i) North Atlantic climate is linked to North Atlantic oceanic heat transport into the high-latitude Greenland, Iceland and Norwegian Seas (the Nordic or GIN Seas), with colder climate and especially colder winters tied to slower heat transport;

ii) Oceanic heat transport can vary both abruptly and gradually, and slow reductions typically have preceded abrupt reductions;

iii) Thermohaline (density) processes are important in North Atlantic oceanic heat transport;

iv) Many factors both in and beyond the North Atlantic can affect the thermohaline circulation; in particular, lower salinity in the North Atlantic correlates with reduced oceanic heat transport into the North Atlantic.

Here, we infer from Greenlandic ice-core data and other evidence that North Atlantic oceanic heat transport into the Nordic Seas has decreased during the last few thousand years of the current, or Holocene, warm period. This inference certainly does not lead to a prediction of abrupt climate

Mechanisms of Global Climate Change at Millennial Time Scales
Geophysical Monograph 112

change in the future, but it does suggest that the climate system has been drifting toward the threshold for an abrupt change.

ICE-CORE DATA

The Greenland Ice Sheet Project II (GISP2) deep ice core was drilled between 1989 and 1993 [*Mayewski et al., 1994*]. The site is at approximately 72.6°N latitude, 38.5°W longitude and 3200 m elevation, 28 km west of the summit of the Greenland ice sheet. Mean-annual (20 m) temperature is approximately -31°C, and accumulation is about 240 mm ice/year [*Alley and Anandakrishnan, 1995*]. High-resolution analyses were conducted of chemical, physical, isotopic, particulate, and gaseous characteristics of the core, for dating, flow information, and paleoclimatic reconstructions. The Greenland Icecore Project (GRIP) deep core, collected at the summit of the ice sheet during the same period (1989-1992), has been analyzed in a similar way, and results from the two projects have provided an exceptionally clear picture of climate changes over the most recent ice-age cycle [*Hammer et al., 1997*].

Especially prominent in the GRIP and GISP2 records are the millennial or few-millennial Dansgaard/Oeschger and Heinrich/Bond oscillations, and the tens-of-millennial Milankovitch cycles [*e.g., Dansgaard et al, 1993; Mayewski et al., 1997*]. The millennial oscillations almost certainly were caused by, or amplified by, changes in North Atlantic oceanic heat transport to high-latitude seas, linked to changes in strength or site of formation of North Atlantic Deepwater (NADW) [e.g., *Lehman and Keigwin, 1992; Keigwin and Lehman, 1994*]. The Milankovitch cycles almost certainly were caused by changes in incoming solar radiation (insolation) related to features of the Earth's orbit [*Imbrie et al., 1992; 1993*]. We explore relations between these two frequency bands.

We first consider Holocene temperature changes at GISP2. Summer and mean-annual insolation have been decreasing for millennia at high northern latitudes, whereas winter insolation has been rising (or is zero sufficiently far north) [*e.g., Berger, 1979; COHMAP, 1988*]. Reconstructed late-Holocene cooling may appear to be consistent with this Milankovitch forcing. However, estimated summertime cooling is smaller than that for wintertime, in contradiction to expectations from the Milankovitch forcing. A role for oceanic heat transport is suggested. The time trends in the ice-isotopic thermometer, in the total gas content of the GRIP core [*Raynaud et al., 1997*], and in climatic variability are consistent with a role for decreasing oceanic heat transport in the late-Holocene cooling.

Holocene Paleothermometry

The stable-isotopic composition of accumulated snow and ice has a clear and direct dependence on site temperature [*e.g., Robin, 1983; Jouzel et al., 1997*]. Ice-isotopic composition is thus widely used as a paleothermometer. However, the ice-isotopic composition is also sensitive to many other factors related to conditions at the moisture source, during transport and deposition, and after deposition. These "correction terms" may be trivial, or may in aggregate be as important as the temperature effect. In most places at most times, the ice-isotopic paleothermometer is useful and reasonably accurate, but its application always requires care [*Jouzel et al., 1997*]. In central Greenland, ice-isotopic data initially have individual-storm resolution, but diffusional processes progressively remove that signal. Seasonal temperature information is preserved for at least a few years [*Shuman et al., 1995*], but does not reliably survive processes in the firn (old snow during its transformation to impermeable ice) during the first century or two following deposition [e.g., *Cuffey and Steig, 1998*].

The physical temperature of the ice sheet provides a second paleothermometer. In central Greenland, for example, ice about halfway through the ice sheet is colder than ice either above or below. The cold ice was deposited at a temperature much lower than modern (by as much as 21°C) [*Cuffey et al., 1995; Johnsen et al., 1995; Cuffey and Clow, 1997; Clow et al., 1997; Dahl-Jensen et al., 1997; 1998*] during the most recent ice age, and has not had time to warm to a steady profile. Borehole paleothermometry provides low time resolution because of diffusive information loss. However, the borehole temperature is almost entirely a record of surface temperature for sites such as central Greenland where melting has been sparse and ice flow is understood reasonably well [e.g., *Alley and Koci, 1990; MacAyeal et al., 1991; Cuffey et al., 1994*].

Joint interpretation of the borehole-temperature and ice-isotopic records [*Cuffey et al., 1992; 1994; 1995; Johnsen et al., 1995; Cuffey and Clow, 1997*] has been used as a shortcut for inverting the borehole-temperature profile to determine surface-temperature histories. Subsequent, independent inversions of borehole-temperature records using more-traditional techniques [*Clow et al., 1997; Dahl-Jensen et al., 1997; 1998*] have verified the utility of this ice-isotope/borehole-temperature technique, as have independent techniques based on the stable-isotopic composition of trapped gases and the physical transformation of snow to ice [*Severinghaus et al., 1998*].

Techniques based on borehole thermometry provide paleotemperature information with time resolution that decreases as age increases, and that typically falls between 1/10 of the age and the age (so a thousand-year-long event ten-thousand years ago is near the practical resolution using typical techniques [*Firestone 1995*]). The joint use of ice-isotopes and borehole temperatures improves the time resolution of borehole paleothermometry, but in a peculiar way: one can demonstrate that high-frequency ice-isotopic data contain paleotemperature information, but the calibration of ice-isotopic ratios to temperature may be frequency-dependent, so the magnitudes of temperature deviations from a long-term trend may not be calibrated well. The Severinghaus et al. [*1998*] techniques and related techniques provide specific temperature estimates at certain times with high (decadal) time-resolution, and so allow the high-frequency oscillations to be calibrated at those times.

Joint interpretation of borehole temperatures and ice-isotopic ratios requires use of a flow model, which can produce estimates of changes in surface elevation [e.g., *Cuffey et al., 1995; Johnsen et al., 1995; Cuffey and Clow, 1997*]. Workers reconstruct temperatures at the ice-sheet surface, which can move up or down through the atmosphere, but some workers also correct the data to estimate temperature changes at a constant elevation. Changes in ice-isotopic ratio certainly include any changes in source-water isotopic ratio, which in turn may reflect changes in the mean composition of the ocean related to changes in volume or composition of isotopically light ice sheets. Different authors either have or have not chosen to make the mean-ocean correction. In interpreting published data, care is required to ensure that comparisons involve reconstructions made using the same assumptions. Note that since the middle Holocene, little surface-elevation change is modeled to have occurred in Greenland [*Cuffey and Clow, 1997; Raynaud et al., 1997*], so surface temperatures and temperatures corrected to constant elevation are nearly equivalent.

Techniques that utilize borehole thermometry and gas isotopes provide information on mean-annual temperatures averaged over a few years or longer, with seasonal temperature signals lost during rapid diffusion in the firn and not preserved in ice [*Cuffey et al., 1995; Severinghaus et al., 1998*]. In contrast, melt layers can be interpreted as recorders of summertime temperatures [*e.g., Herron et al., 1981; Koerner and Fisher, 1990*].

Melting on ice sheets is a clear indication of warmth. The strong seasonality of temperature on ice sheets dictates that, in regions where melting is rare, the high temperatures conducive to melting occur almost solely in summertime [*reviewed by Alley and Anandakrishnan, 1995*].

Paleothermometry from melt-layer frequency is not as well-characterized as are the other techniques discussed above. Traditional study of melt layers involves seeing bubble-free zones in bubbly ice [e.g., *Langway, 1967*]. Hence, formation of a melt layer requires both that enough meltwater be produced, and that enough of that meltwater be refrozen in a restricted zone to make a recognizable feature [cf. *Pfeffer and Humphrey, 1996*]. Melt-layer occurrence thus may be sensitive to temperature gradients in firn as well as to air temperature or insolation. Furthermore, a change in melt-layer frequency in a region where melting is sparse may indicate a change in variability of summer temperatures rather than a change in mean summer temperatures [*Alley and Anandakrishnan, 1995*]. The time-resolution of a melt-layer history is related to the frequency with which melt layers occur; at GISP2, the sparse occurrence (typically one/century) restricts histories to low time resolution (typically millennial).

Calibration of the melt-layer thermometer also involves difficulties. We have used both spatial variability and temporal variability of conditions in calibrating the GISP2 record [*Alley and Anandakrishnan, 1995*]. Close agreement between two independent calibrations improves our confidence in the results.

In many ways, the interpretation of melt thermometry today is at the same stage as for ice-isotopic thermometry before the introduction of joint interpretation with borehole-temperature, gas-isotopic, and general-circulation-modeling techniques [*Cuffey et al., 1992; Jouzel et al., 1997; Krinner et al., 1997*]. We note, however, that most paleoclimatic proxies are calibrated in essentially the same way as we have calibrated the melt paleothermometer, by assuming that spatial dependence of sediment properties on climatic variables can be applied at one point over time, or by assuming that some short-term calibration of sediment properties to instrumental records can be applied over longer times. The melt-layer thermometer probably is useful and accurate, but further work to validate it is desirable.

Comparison of summertime to mean-annual temperature histories can provide further information (Figure 1). We use surface-temperature rather than constant-elevation reconstructions because melting responds to surface temperature. Millennial averaging has been adopted in Figure 1 in light of the statistical uncertainties in the melt-layer data for shorter times [*Alley and Anandakrishnan, 1995*]. The mean-annual GISP2 temperature data used in Figure 1 were provided by K. Cuffey and are based on joint interpretation of ice-isotopic and borehole-temperature data [*Cuffey et al., 1995; Cuffey and Clow, 1997*]. These reconstructions did not include the small correction for oceanic-composition changes, for the reasons discussed in Cuffey and Clow [*1997*]. The mean-annual trend reconstructed using the

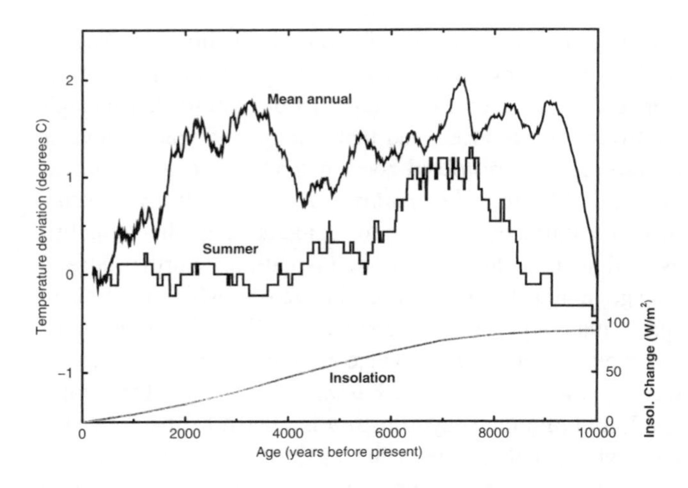

Figure 1. Thousand-year-average Holocene surface-temperature histories for GISP2, central Greenland. The summertime temperatures are from Alley and Anandakrishnan [*1995*]. The cooling since about 7500 years ago is well-resolved; the warming to 7500 years ago is somewhat uncertain owing to the low bubble content of the ice when recovered, as explained by Alley and Anandakrishnan [*1995*]. The mean-annual temperatures are from the joint interpretation of ice-isotopic ratios and borehole temperatures in Cuffey et al. [*1995*] and Cuffey and Clow [*1997*]. Uncertainties are approximately ± 0.5°C, as described in the text. Also shown is the deviation in GISP2 July insolation from the 1950 value, following Berger [*1979*].

Cuffey et al. [*1995*] and Cuffey and Clow [*1997*] results is similar to that we calculated from Johnsen et al. [*1995*], after we removed the corrections for thickness changes and isostatic response as described by Johnsen et al. [*1995*] to obtain surface-temperature change. (Note that Johnsen et al. [1995] did make a small correction for mean-oceanic isotopic changes.)

Figure 1 shows a mean-annual cooling of about 2°C from middle to late Holocene. Cuffey and Marshall [*in prep., personal communication from K. Cuffey, December, 1998*] obtain 1.9 ± 0.6°C for this cooling. This is generally consistant with our interpretation of Johnsen et al. [*1995*]. In comparison, we obtain a summertime cooling of 1.3°C, with statistical uncertainties of about ± *0.5°C* [*Alley and Anandakrishnan, 1995*].

Our best estimate of the mean-annual cooling since the middle Holocene is larger than the summertime change by 0.6 ± 0.8°C. The uncertainties allow summertime and mean-annual changes to have been similar, but a summertime cooling much larger than mean-annual cooling is highly unlikely. This implies that winter (non-summer) cooling has been larger than, or perhaps similar to, the summertime cooling. Also of interest in Figure 1 is that the mean-annual cooling seems to have been delayed relative to the drop in summertime insolation and relative to the summertime cooling.

Non-summer cooling greater than summer cooling is surprising. The summertime insolation has been dropping since before the middle Holocene [*e.g., Berger, 1979*], which should have produced summertime cooling, as observed. However, the wintertime insolation is zero at sufficiently high latitudes and has been rising at lower northern latitudes, which might lead one to expect little wintertime change or wintertime warming since the middle Holocene. This expectation is generally supported by results from atmospheric general circulation models executed using appropriate orbital forcing, greenhouse-gas concentrations and ice-sheet configurations, but without changing deep-ocean circulation.

Climate change from middle to late Holocene in central Greenland simulated in general circulation models using fixed sea-surface temperatures typically shows summertime cooling several degrees larger than wintertime or mean-annual cooling [*e.g., Kutzbach, 1987; Vettoretti et al., 1998*]. A smaller difference in seasonal trends is obtained with a mixed-layer ocean, because some of the excess summer insolation of the mid-Holocene is absorbed by the North Atlantic, which delays sea-ice formation and warms surrounding regions during autumn and early winter (*e.g. Kutzbach et al., 1991; Vettoretti et al., 1998; and the Mitchell et al., 1988* simulations with realistic Laurentide ice sheet specified). Nonetheless, most simulations we have examined show late-Holocene cooling in central Greenland to have been larger for the summer than for non-summer conditions. We have obtained a similar result in our numerical simulations.

Our simulations using the GENESIS General Circulation Model (versions 1.02 and 2) have examined the climatic response to specified changes in the North Atlantic oceanic heat transport at various time slices [*Fawcett et al., 1997; Ágústsdóttir, 1998*]. Of relevance here are simulations with GENESIS v. 1.02A for 8 ka and modern conditions, which show that Milankovitch forcing with unchanged ocean heat transport causes larger summertime than wintertime changes, but that ocean-heat-transport changes can allow a better match to the observations.

Our GENESIS runs used the model atmosphere coupled to a 50 m slab mixed-layer ocean in which poleward heat transport is included as a zonally symmetric function of latitude, tuned to fit modern zonal mean sea-surface temperatures [*Thompson and Pollard, 1995*]. To prevent unrealistic build-up of sea ice in modern simulations of the North Atlantic, a region of enhanced oceanic heat flux during winter has been added to the Nordic Seas in the model.

Table 1. Mean-annual (Ann.) and summer (JJA) climate results for central Greenland obtained from simulations using the GENESIS climate model v. 1.02A [*Ágústsdóttir, 1998*]. In all cases, values are averages for the four grid cells surrounding the summit of the ice sheet, corrected for the elevation difference between the smoothed topography of the model and observed topography following Fawcett et al. [*1997*]. All simulations use a mixed-layer ocean, which can have an additional heat flux from the North Atlantic/Nordic Seas during winter. The Modern simulation and the 8 ka simulation with "modern" oceanic heat flux use the value of this extra heat flux that reproduced the modern climate best in the model tuning of Thompson and Pollard [*1995*]. For the 8 ka "reduced", the extra heat flux in the Nordic Seas was set to zero, and for the 8 ka "enhanced", the extra heat flux was set 60% higher than in the "modern" case. Average values and standard deviations for the last five years of each model run are shown. Interannual variability is seen to be highest in the "reduced" run.

		Temperature		Accumulation (cm/mo)	
Age	Ocean Heat Flux	Ann.	JJA	Ann.	JJA
Modern	Modern	-27.0±1.3	-11.8±0.5	2.62±0.18	3.62±0.49
8 ka	Reduced	-35.7±1.6	-15.0±0.8	1.74±0.25	2.66±0.21
8 ka	Modern	-32.7±0.7	-14.8±0.9	2.21±0.15	3.25±0.13
8 ka	Enhanced	-31.6±1.2	-14.1±0.5	2.40±0.20	3.47±0.24

This enhanced heat flux warms the mixed layer whenever the surface-water temperature falls below 1.04°C in a rectangular region between -10°and 56°E longitude, and 66°and 78°N latitude. The flux increases linearly to 500 W/m^2 as the ocean cools to its freezing point of -1.96°C [*Thompson and Pollard, 1995*]. This simulates the advection of heat by warm ocean currents, and the deepening of the mixed layer in winter. To maintain global energy balance in the model, the extra heat released to the atmosphere by this Nordic Sea adjustment is subtracted uniformly from all oceanic grid cells between 55°S and 55°N latitude. Note that this Nordic-Sea adjustment is not a dynamic oceanic heat flux that responds to changes in fields such as oceanic salinity; rather, this model parameterization allows the atmosphere to extract heat from the model ocean in a vaguely similar way, and in a similar region, to where the modern atmosphere extracts heat from the North Atlantic today.

This GENESIS model parameterization in the wintertime Nordic Seas has allowed us to explore possible atmospheric response to changes in North Atlantic oceanic heat transport [*Fawcett et al., 1997; Ágústsdóttir, 1998; Ágústsdóttir et al., in press*]. We have changed the strength of this extra heat source under boundary conditions suitable for different times, and assessed the sensitivity of the atmosphere to these altered boundary conditions. Three simulations for 8 ka [*Ágústsdóttir, 1998*] set the maximum value of this extra heat flux to 0, 500, and 800 W/m^2 (which we refer to as "reduced", "modern", and "enhanced" oceanic heat flux, respectively). The magnitude of the "enhancement" was chosen to provide significant contrast to the "modern" situation without being physically implausible, and was not chosen to match any particular paleoclimatic data.

The modern control simulation is warmer than any of the 8 ka simulations (Table 1). The control run used modern

Table 2. Summer (ΔT_{jja}) and mean-annual (ΔT_{ann}) temperature changes from mid-Holocene to recent, $\Delta T \equiv T_{mid_Holocene} - T_{recent}$. The 8 ka simulations are GENESIS runs compared to a modern control as shown in Table 1 and described in the text, and include the uncertainties from propagation of model standard deviations over the last five years of the runs. The uncertainties are rather large so statistical confidence is not high, but the 8 ka simulation with "enhanced" heat flux comes closer to the observations than do the other cases, and the 8 ka simulation with "reduced" heat flux provides a poor fit.

Case	ΔT_{jja}(°C)	ΔT_{ann}(°C)	ΔT_{ann}-ΔT_{jja}(°C)
Observed	1.3±0.5	1.9±0.6	-0.6±0.8
8 ka Reduced	-3.2±1.0	-8.7±2.1	-5.5±2.3
8 ka Modern	-3.0±1.0	-5.7±1.5	-2.7±1.8
8 ka Enhanced	-2.3±0.7	-4.6±1.8	-2.3±1.9

CO_2 concentrations; owing to its lack of a model deep ocean, GENESIS is largely equilibrated with the anthropogenically high CO_2 levels during 25 simulated years in our experiments, whereas the Earth probably has not equilibrated with anthropogenically increased CO_2 [*cf. Pollard and Thompson, 1997*]. Comparing the modern and 8 ka simulations, both executed using "modern" North Atlantic oceanic heat transport (Table 2), the seasonal trends (more mean-annual and winter warming from 8 ka to today than summer warming) are more consistent with expectations for the seasonal effects of the Milankovitch forcing and less consistent with the observed trends. (The modeled late-Holocene cooling is larger for the summer than for mean-annual and winter conditions, even though in this case both coolings are negative.) The seasonal differences between the "enhanced" 8 ka run and the "modern" control run come closer to the observed trends than do the 8 ka versus modern runs with "modern" heat flux, although close matches are

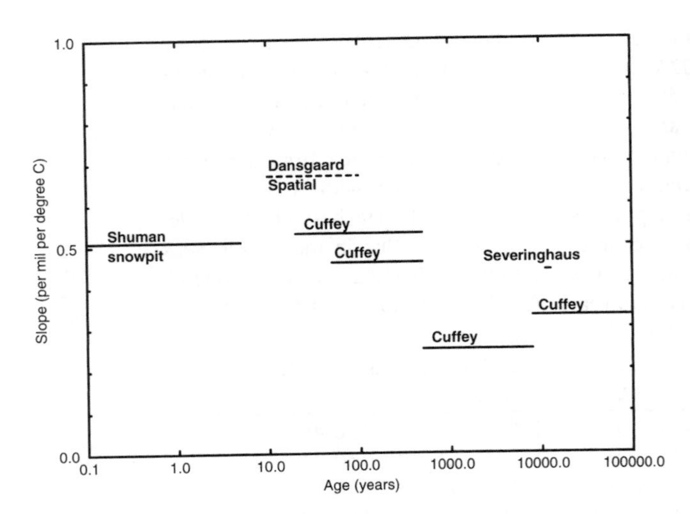

Figure 2. Relation between oxygen-isotopic composition of ice and temperature as a function of age. Data are from Shuman et al. [*1995*], Cuffey et al. [*1992; 1994* for the youngest results labeled "Cuffey"], Cuffey et al. [*1995*] and Cuffey and Clow [*1997*] for the other "Cuffey" results, Severinghaus et al. [*1998*] using the gas-isotope-based temperature change at the end of the Younger Dryas, and Dansgaard [*1964*] plotted against averaging length.

not obtained and considerable uncertainty remains (Table 2).

Our data (Figure 1), if taken at face value, thus show central Greenland cooling since the middle Holocene to have been larger in winter than in summer, yet models driven by the atmospheric effects of Milankovitch forcing alone lead us to expect cooling to have been larger in summer than in winter. A possible explanation of the observed seasonal temperature trends, which is supported by our modeling, is that oceanic heat transport to the Nordic Seas was larger during the middle Holocene than recently. The greater effect of North Atlantic oceanic heat transport on wintertime than on summertime conditions can offset the seasonality of the Milankovitch forcing.

Ice-Isotopic Calibration to Temperature

The calibration method of Cuffey et al. [*1992; 1994; 1995*] produces an estimate of mean-annual temperature, and also of the relation between temperature and ice-isotopic composition. By allowing the calibration to vary across four time windows, Cuffey et al. [*1995*] and Cuffey and Clow [*1997*] produced a new paleoclimatic indicator, the slope of this isotope-temperature relation, which we report as the per-mil change in oxygen-isotopic ratio per degree change in temperature.

The results of the Cuffey et al. [*1994*] Little Ice Age calibration, and the Cuffey et al. [*1995*] Holocene and ice-age calibrations, are shown in Figure 2, together with the snowpit calibrations of Shuman et al. [*1995*], the preliminary calibration based on gas-isotope fractionation of Severinghaus et al. [*1998*], and the spatial-gradient results of Dansgaard [*1964*]. Because of its preliminary nature, the calibration of Severinghaus et al. [*1998*] has large error bars that just include the Cuffey et al. [*1995*] ice-age values and the Dansgaard [*1964*] value. The Shuman et al. [*1995*] seasonal data have been smoothed somewhat by diffusion; the undiffused seasonal slope could be somewhat higher than shown here. For the various calibrations involving comparison of ice-isotopic and borehole temperatures, uncertainties are different for the different ages and techniques plotted. Values obtained are somewhat different for calculations at constant elevation versus those for the ice-sheet surface, and for calculations made using a correction for changing mean ocean composition versus those made without changing the ocean composition. Robust results regardless of computation method chosen are: (i) the time-calibrated slope of the ice-isotopic thermometer is lower than the spatial-gradient results of Dansgaard [1964] for all times considered except possibly the seasonal data; (ii) the slope for the glacial-interglacial transition is lower than for the Little Ice Age; and (iii) the slope for the several-millennial late-Holocene trend is lower than or similar to the slope for the glacial-interglacial transition [*Cuffey et al., 1994; 1995; Cuffey and Clow, 1997; Cuffey and Marshall, in prep.; personal communications from K.M. Cuffey, December 1998 and January, 1999*].

An isotopic calibration to temperature can depend on many factors. As noted above, isotopic ratios in accumulated snow depend on conditions at the moisture source, along the moisture path, and during and following deposition. Changes in any of these that correlate with changes in site temperature can cause the calibration to deviate systematically from the expected value.

For central Greenland, two hypotheses have been advanced as being especially likely to explain the observed calibrations: changes in seasonality of precipitation, and correlation of source and site temperatures. (As discussed by Jouzel et al. [*1997*], changes in source region of the moisture [*Charles et al., 1994*] may have occurred but probably have not exerted a major control on the isotopic composition at the site.)

As described above, Fawcett et al. [*1997*] used the GENESIS atmospheric general circulation model to simulate the effects of reduction in North Atlantic ocean heat transport to the Nordic Seas on the atmosphere during the

Younger Dryas interval. In agreement with a range of climatologies and other climate-model results, Fawcett et al. [*1997*] found that the main effects of North Atlantic ocean heat transport on the atmosphere occur in the winter, with a warm ocean producing warm and wet conditions in surrounding regions. Reduction in oceanic heat transport was simulated to cause wintertime cooling and drying larger than in the summertime for central Greenland and other regions around the North Atlantic. Because the isotopic thermometer only records the temperature when snow is falling, the reduction in wintertime snowfall means that cold intervals would not produce ice as isotopically light ("cold") as one would expect from the mean-annual temperature; hence, a small isotopic shift corresponds to a large temperature shift.

Boyle [*1997*] noted that high-latitude North Atlantic cooling has been correlated with cooling in lower-latitude North Atlantic moisture sources. Temperature changes in source regions and in Greenland are both plausibly linked to changes in North Atlantic circulation. The isotopic composition of precipitation depends on the temperature difference between source and precipitation site as well as on the site temperature [*Boyle, 1997*], so correlated temperature changes between site and source would cause isotopic changes to appear to underestimate temperature changes.

Krinner et al. [*1997*] presented model evidence supporting the seasonality explanation of Fawcett et al. [*1997*] as probably explaining at least much of the observed signal. However, even without choosing between these two models, we note that both can be linked to oceanic processes involving North Atlantic heat transport. If either of these explanations is correct, or if they jointly are correct, and if they apply in similar ways over all times considered, then one can interpret the deviation of the isotopic calibration from the canonical Dansgaard [*1964*] result as an indication of the importance of changes in oceanic heat transport to the climate changes recorded by the isotopic shifts.

In such an interpretation, Little-Ice-Age and perhaps annual shifts involve oceanic processes, but the late-Holocene trend, the Younger Dryas signal, and the glacial-interglacial shift were more strongly affected by the ocean. One could even speculate that the late-Holocene trend and the glacial-interglacial shift were at least in part forced by the ocean (which in turn may have been responding to other forcing), and that the Little-Ice-Age changes involved oceanic feedbacks but not oceanic forcing [*cf. Crowley and Kim, 1996*]. Whether the forcing mechanism is primarily oceanic or atmospheric, a consistent interpretation is that the Holocene calibration differs strongly from the expected value because the late-Holocene cooling was forced by reduction in oceanic heat transport to the Nordic Seas.

Total-Gas-Content Data

Consistent results are provided by total-gas-content measurements. As discussed by Raynaud et al. [*1997*], air content of ice formed from firn depends on several factors, including site elevation, temperature, and local air-pressure conditions. Depositional and near-surface diagenetic processes cause layers deposited during winters to have higher density than layers deposited during summers. The interconnected pore spaces of firn are pinched off to form bubbles at a shallower depth in winter layers than in summer layers. When enough pore spaces are pinched off in a winter layer, it becomes impermeable. Excess air is then trapped in the summer layer beneath, which remains permeable laterally but cannot expel that air through the overlying, impermeable winter layer. Summer layers in an ice core thus contain more air than do winter layers. A decrease in the winter:summer ratio of annual snow accumulation would produce an increase in the air content of an ice core, provided that sufficient winter accumulation remained to create impermeable layers [*Raynaud et al., 1997*].

The total-gas data from the GRIP core, after correction for other known effects, show a significant increase in air content through the Holocene toward the present [*Raynaud et al. 1997*]. If this total-gas increase were caused by an increase in air pressure associated with thinning of the ice sheet, significantly more ice-sheet thinning would be required than is allowed by several model results, as reviewed by Raynaud et al. [*1997*]. A likely interpretation is that the winter:summer ratio of snow accumulation has decreased toward the present [*Raynaud et al., 1997*]. Qualitatively, this is the same result obtained from the calibration of the ice-isotopic thermometer.

Climatic Variability

It is common to associate climatic variability with continentality--extremes result when the atmosphere is not buffered by the longer-response-time ocean. A reduction in North Atlantic oceanic heat transport to the Greenlandic region is in some sense analogous to an increase in the continentality of the island, and might be expected to increase variability. Indeed, glacial times associated with reduced oceanic heat transport exhibit strong variability [*e.g., Ditlevsen et al., 1996; Mayewski et al., 1997*].

The climate-model simulations of Ágústsdóttir [*1998*] show that reducing the oceanic heat transport to the Nordic Seas under other conditions appropriate for 8 ka produces an increase in interannual variability for both precipitation (significant at just less than the 90% level) and temperature (significant at greater than the 90% level) (Table 1). However, no highly significant changes are found in comparing the 8 ka and modern simulations, using either "modern" or "enhanced" heat transport at 8 ka. We take these results as tentative support for the "common-sense" expectation that reduction in oceanic heat transport to the Nordic Seas would increase variability in Greenland.

We have examined the interannual variability of several of the paleoclimatic signals in the GISP2 ice core through the Holocene [*Alley et al., 1996*]. In Figure 3, we plot the accumulation-rate variability for the Holocene, from Spinelli [*1996*]. The coefficient of variation (standard deviation divided by the mean) is shown, but the figure with standard deviations is similar. The accumulation-rate measurements are based on the visible-only stratigraphic dating by Alley et al. [*1997b*], collected in a consistent manner throughout the Holocene to avoid observational biases. (The "official" dating of the GISP2 ice core [*Meese et al., 1997*] used data from different techniques and different investigators for different age ranges, which may have affected the variability, as described in Alley et al [*1997b*]. The data used for Figure 3 are probably more-consistent but less-accurate than the official data, although still with typical accuracy of 1% [Alley et al., 1997b].)

For accumulation rate, and for other climatic indicators studied, variability is higher recently than earlier in the Holocene [*Alley et al., 1996*]. The late-Holocene trend to increasing variability then is consistent with a late-Holocene trend to reduced ocean heat transport to the Nordic Seas.

DISCUSSION

Many modeling studies of Holocene climatic trends have assessed the roles of shrinking ice sheets, varying insolation from orbital processes, evolving vegetation, and changing greenhouse-gas concentrations [*e.g., Kutzbach, 1987; COHMAP, 1988; Kutzbach et al., 1991*]. Such studies are gratifyingly successful in simulating patterns of changes reconstructed from paleoclimatic indicators [*e.g., COHMAP, 1988*]. However, reconstructed changes have sometimes been larger than those produced by models [*e.g., Kutzbach et al., 1996*].

Studies of abrupt climate changes, and of ice-age cycles, have placed greater emphasis on the role of oceanic heat transport in affecting climate, especially at high northern latitudes. Imbrie et al. [*1992; 1993*] argued that orbital forcing produces changes in the dynamics of the high-latitude North Atlantic, affecting formation of North Atlantic Deepwater (NADW). Many workers including Broecker [*1997*] have argued that abrupt, millennial-scale climate changes involve steplike changes in oceanic heat transport, in the same North Atlantic regions suggested to be involved in the changes at orbital timescales. An oceanic role is indicated for the abrupt North Atlantic cooling event about 8200 years ago [*Alley et al., 1997a; Klitgaard-Kristensen et al., 1998; Barber et al., 1998*] as well as for older events [*Lehman and Keigwin, 1992; Keigwin and Lehman, 1994*].

Paleoclimatic reconstructions from North Atlantic sediment cores typically reveal the complexity of the North Atlantic system. Some records, such as those of Koc et al. [*1993*] for Holocene surface-water conditions in the Nordic Seas, and those of Boyle and Keigwin [*1987*] ($\delta^{13}C$ in their Fig. 1 and some other curves) for deepwater conditions on the Bermuda Rise, are consistent with a late-Holocene trend to reduced North Atlantic oceanic heat transport linked to reduced NADW formation; however, not all records we have examined show such a signal.

Events of reduced North Atlantic oceanic heat transport to the Nordic Seas, such as the Younger Dryas, are linked to colder conditions in Greenland characterized by larger climate variability [*e.g., Ditlevsen et al., 1996; Mayewski et al., 1997*], greater reduction in wintertime than in summertime temperature and snow accumulation [*Fawcett et al., 1997; Krinner et al., 1997*], hence ice-isotopic ratios that have not changed as much as one would expect based on the change in mean-annual temperature [*Cuffey et al., 1995; Severinghaus et al., 1998*]. Reduced North Atlantic oceanic heat transport is also linked to climate changes well beyond Greenland, including dry conditions in monsoonal regions of Africa [*Street-Perrott and Perrott, 1990; Alley et al., 1997a*].

Data from African lakes show a clear drying trend through the late Holocene [*Street-Perrott and Perrott, 1990*]. The sign of this trend is well-simulated by models forced with Milankovitch insolation variations, but the changes have been somewhat larger than simulated [*Kutzbach et al., 1996; Jolly et al., 1998; Qin et al., 1998*]. Assuming both that models and data are accurate, a possible explanation is that the North Atlantic has been cooling since the middle Holocene, perhaps linked to reduction in North Atlantic oceanic heat transport that in turn may be linked to Milankovitch forcing.

Holocene data from the GISP2 ice core are most directly interpreted as indicating greater wintertime than summertime reduction in temperature and snow accumulation since the middle Holocene. The inferred reduction in the winter:summer snow-accumulation ratio may explain both the small changes in ice-isotopic ratios relative to the significant reconstructed temperature changes, and the large changes in total-gas content of the GRIP core [*Raynaud et*

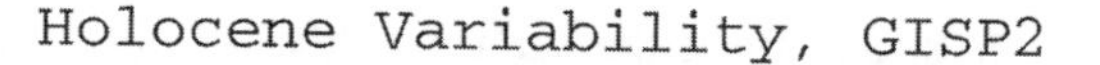

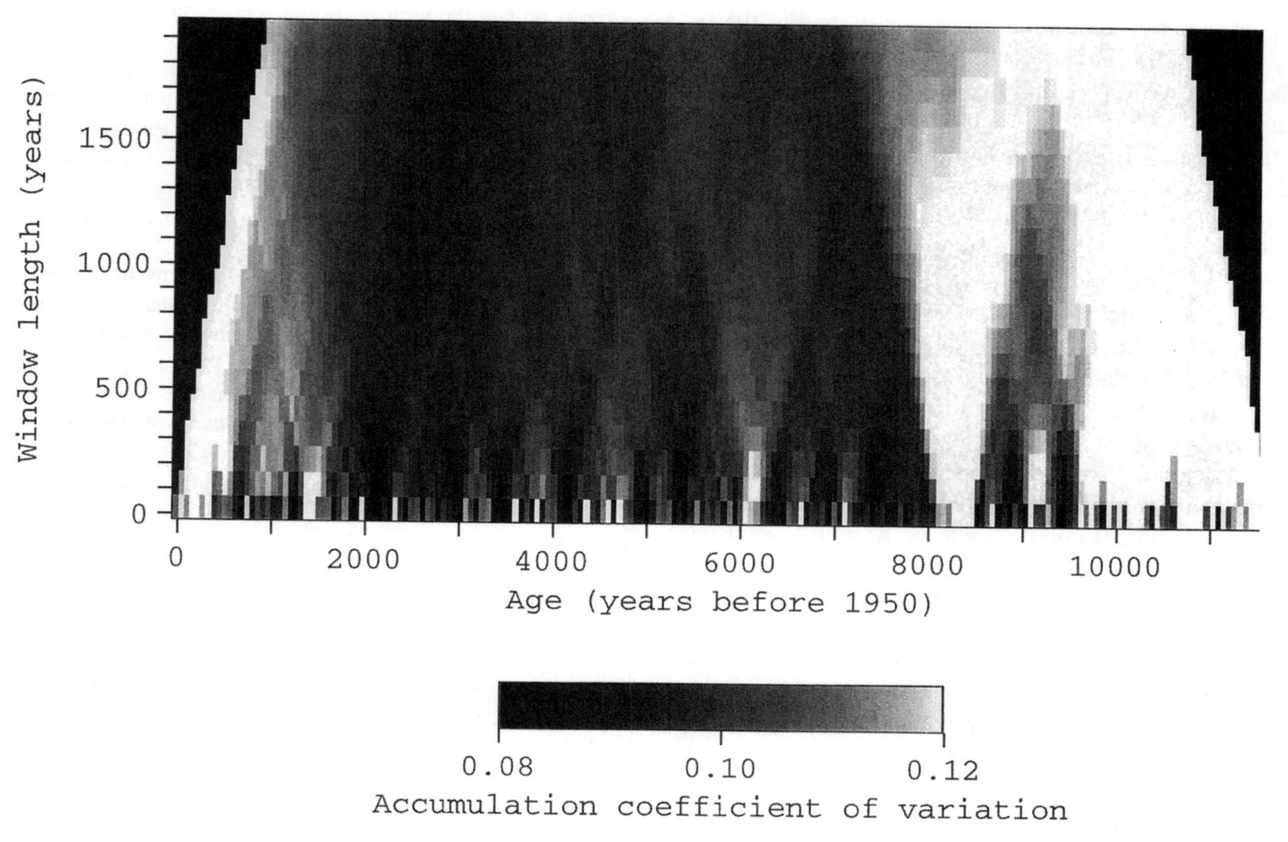

Figure 3. Interannual accumulation-rate variability, from Spinelli [*1996*] using the visible-only data set described in Alley et al. [*1997b*]. Different window lengths (vertical axis) reveal similar time-trends (horizontal axis) in variability (gray scale). Events such as the cooling just prior to 8 ka show up clearly. The general trend is for decreasing variability through the early Holocene with increasing variability more recently. Variability was significantly higher during the Younger Dryas and other, older cold times than during the Holocene.

al., 1997] relative to the small modeled elevation change at the site. A trend to greater interannual climate variability since the middle Holocene also has occurred. Model results indicate that such trends could be caused by a trend to reduced oceanic heat transport to the Nordic Seas, but that these trends are not expected based solely on the atmospheric effects of Milankovitch forcing.

We fully recognize that the interpretations presented here for the paleoclimatic data are nonunique. The quantities interpreted (and especially the variability and the ice-isotopic thermometer calibration) are derived quantities, and their interpretations involve several assumptions that are difficult to test. Errors thus are hard to estimate, but probably are not insignificant. We nonetheless believe that the results presented here show that oceanic feedbacks have occurred, almost certainly involving mixed-layer processes and probably involving dynamic processes that transport heat to the Nordic Seas. If these changes had been larger and more rapid, their effects on climate may have mimicked the abrupt changes of the past.

The Imbrie et al. [*1992; 1993*] interpretation of ice-age cycles suggests that a very early response to insolation changes occurs in the high-latitude seas of the North Atlantic. Our reconstructions support this idea.

There is an apparent dichotomy between our reconstruction of a warmer mid-Holocene occurring with stronger North Atlantic oceanic heat transport to the high latitudes, and some simulations of the future in which greenhouse-gas warming reduces the North Atlantic overturning circulation [*e.g., Stocker and Schmittner, 1997*]. A possible explanation is given by the new model results of Wang et al. [*1999*], who show that the steady-state strength of the North Atlantic thermohaline circulation increases with the strength of global hydrological transports, but that significant transient increases in global hydrological transports over decades to centuries can cause collapse of the North

Atlantic overturning circulation. In steady state, the salinity in the deep convection regions is closely tied to the salinity of the deep ocean, which is nearly constant, but the surface salinity in upwelling regions at southern high latitudes affects the rate of upwelling and thus the rate of the overturning circulation. Transiently, however, fresh-water influx can cap the deep convection regions in the north [*Wang et al., 1999*]. It is possible that the changes associated with orbital forcing are slow enough that they allow the circulation to approach steady state, whereas faster greenhouse warming in the future is more consistent with the transient conditions.

Note that if we relate the strength of the middle-Holocene northern overturning circulation to the Wang et al. [*1999*] steady results, then southern rather than northern high-latitude conditions are most important in controlling the strength of the northern circulation. Many records do show southern high-latitude middle-Holocene warmth [*e.g., Ciais et al., 1992*] which might have produced a stronger hydrological cycle, more-buoyant surface waters in southern upwelling regions, and a stronger overturning circulation than recently. However, a detailed consideration of this possibility is clearly beyond the scope of our paper.

Abrupt climate changes of the past have had a typical pattern in which slow cooling was followed by abrupt cooling, a cold interval, and then abrupt warming [*e.g., Bond et al., 1993*]. Such a pattern can be interpreted to indicate a gradual slowdown and then a sudden slowdown of oceanic heat transport [*Broecker et al., 1990*]. Such behavior is consistent with some models of the thermohaline circulation, in which abrupt changes occur once the system is brought to an appropriate threshold [*e.g., Stocker and Schmittner, 1997*]. These results suggest the hypothesis that the North Atlantic system is trending towards an abrupt shift. However, they provide no information on how close the system is to such a shift, and only suggest that such a shift is closer than it was during the middle Holocene.

Acknowledgements. We thank the GISP2 Science Management Office, the 109th New York Air National Guard, the Polar Ice Coring Office, the National Ice Core Laboratory, the National Science Foundation Office of Polar Programs, two anonymous reviewers, and numerous colleagues at GISP2, GRIP, and elsewhere. We thank Peter Clark and Robin Webb for organizing the meeting. RBA also thanks the D.&L. Packard Foundation for partial funding.

REFERENCES

Ágústsdóttir, A.M., Ph.D. Abrupt climate changes and the effects of North Atlantic deepwater formation: results from the GENESIS global climate model and comparison with data from the Younger Dryas event and the event at 8200 years bp and the present, *Thesis*, The Pennsylvania State University, University Park, PA, 1998.

Ágústsdóttir, A.M., R.B. Alley, D. Pollard and W.H Peterson, Ekman transport and upwelling during Younger Dryas estimated from wind stress from GENESIS climate model experiments with variable North Atlantic heat convergence, *Geophysical Research Letters*, in press, 1999.

Alley, R.B., Icing the North Atlantic, *Nature, 392*, 335-336, 1998.

Alley, R.B., and S. Anandakrishnan, Variations in melt-layer frequency in the GISP2 ice core: implications for Holocene summer temperatures in central Greenland, *Annals of Glaciology, 21*, 64-70, 1995.

Alley, R.B., and P.U. Clark, The deglaciation of the northern hemisphere: a global perspective, *Annual Reviews of Earth and Planetary Sciences, 27*, 149-182, 1999

Alley, R.B., and B.R. Koci, Recent warming in central Greenland?, *Annals of Glaciology, 14*, 6-8, 1990.

Alley, R.B., P.A. Mayewski, and E.S. Saltzman, Holocene climatic-variability trends in central Greenland, *Eos (Transactions of the American Geophysical Union), 77*, S150, 1996.

Alley, R.B., P.A. Mayewski, T. Sowers, M. Stuiver, K.C. Taylor and P.U. Clark, Holocene climatic instability: A prominent, widespread event 8200 years ago, *Geology, 25*, 483-486, 1997a.

Alley, R.B., C.A. Shuman, D.A. Meese, A.J. Gow, K.C. Taylor, K.M. Cuffey, J.J. Fitzpatrick, P.M. Grootes, G.A. Zielinski, M. Ram, G. Spinelli and B. Elder, Visual-stratigraphic dating of the GISP2 ice core: basis, reproducibility, and application, *Journal of Geophysical Research, 102(C12)*, 26,367-26,381, 1997b.

Barber, D.C., A.E. Jennings and J.T. Andrews, Did meltwater trigger the cold event 8,200 cal. yrs ago? Revised 14C ages for drainage of glacial Lake Ojibway, in Clark, P.U., and R.S. Webb, eds., Mechanisms of Millennial-Scale Global Climate Change, *Abstracts with Program*, American Geophysical Union, Washington, D.C., 1998, p. 15.

Berger, A., Insolation signatures of Quaternary climatic changes, *Nuovo Cimento, 2C*, 63-87, 1979.

Bond, G., W. Broecker, S. Johnsen, J. McManus, L. Labeyrie, J. Jouzel, and G. Bonani, Correlations between climate records from North Atlantic sediments and Greenland ice, *Nature, 365*, 143-147, 1993.

Boyle, E.A., Cool tropical temperatures shift the global δ^{18}O-T relationship: An explanation for the ice core δ^{18}O-borehole thermometry conflict?, *Geophysical Research Letters, 24*, 273-276, 1997.

Boyle, E.A. and L. Keigwin, North Atlantic thermohaline circulation during the past 20,000 years linked to high-latitude surface temperature, *Nature, 330*, 35-40, 1987.

Broecker, W.S., Thermohaline circulation, the Achilles Heel of our climate system: will man-made CO_2 upset the current balance?, *Science, 278*, 1582-1588, 1997.

Broecker, W.S., G. Bond, M. Klas, G. Bonani, and W. Wolfli,

A salt oscillator in the glacial Atlantic? 1. The concept, *Paleoceanography, 5*, 469-477, 1990.
Charles, C., D. Rind, J. Jouzel, R. Koster, and R. Fairbanks, Glacial-interglacial changes in moisture sources for Greenland: influences on the ice core record of climate, *Science, 261*, 508-511, 1994.
Ciais, P., J.R. Petit, J. Jouzel, C. Lorius, N.I. Barkov, V. Lipenkov, and V. Nicolaïev, Evidence for an early Holocene climatic optimum in the Antarctic deep ice-core record, *Climate Dynamics, 6*, 169-177, 1992.
Clow, G.D., E.D. Waddington, and N.S. Gundestrup, Reconstruction of past climatic changes in central Greenland from the GISP2 high-precision temperature profile using spectral expansion [abs.]: Eos *(Transactions of the American Geophysical Union), 78*, F6, 1997.
COHMAP Members, Climatic changes of the last 18,000 years: observations and model simulations, *Science, 241*, 1043-1052, 1988.
Crowley, T.J., and K.Y. Kim, Comparison of proxy records of climate change and solar forcing, *Geophysical Research Letters, 23*, 359-362.
Cuffey, K.M., and E.J. Steig, Isotopic diffusion in polar firn: implications for interpretation of seasonal climate parameters in ice core records, with emphasis on central Greenland, *Journal of Glaciology, 44*, 273-284, 1998.
Cuffey, K.M., R.B. Alley, P.M. Grootes, and S. Anandakrishnan, Toward using borehole temperatures to calibrate an isotopic paleothermometer in central Greenland, *Global and Planetary Change, 98*, 265-268, 1992.
Cuffey, K.M., R.B. Alley, P.M. Grootes, J.F. Bolzan, and S. Anandakrishnan, Calibration of the $\delta^{18}O$ isotopic paleothermometer for central Greenland, using borehole temperatures, *Journal of Glaciology, 40(135)*, 341-349, 1994.
Cuffey, K.M., G.D. Clow, R.B. Alley, M. Stuiver, E.D. Waddington, and R.W. Saltus, Large Arctic temperature change at the glacial-Holocene transition, *Science, 270*, 455-458, 1995.
Cuffey, K.M., and G.D. Clow, Temperature, accumulation, and ice sheet elevation in central Greenland through the last deglacial transition, *Journal of Geophysical Research, 102(C12)*, 26,383-26,396, 1997.
Dahl-Jensen, D, K. Mosegaard, N. Gundestrup, G.D. Clow, S.J. Johnsen, A.W. Hansen, and N. Balling, Past temperatures directly from the Greenland Ice Sheet, *Science, 282*, 268-271, 1998.
Dahl-Jensen, D.J., N.S. Gundestrup, K. Mosegaard, and G.D. Clow, Reconstruction of the past climate from the GRIP temperature profile by Monte Carlo inversion [abs.]: *Eos (Transactions of the American Geophysical Union), 78*, F6, 1997.
Dansgaard, W., Stable isotopes in precipitation, *Tellus, 16*, 436-468, 1964.
Dansgaard, W., S.J. Johnsen, H.B. Clausen, D. Dahl-Jensen, N.S. Gundestrup, C.U. Hammer, C.S. Hvidberg, J.P. Steffensen, A.E. Sveinbjörnsdottir, J. Jouzel, and G. Bond, Evidence for general instability of past climate from a 250-kyr ice-core record, *Nature, 364*, 218-220, 1993.
Ditlevsen, P.D., H. Svensmark, and S. Johnsen, Contrasting atmospheric and climate dynamics of the last-glacial and Holocene periods, *Nature, 379*, 810-812, 1996.
Fawcett, P.J., A.M. Ágústsdóttir, R.B. Alley, and C.A. Shuman, The Younger Dryas termination and North Atlantic deepwater formation: insights from climate model simulations and Greenland ice core data, *Paleoceanography, 12*, 23-38, 1997.
Firestone, J., Resolving the Younger Dryas event through borehole thermometry, *Journal of Glaciology*, 41(137), 39-50, 1995.
Hammer, C., P.A. Mayewski, D. Peel and M. Stuiver, eds., Greenland Summit Ice Cores, reprinted from *Journal of Geophysical Research, 102(C12)*, 26,315-26,886, 1997.
Herron, M.M., S.L. Herron, and C.C. Langway, Jr., Climatic signal of ice melt features in southern Greenland, *Nature, 293*, 389-391, 1981.
Imbrie, J., E.A. Boyle, S.C. Clemens, A. Duffy, W.R. Howard, G. Kukla, J. Kutzbach, D.G. Martinson, A. McIntyre, A.C. Mix, B. Molfino, J.J. Morley, L.C. Peterson, N.G. Pisias, W.L. Prell, M.E. Raymo, N.J. Shackleton, and J.R. Toggweiler, On the structure and origin of major glaciation cycles: 1. Linear responses to Milankovitch forcing, *Paleoceanography, 7*, 701-738, 1992.
Imbrie, J., A. Berger, E.A. Boyle, S.C. Clemens, A. Duffy, W.R. Howard, G. Kukla, J. Kutzbach, D.G. Martinson, A. McIntyre, A.C. Mix, B. Molfino, J.J. Morley, L.C. Peterson, N.G. Pisias, W.L. Prell, M.E. Raymo, N.J. Shackleton and J.R. Toggweiler, On the structure and origin of major glaciation cycles: 2. The 100,000-year cycle, *Paleoceanography, 8*, 699-735, 1993.
Johnsen, S.J., D. Dahl-Jensen, W. Dansgaard, and N. Gundestrup, Greenland paleotemperatures derived from GRIP bore hole temperature and ice core isotope profiles, *Tellus Ser. B, 47*, 624-629, 1995.
Jolly, D., S.P. Harrison, B. Damnati and R. Bonnefille, Simulated climate and biomes of Africa during the Late Quaternary: Comparison with pollen and lake status data, *Quaternary Science Reviews, 17*, 629-657, 1998.
Jouzel, J., R.B. Alley, K.M. Cuffey, W. Dansgaard, P. Grootes, G. Hoffmann, S.J. Johnsen, R.D. Koster, D. Peel, C.A. Shuman, M. Stievenard, M. Stuiver and J. White, Validity of the temperature reconstruction from water isotopes in ice cores, *Journal of Geophysical Research, 102(C12)*, 26,471-26,487, 1997.
Keigwin, L.D., and S.J. Lehman, Deep circulation change linked to Heinrich event 1 and Younger Dryas in a middepth North Atlantic core, *Paleoceanography, 9*, 185-194, 1994.
Klitgaard-Kristensen, D., H.P. Sejrup, J. Jaflidason, S. Johnsen, and M. Spurk, A regional 8200 cal yr BP cooling-event in NW-Europe, induced by final stages of the Laurentide ice-sheet deglaciation?, *Journal of Quaternary Science, 13*, 165-169, 1998.
Koc, M., E. Jansen, and H. Haflidason, Paleoceanographic reconstructions of surface ocean conditions in the Greenland, Iceland and Norwegian Seas through the last 14 ka based on diatoms, *Quaternary Science Reviews, 12*, 115-140, 1993.

Koerner, R.M., and D.A. Fisher, A record of Holocene summer climate from Canadian high-Arctic ice core, *Nature, 343*, 630-631, 1990.

Krinner, G., C. Genthon, and J. Jouzel, GCM analysis of local influences on ice core δ signals, *Geophysical Research Letters, 24*, 2825-2828, 1997.

Kutzbach, J.E., Model simulations of the climatic patterns during the deglaciation of North America, in Ruddiman, W.F., and H.E. Wright, Jr., eds., North America and adjacent oceans during the last deglaciation, *Geological Society of America, Boulder, CO*, 425-446, 1987.

Kutzbach, J.E., R.G. Gallimore and P.J. Guetter, Sensitivity experiments on the effect of orbitally-caused insolation changes on the interglacial climate of high northern latitudes, *Quaternary International, 10-12*, 223-229, 1991.

Kutzbach, J., G. Bonan, J. Foley, and S.P. Harrison, Vegetation and soil feedbacks on the response of the African monsoon to orbital forcing in the early to middle Holocene, *Nature, 384*, 623-626, 1996.

Langway, C.C., Jr., 1967, Stratigraphic analysis of a deep ice core from Greenland, CRREL Res. Rep. 77, Cold Reg. Res. and Eng. Lab., Hanover, N.H., 1967. Lehman, S.J., and L.D. Keigwin, Sudden changes in North Atlantic circulation during the last deglaciation, *Nature, 356*, 757-762, 1992.

Mayewski, P.A., M. Wumkes, J. Klinck, M.S. Twickler, J.S. Putscher, K.C. Taylor, A.J. Gow, E.D. Waddington, R.B. Alley, J.E. Dibb, P.M. Grootes, D.A. Meese, M. Ram, M. Whalen and A.T. Wilson. Record drilling depth struck in Greenland, Eos *(Transactions of the American Geophysical Union) 75(10)*, pages 113, 119 and 124, 1994.

Mayewski, P.A., L.D. Meeker, M.S. Twickler, S. Whitlow, Q. Yang, W.B. Lyons, and M. Prentice, Major features and forcing of high-latitude northern hemisphere atmospheric circulation using a 110,000-year-long glaciochemical series, *Journal of Geophysical Research, 102(C12)*, 26,345-26,366, 1997.

MacAyeal, D.R., J. Firestone, and E.D. Waddington, Paleothermometry by control methods, *Journal of Glaciology, 37(127)*, 326-338, 1991.

McCabe, A.M., and P.U. Clark, Ice-sheet variability around the North Atlantic Ocean during the last deglaciation, *Nature, 392*, 373-377, 1998.

Meese, D.A., A.J. Gow, R.B. Alley, G.A. Zielinski, P.M. Grootes, M. Ram, K.C. Taylor, P.A. Mayewski, and J.F. Bolzan, The Greenland Ice Sheet Project 2 depth-age scale: methods and results, *Journal of Geophysical Research, 102(C12)*, 26,411-26,423, 1997.

Mitchell, J.F.B., N.S. Grahame and K.J. Needham, Climate simulations for 9000 years before present: Seasonal variations and effect of the Laurentide ice sheet, *Journal of Geophysical Research, 93(D7)*, 8283-8303, 1988.

Pfeffer, W.T., and N.F. Humphrey, Determination of timing and location of water movement and ice-layer formation by temperature measurements in subfreezing snow, *Journal of Glaciology, 42(141)*, 292-304, 1996.

Pollard, D., and S.L. Thompson, Climate and Ice-sheet Mass Balance at the Last Glacial Maximum from the Version 2 Global Climate Model, *Quaternary Science Reviews, 16*, 841-863, 1997.

Qin, B., S.P. Harrison and J.E. Kutzbach, Evaluation of modelled regional water balance using lake status data: A comparison of 6 ka simulations with the NCAR CCM, *Quaternary Science Reviews, 17*, 535-548, 1998.

Rahmstorf, S., Bifurcations of the Atlantic thermohaline circulation in response to changes in the hydrological cycle, *Nature, 378*, 145-149, 1995.

Raynaud, D., J. Chappellaz, C. Ritz, and P. Martinerie, Air content along the Greenland Ice Core Project core: A record of surface climatic parameters and elevation in central Greenland. *Journal of Geophysical Research, 102(C12)*, 26,607-26,613, 1997.

Robin, G. de Q., ed., *The climatic record in polar ice sheets*, Cambridge University Press, Cambridge, UK, 1983.

Severinghaus, J.P., T. Sowers, E.J. Brook, R.B. Alley and M.L. Bender, Timing of abrupt climate change at the end of the Younger Dryas interval from thermally fractionated gases in polar ice, *Nature, 391*, 141-146, 1998.

Shuman, C.A., R.B. Alley, S. Anandakrishnan, J.W.C. White, P.M. Grootes and C.R. Stearns, Temperature and accumulation at the Greenland Summit: comparison of high-resolution isotope profiles and satellite passive microwave brightness temperature trends, *Journal of Geophysical Research, 100(D5)*, 9165-9177, 1995.

Spinelli, G., A statistical analysis of ice-accumulation level and variability in the GISP2 ice core and a reexamination of the age of the termination of the Younger Dryas cooling episode, *Earth System Science Center Technical Report No. 96-001*, The Pennsylvania State University, University Park, PA U.S.A., 1996.

Stocker, T.F., and A. Schmittner, Influence of CO_2 emission rates on the stability of the thermohaline circulation, *Nature, 388*, 862-865, 1997.

Street-Perrott, F.A., and Perrott, R.A., Abrupt climate fluctuations in the tropics: the influence of Atlantic Ocean circulation, *Nature, 358*, 607-612, 1990.

Thompson, S.L. and D. Pollard, A global climate model (GENESIS) with a land-surface transfer scheme (LSX). Part I: Present climate simulation, *Journal of Climate, 8*, 732-761, 1995.

Vettoretti, G., W.R. Peltier and N.A. McFarlane, Simulations of Mid-Holocene climate using an atmospheric general circulation model, *Journal of Climate, 11*, 2607-2627, 1998.

Wang, Z., P.H. Stone and J. Marotzke, Global thermohaline circulation. Part I: Sensitivity to atmospheric moisture transport, *Journal of Climate, 12*, 71-82, 1999.

Mailing Addresses

Anna Maria Ágústsdóttir; Soil Conservation Service; Gunnarsholt; 851 Hella; Iceland

Richard B. Alley; Deike Bldg., The Pennsylvania State University, University Park, PA 16802

Peter J. Fawcett; Department of Earth & Planetary Sciences;University of New Mexico; Albuquerque, NM 87131

Simulation of the Potential Responses of Regional Climate and Surface Processes in Western North America to a Canonical Heinrich Event

S.W. Hostetler

U.S. Geological Survey, Corvallis, Oregon

P.J. Bartlein

Department of Geography, University of Oregon

We apply a hierarchy of atmospheric and process models to assess the response in Western North America to a canonical North Atlantic Heinrich event that is characterized by lowering of the Laurentide Ice Sheet (LIS) and subsequent warming of North Atlantic sea surface temperatures (SSTs). Global responses to changes in the LIS and SSTs are simulated by the GENESIS general circulation model. Lateral boundary conditions for a high-resolution regional climate model (RegCM) that is run over Western North America are derived from the GENESIS simulations, and output from the RegCM simulations is used as input to a suit of models that we apply to investigate responses of alpine glaciers, lakes, vegetation, and coastal upwelling. The combined expression of synoptic- and regional-scale circulations yields spatially variable climatologies over WNA in each phase of our H2 event that reinforce, cancel, or reverse climatic patterns (e.g. temperature and precipitation) of previous phases. The response of terrestrial and marine processes is complex, and illustrates the possibility of spatially heterogeneous registration of millennial-scale climatic variations both within a particular phase and among the various phases of those variations.

1. INTRODUCTION

Over the past 50,000 years, the climate of western North America (WNA) has varied on both orbital or Milankovitch time scales and on higher frequency millennial time scales [*Clark and Bartlein*, 1995; *Behl et al*, 1996; *Benson et al.*, 1996; *Philips et al.*, 1996; *Oviatt*, 1997; *Benson et al.*, 1998; *Lund and Mix*, 1998]. Although the mechanisms of orbital-scale forcing of climatic changes in this region are relatively well understood [*Whitlock and Bartlein*, 1997; *Thompson et al.*, 1993], our knowledge of the mechanisms associated with millennial-scale climatic variations is less complete. A prominent question regarding climatic variability on the millennial scale is whether and how the relatively abrupt variations, such as those accompanying North Atlantic Heinrich events, were propagated to WNA. Conceptually, there are three pathways by which such millennial scale variations could become registered in

Mechanisms of Global Climate Change at Millennial Time Scales
Geophysical Monograph 112

WNA: 1) through atmospheric or oceanic transmission from the North Atlantic region, 2) through a joint response of the coupled ocean-atmosphere system, or 3) through some combination of both [*Clark and Bartlein*, 1995]. In an example of the propagation of a North Atlantic "signal" via the first pathway, a climatic variation in the North Atlantic region might perturb the atmospheric general circulation or ocean thermohaline circulation. This perturbation in turn might produce regional climatic changes throughout the globe as effects of the circulation changes propagated along their flow paths. In an example of propagation via the second pathway, climatic variations in both the North Atlantic region and WNA would be viewed as the regionally specific responses to some large- (hemispheric or global) scale forcing.

Insights into the nature of propagation along either of the two main pathways can be obtained through modeling. Features of the regional climatic response consistent with joint dependence or the atmospheric transmission of a Heinrich event can be isolated through a straightforward series of simulations in which appropriate changes to the relevant, large-scale boundary conditions [i.e., height of the Laurentide ice sheet (LIS) and North Atlantic sea surface temperatures (SSTs)] are prescribed [*Hostetler et al.,* 1999], and such an approach forms the core of this paper. Modeling the transmission via ocean circulation, or the joint response to ocean-atmosphere interactions during millennial-scale climate variations, is much more complex and requires an earth system model in which the various subsystems, the atmosphere, terrestrial biosphere, land surface, cryosphere, and oceans, are interactively coupled. The full earth-system model required for such an analysis does not yet exist.

The local- and regional-scale variations in surface processes and environmental systems that are recorded by paleoclimatic indicators are generated by a hierarchy of controls and responses that span multiple temporal and spatial scales. Hemispheric- and synoptic-scale atmospheric and oceanic circulation systems are controlled by global boundary conditions such as latitudinal and seasonal distributions of insolation, patterns of SSTs, and the size and distributions of continental ice sheets. Global circulation systems in turn govern regional or mesoscale circulations (those occurring on a spatial scale on the order of tens of kilometers) which in turn are modified by more localized features such as topographic barriers, large lakes, and coastline orientation. Although climatic features at all scales are important, substantial insights into the climate of WNA can be obtained by extracting and quantifying the relative contributions of synoptic-scale features and the expression of those features at the regional scale. In WNA, the primary regional scale influence on synoptic circulations is exerted by complex topography. Topographic effects on atmospheric circulation are characterized, for example, by blocking and channeling of winds, generation of orographic precipitation, and modification of latitudinal temperature gradients. These topographic effects, in turn, strongly influence the distribution and nature of surface water, vegetation and alpine glaciers, and thus the geologic records of changes in these systems.

Models of the general circulation of the atmosphere (AGCMs) provide simulations of large-scale atmospheric circulation, but many circulation features at the regional scale are not explicitly resolved by AGCMs, which typically are run at a horizontal grid spacing of several degrees for climate simulations (Figure 1). For the purposes of explicitly representing the various scales of controls and responses in WNA involved in the atmospheric transmission of a Heinrich event in the North Atlantic region, we designed a climate-modeling experiment based on a hierarchy of atmospheric and process models. A series of simulations of global climate were conducted with the GENESIS AGCM to evaluate large-scale atmospheric responses to three phases of a canonical representation of Heinrich event 2 (H2) (*Hostetler et al.*, 1999). GENESIS has an effective resolution for atmospheric variables of 3.75 degrees, and for land-surface variables of 2 degrees (or grid-square areas of approximately 12.0×10^4 km^2 and 3.4×10^4 km^2, respectively, at 45°N) (Figure 1). From the GENESIS simulations, we derived time series of the necessary lateral boundary conditions for a regional (mesoscale) climate model (the NCAR RegCM) that was run at a grid spacing of 60 km over WNA. The model hierarchy is completed by using output fields from the RegCM to infer vegetation responses, and as input to a series of models that quantify glacier mass balance, surface water balance, lake levels, and lake thermal response, and coastal upwelling, to the canonical Heinrich event.

2. EXPERIMENTAL DESIGN

In a previous study [*Hostetler et al.*, 1999], we used GENESIS to assess the global atmospheric and surface response to three phases of a canonical Heinrich event patterned after the ice sheet and SST variations that accompanied event H2. In phase one, nominal 21 ka boundary conditions are prescribed that include CLIMAP SSTs [*CLIMAP Project Members*, 1981], continental ice sheets [*Peltier*, 1994], and appropriate orbital parameters and atmospheric composition (CO_2 concentration of 200 ppmv). In the following discussion, this simulation is designated *MaxCold*. Phase one of the canonical event is followed by phase two in which a surge of the LIS into the North Atlantic is represented by lowering the LIS in our model from the maximum 21 ka height to a height ~1500 m lower over Hudson Bay [*Licciardi et al.*, 1998], and is

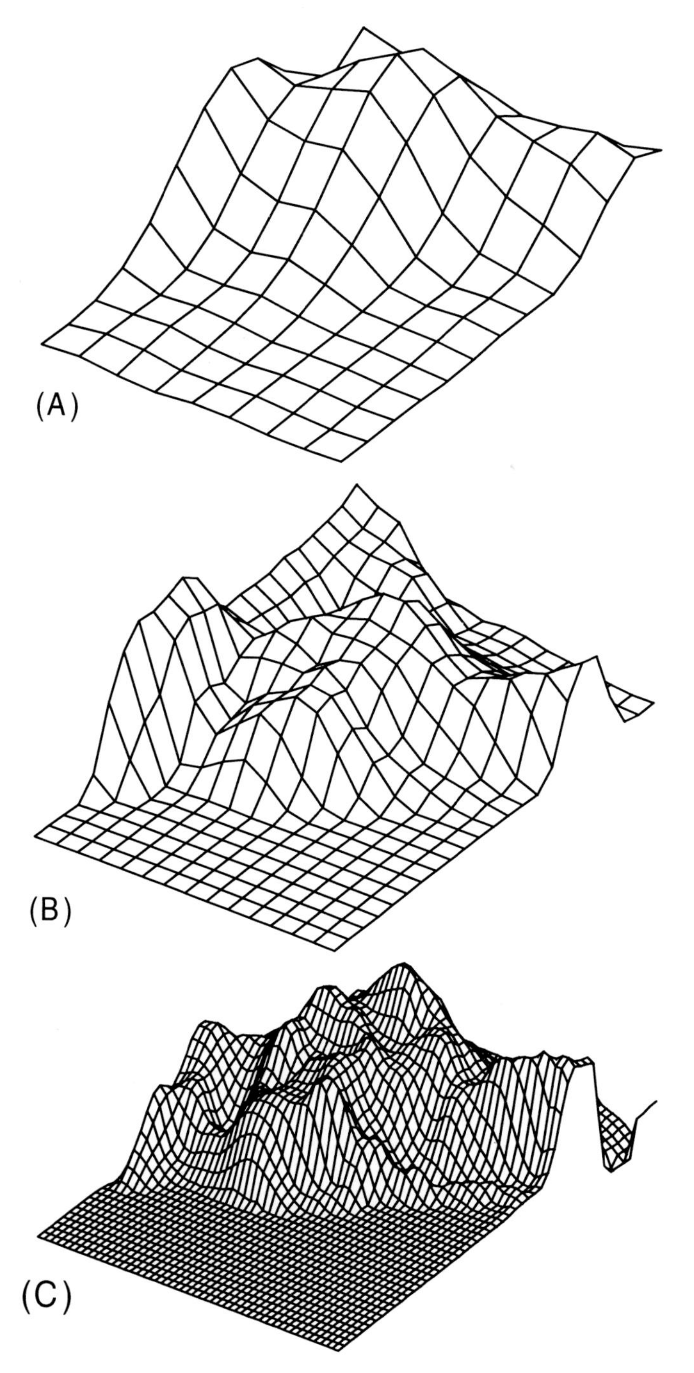

Figure 1. Representation of the topography of western North America at different model resolutions. A: T31 (3.75° × 3.75°) as in GENESIS; B: 2° × 2° as in the land surface scheme LSX in GENESIS; C: ~0.5° × 0.5° (60 km × 60 km) as in RegCM. Areal coverage is approximately the same for each surface map and the vertical scales are exaggerated.

designated *MinCold*. In phase three, designated *MinWarm*, the lower LIS is accompanied by a uniform increase of North Atlantic SSTs north of 30°N to values three fourths of the way between full glacial and modern temperature [*Cortijo et al.*, 1997]. We also include a GENESIS simulation of the present-day climate (designated *Control*) as a reference for comparing the magnitude of Heinrich event responses to those of a glacial-interglacial climate change. Each GENESIS simulation was 15 years long, and the output necessary to derive initial and lateral boundary conditions for the RegCM [surface temperature (including SSTs) and pressure, and vertical profiles of temperature, wind, humidity, and pressure] for 5 years was retained for simulation years 11 through 15.

The version of RegCM used here is based on the standard version that is described by *Giorgi et al.* [1993a, b]. Our version of RegCM includes a) an improved lake submodel [*Small et al.*, in press] for interactive modeling of lake-atmosphere feedbacks, b) a modified method for computing the energy balance and temperature of permanent ice (the mass balance is not computed), c) reduced CO_2 concentration (200 ppmv for 21 ka), and d) calculated solar inputs based on specified values of eccentricity, obliquity and precession. The distributions of vegetation, ice caps and mountain glaciers, the LIS, and pluvial lakes for the regional model domain (Plate 1) are based primarily on published reports [e.g. *Wright et al.*, 1993 and chapters therein], with some modifications (e.g., in the placement of pluvial Lakes Lahontan and Bonneville relative to smoothed model topography). The initial and lateral boundary conditions for the RegCM, which were derived from the GENESIS simulations at 12-hr time steps, begin June 1 and run continuously for 4.5 years. The first half-year of each RegCM run is excluded from our analyses to ensure the model had essentially reached equilibrium.

A suite of process models and computations are used to evaluate the response of surface systems in WNA during the various phases of our H2 event. Glacier mass balance is computed using the temperature and precipitation fields from the RegCM as input to an empirical mass-balance model [*Hostetler and Clark*, 1997]. Lake-level responses are estimated from anomalies of net moisture (precipitation minus evaporation, P-E) from the RegCM and from simulated changes in lake evaporation. Changes in coastal upwelling are computed using the surface wind fields from the RegCM simulations [e.g., *Ortiz et al.*, 1997]. We infer vegetation changes from anomalies of RegCM temperature and effective moisture.

3. RESULTS

We present our results in a "top down" order that parallels the modeling sequence. Discussion of global responses in GENESIS are followed by those for the

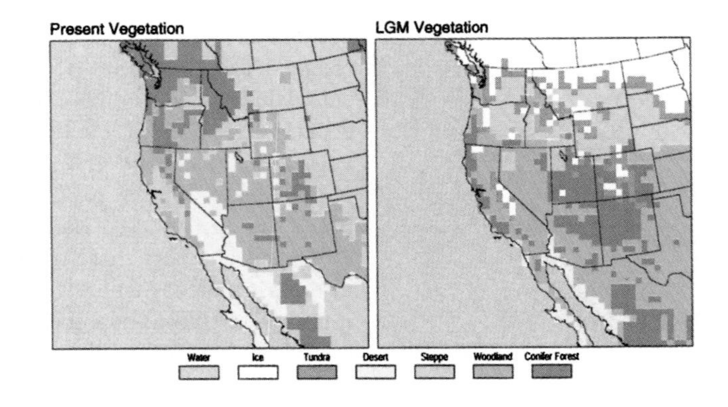

Plate 1. Present and LGM land surface types used in the RegCM simulations.

RegCM and higher resolution process models and computations.

3.1 Global Model

3.1.1 Glacial-Interglacial responses. The glacial-interglacial climate anomalies of our GENESIS experiments (Plate 2, *MaxCold-Control*) are similar to those for other simulations of the LGM conducted with prescribed CLIMAP SSTs in GENESIS [e.g., *Pollard and Thompson, 1997*], and other AGCMs [e.g., *Manabe and Broccoli*, 1985; *Rind*, 1987; *COHMAP Members*, 1988]. The largest glacial/interglacial temperature differences are found in the northern hemisphere over and to the east of the continental ice sheets, and downwind from areas of colder SSTs (Plate 2a). In both seasons, the *MaxCold* simulation produces steeper temperature gradients between the continents and adjacent oceans than those of the *Control* simulation. Air temperatures in the *MaxCold* experiment are colder over WNA than those of *Control* in both winter and summer; however, the magnitude of winter cooling is less than that of summer (Plate 2a). The smaller winter temperature anomalies stem in part from the *Control* simulation being cold relative to modern observations as a consequence of the improper placement of upper level circulation features in GENESIS (and AGCMs of similar resolution). In contrast to actual conditions, persistent high pressure that forms offshore extends inland over the Great Basin, and anchors the simulated longwave circulation pattern. The anomalous pressure pattern, which is attributable to the use of fixed SSTs and to the resolution of the AGCM, produces air flow over the Great Basin predominantly from a northerly direction which sustains the advection of cold Arctic air into the region.

Changes relative to *Control* in pressure patterns and associated wind fields both at the surface and aloft over WNA are also evident in the *MaxCold* simulation. During the boreal winter of the *MaxCold* simulation, stronger-than-present westerlies are simulated over the northern Pacific in response to a steeper temperature gradient set up by colder air temperatures over the Pacific Northwest (PNW) and North Pacific and a subtropical gyre that is as warm as present (Plate 2b). Thermal and topographic blocking effects of the LIS cause anticyclonic flow over the PNW and splitting and southward displacement of the polar jet stream. The split flow is characterized by a stronger subtropical jet, weaker westerlies along the west coast and south of the Cordilleran ice sheet, and stronger southerly flow into Beringia. A split jet was a common feature in earlier simulations of the LGM climate that were run at low resolution (e.g., R15) with the CLIMAP reconstruction of the LIS [eg., *Manabe and Broccoli*, 1985; *Rind*, 1987; *COHMAP Members*, 1988]. Subsequent experiments, in which the lower ICE4-G reconstruction of the Laurentide was used, tend to display an evident, but less well developed split jet [*Bartlein et al.*, 1998]. The recurrence of a split jet in our (and other, e.g., *Pollard and Thompson*, 1997) LGM simulations suggests that the feature may reemerge in simulations conducted with higher resolution AGCMs. During summer, westerlies stronger than those of *Control* are simulated in the *MaxCold* experiment, in response to steep temperature gradients between the colder than present Gulf of Alaska and the warmer (>2°C) subtropical gyre.

In the *MaxCold* experiment there is strong wintertime expression of the glacial anticyclone characterized by weaker surface westerlies south of the LIS, prevailing easterlies over the PNW, and a very well-developed Aleutian Low (Plate 2c). Summer flow patterns in the *MaxCold* simulation reflect the combined influences over WNA of the LIS and surface temperature patterns. A prominent anticyclone over the continent (and extending to the North Atlantic), with a diminished subtropical high in the eastern Pacific and a strengthened Aleutian low produce southwesterly flow along the west coast and easterly flow along the edge of the LIS.

The distribution of precipitation (not shown) and P-E anomalies (Plate 2d) reflects the reorganized wind flow over WNA in the *MaxCold* simulation. Anticyclonic flow from the LIS produces off-shore winds over the PNW resulting in winter conditions at the LGM that were colder and drier than present. Precipitation associated with weaker westerlies and a stronger subtropical jet increase P-E from Central America to the southwestern US (SW). Focusing of the polar jet at the latitude of the Cordilleran ice sheet during summer causes wetter-than-present conditions there, while easterly surface winds along the southern edge of the glacial anticyclone advect Gulf moisture into SW and Rocky Mountains.

3.1.2 Heinrich event responses. Lowering the LIS in the second phase of the canonical Heinrich event (*MinCold*) results in cooling over WNA by as much as 4°C in the winter and 2°C in summer (Plate 2a, *MinCold-MaxCold*). The temperature responses are associated with changes in circulation induced by thermal and topographic effects of the lower ice sheet. Relative to the *MaxCold* phase, in the *MinCold* phase in winter lower pressure forms over the Great Basin and higher pressure forms to the east. Winter of the *MinCold* phase is wetter than the LGM (*MaxCold*) in the PNW and drier than the LGM in the SW. Conditions in the PNW in the *MinCold* phase are wetter than those of the *MaxCold* phase because the strength of the easterlies off the LIS is reduced. During summer in the *MinCold* phase, higher pressure than that of the *MaxCold* phase replaces the wintertime low over the Great Basin and lower pressure forms over the eastern Pacific, causing attendant changes in wind flow and moisture. Consequently, summers in the PNW and the SW are wetter in the *MinCold* phase than

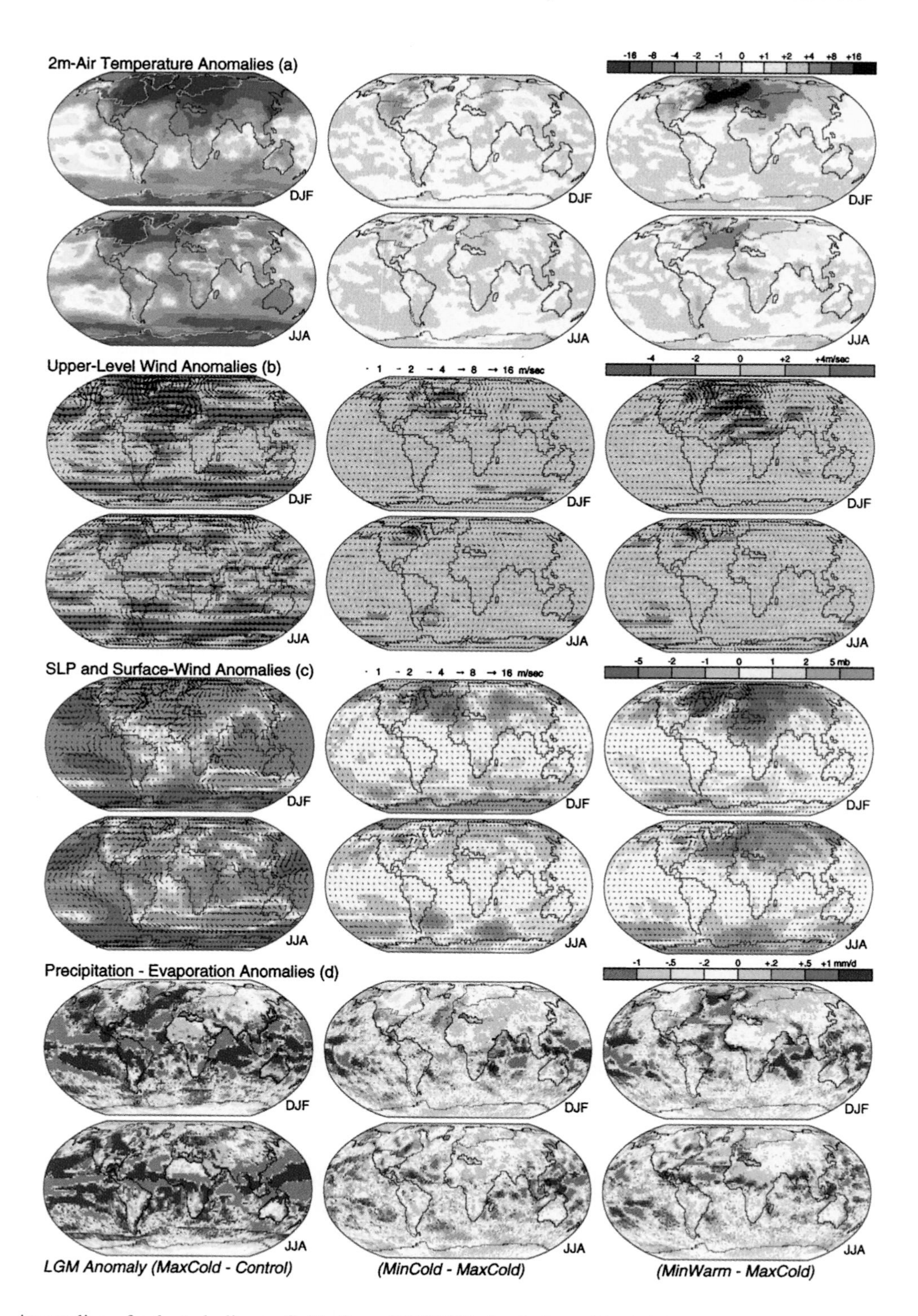

Plate 2. Anomalies of selected climate fields from GENESIS simulations. The left-hand column is glacial/interglacial (*MaxCold* minus *Control*), the middle column is the first phase of the H2 event (*MinCold* minus *MaxCold*), and the right-hand column is the second phase of the H2 event (*MinWarm* minus *MaxCold*). A: Surface (2 m) air temperatures for winter (DJF) and summer (JJA); B: ~500 mb wind vectors and wind magnitudes (shown in color and computed as $\sqrt{u^2+v^2}$, where u and v are the northerly and westerly components of the wind, respectively); C: Sea level pressure (shown in color) and surface wind vectors; D: Precipitation minus evaporation. The anomalies are computed from 12-year averages of the GENESIS simulations.

those of the *MaxCold* phase, with little change elsewhere.

In the third phase of our canonical Heinrich event (*MinWarm* simulation), in which a warm North Atlantic is combined with a low LIS, there is cooling of WNA relative to the *MaxCold* phase, but overall, to a lesser degree than that of the *MinCold* phase (Plate 2a, *MinWarm-MaxCold*). Warmer-than-phase-two temperature anomalies are associated with changes in pressure and circulation patterns (e.g., reduction of the Aleutian Low). Relative to both the *MaxCold* and *MinCold* phases, precipitation and P-E in the *MinWarm* phase decreases over the PNW in winter and increases over the SW during summer.

The climatic response to lowering of the LIS in phase two (*MinCold*) of our H2 event is apparently transmitted to WNA both through changes in the circulation of the northern hemisphere and through circulation changes in North America focused around the LIS. Warming the North Atlantic in phase three (*MinWarm*) causes expected large temperature responses in both the North Atlantic region and downstream over Europe and Asia. Transmission of this response to WNA is primarily accomplished through attendant changes in pressure patterns over the PNW. Temperatures over WNA in the *MinWarm* phase are cooler than those of the *MaxCold* and *MinCold* and P-E is substantially different due to weakening of the Aleutian low and attendant repositioning of wind patterns and storm tracks. Simultaneously changing SSTs in the Pacific and the North Atlantic (i.e., a joint response through changes in oceanic circulation) would likely induce a larger climatic response over WNA than those simulated here.

3.2 Regional Model

3.2.1 Glacial-Interglacial responses. Average January and July temperature, precipitation, and wind fields from the *Control* RegCM simulation (not shown) are consistent with the boundary forcing derived from GENESIS and are in relative agreement with long-term observations [*Thompson et al.*, 1998]. January temperatures over the northern Great Basin, however, are generally colder than observations by up to ~5°C. Simulated January temperatures over the SW are similarly colder than observed temperatures, but by ≤ 2°C. Precipitation in the January *Control* simulation generally agrees with observed values except over the coastal regions of the PNW and the northern Sierra Nevada where the RegCM underestimates precipitation rates by ~2 mm day^{-1}. (In December, simulated precipitation rates agree well with observed values over these regions.) The errors in simulated temperature and precipitation fields are attributed partly to the RegCM and partly to the aforementioned placement of circulation over WNA by GENESIS, which is consequently imposed on the RegCM through the lateral boundary conditions. July temperature and precipitation of the *Control* simulation are in relative agreement with observations over the model domain. Exceptions are found over the Rocky Mountains and the Mexican Plateau where local convective precipitation rates are up to several millimeters per day greater than averaged observed values.

Incorporating the large-scale circulation patterns from GENESIS into the RegCM produces glacial-interglacial (*MaxCold-Control*) climate anomalies over WNA similar in general pattern to those of the AGCM. During January, the flow aloft and at the surface (Plate 3b and 3c, *MaxCold-Control*) combine to generate drier-than-present conditions over the PNW and wetter-than-present conditions over the desert SW. Between the two extremes, a well-defined precipitation gradient is evident across the Great Basin. The precipitation gradient suggests a mechanism whereby lakes to the south (e.g., Estancia, Owens) were at their maximum late-Pleistocene extents during the LGM while lakes in the northern Great Basin (e.g., Lahontan and Bonneville), though much larger than present, were not as large as their post-LGM maxima [*Benson et al.*, 1990; *Hostetler and Benson*, 1990.]. Circulation changes aloft and at the surface in the *MaxCold* simulation also contribute to wetter-than-present conditions over much of the domain during July.

Glacial/interglacial temperature anomalies over the northern part of the domain are largest during January because anticyclonic flow off of the LIS sustains easterly winds across the PNW while farther to the east, northwesterly winds advect cold air off the LIS (Plate 3a *MaxCold-Control*). In the vicinity of Lakes Lahontan and Bonneville, lake heat storage increases air temperatures relative to those of the *Control* [*Hostetler et al.*, 1994]. In the SW, January LGM temperatures range from slightly lower to ~3°C higher than *Control* values. This apparent LGM warming reflects a cold bias in the *Control* simulation and warm air advection associated with the subtropical jet in the *MaxCold* simulation.

In contrast to January, July air temperatures over much of the domain are 4-8°C colder than those in the control, with some areas along the margin of the LIS and over mountain glaciers being up to 16°C colder (Plate 3a). Slightly less cooling also occurs during spring and autumn (figures not shown). Mean annual temperatures of the *MaxCold* simulation are colder than the control by 4-5°C or more over the PNW and northern part of the domain, 1-2°C over the Sierra and the northern Great Basin, and 2-3°C over the SW.

The simulated glacial/interglacial cooling over WNA is generally less than what is inferred from paleo-environmental indicators [*Thompson et al.*, 1993], particularly in winter. The lack of cooling in part is attributable to our use of CLIMAP SSTs which may be too warm in the midlatitudes of the eastern Pacific. Additional

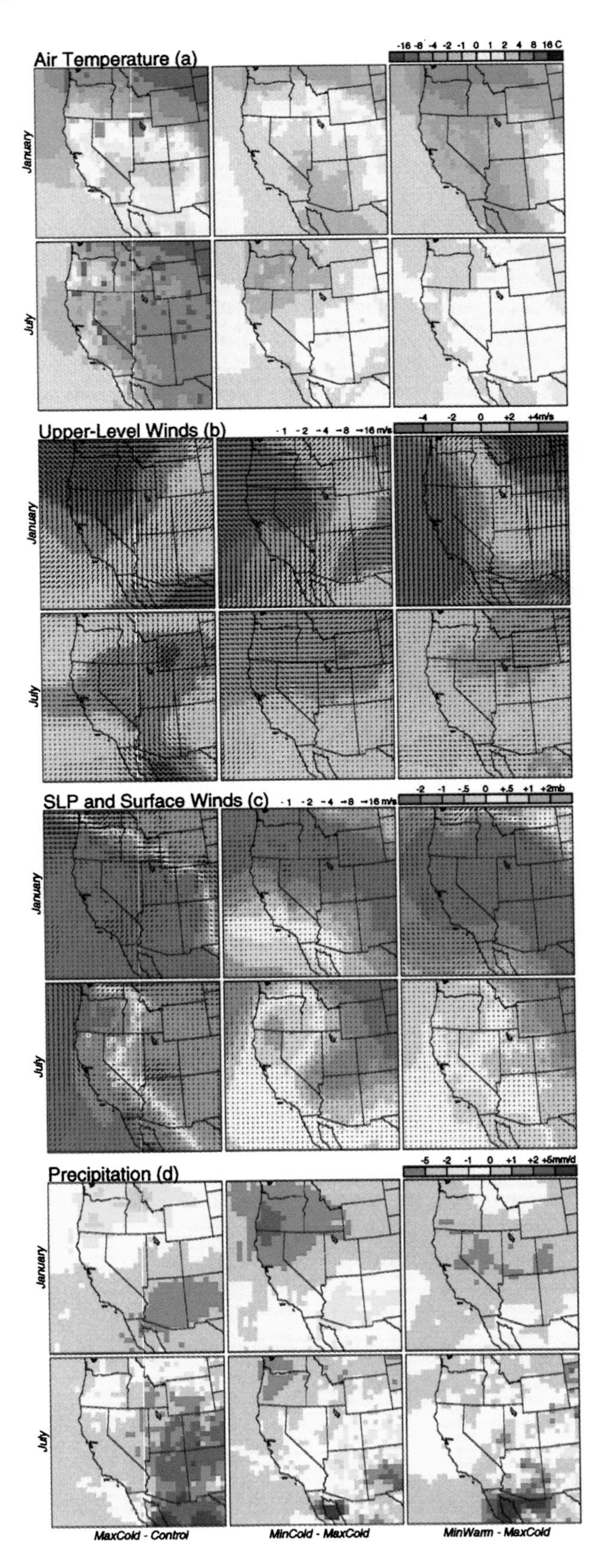

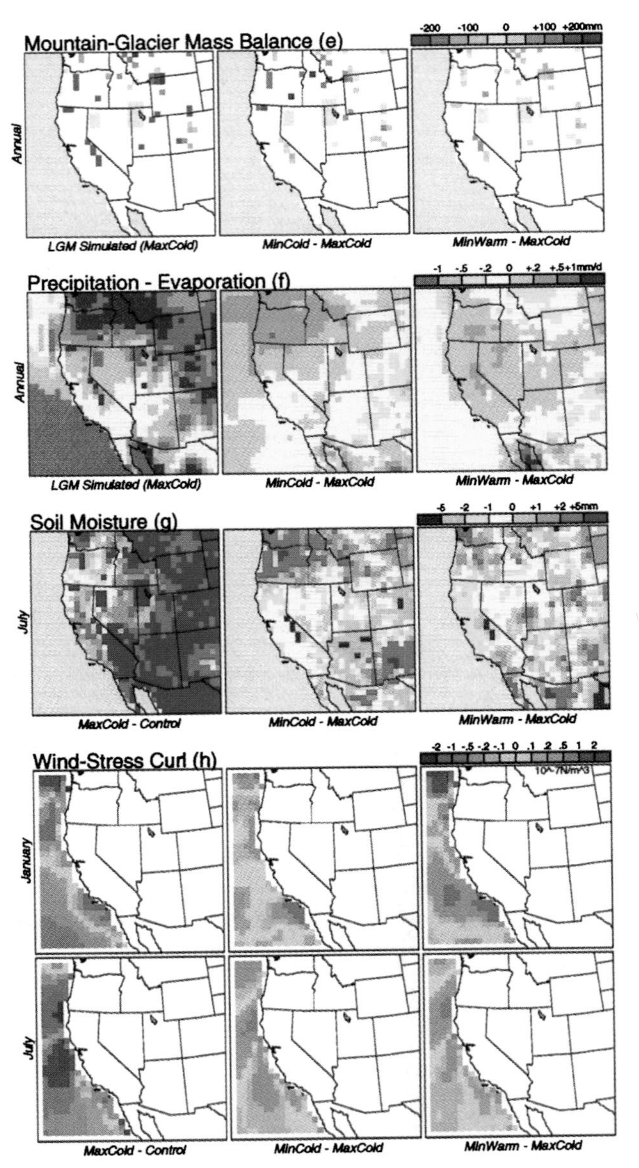

Plate 3. Anomalies of climate fields from the RegCM simulations (left-hand column) and related surface process responses (right-hand column). The order of presentation within both columns is glacial/interglacial (*MaxCold* minus *Control*), the first phase of the H2 event (*MinCold* minus *MaxCold*), and the second phase of the H2 event (*MinWarm* minus *MaxCold*). A: Surface (2 m) air temperatures for winter (DJF) and summer (JJA); B: ~500 mb wind vectors and wind magnitudes (shown in color and computed as $\sqrt{u^2+v^2}$, where u and v are the northerly and westerly components of the wind, respectively); C: Sea level pressure (shown in color) and surface wind vectors; D: Precipitation; E: Annual mass balance of alpine glaciers; F: Annual precipitation minus evaporation from the RegCM; G: July soil moisture from the RegCM; H: January and July wind-stress curl. The anomalies are based on 4-year averages from the RegCM simulations.

LGM cooling over WNA of 2-4°C was simulated by *Pollard and Thompson* [1997] when fixed CLIMAP SSTs were replaced by a slab mixed layer ocean in GENESIS, (perhaps indicating improvement in the simulated wave pattern). We chose to begin our Heinrich event experiments with fixed SSTs so we could isolate and quantify the atmospheric responses associated with prescribed changes in the North Atlantic region, and it is likely that the use of lateral boundary conditions from a mixed-layer (or fully interactive) ocean simulation would produce further winter cooling in the *MaxCold* simulations.

The magnitude and distribution of glacial/interglacial precipitation anomalies, in general, agree with paleoenvironmental reconstructions [*Thompson et al.*, 1993]. The pattern of opposing changes in wetness between the PNW and SW in particular is well simulated. Although there are shortcomings in our simulated glacial/interglacial climate changes owing to limitations in the RegCM, and placement of synoptic-scale circulation features in the AGCM, the simulations provide a sufficient baseline against which the climatic variations of the canonical Heinrich event simulations can meaningfully be compared.

3.2.2 Heinrich event responses. Lowering the Laurentide ice sheet in the second (*MinCold*) phase of the Heinrich event experiments induces changes in the magnitude and distribution of pressure patterns aloft and at the surface, with attendant changes in upper level and surface winds over WNA (Plate 3b and 3c, *MinCold-MaxCold*). During January, there is a breakdown aloft in the effect of the glacial anticyclone and a reduction in the prominence of the split jet (in the AGCM), consequently intensifying the westerlies over the west coast and into the Great Basin while reducing the influence of the subtropical jet in the SW. At the surface there is also more westerly flow into the west coast in the *MinCold* simulation than in the *MaxCold*. In July, higher SLP is centered over Oregon and northern California in the *MinCold* simulation than in the *MaxCold*, and relative to the glacial/interglacial anomalies, northerly shore-parallel winds are further reduced along the PNW coastline. Near-shore winds are also reduced to a lesser degree along coast of southern California.

In contrast to glacial/interglacial changes in moisture patterns, precipitation increases by over 2 mm d^{-1} over the PNW, northern Sierras, and the northern Great Basin and decreases by up to 5 mm d^{-1} over the SW (Plate 3d, *MinCold - MaxCold*). Wetter conditions in the PNW in the *MinCold* simulation reverse the glacial/ interglacial gradient and shift the line separating wetter from drier anomalies to the southwest. Differences between *MinCold* and *MaxCold* July precipitation rates are more heterogeneous than those of January, although the winter pattern of wetter conditions in the PNW and drier conditions in the SW is broadly maintained. The most obvious exception is a shift to wetter conditions over New Mexico which is associated with focused moisture advection from the Gulf of Mexico.

January temperatures in the second (*MinCold*) phase are up to 2-4°C colder (than *MaxCold*) over parts of the PNW, the LIS, southward to the Great Basin, and much of the SW (Plate 3a). These areas of cooling are separated by an area of roughly equivalent warming extending eastward from eastern Oregon to the Yellowstone Plateau and southward into eastern New Mexico. Colder temperatures in the north are associated with increased cloud cover whereas cooling to the south is associated with advection of cold air from the north. The area of warming is produced by increased advection of warm air from the south and east, and to a lesser degree by elevated solar radiation levels from reduced cloud cover. July air temperatures are also up to 4°C lower over the PNW, reflecting strengthening of westerly flow and associated cloudiness. Warming extending in a northwesterly direction from the SW is caused both by increased southerly advection and, as in January, to a lesser degree by elevated solar radiation levels.

In the third (*MinWarm*) stage of the canonical Heinrich event, in which increased North Atlantic SSTs are combined with a lower LIS, lower pressure forms aloft over the PNW (Plate 3b; *MinWarm-MaxCold*) in January causing upper-level wind flow to increase from a northwesterly direction over the eastern Pacific and west coast and to decrease to the east. At the surface, there is a general deepening of low pressure relative to both the *MaxCold* and *MinCold* simulations (Plate 3c; *MinWarm-MaxCold*). Lower surface pressure enhances shore-parallel flow along the northern and central coast and enhances westerly flow along the southern coast. In July, wind flow aloft in the *MinWarm* experiment is little changed relative to the *MaxCold* phase along the northern coast, but there is enhanced westerly flow that extends inland from the PNW across the northern Rocky Mountains. To the south there is strengthened anticyclonic flow aloft in the eastern Pacific and enhanced easterly flow over the southern SW and northern Mexico. At the surface, there are relatively minor changes in SLP and wind flow, except over and to the east of the northern Rocky Mountains.

January circulation changes in the *MinWarm* experiment (relative to the *MaxCold* experiment) generally result in slight redistribution of precipitation over WNA (Plate 3d). The exception is the wet area located along the southern edge of the low pressure cell centered over the PNW. In July, there is a mixed, and generally small, precipitation response with the largest change being wetter conditions over northern Mexico and the SW in response to the increased easterly flow aloft in the *MinWarm* experiment relative to the *MaxCold*.

Advection of cold air from the north by winds associated with low pressure aloft in the *MinWarm* experiment yields generally colder temperatures over WNA than those of the *MaxCold* experiment. Areas of cooling of up to 8°C are located over the northern portion of the domain. Cooling of 1-4°C occurs over the Sierras, the Great Basin and the SW (Plate 3a). July temperature anomalies are not as large as those of January, with cooling of up to 2°C over the PNW and about the same magnitude of warming east of the Northern Rocky Mountains.

3.3 Summary of Regional Model Results

As with the global climate simulations, the responses in WNA to the different phases of the canonical H2 event are spatially heterogeneous. A combination of large-scale controls and regional forcing yield regional responses that are reinforced, canceled, or reversed in the various phases of our H2 event. For example, the LGM pattern of relatively dry conditions in the PNW and wet conditions in the SW is attenuated or reversed in the *MinCold* phase. Subsequent warming of the North Atlantic in the *MinWarm* phase leads to attenuated precipitation responses in some areas (PNW, Yellowstone Plateau, and SW), and produces different precipitation patterns in other areas (e.g., wetter conditions across Nevada, Utah, and eastern Colorado). The magnitudes of the climate anomalies in the *MinCold* and *MinWarm* phases range in value up to those of the glacial/interglacial, suggesting that millennial-scale climate changes such as Heinrich events may have been recorded by surface systems and processes (e.g., lakes, vegetation, glaciers, coastal upwelling). We explore the sensitivity of some of these systems in the following sections.

4. RESPONSE OF SURFACE SYSTEMS AND PROCESSES

4.1 Alpine glaciers and ice caps

Alpine glaciers and ice caps are sensitive and rapidly responding (response time on the order of $10\text{-}10^2$ yr) indicators of climate change [*Paterson,* 1994]. Geologic records from WNA suggest the possibility that glaciers responded not only to glacial/interglacial climate changes but also to millennial-scale climate variations [*Clark and Bartlein,* 1995]. The mass balance of a glacier is the difference between accumulation (precipitation falling as snow), and ablation which is governed by the surface energy balance (for which temperature is a proxy). For a particular glacier, the mass balance is thus a function of winter precipitation and summer air temperature. Empirical equations for computing glacier mass balance from temperature and precipitation have been developed from modern data sets [e.g., *Leonard*, 1989; *Ohmura et al.*, 1992], and we have applied such a relation using output from RegCM to quantify the relative contribution of precipitation and temperature in maintaining LGM glaciers in WNA [*Hostetler and Clark,* 1997]. We follow that approach here to assess the sensitivity of glaciers to the climate variations of the H2 event.

At each model grid cell, total accumulation or snowfall is computed as the sum of precipitation (water equivalent) in the months in which the near-surface (2 m) air temperature is ≤ 0°C. Using average July-August air temperature values from the RegCM, we estimate ablation $A(T)$ from the empirical equation $A(T) = (T + 8.16)^{2.85}$ of *Leonard* [1989]. The mass balance at each grid cell is determined as the algebraic difference of accumulation and ablation. Based on an analysis of the interannual variability of mass balances of high-latitude glaciers at present, computed mass balances within the range of ± 750 mm yr^{-1} are assumed to be in approximate equilibrium [*Hostetler and Clark,* 1997]. (Necessary topographic smoothing in the RegCM causes the elevations of mountains to be lower than actual elevations. We did not apply a lapse-rate correction to model air temperatures to account for differences in elevations, but such a correction would, in general, reduce estimated ablation and thus result in more positive mass balances.)

The distribution of mass balance values for the *MaxCold* or LGM simulation (Plate 3e, *MaxCold*) are similar to those reported by *Hostetler and Clark* [1997]. Glaciers in the Northern Rocky Mountains and the Yellowstone Plateau that are generally controlled by temperature tend to have more positive mass balances than their precipitation-controlled counterparts located to the south and west. During the *MinCold* phase, the increase in precipitation (relative to the *MaxCold* phase) extending from the PNW to the eastern Yellowstone Plateau and southward into the Sierra Nevada and Wasatch Mountains in Utah drives the mass balance of glaciers there more positive, suggesting that these glaciers would have advanced during that phase of our canonical Heinrich event. Increased precipitation (accumulation) over the eastern Yellowstone Plateau during the *MinCold* phase is offset by warmer summer temperatures (ablation), subsequently yielding a more negative mass balance, suggesting possible glacier retreat. A common, large-scale climate event induced by lowering the LIS may thus have caused opposing responses over relatively short distances in this region (including northern Montana). .

During the *MinWarm* phase, a different pattern of glacier response emerges (Plate 3e, *MinWarm-MaxCold*). Colder summer temperatures lead to reduced ablation and, combined with accumulation that is somewhat greater than the *MaxCold* experiment but less than that of the *MinCold* experiment, the mass balances of glaciers in the Cascades become less negative than in the *MaxCold* simulation. At the same time, reduced accumulation to the east causes a general trend toward more negative mass balances there. In the Sierra Nevada, Utah, and the central Rocky Mountains, increased accumulation and decreased ablation also yield more positive mass balances.

The sensitivity of alpine glaciers to millennial-scale climate variations simulated here suggests a complex response on two levels. First, over relatively short distances it might have been possible to record opposing glacier responses within each of the phases of a Heinrich event. Second, the direction of change to a subsequent phase may be of the same sign for the combinations of values of temperature and precipitation that lead to similar mass balances, or the direction of change can oppose that of the previous phase. This spectrum of possible responses suggests that geologic records of glacial advance and retreat during millennial-scale climate events may be difficult to correlate in a simple fashion, while remaining consistent with regional-scale climatic controls.

4.2 Lakes

Physical lake responses to climate change, including thermal structure (e.g., temperature, mixing, surface fluxes, ice cover) and area (level) also can be quantified by modeling. To evaluate thermal responses and interactions with the atmosphere, an interactive lake model was used in RegCM to represent Lakes Bonneville and Lahontan, the largest of the pluvial lakes at the LGM. We can also assess the thermal response of other representative lakes not included in the RegCM by driving the lake model [*Hostetler and Bartlein*, 1990] "off line" with meteorological data from the RegCM simulations. Changes in levels are not explicitly computed in either case; rather we infer level changes from distribution of P-E as simulated by RegCM and the lake model.

Relative to the *MaxCold* simulation, the mean annual surface temperature of Lakes Lahontan and Bonneville are slightly colder ($\leq 0.5^{\circ}C$) in both the *MinCold* and *MinWarm* experiments. Changes in the water balance of the lake basins, which are dominated by changes in precipitation, are more substantial (Plate 3f). In the *MinCold* experiment, runoff from the land surface of the Lahontan basin increases by an average of 0.11 mm day^{-1} (relative to .25 mm day^{-1} in the *MaxCold* experiment) and P-E over the entire basin increases by an average of 0.18 mm day^{-1} (relative to a value of -0.32 mm day^{-1} in the *MaxCold* experiment). Over the Bonneville basin, in the *MinCold* phase, changes in P-E are smaller, with runoff increasing slightly by 0.06 mm day^{-1} (relative to a value of 0.48 mm day^{-1} in the *MaxCold* experiment) and P-E increasing by 0.11 mm day^{-1} (relative to a value of -0.07 mm day^{-1} in the *MaxCold* experiment). This tendency for rising lake levels in the Lahontan basin is sustained during the *MinWarm* stage of the Heinrich event with runoff and P-E both increasing by 0.15 mm day^{-1} relative to the *MaxCold* experiment. In the Bonneville basin, however, the tendency for rising lake levels in the *MinCold* phase is reversed in the *MinWarm* phase wherein the average runoff anomaly remains slightly positive (0.06 mm day^{-1}) but the average P-E anomaly is -0.10 mm day^{-1}.

To provide some insight into the relative sensitivity of lakes of different sizes, climatic settings, and geographic locations, we obtained from the RegCM model the time series of daily values of the meteorological variables (air temperature, wind speed, humidity, longwave radiation, shortwave (solar) radiation) required to drive the lake model [*Hostetler and Bartlein*, 1990]. Sites were selected to represent Carp Lake (modeled depth of 4 m), Owens Lake (modeled depth of 60 m), and Lake Estancia (modeled depth of 10 m). The grid cell nearest to the actual location of the lakes was used as the target for the meteorological data. The simulated lakes are therefore not exact representations of the real lakes, but are sufficient to characterize first-order sensitivities.

The glacial/interglacial changes in the mean annual surface temperature of Owens and Estancia lakes are small ($-1.4^{\circ}C$ and $+1.4^{\circ}C$, respectively). Carp Lake has the largest temperature response of any of the simulated lakes. Because the lake is located in the area of prevailing, cold easterlies off the ice sheet, summer maximum and winter minimum temperatures at the LGM (*MaxCold*) are $\sim 15^{\circ}C$ colder than present, and the annual average temperature is $11^{\circ}C$ colder than the control. Changes in mean annual surface temperatures for all lakes are $\leq 1^{\circ}C$ for both the *MinCold* and *MinWarm* experiments relative to the *MaxCold* experiment.

As is the case for the large lakes, changes in the water balances of the smaller lakes are greater in relative magnitude than are temperature changes. During both the *MinCold* and *MinWarm* phases, simulated lake levels at Carp lake have a tendency to rise, with a greater rise occurring during the *MinCold* phase (Table 1). Lake Estancia, in contrast, tends to fall during both the *MinCold* and *MinWarm* phases. Owens Lake tends to fall during the *MinCold* phase and rise during the subsequent *MinWarm* phase. These lake responses reflect the broad patterns of a wet PNW and dry SW, along with shifts in location and steepness of the gradient between these extremes from one phase of the canonical Heinrich event to another.

Like alpine glaciers, the simulated response of lakes during different phases of our H2 event again suggest variable responses that are characterized primarily by changes in the water balance. Both Lakes Lahontan and Bonneville show increases in basin moisture during the *MinCold* phase (relative to the *MaxCold* phase), suggesting that over decades and centuries the lakes would expand. During the *MinWarm* phase, however, Lake Lahontan

Table 1. Changes in annual water balance components of selected lake locations. The changes are relative to the *MaxCold* simulation. Δr: change in runoff; $\Delta P\text{-}E$: change in net basin moisture; ΔE_{lake}: change in lake evaporation, Lake Level: tendency to rise (+) or fall (-). Units of Δr, $\Delta P\text{-}E$, and ΔE_{lake} mm day^{-1}.

	MinCold		MinWarm		ΔE_{lake}		Lake	
Lake	Δr	$\Delta P\text{-}E$	Δr	$\Delta P\text{-}E$			Level	
Carp	0.3	0.3	0.1	0.1	0.0	0.0	++	+
Owens	-0.1	-0.2	0.2	0.0	0.1	0.0	-	+
Estancia	-0.2	-0.1	-0.1	-0.2	0.0	0.1	-	-

continues to support a lake larger than the *MaxCold* experiment, but a reduction in the moisture in the Bonneville basin would lead to a lake smaller than that of the *MaxCold* phase, and in opposition to the direction of change of Lake Lahontan. Similar pattern of variable response are demonstrated by the other lakes considered.

4.3 Vegetation

We use July soil moisture and seasonal temperature anomalies from the RegCM simulations to infer possible changes in vegetation during the various phases of our H2 event. Relative to the control, the *MaxCold* changes in July soil moisture are substantial and widespread over the model domain (Plate 3g). With the exception of coastal PNW, *MaxCold* soil moisture levels are higher than those of the control, particularly over the SW and into the Rocky Mountains. In the *MinCold* experiment, changes in precipitation result in a general reversal of the glacial/interglacial pattern of soil moisture anomalies, with moisture levels increasing in the PNW and decreasing in the SW, California, and the Great Basin. The pattern of soil moisture anomalies in the *MinWarm* experiment is more heterogeneous than that of the *MinCold* experiment. Soil moisture anomalies are positive, but of lower magnitude over the PNW, and a mix of positive and negative anomalies is found elsewhere. Northern and central California and the Sierra are drier, and a mix of drier and wetter areas is found in the Great Basin and SW.

The regional-scale patterns of the LGM-present (*MaxCold-Control*) anomalies of soil moisture and temperature are consistent with the broadscale patterns of paleoecological data [*Thompson et al.*, 1993] that show widespread replacement of forest by steppe across the northern half of WNA, and the reverse across the southern part of the region. Clear correlations between the paleoecological record and the simulated *MinCold-MaxCold* and *MinWarm-MaxCold* climate anomalies are more difficult to detect, even allowing for the large spatial heterogeneity evident in those anomaly patterns [see *Whitlock and Grigg*, this volume]. It is possible, however, that when the soil moisture anomalies are interpreted in light of other landscape scale controls of vegetation (e.g. soil type, proximity to physiographic barriers, see *Whitlock and Bartlein*, 1993), that consistency will emerge for the millennial-scale vegetation variations.

4.4 Coastal upwelling

A challenging aspect of understanding millennial scale climate changes is linking terrestrial and marine responses. Such a linkage is relatively straightforward in the North Atlantic region where strong (and consistent) marine and terrestrial responses are found for events such as the Younger Dryas climate reversal (H0). There are ample indications of millennial-scale climate events in marine and terrestrial records from the eastern Pacific and WNA, however, correlating these responses in time and space to events in the North Atlantic is more difficult in this distant region. Although we did not model oceanic circulation changes, we can quantify the relative sensitivity of wind-driven coastal upwelling to the various phases of the H2 event. Upwelling is a dominant control of marine productivity along the coast of WNA and changes in productivity recorded in marine cores potentially offer a path for associating terrestrial and marine responses to millennial-scale climate changes..

We use the method of *Bakun and Nelson* [1991] as modified by *Ortiz et al.* [1997] to compute curl of the wind stress on the 60-km grid of the RegCM. The magnitude and direction of the curl determines upwelling (cyclonic direction in the northern hemisphere) and downwelling (anticyclonic direction in the northern hemisphere). Monthly values of the northerly and westerly components of the surface wind (Plate 3c) for the summer (June through August) and winter (December through February) are used in the computation. Seasonal computations are made because there is a distinct seasonal cycle in upwelling along the coast of WNA. The spatial patterns of curl and thus of upwelling and downwelling for our *Control* simulation (not shown) compare favorably in both summer and winter with values computed from observed winds by *Bakun and Nelson* [1991]. The magnitude of the curl fields also compares well with those of *Bakun and Nelson* [1991] in summer, but we found the magnitude of our winter curl values to be generally underestimated. We therefore corrected this underestimate, by scaling the *Control* RegCM winter wind fields by +20% to achieve better average agreement with the observed data. The scaling was applied to all subsequent winter curl computations, and thus does not affect the sign of anomalies among the various phases.

Along the coast and in the open ocean, substantial glacial/interglacial changes in curl of the wind stress and hence upwelling occur in winter and summer (Plate 3h).

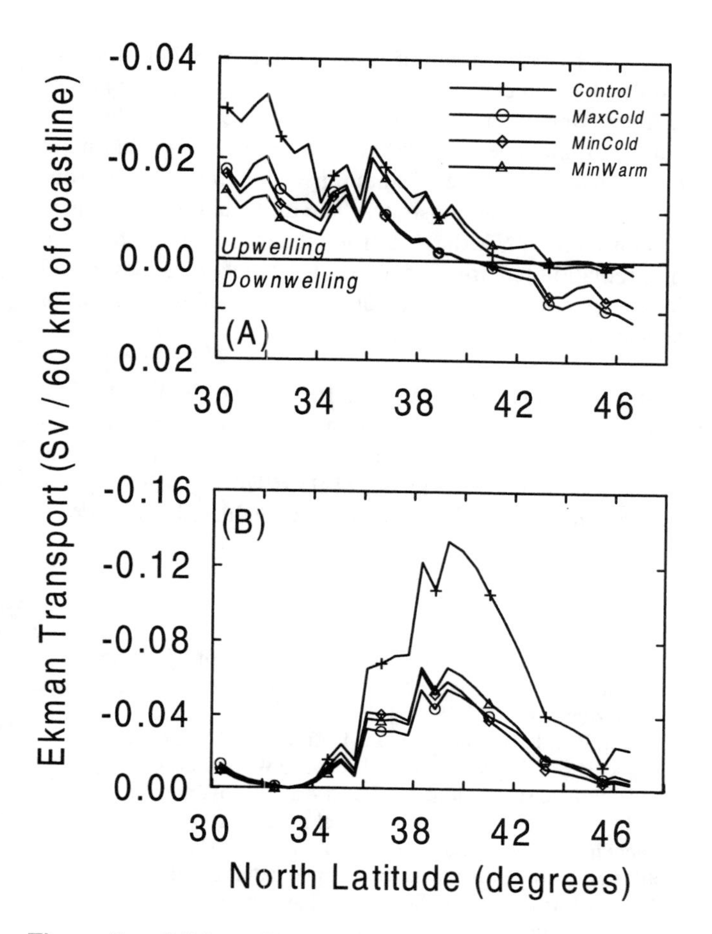

Figure 2. Offshore Ekman transport for *Control*, *MaxCold* (LGM), *MinCold*, and *MinWarm* phases of the H2 event. The transport values are computed using the northerly and westerly components of the surface winds from the RegCM simulations as an approximation to actual shore-parallel winds.

The patterns and magnitude of the changes compare favorably with those of *Ortiz et al.* [1997]. Off the PNW coast in January, upwelling is enhanced where easterly winds from the LIS dominate the *MaxCold* wind flow. To the south, there are areas of reduced upwelling off the coast of southern Oregon and northern California and along the southern California coast. In July, an area of reduced upwelling is located off the PNW and northern California coast, and farther to the south along southern California and the Santa Barbara basin. Enhanced upwelling occurs along and off shore of central California. Coastal upwelling changes are further quantified by the computed Ekman transport, which reflects the volume of offshore flow (Figure 2). In agreement with *Ortiz et al.* [1997], glacial/interglacial Ekman transport is reduced along the entire coastline, with the largest reductions occurring during summer.

In the *MinCold* experiment there is a mixed pattern of upwelling changes in winter, with the largest anomalies being located off the PNW coast and in the vicinity of the Santa Barbara basin (Plate 3h). Summer anomalies tend to reinforce the glacial/interglacial increase in the PNW, and reverse the pattern off California where increased upwelling is indicated. In the *MinWarm* phase, winter upwelling is reduced in the PNW and strengthened off the California coast and in the Santa Barbara basin, whereas summer anomalies are similar to those of the *MinCold* experiment. Winter Ekman transport in the *MinCold* experiment is reduced from 30°N to 36°N and increases to values exceeding the control, north of ~40°N, whereas summer transport exceeds the *MaxCold* from ~36°N to 43°N. In the *MinWarm* phase, Ekman transport is greater than that of the *MinCold* phase in the south, and about the same as the *MinCold* north of 36°. Summer transport values are greater than the *MaxCold* and *MinCold* phases from 34°N to 38N°, and less than previous values northward of 39°N.

Our curl computations suggest the possibility that changes in upwelling during a Heinrich event may be of sufficient magnitude to be recorded as changes in productivity in marine cores. The areas most likely to reflect the changes are the Santa Barbara basin, where winter upwelling is suppressed in both the *MinCold* and *MinWarm* phases, and off the PNW coast where an alternating pattern of increased and decreased upwelling is simulated. Interpretation of the results is not straightforward, however, because we cannot isolate the contributions of changes in the location of the Alaskan gyre and the interplay between offshore and coastal upwelling. Nonetheless, our computed upwelling changes coupled, for example, with changes inferred changes in surface runoff suggest the existence of joint marine and terrestrial responses (e.g., the Columbia, Elk, and Santa Clara Rivers) during various phases of a Heinrich event.

5. DISCUSSION AND CONCLUSIONS

Our modeling study allows us to decompose global, millennial-scale climate variations into synoptic and regional components by using a suite of models to assess the response of global climate changes on regional and local processes in WNA. At the global scale, our canonical Heinrich event is transmitted to WNA primarily by the effects of topographic and thermal forcing of the LIS and secondarily by transmission from the North Atlantic region through circum-polar circulations. In WNA, synoptic-scale responses of precipitation and temperature are induced by changes in the location and strength of large-scale atmospheric pressure patterns and their influence on wind flow at the surface and aloft. Changes in the synoptic-scale climatology simulated by the GCM in the *MinCold* and *MinWarm* phases of the Heinrich event are downscaled by the higher resolution regional climate model that portrays more variable topographic complexity.

Simulated millennial-scale climate responses in WNA are characterized primarily by changes in precipitation patterns, and secondarily by changes in temperature and wind flow both within a particular phase and among the various phases. Spatially, the responses within a particular phase can be of opposite sign (e.g., the pattern of wet PNW and dry SW in the *MinCold* simulation), and these responses can reinforce, cancel, or reverse responses of previous or subsequent phases (e.g., the wet PNW-dry SW pattern of the *MinCold* simulation is opposite that of the *MaxCold* simulation). The potentially complex pattern of regional climate change that ensues from changes in global boundary conditions suggests caution when attempting to infer millennial-scale climate variations from geologic records of WNA. Our process modeling underscores this cautionary statement.

Relative to the LGM, during the *MinCold* phase simulated alpine glaciers would advance in the PNW, and retreat in the Sierra and Rocky Mountains. Moreover, there is an indication that some areas such as the Yellowstone region could have registered opposing responses. In the *MinWarm* phase, simulated glaciers would retreat in the PNW and Rocky Mountains while advances would occur in the Sierra. These variable responses are also evident in changes in simulated lake level, vegetation and upwelling. In addition, during some phases, different systems register opposite responses. For example, in the *MinWarm* phase, glacier advances in the Sierra are indicated, yet negative soil moisture anomalies west of the Sierra suggest a possible vegetation shift to more xeric taxa. Similarly, opposing responses are evident in the PNW and northern Rocky Mountains.

Our modeling study has limitations. First, the canonical Heinrich event patterned after H2 was prescribed as a linear progression of what are thought to be primary components of Heinrich events in general, and it is likely that a collapse of the LIS and changes in thermohaline circulation were not discrete. In addition, we did not change SSTs in the Pacific, although it is likely that some oceanic teleconnection existed through which SSTs in the Pacific did change in response to variations in North Atlantic thermohaline circulation. Changing SSTs in the Pacific in general, or along the coast of WNA in particular, would substantially affect our simulated regional climate. Finally, our RegCM simulations are 4 years long. Over this relatively short period, interannual variability in the climate (or, actually, in the weather) can be large relative to changes among the simulations. For the domain of the RegCM, the envelope of interannual variability is ultimately limited by the driving AGCM, and any given 4-yr period in an AGCM simulation usually differs from other 4-yr periods. In our GENESIS simulations, we achieved significant changes in climate for each phase of the H2 event in 15-yr. simulations. It is therefore likely that we would also achieve significance in our RegCM simulations if they were run out for the same 15 years, but the computational needs for 60 years of RegCM simulation are prohibitive. These limitations notwithstanding, our modeling study demonstrates that millennial-scale climate responses in WNA derived from changes in a combination of circulation features occurring at a variety of scales, could have left a complex signature in the geologic record of terrestrial and marine systems.

Acknowledgments. We thank P. Clark, P. Fawcett, D. Zahnle for their reviews and comments. This research was supported by the U.S. Geological Survey and National Science Foundation grant ATM-9532974 to the University of Oregon. Model simulations were conducted at NCAR.

REFERENCES

Bakun, A. and C.S. Nelson, The seasonal cycle of wind-stress curl in subtropical eastern boundary current regions, *J. Phys. Oceanogr., 21,* 1815-1834, 1991.

Bartlein, P.J., K.H. Anderson, P.M. Anderson, M.E. Edwards, C.J. Mock, R.S. Thompson, R.S. Webb, T.W. Webb, C. Whitlock, Paleoclimate simulations for North America over the past 21,000 years: Features of the simulated climate and comparisons with paleoenvironmental data, *Quat. Sci. Rev., 17,* 549-586, 1998.

Behl, R.J., and J.P. Kennett, Brief interstadial events in the Santa Barbara Basin, NE Pacific, during the last 60 kyr, *Nature, 379,* 243-246, 1996.

Benson, L.V., J.W. Burdett, M. Kashgarian, S.P. Lund, F.M. Philipps, and R.O. Rye, Climatic and hydrologic oscillations in the Owens Lake basin and adjacent Sierra Nevada, California, *Science, 274,* 746-749, 1996.

Benson, L.V., S.P. Lund, J.W. Burdett, M. Kashgarian, T.P. Rose, J.P. Smoot, and M. Schwartz, Correlation of late-Pleistocene lake-level oscillations in Mono Lake, California, with North Atlantic climate events, *Quat. Res., 49,* 1-10, 1998.

Benson, LV., D.R. Currey, R.I. Dorn, K.R. Lajoie, C.G. Oviatt, S.W. Robinson, G.I. Smith, S. Stine, Chronology of expansion and contraction of four Great Basin lakes, *Paleogeog., Paleoclim., Paleoecol., 78,* 241-286, 1990.

Berger, A. Long-term variations of caloric solar radiation resulting from the earth's orbital elements, *Quat. Res., 9,* 139-167, 1978.

Clark, P.U. and P.J. Bartlein, Correlation of late-Pleistocene glaciation in the western United States with North Atlantic Heinrich events, *Geology, 23,* 483-486, 1995.

COHMAP Members, Climatic changes of the last 18,000 years: Observations and model simulations, *Science*, 241:1043-1052, 1988.

Cortijo, E., L. Labeyrie, L. Vidal, M. Vautravers, M. Chapman, J.-C. Duplessy, M. Elliot, M. Arnold, J.-L. Turon, and G. Auffret, Changes in sea surface hydrology associated with Heinrich event 4 in the North Atlantic Ocean between 40° and 60°N, *Earth Planet. Sci. Lett., 146,* 29-45, 1997.

CLIMAP Project Members, Seasonal reconstruction of the earth's surface at the last glacial maximum. *Geol. Soc. Amer. Map & Chart Series* MC-3, 1981.

Giorgi, F., M.R. Marinucci, G.T. Bates, Development of a second-generation regional climate model (RegCM2). Part I: Boundary-layer and radiative transfer processes, *Monthly Weath. Rev., 121*, 2794-2813, 1993a.

Giorgi, F., M.R. Marinucci, G.T. Bates, Development of a second-generation regional climate model (RegCM2). Part II: Convective processes and assimilation of lateral boundary conditions, *Monthly Weath. Rev., 121*, 2814-2832, 1993b

Hostetler, S. W. and P.U. Clark, Climatic controls of western U.S. glaciers at the last glacial maximum, *Quat. Sci. Rev., 16*, 505-511, 1997.

Hostetler, S. W. and P.J. Bartlein, Modeling climatically determined lake evaporation with application to simulating lake-level variations of Harney-Malheur Lake, Oregon. *Water Res. Res. 26*, 2603-2612, 1990.

Hostetler, S. W., P.U. Clark, P.J. Bartlein, A.C. Mix, and N.J. Pisias, Atmospheric transmission of North Atlantic Heinrich events, *J. Geophys Res., 104*, 3947-3952, 1999.

Hostetler, S.W., F. Giorgi, G.T. Bates, and P.J. Bartlein, Lake-atmosphere feedbacks associated with paleolake Bonneville and Lahontan, *Science, 263,* 665-668, 1994.

Hostetler S. W. and L.V. Benson, Paleoclimtic implications of the high stand of Lake Lahontan derived from models of evaporation and lake level, *Clim. Dyn., 4*, 207-217, 1990.

Hostetler S. W. and L. V. Benson, Stable isotopes of oxygen and hydrogen in the Truckee River-Pyramid Lake surface-water system. 2. A predictive model of $\delta^{18}O$ and $\delta^{2}H$ in Pyramid Lake, *Limnol. Oceanogr., 39,* 356-364, 1994.

Leonard, E.M., Climatic change in the Colorado Rocky Mountains: estimates based on modern climate at late Pleistocene equilibrium lines, *Arc. Alp. Res., 21*, 245-255, 1989.

Licciardi, J.M., P.U. Clark, J.W. Jenson, and D.R. MacAyeal, Deglaciation of a soft-bedded Laurentide Ice Sheet, *Quat. Sci. Rev., 17,* 427-448, 1998.

Lund, D.C., and A.C. Mix, Millennial-scale deep water oscillations: Reflections of the North Atlantic in the deep Pacific from 10 to 60 ka, *Paleoceanography, 13,* 1-19, 1998.

Manabe, S. and A.J. Broccoli, The influence of continental ice sheets on the climate of an ice age, *J. Geophys. Res. 90, 2167-2190*, 1985.

Ohmura, A., P. Kasser, and M. Funk, Climate at the equilibrium line of glaciers, *J.. Glaciol., 38*, 397-411, 1992.

Ortiz, J., A. C. Mix, S. W. Hostetler, and M. Kashgarian, The California Current of the last glacial maximum: Reconstruction at 42°N based on multiple proxies, *Paleoceanography, 12*, 191-205, 1997.

Oviatt, C.G., Lake Bonneville fluctuations and global climate change, *Geology, 25*, 155-158, 1997.

Paterson, W.S.B., The Physics of Glaciers. Pergamon Press, Oxford, 480 pp., 1994.

Peltier, W.R., Ice age paleotopography, *Science, 265,* 195-201, 1994.

Phillips, F.M., M.G. Zreda, L.V. Benson, M.A. Plummer, D. Elmore, and P. Sharma, Chronology for fluctuations in late Pleistocene Sierra Nevada glaciers and lakes, *Science, 274,* 1996.

Pollard, D. and S.L. Thompson, Climate and ice-sheet mass balance at the Last Glacial Maximum from GENESIS version 2 global climate model, *Quat.. Sci. Rev., 16*, 841-863, 1997.

Rind, D., Components of Ice Age circulation, *J. Geophys. Res., 92*, 4241-4281, 1987.

Small E.E., L.C. Sloan, F. Giorgi, and S. W. Hostetler, Simulating the regional water balance of the Aral Sea with a coupled regional climate-lake model, *J. Geophys. Res.*, in press.

Thompson, R.S., C. Whitlock, P.J. Bartlein, S.P. Harrison, W.G. Spaulding, Climatic changes in the western United States since 18,000 yr B.P., in *Global Climates Since the Last Glacial Maximum,* edited by H.E. Wright, Jr., J.E. Kutzbach, T. Webb, W.F. Ruddiman, F.A. Street-Perrott, and P.J. Bartlein, pp. 468-513, University of Minnesota Press, Minneapolis, 1993.

Thompson, R.S., S.W. Hostetler, P.J. Bartlein, and K.H. Anderson, A strategy for assessing potential future changes in climate, hydrology, and vegetation in the Western United States, *U.S. Geological Survey Circular 1153*, 20 pp., 1998.

Weaver, A.J., Eby, M., Fanning, A.F., and Wiebe, E.C., Simulated influence of carbon dioxide, orbital forcing, and ice sheets on the climate of the last glacial maximum, *Nature*, *394*, 847-853, 1998.

Whitlock, C. and Bartlein, P.J., Spatial variations of Holocene climatic change in the Yellowstone region. *Quat. Res., 39,* 231-238, 1993.

Whitlock, C. and P.J. Bartlein, Vegetation and climate change in northwest America during the past 125 kyr, *Nature*, 388, 57-61, 1997.

Whitlock, C. and L.D. Grigg, Paleoecological evidence of Milankovitch and sub-Milankovitch climate variations in the Western US during the late Quaternary, in *Mechanisms of Millennial-Scale Global Climate Change*, edited by P.U. Clark, R.S. Webb, L.D. Keigwin, this volume.

Wright, H.E., Jr., J.E. Kutzbach, T. Webb, W.F. Ruddiman, F.A. Street-Perrott, and P.J. Bartlein, editors, *Global Climates Since the Last Glacial Maximum,* University of Minnesota Press, Minneapolis, 569 pp, 1993.

P.J. Bartlein, Department of Geography, University of Oregon, Eugene, OR 97403 (bartlein@oregon.uoregon.edu)

S.W. Hostetler, U.S. Geological Survey, 200 SW 35th St, Corvallis, OR 97333 (steve@ucar.edu)

Sensitivity of Stationary Wave Amplitude to Regional Changes in Laurentide Ice Sheet Topography

Charles Jackson

Atmospheric and Oceanic Sciences Program, Princeton University, New Jersey

The maximum effect of regional changes in Laurentide ice sheet (LIS) topography on the atmosphere's stationary wave circulation is determined using single-layer models of the atmosphere. Model experiments measure the individual contribution of each section of the LIS and Greenland topography to the stationary wave circulation. Results show the possibility of a select volume of topography to control a disproportionate amount of atmosphere's total response to topography. Moreover, the possibility exists that stationary wave amplitude can increase due to a reduction in topographic forcing. These results can be understood by considering how the mean flow controls the horizontal propagation of wave energy and superposition of wave amplitude. Stationary wave sensitivity to topography within the single-layer models suggests that variations in the size or shape of the LIS can be one factor in modulating millennial-scale climate variability.

1. INTRODUCTION

The Laurentide ice sheet (LIS) is thought to participate in millennial-scale climate variability by an instability created by the thawing of basal sediment and the ensuing collapse of one of LIS's ice domes [*MacAyeal,* 1993a, 1993b]. One of the domes comprising the LIS is the Hudson Bay ice dome. Evidence of the collapse of the Hudson Bay ice dome is marked in the North Atlantic sediment by the appearance of layers of detrital carbonate ice rafted debris. These layers, known as Heinrich events, were deposited prior to prominent warmings of Greenland climate and are the primary clues of a connection between the LIS and millennial-scale climate variability [*Bond et al.,* 1993]. One way the LIS can affect climate change is through the increased supply of iceberg meltwater to the North Atlantic. Ocean models have shown that an increased fresh water supply to the North Atlantic reduces the thermohaline circulation component of the ocean's poleward heat transport [*Paillard and Labeyrie,* 1994; *Rahmstorf,* 1994, 1995; *Manabe and Stouffer,* 1995, 1997; *Fanning and Weaver,* 1997; *Cai et al.,* 1997]. Another way the LIS can affect climate change is through the influence of an altered LIS topography on the atmosphere's stationary wave circulation. The effects of changes in ice sheet topography on the atmosphere's stationary waves, the primary circulation feature affected by topography, are not well known. Because the atmosphere's stationary waves affect poleward heat transport as well as precipitation patterns, one might expect that the observed changes in climate were also influenced by the evolution of LIS topography created by ice sheet instability.

The climate effects of a topographically varying LIS during deglaciation are considered by *Felzer et al.* [1998] and *Kutzbach and Guetter* [1986]. In addition, the climate effects of two alternate reconstructions of Glacial Maximum ice sheets are considered by *Shinn and Barron* [1989]. These studies determine that the primary means by which an ice sheet affects climate is through

Mechanisms of Global Climate Change at Millennial Time Scales
Geophysical Monograph 112

the amount of area covered by an ice sheet's high albedo surface. The changing areal coverage of the ice sheets during deglaciation obscure the independent effects of an ice sheet's topography on climate. The change in ice sheet forcing thought to have occurred during the collapse of the Hudson Bay ice dome would likely only involve changes in topography with little change in albedo. A more clear determination of the climate effects of ice sheet topography is considered by *Felzer et al.* [1996] and *Rind* [1987]. *Felzer et al.* [1996] use the Community Climate Model version CCM1 coupled to a mixed-layer ocean to determine that the topography of the Laurentide and European ice sheets warmed limited areas of the surface as much as 10°C in winter and summer. *Rind* [1987] finds qualitatively similar results to *Felzer et al.* [1996] using GISS model II with prescribed sea surface temperatures according to CLIMAP. Although the effects of ice sheet topography are limited in area, *Felzer et al.* [1996] and *Rind* [1987] do show the potential of changes in topography to effect climate. When using these studies to interpret Heinrich events, one limitation is that their prescribed changes in topography involve entire ice sheets rather than the elimination of particular regions.

In order to argue that changes in LIS topography can be a significant part of the climatic response to ice sheet instability, two points need to be addressed: The first point is whether or not the LIS had dominate control over the atmosphere's stationary wave field. The problem here is that although a stationary wave is the primary atmospheric response to topography, its importance can be lessened by the relative importance of other potential stationary wave forcings. The second point is whether or not a volume of the LIS equivalent to 3.5 meters of sea level ascribed to the volume of iceberg discharge during a Heinrich event [*Alley and MacAyeal,* 1994] can significantly alter the atmosphere's stationary field. The subject of this paper mainly concerns the second of these two points. However, the relative importance of topographic forcing over other potential stationary wave forcings is important to the experimental design and will be discussed next.

Besides topography, stationary waves can be forced by diabatic heating sources such as land/sea contrasts as well as transient eddies (storms). Only a few studies have directly calculated the topographic component of stationary wave generation independent of other possible wave sources. *Cook and Held* [1988] use a linearized model of the atmosphere's circulation to dissect those aspects of Last Glacial Maximum climate that generate its stationary wave field. They find that the mechanical effect of LIS topography, its ability to block the flow of the atmosphere, is the main source of stationary wave energy. Their analysis contrasts with modeling studies of modern-climate stationary waves which are found to be forced by equal contributions from topography, transient eddies, and other diabatic heating sources [*Nigam et al.,* 1988]. The relative importance of topography to other possible sources of stationary wave energy can be model specific. For example, *Dong and Valdes* [1998] conduct a series of sensitivity experiments of the Last Glacial Maximum atmosphere and find their model's stationary wave field to be primarily forced by transient eddies. What determines the partitioning between possible stationary wave forcings is not known with great certainty. One factor could be the strength of the low level winds. Strong low level winds favor topographic forcing over diabatic surface forcings [*Held and Ting,* 1990]. Unless more confidence can be placed on the exact partitioning of the means by which stationary waves are forced before, during, and after Heinrich events, there can be no certainty what the effect of changing LIS topography has on climate.

For purposes of argument, the present model experiments assume that topography has exclusive control over the atmosphere's stationary wave field. This assumption can be achieved using models that do not include surface diabatic heating sources or transient eddies. The most simple models that can capture the atmosphere's response to topography without other potential sources of stationary wave energy are single-layer (barotropic) models. Single-layer models have been used to study the dynamics of the way stationary wave energy propagates around the globe as well as the effects of topography on the atmosphere's circulation [*Hoskins et al.,* 1977; *Grose and Hoskins,* 1979; *Branstator,* 1983; *Karoly,* 1983; *Hoskins and Ambrizzi,* 1993]. The present model experiments use a linearized one-dimensional, single-layer model as well as a nonlinear, single-layer model on a sphere to illustrate how variations in LIS topography affect the atmosphere's stationary waves. Model experiments involve systematic alterations to its topography to derive a measure of the sensitivity of stationary wave amplitude to each section of topography. These maps will help evaluate whether or not small changes in ice sheet topography could exert a significant influence over the atmosphere's stationary wave field.

2. ONE-DIMENSIONAL SINGLE-LAYER MODEL

The one-dimensional dynamics of the midlatitude atmosphere can be approximated with a quasi-geostrophic potential vorticity equation linearized about a mean zonal flow. In general, stationary waves are defined in terms of a perturbation streamfunction,

$$\psi' = \overline{\psi}^t - \overline{\psi}^{t\theta} \tag{1}$$

where the overline with a t or θ denotes a time or zonal average. The stationary wave response of the linear model to topography of arbitrary shape η_B that exists for zonal coordinate $x > 0$ can by found by way of Green's functions. Its solution for $x \leq 0$ is

$$\psi'(x) = 0. \tag{2}$$

For $x > 0$,

$$\psi'(x) = -\int_0^x \frac{f_\circ}{Dk_s} \sin(k_s[x - x'])\eta_B(x')dx' \tag{3}$$

where $f_\circ$ is the value of the Coriolis parameter at latitude $\varphi_\circ$, D is the fluid layer depth, k_s is the stationary wave number equal to $\sqrt{\beta/U_\circ}$ where β is the meridional gradient of f at $\varphi_\circ$ and $U_\circ$ is the zonal mean wind.

The solution expressed in Equation 3 depends on the product between topography $\eta_B(x)$ and the atmosphere's stationary wave $\sin(k_s[x])$. One half the wavelength of the stationary wave $R_\circ = \pi/k_s$ can vary depending on the background flow $U_\circ$ but is independent of the details of the topography. Therefore, for a given flow, the response of the one-dimensional atmosphere to topographic forcing depends solely on the relative length scales between $R_\circ$ and $[\eta_B]$, where square brackets around η_B denote horizontal length scale. The sensitivity of the stationary wave solution to small changes in topography η_B in cases where $[\eta_B] \approx R_\circ$ is to be investigated.

The relative influence of various portions of the LIS topography on the atmosphere's stationary wave amplitude will be illustrated with the aid of sensitivity maps. The sensitivity S is defined to be the ratio of the change in the globally averaged stationary wave amplitude to a given change in topography,

$$S = \frac{\Delta|\psi'|}{\Delta\eta_B}. \tag{4}$$

The change in stationary wave amplitude is defined by

$$\Delta|\psi'| = \frac{\int_{-\infty}^{+\infty} \widehat{\psi}_{[a,b]} \cdot \psi' dx}{\int_{-\infty}^{+\infty} \psi' \cdot \psi' dx} \tag{5}$$

where $\widehat{\psi}_{[a,b]}$ gives the influence of topography in region $x \in$[a,b] to the model domain. The influence function is defined as

$$\widehat{\psi}_{[a,b]}(x) = \psi'_{[a,b]}(x) - \psi'(x). \tag{6}$$

The influence function $\widehat{\psi}_{[a,b]}$ is given as the difference between the stationary wave solution with a change in surface elevation of η_B in section $x \in$[a,b] and the stationary wave solution to unaffected topography, $\psi'(x)$. The change in topographic forcing, $\Delta\eta_B$, is found by dividing the integral of topographic perturbation

$$\int_a^b \eta'_B(x)dx \tag{7}$$

where $\eta'_B(x)$ is the change in surface elevation of the topography in section $x \in$[a,b], by the integral of the entire topography

$$\int_{-\infty}^{+\infty} \eta_B(x)dx. \tag{8}$$

That is,

$$\Delta\eta_B = \frac{\int_a^b \eta'_B(x)dx}{\int_{-\infty}^{+\infty} \eta_B(x)dx}. \tag{9}$$

If stationary wave amplitude grows in proportion to topographic volume then $S(x)$ for section $x \in$[a,b] of the topography is equal to 1. For the one-dimensional model at hand that is true for η_B whose zonal extent is much less than the atmosphere's stationary wavelength, $[\eta_B] << R_\circ \approx 4500$ km. The longitudinal extent of the LIS, however, exceeds the stationary wave length scale. Figure 1 shows the sensitivity of the one dimensional model at 45° N with $U_\circ = 30$ ms^{-1} to a rectangular topography of uniform 1 km height. Figure 1 shows a spread of sensitivity values ranging from near 0 to over 1.5 sensitivity units. Changing the height of the topography in regions with $S \approx 0$, near $x = 0$ and $x = 4500$ km, would have negligible effect on stationary wave amplitude. Alternatively changing the topography near $x = 2500$ km would have maximum effect on stationary wave amplitude.

The one-dimensional model shows that topographic features that have a zonal extent similar to the LIS have regions of varying sensitivity to topographic alteration. Because this model is linear, one may conclude that this effect is due to the superposition of wave amplitude

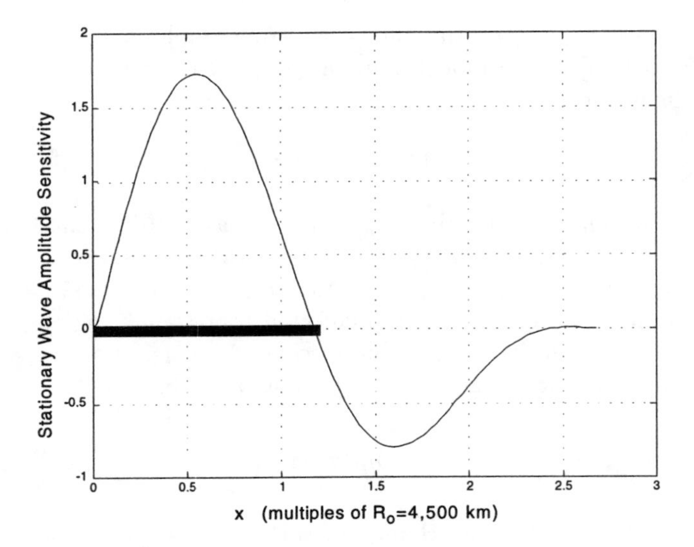

Figure 1. Stationary wave amplitude sensitivity to topography extending 5400 km from $x = 0$ (location indicated by thick line segment starting at $x = 0$). Stationary waves are most sensitive to changes in topography near $x = .6$.

from different wave sources (regions of the topography). In other words, wave superposition causes 'sweet spots' which are defined to be geographic regions whose topographic alteration leads to changes in stationary wave amplitude in excess of what may be expected if stationary wave amplitude were to scale proportionally with forcing.

3. SINGLE-LAYER MODEL ON A SPHERE

The second model to be examined is the single-layer model on a sphere. The 10 km thick layer of constant density is governed by the shallow water equations which are solved numerically using the spectral transform technique. The model is truncated at R15 with corresponding resolution of 7.5° longitude and 4.5° latitude grid spacing. The model's mean zonal winds are forced by relaxing the model's vorticity, divergence, and mass fields to a balanced reference state. The relaxation time scales are shown in Figure 2. In the high latitudes, the vorticity relaxation time scale is set around 14 days, whereas for divergence and mass, their relaxation time scale is set to a more restrictive 10 days. The shorter time scales in the tropics enhance the local absorption of wave activity.

Mean winds that resemble the observed atmospheric flow at 50 kPa pressure surface (~5 km above sea level) is chosen for model experiments. This flow is characterized by midlatitude jets flowing to the east (westerly winds) with diminished amplitude at the poles and subtropics. Westerly-only winds are imposed along the equator instead of the observed easterly trade winds (Figure 3). This choice is motivated by the need to minimize unrealistic wave reflection that occurs off easterly winds in single-layer models. The choice to combine westerly winds with enhanced tropical damping minimizes the distortion of the mid- to high latitude stationary wave field.

A topography reduction factor h is included in the model to enable an approximate matching between observations and the single-layer model's response to modern topography. Its value is determined through a comparison of the atmosphere's stationary wave field at 46°N to the model's steady response to the topography of the Himalaya and Rocky Mountains. An h value of .6 is found to be adequate.

The single-layer model uses *Peltier's* [1994] Ice-4G topography reconstruction of North American and Greenland topography at 21 ka before present (Figure 4). Other topographic features, such as the European ice sheet and Himalaya, are not included in order to simplify the interpretation of model results. Specific model predictions may be sensitive to the presence of these topographic features. It should be kept in mind that the model is useful for the examination of general sensitivity relationships rather than for specific predictions which can depend on processes and topographic features absent from the model.

The change in stationary wave amplitude to a given change in topography is determined through an influence function similar to that of the one dimensional model of Section 2. Influence functions are defined as follows. Thinking of the streamfunction ψ as a solution of a single-layer model function F to a topographic forcing η_B,

$$F(\psi) = \eta_B, \tag{10}$$

Symbol ψ_{ji} is denoted as the solution to Equation 10 but with a 100 meter increase in surface elevation at a grid box at latitude-longitude coordinate (φ_j, θ_i). So

$$F(\psi_{ji}) = \eta_B + \widehat{\eta}_{B_{ji}} \tag{11}$$

where $\widehat{\eta}_{B_{ji}}$ is the 100 meter increase in surface elevation at (φ_j, θ_i). Accordingly, the influence function is defined to be

$$\widehat{\psi}_{ji}(\varphi, \theta) \equiv \overline{\psi}^t_{ji}(\varphi, \theta) - \overline{\psi}^t(\varphi, \theta) \tag{12}$$

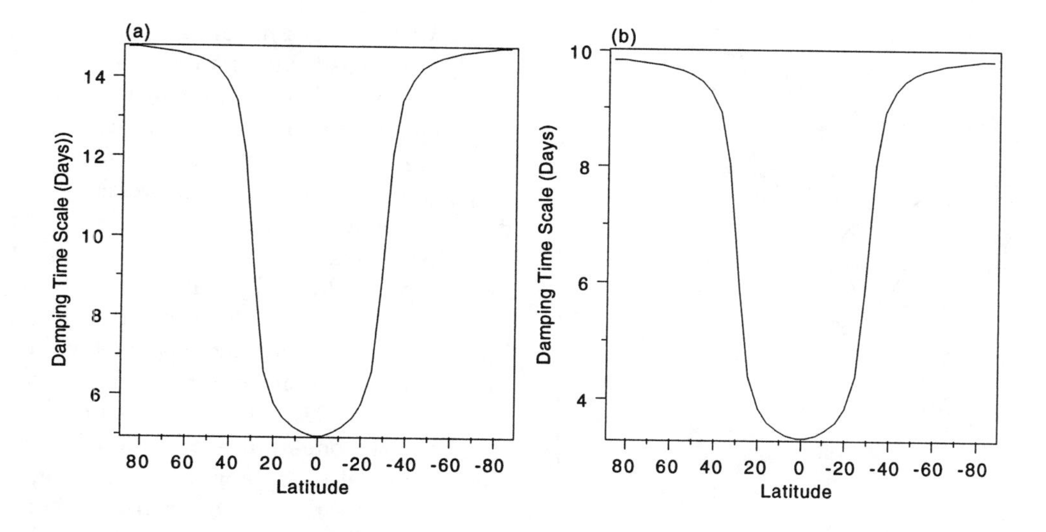

Figure 2. Graph (a) shows latitudinal profile of the vorticity damping time scale. Graph (b) shows latitudinal profile of the divergence and mass damping time scales.

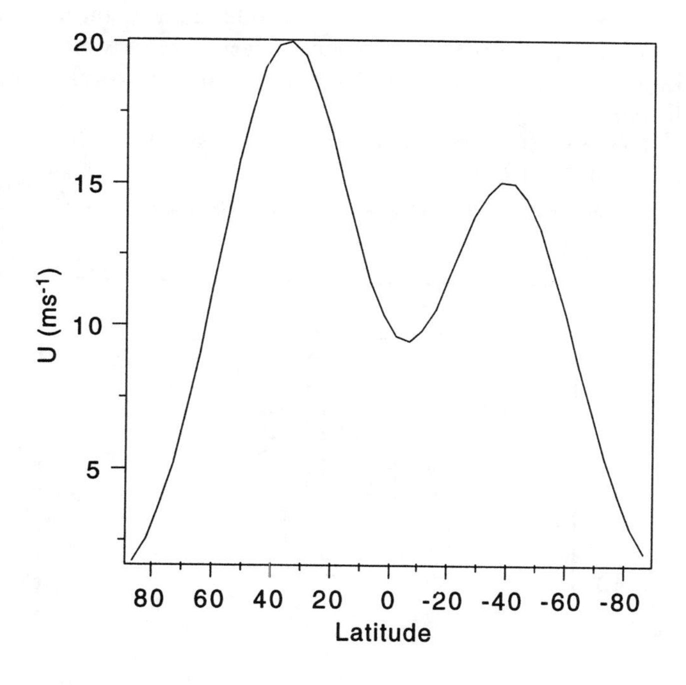

Figure 3. Zonal wind profile used in single-layer model on a sphere.

The influence function $\widehat{\psi}_{ji}$ is defined by the difference between the time averaged flow (time averaged streamfunction) over topography with a 100 meter addition to a grid box centered at (φ_j, θ_i) and the time averaged flow over normal topography. One may interpret the influence function $\widehat{\psi}_{ji}$ as the influence of region (φ_j, θ_i) on the global stationary wave field.

The change in stationary wave amplitude $\Delta|\psi'|$ in a two-dimensional single layer model on a sphere is found by projecting the influence function into the stationary wave field divided by the projection of the stationary wave field into itself,

$$\Delta|\psi'| = \frac{\int_0^{2\pi} \int_{-\pi}^{+\pi} \widehat{\psi}_{ji} \cdot \psi' R^2 \cos(\varphi) \, d\varphi \, d\theta}{\int_0^{2\pi} \int_{-\pi}^{+\pi} \psi' \cdot \psi' R^2 \cos(\varphi) \, d\varphi \, d\theta} \tag{13}$$

In Equation 13, R is the radius of the Earth, φ is the latitude, and θ is the longitude. The change in volume of topography $\Delta\eta_B$ is calculated as the volume of $\widehat{\eta}_{B_{ji}}$ divided by the total volume of topography in the model,

$$\Delta\eta_B = \frac{\int_0^{2\pi} \int_{-\pi}^{+\pi} \widehat{\eta}_{B_{ji}} R^2 \cos(\varphi) \, d\varphi \, d\theta}{\int_0^{2\pi} \int_{-\pi}^{+\pi} \eta_B R^2 \cos(\varphi) \, d\varphi \, d\theta} \tag{14}$$

A sensitivity measurement of 1 at (φ_j, θ_i) means, for example, that a 5% reduction of topographic volume taken from location (φ_j, θ_i) would reduce the model's globally averaged stationary wave amplitude by 5%.

A sensitivity map will be constructed from a 16 by 9 grid covering the Laurentide ice sheet and part of Greenland (Figure 5). Each grid box of Figure 5 has its own influence function that is obtained by the following procedure. First, the model is run long enough for the one-layer streamfunction to reach steady state.

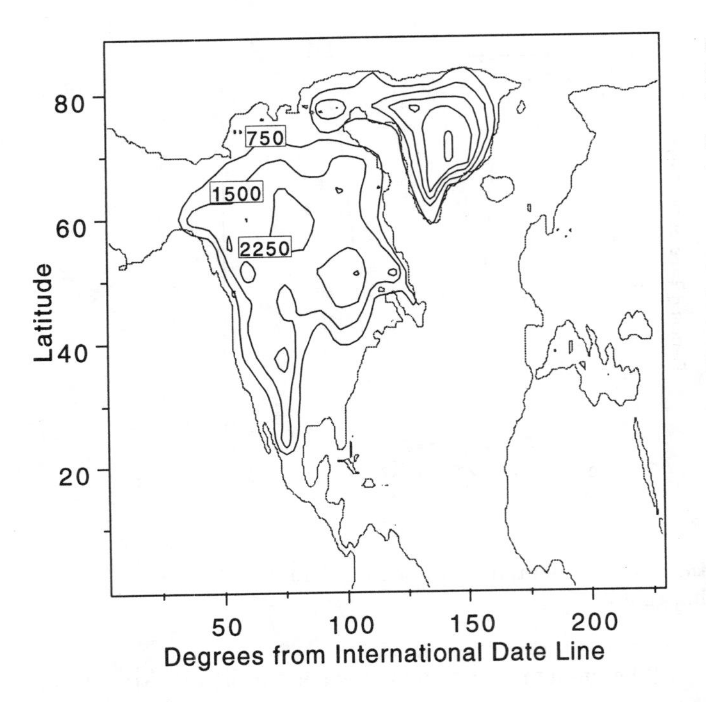

Figure 4. Ice-4G reconstruction of the Laurentide and Greenland ice sheets at Glacial Maximum, 21 ka before present (contour interval is 750 m). Reconstruction includes Rocky Mountains.

A 2000 day run is more than sufficient. Steady state model fields are saved and used as initial conditions for subsequent sensitivity measurements. The influence of grid box at coordinate (φ_j, θ_i) is found by raising the elevation of that grid box by 100 meters and recalculating the steady state streamfunction. For any one grid box, a steady state solution can be found in a 120 day model integration using the last 30 days for the time average. Finally, the procedure is repeated for the remaining grid boxes. Once all the 140 steady state streamfunctions are obtained, a sensitivity map can be generated from Equation 4.

The map of stationary wave sensitivity to the Laurentide and Greenland ice sheets is shown in Figure 6. The results demonstrate three points: First, the sensitivity map of Figure 6 shows the possibility of negative sensitivity. That is, the amplitude of the stationary wave increases with the lowering of the surface elevation of the topography. Second, the sensitivity map shows sensitivity values as high as 2 and as low as -2.5. Recall that a sensitivity of 1 means that the change in stationary wave amplitude scales with the volume of topographic perturbation. A sensitivity value of 2 means that stationary wave amplitude is two times more sensitive to topographic perturbations than if the amplitude scaled with changes in topographic volume. Third and finally, the sensitivity map is subdivided into a number of regions of varying sensitivity. The diversity of sensitivity values means that topographic depressions that develop in one region could affect stationary wave amplitude differently if the depressions happened to be displaced by a small distance across the ice sheet.

Additional experiments were conducted with varying mean topographic height as controlled by topography reduction factor h. These experiments are meant to illustrate how the average size of the LIS affects stationary wave sensitivity to regional changes in topography. In the limit $h \rightarrow 0$, winds become zonally uniform, similar to the one-dimensional model's winds. As h increases, model topography grows in height and begins to force zonal variations in the mean winds. These zonal variations influence the horizontal propagation of wave energy and can have a strong influence on stationary wave sensitivity to topography. These experiments are not meant to simulate Heinrich events since Heinrich events are thought to affect regional variations in topography rather than universal reductions in topographic height.

The sensitivity map was recalculated using two alternate values of h: a value of h that gives a 30% thinner ice sheet (Figure 7), and a value of h that gives a 30%

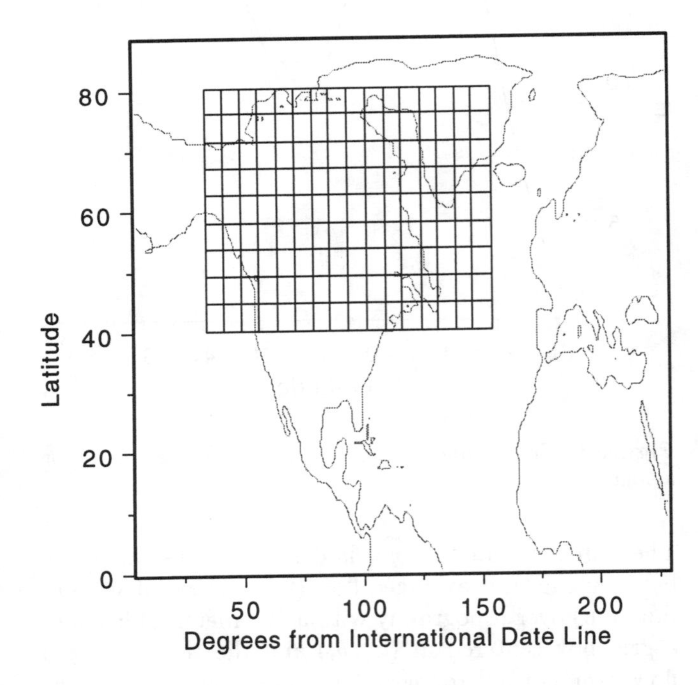

Figure 5. Map describing grid box locations where topography is modified to determine sensitivity of stationary wave amplitude to changes in topography.

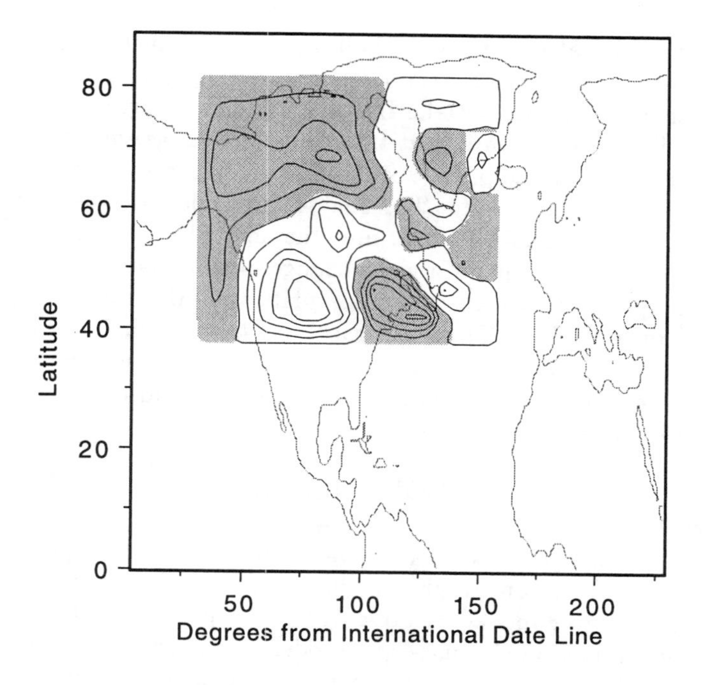

Figure 6. Sensitivity of global stationary wave amplitude (contour interval is 0.5 sensitivity units, negative values shaded) to Laurentide and Greenland ice sheet topography at Glacial Maximum.

thicker ice sheet (Figure 8). The range of sensitivity values for the thick ice sheet sensitivity map of Figure 8 is 3 to -3. This range is greater than the 2 to -2.5 range in sensitivity values that were calculated for Figure 6, the case with the standard topography thickness. In contrast, the thin ice sheet sensitivity map, Figure 7, shows a range of sensitivity values between 2 and -0.5. From the comparisons of the three sensitivity maps of Figures 6, 7, and 8, a trend emerges- Thickening an ice sheet enhances the range of stationary wave sensitivity to topography.

The three sensitivity maps also show the trend that increasing ice sheet thickness increases the mixture of areas with positive and negative sensitivity. Having mostly a single region of positive sensitivity, the sensitivity map for the thin ice sheet, Figure 7, shows the least amount of complexity. A latitudinal slice of Figure 7 around 45° North looks striking familiar to Figure 1 of the linear one-dimensional model which does not allow meridional propagation of wave energy. Longitudinal variations in the zonal wind have a strong influence on the horizontal propagation of wave energy [*Hoskins and Ambrizzi,* 1993]. These zonal variations scale with ice volume and likely cause the increase in the complexity of the sensitivity pattern seen in Figure 8.

The correspondence between the thin topography sensitivity map (Figure 7) and the linear one-dimensional sensitivity map (Figure 1) could be from a similar lack in zonal variations in each model's mean winds.

4. DISCUSSION

One argument for the potential relevance of changes in LIS topography to Heinrich event climate change is that a given volume of topography can control a disproportionate amount of the global stationary wave field. The volume of ice associated with a Heinrich event is estimated to be 1.25×10^{15} m^3 [*Alley and MacAyeal,* 1994]. This represents approximately 3.5 meters of sea level or 3% of the model topography. Sensitivity map of Figure 6 indicates that this volume can alter the globally averaged stationary wave amplitude by as much as ±6% depending on where ice is removed. The change in the stationary wave amplitude over the North Atlantic may be even greater than 6% given its close proximity to the LIS. The present model experiments focus on the LIS's potential to alter stationary wave amplitude. A more complete climate model is needed to evaluate how specific changes in the stationary wave field are

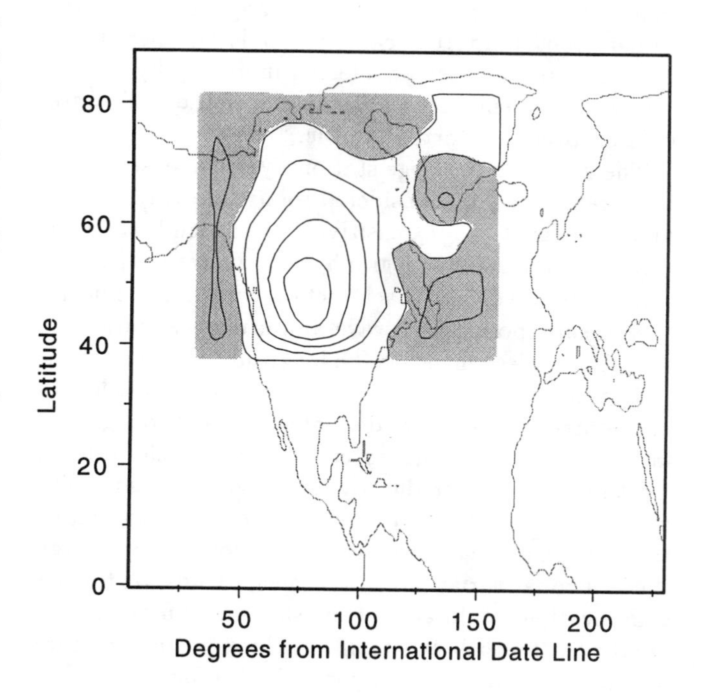

Figure 7. Sensitivity map of global stationary wave amplitude (contour interval is 0.5 sensitivity units, negative values shaded) to 30% thinner Glacial Maximum topography.

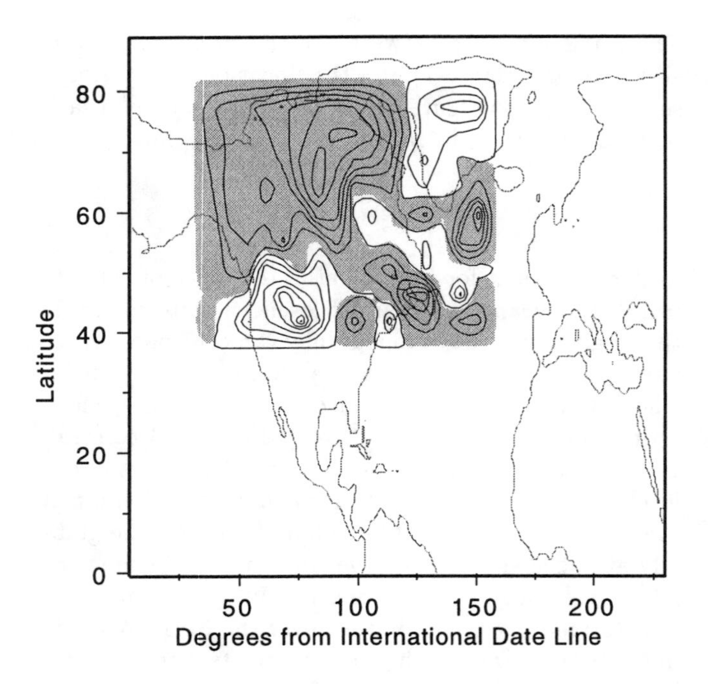

Figure 8. Sensitivity map of global stationary wave amplitude (contour interval is 0.5 sensitivity units, negative values shaded) to 30% thicker Glacial Maximum topography.

manifested in climate. Besides the way a change in stationary wave amplitude can affect poleward heat transport, climate can also be affected indirectly by stationary wave influence on sea ice extent or the fresh water balance over the North Atlantic.

The potential for large stationary wave sensitivity to topography may be considered robust in the sense that modeled variations in sensitivity are dependent on the way each section of topography contributes to the total stationary wave field. Each section contributes linearly by the superposition of wave amplitude. Variations in sensitivity arise whenever the longitudinal extent of the topography is larger than the stationary wave length and contributions of two distant sections of topography interfere. Less confidence can be put on the present model's ability to predict the sensitivity extrema. The position of sensitivity extrema is sensitive to factors that influence the horizontal propagation of wave energy such as zonal variations in the mean winds. A further complication would arise if the shifts in climate associated with Heinrich events affect, through the structure of the mean winds, the regions of LIS topography that control the atmosphere's stationary wave field.

One possible mechanism by which ice sheet instability could lead to warming over Greenland is suggested by the sensitivity maps. Those areas of the maps that show negative sensitivity indicate that a reduction in the height of the ice sheet leads to an increase in stationary wave amplitude. Because stationary waves contribute to poleward heat transport, an increase in stationary wave amplitude could lead to warmer temperatures over Greenland. The single-layer models used here are not adequate to determine the amount of warming that could occur for a given change in stationary wave amplitude. One may also speculate that Bond cycles [*Bond et al.*, 1993], which track long term cooling over Greenland after Heinrich events, represent the slow recovery of the LIS from the collapse of the Hudson Bay ice dome.

5. CONCLUSIONS

An analysis of stationary wave sensitivity to regional changes in topography in one- and two-dimensional single-layer models is presented. The purpose of the one-dimensional single-layer model is to illustrate, in a simple linear context, the effect of regional variations of large (longitudinally extended) topography on stationary wave amplitude. Each section of topography contributes to the total stationary wave field by the superposition of wave amplitude. The model demonstrates that through this process it is possible for large variations in sensitivity to exist. Wave superposition can explain the development of 'sweet spots' defined to be areas whose topographic alteration leads to significant changes in stationary wave amplitude over what may be expected if stationary wave amplitude scales with ice volume.

The purpose of the single-layer model on a sphere is to increase the realism of the one-dimensional model. It includes the influence of a spherical shell domain, zonally varying mean winds, and a realistic topography reconstruction at the Last Glacial Maximum on stationary wave sensitivity to topography. Sensitivity maps are developed to illustrate the influence of various topographic regions on globally averaged stationary wave amplitude. A wide range of sensitivity is found, including an increase of stationary wave amplitude with the removal of some amount of mass from regions with negative sensitivity. It is also speculated that the warming which occurs over Greenland after Heinrich events could be caused by an increase in atmospheric heat transport which is forced by the removal of ice mass from a section of LIS topography with negative stationary wave sensitivity to topography.

Substantial uncertainties remain concerning the importance of topography relative to other potential sources of stationary wave energy as well as specific predictions of which regions of the LIS contain sensitivity extrema. Despite these uncertainties, much can be learned through single-layer models of the potential for variations in LIS topography to be a factor in millennial-scale climate variability.

Acknowledgments. I am grateful to D. MacAyeal, N. Nakamura, and R. Pierrehumbert for their guidance on model development and interpretation. I also thank R. Webb and two anonymous reviewers for their suggestions to improve the manuscript. This research was supported by NSF Grant OPP-9321457.

REFERENCES

Alley, R. B., and D. R. MacAyeal, Ice-rafted debris associated with binge/purge oscillations of the Laurentide ice sheet, *Paleoceanography, 9*(4) 503-511, 1994.

Bond, G., W. Broecker, S. Johnsen, J. McManus, L. Labeyrie, J. Jouzel, and G. Bonani, Correlations between climate records from the North Atlantic sediments and Greenland ice, *Nature, 365* 143-147, 1993.

Branstator, G. W., Horizontal energy propagation in a barotropic atmosphere with meridional and zonal structure, *J. Atmos. Sci., 40* 1689-1708, 1983.

Cai, W., J. Syktus, H. Gordon, and S. O'Farrell, Response of a global coupled ocean-atmosphere-sea ice climate model to an imposed North Atlantic high-latitude freshening, *J. Climate, 10*(5) 929-948, 1997.

Cook, K. H., and I. M. Held, Stationary waves of the Ice Age climate, *J. Climate, 1*(8) 807-819, 1988.

Dong, B., and P. J. Valdes, Simulations of the Last Glacial Maximum climates using a general circulation model: prescribed versus computed sea surface temperatures, *Climate Dyn.*, (14) 571-591, 1998.

Fanning, A. F., and A. Weaver, Temporal-geographical meltwater influences on the North Atlantic Conveyor: Implications for the Younger Dryas, *Paleoceanography, 12*(2) 307-320, 1997.

Felzer, B., R. Oglesby, T. Webb III, and D. Hyman, Sensitivity of a general circulation model to changes in northern hemisphere ice sheets, *J. Geophys. Res., 101*(D14) 19077-19092, 1996.

Felzer, B., T. Webb III, and R. Oglesby, The impact of ice sheets, CO2, and orbital insolation on late Quaternary climates: sensitivity experiments with a general circulation model, *Quaternary Science Reviews, 17* 507-534, 1998.

Grose, W. L., and B. J. Hoskins, On the influence of orography on large-scale atmospheric flow, *J. Atmos. Sci., 36* 223-324, 1979.

Held, I. M., and M. Ting, Orographic versus thermal forcing of stationary waves: The importance of the mean low-level wind, *J. Atmos. Sci., 47* (4) 495-500, 1990.

Hoskins, B. J., and T. Ambrizzi, Rossby wave propagation on a realistic longitudinally varying flow, *J. Atmos. Sci., 50*(12) 1661-1671, 1993.

Hoskins, B. J., A. J. Simmons, and D. G. Andrews, Energy dispersion in a barotropic atmosphere, *Quart. J. Roy. Meteor. Soc., 103* 553-567, 1977.

Karoly, D. J., Rossby wave propagation in a barotropic atmosphere, *Dyn. of Atmos. Oceans, 7* 111-125, 1983.

Kutzbach, J., and P. J. Guetter, The influence of changing orbital parameters and surface boundary conditions on climate simulations for the past 18000 years, *J. Atmos. Sci., 43*(16) 1726-1759, 1986.

MacAyeal, D. R., A low-order model of the Heinrich event cycle, *Paleoceanography, 8*(6) 767-773, 1993a.

MacAyeal, D. R., Binge/purge oscillations of the Laurentide ice sheet as a cause of the North Atlantic's Heinrich events, *Paleoceanography, 8*(6) 775-784, 1993b.

Manabe, S., and R. J. Stouffer, Simulation of abrupt climatic change induced by freshwater input to the North Atlantic ocean, *Nature, 378* 165-167, 1995.

Manabe, S., and R. J. Stouffer, Coupled ocean-atmosphere model response to freshwater input: Comparison to Younger Dryas event, *Paleoceanography, 12*(2) 321-336, 1997.

Nigam, S., I. M. Held, and S. W. Lyons, Linear simulation of the stationary eddies in a GCM. Part II: The 'mountain' model, *J. Atmos. Sci., 45*(9) 1433-1452, 1988.

Paillard, D., and L. Labeyrie, Role of the thermohaline circulation in the abrupt warming after Heinrich events, *Nature, 372* 162-164, 1994.

Peltier, W. R., Ice Age paleotopography, *Science, 265* 195-201, 1994.

Rahmstorf, S., Rapid climate transitions in a coupled ocean-atmosphere model, *Nature, 372* 82-85, 1994.

Rahmstorf, S., Bifurcations of the Atlantic thermohaline circulation in response to changes in the hydrological cycle, *Nature, 378* 145-149, 1995.

Rind, D., Components of the Ice Age circulation, *J. Geophys. Res., 92*(D4) 4241-4281, 1987.

Shinn, R. and E. J. Barron, Climate sensitivity to continental ice sheet size and configuration, *J. Climate, 2*(12) 1517-1537, 1989.

C. Jackson, Atmospheric and Oceanic Sciences Program, Princeton University, Forrestal Campus, US Route 1, Princeton, New Jersey, 08542 (e-mail: csj@gfdl.gov)

Subtropical Water Vapor As a Mediator of Rapid Global Climate Change

Raymond T. Pierrehumbert

University of Chicago, Chicago, IL

This article surveys the essential features of atmospheric water vapor dynamics needed to address current issues regarding the possible role of water vapor changes in mediating climate fluctuations on the millennial to Milankovic time scales. The focus is on the subtropics, which afford the most interesting possibilities for significant feedbacks. The observed distribution of water vapor, the amount by which water vapor must change in order to cause a significant temperature change, and the physical factors that determine the water vapor content of the subtropical atmosphere are discussed. It is shown that halving the subtropical relative humidity would lead to a 2.5K cooling of the tropics, while doubling it would lead to a 3K warming. The humidity content of the subtropics could be reduced by enhancing subsidence, reducing transient eddy activity, or contracting the convective region. Further work is needed to determine which, if any, of these changes occur in concert with the observed millennial and longer scale climate fluctions.

1. INTRODUCTION

High resolution climate proxy records covering the past few hundred thousand years have revealed that, embedded within the long-period Milankovic scale fluctuations are a bewildering variety of persistent shorter period fluctuations without any obvious tie to an external forcing mechanism of the climate system. Climate variations of this sort have shown up in polar and tropical ice cores, corals, and ocean sediment cores. Geochemical signatures of temperature and precipitation variations have been found in alkenones, planktonic and benthic foraminifera $\delta^{18}O$ and $\delta^{13}C$, glacial $\delta^{18}O$ and δD, foramanifera assemblages, dust, ice-rafted debris and methane, to name a few. Climate variations often involve rapid transitions to vastly different states, and can involve time scales from decades to centuries to millenia and on upward. Indeed, it is not at all clear that there is any real spectral gap between what we think of as "weather" and the ponderous 100,000 year waking and sleeping of the great ice sheets. Certain millennial scale events have names, including the Dansgaard-Oeschger events, the Heinrich events and the Bond cycles; myriad others still await christening. Milankovic-scale cycles exhibit a fair amount of global synchrony, and the globalization effect of CO_2 fluctuations can probably account for a good bit of this synchrony. The shorter period events do not exhibit such a clear and consistent spatial pattern, but there is considerable evidence that they do not represent purely local phenomena. They are evidently part of a loosely coupled global system. Evidence of this nature is presented in numerous contributions elsewhere in the present volume.

The best studied of candidate mechanisms for millennial scale climate fluctuations involves the switching on and off of North Atlantic Deep Water formation. However, as reviewed by [Broecker (1994)], and elsewhere in the present volume, it is not entirely clear that

Mechanisms of Global Climate Change at Millennial Time Scales
Geophysical Monograph 112

the mechanism can yield a sufficiently strong response outside the North Atlantic region to account for the observations. General circulation model evidence suggests that with presently understood physics, even a change as major as introducing a Northern Hemisphere ice sheet may induce only a weak response in the Northern Hemisphere tropics, and hardly any response at all south of the Equator [Manabe and Broccoli (1985)]. It is not certain that the millennial scale fluctuations observed in different parts of the globe are all related to each other, but it is certain that such fluctuations are observed just about everywhere, so one should at least entertain the possibility that the NADW picture alone cannot account for everything that is going on. One either needs independent fluctuation mechanisms in the various regions, or a means of globalizing the impact of the NADW signal, or some combination of the two.

Carbon dioxide, methane and N_2O do not seem to exhibit fluctuations of sufficient amplitude to couple the different regions on millennial time scales. In view of the strong jet streams, atmospheric transports alone are probably sufficient to spread the influence of a strong regional event over the entire extratropical hemisphere in which it occurs. There is evidence that a combination of atmospheric and oceanic heat transports can extend the effect of a Northern extratropical anomaly at least part way into the tropics. Almost any change in the circulation induced will have repercussions for atmospheric water vapor. Water vapor has a rapid response time, and because it it such a good greenhouse gas, it has a large leverage over climate. The tropics sit in a good place to couple climate changes between the Northern and Southern hemispheres. Further, because tropical climate is dynamically constrained to be quite zonally uniform, the tropics provide a natural mechanism for spreading climate effects across longitudes. Therefore, it is natural that thoughts should turn in the direction of tropical water vapor when faced with the necessity of finding mechanisms for globalizing rapid climate change. One is on the lookout especially for "threshold" effects which could switch on or off suddenly, and give rise to rapid changes of climate.

This paper surveys the basic aspects of water vapor and climate that provide essential background for addressing questions regarding the role of the atmosphere in millennial scale climate change. Most of the subject matter is also relevant to the general problem of cooling of the Tropics during glacial times. With regard to mechanisms, we will be content to point out some points of vulnerability of climate, which are candidates for amplifying and globalizing climate variations. Our focus is on the subtropics, which exhibit a rich interplay of dynamics, thermodynamics and radiative effects offering much scope for interesting transitions. We will be concerned with moisture in the free atmosphere rather than boundary layer; free-atmosphere moisture has a sharp gradient between the subtropics and the convective region, whereas boundary layer moisture is very uniform. Further, it is the free-atmosphere water vapor which has the most leverage over the radiation budget of the planet.

Section 2 provides an overview of the observed tropical humidity pattern, and of the basic workings of tropical climate. The impact of water vapor changes on tropical temperature and circulation is discussed in Section 3. Mechanisms determining the subtropical water vapor content, and ways in which they may change in a changing climate, are laid out in Section 4. In Section 5 a few remarks are offered as to where water vapor may fit into the bigger picture of millennial climate fluctuation, and on the rocks that need to be turned over in the search for a theory of these enigmatic events.

2. THE OBSERVED HUMIDITY PATTERN

Dynamically, the Tropics is the region of the Earth that lies under the influence of the large scale overturning Hadley and Walker circulations. Roughly speaking, it extends from latitudes 30S to 30N. The division of the tropics into convective (ascending) and nonconvective (subsiding) regions is crucial to the understanding of tropical climate in general, and tropical water vapor feedbacks in particular. This situation contrasts with the extratropics; there, mobile transient baroclinic eddies dominate the climate, and large scale geographically fixed regions of preferred ascent or descent do not play a prominent role. Throughout the following, we will interchangeably refer to the "non-convective region" of the tropics as the "subsiding region," "the subtropics," or "the dry pool."

Only a small portion of the tropics is subject to direct moistening by nearby deep convection. This can be seen clearly from the climatological precipitation pattern, a typical example of which is shown in Plate 1. In the tropics, heavy precipitation (a marker of deep convection) is confined to three zones: the narrow zonally oriented ITCZ band, the Pacific warm pool region, and continental rainfall regions in S. America and Africa. These zones make up about one third of the area of the tropics. The rest of the tropical free atmosphere must get its moisture by some more indirect means.

The convective region is characterized by mean ascent, with the resulting adiabatic cooling balancing the

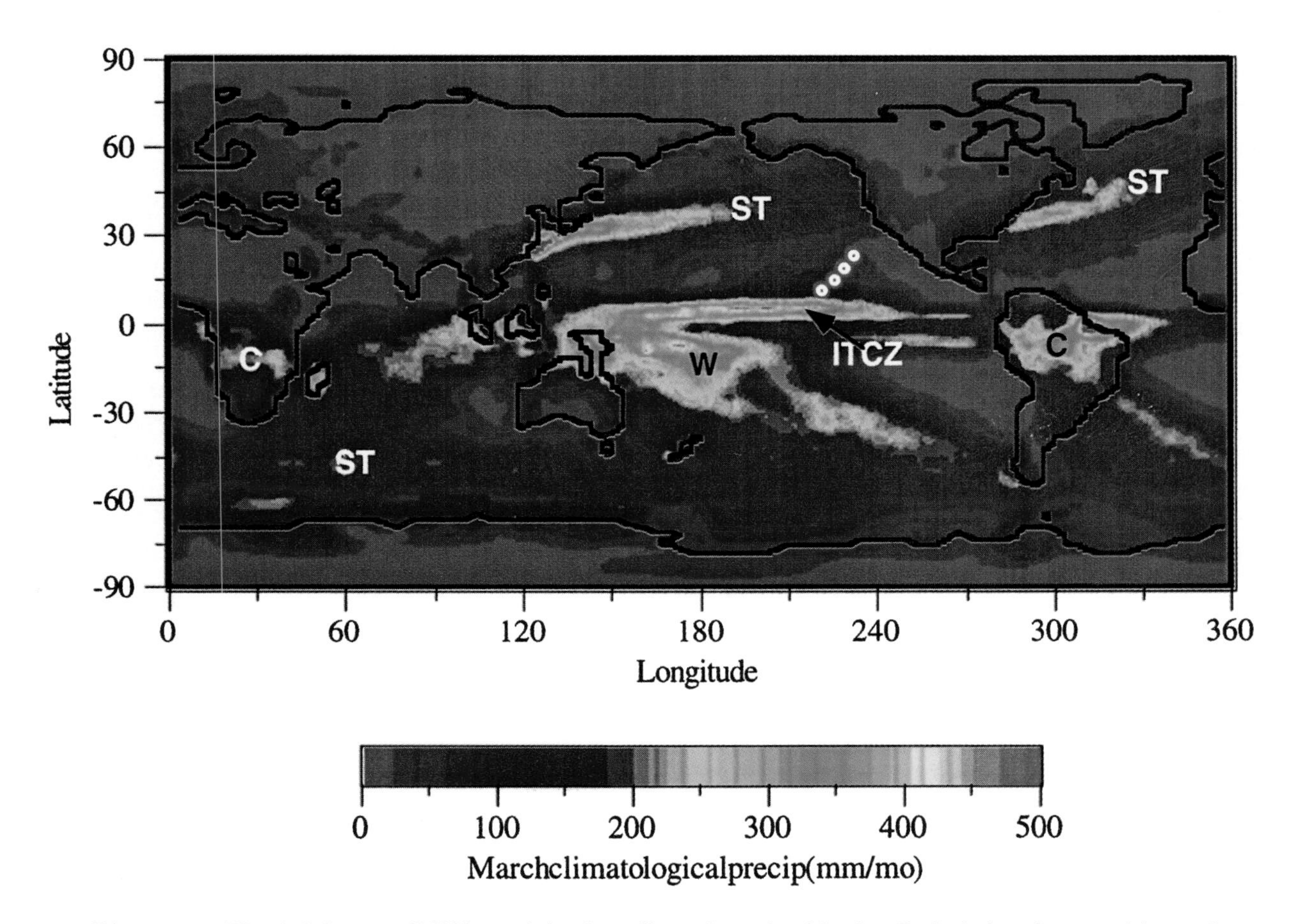

Plate 1. Blended Legates/MSU precipitation climatology for March. Red circles show positions of sondes discussed in the text. The principal convective regions are denoted as follows: ITCZ = Inter Tropical Convergence zone. W = Pacific Warm Pool. C = Tropical Continental. ST = Midlatitude Storm Track.

latent heat release due to precipitation. The nonconvective region is characterized by mean descent, with the resulting adiabatic warming balancing infrared radiative cooling. We will amplify on this picture in Section 3. For now, the key point to keep in mind is that, in the absence of a compensating moisture source, subsidence would cause the subtropical water vapor mixing ratio to relax to the tropopause value - around 10^{-5} - over the course of about 30 days. This relaxation occurs because mixing ratio (also known as specific humidity) is conserved following parcel trajectories. The subsidence thus brings down arid air from the tropopause level, and this air is replaced by outflow from the convective region, which, for consistency with this picture, must also be at the tropopause level. The observed subtropics are dry, but not so dry as would be indicated by this process, so there is evidently a pathway for moist air to reach the subtropics without being wrung out through the cold tropopause.

In Plate 2 we show the pattern of 500-700mb mean moisture mixing ratio for the period March 24-28, 1993. This field was reconstructed from TOVS polar orbiter satellite data. There is a close association between tropical moist regions and regions of high climatological precipitation, even though the moisture data is from a single five-day period. The moist regions generally spread somewhat beyond the high precipitation zones, particularly in the Southern hemisphere tropics. To the north of the ITCZ, there is a very sharp boundary between moist, convective air and the dry pool, both in the Atlantic and in the Pacific. The dry subtropical air in places has specific humidity below .001, but more typically has specific humidity between .001 and .002 at the 500-700mb level. Air with a mixing ratio of .001 would be saturated at the 280mb level, and presumably originated in outflow from the convective region near that level before subsiding to the layer analyzed in Plate 2. Air with a mixing ratio of .0015 would originate from near the 305mb level. The subtropical air is highly undersaturated, with relative humidities in the range of 10%-15%, with occasional values even drier. We have examined the corresponding TOVS data up to 200mb, and found that the mixing ratio is not constant with height in the subtropics, as would be expected if all the outflow were at a single high-altitude level. Instead, the mixing ratio decreases gently with height, in a manner suggestive of moisture injection at multiple levels. The vertical profile of mixing ratio is intermediate between a constant mixing-ratio model and a constant relative-humidity model.

It is important to recognize that the amount of water required to moisten the subtropics is truly tiny. With a mixing ratio of .001 throughout the 500-700mb level, the water content of this layer is equivalent to a layer of only 2 mm of liquid water. Higher layers contain even less water. If we take a typical rainfall rate of 300 mm/month and divide by three to account for the fact that only about a third of the area of the tropics experiences significant precipitation, we find that only 2% of the moisture in the rainfall needs to be retained in the atmosphere over the course of a month in order to achieve the observed moisture level. Clearly, water supply is not the limiting factor in subtropical water vapor. Thus, it would be incorrect to infer any relation between changes in tropical precipitation rate, and changes in the subtropical moisture content.

To keep the discussion firmly grounded in reality, we will make illustrative use of four E. Pacific subtropical soundings obtained during the CEPEX experiment (Williams 1993). CEPEX was a convective-region experiment, but fortunately these four "leftover" sondes were released into the interesting subtropical region on the homeward voyage of the Vickers ship. Their positions are shown in Plates 1 and 2, and they were released at 00Z on March 24, 25, 26 and 27, proceeding from southernmost to northermost. One can not base a comprehensive picture on just four soundings, and what we shall present does not aspire to do so. Nonetheless, the nature of the tropical climate is so clearly drawn that all the "dry pool" concepts that we need to deal with are abundantly revealed in these soundings.

The temperature soundings are shown in Figure 1, and the relative humidity profiles are shown in Figure 2. In accordance with common practice, the relative humidities are reported relative to saturation over liquid , even at altitudes where the dominant condensate would be ice. Because of the limitations of the humidity sensor, the values reported above 300mb are of dubious validity, and should probably be ignored. A comprehensive discussion of the sensor, and of humidity soundings during CEPEX (apart from these four subtropical soundings) can be found in [Kley *et al* (1997)].

Although the soundings span approximately 20 degrees of latitude, the thermal structure shows very little variation amongst them. The farthest north is only slightly cooler than the one closest to the Equator, and all have the tropopause at very nearly the same altitude. This uniformity illustrates the familiar point that, above the boundary layer, the Tropics cannot support large horizontal temperature gradients. The reasons for temperature uniformity in the Tropics are part of well-established tropical lore, which one can find discussed in [Held and Hou (1980)], [Emanuel (1995)], and [Pierrehumbert (1995)]. Because the thermal structure is so

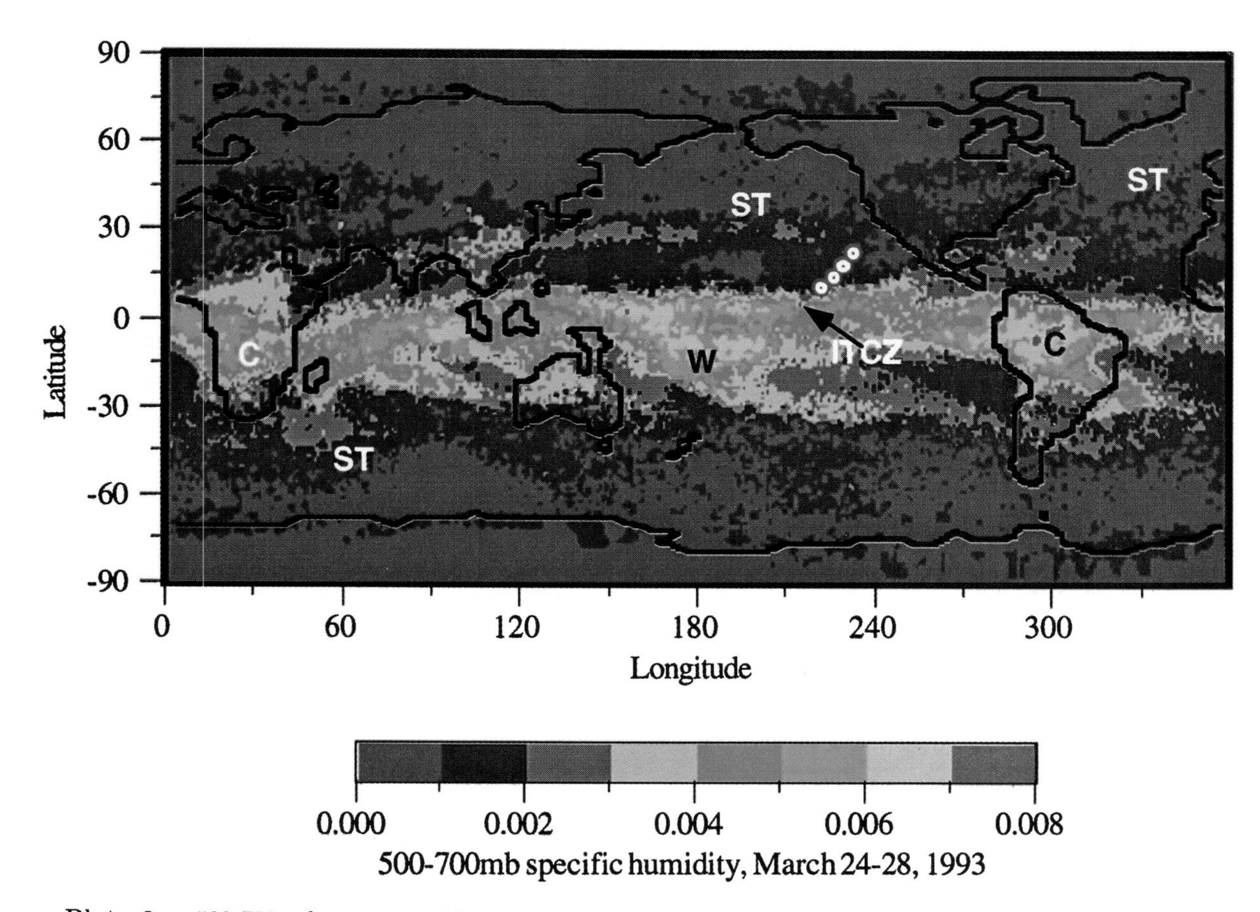

Plate 2. 500-700 mb mean specific humidity from TOVS satellite data, for the mean of the period March 24-29, 1993. Annotations as for Figure 1.

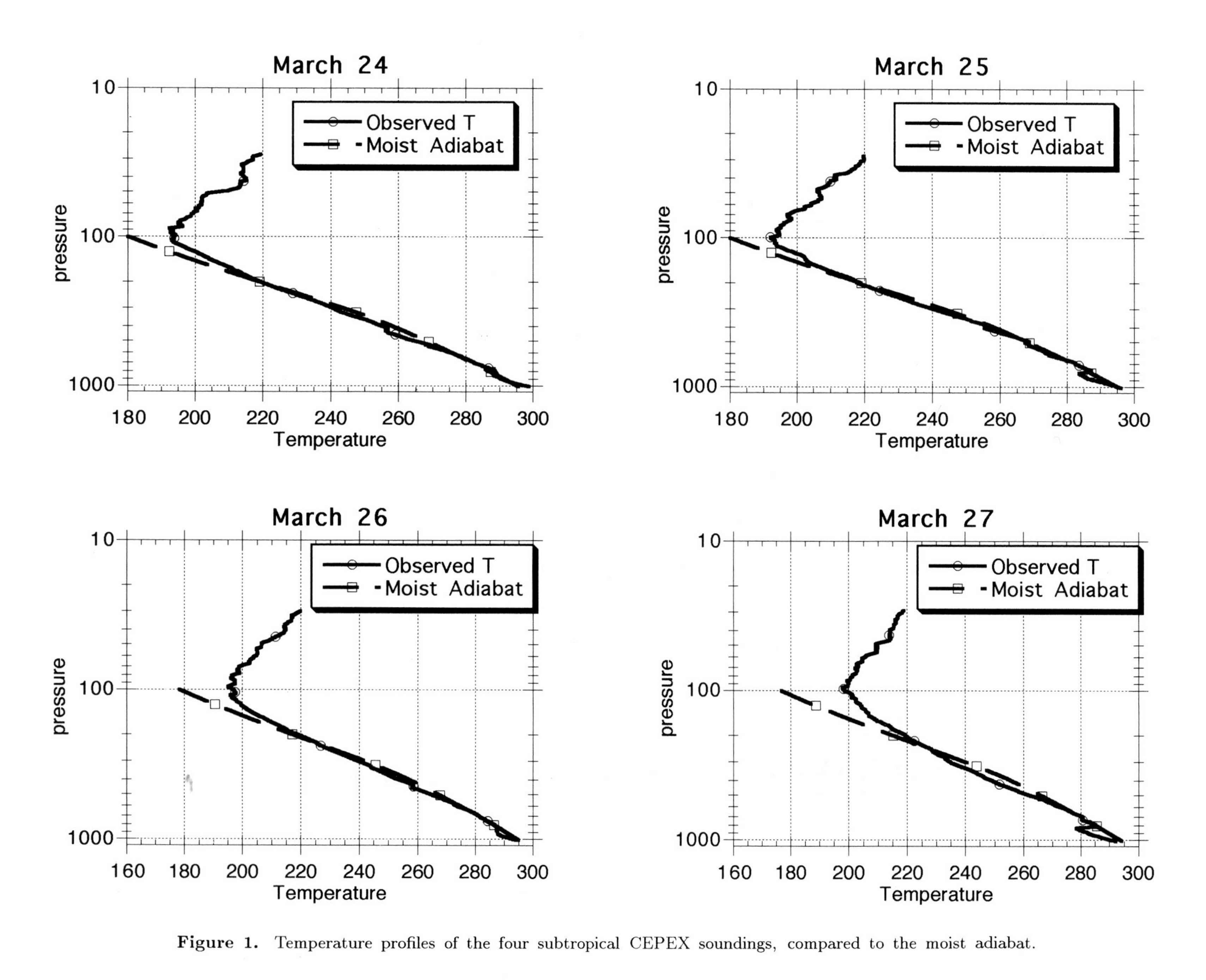

Figure 1. Temperature profiles of the four subtropical CEPEX soundings, compared to the moist adiabat.

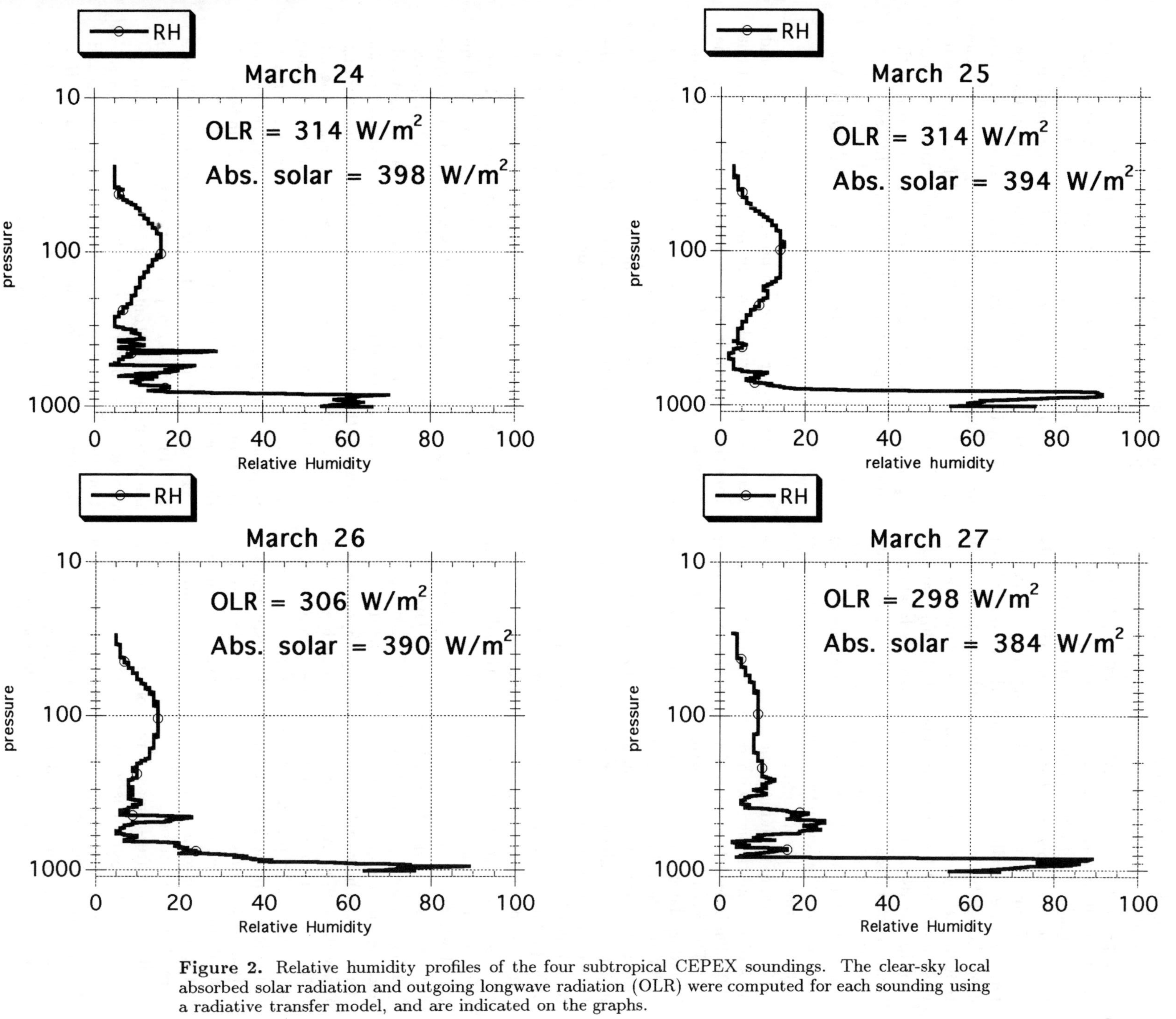

Figure 2. Relative humidity profiles of the four subtropical CEPEX soundings. The clear-sky local absorbed solar radiation and outgoing longwave radiation (OLR) were computed for each sounding using a radiative transfer model, and are indicated on the graphs.

tightly coupled across the whole Tropics, one can make considerable progress in understanding tropical climate by employing radiation budgets averaged over the entire tropics while ignoring regional temperature variability.

Along with each temperature sounding, we also show the saturated moist adiabat corresponding to the observed 700mb temperature. It is significant that the temperature profiles closely follow the moist adiabat even though the air is highly undersaturated, and even though the region does not experience convection. It was established by [Xu and Emanuel (1989)] that this principal holds throughout a comprehensive database of tropical soundings. Ultimately, the heating that maintains the moist adiabat in the subtropics is due to condensation, but not to local condensation; the effects of remote condensation in the convective region are transmitted to the subtropics via their effect on subsidence.

The moisture soundings underscore the general dryness of the subtropics, with relative humidities in the 800-300mb level frequently falling to values as low as 5%. There is a very sharp distinction between the moist boundary layer air (with relative humidity as high as 80%) and the arid air aloft. This generally dry background is interrupted by a number of spikes of relatively moist air. Though some spikes are very thin in the vertical,they can sometimes be thicker, as in the March 27 sounding which has a broad moist intrusion centered on 500mb. The complex vertical structure of the humidity suggests the primacy of horizontal advection as a moisture source for the subtropics. Differential advection in the vertical can create such structures by replacing a thin layer of dry subtropical air with moist air of convective origin, without disturbing the other layers.

This anecdotal picture of the extent of the dry pools, and their degree of dryness, is entirely compatible with the more extensive analysis presented by [Spencer and Braswell (1997)]. A recurrent worry in such endeavors is the reliability of radiosonde moisture data at the low values needed to properly characterize the subtropics. Reliable, comprehensive satellite-based moisture data is only now becoming widely available. The characterization of subtropical moisture will improve in the future, but it will take some years to build up an accurate picture of the climatology.

Although the free-atmosphere subtropical air is arid, the boundary layer air is moist both in the convective and nonconvective regions, with relative humidities in the vicinity of 60-70% (see, e.g. [Broecker (1997)]). Keeping this in mind, we will employ two idealized models of of the subtropical tropospheric humidity profile in subsequent discussion. In both models, the boundary layer air (below 850mb) is kept at fixed and high relative humidity. In the "RH model," we specify the the moisture profile so as to maintain fixed relative humidity in the rest of the troposphere. In the "outflow model" we instead specify a fixed specific humidity q above the boundary layer, with the exception that the humidity is set to saturation for tropospheric altitudes greater than the altitude z_s where the given specific humidity becomes saturated. This model corresponds to saturated convective region outflow at altitudes above z_s only, with moisture communicated to the rest of the subtropical troposphere only through subsidence.

3. CLIMATIC IMPACT OF SUBTROPICAL HUMIDITY

Because water vapor is a potent greenhouse gas, and because the thermal coupling in the tropics is so tight that the temperature changes in concert throughout the tropics, the dryness of the subtropics has a profound cooling effect on the whole tropics. The importance of the subtropics in climate was starkly revealed in the simplified models pursued in [Pierrehumbert (1995)]. Insofar as the tropics are about half the area of the planet, and insofar as they provide a boundary condition for the extratropics, the influence can be expected to extend to the rest of the planet. Curiously, the issue of the water vapor content of the subtropics, and the corresponding climate sensitivity noted in [Pierrehumbert (1995)], has been used in support of the claim that general circulation models are at risk of overpredicting the response to the doubling of CO_2 [Spencer (1997); Spencer and Braswell (1997)]. It is not as widely appreciated that there is an equal or greater scope for errors in the direction of *underestimating* the warming. For example, the water vapor feedback enhancing climate change would be stronger than current thinking suggests, if the dry pools were to contract in area in response to a warming planet. Given that the factors governing the dry pool area are not at all understood, this possibility must give us pause.

Be that as it may, it is the very sensitivity of climate to subtropical water vapor that brings the matter to our attention in the context of millennial scale climate change. The questions we seek to answer in this section are: How do the subtropical humidity content and radiation budget affect the tropical circulation? How much do we have to reduce the subtropical humidity to significantly cool the tropics (all other things being equal)? How much do we need to increase the subtropical humidity to substantially warm the tropics?

3.1. *Radiative cooling and tropical circulations*

Climate is more than just temperature, and it should be kept in mind that extratropical climate forcings could have a more profound impact on tropical circulations than on the tropical temperature itself. This point is especially important to the interpretation of ice-core methane records, which probably reflect changes in precipitation over tropical land areas at least as much as they reflect tropical temperature. Our main preoccupation in this Section will be with the strength of the subtropical subsidence, because the subsidence strength provides a convenient measure of the intensity of tropical circulations, and because it will turn out to be one of the key players in determining the subtropical humidity content.

In order to fully understand the effects of greenhouse gases on climate, it is essential to keep in mind that, under clear sky conditions, the troposphere of a planet generally experiences infrared radiative cooling throughout its depth; in equilibrium this cooling must be balanced by a compensating warming term, and the nature of the balance in large measure determines the character of the climate response. Under some common circumstances, increasing the concentration of a greenhouse gas intensifies the infrared cooling rate of the troposphere. This point is illustrated in Figure 3, where we show the upper tropospheric radiative cooling rate as a function of tropospheric humidity content for the March 24 temperature sounding. These calculations were performed using the NCAR radiation model [Kiehl and Briegleb (1992)] under clear sky conditions, with stratospheric and boundary layer (below 850mb) humidity held fixed at the observed values. For the rest of the troposphere, the observed moisture profile was replaced by a series of idealized profiles corresponding either to the RH model or the outflow model. To make it easier to compare results between the two moisture models, we have plotted the radiative cooling against the 700mb specific humidity for both models. A specific humidity of .001 corresponds to a relative humidity of 11% at 700mb. For both models the radiative cooling becomes stronger as humidity is increased. The two models yield somewhat different patterns of sensitivity to moisture, but in neither case is the sensitivity extreme. In the RH model, it is necessary to vary moisture by a factor of 100, between .0001 and .01, to change the radiative cooling rate from .8K/day

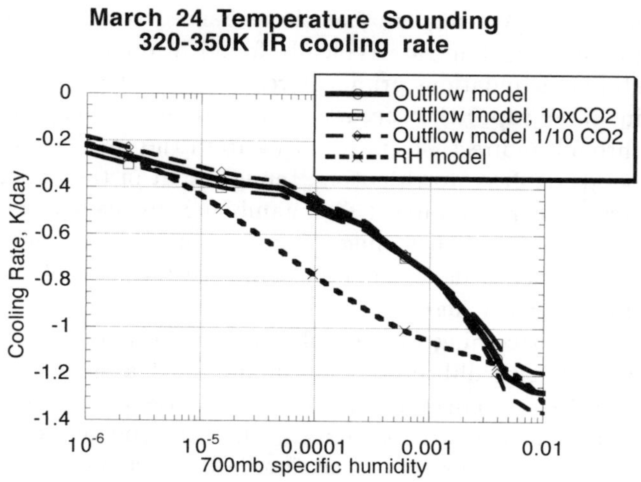

Figure 3. Upper tropospheric infrared radiative cooling for the March 24 temperature sounding, as a function of humidity. Results are shown for the RH model and the outflow model. For the outflow model, the cooling is computed also with 10 times present CO_2 and 1/10 present CO_2. The cooling is averaged over the isentropic layer 320K-350K (about 700-100mb).

to 1.2K/day. For comparison, the cooling rates computed using the actual observed humidity for all four sondes are: .94K/day for the 24th, .91K/day on the 25th, .95K/day on the 26th and .92K/day on the 27th. The uniformity of these radiative cooling rates suggests there may be some profit in making simplified models of the subtropics based on uniform subsidence. The values can be matched with a 700mb specific humidity of .001 in the RH model, or .003 in the outflow model.

The factors governing the cooling rate are subtle, and dependent on the vertical structure of temperature and the greenhouse gas. This importance of the latter is evident from the comparison of the RH with the outflow models. A further appreciation of the subtlety can be obtained from looking at the effects of CO_2 changes on radiative cooling, since this gas unlike water vapor has a constant mixing ratio throughout the whole atmosphere. In Figure 3 we show radiative cooling rates for the RH moisture model with 10x present CO_2 (3300 ppmv) and 1/10 present CO_2 (33ppmv). This is an extreme variation, corresponding at the high end to levels believed to have prevailed during the Cretaceous, and at the low end to values far lower than have ever occurred on Earth (noting that glacial-age CO_2 only falls to 180ppmv). Despite the large change, the upper tropospheric radiative cooling hardly changes at all, and the sign of the effect of CO_2 actually varies with the humidity. We suggest that the enhanced tropospheric radiative cooling rate typically found in doubled CO_2 climate simulations primarily results from the increase in water vapor that occurs as the climate warms, and through indirect effects resulting from the rise in temperature. It is worth noting, though, that in the upper stratosphere, elevated CO_2 unambiguously leads to intensified radiative cooling [Ramaswamy *et al.* (1996)], and in fact leads to an actual temperature drop of the upper stratosphere.

A shortcoming of the calculation shown in Figure 3 is that equilibrium is not maintained as humidity or CO_2 is increased. As the greenhouse gas concentration increases or decreases, the atmosphere would have to warm or cool to accomodate the change in infrared trapping, whereas the calculations were carried out with fixed temperature. Increasing the temperature on the one hand increases the infrared emission, which enhances radiative cooling. On the other hand, it increases the infrared flux upwelling from warm layers nearer the surface, which upon absorption would decrease the radiative cooling. *A priori*, it is not clear which effect wins, and the answer can be subtly dependent on the temperature and greenhouse gas profile. Fortunately, an elementary result can be obtained which links the mean tropospheric cooling rate to the radiation balance at the surface and at the top of the atmosphere. Let $F(p) = I^+ - I^-$ be net net infrared radiation flux (upward positive). Then, the infrared heating rate in degrees K per unit time is

$$H_{IR} = \frac{g}{c_p}\frac{d}{dp}F(p) \tag{1}$$

where c_p is the specific heat of air at constant pressure and g is the acceleration of gravity. Then, integrating over the depth of the troposphere yields

$$\begin{aligned} \bar{H}_{IR} &\equiv \frac{1}{p_s - p_t}\int_{p_t}^{p_s} H_{IR}dp \qquad (2) \\ &= \frac{g}{(p_s - p_t)c_p}F(p_s) - \frac{g}{(p_s - p_t)c_p}F(p_t) \end{aligned}$$

where p_s is the surface pressure and p_t is the tropopause pressure. If the stratosphere is optically thin in the infrared, then $F(p_t)$ is approximately equal to $I^+(p = 0)$, which is also known as "outgoing longwave radiation," abbreviated "OLR". The OLR is positive everywhere, since there is no significant infrared source from space, and its average over the whole planet must equal the absorbed solar radiation, insofar as the planet as a whole is nearly in equilibrium. The remaining term $F(p_s)$ in Eqn. (2) is the net infrared imbalance at the surface. It tends to be small for the following reason. Turbulent fluxes of evaporative and sensible heat keep the surface temperature rather close to the overlying air temperature. Further, the high water vapor content of boundary layer air means that the boundary layer is optically thick in the infrared, and hence radiates back to the surface most of the infrared welling upward from the surface. If this balance were perfect, $F(p_s)$ would vanish, and the mean infrared cooling of the troposphere would be simply the OLR divided by the heat capacity of the troposphere per unit area. Except in the rare circumstances where $F(p_s)$ is not small, the OLR term dominates Eqn. (2) and the troposphere experiences a net infrared radiative cooling, simply because infrared energy drains out the top of the atmosphere faster than it is replenished at the bottom. It is the turbulent coupling of the surface to the overlying air that supplies the necessary ingredient for the latter part of this general argument. As an example of the application of Eqn. (2), an OLR of 240 W/m^2 and a surface cooling of 50 W/m^2 yields a mean tropospheric radiative cooling of 2K/day if p_t is taken to be 200mb. This whole-troposphere average cooling is distinctly greater than the *mid-tropoposheric* cooling directly computed for the CEPEX soundings, because the cooling rate tends to increase sharply toward the ground.

The behavior of the infrared radiative cooling depends intimately on what is going on at the surface of the planet, and so the surface budget bears closer scrutiny. The equilibrium surface budget is

$$F(p_s) + E + kc_p(T_s - T_a) = S(p_s) + F_o \tag{3}$$

where E is the evaporative heat flux out of the surface, k is the wind-dependent sensible heat transfer coefficient, T_s is the surface temperature, T_a is the low-level air temperature, F_o is the implied heat flux due to ocean currents (set to zero over land), and $S(p_s)$ is the solar radiation flux absorbed at the surface. If the right hand side of Eqn.(3) is held fixed, the combination of evaporative and sensible heat flux must go up as $F(p_s)$ goes down, and conversely. Except over dry, hot deserts, the evaporation dominates the sensible heat flux in the warmer parts of the world. As $F(p_s)$ approaches zero, the evaporation dominantly balances the surface solar heating corrected for oceanic internal fluxes; the latter allow the evaporation to greatly exceed locally absorbed solar radiation over the Gulf Stream, for example.

Using the same Legates/MSU precipitation climatology shown for March in Plate 1, it is found that the

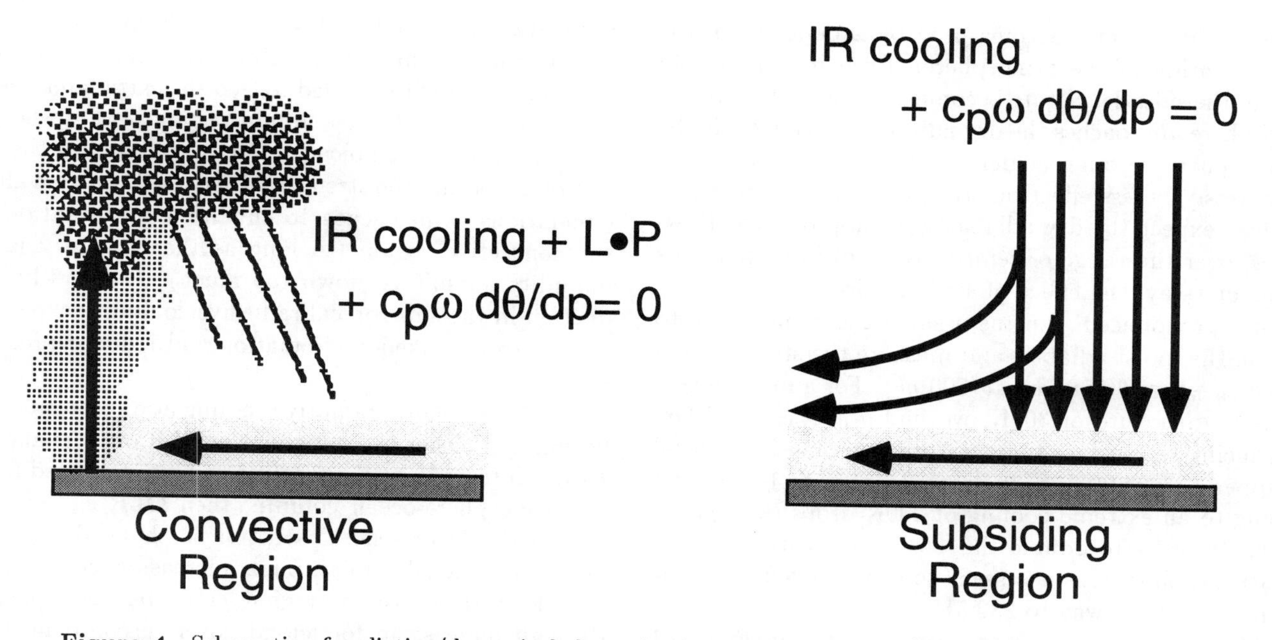

Figure 4. Schematic of radiative/dynamic balance in the convective vs. nonconvective region of the Tropics. In this figure L is the latent heat of condensation, P is the precipitation rate, ω is the pressure velocity and θ is the dry potential temperature.

global mean precipitation at present is .098 meters per month, corresponding to an evaporative flux of 94.5 W/m^2, which is already nearly half of the roughly 200 W/m^2 of solar radiation which is absorbed globally at the surface. In the tropics, the precipitation is equivalent to an evaporative flux of 111 W/m^2, which accounts for an even higher proportion of solar radiation reaching the surface, leaving typical values of $F(p_s)$ of 60 W/m^2 or less. Hence there is rather limited scope for increasing the tropospheric radiative cooling in a warmer world. Given typical subtropical OLR values of 240 W/m^2, the tropospheric radiative cooling would only increase 25% even if $F(p_s)$ were zeroed out completely.

The general increase of precipitation with temperature discussed in the preceding paragraph is at the heart of the CO_2 thermostat regulating the temperature of the planet over geological time scales [Berner (1994)], since removal of atmospheric CO_2 by weathering into carbonate rocks is proportional to precipitation rate over land. In thinking about this thermostat, it would be well to keep in mind the limitations on precipitation increase imposed by Eqn. (3). There is no such intrinsic limitation on the *cooling* side however. In a sufficiently cold (or dry) world, $F(p_s)$ could be increased until the tropospheric radiative cooling were reduced to the point where it balances the sensible heat flux term in Eqn. (3), and precipitation is eliminated entirely.

The preceding should impart some intuition as to the circumstances under which tropospheric radiative cooling intensifies. What are the implications for the circulation of the tropics? The fact that increasing the greenhouse gas concentration increases atmospheric radiative cooling may seem paradoxical. How can the atmosphere warm in the face of a cooling term which increases in magnitude? The atmosphere can warm despite increased radiative cooling, because this cooling is balanced by other heating terms. In the tropics, the compensating heating is provided primarily by adiabatic compression due to subsidence (in the nonconvective regions) or by latent heat release (in the convective regions), as depicted in Figure 4. Though there are additional heating and cooling terms, including heating due to atmospheric absorption of solar radiation, and cooling due to import of cold extratropical air by eddies, the idealization shown in Figure 4 will serve us for a while. The dominant balance in the nonconvective region may be expressed

$$c_p \omega \frac{\partial \theta}{\partial p} = H_{IR} \tag{4}$$

where ω is the rate of change of pressure following a fluid parcel and θ is the dry potential temperature. Hence, the subsidence rate as measured by ω depends on the stratification of the atmosphere. Even if H_{IR} is held fixed, the subsidence rate becomes unbounded as the atmosphere approaches the dry adiabat $\theta = const$. As the atmosphere becomes colder and drier, water vapor has progressively less effect on stratification and one might indeed expect the dry adiabat to be approached. However, over the range of temperatures of interest in the present essay, the effect of stratification on subsidence is not pronounced. Let us assume the atmosphere to lie on the moist adiabat, and measure the stratification by $\theta_{500-700} \equiv \theta(500mb) - \theta(700mb)$. For a modern surface temperature of 300K, we find $\theta_{500-700} = 17.1K$. Reducing the surface temperature to a plausible glacial temperature of 295K only drops $\theta_{500-700}$ to 14.4K, and going to an extreme cooling of 290K drops $\theta_{500-700}$ to 12K. In order to double the subsidence with fixed radiative cooling, it is necessary to drop the surface temperature all the way to 282.5K.

The solar-powered compressor depicted in Figure 4 can operate even if there is no top-of-atmosphere radiative imbalance (i.e. if the absorbed solar radiation S equals the OLR locally), nor does it require a meridional gradient in either S or OLR. Consider the idealized case where the surface sensible heating vanishes and $F(p_s) = 0$, and further suppose the atmosphere to be perfectly transparent to solar radiation. Then, there is subsidence at a rate proportional to the OLR, while S independently penetrates to the surface and is balanced by evaporation; the evaporated moisture is carried away by the current fed by the subsidence, maintaining a steady state. In some sense, the compressor lives off the blackbody temperature difference between the "hot" incoming solar radiation and the "cold" OLR. Yet, for *identical* top-of-atmosphere and surface energy budgets, the system can also support a solution in local radiative-convective equilibrium, with no large scale circulation. It is for this reason that it is difficult to make simple and sweeping generalizations as to the circumstances in which the circulation gets stronger, or weaker.

In the *convective* region the infrared cooling would balance the latent heat release due to precipitation, if there were no mean ascent. With a large scale circulation such as is necessitated by compatibility with the downward mass flux in the subsiding region moisture is imported into the convective region, and the latent heat release more than balances the convective-region radiative cooling, allowing a mean ascent to prevail there. Some of the dry air descending in the subsiding region infiltrates the boundary layer, which keeps the boundary layer unsaturated and sustains the evaporation and lateral moisture flux needed to feed the extra latent heat release needed in the ascending region. The upshot is that increased subtropical radiative cooling has the effect of increasing the strength of the Hadley and Walker circulations, which leads to enhanced precipitation in the convective region (so long as the boundary layer humidity doesn't go down too much). Conversely, reduction in the subtropical radiative cooling is expected to lead to a weakened circulation and reduced precipitation.

The tropospheric radiative cooling at a given site can be changed either by changing the OLR or by changing $F(p_s)$. If there is no change in the energy exported from the atmosphere-ocean column, then OLR can change only if the local solar absorption is changed, e.g. by increasing the albedo through increasing cloud cover (which, in the subsiding region, could be accomplished through increase in low cloud cover, accompanied by a cooling of the subtropics and consequent reduction in OLR). The surface infrared cooling, $F(p_s)$ can be changed by altering conditions at the surface, even if OLR is kept fixed.

In the tropics, $F(p_s)$ is particularly small, because evaporative coupling of the surface to the boundary layer air is strong, and the water vapor content of the boundary layer is high. If the tropics is made warmer by any means whatsoever, then the boundary layer gets moister, $F(p_s)$ tends to get smaller, and evaporative coupling tends to get stronger,. Unless the OLR decreases significantly, the infrared radiative cooling of the troposphere must then increase; according to Eqn. (3) the evaporation, and hence precipitation and latent heat release, increase by a corresponding amount. Conversely, in a colder world $F(p_s)$ would tend to be more competitive with evaporation, and radiative cooling, evaporation, and precipitation would all decrease. This scenario is plausible, but not inevitable. Evaporation could increase in a colder world if surface winds intensified dramatically, or boundary layer relative humidity dropped. The latter has the dual effect of increasing moisture flux for a given surface wind intensity, and also increasing the radiative cooling of the surface. Changes in low level clouds also profoundly affect the tropospheric radiative cooling. Low clouds reduce the local solar absorption, and hence, in equilibrium, reduce the OLR. They also are very effective at making the boundary layer optically thick in the infrared, which reduces $F(p_s)$. If the former effect wins, tropospheric radiative cooling decreases, whereas if the latter effect wins the tropospheric radiative cooling increases (though the influence may show up largely near the boundary layer). Despite the complicating possibilities, the naive picture does seem to have some merit, as global mean precipitation and tropospheric radiative cooling do generally increase in numerical experiments in which the world is made warmer by doubling CO_2. On the other hand, [Ramstein *et al.* (1998)] report simulations in which the Hadley cell strength *increases* for cold LGM conditions, but *decreases* for warm doubled CO_2 conditions. Also [Knutson and Manabe (1995)] report quadrupled CO_2 simulations in which the tropical Pacific precipitation increases despite a slight *weakening* of the tropical Pacific overturning circulation.

The tropospheric radiative cooling can also be affected by oceanic influences: If upwelling or horizontal advection of cold water results in the surface temperature becoming substantially less than the overlying air temperature (as happens in the cold tongue of the tropical Pacific),then $F(p_s)$ is reduced, or even becomes negative, and tropospheric radiative cooling is enhanced.

The tropics are not isolated, but rather export excess solar energy to the extratropics. How does the atmospheric part of this transport affect the subsidence? The effect is important for the purposes of this essay, because climate forcings indigenous to the Northern Hemisphere, such as ice sheets or North Atlantic temperature changes, communicate their influence to the tropics in part through changes in horizontal atmospheric transports. Atmospheric *sensible* heat transport mixes cold extratropical air into the subtropics, and acts in concert with the radiative cooling shown in Figure 4 to maintain the subsidence. However, if the sensible heat transport were increased, the OLR the tropics needs to radiate decreases by a like amount, which by Eqn. (2) reduces the subsidence so as to exactly offset the effects of increased sensible heat transport. An increase in sensible heat transport could still enhance subsidence if it were concentrated near the edge of the subtropics, as in the ice-sheet experiments of [Manabe and Broccoli (1985)]. In that case, the subtropical cooling due to admixture of cold air can locally dominate the required reduction of OLR, which is spread over the whole low latitude band. An increase in horizontal latent heat transport, on the other hand, does not lead directly to a local cooling which affects subsidence; it still reduces the tropical OLR, in equilibrium, and hence reduces subsidence. Finally, if the net lateral atmospheric heat transport is held fixed while its composition shifts from latent to sensible fluxes, then the subsidence should intensify since OLR stays fixed but subtropical internal cooling increases. In the experiments on the influence of a Northern Hemisphere ice sheet by [Manabe and Broccoli (1985)], the net atmospheric heat transport out of the tropics increases only slightly, while there is a large shift from latent to sensible heat transport, in accord with the general reduction of water vapor content in the colder atmosphere.

To the extent that the subsiding regions are zonally symmetric, increasing the intensity of the subsidence can be taken as equivalent to "increasing the strength of the Hadley circulation." The very presence of a Hadley circulation of any type redistributes heat within the low latitudes, making the equatorial regions cooler, and the subtropical regions warmer, than they would be if the low latitudes were in local radiative-convective equilibrium. However, increasing the strength of the Hadley circulation does not have a marked effect on the tropical temperature distribution. So long as the circulation is strong enough to approximately conserve angular momentum aloft in the face of frictional dissipation, the temperature converges to a universal meridional profile independent of the strength of the circulation [Held and Hou (1980)]. For related reasons, it is not easy to change the meridional extent of the Hadley circulation appreciably. There is some leeway for the circulation strength to alter the temperature pattern, since the observed subtropical jet is only about half as strong as it would be if angular momentum were conserved; hence dissipation of angular momentum is evidently not completely overwhelmed, at least toward the poleward boundaries of the subtropics. Increasing the strength of the circulation does, in general, have the following effects: (1) It strengthens the surface easterlies, since the easterlies are determined by a balance between the zonal Coriolis acceleration and surface friction, and the Coriolis acceleration is proportional to the low level meridional flow. (2) It increases the strength of the subtropical jets aloft, especially poleward of 15N or 15S, since friction has less time to dissipate angular momentum if the meridional transport is faster. (3) Thermal wind balance requires that, to the extent the subtropical jets strengthen, the meridional temperature gradient between the equator and the poleward edges of the subtropics increases. This effect would retard the cooling of the subtropics compared to that of the extratropics, and would act conversely in the event of a weakening Hadley circulation. (4) It increases the precipitation in the Intertropical Convergence Zone.

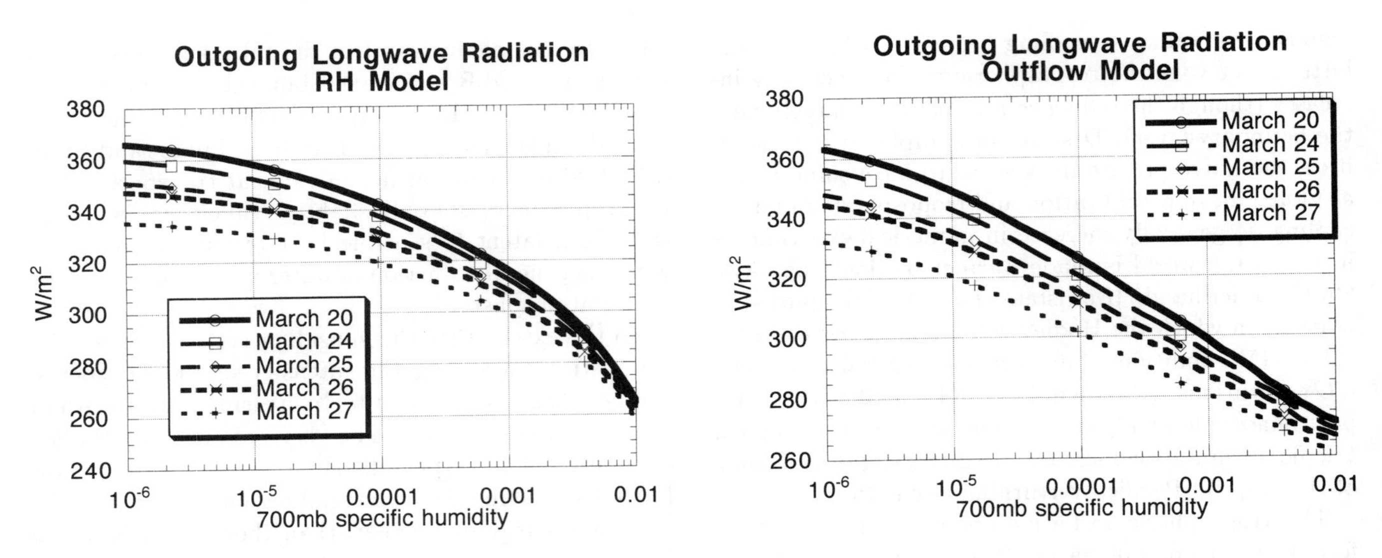

Figure 5. Outgoing longwave radiation as a function of 700mb specific humidity for 5 tropical temperature sounding. Results are shown for the RH model (a) and the outflow model (b).

Our preoccupation with subsidence should not be construed to mean that the the Hadley circulation is "driven by" subtropical cooling, though this point of view is probably closer to the truth than the more conventional one that the circulation is "driven by" latent heat release. In fact, the whole system is coupled, and while the subtropical cooling must somehow adjust itself so as to be compatible with the changes in the circulation, it is not necessarily the prime mover in such changes. For example [Lindzen and Hou (1988)] and [Hou and Lindzen (1992)] found that concentrating heating or moving its center off the equator leads to an intensified circulation; the subtropical radiative cooling adjusts to accomodate these changes, but cannot be regarded as the *cause*. The Hadley circulation is a bit like a flywheel, which tends to stay in motion once it is set in motion. Its ultimate speed is determined by the balance between the aggregate of small dissipating and accelerating mechanisms.

3.2. Influence of water vapor on the radiation budget

To determine the atmospheric temperature change, we must look to the top-of-atmosphere radiation budget. The temperature of a planet is determined by a balance between infrared cooling to space (the "OLR"), and heating by absorption of solar radiation. Any factor that reduces OLR for fixed temperature will have a warming effect, since the planet will then have to heat up in order to bring the OLR back to the point where it can balance the solar radiation. For example, doubling the CO_2 concentration from pre-industrial values reduces the Earth's OLR for fixed temperature by about 4 W/m^2, which can be compensated by a rise in temperature of 2-4C,depending on water vapor and cloud feedbacks.

The potent influence of moisture on the subtropical part of the Earth's radiation budget is seen in Figure 5, where we show the OLR as a function of moisture for typical tropical temperature profiles. The temperature profiles are from the four soundings of Figure 1, plus one additional sounding from 4S near the warm pool region. The results were computed using both the outflow-model idealized moisture profiles and the RH model. Note that for fixed humidity, the OLR decreases gently with distance away from the Equator, but that the effect of moisture variation is far more pronounced than the temperature effect. From these results we can conclude that the OLR is highly sensitive to the small amounts of water vapor contained in the dry pools; dry though they be, making them yet drier would substantially change the OLR. For example, in the outflow model, changing the specific humidity from 10^{-5} (about as dry as it could plausibly get) to 10^{-4} drops the OLR by 25 W/m^2. Increasing the moisture by another factor of 10 drops the OLR by another 25 W/m^2. In fact, except for very dry values, the OLR in the outflow model responds logarithmically to moisture, much as it is known to do for CO_2. The associated elevated sensitivity of OLR to moisture at very dry values — in the sense that adding one molecule of water to a dry atmosphere drops OLR more in absolute terms than adding the same molecule to a moist atmosphere — has been noted by [Spencer and Braswell (1997)], among others. The RH model shows an increase of sensitivity (in the logarithmic sense) for specific humidities above .001, but, recognizing that the results in Figure 5 are presented with a logarithmic humidity axis, this model too indicates high sensitivity to small changes in the humidity of dry air.

The extreme cases in [Pierrehumbert (1995)] in which increasing the dry pool greenhouse gas concentration led to reduced atmospheric temperatures were fundamentally different from the realistic situation treated above, in that the former atmospheres had temperatures far exceeding that of the surface underlying the dry pool. In this case, increasing the infrared optical thickness of the atmosphere actually increased the OLR, by replacing the cold radiating surface with a warmer one in the atmosphere. The atmosphere must then cool down in order to restore balance. This is a classic "anti-greenhouse" effect, the conditions for which cannot easily be met in any plausible climate change scenario on Earth. This does not mean that the "radiator fin" effect described in [Pierrehumbert (1995)] is irrelevant. The dryness of the subtropics still make the climate cooler than it would otherwise be, and in this sense the subtropics indeed act to help the deep tropics radiate away its excess heat. However, the Earth is almost invariably in a regime where increasing the subtropical moisture inhibits this radiation, and thus has a warming effect.

For our purposes, it can safely be assumed that adding moisture to the dry pool will make the atmosphere warmer, and removing it would make the atmosphere cooler But how much warmer or cooler? In order to answer this question, we turn to a simple model of the radiation balance of the whole tropics, comprising the convective and dry-pool regions. The assumptions are as follows: (1) The temperature is horizontally uniform over the entire tropics. (2) The vertical temperature profile is on the moist adiabat corresponding to the surface temperature (which, in turn, is continuous with the low level air temperature). (3) The convective region has 70% relative humidity throughout the troposphere. (4) The dry pool humidity is specified according to the RH model, with various specified free-tropospheric humidities. (5) The convective region occupies $\frac{1}{3}$ of the area of the tropics, while the dry pool occupies the remaining $\frac{2}{3}$ of the tropics. We will neglect the effect of clouds on the radiation budget, on the grounds that the high clouds of the present tropics have a nearly zero net effect on the radiation budget [Ramanathan *et al.* (1989)]. Whether this cancellation would continue to hold in an altered climate is anybody's guess. As discussed by [Miller (1997)], there is a possible important

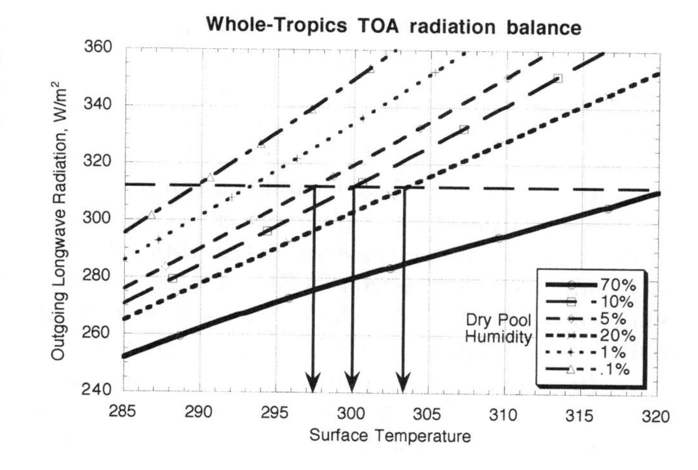

Figure 6. Outgoing longwave radiation averaged over the whole tropics, as a function of surface temperature. Curves are computed for various dry-pool humidities. The vertical profile of humidity was taken according to the RH model. See text for details of the calculation.

role for low marine stratus clouds in affecting climate change. This is an important consideration, but one which we will not take up here as our main interest is in the leverage water vapor has over climate.

We adopted the RH model for the dry pool on the grounds that if the dry pool moisture is ultimately drawn from the convection region, the dry pool humidity ought to go up with temperature roughly according to the way moisture in the convective region behaves. One could achieve a similar behavior in the outflow model by pegging the specific humidity to the value attained at a given altitude in the convective region. We present here results only for the RH model dry pool.

With the above assumptions, one can compute the average tropical OLR as a function of the surface temperature. This curve is shown in Figure 6, for various values of the dry pool relative humidity. These results were computed with CO_2 set to its pre-industrial value of 280 ppmv. If the tropics were energetically closed, then one would obtain the surface temperature by balancing OLR against the absorbed solar radiation. However, this would give an excessively high temperature, since the tropics in fact export energy to the extratropics, cooling the tropics and warming the extratropics. Since we are mainly interested in finding how temperature changes with changes in dry pool humidity, we will adopt the expedient of adjusting the energy input for a "base case" such that the surface temperature comes out to a reasonable value. The required energy input is the true absorbed solar radiation minus the energy (per square meter of tropics) exported to the extratropics. Changes can then be assessed with respect to this base case. This procedure implicitly assumes that the dynamical heat export from the tropics which occurs both in the atmosphere and ocean does not change with changing climate. This is a debatable assumption, but doing better requires a full general circulation model.

Let us take as our base case a situation with 10% drypool humidity, which is consistent with the observed values shown in Plate 2. The tropics is in balance at 300K surface temperature if the net absorbed energy is 312 W/m^2. With an estimated absorbed solar radiation of 360 W/m^2 over 30N-30S (adjusted for the joint effects of clouds on solar reflection and OLR), this implies that the tropics must export about 48 W/m^2 to the extratropics to come into balance. In view of the large role of the dynamical heat transport, it is to be expected that climate changes would alter the heat export, which in turn would have a strong affect on tropical climate. Although the atmosphere accounts for a large part of the required energy transfer (both in models and observations), there is evidence that the globe can tolerate quite large climate changes without incurring much change in the heat exported from the tropics. Notably, in the glacial maximum simulations of [Manabe and Broccoli (1985)] and [Broccoli and Manabe (1987)] it was found that cooling the extratropics by introducing an ice sheet (without reducing CO_2) had a relatively small effect on tropical temperatures, such effect as there was being confined to the north of the Equator. The smallness of the effect indicates that in the simulation the extratropics was drawing only a little more energy out of the tropics than in the present, despite the greatly increased pole-equator temperature gradient. A similar state of affairs holds in the simplified coupled atmosphere-ocean climate model of [Ganopolski *et al.* (1998)]. This result is quite counter-intuitive, given that in glacial times one has a greater temperature gradient and stronger storms,which together constitute all one needs for elevated heat fluxes. Evidently, the glacial-era atmosphere had a strengthened barrier to tropical/extratropical transport, perhaps in the form of a stronger subtropical jet. The heat export from the tropics increases, but not in proportion to the increase in temperature gradient. In [Manabe and Broccoli (1985)] it was found that there was indeed a substantial increase in *sensible* heat export across 30N, but that this was largely compensated by a decrease in *latent* heat transport. In accordance with the discussion of the preceding subsection, this could lead to changes in the strength of the subsidence, even in the absence of appreciable tropical temperature changes. [Hewitt and Mitchell (1997)] repeated the ice-sheet experiments of [Manabe and Broccoli (1985)] with a different GCM, incorporating cloud feedbacks, and also found that Northern Hemisphere extratropical cooling influences lead to only weak tropical cooling and even weaker Southern Hemisphere cooling.

The GCM results of [Webb *et al.* (1997)] differ from the simulations cited above, in that they show a very large tropical and Southern Hemisphere in response to the introduction of Northern Hemisphere extratropical ice sheets, even while holding ocean heat transports fixed. It is noted by [Webb *et al.* (1997)] that the cloud parameterization employed makes the model more sensitive to changes in forcing than some other models. As in [Manabe and Broccoli (1985)], the heat exported from the tropics in [Webb *et al.* (1997)] increases slightly in the LGM climate, as compared to the present. The contrary was stated in [Webb *et al.* (1997)], but this turns out to have been due to a minus-sign error in the analysis of the results (R. Webb 1999, personal communication). On its own, this increased heat export would lead to some tropical cooling; the large response in [Webb *et al.* (1997)] can be viewed as an amplification of this cooling by the model's unusually strong cloud feedbacks. Given the very low resolution of the GCM used in [Webb *et al.* (1997)], which only has six full grid boxes covering the latitudes from 24N to 24S , the results of the simulation should be interpreted with caution, pending confirmation with higher resolution models with alternate cloud treatments. Cloud and convection patterns are quite sensitive to the tropical circulation, and the circulation response may be different in models which better resolve the tropical dynamics.

[Ganopolski *et al.* (1998)] did find an elevated energy export from the tropics in glacial times. The atmosphere accounts for most of the change across 30N, but by 20N the additional heat transport is mostly carried by the Northern subtropical ocean. It is an interesting question whether the atmosphere alone would be able to take up the slack if the oceanic heat transport were absent (as in a mixed layer ocean model). [Ganopolski *et al.* (1998)] found an increase of $.7x10^{15}$ watts in the oceanic heat flux across 20N. Averaged over the band 20N-20S, this yields an average heat export of 4 W/m^2 . On the basis of the OLR curve for the base case in Figure 6, this would lead to a tropical cooling of 1.5C, which is somewhat less than the mean oceanic cooling found in the simulation. We emphasize that this cooling includes the "standard" water vapor feedback, i.e. that which goes along with keeping relative humidity fixed as temperature changes. In the following, one ought to keep in mind that plausible increases in the oceanic heat transport can account for a substantial portion of the tropical cooling, even in the absence of fundamental changes in the nature of the water vapor feedback.

We are now in a position to consider the sensitivity of climate to changes in the subtropical humidity. If the dry pool humidity is halved to 5%, then the temperature drops to 297.4K. On the other hand, if the dry pool humidity is doubled to 20%, the tropics warms to 303.2K. Compared to the tropical temperature variations between interglacial and glacial times (estimated at 2-5K) these are highly significant figures. Still, the bottom line is that it is not enough to make small relative changes in the dry pool humidity in order to gain leverage over climate. One must have in hand mechanisms that can halve or double the humidity. To take more extreme cases, a drop in dry pool humidity to 1% cools the tropics to 293K, and a further drop to .1% cools the tropics to 290K. Clearly, the small amounts of humidity that exist in the subtropics are playing a significant role in keeping the planet warm, so it is important to know what makes the observed humidity hover around, say, 10% rather than 1%. In the other extreme, if the dry pools broke down completely, so that the convective region expanded to fill the entire tropics, the tropical temperature would shoot up to 320K. The geological record suggests strongly that mean tropical temperatures this high have never been achieved during the entire span of the fossil record, providing indirect evidence that the dry pools have never, in fact, collapsed.

These moisture effects occur jointly with the effect of CO_2 fluctuation. A recomputation of the values in Figure 6 shows that reducing CO_2 to the glacial value of 180ppm drops the OLR for a given temperature between 2 and 3 W/m^2, with the higher values obtaining for the very dry cases. Using the OLR curve for the 10% humidity base case, this decline would lead to a temperature drop of .75K to 1K, which is consistent with the GCM results of [Broccoli and Manabe (1987)]. These values are not negligable compared to the moisture effect, but the moisture effect is distinctly greater, and moreover moisture can change nearly instantaneously, whereas CO_2 cannot, and is moreover known not to have changed appreciably during most millennial scale fluctuations.

This model embodies a very crude picture of the way the tropical climate works, but it has the virtue of a more realistic radiative model of the dry pool than that adopted (for reasons of analytic simplicity) in [Pierrehumbert (1995)]. Two-box models of tropical climate have become popular recently, and even more realism has been pursued [Miller (1997); Clement and Seager (1998); Larson and Hartmann (1998)]. Much has been learned from these models, and there has even been the beginning of at attempt to take ocean dynamical transports into account. A major shortcoming of such models remains that none has been able to give an account of what determines the relative areas of convective and subsiding regions. A further need is to replace the various *ad hoc* assumptions regarding subtropical moisture content with something more dynamically based. On the latter score, at least, there is grounds to hope for rapid progress in the near future, given what has been learned about the dynamics of subtropical moisture.

4. WHAT CONTROLS SUBTROPICAL HUMIDITY?

Having established the sensitivity of climate to subtropical water vapor, we must now look to the question of what processes could change the dry pool humidity. Since there is no obvious moisture source *within* the subsiding region, the moisture of an air parcel there must be determined by tracking it back to its origins within the convection region.

Consider an air parcel moving around in the atmosphere. Its motion is characterized by its trajectory in three dimensional space, $\vec{r}(t)$. Along the trajectory, one can obtain the time series of temperature, $T(t)$, and of saturation mixing ratio $q_s(t)$. At this point, the trajectory can be defined by whatever process one likes, be it winds from observations, a general circulation model or some simplified stochastic scheme. We suppose that the specific humidity q of the air parcel is conserved except for two possible kinds of events: If the parcel wanders into a source region (e.g. the boundary layer), its moisture is reset to saturation corresponding to the value of $q_s(t)$ where the parcel encounters the region. Further, if the parcel wanders into a region so cold that its current humidity q exceeds the current saturation value $q_s(t)$, then q is reset to $q_s(t)$, the balance of the moisture being rained out. This model is a simplification of the real moisture dynamics, in that moisture changes due to mixing amongst moist and dry air parcels, and due to evaporation of precipitation falling into the parcel from above, are neglected. One may hope that both processes are not crucial in the non-convective region. With the above simplifications, the moisture at a given point is given by the minimum q_s encountered going

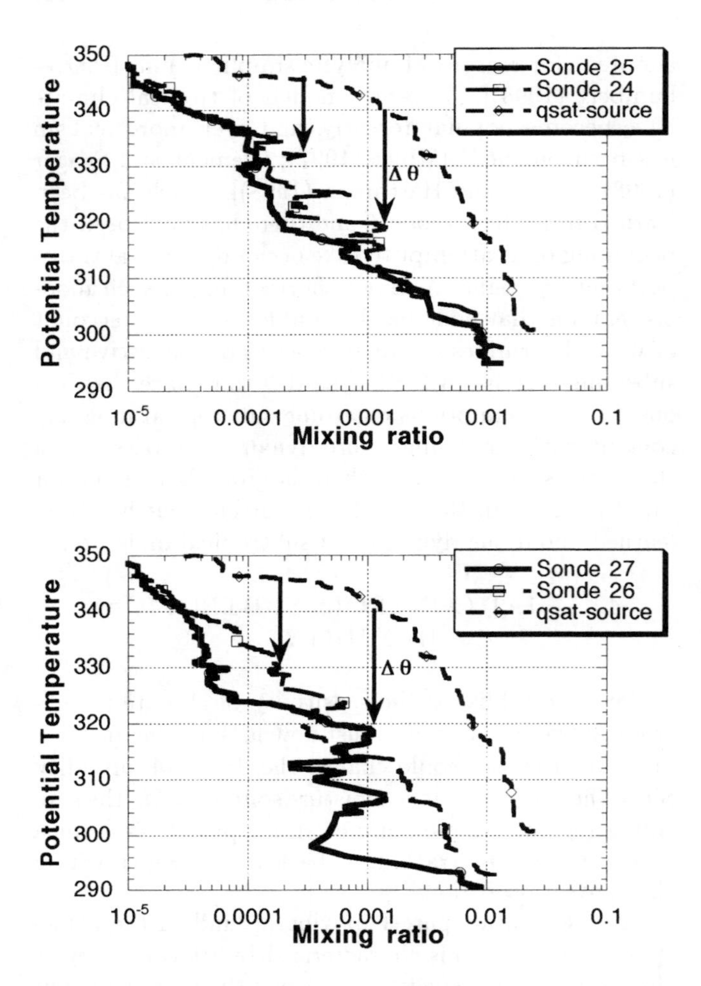

Figure 7. A typical back-trajectory in the advection/subsidence model, showing how p_{min} is determined by the time taken for air to encounter the convective region

backwards in time along the back-trajectory from the point, up to the most recent encounter with the source region. For any given model of the trajectory, this quantity can be computed by tracking q_s along the back-trajectory until the trajectory hits the source region, and taking the minimum over this finite stretch of the back-trajectory. Each back-trajectory is characterized by a unique value of $min(q_s)$, and it is the probability distribution of this quantity over the ensemble of trajectories that one must come to understand. Since q_s is primarily dependent on temperature, this quantity is primarily determined by the minimum temperature encountered along back-trajectories. Many interesting models of this statistic can be made, using either realistic winds or random walk models of the trajectories. Here, we shall confine ourselves to the most basic inferences regarding the statistic that can be made using qualitative reasoning.

A typical trajectory illustrating the preceding idea is shown in Figure 7. The chief consequence of this picture is that the longer it takes for a parcel to reach a given point in the subtropics, the drier it is. This is so because "older" air parcels have subsided more (and hence originated from colder and drier parts of the convective region); also because "older" air parcels have a higher probability of having been dried out by processing through cold extratropical regions [Yang and Pierrehumbert (1994)]. Weakened lateral mixing leads to a drier subtropics, as does enhanced subsidence. A larger temperature gradient between subtropics and extratropics also has a drying effect, since an air parcel does not have to be moved as far to be processed through a cold region.

Returning to the sondes of Figure 1, we can get an idea of the amount of subsidence needed to account for the observed dryness. In Figure 8 we plot the specific humidity q of the four sondes against the dry potential temperature θ. The potential temperature is used as the vertical coordinate because it would be conserved if there were no diabatic cooling (i.e. subsidence). In Figure 1 we also show the humidity profile $q_s(\theta)$ for the nominal "source air," which we model as the saturated mixing ratio for an equatorial sonde. If the moisture were mixed into the subtropics without subsidence, all the observed $q(\theta)$ profiles would relax to $q_s(\theta)$. Subsidence causes the air to descend to lower values of θ, which are closer to the surface. By comparing the observed $q(\theta)$ with $q_s(\theta)$, we can estimate the subsidence if the moisture originated in an air mass with humidity profile $q_s(\theta)$. For example, the moist spike at 320K in the March 24 sounding would have to have subsided

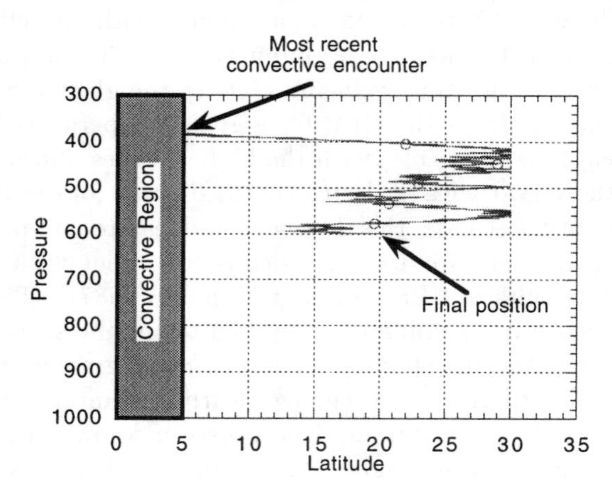

Figure 8. Specific humidity profiles for the four CEPEX subtropical soundings plotted as a function of dry potential temperature. The "source" region sounding is the saturation specific humidity for a sounding taken on the Equator.

from the 340K level to the 320K level, if it began life as a saturated parcel in the convective region. A similar conclusion applies to the spike at 320K in the March 27 sounding. Based on the typical cooling rates of about 1K per day indicated in Figure 3, the profiles suggest that this air took 20 days to arrive at its ultimate subtropical location. These mixing times are compatible with estimates based on direct trajectory calculations [Pierrehumbert (1998)]. The moisture variability in the soundings can be ascribed to the differences in the time it takes for the convective air to arrive at the sonde locations. "Older" air will be drier, because it has subsided more.

There is a great deal of observational support for the notion that the subtropical humidity is governed by a balance between subsidence and lateral mixing due to large-scale wind fields, where "large-scale" in this context means winds of the scales typically resolved by general circulation models [Sherwood (1996); Salathe and Hartmann (1997); Soden (1998); Pierrehumbert (1998)]. Recently, [Pierrehumbert and Roca (1998)] showed that the observed subtropical humidity pattern can be reproduced in great detail using the large scale advection-subsidence model, but that the humidity is crucially dependent on the amount of transient eddy activity. In the extreme case where transients are suppressed altogether, the subtropics become essentially completely dry. Thus, changes in the subtropical transient eddy activity, or changes in the mean flow environment which modulates the mixing action of the transients, can lead to profound changes in the subtropical moisture. Further, general circulation models must be able to faithfully reproduce the mixing action of low latitude transient eddies, if they are to accurately simulate water vapor changes.

Another way to change the water vapor content of the low latitudes is to change the area covered by the highly saturated convective region. The factors that govern the configuration of the Pacific warm pool and cold tongue are subtle and sensitive to details of the interaction between the tropical oceanic transports and atmospheric convection [Dijkstra and Neelin (1995)]. Because a small change in surface temperature relative to the air temperature can change a region from convective to nonconvective, the system governing the warm pool provides good scope for threshold phenomena, and possible drastic reorganization of convection in response to gradual climate forcings like the precessional insolation cycle.

Contemporary fluctuations in tropical convective area provide some indication of the extent to which convection can expand or contract. The convective region area, shown in Figure 9 undergoes a pronounced seasonal cycle, with peaks usually occurring in January, during the Southern Hemisphere summer. The effect of El Niño on this cycle is not completely consistent, as can be seen by comparing the convection time series with the Niño3 index time series in Figure 9. Strong El Niño's, like 1983 and 1998, tend to contract the convective region, whereas strong La Niña's,notably 1999, expand the convective area. The association is disrupted during the series of amorphous events in 1990-1995. Overall, the variations in the convective region are quite significant, ranging from 27% convective region coverage in 1998 to 34% in 1999. The indication is that a "permanent El Niño," corresponding to a breakdown of the warm pool, would yield a more zonally symmetric circulation with a smaller convective area and larger dry pool, and should lead to a cooler tropics once equilibrium is established. This hypothesized behavior conflicts with the observation that El Niño years tend to have warmer global mean surface temperatures; the discrepancy may be a matter of time scale for equilibration, or may be a matter of competing effects which overwhelm the cooling tendency of the contraction of the convective region during El Niño years.

The supposition that increasing the area of the convective region would have a warming effect (and conversely for contraction) would be straightforward if one only had to worry about water-vapor feedbacks and not cloud feedbacks. High clouds in the convective region reduce OLR beyond the water vapor effects, but also reduce solar accumulation through their albedo effects. At present, the state of affairs is such that the convective region is a strong net accumulator of energy, compared to the subsiding regions, because the cloud and water vapor OLR effects combine to dominate the albedo effects in the convective region. Any increase in the cloud albedo effect relative to the cloud OLR effect could alter this picture, and make the convective regions a cooling influence rather than a warming influence.

The four chief influences determining subtropical water vapor are thus:

Convective region humidity. All other things being equal, the subtropical moisture mixing ratio is proportional to the mixing ratio in the convective region, which is the proximate moisture source for the subtropics. It is a reasonable working hypothesis that vigorous deep convection maintains the convective region humidity at a roughly fixed fraction of its saturation value at each altitude. In consequence, a cooler climate will have a drier subtropics, in proportion to the reduction of saturation vapor pressure. This is the *conventional water vapor feedback*.

Subtropical subsidence rate. All other things being equal, a more rapid subidence rate brings down moisture from higher levels in the convective source region, and results in a drier subtropics. Since the source region mixing ratio decreases nearly exponentially with altitude, the subtropical humidity content decreases approximately exponentially with increasing subsidence.

Large scale horizontal mixing. Reduction in horizontal mixing, either by reducing transient eddy intensity, or by increasing mixing barriers by intensifying subtropical jets, reduces subtropical moisture, and leads to a cooling.

Change in area of convective region. Contracting the area of the convective regions will generally have a cooling effect, whereas expanding them would have a warming effect, provided cloud feedbacks do not change. However, changes in the balance between high cloud effects on OLR and albedo have the potential to fundamentally alter the impact of the convective regions on the radiation budget, either in the direction of warming or cooling.

In discussions of water vapor feedback, it is both appealing and common to assume that increases in evaporation imply increased atmospheric water vapor content. This assumption is a fallacy. In equilbrium, the evaporation equals the precipitation, and both represent the rate of flux of water through the system. The rapidity with which water cycles through the system has little bearing on how much vapor is retained in the system. A steady drip can fill up a bucket if the bucket is tightly caulked, but a torrent won't suffice to fill up a sieve. Likewise, a low evaporation rate can keep an atmosphere near saturation in equilibrium if the factors producing dry air are weak, and conversely a high evaporation rate can coexist with a dry atmosphere if dry air production is strong. As a specific example of the latter possibility, consider the situation of Figure 4, but with the additional assumptions that air is mixed from the convective region to the subsiding region at an invariable rate, that the temperature is held fixed, and that convective region is saturated with respect to water. Then, if the intensity of the Hadley circulation is increased, evaporation goes up, as does precipitation in the convective region. However, the convective region doesn't get any moister since it is already saturated, while the increased subsidence dries the subtropics. More generally, recall that the "radiatively active" water vapor aloft is a small fraction of the total atmospheric vapor, so that vapor *supply* is not in any event a limiting factor in the water vapor feedback. Only a tiny part of the water evaporated into the atmosphere needs to be diverted into the dry subsiding regions in order to create a large radiative feedback.

5. DISCUSSION

The ultimate trigger for millennial scale climate variations has not yet been identified, but subtropical water vapor changes can be involved in a number of ways. Always lurking in the background is what might be called the "conventional" water vapor feedback, in which a cooler atmosphere contains less moisture in proportion to the reduction in saturation vapor pressure. We have argued that this kind of feedback should apply also to the subtropics, even though the water vapor source is remote, in the convective region, rather than local. This water vapor feedback amplifies the cooling or warming tendencies from any source whatsoever. With regard to the tropical cooling of the Last Glacial Maximum, it would appear that no other form of water vapor feedback is called for; a reduction in CO_2, combined with a modest and entirely plausible increase in oceanic heat export from the tropics appears nearly sufficient to account for the data, when amplified by the conventional water vapor feedback.

Additional coolings or warmings on the order of 2K can be obtained if the subtropical water vapor content is halved or doubled, as compared to the predictions of the "conventional" feedback. Any agent that increases the subtropical subsidence rate will dry the atmosphere beyond the conventional water vapor feedback. An increase in subsidence could be mediated by an increase in sensible heat transported to the extratropics by atmospheric eddies, or an ocean surface cooling induced by enhanced ocean heat transports. Reorganization of convection can also increase the subsidence, as discussed by [Lindzen and Hou (1988); Hou and Lindzen (1992)]. This seems like an especially fruitful possibility to pursue, given that such reorganizations could perhaps be driven by precessional insolation changes indigenous to the tropics. Aside from changes in subsidence, a decrease in large scale mid-tropospheric lateral mixing of moisture could dry the subtropics. The mixing could decrease in response to intensified subtropical jets, a poleward retreat of the extratropical storm tracks, a change in the Northern Hemisphere planetary wave configuration, or a shut-off of some as yet unidentified eddy producing instability indigenous to the tropics. In the domain of ocean-atmosphere collective phenomena, it is also possible that subtle changes in tropical or extratropical conditions could lead to a contraction or

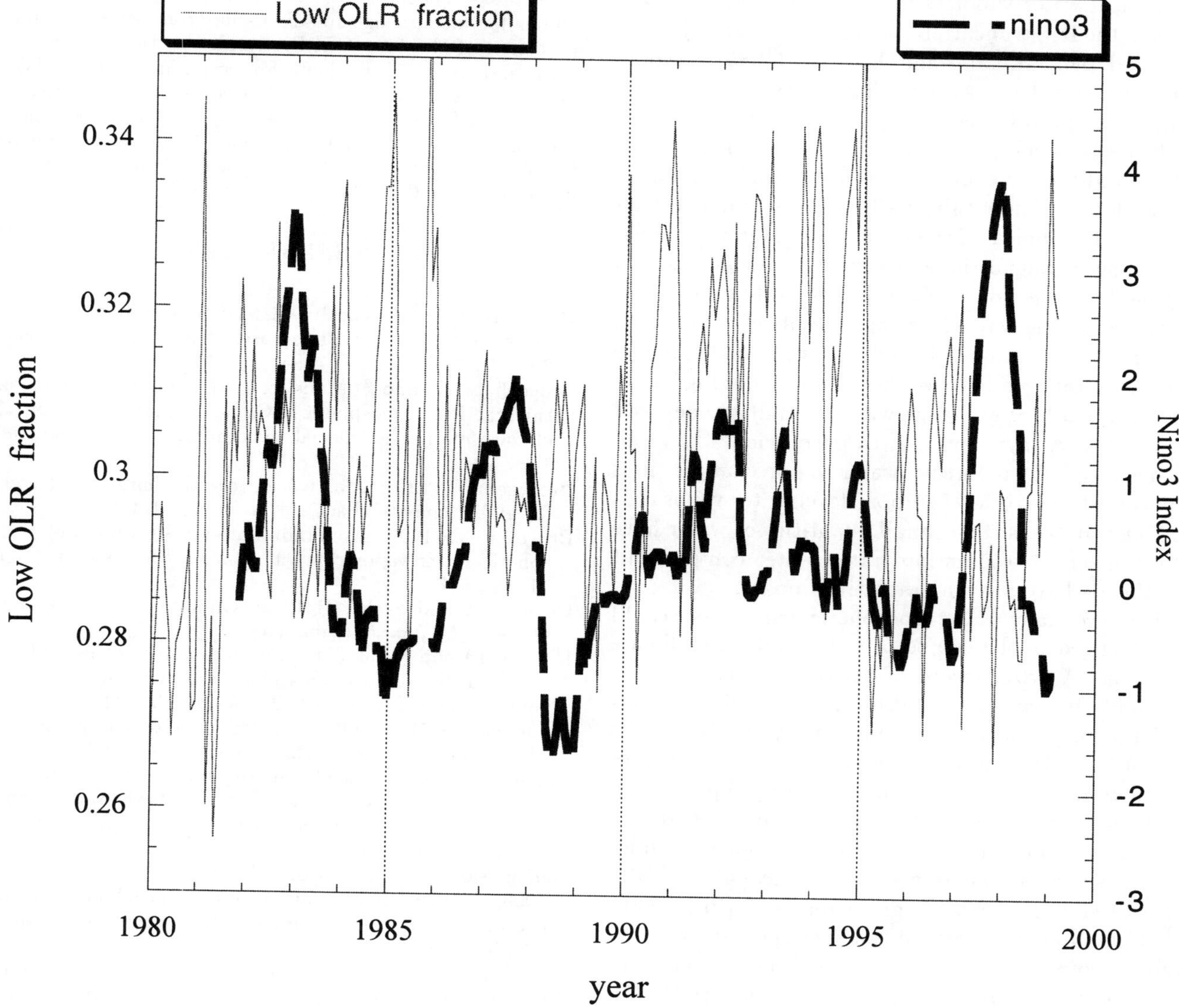

Figure 9. Time series of the Niño3 index, and of a measure of the fractional area covered by the tropical convective region. A high Niño3 index denotes a strong El Niño pattern. The tropical convective index is formed by counting the proportion of the tropical area (25N-25S) for which the monthly-mean OLR is less than 240 W/m^2.

expansion of the Pacific warm pool region, with associated changes in low latitude humidity content. To date, extratropical triggers (such as shut off of North Atlantic Deep Water formation) do not seem able to create enough water vapor change in the tropics to produce a global response in climate simulations, but the possibility remains that this failure could be due to inadequate resolution of tropical atmospheric or oceanic dynamics (keeping in mind also the possibility that improvements in dynamic sea ice treatment may enhance the extratropical response itself).

The preceding conclusions regarding water vapor and millennial scale variability are unsatisfyingly vague, and a further examination of the possibilities must rely heavily on general circulation models. There are a few lessons for modelers to be found in this essay. First, a good resolution of tropical dynamics, with regard to transient eddies, jets and subsidence, is essential if one is to accurately capture the subtropical water vapor feedback. The low resolutions employed to date in models of the LGM and subsequent climate variations is a source of concern, and much more will be learned about the subtropical water vapor as models with resolution better than the R15 truncation (about 4 degrees resolution) employed by [Manabe and Broccoli (1985)] become common. Another lesson is that there is a lot going on in the tropical boundary layer that ought to concern us. Processes here govern the boundary layer humidity, surface radiative cooling, evaporation, and generation of low clouds. These things all have a strong effect on the way the tropical circulation and water vapor content change. Diagnostics addressing the issues raised in Section 3 and Section 4 of this essay would be valuable in comparing the way water vapor feedbacks operate in the various simulations. Such diagnostics might include subsidence rate, mid-tropospheric lateral mixing, sensible and latent heat exchange with the midlatitudes, and both boundary layer and mid-tropospheric water vapor mixing ratio. In reporting water vapor changes, it is important to diagnose the upper level water vapor content separately, as these small quantities of water have little impact on column-integrated water content, but nonetheless have a strong radiative impact.

There are many ways tropospheric water vapor could have changed in the course of millennial scale climate variations. A proxy record of paleo-humidity would be invaluable in narrowing down the possibilities. Certainly, boundary layer relative humidity affects the levels of $\delta D$and $\delta^{18}O$ in the boundary layer water vapor, and there may be some way of exploiting this effect in the Tropics. However, what is most of interest for the tropical radiation budget is the free-atmosphere humidity. [Broecker (1997)] argued that tropical glacier $\delta^{18}O$ was indeed such a paleo-hygrometer, but [Pierrehumbert (1999)] pointed out why the isotope data was unlikely to contain much information about the ambient free-troposphere humidity. In [Pierrehumbert (1999)] it was argued that the $\delta^{18}O$ instead is telling us something about the degree to which rainfall over tropical continents (specifically over the Amazon Basin) is lost to runoff. The results indicate that a greater proportion of rainfall was lost to runoff during the Last Glacial Maximum than is lost at present, and argue for a reduction in forest cover during the LGM. Owing to lack of snowline data for later times, it is not known whether similar changes in hydrology are implied in concert with millennial scale variability. The issue is interesting, but its resolution is unlikely to shed much light on the behavior of free-tropospheric water vapor. For the forseeable future, the chief tools one can bring to bear on the problem are likely to be simulation, theory, and development of analogies with presently observable climate variations.

Acknowledgments. We are grateful to Mr. Matthew Huber for much insightful discussion concerning the interpretation of the four CEPEX sondes in the subsiding region, and to Mr. Hui Zhang for preparation of the TOVS moisture data map. The research upon which this essay is based was funded in part by the National Science Foundation, under grant ATM 9505190, and by the National Oceanographic and Atmospheric Administration, under grant NA56GP0436.

REFERENCES

Berner R. A. 1994: GEOCARB II: A revised model of atmospheric CO2 over Phanerozoic time. *Amer. J. Science* 294 , 56-91.

Broccoli, A. J., and S. Manabe, 1987: The influence of continental ice, atmospheric CO2, and land albedo on the climate of the last glacial maximum. *Climate Dynamics* 1, 87-99.

Broecker WS 1994: Massive iceberg discharges as triggers for global climate change. *Nature* 372, 421-424.

Broecker WS 1997: Mountain glaciers: Recorders of atmospheric water vapor content?. *Global Biogeochem. Cycles* 11, 589-597.

Clement A and Seager R 1998: Climate and the Tropical Oceans. *J. Climate* (submitted).

Dijksra HA and Neelin JD 1995: Ocean-Atmosphere interaction and the tropical climatology Part II: Why the Pacific cold tongue is in the east. *J. Climate* 5. 1344-1359.

Emanuel KA 1995: On Thermally Direct Circulations In Moist Atmospheres. *J Atmos Sci* 52, 1529-1534.

Emanuel, K and Pierrehumbert, RT 1996: Microphysical and dynamical control of tropospheric water vapor. in *Clouds, Chemistry and Climate*, Nato ASI Series 35. Springer:Berlin, 260pp.

Ganopolski F, Rahmstorf S, Petouikhov V and Claussen M 1998: Simulation of modern and glacial climates with a coupled global model of intermediate complexity. *Nature* 391, 351-356.

Held IM, and Hou AY, 1980: Nonlinear axially symmetric circulations in a nearly inviscid atmosphere.*J. Atmos. Sci.*, 37 515-533.

Hewitt CD and Mitchell JFB 1997: Radiative forcing and response of a GCM to ice-age boundary conditions: Cloud feedback and climate sensitivity. *Climate Dynamics*, 13 821-834.

Hou AY and Lindzen RS 1992: The Influence of concentrated heating on the Hadley circulation. *J. Atmos. Sci.* 49, 1233-1241.

Kalnay E *et al* 1996: The NCEP/NCAR 40-Year Reanalysis Project. *Bull. Am. Meteorological Soc.* 77, 437-471.

Kiehl, J. T. and Briegleb, B. P. 1992: Comparison of the observed and calculated clear sky greenhouse effect: Implications for climate studies. *J. Geophys. Res.* 97, 10037-10049.

Kley D, Smit HGJ, Vömel H, Grassl H, Ramanathan V, Crutzen PJ, Williams S, Meywerk J and Oltmans SJ 1997: Tropospheric water-vapour and ozone cross-sections in a zonal plane over the central equatorial Pacific Ocean. *Q. J. R. Meteorol. Soc.* 123, 2009-2040.

Knutson TR and Manabe S 1995: Time-mean response over the tropical Pacific to increased CO_2 in a coupled ocean-atmosphere model. *J. Climate* 8, 2181-2199.

Larson K, Hartmann DL, and Klein SA 1998: Climate sensitivity in a two-box model of the tropics. *J. Climate* (submitted).

Lindzen RS and Hou AY 1988: Hadley circulations for zonally averaged heating centered off the Equator. *J. Atmos. Sci.* 45, 2416-2427.

Manabe, S. and Wetherald, R. T. 1967: Thermal equilibrium of the atmosphere with a given distribution of relative humidity. *J. Atmos. Sci* 24, 241 - 259.

Manabe, S., and A. J. Broccoli, 1985: The influence of continental ice sheets on the climate of an ice age. *J. Geophys. Res.*, 90(C2), 2167-2190.

Miller, R. L. 1997: Tropical Thermostats and Low Cloud Cover. *J. Climate* 10, 409-440.

Pierrehumbert RT 1998 Lateral mixing as a source of subtropical water vapor. *Geophys. Res. Lett* 25, 151-154.

Pierrehumbert RT 1995: Thermostats, Radiator Fins, and the Local Runaway Greenhouse. *J. Atmos. Sci.* 52, 1784-1806.

Pierrehumbert, RT and Roca R 1998: Evidence for Control of Atlantic Subtropical Humidity by Large Scale Advection. *Geophys. Res. Letters* 25, 4537-4540.

Pierrehumbert RT 1999: Huascaran $\delta^{18}O$ as an indicator of tropical climate during the last glacial maximum. *Geophys. Res. Letters* 26,1345-1348.

Ramanathan, V., Cess, R.D., Harrison, E.F., Minnis, P., Barkstrom, B.R., Ahmad, E. and Hartman,D. 1989: Cloud-radiative forcing and the climate: Results from the Earth Radiation Budget Experiment. *Science* 243,57-63.

Ramaswamy, V., and M. D. Schwarzkopf, W. J. Randel, 1996: Fingerprint of ozone depletion in the spatial and temporal pattern of recent lower-stratospheric cooling. *Nature*, 382, 616-618.

Ramstein G, Serafini-Le Treut Y, Le Treut H, Forichon M, Joussaume S 1998: Cloud processes associated with past and future climate changes. *Climate Dynamics* 14, 233-247.

Salathe EP and Hartmann DL 1997: A trajectory analysis of tropical upper-tropospheric moisture and convection. *Journal of Climate* 10,2533-2547.

Sherwood SC 1996: Maintenance of the free-tropospheric tropical water vapor distribution, Part II: Simulation by large-scale advection. *Journal of Climate* 9 pp 2919-2934.

Soden BJ 1998: Tracking upper tropospheric water vapor. *J. Geophys. Research* 103D14,17069-17081.

Spencer RW, Braswell WD 1997: How dry is the tropical free troposphere? Implications for global warming theory B Am Meteorol Soc 78, 1097-1106 (1997).

Spencer Roy W 1997: Statement Concerning the Role of Water Vapor Feedback in Global Warming. Statement presented to the U.S. House of Representatives Science Committee, subcommittee on Energy and the Environment, 7 October, 1997.

Sun D-Z, and Lindzen RS 1993a: Distribution of tropical tropospheric water vapor.*J. Atmos. Sci* 50, 1643 - 1660.

Sun D-Z, and Lindzen RS 1993b: Water vapor feedback and the ice age snowline record. *Ann. Geophysicae* 11, 204-215.

Webb RS, Rind DH, Lehman SJ, Healy RJ, and Sigman D 1997: Influence of ocean heat transport on the climate of the Last Glacial Maximum. *Nature* 385 695-699.

Williams, S. F. 1993: *Central Equatorial Pacific Experiment (CEPEX) Operations Summary*, UCAR Office of Field Project Support. Boulder, Colorado. 321pp.

Xu KM and Emanuel KA 1989: Is the tropical atmosphere conditional unstable? *Mon Weather Rev* 117, 1471-1479.

Yang, H. and Pierrehumbert, R. T. 1994: Production of dry air by isentropic mixing. *J. Atmos. Sci.* 51, 3437-3454.

R. T. Pierrehumbert, Department of Geophysical Sciences, University of Chicago, Chicago, IL 60637. (e-mail: rtp1@geosci.uchicago.edu)

A Role for the Tropical Pacific Coupled Ocean-Atmosphere System on Milankovitch and Millennial Timescales. Part I: A Modeling Study of Tropical Pacific Variability

Amy C. Clement [1] and Mark Cane

Lamont-Doherty Earth Observatory of Columbia University, Palisades, NY

Climate records from over much of the world show variability on both Milankovitch and millennial timescales. However, mechanisms in the climate system that have a global scale are lacking. In this two-part paper, we turn attention to a part of the system that is known from the modern climate record to be capable of organizing global scale climate events: the tropics. In the first part, we isolate this system using a coupled ocean-atmosphere model of the tropical Pacific. The model results demonstrate a tropical mechanism which can amplify Milankovitch forcing and generate a mean climate response. The model is also run for 150,000 years with no forcing. The results raise the possibility that millennial timescale variability can be generated within the tropics through non-linear interactions. In that case, power is to be expected over a range of frequencies, rather than in any particular narrow band. The second part of our study, presented in the following paper, will discuss the potential impact on the global climate of these mechanisms of tropical climate variability.

1. INTRODUCTION

The El Nino/Southern Oscillation (ENSO) is perhaps the most well-studied modern climate phenomenon. The observational record indicates that ENSO tends to have most power at 2-7 year periods. At the heart of ENSO is a positive feedback between the ocean and atmosphere. A reduced equatorial sea surface temperature (SST) gradient, as in the warm phase of ENSO, leads to a slackening of the trade winds, reduced equatorial upwelling, and a deepening of the thermocline in the eastern equatorial Pacific, all of which further weaken the SST gradient [*Bjerknes*, 1969; *Cane*, 1986]. In the cold phase, the feedbacks push the system in the opposite direction. An increased equatorial SST gradient is enhanced by stronger trades, more equatorial upwelling, and a steeper thermocline tilt. These interactions are not restricted to the 2-7 year timescale. The modern climate record indicates that ENSO may have variability on decadal timescales [*Trenberth and Hurrell*, 1994]. Whether this variability is internal to the tropics [i.e., *Zebiak and Cane*, 1991], or is the result of extra-tropical forcing of ENSO [i.e., *Kleeman et al.*, 1999], the same positive feedbacks will operate on this longer timescale.

In addition to the potential for natural, perhaps internal low-frequency tropical variability, there is also the great likelihood that ENSO physics play an important

[1] Now at LODYC, Université de Paris 6, Paris, France.

Mechanisms of Global Climate Change at Millennial Time Scales
Geophysical Monograph 112

role in the response of the climate to an external forcing [*Clement et al.*, [1996]; *Dijkstra and Neelin*, 1995; *Cane et al.*, 1997]. There is also the possibility that the ENSO behavior will be altered by such a forcing. Whether ENSO has changed in response to greenhouse forcing is currently an issue of debate in the climate community [*Trenberth and Hoar*, 1996; *Rajagopalan et al.*, 1997; *Latif et al.*, 1997; *Goddard and Graham*, 1997]. Modeling studies have shown that the temporal characteristics of ENSO may be altered by greenhouse forcing [*Zebiak and Cane*, 1991; *Clement et al.*, 1996; *Timmermann et al.*, 1998].

Separating natural and forced variability embedded in the tropical Pacific in the modern climate record is difficult because the same coupled physics can be instrumental in the response to a forcing and in generating natural low-frequency variability. The result is a spatial pattern of tropical Pacific SST variability in the 20th century that can look remarkably similar over a range of timescales [*IPCC*, 1995; *Zhang et al.*, 1997; *Latif et al.*, 1997]. Is this low-frequency variability of ENSO or a change in the mean state? Is it forced or natural? The simplest answer is that the tropical Pacific climatology must be thought of as the result of similar physics, whether forced or arising from internal instabilities, operating on a variety of timescales that interact with each other, and cannot be separated.

In this paper, we extend the study of ENSO-like variability to Milankovitch and millennial timescales. The approach we take is to isolate the tropical Pacific using a simple coupled ocean-atmosphere model. The focus of this work is on determining to what extent the tropical Pacific climate can change *on its own* with no influence from higher latitudes. We will explore the mechanisms of climate change on these timescales and make some tentative statements about what kind of temporal variability to expect from this system alone.

2. MODELLING EXPERIMENTS

We perform experiments with the Zebiak-Cane ENSO model [*Zebiak and Cane*, 1987]. This is a coupled model of the tropical Pacific which computes anomalies in the circulation and SST about a mean climatology that is specified from observations. The model domain is 124°E to 80°W and 29°N to 29°S. The dynamics in the atmosphere and ocean are described by linear shallow water equations on an equatorial beta-plane. In the ocean, an additional shallow frictional layer of constant depth (50 m) is included to account for the intensification of wind driven currents near the surface. An atmospheric heating anomaly is computed from the SST anomaly, and the specified background wind divergence field. Wind field anomalies are computed from this heating, and are used to drive the ocean model. The anomalous ocean circulation and thermocline depth are used to compute a new SST anomaly. The subsurface temperature anomaly is a non-linear function of the thermocline depth anomaly. This coupled model produces self-sustained interannual oscillations and contains the main physics thought to be relevant for ENSO.

The model is run for the past 150,000 years and forced with variations in solar radiation due to changes in eccentricity, obliquity, and precession of the equinoxes [*Berger*, 1978]. The Milankovitch solar forcing is implemented as an anomalous heat flux into the ocean surface. The solar radiation anomaly relative to today is computed as a function of time in the past (Kyr), season (τ), and latitude (θ), and converted to a surface heat flux by:

$$F_o'(Kyr, \tau, \theta) = Q_s(T_a + 0.0019\theta_z) - F_o(0, \tau, \theta) \quad (1)$$

where $T_a = 0.7$ which is the transmissivity of the atmosphere meant to represent the bulk effect of absorbtion and reflection of solar radiation by all atmosphere constituents, Q_s is the derived solar radiation at the top of the atmosphere, θ_z is the solar zenith angle [*Berger*, 1978], and $F_o(0, \tau, \theta)$ is the modern distribution of surface solar radiation similarly computed. In addition, a 150,000 year control run is performed where there is no external forcing.

2.1. *Milankovitch Forcing*

Figure 1a shows 500 year averages of the NINO3 index (the SST anomaly averaged over 150°, 90°W and 5°S, 5°N) for the model experiment with Milankovitch forcing. NINO3 is generally taken to be an index of ENSO. The mean NINO3 value for the control run is about $0.4K$ [*Zebiak and Cane*, 1987], and is subtracted from the NINO3 index of the forced run. This 500 year average index has a large precessional peak (about 21 kyr) as well as significant power at about 11 kyr. The 11 kyr cycle will be discussed elsewhere. The obliquity cycle (41 kyr) is not present. Warm periods generally occur when perihelion (anomalous heating) occurs between December and June, and cold periods occur when perihelion occurs between July and November. The spatial pattern that arises on the precessional timescale is shown in Figure 2. This is the first empirical orthogonal function (EOF) which describes 75% of the variance of the lowpass filtered (> 20 kyr periods) SST field. The pattern looks much like ENSO with the largest signal occuring in the eastern equatorial Pacific.

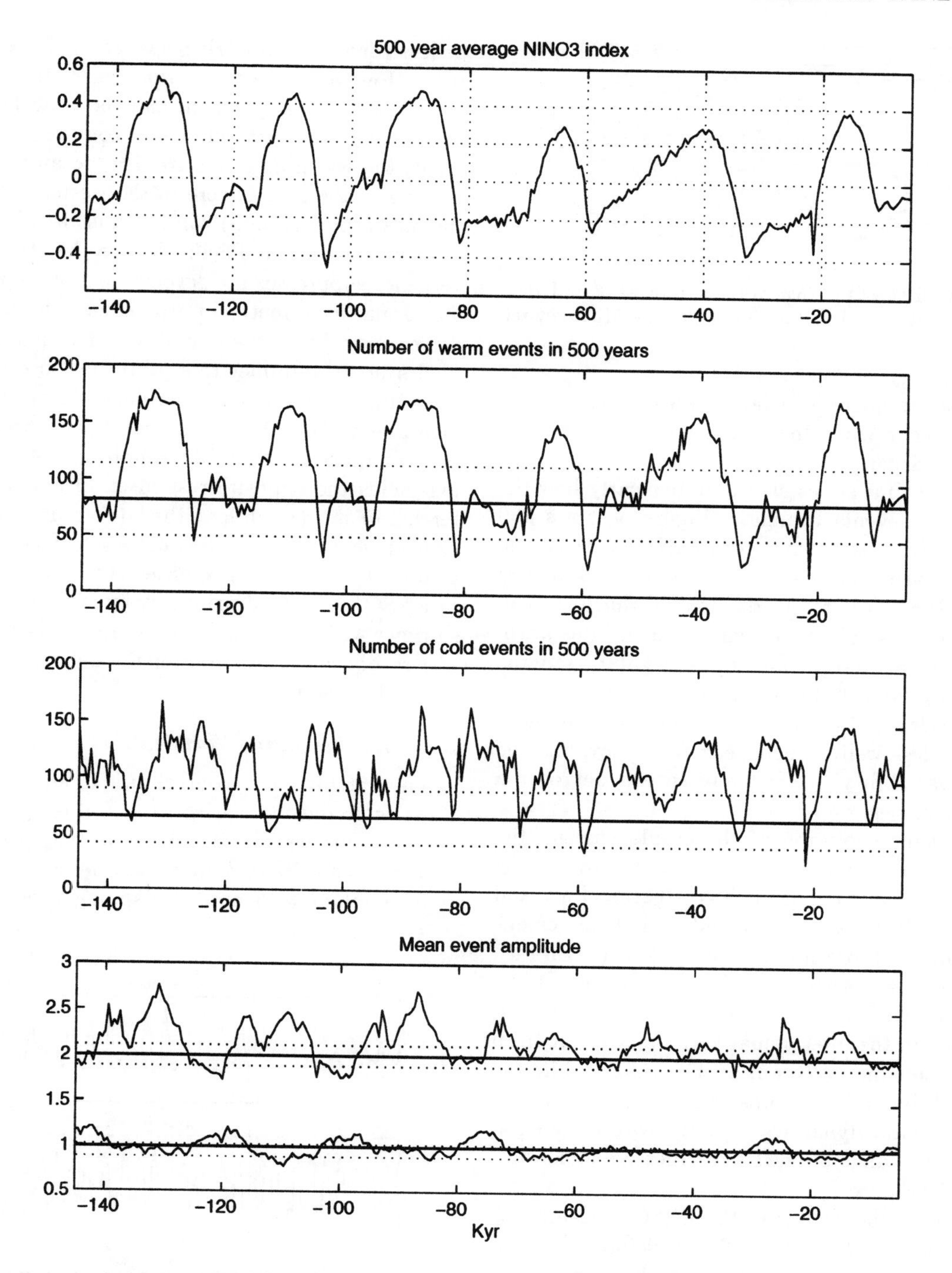

Figure 1. (a) 500 year average NINO3 (degrees Celsius) index from Zebiak-Cane model forced with Milankovitch solar forcing, (b) number of warm events defined as in text for 500 year non-overlapping periods, (c) number of cold events (d) mean amplitude of warm (solid) and cold (dashed) events. The bold line shows the mean values for each of these statistics for the control run with 95% confidence limits plotted (dotted).

What causes this annual mean response to a forcing that has approximately zero annual mean? Figures 1b and 1c show the number of warm and cold events in non-overlapping 500 year windows. A warm event is defined here to occur when the annual mean value of NINO3 exceeds 1 K, and a cold event when the index is less than -1 K. The mean amplitude of events is also shown (Figure 1d). Generally, warm periods occur when

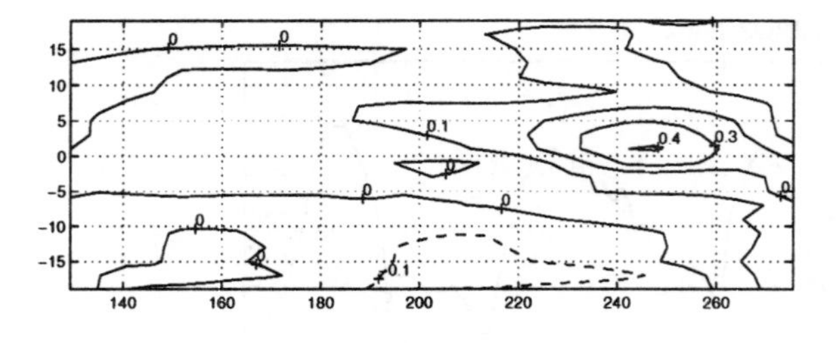

Figure 2. The first EOF (75% of the variance) of SST data with a lowpass filter applied (> 20 kyr) for the Milankovitch forcing run.

there are more frequent and larger warm events while cold periods occur when the cold event frequency and amplitude are larger.

A look at the time series shows more clearly how the character of the events changes. Figure 3 shows 100 year segments from the control run NINO3 time series, from typical warm periods (i.e. 135 kyr), and typical cold periods (i.e. 75 kyr). In the control run, the peak frequency occurs at about 4 years. During the warm periods, events are larger and more regular, and the peak frequency shifts to approximately 3 years. During the cold periods, cold events tend to be more frequent but there is a less well defined peak frequency, and the interannual variability is effectively spread out over a range of frequencies.

While the time evolution of the events change, the spatial structure remains essentially the same. The first EOF of the annual mean SST, thermocline, and wind field during the warm and cold periods are essentially the same as those for the control run (Figure 4). These patterns are described by *Zebiak and Cane* [1991] as the mature ENSO signal. Thus, the orbitally induced changes in the interannual variability result from a change in the time evolution of the events, while the basic coupled dynamics are unaltered. The persistence of the fundamental dynamics explains why the spatial pattern of SST change on a 21 kyr timescale looks similar to ENSO (Figure 2).

The reason for the change in the ENSO variability is discussed in *Clement et al.* [1999] in detail. In brief, it is the result of a seasonal cycle in the response of the system to the forcing. Consider a uniform heating of the tropical Pacific, which approximates the precessional forcing. When a uniform heating is applied, it will initially generate a warm SST anomaly which will affect the atmosphere differently in different seasons. For example, in spring, the ITCZ is near the equator and the wind field is convergent all across the equator. Thus, the uniform SST anomaly will generate a more or less zonally symmetric response in atmospheric heating. However, in late summer/early fall, the eastern Pacific ITCZ moves north, and the wind field becomes divergent in the east, while still convergent in the west. Thus, the warming of the tropical ocean yields a larger heating of the atmosphere in the western Pacific where the mean (background) wind field is already convergent, than in the eastern Pacific, where the strong mean divergence suppresses the development of deep convection. The zonal asymmetry in atmospheric heating anomaly drives easterly wind anomalies at the equator. The coupled system amplifies this perturbation on an interannual timescale via the same unstable interactions that give rise to ENSO, and the result is a cooler east Pacific, or more La Niña-like conditions. A uniform cooling yields the opposite response: little response to the forcing in spring, while in the late summer/early fall the result is *westerlies* on the equator which can develop into a more El Niño-like response. Thus, the mean change in ENSO is dictated primarily by the forcing in the late summer with warming in the summer giving a La Niña-like response and cooling in the summer giving an El Niño-like response.

2.2. Low Frequency Variability in the Control run

The 150,000 year control run is analyzed to assess the low frequency variability of this system with no external forcing. Figure 5 shows the spectrum of the annual mean NINO3 index computed using a multitaper method with non-overlapping windows of length

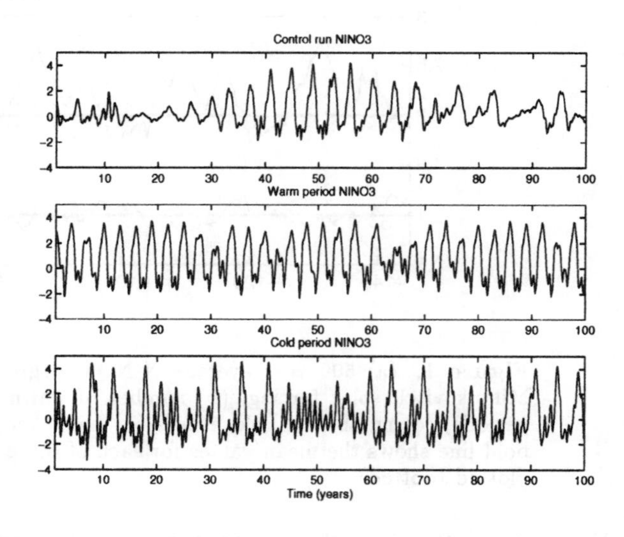

Figure 3. NINO3 time series segments from (a) control run (b) warm period of the forced model run and (c) cold period from the forced model run.

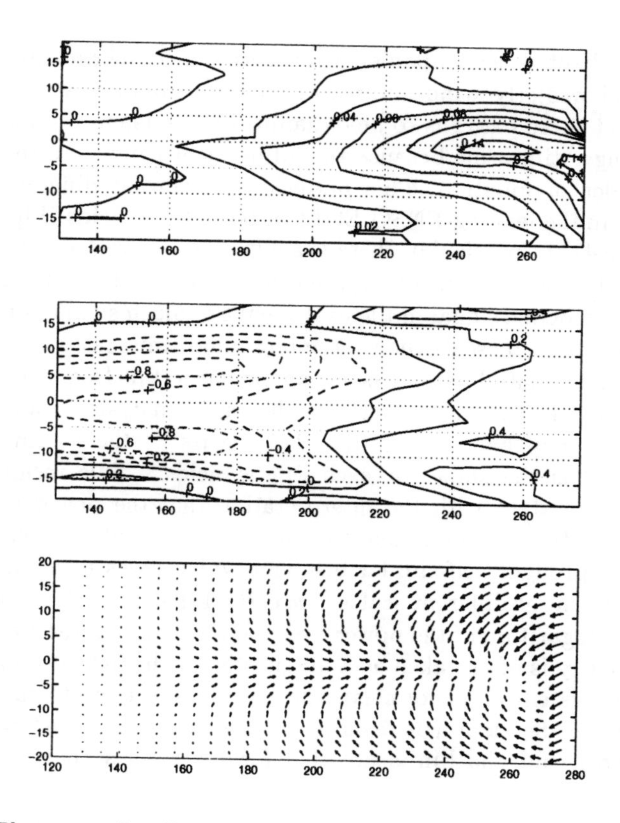

Figure 4. Leading EOF of (a) SST (85% of the variance) (b) thermocline (55% of the variance) and (c) wind field vectors (80% of the variance) for annual mean values from the control run.

2^{14} years [*Mann and Lees*, 1996 and references therein]. This system has a known peak at 4 years [*Zebiak and Cane*, 1987]. The power drops off significantly towards longer periods reaching a minimum around 40 years. In the range of 40 to 400 year periods, the power ramps up, and then levels off at periods longer than 400 years. These properties of the spectrum are independent of the spectral estimation technique. The power around 4 years is ENSO, and there is a vast literature on the theory for this phenomenon [*Neelin*, 1998]. At lower frequencies, however, there is no theory. How do we assess the significance of the power at these frequencies?

The general approach in interpreting the spectra of climate records is to look for frequency bands in which the power rises above the level of a random process. Typically, the process used to define this level is an autoregressive model of order 1 (AR(1)) [*Crowley and North*, 1991]. This AR (1) model represents a linear system in which high-frequency variability is smoothed, and power at low-frequencies is emphasized. *Hasselman* [1976] described this process as integration of "weather" by the climate system. If peaks in a spectrum are above the level of the AR(1) process, they are taken to be significant at some confidence level, and can, in principle, be attributed to some deterministic process. An AR(1) process, however, is not the appropriate one against which to compare the variability in the Zebiak-Cane model. This model has a peak at a period of 4 years where most of the variance concentrated, while the AR(1) process will not allow any narrow band peaks. The next order model that is appropriate is an AR(2) process, which describes a linear, damped oscillator that is driven by white noise. Coefficients of the AR(2) are found which provide the best fit with the NINO3 index, and a 150,000 year time series is generated with white noise forcing. Comparing the spectra of the Zebiak-Cane model and the AR(2) process then is a test of whether the power at millennial timescales in the Zebiak-Cane model is significantly different than what would arise from random fluctuations of a linear process.

The spectrum of the NINO3 index and that of the AR(2) process with 95% confidence limits are shown in Figure 6. At periods longer than about 100 years, the Zebiak-Cane model has power that is distinguishable from random noise in the linear model at the 95% confidence level. We suggest that power at these frequencies is due to non-linearity in the Zebiak-Cane model.

Lorenz [1991] pointed out that power in non-linear systems can occur at unexpected low-frequencies. He writes that "...in chaotic dynamical systems in general, very-long-period fluctuations, much longer than any obvious time constants appearing in the governing laws,

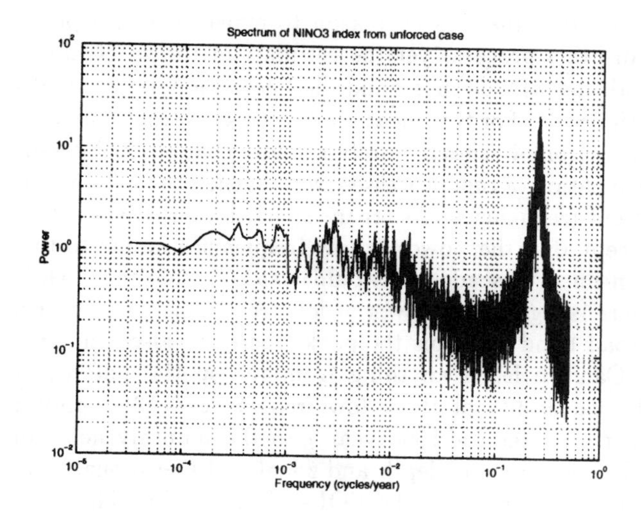

Figure 5. Multi-taper spectrum of NINO3 index from unforced Zebiak Cane model.

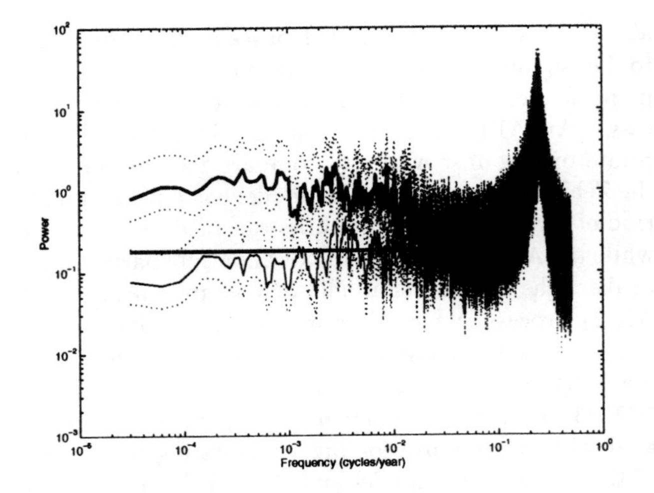

Figure 6. Multi-taper spectrum of NINO3 index from unforced Zebiak Cane model with 95% confidence limits (dotted lines around bold solid line) and multi-taper spectrum of a linear damped oscillator driven by white noise with 95% confidence limits (dotted lines around light solid line), and theoretical spectrum for an AR(2)- solid line.

are capable of developing without the help of any variable external influences." The spectrum of NINO3 can be interpreted in this context. The Zebiak-Cane model is chaotic in some regions of parameter space [*Zebiak and Cane*, 1987; *Tziperman et al.*, 1994; 1997]. The dynamics in both the atmosphere and ocean are linear and damped on a timescale of days to years [*Zebiak and Cane*, 1987]. The non-linearity arises in the coupling between the atmosphere and ocean, and thus it is the coupling that can potentially generate power at low frequencies through a non-linear cascade. It should be noted that there is no reason to expect spectral peaks from this process, yet the estimate of the spectrum of both the NINO3 index and the linear model show peaks (Figure 6). For the linear model, we know that any peak away from 4 years is purely random chance. A different realization of this processes would generate peaks at different frequencies. As for the spectrum on NINO3, if we accept that the power at low frequencies is due to this non-linear cascade, we would also conclude that, while there is significant power at low frequencies, it is over a broad band and thus the peaks there are also random.

Can we learn anything from the spatial pattern of the variability at these low frequencies? We apply a lowpass filter (> 1000 years) to the monthly fields in SST, thermocline depth and zonal and meridional wind fields, and compute the EOFs of each field. The leading EOFs (Figure 7) explain approximately 90% of the variance in each field, and the principal components are almost exactly in phase. A decreased equatorial SST gradient is accompanied by a less sloped thermocline and westerly wind anomalies on the equator. We expect these patterns to be dynamically consistent on this longer timescale because the adjustment time is on the order of months. This pattern is somewhat different from the leading ENSO EOFs of these three fields (Figure 4). In particular, there is a signal in the west Pacific that is not present on an interannual timescale, and the thermocline signal in the east Pacific is much smaller on the long timescale than the interannual timescale.

This spatial pattern is not particular to this frequency band. In fact, if we perform the same analysis on any frequency band lower than 1/50 years^{-1}, the leading EOF is the same for each of the fields. This result is consistent with the interpretation that the power at these frequencies is due to chaos. Chaotic interactions can produce self-similar behavior which gives identical patterns at different scales [*Turcotte*, 1993]. The slope of the logarithm of the power over the logarithm of the frequency is often taken as a diagnostic for the behavior of the system. It is not clear where in the spectrum to calculate the slope (there is no obvious cut-off frequency here), but the limits are -0.8 for the reddest part of

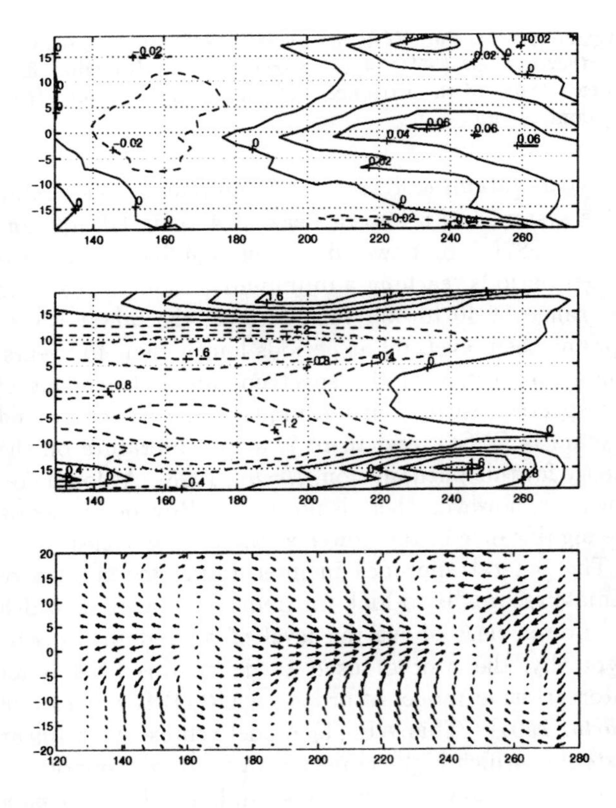

Figure 7. Leading EOF of (a) SST (K) (b) thermocline (m) and (c) wind field vectors for low pass (> 1000 yrs) unforced run.

the spectrum (1/40 - 1/400 years^{-1}), and -0.5 over the entire low frequency part of the spectrum (frequencies lower than 1/40 years^{-1}).

3. DISCUSSION

The model experiments demonstrate a mechanism by which the tropical climate can amplify Milankovitch forcing *on its own*. The mechanism is ENSO. The effect of changes in the Earth's orbital parameters in the tropics is to alter the seasonal cycle of solar radiation. ENSO is thought to be the result of interactions between the seasonal and interannual timescales [*Zebiak and Cane*, 1987; *Münnich et al.*, 1991; *Tziperman et al.*, 1994; 1997; *Jin et al.*, 1994; *Chang et al.*, 1995], and is known to be phase locked to the seasonal cycle [*Rasmusson and Carpenter*, 1982]. Thus, as the seasonal cycle in solar radiation changes, ENSO *must* be affected. The details of the response may be particular to the model, but the mechanism is not.

The results can be considered a demonstration of the idea put forth by *Palmer* [1993] that climate change in a system with strong attractors will manifest as a change in the probability that the system will reside in a particular regime without changing the character of the attractors. The primary mode of variability in the tropical climate is ENSO. The spatial structure of this mode of variability is unaltered by the solar forcing. Rather, the Milankovitch forcing changes the temporal evolution which results in a mean tropical climate change due to a change in the *statistics* of events. Again, while we can not be sure of the details, the model results presented here outline a mechanism of low frequency climate change that fits into Palmer's framework.

The model results for the unforced case raise some general issues about the interpretation of millennial scale variability in the climate record. If the real ENSO system is truly chaotic, we can expect there to be power in the climate system at low frequencies from purely tropical processes. Furthermore, this only involves near surface coupled interactions, and the deep ocean is *not* necessary for generating power at these timescales. The actual numerical results raise some particular issues. The estimate of the spectrum for both the Zebiak-Cane model and for the linear damped oscillator show peaks at low frequencies. These peaks are interpreted here to be purely random. On the basis of theory [*Turcotte*, 1993], we expect chaotic interactions to generate power in a broad band rather than at particular frequencies. We suggest that this is a more appropriate null hypothesis for testing peaks in the spectra of climate records. There may well be processes internal to the climate system that have spectral peaks at millennial timescales, but these peaks should be significant above the level of power that we expect from chaos in the climate system.

The model used in this study is highly idealized, and thus the results are subject to certain caveats. There are many processes that are not included in the model which may influence the way the coupled physics operate on Milankovitch and millenial timescales. It has been suggested that changes in the mean tropical SST and zonal SST gradient influence ENSO behavior [*Sun and Liu*, 1996, *Sun*, 1999], but many of the processes thought to be important in determining tropical SST are not included in this model. These include atmospheric convective boundary layer processes [*Sarachik*, 1978; *Betts and Ridgway*, 1989], atmospheric circulation [*Wallace*, 1992; *Fu et al.*, 1990; *Hartmann and Michelsen*, 1993; *Pierrehumbert*, 1995], ocean circulation [*Clement and Seager*, 1999], stratus clouds [*Miller*, 1997], and tropospheric water vapor content [*Larson and Hartmann*, 1999, *Seager et al.*, 1999]. On glacial-interglacial timescales, atmospheric CO_2 and high latitude climate conditions [*Broccoli and Manabe*, 1987; *Hewitt and Mitchell*, 1998] will likely affect tropical SST through both the ocean [*Bush and Philander*, 1998], and through the atmopshere [*Broccoli*, 1999]. Furthermore, changes in the strength of the Asian monsoons, either forced on a Milankovitch timescale, or unforced on a millennial timescale, are likely to have some impact on the ENSO system [i.e., *Shukla and Paolina* 1983].

The approach we have taken in this modelling study was to investigate the forced and unforced variability of this simplified coupled ocean-atmosphere system. More complete models are needed to further test the ideas presented here. The specific model results may be modified in the presence of more complete physics describing evolving climate conditions on these timescales. Nevertheless, the physical links underlying the model behavior on Milankovitch and millennial timescales are well known and unlikely to fail.

Acknowledgments. We are grateful to the two anonymous reviewers for providing useful insight. Amy Clement was supported by NASA grant NAGW-916. MAC was supported by NOAA grant NA36GP0074-02.

REFERENCES

Berger, A., Long-term variations of daily insolation and Quaternary climate changes, *J. Atmos. Sci.*, *35*, 2362–2367, 1978.

Betts, A. K., and W. Ridgway, Climatic equilibrium of

the atmospheric convective boundary layer over a tropical ocean, *J. Atmos. Sci.*, *46*, 2621–2641, 1989.

Bjerknes, J., Atmospheric teleconnections from the equatorial Pacific, *Mon. Weather. Rev.*, *97*, 163–172, 1969.

Broccoli, A., Extratropical Influences on tropical paleoclimates , *AGU Monograph: Mechanisms of millenial scale global climate change*, 1998, submitted.

Broccoli, A., and S. Manabe, The influence of continental ice, atmospheric CO_2 and land albedo on the climate of the Last Glacial Maximum, *Clim. Dyn.*, *1*, 87–99, 1987.

Bush, A.B.G. and S.G.H. Philander, The role of ocean-atmosphere interactions in tropical cooling during the last glacial maximum, *Science*, *279*, 1341–1344, 1998.

Cane, M., El Niño, *Annu. Rev. Earth Planet Sci.*, *14*, 43–70, 1986.

Cane, M. A., A. C. Clement, A. Kaplan, Y. Kushnir, D. Pozdnyakov, R. Seager, S. E. Zebiak, and R. Murtugudde, Twentieth century sea surface tempature trends, *Science*, *275*, 957–906, 1997.

Chang, P., B. Wang, T. Li, and L. Ji, Interactions between the seasonal cycle and the Southern Oscillation: Frequency entrainment and chaos in a coupled ocean-atmosphere model , *Geophys. Res. Lett.*, *21*, 2817–2820, 1994.

Clement, A. C., and R. Seager, Climate and the tropical oceans, *J. Clim.*, 1999, in press.

Clement, A. C., R. Seager, M. A. Cane, and S. E. Zebiak, An ocean dynamical thermostat, *J. Climate*, *9*, 2190–2196, 1996.

Clement, A. C., R. Seager, and M. A. Cane, Suppression of el niño during the mid-holocene by changes in the earth's orbit, *Science*, 1999, submitted.

Crowley, T., and G. North, *Paleoclimatology* , Oxford Univ. Press, New York., 1991.

Dijkstra, H. A., and J. D. Neelin, Ocean-atmosphere interaction and the tropical climatology, II, Why the Pacific cold tongue is in the east, *J. Clim.*, *8*, 1343–1359, 1995.

Fu, R., A. D. DelGenio, and W. B. Rossow, Behavior of deep convective clouds in the tropical Pacific deduced from ISCCP radiances, *J. Clim.*, *3*, 1129–1152, 1990.

Goddard, L., and N. Graham, El Niño in the 1990s., *J. Geophys. Res.*, *102*, 10,423–10,436, 1997.

Hartmann, D. L., and M. Michelsen, Large scale effects on the regulation of tropical sea surface temperature, *J. Clim.*, *6*, 2049–2062, 1993.

Hasselmann, K., Stochastic climate models. Part I: Theory, *Tellus*, *XXVIII*, 473–485, 1976.

Hewitt, C. D., and J. F. B. Mitchell, A fully coupled GCM simulation of the climate of the mid-Holocene, *Geophys. Res. Lett.*, *25*, 361–364, 1998.

Intergovernmental Panel on Climate Change, *Climate change: The IPCC Scientific Assessment*, Cambridge University Press, Cambridge, England, 365pp, 1995.

Jin, F.-F., J. D. Neelin, and M. Ghil, El Niño on the Devil's staircase: annual subharmonic steps to chaos, *Science*, *264*, 70–72, 1994.

Kleeman, R., J. McCreary, and B. Klinger, A mechanism for generating ENSO decadal variability , *Science*, 1999, submitted.

Larson, K., and D. L. Hartmann, A two box equilibrium model of the tropics, *J. Clim.*, 1999, in press.

Latif, M., R. Kleeman, and C. Eckert, Greenhouse warming, decadal variability or El Niño? An attempt to understand the anomalous 1990s, *J. Clim.*, *10*, 2221–2239, 1997.

Lorenz, E., Chaos, spontaneous climatic variations and detection of the Greenhouse effect, in *Greenhouse-Gas-Induced Climate Change: A Critical Appraisal of Simulations and Observations*, edited by M. E. Schlesinger, pp. 445–453, Elsevier, Amsterdam, the Netherlands, 1991.

Mann, M., and J. Lees, Robust estimation of background noise and signal detection in climatic time series, *Clim. Change*, *33*, 409–445, 1996.

Miller, R. L., Tropical thermostats and low cloud cover, *J. Clim.*, *10*, 409–440, 1997.

Münnich, M., M. A. Cane, and S. E. Zebiak, A study of self-excited oscillations of the tropical ocean atmosphere system, II, Nonlinear cases., *J. Atmos. Sci.*, *48*, 1238–1248, 1991.

Neelin, J. D., D. Battisti, A. Hirst, F. Jin, Y. Wakata, T. Ymamgata, and S. Zebiak, ENSO theory , *J. Geophys. Res.*, *103*, 14,261–14,290, 1998.

Palmer, T. N., Extended-range atmospheric prediction and the Lorenz model, *Bull. Am. Meteorol. Soc.*, *74*, 49–65, 1993.

Pierrehumbert, R. T., Thermostats, radiator fins, and the runaway greenhouse, *J. Atmos. Sci.*, *52*, 1784–1806, 1995.

Rajagopalan, B. and U. Lall and M.A. Cane, Anomalous ENSO occurences: an alternate view , *J. Climate.*, *10*, 2351–2357, 1997.

Rasmusson, E., and T. Carpenter, Variations in tropical sea surface temperature and surface wind fields associated with the Southern Oscillation/El Niño, *Mon. Weather. Rev.*, *110*, 354–384, 1982.

Sarachik, E. S., Tropical sea surface temperature: An interactive one-dimensional model., *Dyn. Atmos. Oceans*, *2*, 455–469, 1978.

Seager, R., A. Clement, and M. Cane, Glacial cooling in the tropics: Exploring the roles of tropospheric water vapor, surface wind speed and boundary layer processes, *J. Atmos. Sci.*, 1999, submitted.

Shukla, J., and D. Paolina, The Southern Oscillation and long range forecasting of the summer monsoon rainfall over India, *Mon. Weather. Rev.*, *111*, 1830–1837, 1983.

Sun, D.-Z., Global climate change and El Niño: A theoretical framework , in *El Niño and the Southern Oscillation: Multiscale variability, global and regional impacts*, edited by E. by H.F. Diaz and V. Margraf, p. in press, Cambridge Univ. Press, New York, 1999.

Sun, D.-Z., and Z. Liu, Dynamic ocean-atmosphere coupling: a thermostat for the tropics, *Science*, *272*, 1148–1150, 1996.

Timmermann, A., J. Oberhuber, A. Bacher, M. Esch, M. Latif, and E. Roeckner, ENSO response to greenhouse warming, *Geophys. Res. Letts.*, 1998, submitted.

Trenberth, K., and T. J. Hoar, The 1990-1995 El Niño-Southern Oscillation event: Longest on record, *Geophys. Res. Lett.*, *10*, 2221–2239, 1996.

Trenberth, K., and J. W. Hurrell, Decadal atmosphere-ocean variations in the pacific, *Clim. Dyn.*, *9*, 303–319, 1994.

Turcotte, D., *Fractals and chaos in geology and geophysics*, Cambridge University Press, 1993.

Tziperman, E., M. A. Cane, and S. E. Zebiak, Irregularity and locking to the seasonal cycle in an ENSO prediction model as explained by the quasi-periodicity route to chaos, *J. Atmos. Sci.*, *52*, 293–306, 1994.

Tziperman, E., S. E. Zebiak, and M. Cane, Mechanisms of seasonal-ENSO interaction, *J. Atmos. Sci.*, *54*, 61–71, 1997.

Wallace, J. M., Effect of deep convection on the regulation of tropical sea surface temperature, *Nature*, *357*, 230–231, 1992.

Zebiak, S. E., and M. A. Cane, A model El Niño-Southern Oscillation, *Mon. Weather. Rev.*, *115*, 2262–2278, 1987.

Zebiak, S. E., and M. A. Cane, Natural climate variability in a coupled model., in *Greenhouse-Gas-Induced Climate Change: A Critical Appraisal of Simulations and Observations*, edited by M. E. Schlesinger, pp. 457–469, Elsevier, New York, 1991.

Zhang, Y., J. M. Wallace, and D. S. Battisti, ENSO-like decade-to-century scale variability: 1900-93, *J. Clim.*, *10*, 1004–1020, 1997.

M. Cane, Lamont-Doherty Earth Observatory, Palisades, New York 10964 (e-mail: mcane@ldeo.columbia.edu)

A. Clement, LODYC, Tour 26, étage 4, CASE100, 4 Place Jussieu, 75252 Paris, Cedex 05, France (e-mail:Amy.Clement@ipsl.jussieu.fr)

A Role for the Tropical Pacific Coupled Ocean-Atmosphere System on Milankovitch and Millennial Timescales. Part II: Global Impacts

Mark Cane and Amy C. Clement

Lamont-Doherty Earth Observatory of Columbia University, Palisades, New York

We offer the hypothesis that global scale millennial and glacial cycles may be initiated from the tropical Pacific. Part I used model results to illustrate how nonlinear ocean-atmosphere interactions in the tropical Pacific could generate variations in the field of sea surface temperature (SST) on both orbital and millennial timescales. The physics underlying these variations is essentially the same as that causing ENSO (El Niño - Southern Oscillation) variability in the modern climate. Here we argue that, as with ENSO but on paleoclimatic timescales, these changes in SST distribution will be accompanied by changes in the location of atmospheric convection, which will alter the global climate via atmospheric teleconnections. By analogy with ENSO, it is hypothesized that the cold phase will increase the glaciation over North America, increase low cloud cover, and reduce atmospheric water vapor. All tend to cool the earth either by increasing the planetary albedo or reducing greenhouse trapping. The warm phase has the opposite tendencies.

We also critique the hypothesis that millennial changes are triggered by changes in the production of North Atlantic Deep Water.

1. INTRODUCTION

The discovery of global climate variations with millennial timescales is arguably the most profound paleoclimate surprise of recent decades. Though the observational picture is still quite incomplete, evidence is accumulating rapidly to show that these changes can set on abruptly, impact vast areas of the globe, and have existed through glacials and interglacials over the past 500,000 years. Illustrations abound in the present volume. Theory lags behind: it is fair to say that there are no compelling explanations for the causes of these millennial variations.

Here we propose a mechanism that locates the origin of these changes in ocean-atmosphere interactions in the tropical Pacific. The general idea may be grasped by analogy with the largest cause of climate variability in the modern period, ENSO (El Niño and the Southern Oscillation). ENSO is not a consequence of external forcings, but of coupled instabilities internal to the tropical Pacific ocean- atmosphere system. ENSO has global consequences because variations in the location of tropical convection perturb the global atmospheric circulation.

We hypothesize that the same physics operates on millennial timescales. In Part I of this two part study we presented evidence from a model calculation that the tropical Pacific interactions can induce fluctuations at these much longer timescales. Here we will argue that the global consequences of changes in the state of the tropical Pacific, which depend on relatively fast atmospheric physics, also operate at these longer timescales. Changes in the state of the tropical Pacific also may be forced by orbital variations (Clement et al, 1999). With

Mechanisms of Global Climate Change at Millennial Time Scales
Geophysical Monograph 112

orbital forcings the very large changes in the tropical Pacific are a consequence of its great sensitivity, whereas with millennial variations they are due to its outright instability. With either reason for the tropical Pacific changes, the global consequences follow. Our argument here applies to both millennial and Milankovitch variations. Evidence for or against it in one setting would likely apply to the other. However, our hypothesis is still crude and sketchy, and differences between the two cases will surely emerge as it is refined. For example, feedbacks from the extratropics to tropics should differ because of the direct influence of orbital variations on high latitudes.

It is worthwhile to recall that the state of knowledge a decade ago seemed to rule out a role for the tropics in causing any of the changes on paleoclimate timescales. According to the CLIMAP (1976) reconstruction of global sea surface temperature (SST), tropical temperatures at the Last Glacial Maximum (LGM) hardly differed from their present values. The largest changes were found in the North Atlantic, typically 10^oC and more, rendering much of the surface ice-covered. Further, the North Atlantic is the "source" of the global ocean circulation in that surface water there can become cold and salty enough to sink to the bottom and then spread globally. As it spreads it works its way back to surface, eventually returning to the North Atlantic to sink once again. This picture of a "conveyor" circulation is, of course, a vast simplification of the very complicated and nuanced global plumbing system that is the global ocean's thermohaline circulation, but it is a useful conceptual picture all the same.

A great deal of heat is associated with the conveyor, heat that is given up to the atmosphere when the water in the North Atlantic cools. Noting that deep water does not form in the Pacific, one may take the fact that Europe is several degrees warmer than corresponding latitudes in western North America as a measure of the power of the conveyor. There is strong evidence that the character of the conveyor has varied over the last 800 kyr (Broecker 1995 p171, Raymo et al. 1990), perhaps shutting down altogether at the LGM. It was argued that such a shutdown would surely cool Europe, and the rest of the world as well.

In the last two decades new data has enriched and occasionally changed the CLIMAP picture, providing facts to explain, new constraints on hypotheses for how the earth's climate system operates. Important among these:

(i) climate changes are often (usually?) abrupt;

(ii) there are many climate changes with periods of O(1 kyr) or more (e.g. Dansgaard-Oeschger cycles, Heinrich events) which have no obvious orbital pacing;

(iii) changes are global in extent, not just in the Northern Hemisphere, and not just near the poles;

(iv) many (most?) changes are globally simultaneous;

(v) the tropics cooled by 3^oC - 5^oC at the LGM.

This new information raises 2 new questions: Why are the changes abrupt? What drives the non-orbital (aperiodic) cycles? Both point to the fact that the response of the earth's climate system does not closely follow the orbital forcing.

Before beginning to address these questions, points (i) -(v) require a few comments. Evidence from ice cores (Johnsen et al. 1992, Grootes et al. 1993, Brook et al. 1996) indicates that in Greenland at least, "abrupt" can mean within a few decades. It is possible that changes are equally abrupt elsewhere, but the evidence doesn't allow us to say so with certainty. Though the orbital variations are complex, there is no obvious orbital signal significant enough to account for variations at periods shorter than 11 Kyr. It is particularly relevant to the arguments in this paper to note that the millennial variations are well documented in the tropics: for example, see Curry and Oppo (1997) for the Atlantic Ocean; Sirocko (1993) for the Indian Ocean; Linsley (1996) and Beck et al. (1997) for the Pacific Ocean. Changes over land in the tropics are indicated by methane variations (Chappellaz et al. 1993). Other references may be found thoughout the present volume. By stating that global variations are simultaneous, we only mean to say that many global changes happen at the same time to within the resolution of the age models for the data. We do not contradict such evidence as Blunier et al (1998) by claiming that all simultaneous changes are in the same sense (e.g. that all points on the globe warm or cool synchronously). ENSO provides a modern example of a set of simultaneous climate changes in which some parts of the globe get warmer while others cool, and some experience drought while others flood.

A colder tropics at once makes the North Atlantic a less viable cause and opens the possibility of a role for the tropics. The perspective being presented here would be unsupportable if the tropical SSTs were not appreciably different in glacial times, so we briefly review the evidence for it. There is not yet universal agreement that the CLIMAP version must be abandoned, but the preponderant evidence now favors a substantial tropical cooling in glacial times. The CLIMAP reconstruction is based on finding the relative abundances of different types of forminifera in sediment cores, and then estimating paleo SSTs from statistical relations derived from observations of the modern ocean. The assumption is that the relative abundances are indeed thermometers, and that they have not changed over time as so many other things have changed. The more foolproof paleothermometers, based on geochemical analyses (Sr/Ca, noble gases) indicate a cooling of $3-5^oC$. (Guilderson et al., 1994; Stute et al., 1995). The $\delta^{18}O$ difference, which at first seemed to agree with CLIMAP, implies much colder temperatures if one accepts pore water $\delta^{18}O$ changes as a measure of ice volume (Schrag et al., 1996). The influence of changing pH (Spero et al, 1997) may be another factor helping to reconcile the geochemical evidence. Measurements of $\delta^{18}O$ in ice cores from Peru and Tibet are additional evidence for colder tropical sea surface temperatures (Thompson et al., 1995; 1997).

There is also powerful physical evidence. Mountain snowlines descended in the tropics, suggesting that the whole atmosphere cooled, right down to the surface. The only other possibility is that the atmospheric lapse rate steepened, but it has proven difficult to construct a plausible scenario where the lapse rate changes and the SST hardly changes. The observed increase in large dust particles indicates stronger winds, and stronger winds implies colder SSTs – as in a modern La Niña. Finally, there was less CO2 in the atmosphere, and lower greenhouse gas content imples lower temperatures. An SST field consistent with the dominant evidence would be about 5^oC lower in the E Pacific and Atlantic, and close to 3^oC lower in the W Pacific and Indian Oceans. Among the many newer tropical estimates, only one, the alkenone data is not fully consistent. It indicates a change of about 2^oC all across the tropics, including the eastern Pacific.

2. A RECONSIDERATION OF THE ROLE OF THE NORTH ATLANTIC

At first these new facts regarding millennial timescale variability seemed to promote the importance of the North Atlantic and its conveyor circulation. Many of the observations supporting them come from the North Atlantic. Both theory and models shown that changes in salinity, generated either by glacial melting or by internal oscillations in the ocean, could cause the conveyor circulation to shut down abruptly. The North Atlantic would then cool, the nearby atmosphere would cool, and the climate would change – abruptly. However, it is now seen that on the whole these new facts cast some doubt on the paradigm with the North Atlantic led conveyor as prime mover. In addition to the questions raised at the Chapman Conference, it is noteworthy that in a recent article Broecker (1997), though not abandoning the conveyor circulation as a drier of millenial scale climate variability, argues for a central role for water vapor, a variable with a largely tropical source.

There is no doubt that there is a conveyor circulation in the modern ocean, and that it has varied in the past (e.g. Broecker, 1997). But not all the fluctuations observed in the climate system (e.g. each of the Dansgaard-Oeschger cycles) correspond to major conveyor changes.

How to have the changes be global and simultaneous is a difficult problem to solve via the conveyor (viz Broecker 1995 p. 258, also Broecker 1997]. Broecker (1997) has calculated the conveyor heat loss to be $3x10^{21}calyr^{-1}$, heat that goes to warm the atmosphere. This is a reasonable estimate of heat loss in the North Atlantic, but whether one should interpret it as a loss by the conveyor is open to interpretation. The North Pacific is not a deep water formation site, but nonetheless it gives up a great deal of heat to the atmosphere. Further, the evidence is that its exceptional for the conveyor to shut down totally . In its weak state waters continue to sink, but stop well short of the bottom, returning south at an intermediate depth (Boyle 1995, Broecker 1995 p.264). The change in the heating of the atmosphere is then only a fraction of the $3x10^{21}calyr^{-1}$ it would be for a total collapse of the conveyor.

While the conveyor carries a lot of heat by local standards, making a marked difference in the climate for those of us who live near the North Atlantic, it is not a lot of heat by global ocean measures. For example, the surface heat exchange difference between ENSO warm and cold phases is about $150Wm^{-2}$ over an equatorial Pacific area of about 20x 75 degrees of latitude and longitude, a difference in heat loss of $4.5X10^{19}calday^{-1}$. Each phase persists for about 200 days, yielding a change over the approximately 2 years between extremes of $9x10^{21}cal$.

No matter what it does to the deep ocean, the conveyor must change the temperature at the sea surface if it is to affect the atmosphere. It is difficult to see how changes in the conveyor could have an appreciable impact on the whole globe, as the paleo data now calls for. The easiest way would be to have the conveyor change the tropics, which would then change everything else. In order to give the conveyor an unrealistically favorable chance, let us confine ourselves to the tropical Pacific alone and imagine that all the 10 Sv ($10x10^6m^3s^{-1}$) estimated to leave the North Atlantic is piped directly to the surface layer (the top 50 m) of the tropical Pacific

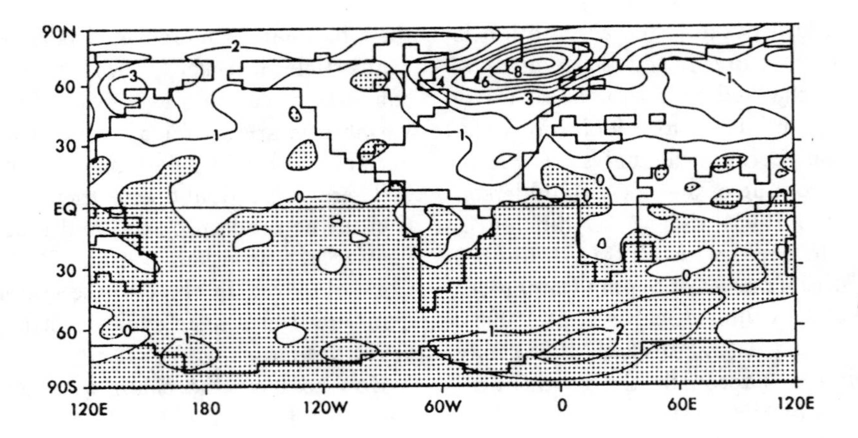

Figure 1. Difference in surface air temperature between a coupled GCM run with a strong thermohaline overturing in the North Atlantic and with no thermohaline overturning (from Manabe and Stouffer 1988).

(10^oNto10^oS, say). The consensus is that the rate of deep water formation is < 20 Sv, with approximately 10 Sv getting out of the N Atlantic across the equator; e.g. Speer and Tziperman (1992) estimate 9 Sv crosses the equator and consistent with that, they find the same amount of water crossing isopycnals. That is, about half of the 20 Sv is entrained locally in the N Atlantic, and recycled there. In fact, most of the goes to the southern ocean, and is much changed before it moves on to the rest of the world. At an input rate of 10 Sv it would take about 5 years to refill this tropical Pacific surface layer. Now suppose the conveyor switches off and this water supply suddenly becomes 7^oC colder. It would surely tend to cool the tropics, but meanwhile the surface heat exchange would be tending to warm the surface back to where it was; the e-folding time for this in the present climate is about 0.5 yrs (Seager et al 1988) – about 10 times faster than the conveyor is changing it. The net temperature change would only be about 0.6^oC in this fantastically favorable case.

This simple calculation is supported and extended by model calculations. There have been many model calculations, including some with comprehensive ocean-atmosphere coupled models, in which the conveyor is shut down and/or the North Atlantic SST is cooled (Rind et al 1986, Manabe and Stouffer 1988, Rahmstorf 1994, Tziperman 1997). A number of them are cited in papers advocating a major role for the conveyor in changing climate. Without exception they all show a strong impact in the areas adjacent to the North Atlantic, and little effect elsewhere on the globe (Figure 1). Some do show significant effects in polar latitudes of the Southern Ocean, and in the Kuroshio region. Thus these models, which include pathways through the atmosphere as well as the ocean, are unable to create a global impact.

One may also look to the modern data record, though with the caveat that it contains nothing in the North Atlantic as strong as the millennial oscillations. Still, the many data studies of the principle mode of climate variability there, the North Atlantic Oscillation, show strong signals around the North Atlantic and downstream of it, but nothing over the rest of the globe (e.g. Hurrell and van Loon, 1997). The pattern is consistent with model calculations, and may be taken as evidence that the atmospheric models do a credible job of simulating the response to SST variations in the North Atlantic.

A final problem with the North Atlantic centered view is the evidence that the global exchange between the surface and deep ocean goes on more or less unchanged even as the formation rates of NADW vary (cf. Broecker 1995) It appears that the other deep water sources in operation today, those around Antarctica, step up production and make up the difference. Its hard to see how the change in source could matter for the delivery of heat to the global surface layers.

3. A ROLE FOR THE TROPICAL PACIFIC

None of the foregoing denies that there are changes in the conveyor. We view them as part of the response to whatever is driving the climate system into new states. They are not the prime mover, but part of the chain of consequences. We now explore the possibility that this chain is set in motion from the tropical Pacific for both forced (orbital) and internal (Dansgaard-Oeschger) climate cycles.

Orbital or not, the incident global annual insolation is almost unchanged through these cycles. Since all these cycles last long enough to rule out significant heat storage, the net heat exchange to space must be close to zero. Even with orbital cycles, only the latitudinal and seasonal distributions of insolation at the top of the atmosphere change appreciably. Globally and annually averaged, the planet receives about the same energy input from the sun. Thus the heat leaving at the top of the earth's atmosphere also must be nearly unchanging. In order for this to occur at a cooler planetary temperature, the earth must either reflect more short wave (solar) radiation back to space via greater ice and snow extent, or more cloud cover; or trap less infrared radiation via less greenhouse gas (water vapor, CO2) or less cloud. Clouds have potentially conflicting roles, and, indeed, for reasons still mysterious, high clouds seem to exactly compensate the heat they reflect back to space by blocking the escape of longwave energy from below (Hartmann et al. 1992). Low clouds, however, yield a net cooling.

Our hypothesis is that both orbital variations and internal instabilities change tropical Pacific SST distributions. This change in the thermal boundary condition on the atmosphere alters the distribution of atmospheric heating in the tropics, which in turn alters the global climate. ENSO is the most familiar instance of such a mechanistic chain, but the physical links are quite general. Given a change in tropical Pacific SST distribution, the rest of the sequence works through the atmosphere and applies on all time scales intraseasonal and longer: centers of tropical convection tend to lie over the warmest water, so moving the locus of that water moves the convection; changing the location of the convection changes the teleconnection pattern – the impacts on distant locations (Hoerling et al 1997). ENSO provides clear examples. In a warm ENSO event (El Niño) the warm waters and the center of atmospheric convection move eastward from near Indonesia to the vicinity of the dateline. As a consequence northern North America warms. In a cold event (La Niña) the tropical Pacific warm pool contracts back toward Australasia, the convection follows, and northern North America typically is colder than average. Figure 2 illustrates for the recent warm event. Again, this chain of effects is not particular to ENSO. The atmospheric response time is short, and the same physics is known to work on intraseasonal timescales (Higgins and Mo 1997). Nor is there any reason why it would fail for longer time periods: numerical experiments with permanent anomalies show a very similar response (Kumar et al 1994).

Nor is the "ENSO physics" leading to tropical SST changes restricted to the ENSO cycle. As first suggested by Bjerknes (1969; for a succinct account see Cane 1991; for an update see Neelin et al 1998), a positive feedback operates in the coupled tropical Pacific ocean-atmosphere system. In the cold phase a stronger east-west SST gradient along the equator drives stronger easterly winds, which increase upwelling, thermocline tilt, and zonal currents, resulting in still stronger SST gradients. In the warm phase, a weaker SST gradient causes the winds to slacken which changes the ocean to further weaken the SST gradient. The ocean is too slow for this sequence to operate on intraseasonal timescales, but the equatorial ocean is fast enough to participate in variability on all timescales longer than the 4 years or so characteristic of the ENSO cycle. There is no obvious reason why this positive feedback should fail at millennial or orbital periods.

According to our present understanding ENSO is an internal instability of the tropical Pacific coupled ocean-atmosphere system (Neelin et al 1998). In this theory, ENSO variability in the modern climate record can be explained in terms internal to the climate system. Nonetheless, the same theory indicates that because of its sensitivity to the seasonal cycle in the tropics (Zebiak and Cane 1987, Tziperman et al. 1994) ENSO should be altered by orbital variations, especially precessional changes. The same physical links imply that the mean state of the tropical ocean-atmosphere also should change as the seasonal solar heating varies. This idea has been verified in a number of model studies (Dijkstra and Neelin 1995, Clement et al 1996).

Thus there is a reasonable expectation for tropical Pacific variability on orbital timescales. The model runs reported in Part I were intended to check this expectation. Briefly, the same Zebiak and Cane (1987) model in use for more than a decade to study and predict ENSO was run for 150,000 years, imposing the anomalies in solar heating due to orbital variations (see Clement et al. 1999 in addition to Part I). In response, the model shows increases or decreases in the frequency of ENSO warm or cold events, and changes in their average amplitudes. In some orbital states there are more warm events, in some more cold events, in some fewer extremes of either sign. Along with such variations go changes in average SST and thus in the mean position of the warmest SSTs. The changes in SST regime, which often take less than

1000 years, are more abrupt than the orbital variations, though not so abrupt as the changes observed in the Greenland ice core.

Surprisingly, this model run also showed more power at millennial timescales than would be expected just from random fluctuations of a process with its power concentrated at typical ENSO timescales (see Part I). This excess power must arise from nonlinear interactions within the model system, not as the red noise extension of the subdecadal ENSO cycle. Nor can it arise from external causes: the excess millennial power is even more impressive in a 150,000 year run with orbital parameters fixed at present day values (see Fig. 5 of Part I).

These runs do show peaks at periods near 1500 years, similar to those in the paleoclimate record, (viz Fig. 5 of Part I). However, as discussed in Part I (viz Fig. 6 there), the most reasonable interpretation of the model run is that while there is broad band power at millennial periods, any peaks are just artifacts from too short a record. We suspect that the same is true of the paleoclimate record, which supports the idea of these fluctuations stemming from internal instabilities rather than external forcings. The observations showing marked peaks extend for no more than a few 10s of kyrs, too short to resolve peaks around 1500 years. Moreover, different paleoproxies show somewhat different peaks, and the one record extending for several 100 ky, long enough to isolate millennial peaks, shows the very many peaks one would expect if the true process is broad band (Oppo et al 1998). On the other hand, we cannot absolutely rule out the possibility that there is truly a single strong peak or a few strong peaks, split into many peaks in the observational record by errors in the age model used in converting from sediment core depth to time.

The highly simplified Zebiak and Cane model was not designed with paleoclimate variations in mind, and we do not interpret these runs in any detail. (Part I discusses model limitations.) We use the model results here solely to illustrate two ideas that may be derived from a consideration of the physics of the coupled ocean-atmosphere system. First, the orbital variations alter the seasonal cycle which exerts such a powerful influence on tropical Pacific interannual variability. Consequently, the tropical Pacific may vary on orbital timescales independent of influences from the rest of the climate system. Second, in common with other nonlinear systems (cf Lorenz, 1991), the tropical Pacific ocean-atmosphere may exhibit regime like behaviors which persist far longer than any obvious intrinsic physical timescale. Consequently, the tropical Pacific may vary on millennial timescales independent of influences either from within the rest of the climate system or from external causes.

The crucial issue for the rest of the climate system is the locus of atmospheric convection in the tropics. In the present climate we know that different locations have different implications for the impact on higher latitudes, and these would presumably be influenced by other differences between the glacial atmospheric circulation and the modern. We are not at all confident that our simple model yields a trustworthy notion of where the center of convection moves as the orbital configuration varies, or where it is during the stadial and interstadial phases of the millennial cycles. Currently, the best documented description is that for the ENSO cycle. In the warm (El Niño) phase, convection moves with the warmest water into the central Pacific, in the vicinity of the dateline. In the cold (La Niña) phase the warm pool contracts and the convection moves back over the Indonesian region. We describe some of the warm phase response; the cold phase response is roughly (sometimes very roughly) opposite (see Hoerling et al 1997 for an account of the differences).

Figure 2 illustrates one well known influence of El Niño, a warming of northern North America. An interesting if somewhat anecdotal confirmation is based on Hudson Bay Trading Company records of the date that the ice goes out on Hudson Bay (Moodie and Catchpole, 1975). Hamilton and Garcia (1986) showed that in El Niño years the ice goes out early. In view of the broad spatial scale of the teleconnection pattern, the implication is that moving the convection to the central Pacific will help to melt ice and snow in Canada, while positioning it in the far western Pacific will favor ice sheet growth. Extrapolating, moving the convection to the far west for a long enough period will help to grow a Laurentide Ice Sheet, while the state resembling El Niño will tend to melt it. If the pattern of ENSO teleconnections resembled the present one, variations in the southern edge of the ice sheet would accompany millennial cycles.

Something of a case can be made for the ENSO cycle having the proper impacts on the other albedo/greenhouse influences listed above. Satellite based cloud climatologies (Rossow and Schiffer 1991) show a tendency toward less low cloud during warm events (Figure 3), when the tropical area covered by cold waters is at a minimum. This tendency is consistent with the results of Klein and Hartmann (1993), who suggest that there will be more low cloud with a strengthening of the inversion at the top of the atmospheric boundary layer. As illustrated in

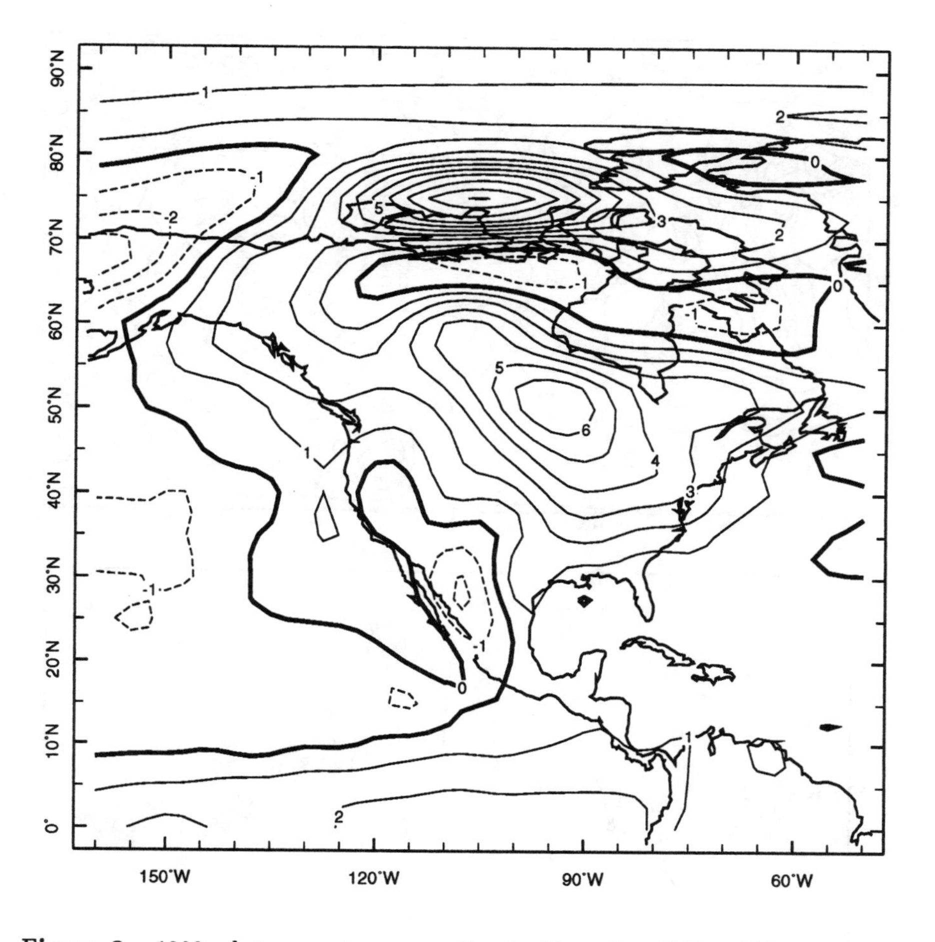

Figure 2. 1000 mb temperature anomalies for December 1997 to February 1998.

Figure 4 from Soden (1997) there is also some evidence for an increase in atmospheric water vapor during warm events (also see Sun and Oort 1995). These changes are small in amplitude, but they all line up in the right sense for El Niño like heating to favor interglacial conditions and La Niña like heating to favor glacial conditions. In the modern record, only atmospheric CO_2, which decreases during El Niño events, appears to be out of line. However, it may be that the decrease is short-lived, and that the long term impact of El Niño events is a net increase (Keeling et al, 1995; Rayner et al, 1999). (Also see Anderson et al. (1990), whose study of pleistocine marine varves off California led them to suggest that more CO_2, may be sequestered in the ocean when cold event conditions prevail.) A more complete model of the global climate system will be needed to see if these effects can trigger the substantial changes observed in millennial and orbital timescale climate records.

The limitations of the model render it silent on the question of how the rest of the world will feed back on the tropical Pacific. Earlier studies of the role of the tropics in paleoclimate have emphasized the changes in the monsoons (Kutzbach and Liu 1997, Lindzen 1993, Lindzen and Pan 1994). The evidence that the monsoons change with orbital variations is compelling, and it is highly likely that these changes would influence the tropical oceans. Once the higher latitudes have cooled appreciably, there is likely to be an impact on the tropical Pacific, either through the ocean (Bush and Philander 1998) or the atmosphere (Broccoli 1998). Thus, we must remain open to the possibility that feedbacks onto the tropical Pacific will alter the picture substantially. At present we cannot say whether they are more likely to reinforce or to interfere with the local feedbacks.

In addition to its regional limits, the model is an anomaly model, calculating only changes from the modern climate. By construction it is constrained to stay near the present climate, so its responses to orbital changes and its millennial variations are likely to be understated versions of reality. They are best taken as

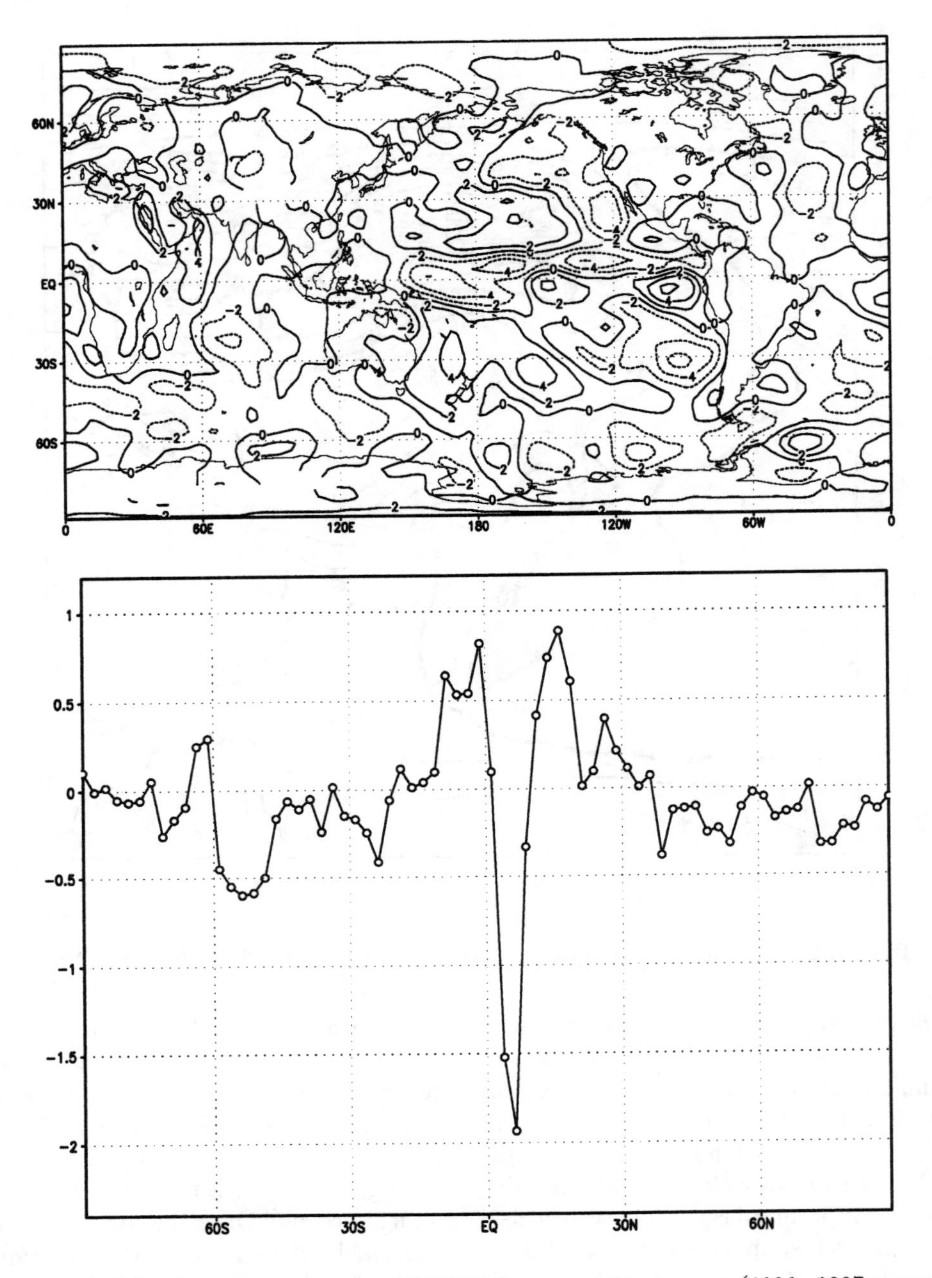

Figure 3. (a) Low cloud frequency from ISCCP for warm event years (1986, 1987 averaged from July-June) minus cold events years (1984,1988). (b) Zonal average of (a).

illustrations of possible behaviors rather than as literal predictions of the paleoclimate record. Further experiments with more complete models will be needed to build confidence in the idea that the tropical Pacific is a substantial driver of paleoclimate variations.

It will be important to develop a solid notion of where the center of convection moves as the orbital configuration varies, where it is during the stadial and interstadial phases of the millennial cycles. Different locations have different implications for the impact on higher latitudes, though these would presumably be influenced by other differences between the glacial atmospheric circulation and the modern.

In addition to more modeling work, there is a need for more observational studies. Most of the paleoclimate record comes from higher latitudes, especially the North Atlantic. There is a strong need for a more complete picture of the tropical Pacific. What can be said

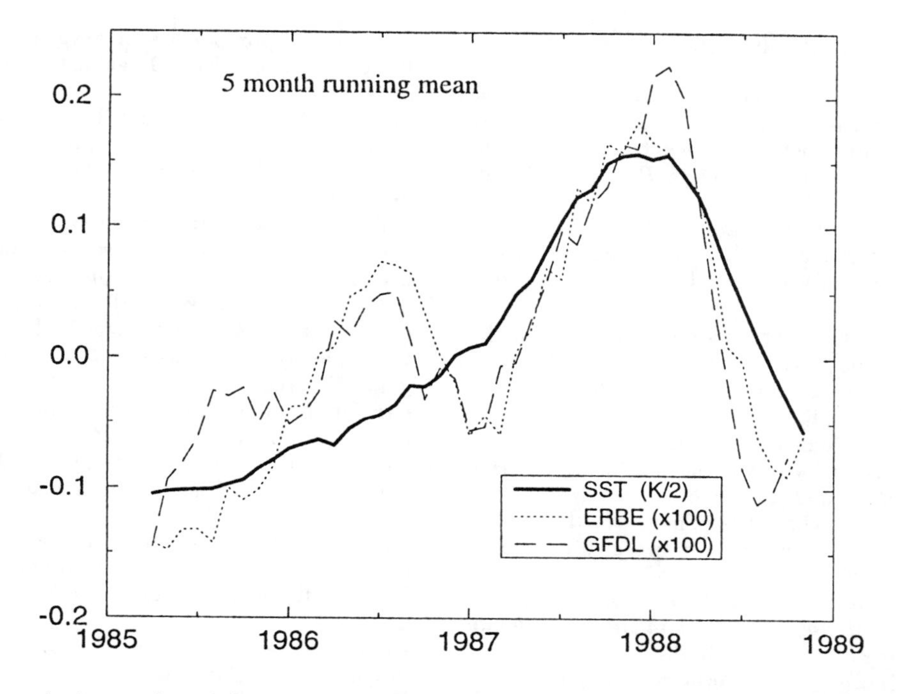

Figure 4. A time series of the tropical-mean interannual anomaly of fractional greenhouse trapping (which removes the effect of the change in surface temperature on greenhouse trapping) for ERBE observations dotted, for the GFDL GCM simlulations, dashed, and observed SST anomaly, solid (from Soden 1997).

about the changes in SST, east and west? How did the thermocline depth and slope vary over time? Can these changes be synchronized with changes in the tropical Atlantic and Indian Oceans, as well as with high latitude observations?

As with the North Atlantic hypotheses, the tropical scenario is neither complete and nor problem free. Interactions with the rest of the global climate system must be spelled out. Even granting that the tropical Pacific influences glacial growth and decay in North America, variations in Northern Hemisphere summer insolation must have an influence as well, and the proper accounting of the combined impact of the two is needed. Even if the North Atlantic is not the prime mover, there is no doubt that proxies in that region have recorded large changes during millennial and orbital cycles, changes that any theory must explain. While it is true that ENSO's impacts are nearly global, the North Atlantic and Europe are among the the places it seems to largely miss. It is not a complete miss (Davies et al. 1997; Rogers 1984), and recently Huang et al (1998) have suggested that there is a coherent impact on the NAO when the Pacific anomalies are strong. There is no obvious strong direct connection from the tropical Pacific, but an indirect mechanism is a possibility For example, one may argue from the well established fact that there is an ENSO impact on the tropical Atlantic, and the increasing evidence that the North Atlantic Oscillation, which has strong signals in Europe as well as its eponymous region, can be driven from the low latitude Atlantic (Robertson et al. 1999).

It is premature to make strong claims for the tropical Pacific hypothesis sketched here. The causal chain from the tropical Pacific to millennial scale variability still has serious gaps and weaknesses, though arguably no worse than other proposed explanations for the expanded inventory of paleoclimate cycles. However, paleoclimate evidence indicates that these cycles are global, including the tropics, and quite often globally synchronous, so our understanding and observations of modern climate make it hard to see how an adequate scenario for paleoclimate variations can be constructed without the tropical Pacific as a player. Even at this early stage, the perspective presented here points in a new direction for modeling and observational studies, one that uses our ideas of modern climate variability as a guidepost.

Acknowledgments. We thank two anonymous reviewers and our editor, Robin Webb, for constuctive suggestions that substantially improved the paper. This work was supported by National Science Foundation Grant No. ATM 92-24915.

REFERENCES

Anderson, R., Linsely, B., and Gardner, J. (1990). Expression of seasonal and enso forcing in climate variability at lower than enso frequencies: evidence from pleistocene marine varves off california. *Paleogeo. Paleoclim. Paleocean.*, *78*, 287–300.

Beck, J., Récy, J., Taylor, F., Edwards, R. L., and Cabioch, G. (1997). Abrupt changes in the early Holocene tropical sea surface temperature derived from coral records. *Nature*, *385*, 705–707.

Bjerknes, J. (1969). Atmospheric teleconnections from the equatorial Pacific. *Mon. Wea. Rev.*, *97*, 163–72.

Blunier, T., Chappellaz, J., Schwander, J., Dallenbach, A., Stauffer, B., Stocker, T., Raynaud, D., Jouzel, J., Clausen, H., Hammer, C., and Johnsen, S. (1998). Asynchrony of Antarctic and Greenland climate change during the last glacial period. *Nature*, *394*, 739–743.

Boyle, E. A. (1995). Last Glacial Maximum North Atlantic deep water- on, off, or somewhere in between. *Phil. Trans. Roy. Soc.*, *348*, 243–253.

Broccoli, A. (1998). Extratropical Influences on tropical paleoclimates . *AGU Monograph: Mechanisms of millenial scale global climate change*. submitted.

Broecker, W. (1995). *The glacial world according to Wally*. Eldigio, Palisades, NY. p.318.

Broecker, W. (1997). Thermohaline circulation, the Achilles heel of our climate system: will man made CO_2 upset the current balance? *Science*, *278*, 1582–1588.

Brook, E., Sowers, T., and Orchardo, J. (1996). Rapid variations in atmospheric methan concentration during the past 110,000 years. *Science*, *273*, 1087–1091.

Bush, A. and Philander, S. (1998). The role of ocean-atmosphere interactions in tropical cooling during the last glacial maximum. *Science*, *279*, 1341–1344.

Cane, M. A. (1991). Forecasting El Niño with a Geophysical Model. In Glantz, M., Katz, R., and Nicholls, N., editors, *Teleconnections Linking Worldwide Climate Anomalies.*, chapter 11, pp. 345–369. Cambridge U. Press, Cambridge, England.

Chappellaz, J., Blunier, T., Raynaud, D., Barnola, J., Schwander, J., and Stauffer, B. (1993). Synchronous changes in atmospheric CH_4 and Greenland climate between 40 and 8 kyr BP. *Nature*, *366*, 443–445.

Clement, A. C., Seager, R., and Cane, M. A. (1999). Orbital controls on the tropical climate. *Paleoceanography*. accepted.

Clement, A. C., Seager, R., Cane, M. A., and Zebiak, S. E. (1996). An ocean dynamical thermostat. *J. Climate*, *9*, 2190–2196.

Curry, W. B. and Oppo, D. (1997). Synchronous, high-frequency oscillations in the tropical sea surface temperatures and North Atlantic Deep Water production during the last glacial cycle. *Paleoceanography*, *12*, 1–14.

Davies, J., Rowell, D., and Folland, C. (1997). North Atlantic and European seasonal predictability using an ensemble of multidecadal atmospheric GCM simulations. *Int. J. Clim.*, *17*, 1263–1284.

Dijkstra, H. A. and Neelin, J. D. (1995). Ocean-atmosphere interaction and the tropical climatology. Part II: Why the Pacific cold tongue is in the east. *J. Climate*, *8*, 1343–1359.

Grootes, P., Stuiver, M., White, J., Johnsen, S., and Jouzel, J. (1993). Comparison of oxygen-isotope records from the GISP2 and GRIP Greenland ice cores. *Nature*, *366*, 552–554.

Guilderson, T. P., Fairbanks, R. G., and Rubenstone, J. L. (1994). Tropical temperature variations since 20 000 years ago: Modulating interhemispheric climate change. *Science*, *263*, 663–665.

Hamilton, K. and Garcia, R. (1986). El Nino Southern Oscillation Events and their associated mid latitude teleconnections 1532-1828 . *Bull. Am. Soc.*, *67*, 1354–1361.

Hartmann, D. L., Ockert-Bell, M. E., and Michelsen, M. (1992). The effect of cloud type on Earth's energy balance: Global analysis. *J. Climate*, *5*, 1281–1304.

Higgins, R. and Mo, K. (1997). Persistent North Pacific circulation anomalies and the tropical intraseasonal oscillation. *J. Climate*, *10*, 223–244.

Hoerling, M.P., A. K. and Zhong, M. (1997). El Niño, La Niña, and the nonlinearity of their teleconnections. *J. Climate*, *10*, 1769–1786.

Huang, J., Higuchi, K., and Shabbar, A. (1998). The relationship between the North Atlantic Oscillation and El Nino Southern Oscillation. *Geophys. Res. Lett.*, *25*, 2707–2710.

Hurrell, J. W. and van Loon, H. (1997). Decadal Variations in Climate associated with the North Atlantic Oscillation. *Climatic Change*. In press.

Johnsen, S., Clausen, H., Dansgaard, W., Fuhrer, K., Gundestrup, N., Hammer, C., Iversen, P., Jouzel, J., Stauffer, B., and Steffensen, J. (1992). Irregular glacial interstadials recorded in a new Greenland ice core. *Nature*, *359*, 311–313.

Keeling, C.D. and Whorf, T.P. and Wahlen, M. and Vanderplicht, J. (1995). Interannual extremes in the rate of rise of atmospheric carbon-dioxide since 1980. *Nature*, *375*, 666–670.

Klein, S. A. and Hartmann, D. L. (1993). The seasonal cycle of low stratiform clouds. *J. Climate*, *6*, 1587–1606.

Kumar, A., Leetmaa, A., and Ji, M. (1994). Simulations of atmospheric variability induced by sea-surface temperatures and implications for global warming. *Science*, *266*, 632–634.

Kutzbach, J. and Liu, Z. (1997). Response of the African monsoon to orbital forcing and ocean feedbacks in the Middle Holocene. *Science*, *278*, 440–443.

Lindzen, R. S. (1993). Climate dynamics and global change. *Ann. Rev. Fluid Mech.*, *26*, 353–278.

Lindzen, R. S. and Pan, W. (1994). A note on orbital control of equator-pole heat fluxes. *Clim. Dyn.*, *10*, 49–57.

Linsley, B. (1996). Oxygen isotope evidence of sea level and climatic variations in the Sulu Sea over the past 150,000 years. *Nature*, *380*, 234–237.

Lorenz, E. (1991). Chaos, spontaneous climatic variations and detection of the Greenhouse effect. In Schlesinger, M. E., editor, *Greenhouse-Gas-Induced Climate Change: A Critical Appraisal of Simulations and Observations*, pp. 445–453. Elsevier, Amsterdam, the Netherlands.

Manabe, S. and Stouffer, R. (1988). Two stable equilibria of a coupled ocean-atmosphere model. *J. Climate*, *1*, 841–866.

Moodie, D. and Catchpole, A. (1975). Environmental data from historical documents by content analysis . *Manitoba Geographical Studies*, *5*, 1–119.

Neelin, J. D., Battisti, D., Hirst, A., Jin, F., Wakata, Y., Ymamgata, T., and Zebiak, S. (1998). ENSO Theory . *J. Geophys. Res.*, *103*, 14261–14290.

Oppo, D., McManus, J., and Cullen, J. (1998). Abrupt climate events 500,000 to 340,000 years ago: evidence from subpolar North Atlantic sediments. *Science*, *279*, 1335–1338.

Rahmstorf, S. (1994). Rapid climate transitions in a coupled ocean-atmosphere model. *Nature*, *372*, 82.

Raymo, M., Ruddiman, W., Shackleton, N., and Oppo, D. (1990). Atlantic-Pacific carbon isotope differences over the last 2.5 million years. *Earth Planet. Sci. Lett.*, *97*, 353–368.

Rayner, P.J., Law, R.M., and Dargaville, R. (1999). The relationship between tropical CO2 fluxes and the El Niño-Southern Oscillation. *Geophysical Research Letters*, *26*, 493–496.

Rind, D., Peteet, D., Broecker, W., McIntyre, A., and Ruddiman, W. R. (1986). The impact of cold North Atlantic sea surface temperature on climate: implication for Younger Dryas cooling (11-10 k). *Climate Dyn.*, *1*, 3–33.

Robertson, A. W., Mechoso, C. R., and Kim, Y. (1999). The influence of Atlantic sea surface temperature anomalies on the North Atlantic oscillation. *J. Climate.* submitted.

Rogers, J. (1984). The association between the North Atlantic Oscillation and the Southern Oscillation in the Northern Hemisphere. *Mon. Wea. Rev.*, *112*, 1999–2015.

Rossow, W. B. and Schiffer, R. A. (1991). ISCCP cloud data products. *Bull. Am. Meteor. Soc.*, *72*, 2–20.

Schrag, D., Hampt, G., and Murray, D. (1996). Pore fluid constraints on the temperature and oxygen isotopic composition of the glacial ocean. *Science*, *272*, 1930–1931.

Seager, R., Zebiak, S. E., and Cane, M. A. (1988). A model of the tropical Pacific sea surface temperature climatology. *J. Geophys. Res.*, *93*, 1265–1280.

Sirocko, F., Sarnthein, M., Erlenkeuser, H., Lange, H., Arnold, M, and Duplessy, J.C (1993). Century-scale events in monsoonal climate over the past 24,000 years. *Nature*, *364*, 322–324.

Soden, B. J. (1997). Variations in the tropical greenhouse effect during El Niño. *J. Climate*, *10*, 1050–1055.

Speer, K. and Tziperman, E. (1992). Rates of water mass formation in the North-Atlantic ocean. *J. Phys. Oceanogr.*, *22*, 93–104.

Spero, H., Bijma, J., Lea, D., and Bemis, B. (1997). Effect of seawater carbonate concentration on foraminiferal carbon and oxygen isotopes . *Nature*, *390*, 497–500.

Stute, M., Forster, M., Frischkorn, H., Serejo, A., Clark, J., Schlosser, P., Broecker, W., and Bonani, G. (1995). Cooling of tropical Brazil during the last glacial maximum. *Science*, *269*, 379–383.

Sun, D.-Z. and Oort, A. H. (1995). Humidity-temperature relationships in the tropical troposhere . *J. Climate*, *8*, 1974–1987.

Thompson, L., Mosley-Thompson, E., Davis, M., Lin, P., Cole-Dai, J., and Bolzan, J. (1995). A 1000 year climate ice-core record from the Guliya ice cap, China: its relationship to global climate variability . *Ann. Glaciol.*, *21*, 175–185.

Thompson, L., Tao, T., Davis, M., Henderson, K., Mosley-Thompson, E., Lin, P., Beer, J., Synal, H., Cole-Dai, J., and Bolzan, J. (1997). Tropical climate instability: The last glacial cycle from a Qinghai- Tibetan ice core. *Science*, *276*, 1821–1825.

Tziperman, E. (1997). Inherently unstable climate behavior due to weak thermohaline circulation. *Nature*, *386*, 592–595.

Tziperman, E., Cane, M. A., and Zebiak, S. E. (1994). Irregularity and locking to the seasonal cycle in an ENSO prediction model as explained by the quasi-periodicity route to chaos. *J. Atmos. Sci.*, *52*, 293–306.

Zebiak, S. E. and Cane, M. A. (1987). A model El Niño-Southern Oscillation. *Mon. Wea. Rev.*, *115*, 2262–2278.

M.A. Cane and A.C. Clement, Lamont-Doherty Earth Observatory, Columbia University, P.O. Box 1000, Palisades, NY 10964-8000. (e-mail: mcane@ldeo.columbia.edu; clement@ldeo.columbia.edu)

Making Sense of Millennial-Scale Climate Change

R.B. Alley[1], P.U. Clark[2], L.D. Keigwin[3], R.S. Webb[4]

Recent results, and especially those presented at the AGU Chapman Conference on Mechanisms of Millennial-Scale Global Climate Change, allow formulation of a consistent hypothesis for millennial-scale climate change. The observed large, abrupt, widespread millennial-scale climate changes of the last glaciation are hypothesized to have been forced by North Atlantic atmosphere-ocean-ice interactions. Shifts in ocean circulation (oceanic jumps) between modern and glacial modes caused the Dansgaard/Oeschger oscillations, in which the cold, dry and windy signal of the glacial mode was transmitted through the atmosphere to hemispheric or broader regions. These jumps were triggered by changes in freshwater delivery to the North Atlantic, and possibly by other causes extending beyond the North Atlantic. The additional jumps to the Heinrich mode caused atmospheric anomalies similar to but somewhat stronger and more-widespread than during the cold phases of the Dansgaard/Oeschger oscillation. Heinrich-mode effects also were transmitted through the ocean, with antiphase behavior between much of the world and some southern regions centered on and downwind of the South Atlantic. Surging of the Laurentide Ice Sheet supplied extra freshwater to force jumps to the Heinrich mode; surging probably was triggered by Dansgaard/Oeschger cooling after Laurentide growth exceeded a MacAyeal threshold for thawing the ice-sheet bed.

INTRODUCTION

Decades of research have established the importance of millennial-scale change in the paleoclimatic record [e.g., *Denton and Karlen*, 1973; *Pisias et al.*, 1973; *Mangerud et al.*, 1974]. A global network of high-resolution records, anchored by the results from the Greenland ice-coring projects at Summit [*Hammer et al.*, 1997], has demonstrated that abrupt, widespread, (nearly-) synchronous changes have been important or dominant in paleoclimatic variability at many times and places. These data motivated the AGU Chapman Conference on Mechanisms of Millennial-Scale Global Climate Change (June, 1998).

During the conference, the Program Committee attempted to formulate a consistent hypothesis for millennial-scale climate change, which we expand upon here. This hypothesis is based on the assumption that ice sheet-ocean-atmosphere interactions in the North Atlantic basin drive climate change elsewhere. These North Atlantic changes are postulated to be transmitted through the atmosphere and the ocean with various feedbacks amplifying and further broadcasting the signal hemispherically to globally. We have not attempted to present a consensus document for the meeting, for the Program Committee, or even the four of us, although we certainly acknowledge insight and inspiration

[1]Environment Institute and Department of Geosciences, The Pennsylvania State University, University Park, PA
[2]Department of Geosciences, Oregon State University, Corvallis, OR
[3]Woods Hole Oceanographic Institution, Woods Hole, MA
[4]NOAA-National Geophysical Data Center, Paleoclimatology Program, Boulder, CO

Mechanisms of Global Climate Change at Millennial Time Scales
Geophysical Monograph 112

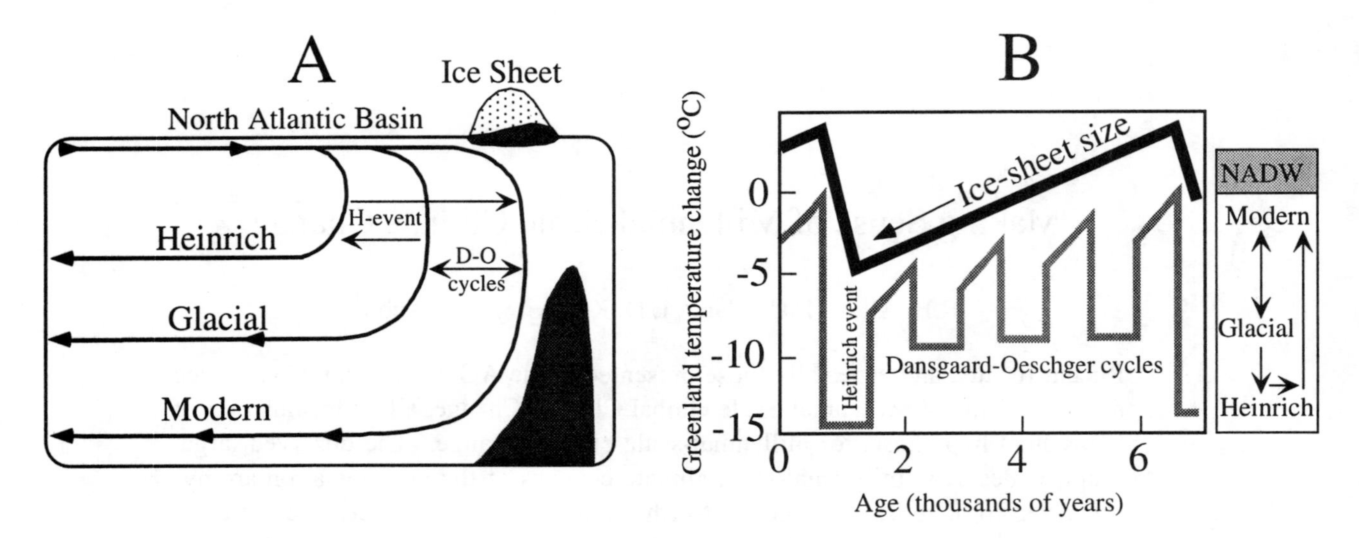

Figure 1. (A) Conceptual cartoon of three modes of North Atlantic circulation (*modern, glacial*, and *Heinrich*). Oceanic jumps between modern and glacial modes caused the Dansgaard/Oeschger oscillations, in which the cold, dry and windy signal of the glacial mode were transmitted elsewhere largely through the atmosphere. D/O oscillations were amplified during times of intermediate sized circum-North Atlantic ice sheets. The additional jumps to the Heinrich mode occurred in response to surging of the Laurentide Ice Sheet, and caused atmospheric anomalies similar to the cold phases of the D/O oscillations, but also transmitted climate change through the ocean. (B) An idealized time series of changes in climate between Heinrich events, during which several D/O oscillations occur (after *Alley* [1998]). Changes in ocean circulation associated with D/O oscillations and Heinrich events are shown on the right. Note that not all Heinrich events are colder than D/O oscillations in the Greenland ice cores.

both from the Program Committee and from other meeting attendees. Other hypotheses for mechanisms of millennial-scale climate change were discussed at the conference and are presented elsewhere in this volume [e.g., *Bond et al., Clement and Cane, Cane and Clement, Ninnemann and Charles*]. Nevertheless, by presenting a unified hypothesis for millennial-scale climate change that strives to reconcile the data and paleoclimate modeling results, we hope to show strengths, weaknesses, and interrelationships that may help guide further research.

MODES OF NORTH ATLANTIC DEEP WATER FORMATION

Data and models suggest that the North Atlantic Ocean usually operates in one of three distinct modes or bands of circulation; small changes may occur within a mode, but jumps in conditions separate the modes (Figure 1, 2, 3) [cf. *Rahmstorf*, 1994, 1995; *Sarnthein et al.*, 1994; *Oppo and Lehman*, 1995; *Maslin et al.*, 1995; *Manabe and Stouffer*, 1997; *Vidal et al.*, 1997; *Ganopolski et al.*, 1998; *Seidov and Maslin*, 1999; *Weaver*, this volume]. The first mode is characterized by deep-water formation in both the Nordic Seas and farther south in the North Atlantic (Figure 1A, 2A) (the two-pump ocean of *Imbrie et al.* [1993]; Holocene or interglacial mode of *Sarnthein et al.* [1994]; mode 1G of *Stocker* [Chapman Conference Abstracts, 1998, hereafter designated C*]; modern mode of *Alley and Clark* [1999]). In the second mode, deep-water formation largely stopped in the Nordic Seas but with continued vigorous deep-intermediate water formation farther south in the North Atlantic (Figure 1B, 2B) (the one-pump ocean of *Imbrie et al.* [1993]; glacial mode of *Sarnthein et al.* [1994] and *Alley and Clark* [1999]; mode 1L of *Stocker* [C*]). In the third mode, deep-intermediate water formation are greatly reduced in both locations (Figure 1C, 2C) (a no-pump mode not discussed by *Imbrie et al.* [1993]; meltwater mode of *Sarnthein et al.* [1994]; mode 0 of *Stocker* [C*]; Heinrich (H) mode of *Alley and Clark* [1999]; Imbrie et al. [1993] did not discuss the possibility of a "no-pump mode"). Although debate remains within the community on whether there was complete shutdown of circulation in the North Atlantic at least for a part of the time during each Heinrich event, within our hypothesis we chose to treat the greatly reduced deep and intermediate water formation within Heinrich as an extreme amplification of the partial shutdown during glacial maximum. Throughout the rest of the paper, we refer to these three distinct modes of North Atlantic Ocean circulation as *modern, glacial*, and *H* modes, respectively (Figures 1, 2, 3).

The modern mode of circulation is associated with warm times in the North Atlantic. The glacial mode is reached at

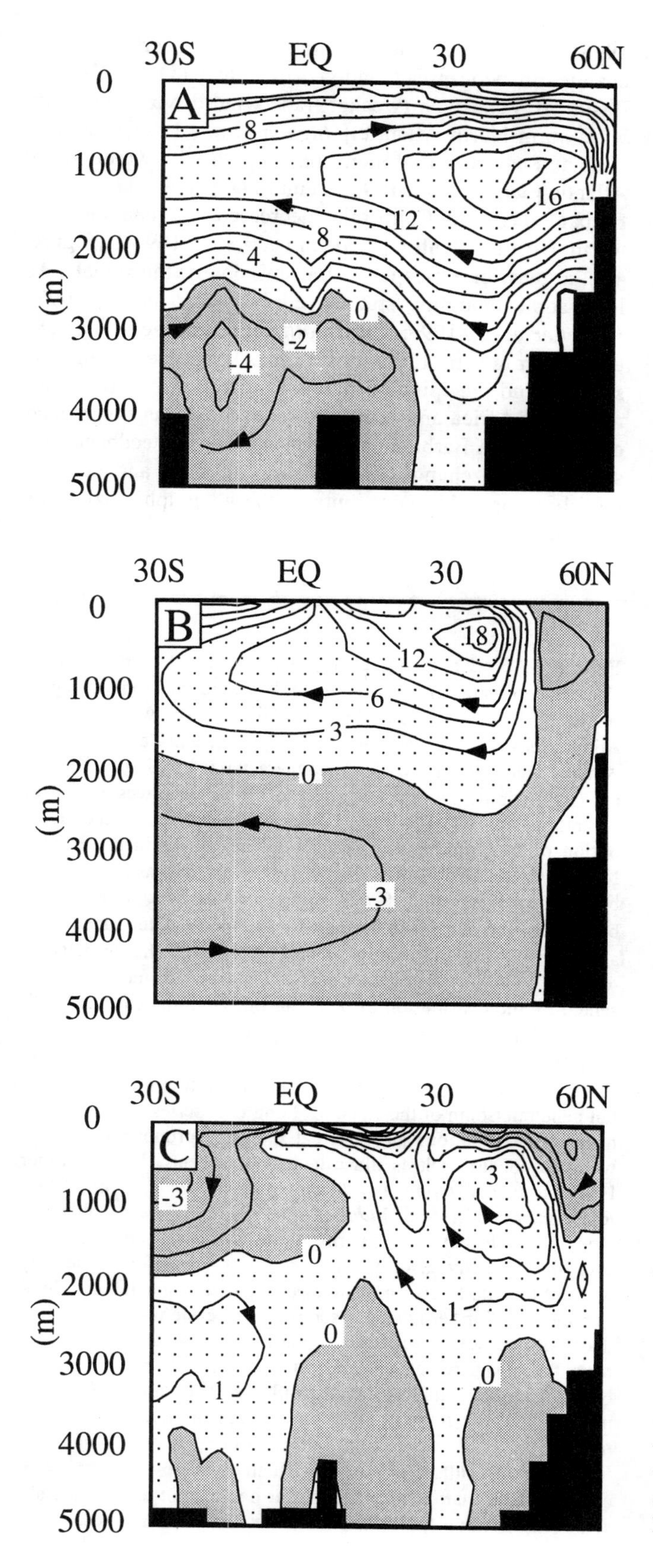

glacial maximum and during the cold or stadial times of the ~1500 year Dansgaard/Oeschger (D/O) oscillation. The H mode is achieved during especially large meltwater inputs, such as are associated with some Heinrich (H) events. Milankovitch-linked glacial-interglacial cycles cause the ocean to remain primarily in the glacial mode near glacial maximum and the modern mode during interglacials, but even at these times mode-jumps are possible with sufficient forcing.

DANSGAARD-OESCHGER OSCILLATIONS

The D/O oscillation is the dominant millennial-scale climate-change signal (Figure 4). Prominent events have had an approximate spacing of 1500 years [*Bond and Lotti*, 1995; *Bond et al.*, 1997; *Grootes and Stuiver*, 1997; *Mayewski et al.*, 1997]. D/O changes are centered on the North Atlantic (Figure 4A), and on regions with strong atmospheric response to changes in the North Atlantic, including tropical and subtropical monsoonal areas of Africa and to a lesser extent Asia [e.g., *Street-Perrott and Perrott*, 1990; *Gasse and van Campo*, 1994; *Hostetler et al.*, 1999]. Warm North Atlantic conditions associated with the modern mode of circulation produce enhanced monsoonal precipitation, contributing to the increased methane (Figure 4B) that apparently is an almost-instantaneous (<30 years) feedback on North Atlantic warming [*Severinghaus et al.*,

---

Figure 2. Model results suggesting three modes of North Atlantic deep-water circulation. (A) The modern mode is suggested by the meridional transport stream function of the thermohaline circulation (THC) [in units of sverdrups (Sv = $10^6 m^{-3}$)] from the control experiment of *Manabe and Stouffer* [1997]. The model is a coupled atmosphere-ocean general circulation model. Model results show sinking of North Atlantic Deep Water (NADW) to fill the northern North Atlantic basin (light stipple pattern). (B) The glacial mode is represented by the stream function of THC from the model of the last glacial maximum by *Ganopolski et al.* [1998]. The model couples a dynamical-statistical atmosphere model to a zonally averaged ocean model with three separate basins and is forced by insolation, CO_2, ice sheets, and sea level consistent with boundary conditions of the last glacial maximum. Model results show a shoaling of NADW and a southward shift in convection sites, but indicate that the overall rate of NADW formation was only slightly reduced from modern. (C) The Heinrich mode is represented by the stream function of THC from the model of a Heinrich event by *Weaver* [this volume]. The model is an energy-moisture balance model coupled to an ocean general circulation model. Model results show that the response to a 105-year freshwater forcing with an average flux of 0.013 Sv results in shoaling to ~2000 m as well as a significant reduction in rate of formation of NAIW (light stipple pattern) compared to NADW (e.g., A or B).

1998]. The synchroneity (within dating uncertainties) in changes in widespread regions, including (1) the Cariaco Basin off Venezuela (weaker trade-wind-driven upwelling with warm North Atlantic [*Hughen et al.*, 1998]), (2) the Arabian Sea (increased monsoon strength with warm North Atlantic [*Schulz et al.*, 1998]) (Figure 4C), and (3) the Santa Barbara Basin off California (decreased oxygenation with warm North Atlantic [*Behl and Kennett*, 1996]) (Figure 4D), argues for atmospheric transmission of the signal. At least some modeling studies [e.g., *Fawcett et al.*, 1997; *Hostetler et al.*, 1999; *Agustsdottir et al.*, in press] indicate that warming in the Nordic Seas increases monsoon strength and precipitation-minus-evaporation in monsoonal regions of Africa and Asia, and reduces Ekman divergence over the Santa Barbara and Cariaco Basins. Feedbacks associated with changes in monsoon strength and oceanic upwelling probably contribute to the geographic extent of D/O impacts.

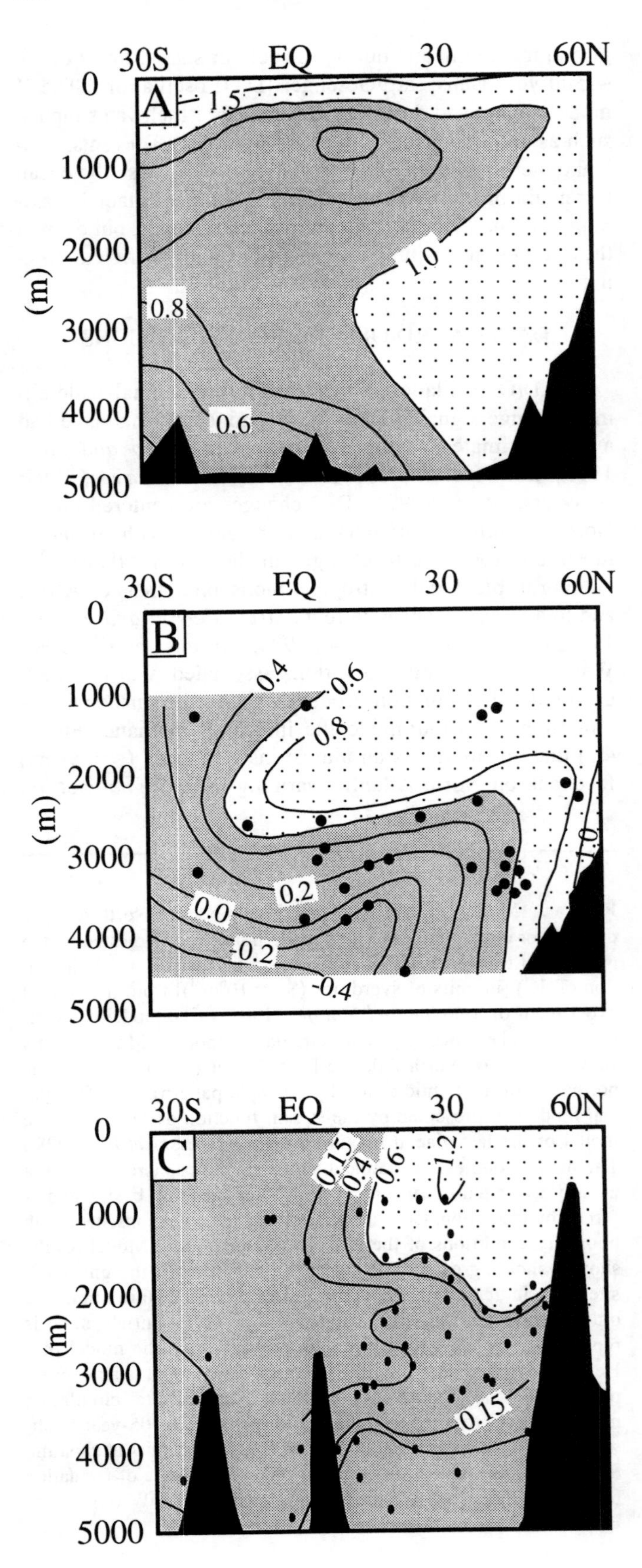

Figure 3. $\delta^{13}C$ data suggesting three modes of North Atlantic Ocean deep-water circulation. (A) The modern mode is represented by the modern distribution of $\delta^{13}C$ in the western North Atlantic basin [after *Kroopnick*, 1985]. Nutrient-depleted North Atlantic Deep Water (NADW) is characterized by high $\delta^{13}C$ values (light stipple pattern) and sinks to fill the deep western North Atlantic basin. (B) The glacial mode is represented by the distribution of $\delta^{13}C$ during the last glacial maximum in the eastern North Atlantic basin (after *Duplessy et al.* [1988], as recontoured by *Broecker* [1989]). $\delta^{13}C$ values indicate that NADW (identified by values >0.6 per mil to account for the glacial-interglacial $\delta^{13}C$ shift of ~0.4 per mil; *Curry et al.* [1988]) was shallower (to a depth of ~3000 m) than modern. (C) The Heinrich mode is represented by the distribution of $\delta^{13}C$ during H1 in the eastern North Atlantic [after *Sarnthein et al.*, 1994]. $\delta^{13}C$ values indicate that NADW shoaled to ~2000 m to form North Atlantic Intermediate Water (NAIW) (light stipple pattern) and that this water mass did not penetrate south of the equator. Long time series of surface and deep-water variations suggest that the most pronounced deep-water responses such as illustrated in (C) occurred during other Heinrich events [*Oppo and Lehman*, 1995; *Maslin et al.*, 1995; *Rasmussen et al.*, 1996; *Vidal et al.*, 1997; *Curry et al.*, this volume]. It is clear from comparing timeslices in (B) and (C), however, that in many cases a single record of deep-water variability will not be able to discriminate clearly between glacial and Heinrich modes. Furthermore, a response in the eastern North Atlantic does not necessarily reflect a strong effect in what was exported from the basin, which can only be measured from the western North Atlantic. Finally, model results by *Marchal et al.* [1998] indicate that the relation between changes in NADW formation rate and in $\delta^{13}C$ values is not linear. Model (Figure 2) and data (this figure) results are thus consistent in suggesting significant changes in formation of NADW, but one cannot extrapolate absolute changes in NADW formation rates from $\delta^{13}C$ records.

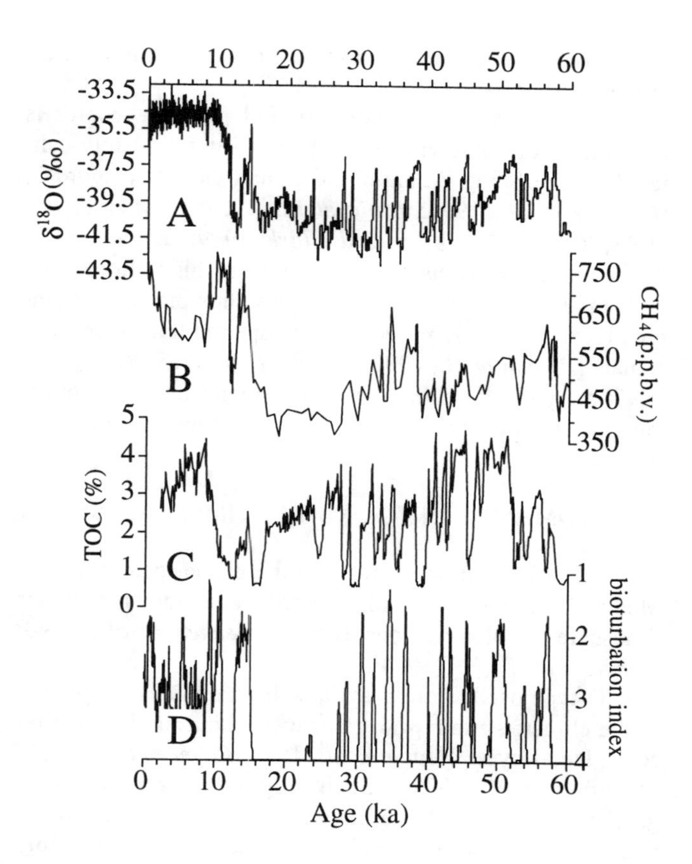

Figure 4. Records showing Dansgaard/Oeschger oscillations over the last 60 ka from: (A) the GISP2 $\delta^{18}O$ record (in per mil) [*Grootes et al.*, 1993]; (B) the GISP2 methane record [*Brook et al.*, 1996]; (C) total organic carbon (TOC) (in %) from Arabian Sea sediments [*Schulz et al., 1998*], and (D) changes in the bioturbation index from ODP Site 893 (Santa Barbara basin) (1 = laminated sediments, 4 = massive sediments) [*Behl and Kennett*, 1996].

Intermediate to deep oceanic impacts of the modern-to-glacial D/O jumps are probably small beyond the high-latitude North Atlantic. These impacts are limited to northward cross-equatorial surface-water flow in the Atlantic that continues during cold and warm phases, as does ventilation to at least mid-depth in the Atlantic [*Rahmstorf*, 1995; *Yu et al.*, 1996; *Ganopolski et al.*, 1998; *Stocker*, C*] (Figure 2) and supply of North Atlantic-origin waters to the Antarctic thermocline. Some model results suggest that glacial-mode conditions are accompanied by increased formation of intermediate water in the North Pacific [*Ganopolski et al.*, 1998; *Keigwin*, 1998], causing Pacific and Atlantic glacial circulations to be similar.

The D/O oscillation is an oceanic process, often triggered by meltwater changes but possibly oscillating freely in response to some as-yet-unidentified process(es). Although the oscillations appear periodic, close inspection indicates that there is much variability about the mean spacing [*Bond and Lotti*, 1995], and our impression is that during certain times (especially during the deglaciation) there is some uncertainty in identifying D/O-type events and in determining their temporal spacing. Persistence of an apparent D/O-type oscillation during the Holocene (at greatly reduced amplitude) [*Denton and Karlen*, 1973; *Keigwin and Jones*, 1989; *Bond et al.*, 1997; *Bianchi and McCave*, 1999] also involves much deviation in event spacing. One cannot exclude great variability in spacing of tones off Milankovitch or of solar or other forcing [*Mayewski et al.*, 1997], but a more-periodic behavior might be expected by analogy to known shorter solar cycles and longer Milankovitch cycles [e.g., *Imbrie et al.*, 1992]. A solar forcing is further questioned by the lack of a cosmogenic-nuclide signal of solar-wind changes in Holocene ice-core records [e.g., *Finkel and Nishiizumi*, 1997]; the ice cores record both solar forcing and climatic response at higher frequencies [*Finkel and Nishiizumi*, 1997; *Grootes and Stuiver*, 1997], but thus far we know of no credible evidence for significant solar forcing at the millennial scale.

Internal processes are strongly supported by the observation that all of the events during the most recent deglaciation are plausibly related to changes in meltwater delivery to the North Atlantic [e.g., *Broecker et al.*, 1988; *Teller*, 1990; *Keigwin et al.*, 1991; *Clark et al.*, 1996; *Barber et al.*, C*; *Liccardi et al.*, this volume], a process which has been shown in numerous models to force oceanic changes [e.g., *Rahmstorf*, 1995; *Fanning and Weaver*, 1997; *Stocker*, C*]. (Figure 5A) probably reflects the growth of the Laurentide ice sheet--surges are only possible once a growth threshold is reached, which may require several D/O cycles but which is easier/faster during colder times near glacial maximum [*McManus et al.*, 1999]. The overall cooling trend associated with the H events [*Bond et al.*, 1993] is plausibly the effect on the atmosphere of changing ice-sheet as well as oceanic conditions [*MacAyeal*, 1993; *Jackson*, this volume; *Hostetler et al.*, 1999].

In addition to an atmospheric transmission as seen for D/O oscillations, H events are also transmitted elsewhere through the ocean. Meltwater associated with H events forces (or reinforces) the H mode or band of ocean circulation [*Keigwin and Lehman*, 1994; *Curry and Oppo*, 1997; *Vidal et al.*, 1997; *Curry et al.*, this volume]. Loss of the second site of deep-water formation has only a small additional direct effect on the atmosphere (oceanic heat being more important to the total energy budget in wintertime high latitudes than at lower latitudes), but affects the oceanic circulation more prominently than does loss of higher-latitude deep-water formation. The main effect of switching to the H mode is to reduce the cross-equatorial flow of Atlantic surface waters (Figure 2), which leaves heat in the South Atlantic, warming high southern latitudes [*Manabe*

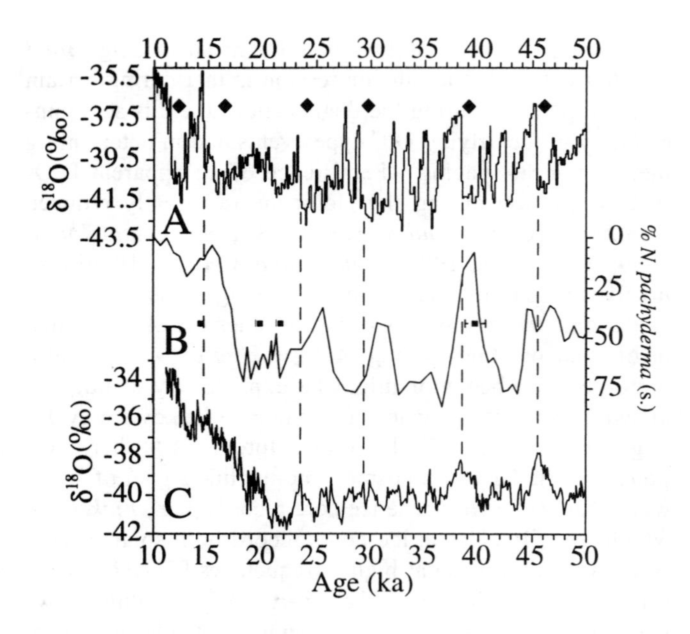

Figure 5. Records showing "bipolar seesaw" relationship between the North Atlantic (A: GISP2 $\delta^{18}O$ record (in per mil) [*Grootes et al.*, 1993]), the South Atlantic (B: percentage of *N. pachyderma* (s.) from core GeoB 1711 [*Little et al.*, 1997]; dating control indicated by squares), and Antarctica (C: Byrd $\delta^{18}O$ record (in per mil) [*Johnsen et al.*, 1972]). Heinrich Events are indicated by diamonds along the top of the figure (with increasing age, the Younger Dryas or H0, and then H1-H5). Ed Brook (personal communication, October, 1998) correlated the Byrd $\delta^{18}O$ chronology to the GISP2 $\delta^{18}O$ chronology using methane data from the GISP2 ice core [*Brook et al.*, 1996] and methane data from the Byrd ice core [*Blunier et al.*, 1998]. Specifically, GISP2 gas (methane) ages were assigned to depths in the Byrd record where there are corresponding rapid changes in methane, allowing interpolation of a gas-age timescale for the Byrd record. An ice-age timescale for the Byrd $\delta^{18}O$ record was then calculated using estimated delta age (ice-age/gas-age offset). Note that there are no control points for the interval 17-30 ka.

and Stouffer, 1988, 1997; *Stocker et al.*, 1992; *Crowley*, 1992; *Rahmstorf*, 1994; *Schiller et al.*, 1997; *Weaver*, this volume]. Loss of North Atlantic Deep Water formation probably also stimulates additional deep-water formation in specific places in the south [*Schiller et al.*, 1997; *Broecker*, 1998; *Stocker*, C*] and enhanced intermediate water formation in the North Pacific [*Rahmstorf*, 1995; *Keigwin*, 1998; *Lund and Mix*, 1998], resulting in an oceanic "seesaw" [*Broecker*, 1998; *Lund and Mix*, 1998]. The warming associated with vigorous deep-water formation is much smaller in the south than in the north, however, because the Coriolis effect prevents warm surface waters from crossing the Antarctic Circumpolar Current at the latitude of Drake Passage [*Toggweiler and Samuel*, C*] so that southern deep waters must be formed from upwelling of already-cool intermediate waters.

Antiphase behavior of northern and southern records is prominent for some H events at some sites probably centered on the South Atlantic and Indian Oceans (Figure 5) [*Charles et al.*, 1996; *Little et al.*, 1997; *Blunier et al.*, 1998; *Stocker*, C*; *Alley and Clark*, 1999; *Bender et al.*, this volume; *Niennemann and Charles*, this volume], but with other southern-hemisphere sites showing a northern-type record probably reflecting atmospheric transmission of signals [*Pichon et al.*, 1992; *Lowell et al.*, 1995; *Bard et al.*, 1997; *Steig et al.*, 1998]. Southern Hemisphere signals are muted compared to changes in the Northern Hemisphere (Figure 5).

COMPARISON TO ORBITAL TIMESCALES

The North Atlantic operates in the glacial band for both Modeled response to such forcing often has shown a similar timescale to observations [e.g., *Broecker et al.*, 1990; *Weaver and Hughes*, 1994].

D/O-type oscillations may have been a persistent feature of the climate system prior to as well as during the most recent glacial cycle [*Oppo et al.*, 1998; *Raymo et al.*, 1998; *McManus et al.*, 1999]. Such long-term persistence suggests that some D/O oscillations were not forced. If so, then a free oscillation of unknown cause is active, possibly related to ENSO [*Clement and Cane*, this volume; *Cane and Clement*, this volume], tidal [*Keeling*, C*], or other processes including stochastic variability [*Weaver and Hughes*, 1994; *Weaver*, this volume]. The millennial frequency points to a major role for the ocean; some ice sheets have longer response times, and the atmosphere has a much shorter response time. The increased ice-rafted debris from multiple sources in North Atlantic sediments from the cold phases of the D/O oscillations likely reflects the normal response of ice sheets to the cooling associated with the events rather than any ice-dynamical forcing of the coolings [*Bond and Lotti*, 1995; *McCabe and Clark*, 1998; *van Kreveld et al.*, C*].

D/O oscillations have been largest at times of intermediate temperatures and ice extent, often when orbital forcing and CO_2 were changing rapidly, rather than during the coldest or warmest times near insolation and CO_2 maxima or minima [*Alley and Clark*, 1999; *McManus et al.*, 1999]. This may indicate that the system can remain in one mode of operation with appropriate forcing, but that oscillations result when forcing would place the climate system in the gap between two modes. Alternatively, the stability of the system may be sensitive to the rate of change of forcing variables, such that climate response is analogous to that of a drunken human: when left alone, it sits; when forced to move, it staggers with abrupt changes in direction.

HEINRICH EVENTS

The H events (or at least most of them) involved surging of the Laurentide Ice Sheet through Hudson Strait apparently triggered by D/O cooling [*Bond et al.*, 1993]. The tremendous sediment volume [e.g., *Bond et al.*, 1993], dominant Hudson-Bay source [e.g., *Gwiazda et al.*, 1996], and rapid rate of sediment supply [e.g., *McManus et al.*, 1998] of H events set them apart from other stadials (millennial-scale cold events), and seem to require an ice-dynamics instability [*Alley and MacAyeal*, 1994]. In most marine sediment records, cooling preceded the sharp-based, carbonate-rich, rapidly deposited debris of H layers [e.g., *Bond et al.*, 1993; *Bond and Lotti*, 1995]. A spectrum of mechanisms has been proposed to explain the triggering of H events [*Alley et al.*, 1996; *Parizek et al.*, 1997; *Clarke et al.*, this volume; *MacAyeal*, C*], but triggering by cooling is at least plausible provided the key region is ice-marginal and so can respond rapidly to changing climate. The long (and variable) spacing of H events compared to D/O timing glacial-maximum and stadial conditions of the D/O oscillation, which appear similar in many North Atlantic records. Far from the North Atlantic, climate anomalies associated with glacial maximum are much more prominent than those during stadial conditions, because CO_2, ice sheets, and associated ice-albedo, vegetation and other feedbacks, change greatly from glacial to interglacial but not from stadial to interstadial [*Alley and Clark*, 1999; *Hostetler et al.*, 1999]. Controls on CO_2 are poorly understood. Enhanced polar cooling associated with ice-albedo and other feedbacks in response to CO_2 reduction and orbitally induced ice-sheet growth steepens the equator-to-pole temperature gradient. This steepened gradient increases wind strength, cooling the tropics through upwelling of colder waters or entrainment of colder extratropical waters [e.g., *Bush and Philander*, 1998; *Ganopolski et al.* 1998] and further cooling the tropics and extratropics through water-vapor feedbacks [*Pierrehumbert*, this volume]. Glacial and stadial modes thus differ primarily in ways that do not directly involve the deep ocean. In contrast with the glacial climate changes, the stadials do not persist long enough for major CO_2 and ice-sheet feedbacks, and may lack forcing in appropriate regions to trigger those feedbacks.

DISCUSSION

Based on current data and model results, we hypothesize that millennial-scale climate change during the last glaciation: (1) consists of quasi-periodic D/O oscillations and aperiodic H events; (2) is amplified during times of intermediate ice cover and rapidly changing forcing; (3) originates by jumps between three modes of NADW formation (modern, glacial, and H modes); (4) is transmitted from the North Atlantic elsewhere through the atmosphere (modern to glacial mode switch) and oceans (glacial to H mode switch); and (5) is further propagated through various feedbacks (water vapor, upwelling). Switches to at least some glacial modes and all H modes are forced by changes in meltwater delivery.

Although our hypothesis focuses on ice sheet-ocean-atmosphere interactions in the North Atlantic region as the primary mechanism driving millennial-scale climate variability during the last glaciation, atmospheric and oceanic transmission of North Atlantic changes to other regions played key roles in constructing the overall climate-system circuitry that resulted in a complex global response. Global transmission of a signal originating (or amplified) in a small region of the Earth (ENSO variability in the western Pacific) is a well known characteristic of the climate system at decadal timescales [*Clement and Cane*, this volume]. We believe that the data of millennial-scale variability (e.g., Figure 4, 5) have similarly identified important pathways (monsoon, Southern Ocean) by which climate change in the North Atlantic is transmitted outside of the region. Nevertheless, additional well-dated high-resolution records from terrestrial and marine sites in the southern hemisphere are clearly needed to determine the extent and timing of H and non-H D/O oscillations. Also needed are well-dated, high-resolution records from all latitudes of the Pacific--the largest ocean looms as a large uncertainty in our hypothesis. Our understanding of teleconnections at decadal [e.g., *White et al.*, 1996; *Huang et al.*, 1998; *Mysak and Venegas*, 1998; *Barnett et al.*, 1999] and orbital [*Imbrie et al.*, 1992, 1993] timescales allows us to speculate that a number of a pathways in addition to those suggested by existing data served to further transmit millennial-scale climate change originating in the North Atlantic. Only further high-resolution records will be able to decipher these additional pathways. Significant ambiguities remain in the North Atlantic as well. Further understanding of the controls on CO_2 and ice sheets would be valuable, as would an understanding of millennial processes in ocean circulation.

We are impressed with the recent paleoclimate research and modeling results that have contributed to a rapid progress in our knowledge of millennial-scale climate change. Nevertheless, complete understanding of the complex interactions of mechanisms and teleconnections of millennial-scale changes in the climate system remains elusive. Our current lack of complete understanding of the millennial-scale climate processes, their relative timing, and whether these are internally or externally forced, limits our ability to unequivocally attribute these abrupt climate shifts to known forcings. The all-encompassing hypothesis we have presented to explain millennial-scale climate change represents an attempt to summarize a series of complex interactions within the climate system, and should be recognized as only a cartoon providing a useful but certainly not yet complete description.

Acknowledgments. We thank numerous colleagues who attended the 1998 AGU Chapman Conference for discussions, T. Blunier and H. Schulz for providing data, E. Brook for correlating the Byrd record to the GISP2 record, and the National Science Foundation for funding.

REFERENCES

Alley, R.B., Icing the North Atlantic, *Nature, 392*, 335-336, 1998.

Alley, R.B., and D.R. MacAyeal, Ice-rafted debris associated with binge/purge oscillations of the Laurentide Ice Sheet, *Paleoceanography, 9*, 503-511, 1994.

Alley, R.B., and P.U. Clark, The deglaciation of the northern hemisphere: a global perspective, *Ann. Rev. Earth Planet. Sci., 27*, 149-182, 1999.

Alley, R.B., S. Anandakrishnan, and K.M. Cuffey, Subglacial sediment transport and ice-stream behavior, *Ant. J. U.S., 31*, 81-82, 1996.

Agustsdottir, A.M., R.B. Alley, D. Pollard and W.H. Peterson, Ekman transport and upwelling during Younger Dryas estimated from wind stress from GENESIS climate model experiments with variable North Atlantic heat convergence, *Geophys. Res. Let.*, in press.

Barber, D.C., A.E. Jennings, and J.T. Andrews, Did meltwater trigger the cold event 8,200 cal. yrs ago? Revised 14C ages for drainage of glacial Lake Ojibway, *C**, 15.

Bard, E., F. Rostek, and C. Sonzogni, Interhemispheric synchrony of the last deglaciation inferred from alkenone palaeothermometry, *Nature, 385*, 707-710, 1997.

Barnett, T.P., D.W. Pierce, M. Latif, D. Dommenget, and R. Saravanan, Interdecadal interactions between the tropics and midlatitudes in the Pacific basin, *Geophys. Res. Let., 26*, 615-618, 1999.

Behl, R.J., and J.P. Kennett, Brief interstadial events in the Santa Barbara basin, NE Pacific, during the past 60 kyr, *Nature, 379*, 243-246, 1996.

Bender, M.L., Interhemispheric phasing of millennial-duration climate events during the last 100 ka, *C**, 10.

Bianchi, G.G., and I.N. McCave, Holocene periodicity in North Atlantic climate and deep-ocean flow south of Iceland, *Nature, 397*, 515-517, 1999

Blunier, T., J. Chappellaz, J. Schwander, A. Dallenbach, B. Stauffer, T.F. Stocker, D. Raynaud, J. Jouzel, H.B. Clausen, C.U. Hammer, and S.J. Johnsen, Asynchrony of Antarctic and Greenland climate change during the last glacial period, *Nature, 394*, 739-743, 1998.

Bond, G.C., and R. Lotti, 1995, Iceberg discharges into the North Atlantic on millennial time scales during the last deglaciation, *Science, 267*, 1005-1010, 1995.

Bond, G., W. Broecker, S. Johnsen, J. McManus, L. Labeyrie, J. Jouzel, and G. Bonani, 1993, Correlations between climate records from North Atlantic sediments and Greenland ice, *Nature, 365*, 143-147, 1993.

Bond, G., W. Showers, M. Cheseby, R. Lotti, P. Almasi, P. deMenocal, P. Priore, H. Cullen, I. Hajdas, and G. Bonani, 1997, A pervasive millennial-scale cycle in North Atlantic Holocene and glacial climates, *Science, 278*, 1257-1266, 1997.

Broecker, W.S., Some thoughts about the radiocarbon budget for the glacial Atlantic, *Paleoceanography, 4*, 213-220, 1989.

Broecker, W.S., Paleocean circulation during the last deglaciation: A bipolar seesaw?, *Paleoceanography, 13*, 119-121, 1998.

Broecker, S.S., M. Andree, W. Wolfi, H. Oeschger, G. Bonani, J. Kennett, And D. Peteet, The chronology of the last deglaciation: Implications to the cause of the Younger Dryas event, *Paleoceanography 3*, 1-19, 1988.

Broecker, W.S., G. Bond, and M. Klas, A salt oscillator in the glacial Atlantic? 1. The concept, *Paleoceanography, 5*, 469-477, 1990.

Brook, E.J., T. Sowers, and J. Orchardo, Rapid variations in atmospheric methane concentration during the past 110,000 years, *Science 273*, 1087-1091, 1996.

Bush, A.B.G. and S.G.H. Philander, The role of ocean-atmosphere interactions in tropical cooling during the last glacial maximum, *Science, 279*, 1341-1344, 1998.

Clark, P.U., R.B. Alley, L.D. Keigwin, J.M. Licciardi, S.J. Johnsen, and H. Wang, Origin of the first global meltwater pulse following the last glacial maximum, *Paleoceanography, 11*, 563-577, 1996

C*: Clark, P.U., and R.S. Webb, Mechanisms of Millennial-Scale Global Climate Change, *Abstracts with Program*, American Geophysical Union, Washington, D.C., 1998.

Charles, C.D., J. Lynch-Stieglitz, U.S. Ninnemann, and R.G. Fairbanks, Climate connections between the hemisphere revealed by deep sea sediment core/ice core correlations, *Earth Planet. Sci. Lett., 142*, 19-27, 1996.

Crowley, T.J., North Atlantic deep water cools the southern hemisphere, *Paleoceanography, 7*, 489-497, 1992.

Curry, W.B., and D.W. Oppo, Synchronous, high-frequency oscillations in tropical sea surface temperatures and North Atlantic deep water production during the last glacial cycle, *Paleoceanography, 12*, 1-14, 1997.

Curry, W.B., J.C. Duplessy, L.D. Labeyrie, and N.J. Shackleton, Changes in the distribution of $\delta^{13}C$ of deep water ΣCO_2 between the last glaciation and the Holocene, *Paleoceanography, 3*, 317-341, 1988.

Denton, G.H., and W. Karlen, Holocene climatic variations--their pattern and possible cause, *Quaternary Research, 3*, 155-205, 1973.

Duplessy, J.C., N.J. Shackleton, R.G. Fairbanks, L. Labeyrie, D. Oppo, and N. Kallel, Deepwater source variations during the last climatic cycle and their impact on the global deepwater circulation, *Paleoceanography, 3*, 343-360, 1988.

Fanning, A.F., and A.J. Weaver, Temporal-geographical meltwater influences on the North Atlantic conveyor: Implications for the Younger Dryas, *Paleoceanography, 12*, 307-320, 1997.

Fawcett, P.J., A.M. Agustsdottir, R.B. Alley, and C.A. Shuman, The Younger Dryas termination and North Atlantic deepwater formation: insights from climate model simulations and Greenland ice core data, *Paleoceanography, 12*, 23-38, 1997.

Finkel, R.C., and K. Nishiizumi, Beryllium 10 concentrations in the Greenland Ice Sheet Project 2 ice core from 3-40 ka, *J. Geophys. Res., 102*, 26699-26706, 1997.

Ganopolski, A., S. Rahmstorf, V. Petouknov, and M. Claussen, Simulation of modern and glacial climates with a coupled global model of intermediate complexity, *Nature, 319*, 351-356, 1998.

Gasse, F., and E. van Campo, Abrupt post-glacial climate events in West Asia and North Africa monsoon domains, *Earth Planet. Sci. Lett., 126*, 435-456, 1994.

Grootes, P.M., M. Stuiver, J.W.C. White, S.J. Johnsen, and J. Jouzel, Comparison of oxygen isotope records from the GISP2 and GRIP Greenland ice cores, *Nature, 366*, 552-554, 1993.

Grootes, P.M., and M. Stuiver, Oxygen 18/16 variability in Greenland snow and ice with 10^{-3}- to 10^{5}- year time resolution, *J. Geophys. Res., 102*, 26455-26470, 1997.

Gwiazda, R.H., S.R. Hemming, and W.S. Broecker, Provenance of icebergs during Heinrich event 3 and the contrast to their sources during other Heinrich episodes, *Paleoceanography, 11*, 371-378, 1996.

Hammer, C., P.A. Mayewski, D. Peel and M. Stuiver, eds., Greenland Summit Ice Cores, American Geophysical Union, reprinted from *Journal of Geophysical Research, 102*, 26,315-26,886, 1997.

Hostetler, S., P.U. Clark, P.J. Bartlein, A.C. Mix, and N.G. Pisias, Mechanisms for the global transmission and registration of North Atlantic Heinrich events, *Journal of Geophysical Research, 104*, 3947-3952.

Hostetler, S., P.U. Clark, P.J. Bartlein, A.C. Mix, and N.G. Pisias, Response of western North America surface processes to a canonical Heinrich event, *C***, 31-32.

Huang, J., K. Higuchi, and A. Shabbar, The relationship between the North Atlantic Oscillation and El Nino Southern Oscillation, *Geophys. Res. Let., 25*, 2707-2710, 1998.

Hughen, K.A., J.T. Overpeck, S.J. Lehman, M. Kashgarian, J. Southon, L.C. Peterson, R. Alley and D.M. Sigman, Deglacial changes in ocean circulation from an extended radiocarbon calibration, *Nature, 391*, 65-68, 1998.

Imbrie, J., E.A. Boyle, S.C. Clemens, A. Duffy, W.R. Howard, G. Kukla, J. Kutzbach, D.G. Martinson, A. McIntyre, A.C. Mix, B. Molfino, J.J. Morley, L.C. Peterson, N.G. Pisias, W.L. Prell, M.E. Raymo, N.J. Shackleton, and J.R. Toggweiler, On the structure and origin of major glaciation cycles: 1. Linear responses to Milankovitch forcing, *Paleoceanography, 7*, 701-738.

Imbrie, J., A. Berger, E.A. Boyle, S.C. Clemens, A. Duffy, W.R. Howard, G. Kukla, J. Kutzbach, D.G. Martinson, A. McIntyre, A.C. Mix, B. Molfino, J.J. Morley, L.C. Peterson, N.G. Pisias, W.L. Prell, M.E. Raymo, N.J. Shackleton and J.R. Toggweiler, On the structure and origin of major glaciation cycles: 2. The 100,000-year cycle, *Paleoceanography, 8*, 699-735, 1993.

Johnsen, S.J., W. Dansgaard, H.B. Clausen and C.C. Langway, Jr., Oxygen isotope profiles through the Antarctic and Greenland ice sheets, *Nature 235*, 429-434, 1972.

Keeling, C.D., An 1800-year periodicity in oceanic tidal dissipation as a possible contributor to millennial-scale global climate change, *C***, 20.

Keigwin, L.D., Glacial-age hydrography of the far northwest Pacific Ocean, *Paleoceanography, 13*, 323-339, 1998.

Keigwin, L.D., and G.A. Jones, Glacial-Holocene stratigraphy, chronology, and paleoceanographic observations on some North Atlantic sediment drifts, *Deep-Sea Research, 36*, 845-867, 1989.

Keigwin, L.D., and S.J. Lehman, Deep circulations change linked to Heinrich event 1 and Younger Dryas in a middepth North Atlantic core, *Paleoceanography, 9*, 185-194, 1994.

Keigwin, L.D., G.A. Jones, and S.J. Lehman, Deglacial meltwater discharge, North Atlantic deep circulation, and abrupt climate change, *Journal of Geophysical Research, 96(C9)*, 16,811-16,826, 1991.

Kroopnick, P.M., The distribution of ^{13}C of ΣCO_2 in the world oceans: Deep-Sea Research, v. 32, p. 57-84, 1985.

Little, M.G., R.R. Schneider, D. Kroon, B. Price, T. Bickert, and G. Weger, Rapid paleoceanographic changes in the Benguela upwelling system for the last 160,000 years as indicated by abundances of planktonic foraminifera, *Palaeogeog., Palaeoclim., Palaeoecol., 130*, 135-161, 1997.

Lowell, T.V., C.J. Heusser, B.G. And`ersen, P.I. Moreno, A. Hauser, L.E. Heusser, C. Schluchter, D.R. Marchant, and G.H. Denton, Interhemispheric correlation of Late Pleistocene glacial events. *Science, 269*, 1541-1549, 1995.

Lund, D.C., and A.C. Mix, Millennial-scale deep water oscillations: reflections of the North Atlantic in the deep Pacific from 10 to 60 ka, *Paleoceanography, 13*, 1-19, 1998.

MacAyeal, D.R., A low-order model of growth/purge oscillations of the Laurentide Ice Sheet, *Paleoceanography, 8*, 767-773, 1993.

MacAyeal, D.R., Ice sheet model simulations of North Atlantic ice rafting events: rogue ice streams or submissive calving margins? *C***, 31.

Manabe, S., and R.J. Stouffer, Two stable equilibria of a coupled ocean-atmosphere model, *J. Climate, 1*, 841-866, 1988.

Manabe, S., and R.J. Stouffer, Coupled ocean-atmosphere model response to freshwater input: Comparison to Younger Dryas event, *Paleoceanography, 12*, 321-336, 1997.

Mangerud, J., S.T. Andersen, B.E. Berglund, and J.J. Donner, Quaternary stratigraphy of Norder, A proposal for terminology and classification, *Boreas 3*, 110-127, 1974.

Marchal, O., T.F. Stocker, and F. Joos, Impact of oceanic reorganizations on the ocean carbon cycle and atmospheric carbon dioxide content, *Paleoceanography, 13*, 225-244, 1998.

Maslin, M.A., N.J. Shackleton, and U. Pflaumann, Surface water temperature, salinity, and density changes in the northeast Atlantic during the last 45,000 years: Heinrich events, deep water formation, and climatic rebounds, *Paleoceanography, 10*, 527-544, 1995.

Mayewski, P.A., L.D. Meeker, M.S. Twickler, S. Whitlow, Q. Yang, W.B. Lyons, and M. Prentice, 1997, Major features and forcing of high-latitude northern hemisphere atmospheric circulation using a 110,000-year-long glaciochemical series, *Journal of Geophysical Research, 102(C12)*, 26, 345-26,366.

McCabe, A.M., and P.U. Clark, Ice-sheet variability around the North Atlantic Ocean during the last deglaciation, *Nature, 392*, 373-377, 1998.

McManus, J.F., R.F. Anderson, W.S. Broecker, M.Q. Fleisher, and S.M. Higgins, Radiometrically determined sedimentary fluxes in the sub-polar North Atlantic during the last 140,000 years, *Earth and Planetary Science Letters, 55*, 29-43, 1998.

McManus, J.F., D.W. Oppo, and J. Cullen, A 0.5-million-year record of millennial-scale climate variability in the North Atlantic, *Science, 283*, 971-975, 1999.

Mysak, L.A., and S.A. Venegas, Decadal climate oscillations in the Arctic: A new feedback loop for atmosphere-ice-ocean interactions, *Geophys. Res. Let., 25*, 3607-3610, 1998.

Oppo, D.W., and S.J. Lehman, Suborbital timescale variability of North Atlantic Deep Water formation during the last 200,000 years, *Paleoceanography, 12*, 191-205, 1995.

Oppo, D.W., J.F. McManus, and J.L. Cullen, Abrupt climate events 500,000 to 340,000 years ago:evidence from subpolar North Atlantic sediments, *Science, 279*, 1335-1338, 1998.

Parizek, B.R., D.R. MacAyeal, and R.B. Alley, Fast 2-D thermomechanical ice flow model for Heinrich event simulations, *Eos (Trans. Am. Geophys. Union), 78*, F228 (abstract), 1997.

Pichon, J.J., L.D. Labeyrie, G. Bareille, M. Labracherie, J. Duprat, and J. Jouzel, Surface water temperature changes in the high latitudes of the southern hemisphere over the last glacial-interglacial cycle, *Paleoceanography, 7*, 289-318,1992.

Pisias, N.G., C. Sancetta, and P. Dauphin, Spectral analysis of late Pleistocene-Holocene sediments, *Quaternary Research, 3*, 3-9, 1973.

Rahmstorf, S., Rapid climate transitions in a coupled ocean-atmosphere model, *Nature, 372*, 82-85, 1994.

Rahmstorf, S., Bifurcations of the Atlantic thermohaline circulation in response to changes in the hydrological cycle, *Nature, 378*, 145-149, 1995.

Rasmussen, T.L., E. Thomsen, T.C.E. van Weering, and L. Labeyrie, Rapid changes in surface and deep water conditions at the Faeroe Margin during the last 58,000 years, *Paleoceanography, 11*, 757-772, 1996.

Raymo, M.E., K. Ganley, S. Carter, D.W. Oppo, and J. McManus, Millennial-scale climate instability during the early Pleistocene epoch, *Nature, 392*, 699-702.

Sarnthein M, K. Winn, S.J.A. Jung, J.C. Duplessy, L. Labeyrie, H. Erlenkeuser, and G. Ganssen, Changes in east Atlantic deepwater circulation over the last 30,000 years: Eight time slice reconstructions, *Paleoceanography, 9*, 209-267, 1994.

Schiller, A., U. Mikolajewicz, and R. Voss, The stability of the North Atlantic thermohaline circulation in a coupled ocean-atmosphere general circulation model, *Clim. Dyn., 13*, 325-347, 1997.

Schulz, H., U. von Rad, and H. Erlenkeuser, Correlations between Arabian Sea and Greenland climate oscillations of the past 110,000 years, *Nature, 393*, 54-57, 1998.

Severinghaus, J.P., T. Sowers, E.J. Brook, R.B. Alley and M.L. Bender, Timing of abrupt climate change at the end of the Younger Dryas interval from thermally fractionated gases in polar ice, *Nature, 391*, 141-146, 1998.

Seidov, D. and M. Maslin, North Atlantic deep water circulation collapse during Heinrich events, *Geology, 27*, 23-26, 1999.

Steig, E.J., E.J. Brook, J.W.C. White, C.M. Sucher, M.L. Bender, S.J. Lehman, D.L. Morse, E.D. Waddington, and G. D. Clow, Synchronous climate changes in Antarctica and the North Atlantic, *Science, 282*, 92-95, 1998.

Stocker, T.F., Is there a unique mechanism for abrupt changes in the climate system? *C**, 30.

Stocker, T.F., D.G. Wright, and W.S. Broecker, The Influence of High-Latitude Surface Forcing on the Global Thermohaline Circulation, *Paleoceanography, 7*, 529-541, 1992.

Street-Perrott, F.A., and R.A. Perrott, Abrupt climate fluctuations in the tropics: the influence of Atlantic Ocean circulation: *Nature, 358*, 607-612, 1990.

Teller, J.T., Meltwater and precipitation runoff to the North Atlantic, Arctic, and Gulf of Mexico from the Laurentide ice sheet and adjacent regions during the Younger Dryas, *Paleoceanography, 5*, 897-905, 1990.

Toggweiler, J.R., and B. Samuel, Energizing the ocean's large-scale circulation for climate change. *C**, 31.

van Kreveld, S., U. Pflaumann, M. Sarnthein, P. Grootes, M. Eythstrasse, M.J. Nadeau, and H Erlenkeuser, Correlation between temperature records from Reykjanes Ridge sediment and GISP2 ice-core, *C**, 21.

Vidal, L., L. Labeyrie, E. Cortijo, M. Arnold, J.C. Duplessy, E. Michel, S. Becque, and T.C.E. van Weering, Evidence for changes in the North Atlantic Deep Water linked to meltwater surges during the Heinrich events. *Earth Planet. Sci. Lett., 146*, 13-27, 1997.

Weaver, A.J., and T.M.C. Hughes, Rapid interglacial climate fluctuations driven by North Atlantic ocean circulation, *Nature, 367*, 447-450, 1994.

White, W.B., and R.G. Peterson, An Antarctic circumpolar wave in surface pressure, wind, temperature and sea-ice extent, *Nature, 380*, 699-702, 1996.

Yu, E.F., R. Francois, and M.P. Bacon, Similar rates of modern and last-glacial ocean thermohaline circulation inferred from radiochemical data, *Nature, 379*, 689-694, 1996.

R.B. Alley, Environment Institute and Department of Geosciences, The Pennsylvania State University, University Park, PA (ralley@essc.psu.edu)

P.U. Clark, Department of Geosciences, Oregon State University, Corvallis, OR 97331 (clarkp@geo.orst.edu)

L.D. Keigwin, Woods Hole Oceanographic Institution, Woods Hole, MA 02543 (lkeigwin@whoi.edu)

R.S. Webb, NOAA-National Geophysical Data Center, Paleoclimatology Program, Boulder, CO 80303 (rwebb@ngdc.noaa.gov)